NATO ASI Series

Advanced Science Institutes Series

A series presenting the results of activities sponsored by the NATO Science Committee, which aims at the dissemination of advanced scientific and technological knowledge, with a view to strengthening links between scientific communities.

The Series is published by an international board of publishers in conjunction with the NATO Scientific Affairs Division

A Life Sciences B Physics	Plenum Publishing Corporation London and New York
C Mathematical and Physical Sciences D Behavioural and Social Sciences E Applied Sciences	Kluwer Academic Publishers Dordrecht, Boston and London
F Computer and Systems Sciences G Ecological Sciences H Cell Biology I Global Environmental Change	Springer-Verlag Berlin Heidelberg New York Barcelona Budapest Hong Kong London Milan Paris Santa Clara Singapore Tokyo

PARTNERSHIP SUB-SERIES

1. Disarmament Technologies	Kluwer Academic Publishers
2. Environment	Springer-Verlag
3. High Technology	Kluwer Academic Publishers
4. Science and Technology Policy	Kluwer Academic Publishers
5. Computer Networking	Kluwer Academic Publishers

The Partnership Sub-Series incorporates activities undertaken in collaboration with NATO's Cooperation Partners, the countries of the CIS and Central and Eastern Europe, in Priority Areas of concern to those countries.

NATO-PCO DATABASE

The electronic index to the NATO ASI Series provides full bibliographical references (with keywords and/or abstracts) to about 50000 contributions from international scientists published in all sections of the NATO ASI Series. Access to the NATO-PCO DATABASE compiled by the NATO Publication Coordination Office is possible in two ways:

- via online FILE 128 (NATO-PCO DATABASE) hosted by ESRIN, Via Galileo Galilei, I-00044 Frascati, Italy.
- via CD-ROM "NATO Science & Technology Disk" with user-friendly retrieval software in English, French and German (© WTV GmbH and DATAWARE Technologies Inc. 1992).

The CD-ROM can be ordered through any member of the Board of Publishers or through NATO-PCO, Overijse, Belgium.

Series F: Computer and Systems Sciences, Vol. 143

Springer
Berlin
Heidelberg
New York
Barcelona
Budapest
Hong Kong
London
Milan
Paris
Santa Clara
Singapore
Tokyo

Batch Processing Systems Engineering

Fundamentals and Applications for Chemical Engineering

Edited by

Gintaras V. Reklaitis

School of Chemical Engineering, Purdue University
West Lafayette, IN 47907, USA

Aydın K. Sunol

Department of Chemical Engineering, College of Engineering
University of South Florida, 4202 East Fowler Avenue, ENG 118
Tampa, FL 33620-5350, USA

David W.T. Rippin†

Laboratory for Technical Chemistry, Eidgenössische
Technische Hochschule (ETH) Zürich, Switzerland

Öner Hortaçsu

Department of Chemical Engineering, Boğaziçi University
TR-80815 Bebek-Istanbul, Turkey

Springer
Published in cooperation with NATO Scientific Affairs Division

Proceedings of the NATO Advanced Study Institute on Batch Processing Systems Engineering: Current Status and Future Directions, held in Antalya, Turkey, May 29 – June 7, 1992

Library of Congress Cataloging-in-Publication data applied for

CR Subject Classification (1991): J.6, I.6, J.2, G.1, J.7, I.2

ISBN-13:978-3-642-64635-5 e-ISBN-13:978-3-642-60972-5
DOI:10.1007/978-3-642-60972-5

Springer-Verlag Berlin Heidelberg New York

Softcover reprint of the hardcover 1st edition 1996

Typesetting: Camera-ready by editors
Printed on acid-free paper
SPIN: 10486088 45/3142 – 5 4 3 2 1 0

Preface

Batch Chemical Processing, that ancient and resilient mode of chemical manufacture, has in the past decade enjoyed a return to respectability as a valuable, effective, and, indeed, in many instances, preferred mode of process operation. Batch processing has been employed in the past in many sectors of chemical processing industries including food, beverage, pharmaceuticals, agricultural chemicals, paints, flavors, polymers, and specialty chemicals. The batch mode is increasingly being rediscovered by sectors that neglected it as the industry is focusing on more specialized, application tailored, small volume but higher margin products. Moreover, as information and control technologies have become both more technically accessible and economically affordable, the operation of batch facilities has become more efficient, gradually shifting from the conservative and simple operating strategies based on dedicated and cyclically operating trains to more sophisticated and complex operating strategies involving flexibly configured production lines using multi-functional equipment and employing just-in-time inventory management strategies.

The effects of these trends on the process systems engineering community has been a renewed intensity of efforts in research and development on computational approaches to modeling, design, scheduling, and control problems which arise in batch processing. The goal of the NATO Advanced Study Institute (ASI), held from May 29 to June 7, 1992, in Antalya, Turkey, was to review state-of-the-art developments in the field of batch chemical process systems engineering and provide a forum for discussion of the future technical challenges which must be met. Included in this discussion was a review of the current state of the enabling computing technologies and a prognosis of how these developments would impact future progress in the batch domain.

The Institute was organized into two interrelated sections. The first part dealt with the presentations on the state of the batch processing in the Chemical Process Industries (CPI), discussion of approaches to design and operation of more complex individual unit operations, followed by the reviews of the enabling sciences. This four-day program served to set the stage for a five-day program of discussions on the central problem areas of batch processing systems engineering. That discussion was preceded by a one-day interlude devoted to software demonstrations, poster sessions, and small group meetings. A unique feature of this ASI was the presence of a computer room at the hotel site equipped with an IBM RISC workstation, terminals, and personal computers which could be used for application software demonstrations and trials.

The Institute opened with industrial and academic perspectives on the role of batch processing systems engineering in the CPI. Two presentations on the status of batch processing systems engineering in Japan and Hungary provided perspectives on developments in the Far East and the former eastern block countries. The Japanese innovations in batch plant organization using moveable vessels offered insights into materials handling arrangements particularly suitable for multiproduct, small-batch production environments. These presentations were followed by a suite of papers describing applications in CPI sectors such as polymer processing, food and beverages, biochemical, specialty chemicals, textile, and leather industries.

The more complex batch unit operations which give rise to special modeling, design, and control problems were given attention in separate lectures. These included batch distillation, reactors with complex reacting systems, and sorptive separation systems. These presentations were complemented by expositions on the estimation and unit control issues for these more complex systems.

The three categories of enabling technologies which were reviewed were simulation, mathematical programming, and knowledge based systems. The simulation component included discussion of solution techniques for differential algebraic systems, the elements of discrete/continuous simulation, and available simulation environments, as well as prospects offered by advanced computer architectures. The mathematical programming review included a critical assessment of progress in nonlinear optimization and mixed integer programming domains. The knowledge based systems program consisted of a review of the field, continued with its elements and closed with more advanced topics such as machine learning including neural networks.

During the fifth day, attendees divided into small discussion groups on specific topics, participated in the software demonstrations and workshops, and participated in the poster sessions. The software demonstrations included the DICOPT MINLP solver from Carnegie Mellon University, the BATCHES simulation system from Batch Process Technologies, and the BATCH-KIT system (a knowledge based support systems for batch operations scheduling) developed at ETH Zurich.

The central problem areas in batch process systems engineering are those of plant and process design and plant operations. One day was devoted to the former topic, focusing especially on retrofit design as well as approaches to incorporating uncertainty in the design of processing systems. The second day was devoted to scheduling and planning, including consideration of the integration issues associated with linking the control, scheduling, and planning levels of operational hierarchy. The Institute concluded with plenary lectures on future of batch processing systems engineering and an open forum on questions which arose or were stimulated during the course of the meeting.

The ASI clearly could not have convened without the financial resources provided by the Scientific and Environmental Affairs Division of NATO. The support, advice, and understanding provided by NATO, especially through the Division Director Dr. L. V. da Cunha, is gratefully acknowledged. The additional financial support for specific attendees provided by the NATO offices of Portugal and Turkey and by the US National Science Foundation is highly appreciated.

The enthusiastic and representative participation of the batch processing systems engineering community was important for the realization of the goals of the ASI. Fortunately, such participation was realized. Indeed, since the participation represented all the main research groups in this domain, at one point the meeting concerns were voiced about the dire fate of the field if some calamity were to visit the conference site. Fortunately, these concerns were abated the next morning when the participants were greeted by maneuvers of NATO naval forces in Antalya bay. Without question, the active participation of the distinguished lecturers, session chairs, reviewers, and participants made this Advanced Study Institute a great success. Thanks are due to all!

Most of the manuscripts were updated considerably beyond the versions made available to attendees during the Institute and we thank the authors for their diligent work. We sincerely appreciate Springer-Verlag's understanding with unforeseeable delays with the manuscript as well as their kind assistance throughout this endeavor. Special thanks are due to Dr. Hans Wössner and J. Andrew Ross.

Finally, the organizers would like to recognize the help of the following individuals and organizations without whom the Institute would have considerably diminished if not ineffective: Sermin Gönenç (now Sunol), Muzaffer Kapanoğlu, Praveen Mogili, Çağatay Özdemir, Alicia Balsera, Shauna Schullo, Nihat Gürmen, C. Chang, and Burak Özyurt for assistance with brochures, program, re-typing, indexing, and correspondence; Dean M. Kovac and Chairman R. Gilbert of University of South Florida for supplementary financial support; Boğaziçi Turizm Inc.

and Tamer Tours for local arrangements in Turkey and social programs; IBM Turkey, especially Münire Arıkol, for the RISC Station and personal computers; Canan Tamerler and Vildan Dinçbaş (ASI's Angels) for tireless help accompanied by perpetual smiles throughout the ASI; and Falez Hotel management and staff, especially Filiz Güney, for making our stay a very pleasant one.

The idea of organizing a NATO ASI on systems engineering goes back to 1988 and was partially motivated by AKS's desire to do something in this domain at home for Turkey. However, its realization was accompanied by personal losses and impeded by unanticipated world events. A week before the proposal was due AKS lost his best friend, mentor, and mother, Mefharet Sunol. The Institute had to be postponed due to the uncertainties arising from the Gulf crisis. A few months before finalization of this volume, our dear friend and esteemed colleague, Prof. David W. T. Rippin passed away. It is fitting that this proceedings volume be dedicated to the memories of ***Mefharet Sunol*** and ***David Rippin.***

Gintaras V. Reklaitis and Aydın K. Sunol
West Lafayette, Indiana and Tampa, Florida
September 1996

List of Contributors and Their Affiliation

Organizing Committee and Director

Öner Hortaçsu, Chemical Engineering Department, Boğaziçi Üniversitesi, Istanbul, Turkey

Gintaras. V. Reklaitis, School of Chemical Engineering, Purdue University, USA

David W.T. Rippin, Technical Chemistry Lab, ETH Zurich, Switzerland

Director: Aydın K. Sunol, Chemical Engineering Department, University of South Florida, USA

Main Lecturers and Their Current Affiliation

Michel Lucet, Rhone Poulenc, France

Sandro Macchietto, Imperial College, UK

Rodger Sargent, Imperial College, UK

Alirio Rodriguez, University of Porto, Portugal

John F. MacGregor, McMaster University, Canada

Christos Georgakis, Lehigh University, USA

Arthur W. Westerberg, Carnegie Mellon University, USA

Ignacio E. Grossmann, Carnegie Mellon University, USA

Giresh S. Joglekar, Batch Process Technologies, USA

Jack W. Ponton, University of Edinburgh, UK

Kristian M. Lien, Norwegian Institute of Technology, Norway

Luis Puigjaner, Catalunya University, Spain

Special Lecturers

Mukul Agarwal, ETH Zurich, Switzerland

Rıdvan Berber, Ankara Üniversitesi, Turkey

Cristine Bernot, University of Massachusetts, USA

Ali Cınar, IIT, USA

Shinji Hasebe, Kyoto University, Japan

Laszlo Halasz, ETH, Switzerland

Gyula Körtvélyessy, SZEVIKI R&D Institute, Hungary

Joe Pekny, Purdue University, USA

Dag E. Ravemark, ETH, Switzerland

Nilay Shah, Imperial College, University of London, UK

Eva Sorensen, University of Trondheim, Norway

Venkat Venkatasubramanian, Purdue University, USA

Zentner M. G., Purdue University, USA

Denis L.J. Mignon, Université Catholique de Louvain, Belgium

Session Chairs (In Addition to Organizers and Lecturers)

Yaman Arkun, Georgia Tech, USA

Türker Gürkan, METU, Turkey

İlsen Önsan, Boğaziçi Üniversitesi, Turkey

Canan Özden, METU, Turkey

L. H. Garcia-Rubio, University of South Florida, USA

Additional Poster Contributors

Bela Csukas, Vezsprem University, Hungary

Bilgin Kısakürek, METU, Turkey

Table of Contents

Plenary Papers

Status of Batch Processing Systems Engineering in the World

Design of Batch Processes

Control of Batch Processes

Enabling Sciences: Simulation Techniques

Enabling Sciences: Mathematical Programming

Enabling Sciences: Knowledge Based Systems

Current Status and Challenges of Batch Processing Systems Engineering

David W.T. Rippin

Labor für Technische Chemie, E.T.H. Zurich, Switzerland

Abstract: The field of fine and speciality chemicals exhibits an enormous variety in the nature of the products and the character and size of their markets, in the number and type of process operations needed for production, in the scale and nature of equipment items and in the organizational and planning procedures used.

This introductory paper draws attention to the need for a careful analysis of a batch situation to identify the dominant features. Some methods and tools are identified to aid design, planning and operations and some challenges or growing points for future work are suggested. These themes will be taken up in more detail in later papers.

Keywords: Recipe, batch size, cycle time, design, multi product plant, multi plant, multipurpose plant, scheduling, equipment capacity

Factors Characterizing the Batch Production of Chemicals

Any system for producing chemical products has three necessary components (Figure 1)

1. A market
2. A sequence of process tasks whereby raw materials are converted into products
3. A set of equipment items in which the process tasks are carried out

For the production of a single product in a continuous plant, the links between these components are firmly established at the design stage. The process task sequence is designed to serve a specific market capacity for the product and the equipment is selected or specially designed to perform precisely those necessary tasks most effectively.

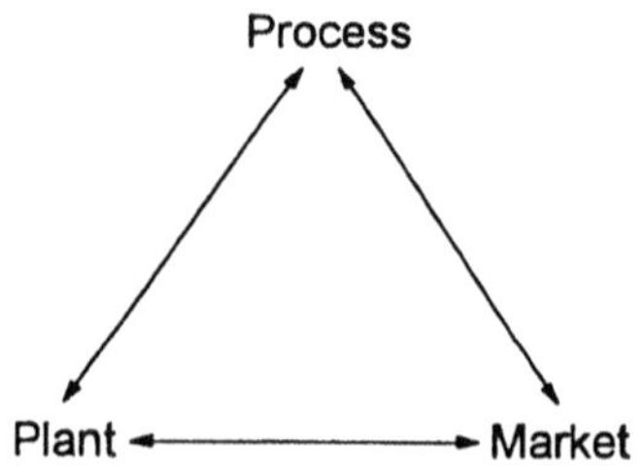

Figure 1: Components of a processing system

In a batch production system, the components themselves are much less rigidly defined and the links between them are subject to change or fuzzy. For example, the market will not be for a defined amount of a single product but may be for a range of products of which the relative demands are likely to vary and to which new products are likely to be added. A variety of processes may have to be considered which cannot be characterized with the precision demanded for a continuous single product plant. Available equipment must be utilized or new equipment selected from a standard range to serve multiple functions rather than specially designed. Similarly, the allocation of process tasks to equipment items must be changed frequently to match the changing requirements of the market.

A much wider range of choices may be available to the operators of batch plants than in a continuous plant environment.

Furthermore, there is an enormous diversity of batch processing situations determined by factors such as:

1. Scale of production
2. Variability of market demand
3. Frequency of "birth" of new products and "death" of old ones
4. Reproducibility of process operations
5. Equipment breakdowns
6. Value of materials in relation to cost of equipment
7. Availability and skill of process workers
8. Skill and experience of planning personnel
9. "Company culture"

Thus, the *dominating* factors may be quite different in one batch environment compared with another. No single sequential procedure can be called upon to solve all problems. A *variety* of tools and practices will be needed which can be matched as necessary to each situation.

The diversity of batch situations makes it important that, before starting on a particular plan of action, an analysis of the situation should be made to assess problems and opportunities, to identify the potential for improvement and to determine where effort and resources can most effectively be invested.

In a busy production environment, the natural tendency, particularly if the overall situation is not fully appreciated, is to treat the most pressing problems first, perhaps missing much more profitable, but less obvious opportunities. Obviously, a right balance should be sought between solving pressing problems, anticipating new problems, and grasping new opportunities. An overall view is needed to do this effectively.

For example, a large chemical company I visited had set itself the task in a 5-year program for its batch processes to reduce labor, cycle time and lost or waste material all by 35%. This certainly provided a focus for activity, although it may not, of course, be the best focus in another environment.

Analyzing Batch Processes

Some directions in which action might be taken to improve batch processing can be illustrated by considering a series of questions:

1. What has to be done? - a breakdown of necessary activities
2. How is achievement measured?
3. How to assess potential for improvement?
4. How to make improvement?
5. What facilities/tools are needed or advantageous?
6. Where to begin?

What Has to Be Done?

1. Batch recipe definition
2. Coordination of process tasks

3. Design of new equipment or realization in existing equipment
4. Allocation of process tasks to equipment items
5. Sequencing of production
6. Timing of production
7. Realization in production with appropriate supervision and control measures
8. Matching the customer's requirements

How Is Achievement Measured?

Measures of performance can be associated with each of the activities that have been defined. They may indicate the need for change in the corresponding activity, or in other activities earlier in the list.

The batch recipe definition may be evaluated in terms of the raw materials and reaction sequence, waste products, requirements for utilities and other resources, proportion of failed batches and more quantitatively in the yield or productivity of reaction stages, the achieved recovery of separations.

Coordination of process tasks at the development stage may be judged by how diverse are the cycle time requirements of the different process tasks (although equipment and organizational measures may be able to counter this diversity, if necessary). Overall optimization of a chain of process tasks may give different results than attention only to isolated units.

Choice of new equipment items may be judged by investment cost. For existing equipment, charges may be made for utilization or additional equipment may be called for. Choice of equipment may also be judged by frequency of breakdown or availability.

Process definition and choice of new equipment will also be judged by whether the plant performs to design specifications and achieves its design capacity.

Allocation of process tasks to equipment items may be judged by the size and frequency of batches which can be produced of a single product or in some cases of a number of different products.

The sequence in which products are produced may influence the total time and cost required to produce a group of products, if the change over time and cost are sequence dependent.

The timing of production combined with the sequencing determines if the delivery requirements and due dates of the customers can be satisfied.

Due to the intrinsic variability of the manufacturing situation, a variety of supervision and control procedures may be required to counter both recognized and unrecognized changes. They will be applied and their effectiveness assessed at different levels:

1. Process control on individual process tasks/equipment items
2. Sequence control on individual items and between interconnecting items
3. Fault detection and methods of statistic process control to detect changes in the process or its environment
4. Quality control to monitor the quality of products delivered to the customer or perhaps also raw material deliveries or environmental discharges

The success of the plant as a whole will be assessed by the overall production costs and the extent to which the market requirements can be met. Customer satisfaction will reflect delivery times and the quantity and quality of products delivered. Production costs will be ascribed to different components or activities of the system and may indicate dominant factors.

How to Assess Potential for Improvement?

First indications of where action can profitably be taken can often be obtained by making extreme or limiting assumptions about the changes which might be effected.

For example:

- At the process definition stage, what benefit would accrue if complete conversion/selectivity could be achieved in the reactor or complete recovery in the separator? - or some less extreme but conceivable assumption
- What consequence would there be for equipment requirement or utilization if the interdependencies between the processing tasks were neglected and each equipment item sized or assigned assuming 100% occupation for those tasks it had to perform?
- What benefits would there be in terms of operating costs, inventory costs, customer satisfaction, if failed batches or equipment breakdowns could be eliminated?
- What benefits in terms of timeliness of production, customer satisfaction, inventory reduction or production capacity by improvement of scheduling?
- What benefits if variability could be reduced or eliminated in terms of the costs of supervision and inspection, product quality give away?

The discipline of assessing how achievement is measured and what potential for improvement may be identified should draw attention to those features of the system where further effort may best be invested. If properly applied, such an analysis should be equally valid for exploring measures which might be taken to solve a current problem, to preempt a potential problem or to exploit a new opportunity.

How to Make Improvement?

Attention should be directed, initially, to the most promising (or threatening!) aspects of the situation. These may vary widely from case to case, so no general recipe can be given. An obvious defect or lack of reliability in the way a particular operation is carried out may claim immediate attention and resources. However, sight should not be lost of the fact that equal or greater benefits may derive from some higher level coordination or planning activity of which the effects may not be immediately obvious to the plant operator or manager.

To appreciate the potential benefits of many of the possible actions, a quantitative characterization is required of the processing requirements of a product, how these can be matched to available or potentially available equipment, and how the requirements of different products can be accommodated.

At the level of individual process tasks, there may be potential for improvement, for example, with respect to reactor conversion, selectivity, separator recovery, waste production, time requirements, in the light of the current objectives. This may be realized by developing a mathematical model to be used for determining the optimal levels for operating variables. In some circumstances there may be further advantage in making optimal choice not only of the levels of operating variables, but also of the time profile of these variables during the course of the batch. Reviews are available of where such benefits may be expected.

Quantitative Characterization of Batch Processing Requirements

Quantitative characterization of the processing requirements at each step of the process allows the determination of the performance which can be achieved (amount of product per unit time) when any step is allocated to an equipment item of appropriate type and a specified size. A process step is any physical or chemical transformation which could be carried out in a separate equipment item. For example heating, dissolving a solid, reacting, cooling, crystallization, are all regarded as separate steps in the process specification, even though, at a later stage, several steps may be carried out in sequence in the same equipment item.

The processing requirement of a step can be represented in different levels of detail by models of widely differing complexity. The minimal specification of capacity is the size factor S_{ij} that is the volume (or other measure of capacity) required in a step j for each unit of product i to be produced at the end of the batch and the batch time T_{ij} required to complete the step.

Since the size factors of the processing steps are expressed relative to the amount of product produced at the end of the batch, their calculation requires a complete balance of material and corresponding capacity requirements encompassing all steps of the process.

If the size factor and cycle time are regarded as fixed, then the selection and allocation of equipment and the determination of the overall production capacity of single and multi product plants are relatively straightforward. Such a specification corresponds to a fixed chemical recipe

Table 1: Suggested Hierarchy of models

Item	Model Type	Function	Derivation	Use
A	Comprehensive model	Performance as function of time and all operating variables	Mechanistic understanding	Effect of individual unit on whole plant
B	Performance model	Performance as function of time	Reduced model or empirical fit	Coordination of cycle time of sequence
C	Model of time and capacity requirement	Fixed performance, requirement may depend on equipment or batch size	Reduced model or external specification	Equipment sizing and task assignment
D	Model of time requirement only		Reduced model or independent specification	Simple sequencing and production
E	Stochastic model		Superposition on any of above	

from which all change in operating conditions is excluded. It may be appropriate when a fixed manufacturing specification is taken over from conditions established, for example, in a development laboratory or in previous production experience.

The allowance for variation in operating conditions calls for more complex models to predict the effects of these variations on the processing requirements and probably iterative recalculation of equipment allocation and overall production capacity determination. A hierarchy of models can be envisaged for different purposes with information being fed from one stage of the hierarchy to another as required (Table 1). However, in many practical applications for the planning and operation of batch facilities the assumption of constant size factor and batch time will be acceptable.

The simple quantitative characterization of batch processing requirements is used to determine production capability when a defined process is allocated to a set of equipment items of given size. No account is taken here of uncertainty or variability in processing requirements.

Production Capability of a Set of Equipment Items

If a process for product i is allocated to a set of equipment items of specified capacity V_j the production capability is determined by:

The Limiting Batch Size B_{li}

An upper limit for the amount of product I which can be manufactured in a batch is imposed by the equipment at the stage j with the smallest value of the ratio

$$\frac{\textit{equipment capacity}}{\textit{size factor (capacity requirement per unit product)}} = \frac{V_j}{S_{ij}}$$

$$B_{Li} = \underset{j}{Min} \frac{V_i}{S_{ij}}$$

The Limiting Cycle Time T_{Li}

A lower limit to the interval between producing successive batches of product i is imposed by the process stage with the largest batch time

$$T_{Li} = \max T_{ij}$$

The maximum production rate of product i per unit time is then

$$\frac{limiting\ batch\ size}{limiting\ cycle\ time} = \frac{B_{Li}}{T_{Li}}$$

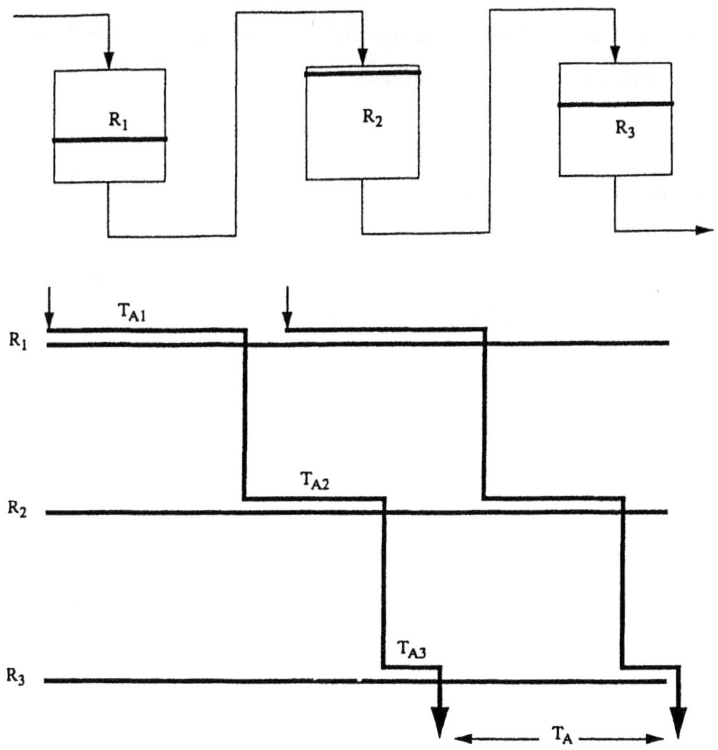

Figure 2: Limiting batch size and cycle time

The effect of the batch size and cycle time limitations are best illustrated graphically by a simple example with three equipment items (Figure 2). The cycle time limitation occurs on the first item and the batch size limitation on the second. If it is desired to increase the production rate of the product, this can be done by addressing either of the two bottlenecks in capacity or cycle time.

The following measures can be taken to increase production rate, or to make underutilized equipment available for other-purposes:

- Add *parallel* equipment items
  which operate in phase to increase batch size
  which operate out of phase to reduce cycle time
- Merge neighboring tasks to carry them out in the same equipment with potential saving of equipment items or split a task, allowing it to be carried out in two consecutive items thus reducing the cycle time
- Insert storage provision between two consecutive tasks allowing different batch sizes and cycle times up and downstream of the storage.

Design of Single Product Plant

If a batch plant is to be designed to meet a specified demand of a single product I with a known process, the number of batches which can be manufactured is given by

$$\frac{\textit{available time}}{\textit{limiting cycle time}}$$

The required batch size is then

$$B_i = \frac{\textit{total product demand}}{\textit{number of batches}}$$

The necessary size of each equipment item is

$$V_j = (\textit{batch size}) \times (\textit{size factor}) = B_i\, S_{ij}$$

The design of a single product plant can be carried out immediately, if the configuration of the plant is specified.

However, if the designer is free to choose different structural alternatives of the type discussed in the previous section, consideration may be given to

- The installation and function of parallel equipment items
- Changing the allocation of process tasks to equipment items
- The installation of intermediate storage

An acceptable design may be arrived at by examination of alternative cases or by optimization over the range of discrete choices available, for example to minimize total equipment cost as is discussed in later papers.

Multiproduct Operation

In multiproduct operation, several products are to be produced in sequence on the same plant with a defined allocation of process tasks to equipment items for each product. For each product, the limiting cycle time can be calculated and hence the production rate. It is then easy to check if specified demands for a given set of products can be met within the available production time.

Bottlenecks to increased production for a particular product can be relaxed in ways already discussed. However, in a multiproduct plant, the situation is likely to be complicated by the fact that bottlenecks for different products will be differently located.

This is illustrated in Figure 3. The first product, as considered previously, is limited in batch size by the equipment item on the second stage and its limiting cycle time is on the first stage. The second product has its capacity limitation on the first stage and the cycle time limitation on the second.

It is not immediately clear what is the most cost effective way of increasing the capacity of such a plant. For example, increasing the size of the second equipment item would increase the batch size of the first product enabling more to be manufactured in the same number of batches, or the same amount to be manufactured in fewer batches leaving more time available for the production of the second product. There are alternative ways, already discussed, for either product by which its production rate can be increased.

The best design, for example, to satisfy specified demands for a set of products at minimum equipment cost can be determined by solving an optimization problem. The optimization can be

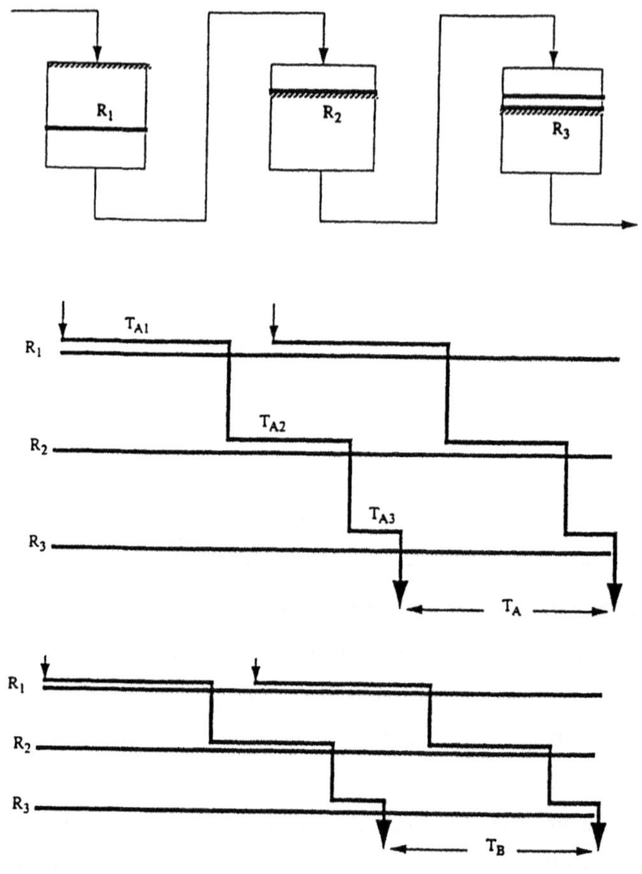

Figure 3: Multi-product production

formulated as choosing the best sizes (and possibly also the configuration) of the equipment items to satisfy the product demands within the available production time, or alternatively it can be viewed as choosing how the available time is divided between production of the different products. With a fixed equipment configuration, when the time allocation to a product has been fixed the equipment sizes to accommodate that product are also determined. For a particular time allocation to the complete set of products the necessary size of any equipment item is determined by scanning over the equipment size requirement at that stage for all products and choosing the largest.

In a practical situation where exact cost optimization is not of major importance, a plausible design can be arrived at by allocating the available time to products or groups of products determining the equipment cost and checking its sensitivity to time allocation or equipment configuration.

Bottlenecks may be removed by changing configuration as for the single product.

Discrete Equipment Sizes

Much batch equipment is not available in continuously variable sizes. For example volumes of available reaction vessels may increase by a factor of ~1.6. A discrete optimization procedure, such as a branch and bound search, may be used to make an optimal selection of equipment items in these circumstances, if required.

Partly Parallel Production (See Figure 4)

Products being produced on the same plant may differ substantially in the number of equipment items needed. In a plant equipped to produce a complex product with many transformations it may be possible at another time to produce several less demanding products in parallel. Capacity evaluation or design can be carried out in ways similar to those already discussed.

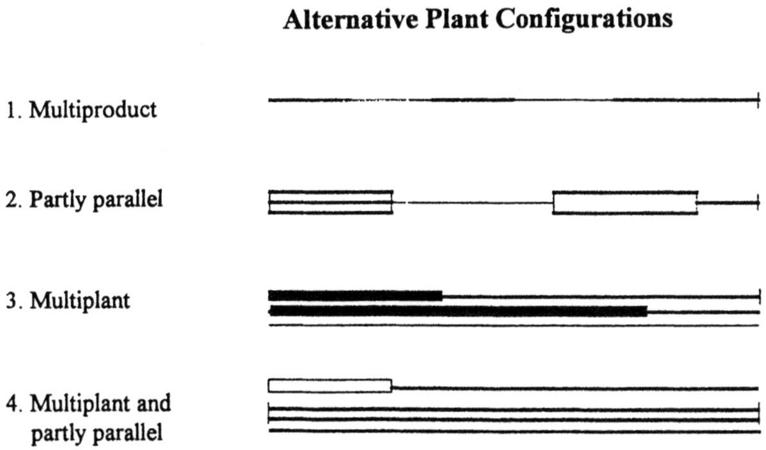

Figure 4: Partly parallel production

Multiplant Production

When a number of products have similar equipment requirements, it may appear attractive to manufacture as many of them as possible in a single large multiproduct plant. The product will be made in large batches and, by economy of scale, the equipment cost will be less than the total cost of a number of smaller plants operating in parallel to produce subgroups of products or even individual products.

However, other factors may militate against the very large multiproduct plant. Manufacturing many products in the same plant will call for frequent product changeovers with associated loss of time and material and other expenses. In addition, if a product is in continuous demand, but is only manufactured for short periods, as will be the case if it shares the plant with many other products, then the inventory level to maintain customer supplies will be much higher than for a plant in which only a few products are manufactured. For high value products the cost of such inventories may be far more significant than "economy of scale" savings on equipment costs. If in addition, purity requirements are high leading to very stringent changeover procedures, the use of parallel plants for single, or a few products, may easily be justified on economic grounds. Discrete optimization procedures can be set up to assist the grouping of products into subsets for production together.

Multipurpose Operation

Many batch products are made in multipurpose plants. A set of equipment items is available in which a group of products is manufactured. The products may change from time to time calling for the plant to be reconfigured. Several products may be manufactured in the plant at one time and the same product may follow different routes through the plant at different times, perhaps depending upon which other products are being produced at the same time.

One way of assessing the capacity of such a plant is to plan its activity in campaigns. A campaign is the assignment of all the equipment in the plant to the production of a subgroup of products for a period. The total demand for a set of products over a planning period can be satisfied by selecting and assigning times to an appropriate set of campaigns chosen from a larger set of possible campaigns. This selection can be made by a linear programming procedure.

The candidate campaigns can be built up by assigning equipment to the manufacture of batches of the same or different products in parallel (Figure 5). Candidates which do not make good use of the equipment can be screened out, leaving a modest number of efficient campaigns, from which the linear programming procedure can make an appropriate selection corresponding to any distribution of demand.

The method described seeks to make the best use of the capacity of an existing multipurpose plant. It could theoretically be extended to consider the selection of equipment items to include in the plant, but at the cost of enormously more computing. It is also not at all clear how to specify the potential product demand for such a plant at the design stage. In fact, many multipurpose plants are built with limited knowledge of the products to be made in them, or even with the intention that as yet unknown products will be produced there.

In those circumstances it is not surprising that the selection of items to include in a multipurpose plant is commonly based on past experience rather than any mathematical procedure. The choice of items may be based on the requirements of one, or a group of representative products, possibly with special provision for other unusual products. Alternatively, a typical mixture of equipment items may be chosen that has given good service in the past and has sufficient flexibility to accommodate a range of types and amounts of products. Of course, the selection of equipment items made in this way may be heavily dependent on the particular application field being considered.

The Choice of Production Configuration and Short Term Scheduling

The choice of a production configuration depends on the character of the processes and the market situation. A group of products with very similar equipment requirements can often be produced conveniently in a multiproduct configuration. Very diverse products may be produced in independent production lines or, depending on the availability of equipment and the level and variability of demand in mixed multipurpose campaigns.

The structuring of campaigns described in the previous section is a device for identifying favorable combinations of products. It may be useful for medium term and capacity planning, but

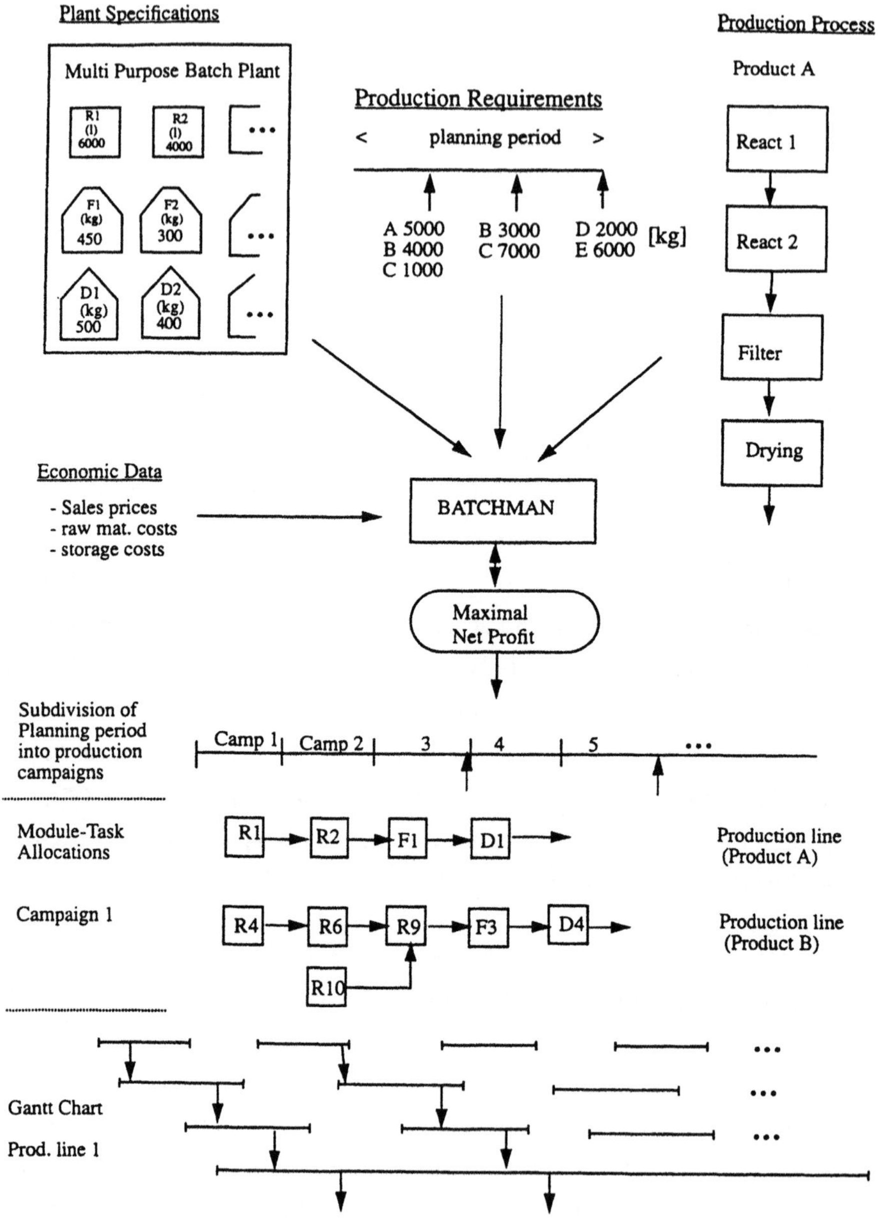

Figure 5: Multi-purpose planning

it will not be rigidly adhered to in day to day production planning. There must be flexibility to adapt, for example, to short-term changes in equipment availability, local working conditions or utility capacity. There will probably be less freedom to assign process tasks to equipment items than in the earlier construction of multipurpose campaigns. However, there must be the possibility to make local adjustment to the timing and allocation of individual batches and to propagate the consequences of these changes.

In practice, this is often done manually on a bar chart. A similar graphical facility can be made available in a computer, or a more comprehensive computer program could not only propagate the consequences of a change, but also examine alternative measures to adapt most profitably to the change. Various aspects of scheduling will be reviewed later.

Accommodating the Variety of Batch Processing

If the manufacture of fine and speciality chemicals is to be undertaken, the following questions should be considered:

1. Which products, in consideration of their process and market characteristics are suitable for batch processing?
2. Which products might be produced together in shared equipment?
3. Which groups of products have sufficiently homogeneous processing requirements that they might be produced consecutively in a single multi-production line?
4. Which products have diverse or fluctuating requirements suggesting that they might be produced in a general multi-purpose facility?
5. On the basis of past experience and anticipated developments, what range of equipment items should be installed in a multipurpose facility?

Whatever decisions are taken about product assignments to particular types of production and choice and configuration of equipment, there will be continual need for monitoring performance, for scheduling and rescheduling production, and for evaluation the effect of introducing new products and deleting old ones.

Harnessing the inherent flexibility of batch processing to deal effectively with change and uncertainty is a problem which is solved routinely in practice. However, the mathematical representation of this problem and how it can be solved in a truly optimal way, are still the subject of study, some of which is reported later.

What Facilities / Tools Are Needed or Advantageous?

- The ability to assess relevant overall data and make rapid order of magnitude estimates of the effect of constraining factors or potential benefits

Hence, to identify elements exercising the dominant constraints on improvements

- The ability to predict the effect of measures to relieve the constraints and hence expected improvements resulting from suggested changes
- In some cases, optimization capability to extract the maximum benefit from certain defined types of changes may be justified
- Packages to perform some of these tasks may be available: overall or detailed simulation, design, scheduling, optimization
- Flexible capability to do a variety of simple calculations and easy access to the necessary basic data may often be important

Some Challenges and Opportunities in Batch Processing

1. Because of the great diversity of batch processing, measures are needed to characterize a batch processing situation to enable computer and other aids to be matched to requirements.
2. Quick estimation procedures to assess whether features of the system are sufficiently significant to be considered in greater detail.
3. Integration of knowledge to make available as appropriate the totality of knowledge about the system including hierarchical model representations.
4. Batch process synthesis including co-ordination of the performance of individual stages with overall requirements of assignment and scheduling, perhaps also coupled with multi-product considerations.
5. Non-linear and integer programming - efficient problem formulation, significance and exploitation of problem structure.
6. Catalogue of potential benefits of profile optimization for different batch operations, reaction kinetics and objective functions.

7. Guidelines for potential usefulness of adaptive control, on-line estimation, optimization as a function of the batch operation, the conditions to which it is exposed and the wider plant environment in which it is situated.
8. Single/multi-product plant design - potential benefits of recipe adjustment. Is the more detailed examination of scheduling at the design stage beneficial?
9. Effect of wide ranging uncertainty on multi-product or multi-purpose design. When is it sensible to drop explicit consideration of product demands and what should be done then?
10. Are there significant advantages in coordinating the dynamic simulation of individual batch units over the whole process and coupling with sequence control?
11. What can be achieved in scheduling, and where will further progress be made, for example with reference to problem size, interaction with the user, relative merits of algorithmic (integer programming) versus heuristic, knowledge-based methods?
12. What is really new - pipeless plants? anything else?

Conclusion

Industrial interest in fine and speciality chemicals has increased substantially in recent years, not least because these often seem to be the most profitable parts of the chemical industry. Over the same period academic work has produced a number of models which have been refined in various ways to express details of how batch processing could be carried out.

There is certainly scope for further interaction between industry and university to match modeling and optimization capabilities to industrial requirements. One benefit of the present institute could be further moves in this direction.

Addendum

For further details and a comprehensive reference list, the reader is directed to D. W. T. Rippin, Batch Process Systems Engineering: A Retrospective and Prospective Review, ESCAPE-2, Supplement to Comput. & Chem. Engineering, 17, Supplement, S1-S13 (1993)

Future Directions for Research and Development in Batch Process Systems Engineering

Gintaras V. Reklaitis

School of Chemical Engineering, Purdue University, West Lafayette, IN 47907-1283, USA

Abstract: The global business and manufacturing environment, to which the specialized and consumer products segments of the CPI are subjected, inexorably drive batch processing into the high tech forefront. In this paper the features of the multipurpose batch plant of the year 2000 are reviewed and the implications on its design and operation summarized. Research directions for batch process systems engineering are proposed, spanning design applications, operations applications and tool developments. Required advances in computer aided design encompass task network definition, preliminary design of multipurpose plants, retrofit design, and plant layout. Needs in computer support for operations include integration of the application levels of the operational hierarchy as well as specific developments in scheduling, monitoring and diagnosis, and control. Advances in tools involve improved capabilities in developing and testing algorithms for solving structured 0-1 decision problems and interpreting their results, further enhancements in capabilities for handling large scale differential algebraic simulation models with implicit discontinuities, and creation of flexible data models for batch operations.

Keywords: Algorithm adversary, computer integrated manufacturing, continuous/discrete simulation, control, data model, heat integration, manufacturing environment, materials handling, mixed integer optimization, monitoring and diagnosis, multiplant coordination, multipurpose plant, plant layout, preliminary design, reactive scheduling, resource constrained scheduling, retrofit design, task networks, uncertainty

Introduction

The lectures and presentations of this Advanced Study Institute have amply demonstrated the vigor and breadth of contemporary systems engineering developments to support batch chemical processing. Much has been accomplished, particularly in the last ten years, to better understand the design, operations, and control issues relevant to this sector of the chemical industry. New computing technologies have been harnessed to solve very challenging and practical engineering problems. Yet, given the explosive growth in hardware capabilities, software engineering and tools, and numerical and symbolic computations, further exciting developments are within our reach. In this presentation, we will attempt to sketch the directions for continuing developments in this domain. Our proposals for future research are based on the conviction that the long term goal for batch process systems engineering should be to fully realize computer integrated manufacturing and computer aided engineering concepts in the batch processing industry in a form which faithfully addresses the characteristics of this mode of manufacturing.

Trends in the Batch Processing Industry

Any projections of process systems engineering developments targeted for the batch processing industry over the next five to ten years necessarily must be based on our anticipation of the manufacturing challenges which that industry will need to face. Thus, the point of departure for our discussion must lie in the assumptions that we make about the directions in which the batch processing industry will evolve. Accordingly, we will first present those assumptions and then launch into our discussion of necessary batch process systems developments.

Chemical Manufacturing Environments 2000

As noted by Loos [12] and Edgerly [6] the chemical industry in the year 2000 will evolve to four basic types of manufacturing environments: the consumer products company, the specialized company, the utility, and the megacompany.

The consumer products company, of which 3M, Procter & Gamble, and Unilever are present day precursors, features captive chemical manufacturing which supports powerful consumer franchises. The manufacture of fine chemicals, polymers, and petrochemicals within this type of company will be driven by consumer demands and as these peak and wane chemical processing will have to change to accommodate. The market life of such products is often two years or less [24]. Many food processing companies are tending in this direction.

The specialized companies, of which Nalco and Lubrizol are illustrative, will be midsized organizations that possess unique technical capabilities and marketing/customer access. These organizations will be involved in continuous technical innovation and intensive focus on customer service and thus their manufacturing functions will subject to continuous product turn-overs, pressures for quick start-up and rapid response to market needs. Pharmaceutical companies are evolving in this direction.

The utility, of which SABIC is a precursor, will be the low cost converter of chemical feedstocks into basic building block chemicals for the other sectors of the CPI. This type of organization will flourish by virtue of its leading-edge technology, world-class scale of production, and advantaged access to raw material or energy sources. Manufacturing in such an organization will be highly automated and optimized for operation with minimum upsets and quality deviations.

The fourth category, the *megacompany* will be the leader in diverse segments of the chemical market, encompassing some or all of the above manufacturing models. These organizations will operate on a global basis with great technical depth, and financial and marketing strength. DuPont, Hoechst, and ICI could be precursors of such companies. Manufacturing in the megacompany will be subject to the same factors as the above three types of companies depending upon the sector in which that particular arm of the organization competes.

It is clear that the specialized and consumer products companies and analogous arms of the megacompanies will be the CPI components in which batch processing will continue to grow and flourish. These sectors of the processing industry will increasingly share in the same business environment as experienced in discrete manufacturing: a high level of continuous change in products and demands, close ties to the customer, whether consumer or other organization, strong emphasis on maintaining quality and consistency, accelerating demands for worker, product, and community safety and prudent environmental stewardship, and relentless competitive pressures to be cost effective.

Consequences of the Changing Environment

The consequences of these factors are that batch processing, the most ancient mode of chemical manufacturing, will be increasingly driven into the high technology forefront. The batch plant of the year 2000 will be a multipurpose operation which uses modularized equipment and novel materials handling methods and is designed using highly sophisticated facilities layout tools. It will feature a high degree of automation and well integrated decision support systems and, consequently, will require significantly lower levels of operating staff than is present practice. The batch processing based firm will employ a high degree of integration of R&D, manufacturing and business functions, with instantaneous links to customers, suppliers, as well as other cooperating plant sites on a global basis. It will employ computer aided process engineering tools to speed the transition from the development of a product to its manufacture, without the protracted learning curves often now encountered.

Design Implications: Short product life and intensely competitive markets will impose major challenges on both the manufacturing and the product development processes. Responsiveness to the customer needs for tailored formulations, generally, will lead to increasing specialization and multiplication of products, resulting in increased focus on very flexible, small batch production. The multipurpose chemical plant will become the workhorse of the industry. At the same time, in order to reduce the operational complexity often associated with multipurpose plants, standardization and modularization of the component units operations will be employed, even at the cost of higher capital requirements and possibly lower capacity utilization levels. As in the manufacture of discrete parts, streamlining of flexible, small batch production will require increased focus on the materials handling aspects of batch operations. With small batch operation, the traditional pipes, pumps, and compressors can, depending upon recipe details, loose their effectiveness as material transfer agents. Other modes of transfer of fluids and powders such as moveable bins, autonomous vehicles, and mobile tanks can become more efficient alternatives. Factors such as efficient materials handling logistics, reduction of in-process inventories, minimization of cross-contamination possibilities and increased operating staff safety will dictate that consideration of physical plant layout details be fully integrated into the design process and given much greater attention than it has in the past.

The industry will also need to give increased emphasis to reducing the product development cycle. This means employing sophisticated computational chemistry tools to guide molecule design and exploiting laboratory automation at the microquantity level to accelerate the search for optimum reaction paths, solvents and conditions. The use of large numbers of automated parallel experimental lines equipped with robotic aids and knowledge based search tools will become quite wide-spread. To insure that recipe decisions take into account available production facilities early in product development, process chemists will need to be supported with process engineering tools such as preliminary design and simulation software. Use of such tools early in the development process will identify the time, resource, and equipment capacity limiting steps in the recipe, allowing process engineering effort to be- focused on steps of greatest manufacturing impact. The models and simulations used in development will need to be transferred to production sites in a consistent and usable form to insure that processing knowledge gained during development is fully retained and exploited.

Operational Implications: Since maintenance of tight product quality standards will be even more of a necessity, sophisticated measurement and sensing devices will be required. The need to control key product quality indices throughout the manufacturing process will put high demands on the capabilities of regulatory and trajectory tracking control systems. Early prediction and correction of recipe deviations will become important in order to reduce creation of off-spec materials and eliminate capacity reducing reprocessing steps. Thus, integrated process monitoring, diagnosis and control systems will be widely employed. The needs to largely eliminate operator exposure to chemical agents and to contain and quickly respond to possible releases of such materials will further drive increased use of automation and robotic devices. Indeed, all routine processing and cleaning steps will be remotely controlled and executed. To allow the reduced number of operating staff to effectively manage processes, intelligent information processing and decision support systems will need to be provided. Effective lateral communications means between corporate staff world-wide will be employed to facilitate sharing of manufacturing problem solving experience, leading to continued manufacturing quality improvements. Rapid and predictable response to customer orders will require development of simple, reliable operating strategies and well integrated scheduling tools. Manufacturing needs for just-in-time arrival of raw materials, key intermediates, and packaging supplies will drive the development of large scale planning tools that can encompass multiple production site and suppliers. The realization of computer integrated process operations will require extensive and realistic operator and management training using high fidelity plant training simulations. These models will further be used in parallel with real plant operation to predict and provide benchmarks for manufacturing performance.

The inescapable conclusion resulting from this view of trends in batch chemical processing is that the needs for information management, automation, and decision support tools will accelerate dramatically over the next decade. The marching orders for the process systems community are thus to deliver the concepts and tools that will increase the cost-effectiveness, safety, and quality of multipurpose batch operations. The greatest challenge will be to use these tools to discover concepts and strategies that will lead to drastic simplifications in the design and operation of batch facilities without loss in efficiency or quality. Simplifications and streamlining of potentially complex manufacturing practices will be the key to maximum payoffs in safety, reliability, and competitiveness.

Research Directions for Batch Process Systems Engineering

In this section, we will outline specific areas in which productive process systems developments should be made. Our projection of research directions will be divided into three areas: design applications, operations applications, and tool development. This division is admittedly artificial since in the batch processing domain design and operation are closely linked and successful developments in both domains depend critically on effective tools for optimization, simulation, and information processing and solution comprehension. However, the division is convenient for discussion purposes.

Advances in Design

While considerable methodological progress has been made since the review of computer aided batch process design given at FOCAPD-89 [20], a number of issues remain unexplored. These can be divided into four categories: task network definition, preliminary design methodology, retrofit design approaches, and plant layout.

Task Network Definition: The key process synthesis decisions made early in the development of a product center on the definition of the precise product recipe and the aggregation of subsets of the contiguous steps of the recipe into tasks which are to be executed in specific equipment types. These decisions define the task network which is the basis for selecting the number and size of the process equipment. Recipe definition is usually made by process chemists in the course of exploring alternative synthesis paths for creating the product or molecule of interest. The decisions involved include the selection of the most effective reaction path which has direct impact on solvents to be employed, reaction conditions, by-product formation, and the types of unit operations which will be required. Task selection involves a range of qualitative and experiential information which incorporates choices of the broad types of equipment which will be selected to execute the tasks. The overall task network definition problem would be greatly facilitated if a knowledge based framework could be developed for task network synthesis which incorporate both the recipe definition and task selection components. To date the proprietary PROVAL package [1] remains the only development which addresses some aspects of this synthesis problem.

Preliminary Design of Multipurpose Plants: The recent work of Papageorgaki [16] and Shah and Pantelides [23] does address the deterministic, long campaign case, while Voudouris and Grossmann [27] offer approaches to incorporating discrete equipment sizes. However, one of the key issues in grass roots design of such facilities is the treatment of uncertainties. Shah and Pantelides [22] do suggest an approach for treating multiple demand scenarios within a deterministic formulation of the problem, an idea which had previously been advanced by Reinhart and Rippin [19] in the multiproduct setting. Moreover, it would appear that the staged expansion concept, initially explored by Wellons [28,29] for longer term demand changes, merits consideration in the multipurpose case, especially in the context of modular plant expansion. Yet missing is a framework for handling likely changes in the product slate, in other words uncertainties in recipe structures, since one of the key reasons for the existence of multipurpose plants is the adaptability of the plant in accommodating not only demand but also product changes. The latter aspect of flexibility needs to be given a quantitative definition.

The increasing interest in alternative material handling modes raises questions of under what recipe conditions these various alternatives are most cost effective. For instance, the vertical stacker crane concept appears to be advantageous for the short campaign, reaction dominated recipe, while the tracked vessel concept is said to be appropriate for mixing/blending type recipes. Clearly, depending upon recipe structure and campaign length, different combinations of these handling modes together with conventional pipe manifold systems might be most economical. The incorporation of these material handling options within an overall process design framework would appear to be highly desirable as a way of allowing quantitatively justified decisions to be made at the preliminary design stage.

While mathematical programming based design formulations are adequate at the preliminary design stage, detailed investigation of designs requires the use of simulation models. Simulation models do allow consideration of the dynamics of key units, step level recipe details, complex operating strategies, as well as stochastic parameter variations. Ideally such simulations should also form the basis for detailed design optimizations. However, while batch process simulation capability does exist (see [3]), the optimization of dynamic plant models with many state and time event discontinuities continues to present a challenging computational problem. Although some interesting developments in the optimization of differential/algebraic systems involving applications such as batch distillation columns [2] have been reported, further investigation of strategies for the optimization of DAE systems with implicit discontinuities are clearly appropriate.

Retrofit Design: A MINLP formulation and decomposition based solution approach for the retrofit design of multipurpose plants operating under the long campaign operating strategy was reported in [15]. This formulation included consideration of changes in product slate and demands as well as addition of new and deletion of old units with the objective of maximization of net profit. An extension of this work to accommodate resource constraints was reported at this conference [17]. Incorporation of the effects of campaign changeover and startup times is straightforward in principle, although it does introduce additional 0-1 variables. Developments which merit investigation include incorporation of continuous units and investigation of key state variable trade-offs during retrofit. In principle, the latter would require inclusion of the functional dependence of key recipe parameters on state and design variables such as temperature, conversion, and recovery. Since these nonlinear dependencies may need to be extracted from simulations of the associated processing tasks, a two level approach analogous to the SQP strategy widely employed in steady state flowsheet optimization may be feasible.

Further key factors not treated in available retrofit design approaches include consideration of space limitations for the addition of new equipment and changes in the materials handling requirements which the addition of new equipment and elimination of old equipment impose. These factors clearly can only be addressed within the framework of the plant layout problem which will be discussed in a later section of this paper.

Heat integration is a feasible retrofit option for batch operations, especially under long campaign operation, and has been investigated in the single product setting [25]. Recent work has led to an MILP formulation which considers stream matches with finite heat exchange times and batch timing modifications to minimize utilities consumption [11]. Interesting extensions which should be pursued include scheduling of multifunctional heat exchange equipment so as to minimize the number of required exchanger units as well as consideration of the integrated use of multiple intermediate heat transfer fluids.

Plant Layout: Once the selection of the number and capacities of the plant equipment items has been made, the next level of design decision involves the physical layout of the process equipment. The physical layout must take into account (1) the sizes/areas/volumes of the process equipment, (2) the unit/task assignments, which together with the recipe fix the materials transfer links between process vessels, (3) the materials transfer mechanisms selected to execute these links, and (4) the geometry of the process structure within which the layout is imbedded. Clearly, safety considerations, maintenance access requirements, cross-contamination prohibitions, and vibrational and structural loading limitations will further serve to limit the placement of process equipment. While these aspects of plant layout have been handled traditionally via rules of thumb and the evolving practices of individual engineering firms, the trend toward truly multipurpose facilities which employ a variety of material handling options will require that the plant layout problem be approached in a more quantitative fashion. This is particularly the case with layouts involving enclosed multilevel process buildings which are increasingly being employed for reasons of esthetics, safety, and containment of possible fugitive emissions.

As discussed in [7], the plant layout problem can be viewed as a two level decision problem involving the partitioning of the equipment among a set of levels and the location of the positions of the equipment assigned to each level. The former subproblem can be treated as a constrained set partitioning problem in which the objective is to minimize the cost of material transfers between units and the constraints involve limitations on additive properties such as areas and weights of the vessels assigned to each level. Because of the effects of gravity, different cost structures must be associated with transfers in the upwards, downwards, and lateral directions. As shown in [7], the problem can be posed as a large MILP and solved using exact or heuristic partial enumeration schemes. The subproblem involving the determination of the actual positions of the equipment assigned to a level is itself a complex decision problem for which only rather primitive heuristic approaches have been reported [8]. The integrated formulation and solution of these subproblems needs to be investigated as such a formulation could form the basis for investigating alternative combinations of material handling strategies. Ultimately, the layout problem solution methodology should be linked to a computer graphics based 3D solids modeling system which would allow display and editing of the resulting layout. Further linkage of the 3D display to a plant simulation model would allow animation of the operation of the multipurpose plant, especially of the material transfer steps. Such virtual models of plant operation could be very effectively used for hazards and operability analysis, operator training, and design validation studies.

Advances in Operations

The effective operation of the batch processing based enterprise of the year 2000 will require full exploitation of computer integrated manufacturing developments and the adaptation of these developments for the specific features of batch operations. While descriptions of the information flows in process oriented CIM systems have been formulated [30] and implementation standards oriented to the chemical industry are being formalized [31], no concrete implementations have actually been realized to date [18]. Given the state of the art, the required research must proceed along two tracks: first, investigation of integrated process management frameworks spanning the planning, scheduling, control, diagnosis and monitoring functions and, second, basic developments in the component applications themselves. In this section we will briefly outline research thrusts in both of these tracks.

Application Integration: From the perspective of a process scheduling enthusiast, the batch processing CIM framework can be viewed as a multilevel integrated scheduling problem, as shown in Figure 1. At the top-most level of the hierarchy is the coordination of the production targets and logistics involving multiple plant sites. This level interfaces with the master schedules for individual plant sites which treat resource assignment and sequencing decisions of a medium term duration. The master schedule in turn must be continuously updated in response to changes on the plant floor and in the business sector. The need for these changes is identified by the process monitoring and diagnosis system which includes key input from the operating staff. The actual changes in the timing and linkage to the required resources are implemented through the control system. The entire framework is of course linked to business information systems, including order entry and inventory tracking systems.

One of the key impediment to assembling such a framework is the difference in level of aggregation of the information and decisions employed at each level of the hierarchy. As noted recently by Macchietto and coworkers, the master schedule operates at the level of tasks, the reactive scheduling function deals with timing of steps and material transfers, while the control system operates at the even more detailed level of valve operations, interlock checks, and regulatory loops. An initial and instructive experiment at integrating three specific functional blocks, namely, master scheduling, reactive scheduling, and sequential control translation blocks has been reported [5] . The principal focus of that work was on reconciling the differences in the procedural information models employed by each of these blocks. Further work is clearly needed to examine the implications of linking the process monitoring and diagnosis functionalities into a comprehensive manufacturing control system, as shown in Figure 2 (after [18]). As envisioned,

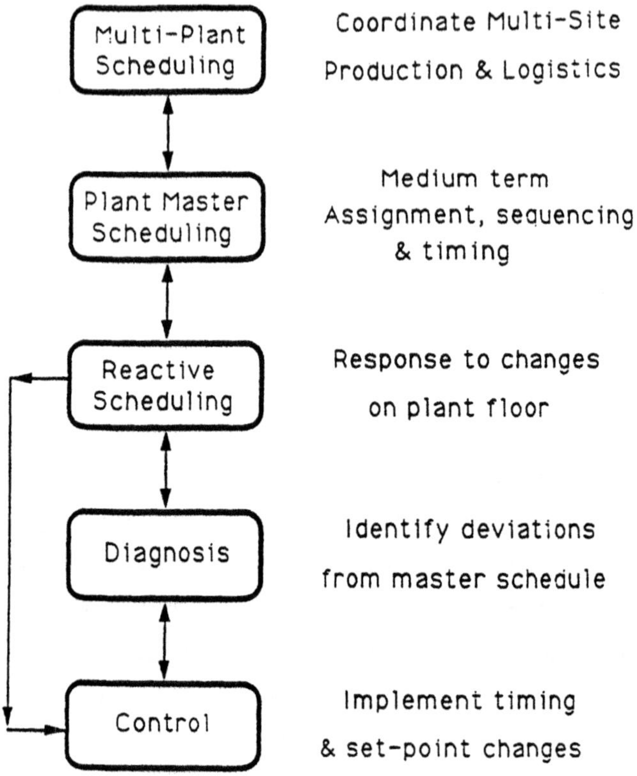

Figure 1. Batch processing CIM levels

the control system generates process information which is filtered and tested to eliminate gross errors. The intelligent monitoring system extracts important qualitative trends, processes these trends to provide near term forecasts, and provides qualitative description of process behavior. The fault detection system detects deviations from expected behavior and recommends corrective actions. These recommendations are offered to the user through a high level interface and, if validated, are presented to the supervisory control system which selects appropriate control configurations, algorithms, and settings, including changes in timing of actions. Neural networks and knowledge based, especially rule based, systems would appear to be the relevant technologies for these functions.

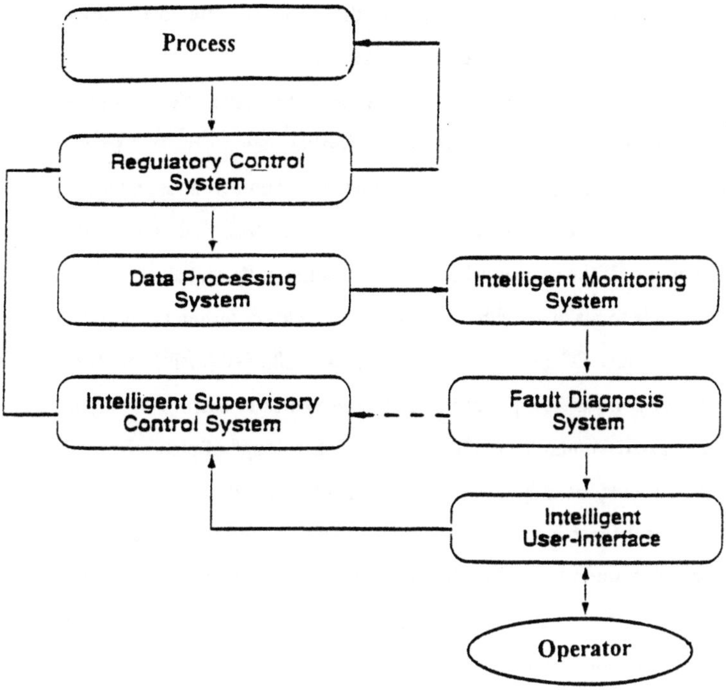

Figure 2. Integrated process monitoring, diagnosis and control

A more fundamental question which underlies the very structure of the above conceptual integration framework is how to quantitatively and rigorously deal with the uncertainty which is inherent to the manufacturing environment. Under present methodology, each component of the hierarchical framework shown in Figure 1, employs a deterministic approach to dealing with uncertainly at its level of aggregation. The multiplant coordinator looks over longer time scales than the plant master scheduler and thus deals with demand uncertainties through the rolling horizon heuristic. The master schedule again operates on representative deterministic information, relies on short term corrections applied by the reactive scheduler to correct infeasibilities and resolve conflicts, and again is employed under the rolling horizon heuristic. In other words, the master schedule is totally recomputed when the departures from the current master schedule become too severe or at some predetermined time interval, whichever arises first. Finally, the monitoring and control systems account for the smallest time scale variations which are encountered between actions of the reactive scheduler.

The key research question which must be addressed is, thus, what is the best way to reflect the uncertainties in demands, order timing and priorities, equipment availabilities, batch quality indices, resource availabilities, and recipe parameter realizations at each level of the hierarchy. Clearly, if the multiplant schedule incorporates sufficient slack time, the plant master schedule gains more flexibility. If the master schedule incorporates sufficient slack time, then the reactive scheduler will be able to make do with less severe corrective actions and may prompt less frequent master scheduling reruns. Of course, if too much slack is allowed, manufacturing capacity will be under-utilized. By contrast, simple use of expected values at each level may lead to infeasible schedules and excessive continual readjustment or "chattering" of plans at each level, disrupting the orderly functioning of shipping, receiving, shift scheduling, and materials preparation activities. At present there is no guidance which can be found in the literature on how "slack" should be distributed among the levels in a way which adequately reflects the underlying degree of uncertainty in the various manufacturing inputs. Exploratory work in this area would be quite valuable in guiding the development of CIM systems for batch operations.

Advances in Operations Applications

While integration of the various decision levels of the CIM hierarchy is the highest priority thrust for research in the operations domain, that integration is only as effective as the methodology which addresses the main application areas of scheduling, monitoring/diagnosis, and control. Important work remains to be carried out in all three of these areas.

Scheduling: The three key issues in scheduling research are investigation of more effective formulations for dealing with resource constraints, approaches to the reactive scheduling problem which address a broader range of decision mechanisms for resolving scheduling conflicts, and formulations and solution strategies for multiplant applications. These application areas are especially important for the short campaign operating mode likely to be favored in multipurpose plants of the future.

The key to effective treatment of globally shared resource constraints is to effectively handle the representation of time. The classical approach of discretizing time in terms of a suitably small time quantum (see [10]) can be effective. However, in order to be able to accommodate problems of practical scope, considerably more work needs to be invested on solution algorithms which can exploit the fine structure of the problem. This is essential in order to have any hope of treating sequence dependent set-ups and clean-outs. Such refinements must go beyond the conventional reformulations and cuts typically employed with MILP's. The interval based approach, explored in [32], has considerable potential but needs further development and large scale testing. Again careful analysis of structure is required in order to generate a robust and practical solution tool. Furthermore, since the interval elimination logic, which is part of that framework, appears to have promise as a preanalysis step for the uniform discretization approach as well as for reactive scheduling approaches, it is worthwhile to investigate this approach in more detail and generality.

While the work of Cott and Macchietto[4] and Kanakamedala et al [9] provide a useful start further research is required to exploit the full range of reactive scheduling decision alternatives, which are shown in Fig. 3. In particular, it is important to investigate interval based mathematical programming formulations which would allow simultaneous adjustments using all of the possible decision alternatives, while permitting treatment of step details and the realistic timing of materials transfers. It would appear that it would also be useful to formulate more realistic scheduling criteria than minimization of deviations from the master schedule. Furthermore, both reactive scheduling and master scheduling formulations need to be expanded to include consideration of

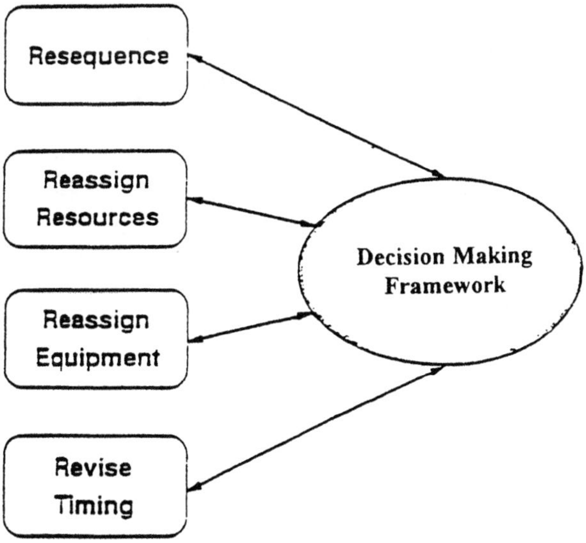

Figure 3. Reactive scheduler structure

alternative material handling modes. While the movable processing vessel/rigid station configuration can be treated as another shared resource, the transfer bin concept appears to introduce some new logistics considerations into the problem. Furthermore, as noted earlier, a theoretical and computational framework also needs to be developed for linking the master scheduling and reactive scheduling functions.

Finally, the upper level of the integrated scheduling hierarchy which deals with the coordinated scheduling of multiple plant sites needs to be investigated. An important consideration at this level is the incorporation of the logistical links between the plants. Thus, the geographical distribution of plant sites, the geographic distribution of inventories, and the associated transport costs and time delays need to be addressed in the scheduling formulation. Moreover, in order to effectively deal with the interchanges of products and feeds which enter and leave a plant at various points in the equipment network, as in the case of the plant of Figure 4, reasonably detailed models of the individual plants must be employed. The conventional lumping of the details of an entire plant into a single black box can not adequately reflect the actual processing rates which are achieved when production units are shared among products. Thus, considerable scope exists for large enterprise scale models and solution methods.

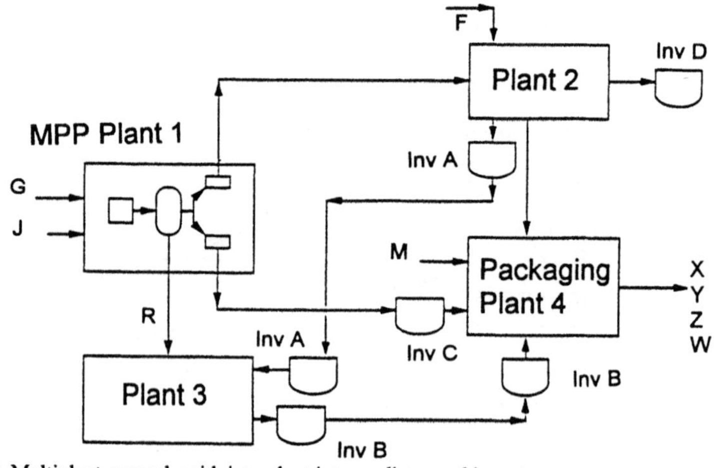

Figure 4. Multiplant example with interplant intermediates and inventory

Monitoring and Diagnosis: A key prerequisite for any identification of process deviations is the ability to identify process trends. In the case of batch operations, such trends will give clues to the progress of a batch and, if available in a timely fashion, can lead to on-line corrections which can save reprocessing steps or wasted batches. Effective knowledge based methods for identifying trends from raw process data need to be developed, directed specifically at the wide dynamic excursions and trajectories found in batch operations.

Although considerable work has been focused on fault diagnosis in the continuous process setting, attention needs to be directed at the special opportunities and needs of batch operations with their step/task structure. For instance, timely forecast of the delayed or early completion of a task can lead to corrective action which minimizes the impact of that delay or exploits the benefits of early completion. For this to occur, the diagnosis system must be able to extract that forecast from the trend information presented to it. As noted earlier, although the monitoring, diagnosis, and control blocks must be integrated to achieve maximum benefit, such integrated frameworks remain to be developed.

Control: The control of nonlinear batch operations such as batch reaction remains a major challenge since typically such batch reaction steps involve complex kinetics, and parameter values which evolve over time and, thus, are not well understood or rigorously modeled, Consequently, the key to effective batch control is to develop nonlinear models which capture the essential elements of the dynamics. Wave propagation approaches which have been investigated in the context of continuous distillation and tubular reactors offer promise in selected batch operations. Schemes for identifying and updating model parameters during the course of regular operations and for inferring properties from indirect measurements are as important in the batch domain as they are in the continuous. The use of neural networks and fuzzy logic approaches appear to offer real promise [26].

Although the trajectory optimization problem has been the subject of research for several decades, the numerics of the optimization of DAE systems with discontinuities remains an area for fruitful research. Recent progress with batch distillation is very encouraging (see [13] in these proceedings) but the routine optimization of the operation of such complex subsystems remains a challenge given that such applications often involve complex mixtures, multiple phases, and poorly characterized vapor liquid properties.

Advances in Tools

The design and scheduling applications discussed in the previous sections rely critically on the availability of state-of-the-art tools for discrete optimization, process simulation, and intensive input/output information processing. Indeed, the scope and complexity of the applications which must eventually be handled in order to fully exploit the potential for computer aided batch process and plant design and computer integrated operations are beyond the capabilities of existing methodology and software implementations. Therefore, the process systems engineering community will need to take a leadership role not only in applications development but also in the design and creation of the enabling tools. In this section, we briefly review essential tool development needs in the areas of optimization, simulation, and information processing.

***Optimization Developments*:** The preliminary design, retrofit, plant layout, scheduling, and trajectory optimization applications all are at root large scale 0-1 decision problems with linear and nonlinear constraints. Indeed, the solution of high dimensionality MINLP and MILP problems with various special structures is a pervasive and key requirement for batch process systems engineering. Unfortunately, the limitations of contemporary general purpose algorithms make the routine solution of problems with over 200 0-1 variables impractical. Indeed as shown in Figure 5, although computing power has grown considerably in the last two decades the capabilities of general purpose solvers for discrete mathematical programming problems have not kept pace. Thus, since applications with hundreds of thousands of 0-1 variables can readily arise in practice, it is clear that general purpose solvers are not the answer. Instead as shown by recent accomplishments within and outside of chemical engineering, a high degree of exploitation of problem structure must be undertaken in order to achieve successful, routine solution. Such enhancements of solution efficiency typically involve not only reformulation techniques, exploration of facets, cut exploitation, and decomposition techniques but also use of special algorithms for key problem components, specialized bounding techniques, primal/dual relationships, graph theoretic constructions and very efficient implementations of key repetitive calculations. Since the software development effort involved in designing and implementing a special purpose solver, tailored for a specific application, which employs all of these enhancements is very large, it is essential that a framework and high level tool kit for algorithm developers be created for efficiently building and verifying tailored algorithms [18]. The core framework of such a solver would consist of the branch and bound structure as this lies at the root of all 0-1 problem solution strategies but integrated within this framework would be a range of algorithmic tools, including features for exploiting parallel and distributed computing, data compression techniques, and caching techniques. In view of the major effort such a development would entail, it is essential that all software components which are available or extractable from the existing commercial and

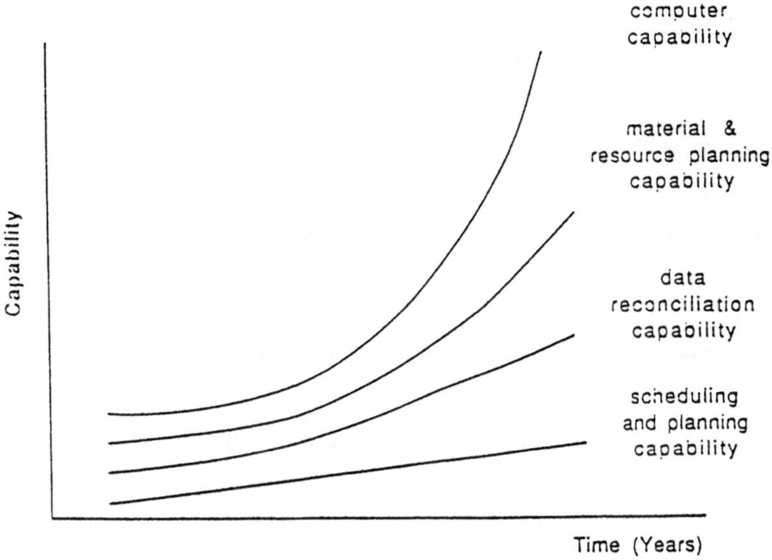

Figure 5. Gap between optimization capability and computer capability

academic inventory be incorporated in the proposed framework.

A key feature of the proposed framework would be the provision of a capability for systematically testing the performance of any particular tailored algorithm and thus discovering and exposing its weak points. In the literature, developers of specialized 0-1 solution algorithms typically only report computational results for a small number of problems, perhaps those which exhibit the most favorable performance, and from these draw broad conclusions about the potential of the strategy for a whole class of problems. Unfortunately, for combinatorial problems such generalizations are in almost all cases invalid. Clearly, in studying an algorithm it is important not only to identify the structural and data features which make it particularly effective but also to identify those for which its performance will substantially deteriorate. This is especially important for industrial application in an operating environment where reliability and predictability are critical for acceptance and continued use of a technology. To facilitate such rigorous testing, Pekny et al [18] propose the creation of an adversary, possibly built using AI methods and genetic algorithms, which would purposely attempt to find data instances that would lead to algorithm performance deterioration. In view of the practical implications of such a capability, its investigation should be accorded the highest priority for future research. It may indeed be an excellent opportunity for collaborative work between several systems research groups.

Finally, in addition to providing a framework for efficient construction of 0-1 solution algorithms for use by an expert, a shell needs to be provided which will allow the application user to employ the tailored algorithm without concern for its fine technical detail. This shell should also provide the user with capabilities for interpretation of the quality and robustness of the solution. Although a general LP type sensitivity analysis is not available for discrete optimization problems, intelligent bounds and procedures for generating approximate solutions should be developed which might generate sensitivity-like information under the control of a rule based system.

Process Simulation Developments: While analytical and algebraic models of the MILP and MINLP form can be extremely powerful tools for design and schedule optimization, such models generally are simplification and approximations of more complex physical phenomena and decision processes. Thus the solutions generated using these models must be viewed as good estimates which ultimately must be refined or at least validated using more detailed models described in terms of differential algebraic equations, stochastic elements, and detailed operating procedures. In the continuous processing domain, process simulation systems have served as vehicles for the creation of such more detailed models. The BATCHES system (see [3] as reported at this ASI) offers such a tool for the simulation of combined continuous/discrete batch operations and recent developments at Imperial College also point in that direction [14]. While BATCHES is an effective, practical tool in its present state, it is limited in three aspects: efficient solution of large scale DAE systems with frequent discontinuities, flexible description of tailored batch operating decision rules, and optimization capability.

BATCHES marches through time by integrating the currently active set of DAE's from one state/time event to the next using a widely available DAE solver. Since in principle the set of active DAE's changes with each event, the previous solution history can not be directly utilized in restarting the integration process at the completion of the logic associated with the current event. The resulting continual restarting of the solver can be quite demanding of computer time, especially for larger scale nonlinear models. Research is thus needed on more efficient ways of taking advantage of previous solution history during restart, say, in the form of suitably modified polynomials, for those equations sets that remain unchanged after an event. Further continuing research is of course also needed in developing more efficient ways of solving large structured DAE's.

One of the key differences between batch process simulation and conventional dynamic simulation is that in the batch case one must model the operational decisions along with the processing phenomena themselves. In BATCHES, a wide range of options is provided under which unit to task allocations, resource allocations, and materials transfers, and batch sequencing choices are defined and executed. However, since any finite choice of options can not encompass all possibilities, the need does arise for either approximating the desired logic using combinations of the available options or developing special purpose decision blocks. Because of the extensive information needs of such decision blocks, creation of such blocks is beyond the scope of a typical casual user. The need thus exists for developing a high level language for describing tailored decision blocks which could be employed in the manner in which "in-line" FORTRAN is now used within several of the flowsheeting systems. A natural language like rule based system would appear to be the most likely direction for such a development.

Finally, once a batch process simulation model is developed and exercised via a set of case studies, a natural next step would be to use it to perform optimization studies for design, retrofit, or operational improvements. This idea is, of course, a natural parallel to developments in the steady state simulation domain. Regrettably, combined continuous discrete simulations do not directly lend themselves to the SQP based strategies now effectively exploited for steady state applications because of three features: the presence of state and time event discontinuities, the frequent model changes which are introduced as the active set of equipment or their modes of operation changes over simulation time, and the discontinuities introduced by the Monte Carlo aspects of the simulation. As a result the optimization of combined continuous-discrete simulations has to date only been performed using direct search methods which treat the simulation model as a black box. The challenge to the optimization community is to develop strategies which would allow more direct exploitation of the structure of the simulation model (the "gray" box approach) for more effective optimization.

Information Processing Developments: One of the key characteristics of batch operations is the large amount of information required to describe a design or scheduling application. This information includes detailed recipe or task network specifications for each product, the equipment specifications, task suitabilities and inter-unit connectivities, the operating decisions and logic, the production requirements and the initial condition of the entire plant. The quantitative description of the operation of a batch plant over a specified time period is also quite information intensive as such a description must cover the activity profile of each processing unit, transfer line or mechanism, and resource over that time period. One of the challenges to the systems community is to develop effective means of generating, validating, maintaining, and displaying this mass of information in a way which enhances understanding of the operation or design. Graphical animation as is provided in BATCHES does help in qualitative assessment that the plant is operating in a reasonable way. The colorful Gantt charts and resource profile charts made available in contemporary scheduling support software such as [21] are certainly helpful. Nonetheless these display tools provide the information about the operation essentially as a huge flat file and, thus, overload the user with detail. Intelligent aids are needed that would help in identifying key operational features, bottlenecks, and constraints and thus focus the users attention on critical problem elements. An object oriented approach which allows one to traverse in the information domain both in extent and in depth, as dictated by analysis needs, may be a useful model.

A further research need is to develop a flexible data model of batch operations which would provide a structured, common information framework for all levels and tools employed in the batch operations CIM hierarchy. A prototype data model for batch scheduling applications was proposed by Zentner [33] and a simulation specific data model implemented using a commercial data base is employed within BATCHES. The need for such a data model which is application independent was recognized in [5] in the course of executing a limited integration study. The key, of course, is application independence. The step level description required in a BATCHES simulation differs only in some further details from the description required for a sequencing control implementation. The step level description could also be employed in a rescheduling application, while a task level aggregation might suffice for master scheduling purposes or for a retrofit design application. This data model should be supported with generalized consistency checking and validation facilities which are now scattered across various applications and tools such as the BATCHES input processor, the input preprocessors developed for various scheduling formulations, and the detailed sequencing control implementation codes provided by control vendors. Such unified treatment of process and plant information clearly is an essential prerequisite for computer aided engineering developments as a whole and CIM implementations in particular.

Summary

In this paper, the future directions of technical developments in batch process systems engineering have been motivated and outlined. In the process design domain, methodology to support the synthesis and definition of task networks, approaches for quantitatively balancing plant flexibility with demand and product uncertainties, retrofit design aspects including heat integration, and quantitative approaches to plant layout were proposed for investigation. In the operations domain, integration of the levels of the CIM hierarchy, especially of the multiplant, individual plant and plant reactive scheduling levels and of the monitoring, diagnosis and control levels were offered as high priority developments. The general problem of explicit treatment of uncertainty in the CIM hierarchy is a highly appropriate subject for basic study at the conceptual and quantitative levels. Operations applications requiring further attention include treatment of time within resource constrained formulations, a broader investigation of reactive scheduling strategies, and the study of multiplant scheduling formulations. Intelligent trend analysis to support diagnosis and further developments in low order nonlinear modeling for control purposes also offer significant promise for batch operations. In the area of tool development, the need for a flexible and well integrated framework for discontinuous optimization was proposed, including provisions for both a developer's and an algorithm user's view of the tool and the provision of an adversary feature for algorithm testing. In the simulation domain, requirements for improvements in the solution of DAE systems, flexible description of operational rules, and optimization capabilities were noted. Finally, in the information processing area, the case was made for intelligent aids for plant and process data analysis, visualization, and interpretation as well as the need for a batch operations data model, which would form the basis for computer aided engineering developments. The scope of these themes is such as to offer challenging and fruitful research opportunities for the process systems engineering community well into the next decade.

Acknowledgment

This presentation benefited considerably from the ideas on these topics which have been developed by my colleagues in the Purdue Computer Integrated Process Operations Center, namely, Profs. Ron Andres, Frank Doyle, Joe Pekny, Venkat Venkatasubramanian, Dr. Mike Zentner, our collective graduate student team, and our supportive industrial partners.

References

1. S. Bacher: Batch and Continuous Process Design. Paper 33d, AIChE National Mtg., Houston (April, 1989)
2. L. Biegler: Tailoring Optimization algorithms to Process Applications. Comput. Chem. Eng., ESCAPE 1 supplemental volume (1992)
3. S. Clark and G. Joglekar: General and Special Purpose Software for Batch Process Engineering. This volume p. 376

4. B.J. Cott and S. Macchietto: A General Completion Time Determination Algorithm for Batch processes. AIChE Annual Meeting, San Francisco, (Nov.1989)
5. C.A. Crooks, K. Kuriyan, and S. Macchietto: Integration of Batch Plant Design, Automation, and Operation Software Tools. Comput Chem. Eng., ESCAPE-1 supplemental volume (1992)
6. J. B. Edgerly: The Top Multinational Chemical Companies. Chemical Processing, pp.23-31, (Dec.1990)
7. S. Jayakumar and G.V.Reklaitis: Graph Partitioning with Multiple Property Constraints for Multifloor Batch Plant Layout. Paper 133d, AIChE Annual Mtg. Los Angeles (Nov., 1991). See also, Comput. Chem. Eng., 18, 441-458 (1994)
8. S. Jayakumar: Chemical Plant Layout via Graph Partitioning. Ph.D. Dissertation, Purdue University, May, 1992
9. K.B. Kanakamedala, V. Venkatasubramanian, and G.V.Reklaitis: Reactive Schedule Modifications in Multipurpose Batch Chemical Plants, Ind. Eng. Chem. Res., 32, 3037-3050 (1993)
10. E. Kondili, C.C. Pantelides, and R.W.H. Sargent: A General Algorithm for Scheduling Batch Operations. Comput. Chem. Eng., 17, 211-229 (1993)
11. I. Lee and G.V.Reklaitis: Optimal Scheduling of Batch Processes for Heat Integration. I: Basic Formulation. Comput. Chem. Eng., 19, 867-882, (1995)
12. K.B. Loos: Models of the Large Chemical Companies of the Future. Chemical Processing, pp. 21-34 (Jan.1990)
13. S. Macchietto and I. M. Mujtaba: Design of Operation Policies for Batch Distillation. This volume, p. 174
14. C.C. Pantelides and P.J. Barton: The Modeling and simulation of Combined Discrete/Continuous Processes. PSE'91, Montebello, Canada (August, 1991)
15. S. Papageorgaki and G.V.Reklaitis: Retrofitting a General Multipurpose Batch Chemical Plant. Ind. Eng. Chem. Res. 32, 345-361 (1993)
16. S. Papageorgaki and G.V.Reklaitis: Optimal Design of Multipurpose Batch Plants: Part 1, Formulation and Part 2, A Decomposition Solution Strategy. Ind. Eng. Chem. Res., 29, 2054-2062, 2062-2073 (1990)
17. S. Papageorgaki, A.G. Tsirukis and G.V.Reklaitis: The Influence of Resource Constraints on the Retrofit Design of Multipurpose Batch Chemical Plants. This volume, p. 150
18. J. Pekny, V. Venkatasubramanian, and G.V.Reklaitis: Prospects for Computer Aided Process Operations in the Process Industries. Proceedings of COPE-91, Barcelona, Spain (Oct, 1991)
19. H. J. Reinhart and D.W.T. Rippin: Design of Flexible Batch Plants. Paper 50e, AIChE Nat'l Mtg, New Orleans (1986)
20. G. V. Reklaitis: Progress and Issues in Computer Aided Batch process Design. In Proceedings of Third Int'l Conference on Foundations of Computer Aided Process Design, CACHE-Elsevier, New York, pp.241-276 (1990)
21. Scheduling Advisor, Stone & Webster Advanced Systems Development Services, Boston, MA 202210. (1992)
22. N. Shah and C. C. Pantelides: Design of Multipurpose Batch Plants with Uncertain Production Requirements. Ind. Eng. Chem. Res., 31, 1325-1337 (1992)
23. N. Shah and C. C. Pantelides: Optimal Long Term Campaign Planning and Design of Batch Plants. Ind. Eng. Chem. Res., 30, 2308-22321(1991)
24. K. Tsuto and T. Ogawa: A Practical Example of Computer Integrated Manufacturing in Chemical Industry Japan. PSE'91, Montebello, Canada (August, 1991)
25. J.A. Vaselnak, I.E. Grossmann, and A.W. Westerberg: Heat Integration in Batch Processing. Ind. Eng. Chem. Process Des. Dev., 25, 367-366(1986)
26. V. Venkatasubramanian: Purdue University, School of Chemical Engineering, private communication (May, 1992)
27. V.T. Voudouris and I.E. Grossmann: Mixed Integer Linear Programming Reformulations for Batch Process Design with Discrete Equipment Sizes. Ind. Eng. Chem. Res., 31, 1315-1325, (1992)
28. H.S. Wellons and G.V.Reklaitis: The Design of Multiproduct Batch Plants under Uncertainty with Staged Expansion. Comput. Chem. Eng., 13, 115-126 (1989)
29. H.S. Wellons: The Design of Multiproduct Batch Plants under Uncertainty with Staged Expansion, PhD Dissertation, Purdue University, School of Chemical Engineering, December, 1989
30. T. J. Williams: A Reference Model for Computer Integrated Manufacturing: A Description from the Viewpoint of Industrial Automation. ISA, Research Triangle Park, N.C.(1989)
31. T. J. Williams: Purdue Laboratory for Applied Industrial Control, private communication (April, 1992)
32. M. Zentner and G.V.Reklaitis: An Interval Based Mathematical Formulation for Resource Constrained Batch Scheduling. This volume, p. 779
33. M. Zentner: An Interval Based Framework for the Scheduling of Resource Constrained Batch Chemical Processes. Ph.D. Dissertation, Purdue University, School of Chemical Engineering, May, 1992

Role of Batch Processing in the Chemical Process Industry

Michel Lucet, Andre Charamel, Alain Chapuis, Gilbert Guido, Jean Loreau

Rhône-Poulenc Industrialisation, 24 Avenue Jean Jaurés, 69151 Décines, France

Abstract: As the importance of batch processing increases in the Chemical Process Industry, plants are becoming more specialized, equipment is standardized, computer aided process operations methods are being improved and more widely used by manufacturers.

In 1980, the management of Rhône-Poulenc decided to develop fine chemical, specialty chemistry, pharmaceutical, and agrochemical products rather than petrochemicals. Twelve years later, Rhône-Poulenc has become an important producer of small volume products and has acquired certain skills in this domain.

Keywords: Batch equipment; standardization; batch plant types; operations sequences; flexibility.

Batch process for low tonnage

A majority of chemical products whose production rates are less than 1000 t/y are unable to support either significant amounts of research and development or major capital investments by themselves. Therefore, they are processed in existing batch plants in a very similar mode to the laboratory experiments involved in their invention.

Different kinds of batch plants

We distinguish four kinds of batch plants depending upon the average amount of each product processed during one year.

1. Pilot batch plants (zero to 30 t/y of each product)
   These plants are devoted to new products: samples are made to test the market.
   Products are ordered in very small quantities.
2. Flexible and polyvalent batch plants (30 to 300 t/y of each product)
   These plants are used to process a large number of products.
   The recipes of the different products may vary significantly from one product to another.
3. Multiproduct batch plants (300 to 700 t/y of each product)
   These plants run a small number of long campaigns. Often, the recipes are very similar from one campaign to another.
4. Specialized batch plants (700 t/y and above)
   The plant processes the same product all year long.

Standardization of equipment

We need a maximum level of flexibility to respond to the random demand for chemical products. We have to be able to process a given recipe in a maximum number of different plants. So, we have defined a number of standard equipment that will be the same in different plants; they are:

- Reactor under pressure
- Reactor for corrosive compounds
- Reactor at atmospheric pressure
- Distillation linked with reactor
- Rectification
- Crystallization
- Forming of solids, casting
- Liquid-liquid extraction

In Figure 1, a schematic diagram of a standard pressure reactor is given while Table 1 exhibits the statistics about the frequency of use of standard equipment types which arises in the processing of 87 of our products.

This standardization of equipment allows us to have uniform control procedures transferable from one plant to another.

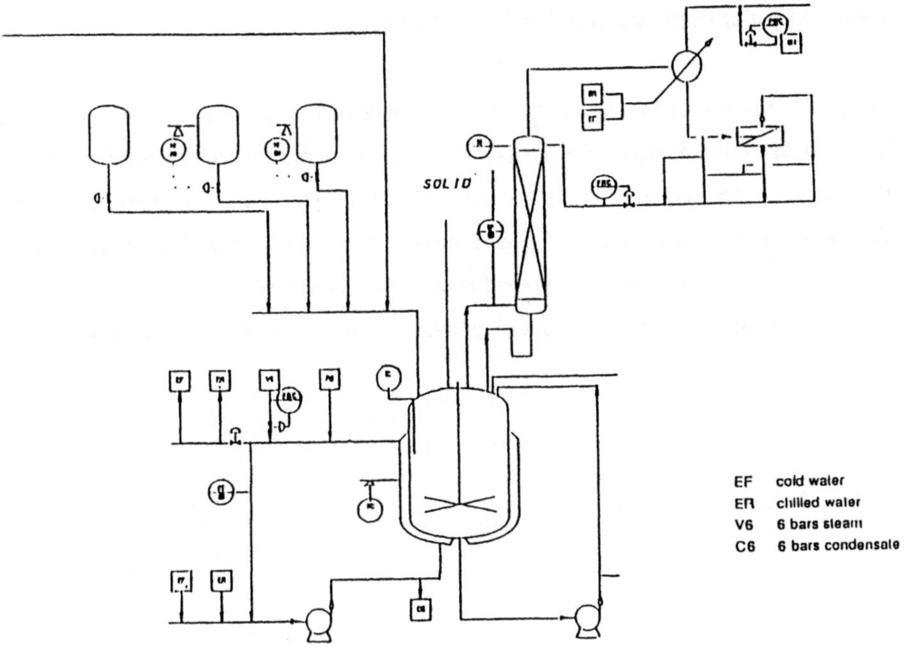

Figure 1: Pressure reactor

Table 1. Statistics for over 87 products - Use of equipments

Pressure reactor	23
Corrosion resistant reactor	25
Atmospheric reactor	62
Distillation linked with reaction	74
Rectification	52
Crystallization	46
Flaking, casting	35
Liquid-liquid extraction	35
Phases	345
Average phases/product	4

Favored links between standard equipment

In Figure 2, a statistical overview of the sequences of equipment use is presented. Some favored links are immediately apparent. The number of standard equipment in batch plants have to reflect these figures, so that the plant is technically adaptable to give a good response to the market demand. Moreover, sometimes, some phases of production of new products are slightly changed to fit correctly the equipment that is present in pilot batch plants. When the product is developed independently, the constraint of adaptability to the equipment becomes less and less effective.

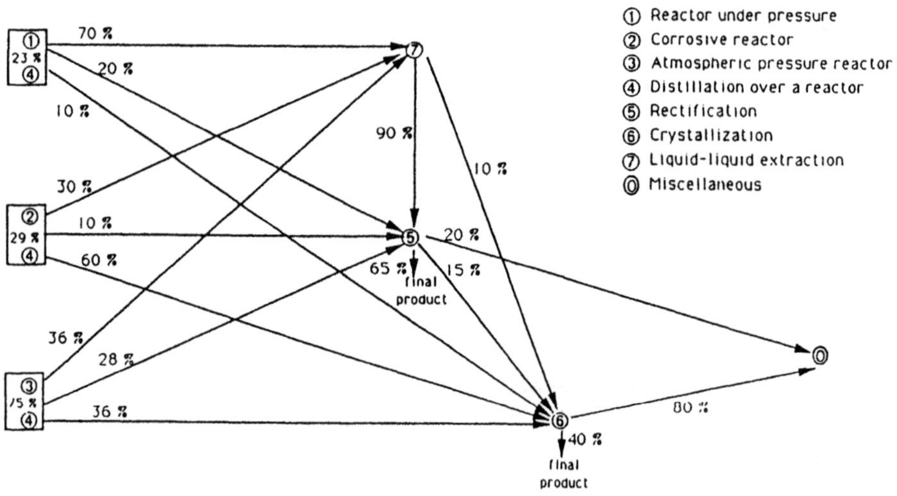

Figure 2. Statistics over 345 processing phases (percentage of consecutive uses of equipments)

The batch diagram logic

The sequence of operations in the production of a product is shown in block diagram form in Figure 3.

A mass balance is made, to yield the quantities transferred from one block to the following one. This information is displayed phase by phase.

Next, a Gantt Chart is drawn showing task by task, subtask by subtask, the occupation time in each equipment. There is also an option to compute the demand on resources such as manpower, electricity, steam, etc ... if these constraints are active in the plant. The program does not handle the constraints on the availability resources. It only shows which subtask requires a given resource and allows the user to slightly modify the chart.

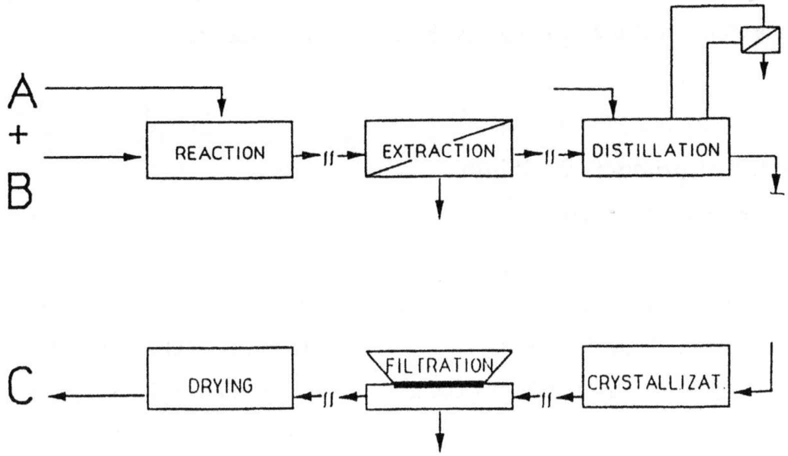

Figure 3. Sequence of operations

The control of batch processes

The sequence of operations is also described at the level of the subtask, for example:

- open the output valve
- wait until the mass of the reactor is less than 1 t
- wait 10 more minutes
- close the output water
- etc...

So the whole operation is logically represented by block diagrams, then tasks, then subtasks, then sequence of operations.

Optimization of batch processes

There are different levels of optimization. For new products, involving small quantities, the optimization involves in the better use of existing plants. For products more mature that are processed in bigger quantities and more often, there is a need to optimize the process by itself. This optimization is mainly obtained by improving the reaction part of the process from one run to the following one. We need automatic data collection from the process and computer assisted analysis of the present and past data to achieve better running parameters.

Industrial life of product - flexibility

Processing in different batch plants

If frequently happens that some phases of processing of a given product are made in different plants through the world. It also happens that some phases are processed in plants of contractors. Thus the planning manager has to take in account a large number of possibilities in processing costs and time delay.

Flexibility in policy storage

There is a real storage policy for products run in long campaigns. The storage of the final products has a cost and this depends upon the storage capacity for this kind of products. For intermediate storage during a short campaign we sometimes use temporary storage - a cart for example.

Conclusion

As chemical products tend to be precisely tailored to sharp specifications. the number of small products is increasing, the processing of these batch products is at the moment far from being optimized as it is for continuous large products. Even if some standardization is made at the moment, each product by itself cannot justify extensive studies. So we have to develop and improve automatic method to optimize these processes.

Present Status of Batch Process Systems Engineering in Japan

Shinji Hasebe and Iori Hashimoto

Department of Chemical Engineering, Kyoto University, Kyoto 606-01, Japan

Abstract: Rapid progress in computer technology has had tremendous effect on batch plant operation. In this paper, the present status of batch plant operation in Japan is reported first by referring to the questionnaires. The main purpose of the introduction of CIM in chemical plants is to produce various kinds of products with a short lead time without increasing inventory. In order to accomplish this purpose, the development of a sophisticated scheduling system is vital. The role of the scheduling system in CIM is discussed next. In addition to the development of computer systems, development of hardware for the batch plant suitable for flexible manufacturing is also important to promote CIM. Recently, a new type of batch plant called a "pipeless batch plant" has received great attention from many Japanese companies. The characteristics of pipeless batch plants and their present status are explained, and a design method and future problems are discussed.

Keywords: Batch plant, computer integrated manufacturing, scheduling, pipeless plant

1. Introduction

An increasing variety of products have been produced in batch plants in order to satisfy diversified customer needs. The deadline requirements for the delivery of products have also become increasingly severe. In order to deliver various kinds of products by a given due date, each product has to be stocked or frequent changeovers of the plant operation are required to produce required products just in time. As a result, the inventory cost and the changeover cost increase and the productivity of the plant decreases.

In the 1980s, the rapid progress of computer technology accelerated the introduction of computer control systems even into the many small- to medium-sized batch plants. And it contributed to the reduction of manpower. In recent years, in order to cope with increases in product types, the development of Computer Integrated Manufacturing (CIM) system is being promoted actively in both continuous and batch plants. The dominant purpose of the development of CIM is to produce various kinds of products under severe time constraints without increasing the amount of inventory or decreasing the plant efficiency.

In this paper, the present status of batch plant operation in Japan is reported first by referring to the questionnaires which were distributed by the Society of Chemical Engineers, Japan. Then the problem of CIM in batch chemical plants is discussed from the viewpoint of the Just-in-Time (JIT) production system which is successfully used in assembly industries. Next, considering the important role a scheduling system plays in CIM, the present status of the study on scheduling problems in Japan and future problems related to scheduling are discussed.

In addition to the development of computer systems, development of hardware for the batch plant suitable for flexible manufacturing is also important to promote CIM. Recently, a new type of batch plant called a "pipeless batch plant" has received great attention as a new-generation production system. The pipeless plant has a structure which is suitable for the production of various kinds of products. The difference between the ordinary batch plant and the pipeless plant, and the present status of the introduction of the pipeless plant in Japanese chemical industries are reported.

2 Present Status of Batch Plant Operation

In order to obtain the information on the present status of batch plant operation, three questionnaires were distributed by the plant operation engineering research group of the Society of Chemical Engineers, Japan. The first questionnaire was distributed in 1981 and the purpose was to obtain information on the present status of batch plant operation and on future trends.[14] Plant operation using cathode-ray tubes (CRT operation) became familiar in the '80s instead of operation using a control panel. The second and third questionnaires were distributed in 1987 and 1990 to obtain information on the present status of CRT operation [15],[16], and the problems and future roles of plant operators. In this chapter, the present status of batch plant operation in Japan is discussed by referring to the results of these questionnaires.

Questionnaire #1 was sent to 51 leading companies that use batch plants; 60 plants from 34 companies replied. The classification of the plants is shown in Fig. 1. Questionnaires #2 and #3 were sent to companies that have continuous and/or batch plants in order to investigate the present status of and future trends in CRT operation. The classification of the plants is shown in Table 1.

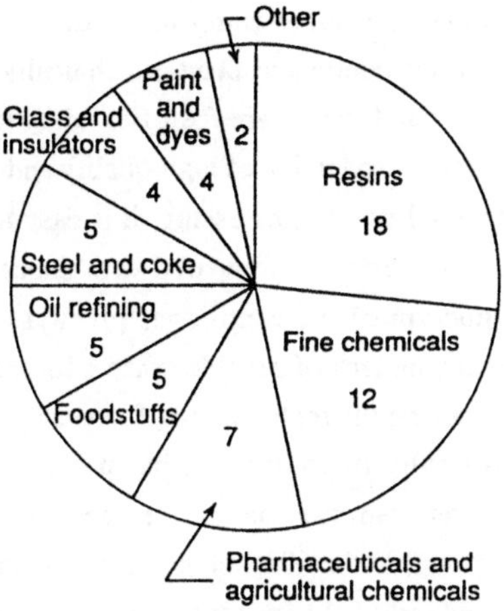

Figure 1. Classification of Batch Plants (Ques.#1) f Batch Plants (Ques.#1)

Table 1. Classification of the Plants (Ques. #2 and #3)

Type of plants	Number of plants	
	Questionnaire #2	Questionnaire #3
(Group 1) continuous chemical plant	18	18
(Group 2) batch chemical, pharmaceutical, or food-processing plant	23	42
(Group 3) oil refining or coal-gas generation plant	22	20

Usually, a product developed at the laboratory is first produced by using a small-size batch plant. Then a continuous plant is used according to increases in the production demand. Should a batch plant be regarded as a transitional plant in the change from a pilot plant to a continuous plant? In order to clarify the present status of batch plants, questionnaire #1 first asked about the possibility of replacing a batch plant with a continuous plant if the current batch plant would be rebuilt in the near future. For only 18% of the plants, replacement by a continuous plant was considered; most of them were pharmaceutical and resin plants.

Figure 2 shows the reasons why batch plants will still be used in the future. Except in cases where the introduction of the continuous plant is technically difficult, all of the reasons show that batch plants have some advantages compared with continuous plants. This means that the batch plant is used even if the continuous plant is technically feasible.

The dominant advantages of batch plants are their flexibility in producing many kinds of products, and their suitability for the production of high-quality and high value-added products. As the material is held in the vessel during processing, it is easy to execute a precise control scheme and perform complicated operations compared with the continuous plant. Technical problems hindering the introduction of the continuous plant are the difficulty of handling materials, especially powders, and the lack of suitable sensors for measuring plant conditions.

The main reasons for considering the replacement of a batch plant with a continuous plant were low productivity and difficulty in automating batch plants. In order to increase the productivity of multiproduct or multipurpose batch plants, generation of an effective operation schedule is indispensable. However, in the 60 batch plants that responded to questionnaire #1, mathematical methods were used only for a quarter of the plants to determine the weekly or monthly production schedule. A Gantt chart was used for a quarter of the plants, and for half of the plants an experienced plant manager generated the schedule by hand. The situation has changed drastically during the last decade due to progress in CIM discussed in chapter 3.

In order to automate batch plant operation, introduction of computer control system is indispensable. In questionnaire #1, computer control systems had been introduced at 56% of the plants. The dominant purposes of the introduction of computers were manpower reduction, quality improvement, and safer and more stable operations, as shown in Fig. 3. Productivity improvement did not gain notice, because the introduction of computers was limited to only

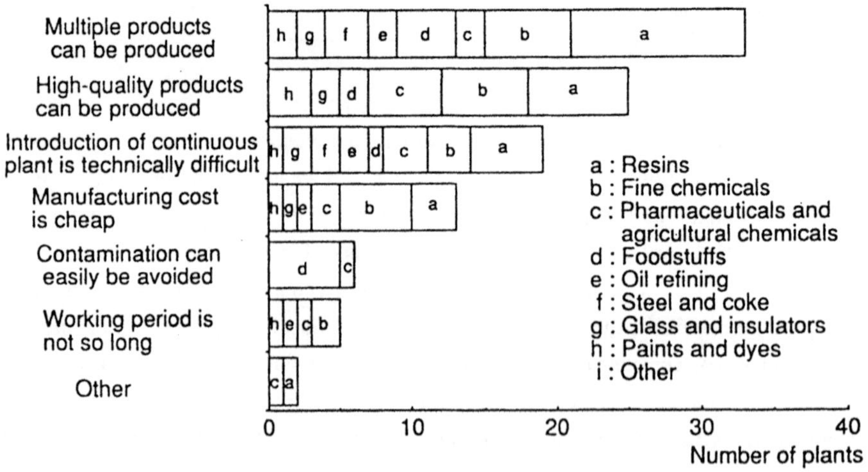

Figure 2. Reasons for Using Batch Plant

batch equipment or batch plants. In order to improve plant productivity by introducing a computer, it is required to develop a factory-wide computer system, which may be upgraded to CIM.

Factors obstructing the introduction of computer control in 1981 were the lack of suitable software, high computer costs, and lack of suitable sensors and actuators as shown in Fig. 4. Due to rapid progress in computer technology, all of the obstacles shown in Fig. 4 may be resolved now except for the lack of suitable sensors. Development of methods which can estimate unmeasurable variables by using many peripheral measurable variables remains as an interesting research area.

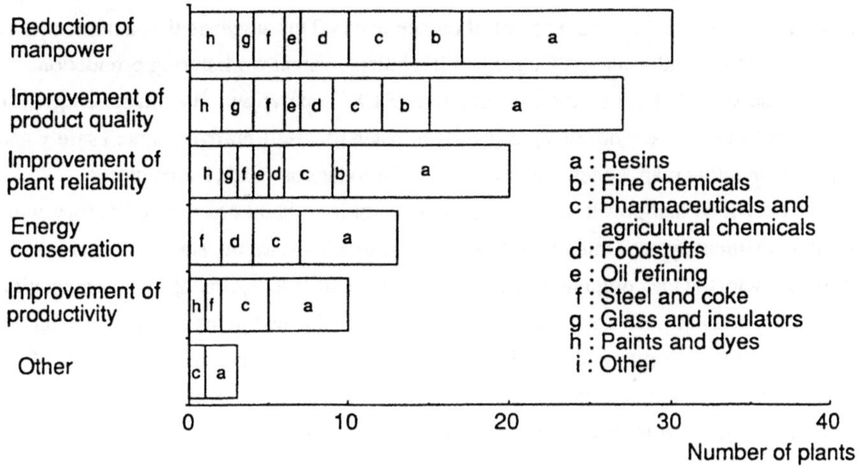

Figure 3. Purpose of Using Computers (Ques.#1)

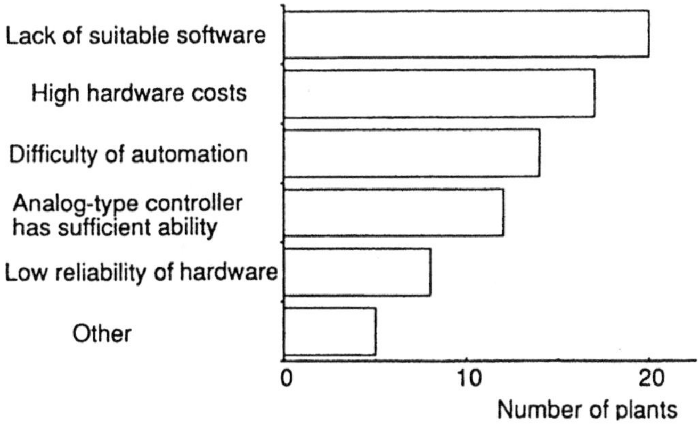

Figure 4. Factors Obstructing the Introduction of Computers (Ques#1)

Questionnaires #2 and #3 were premised on the introduction of computer control system. From Fig. 5, it is clear that the introduction of CRT operation has advanced significantly since 1982. Especially in plants of group B in Table 1 (batch plants), CRT operation was introduced earlier than in plants of other groups.

Due to increases in the number of products, it became a crucial problem to improve sequence control systems so that the addition of new sequences and sequence modification can be easily executed by operators. From Fig. 6, showing the purposes of introduction of CRT operation, it is clear that CRT operation was introduced to many batch plants in order to improve sequence control function.

The other chief purpose of the introduction of the CRT operation was the consolidation of the control rooms in order to reduce the number of operators. However, for batch plants the consolidation of the control rooms was not often achieved. This suggests that automation of the batch plant was very difficult and the plant still requires manual work during production.

Figure 7 shows the effect of the introduction of CRT operation. By introducing CRT operation, manpower can be reduced significantly in batch plants. Improvements in safety and product quality are other main beneficial effects of the introduction of CRT operation.

In order to reduce manpower still further and to improve plant efficiency, factory-wide computer management system must be introduced. Figure 8 shows the state of the art of the introduction of factory-wide computer management systems in 1987. Although implementation of such a system had not been completed yet, many companies had plans to introduce such a system. This system is upgraded to CIM by reinforcing the management function. The purposes of the introduction of factory-wide computer management systems were reduction of manpower and speed-up of information processing.

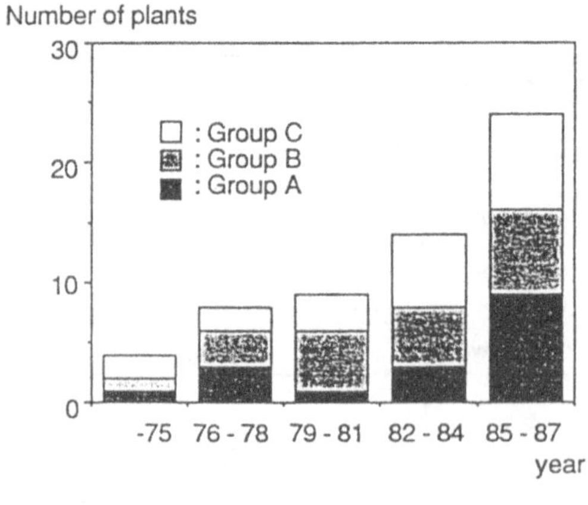

Figure 5. Introduction of CRT Operation

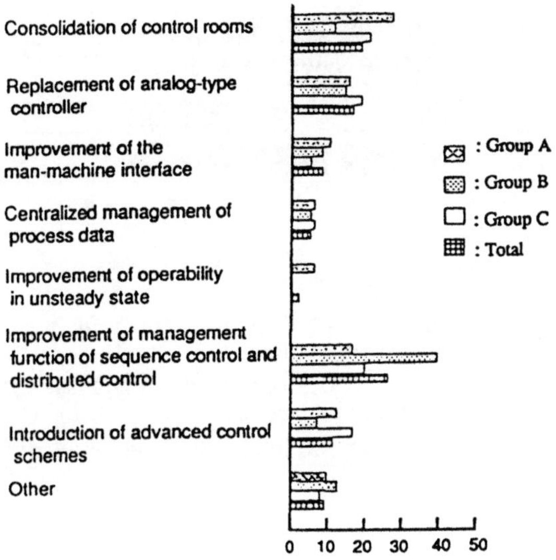

Figure 6. Purpose of CRT Operations

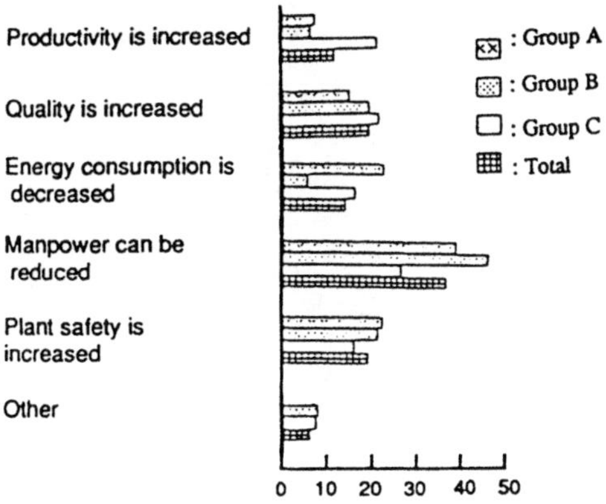

Figure 7. Effect of CRT Operation

Automation of plant operation reduces the number of operators. For a batch plant, because the plant condition is always changing, the possibility of malfunctions occurring is larger than in a continuous plant. Therefore, the contribution of the operator to the plant operation of a batch plant is larger than in a continuous plant. In other words, the role of operator becomes very important.

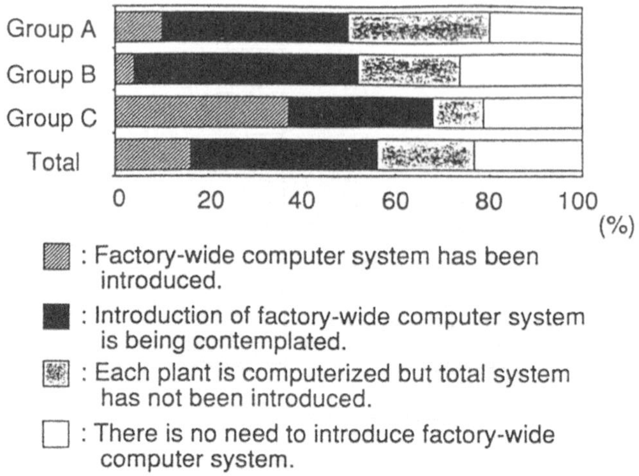

Figure 8. Introduction of Factory-wide Computer System

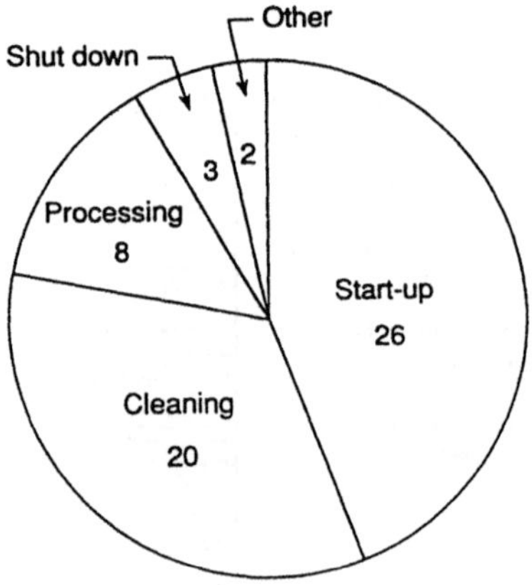

Figure 9. Occurrence of Malfunctions

Figure 9 shows when the malfunctions occurred. It is clear from this figure that malfunctions occurred during unsteady state operations such as start-up and cleaning. Half of these malfunctions were caused by operator errors, and 20 % were caused by trouble in the control systems. The importance of the continued training of operators and the preparation of revised, more precise operation manuals were pointed out by many companies.

In the early days of CRT operation, the necessity of the control panel as a back-up for CRT operation had been discussed widely and often. Many managers feared operators would not be able to adapt to CRT operation. However, they have been happily disappointed, and recently CRT operation without control panels has become common. It is clear that plant operation has become more difficult and complicated, because a sophisticated control scheme has been introduced and interaction between plants strengthened. What is the future role of operators in an advanced chemical plant? Figure 10 shows future plans for plant operation. For 60 to 65% of the plants, it is expected that the operation of the plant will be executed by engineers who are university graduates and have sufficient knowledge of the plant and control systems. For about 20% of the plants, plant operation is expected to become easier as a result of automation and the introduction of operation support systems.

It becomes clear from questionnaire #3 that most of the batch chemical plants are operated 24 hours per day in order to use the plant effectively. And at almost all the plants, the number of operators is the same for day and night operations. Nowadays, the general way of thinking of Japanese young people is changing. Most of the young workers do not want to work at night. Furthermore, young workers have come to dislike entering manufacturing industries, causing labor shortages. By taking these facts into account, much effort should be devoted to reducing night operation. Progress in automation and development of sophisticated scheduling systems may be keys for reducing the number of operators during the night without decreasing productivity.

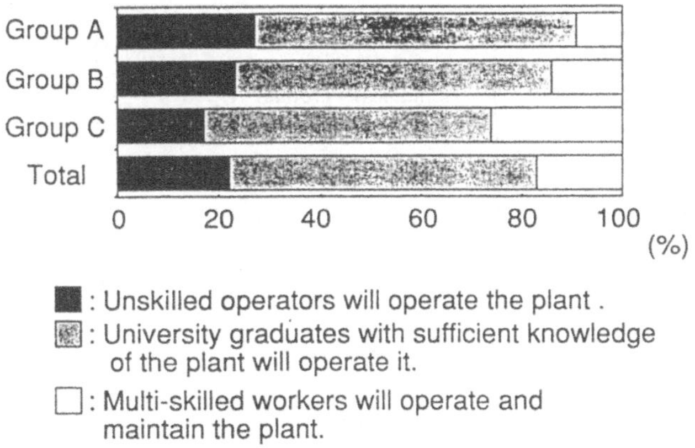

Figure 10. Future Plant Operation

3. Computer Integrated Manufacturing in Batch Plant

In recent years, due to increases in the number of products, inventory costs have become considerably large in order to satisfy rapid changes in production demand. Figure 11 shows an example of the increase of the number of products in the field of foodstuffs [2]. In order to cope with this situation, CIM is being promoted actively in both continuous and batch plants.

Figure 12 shows the purpose of the introduction of CIM in manufacturing industries [4]. From this figure, it may be concluded that the purpose of introducing CIM is to produce more kinds of products with a short lead time without increasing inventory. In order to realize such a system, automation of production is essential, and yet it is not enough. It is equally, or even more important to further promote computerization of the production management system, including the delivery management systems.

First, let us consider the Just-in-Time (JIT) production system, which is actively used in assembly industries and is used successfully to produce various products with a small inventory. The operational strategy of JIT is to produce just the required amount of product at just the required date with a small inventory. In order to produce various products at just the required date, frequent changeovers of operation are required. As a result, the changeover time and the changeover cost increase and the working ratio of machinery decrease. In order not to decrease the productivity of the plant, the following efforts have been undertaken in assembly industries:

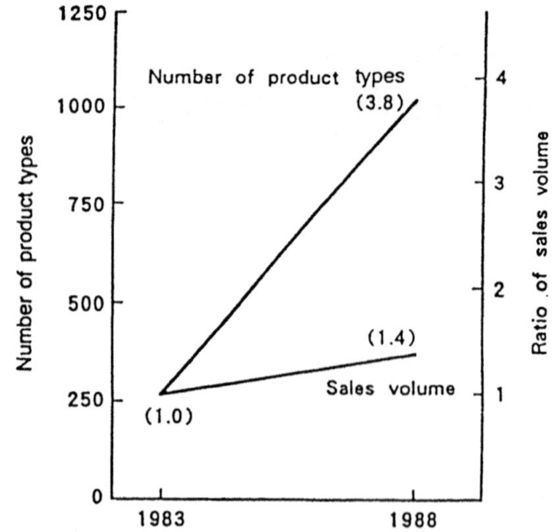

Figure 11. Trend in Number of Products and Sales Volume of Frozen Foods at a Japanese Company

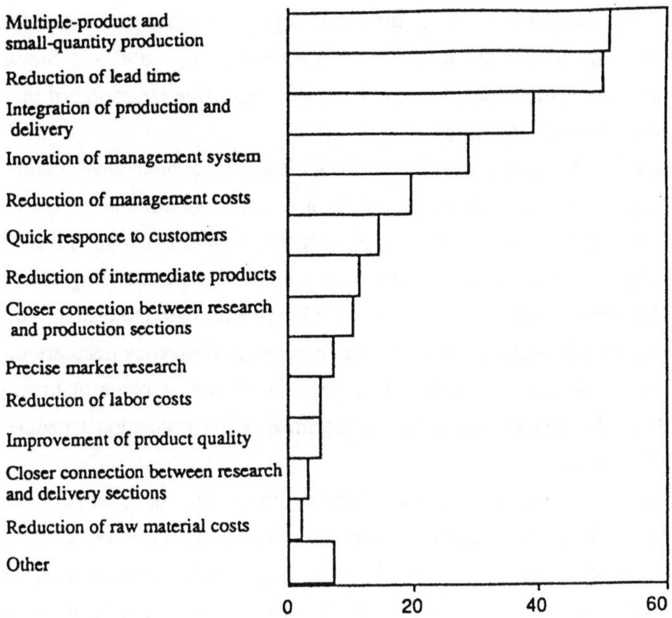

Figure 12. Purpose of the Introduction CIM

1) Improvement of machinery so that changeover time from one product to another is cut greatly,
2) Development of multi-function machines which can execute many different kinds of operations,
3) Training of workers to perform many different kinds of tasks.

In JIT in assembly industries, the reduction of changeover time is realized by introducing or improving hardware, and considerable reduction of inventory is achieved. By introducing multi-function machines, it becomes possible to maintain a high working ratio even if the product type and the amount of products to be produced are changed. It is expected that the benefits obtained from inventory reductions exceed the investment costs necessary for improving hardware.

In assembly plants, a large number of workers are required, and production capacity is usually limited by the amount of manpower and not by insufficient plant capacity. Therefore, in an assembly plant, variations of the product type and the amount of products are adjusted by using the abilities of the multi-skilled workers and by varying the length of the working period. On the other hand, chemical plants require few workers, but a great deal of investment in equipment. Thus having machinery idle is the significant problem. In order to keep high working ratio of machinery, the inventory is used effectively to absorb the variation of the production demand. This is one of the reason why extensive reduction of inventory has not been achieved in chemical industries.

In chemical batch plants, reactors have to be cleaned to avoid contamination when product type is changed. The need for the cleaning operation increases as product specifications become stricter. Furthermore, cleaning of pipelines as well as batch units is required when the product is changed. Efforts for promoting the automation of the cleaning operation have been continued. However, it will be difficult to completely automate the cleaning operation and to reduce cleaning time drastically. Increases in changeover frequency decrease productivity and also increase the required amount of manpower. Therefore, in batch plants much effort has been devoted to reducing changeover time by optimizing the production schedule rather than by improving plant hardware.

The reduction of the amount of inventory increases the changeover time and the changeover cost. Therefore, a reasonable amount of inventory has to be decided by taking into account inventory and changeover costs. In order to accomplish this purpose, the development of a sophisticated scheduling system is vital. And in order to rapidly respond to variations in production demand, inventory status and customer requirements must be transferred to the scheduling system without delay. In other words, systems for scheduling, inventory control, and production requirement management must be integrated. For these reasons, the development of company-wide information systems has been the main issue discussed in the study of CIM in chemical batch plants. The role of the scheduling system in CIM is discussed in the next chapter.

Recently two types of batch plants have been developed to reduce the time and cost of the cleaning operation. One involve the introduction of a "multipurpose batch unit" in which several kinds of unit operations, such as reaction, distillation, crystallization, and filtration can be executed [1]. By introducing multipurpose units, the frequency of material transfer between units can be reduced. However, it should be noted that in a multipurpose unit, only one function is effectively performed during each processing period. For example, the equipment used for distillation and filtration is idle during the reaction period. This means that the actual working periods of many of the components which compose a multipurpose unit are very short even if the entire unit is used without taking any idle time. Therefore, the beneficial characteristics of the multipurpose unit, such as the reduction of pipelines, should be fully exploited in order to compensate for this drawback when a plant using multipurpose units is designed.

The other method of reducing the time and cost of cleaning operation is to reduce pipelines by moving the reactors themselves. Such a plant is called a "pipeless batch plant." The pipeless batch plant consists of a number of movable vessels; many types of stations where feeding, processing, discharging, and cleaning operations are executed; and automated guided vehicles (AGV) to carry vessels from one station to another. Many Japanese engineering companies are paying much attention to pipeless plants from the viewpoint of the flexibility. The characteristics of the pipeless plants and the present status of their development are discussed in chapter 5.

4. Scheduling System of Batch Plants

A scheduling system is one of the dominant subsystems of the production management system. And the computerization of the scheduling system is indispensable to promote CIM. By regarding a scheduling system as an element of CIM, the functions which the scheduling system should provide become clearer. In this chapter, the relationships between the scheduling system and other systems which compose CIM are first considered to make clear the purpose of the scheduling system in CIM. Then, a scheduling system which has sufficient flexibility to cope with changes in various restrictions is briefly explained.

Scheduling System in CIM

When the scheduling system is connected to other systems, what is required of the scheduling system by these other systems? For a plant where customer demands are met by inventory, a production schedule is decided so as to satisfy the production requirement determined by the production planning (long-term scheduling) system. The scheduling system must indicate the feasibility of producing the required amount of product by the due date. The response from the scheduling system is usually used to determine the optimal production plan. Therefore, quick response takes precedence over optimality of the derived schedule.

Information from the scheduling system is also used by the personnel in the product distribution section. They always want to know the exact completion time of production for each product, and the possibility of modifying the schedule each time a customer asks for a change in the due date of a scheduled material or an urgent order arrives.

If information on the condition of a plant can be transferred to the scheduling system directly from the control system, information on unexpected delays occurring at the plant can be taken into the scheduling system and the rescheduling can be executed immediately.

From the above discussion, it becomes clear that the following functions are required of the scheduling system when it is connected with other systems.

One is the full computerization of scheduling. The scheduling system is often required by other systems to generate schedules for many different conditions. Most of these schedules are not used for actual production but rather to analyze the effect of variations in these conditions. It is a very time-consuming and troublesome task to generate all these schedules by hand. Therefore, a fully computerized scheduling system that generates a plausible schedule quickly is required when the scheduling system is connected with many other systems. This does not decrease the importance of the manual scheduling system. The manual scheduling system can be effectively used to modify and improve the schedule, and it increases the flexibility of the scheduling system.

The other function is to generate schedules with varying degrees of precision and speed. In some cases a rough schedule is quickly required. And in some cases a precise schedule is needed that considers, for example, the restriction of the upper bound of utility consumption or of the noon break. Computation time for deriving a schedule depends significantly on the desired preciseness. Therefore, a schedule suitable to the request in terms of precision should be generated.

The objective of scheduling systems is twofold: one is to determine the sequence in which the products should be produced (sequencing problem), and the other is to determine the starting moments of various operations such as charging, processing, and discharging at each unit (simulation problem).

There are two ways to solve the scheduling problem. One is to solve both the sequencing and the simulation problems simultaneously. Kondili, Pantelides, and Sargent [11] formulated the scheduling problem as an MILP and solved both problems simultaneously. They proposed an effective branch-and-bound method but the problems that can be treated by this formulation are still limited because the necessary computations are time-consuming. For cases where many schedules must be generated, the time required for computation becomes very great. The other way is to solve the sequencing and the simulation problems separately.

From the viewpoint of promoting CIM, many scheduling systems have been developed by Japanese companies, and some of them are commercially sold [9],[18],[20]. Most of them take the latter approach. In some systems, backtracking is considered to improve the schedule, but the production sequence is determined mainly by using some heuristic rules. In order to determine the operation starting moments at each batch unit, it is assumed that a batch of product is produced without incurring any waiting time. That is, a zero-wait storage policy is taken in many cases.

In these systems, creation of a user-friendly man-machine interface is thoroughly considered, and the optimality of the schedule is not strongly emphasized. That is, the schedule derived by computer is regarded as an initial schedule to be improved by an experienced plant operator. The performance index for the scheduling is normally multi-objective, and some of the objectives are difficult to express quantitatively. It is also difficult to derive a schedule while considering all types of constraints. For these reasons, the functions that are used to modify the schedule manually (such as drawing a Gantt chart on a CRT and moving part of it by using a mouse), are regarded as the main functions of a scheduling system. However, it is clear that functions to derive a good schedule or to improve the schedule automatically are required when the scheduling system is connected with many other systems, as mentioned above.

In addition to a good man-machine interface, two kinds of flexibility are required for the system. One is ease in schedule modification, and the other is ease in modification of the scheduling system itself.

A generated schedule is regularly modified by considering new production requirements. Furthermore, the schedule would also be modified each time a customer asks for a change in the due date of a scheduled material, an urgent order arrives, or an unexpected delay occurs while the current schedule is being executed. Therefore, a scheduling system must be developed so that the scheduling result can be modified easily.

In a batch plant, it is often the case that a new production line is installed or a part of the plant is rebuilt according to variations in the kinds of products and/or their production rates. As a result, a batch plant undergoes constant modifications, such as installation of recycle flow, splitting of a batch, replacement of a batch unit by a continuous unit, etc. A new storage policy between operations is sometimes introduced, and the operations which can be carried out at night or over the weekend may be changed. It is important that the scheduling algorithm has a structure which can easily be modified so as to cope with changes in the various restrictions imposed on the plant.

Flexible Scheduling System

By taking these facts into account, a flexible scheduling system for multiproduct and multipurpose processes is developed. Figure 13 shows an outline of the proposed scheduling system. In this system, a plausible schedule is derived by the following steps:

First, an initial schedule is generated by using a module-based scheduling algorithm. Each g_i in the figure shows one of the possible processing orders of jobs at every unit, and is called a "production sequence."

Then, a set of production sequences is generated by changing the production orders of some jobs for the production sequence prescribed by g_0. Here, two reordering operations, the insertion of a job and the exchange of two jobs, are used to generate a set of production sequences [8]. For each production sequence g_i, the starting moments of jobs and the performance index are calculated by using the simulation program. The most preferable sequence of the generated production sequences is regarded as the initial sequence of the recalculation, and modification of the production sequence is continued as far as the sequence can be improved.

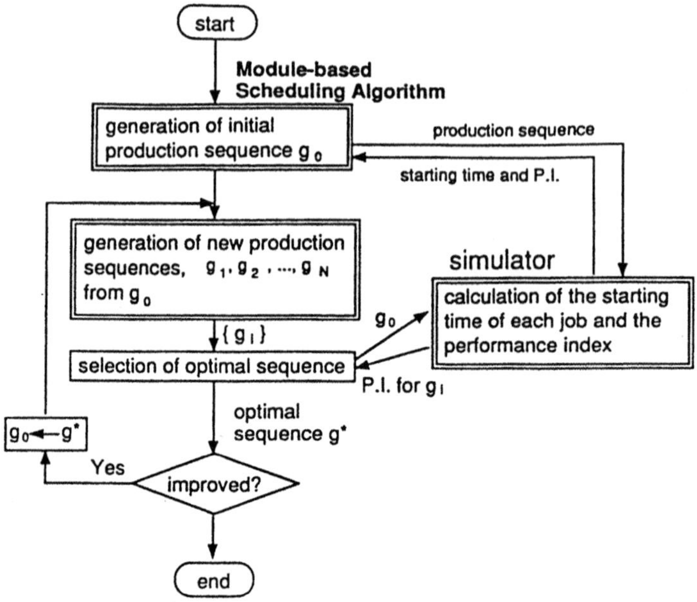

Figure 13. Structure of a Scheduling Algorithm

One feature of this system is that the generation of the initial schedule, the improvement of the schedule, and the calculation of the starting moments of jobs are completely separated. Therefore, we can develop each of these subsystems independently without taking into account the contents of others. The concept of the module-based scheduling algorithm and the constraints which must be considered in the simulation program are explained using the rest of this chapter.

Module-Based Scheduling Algorithm

In order to make the modification of the algorithm easier, the algorithm must be developed so as to be easily understood. That is, a scheduling program should not be developed as a black box. The algorithm explained here is similar to an algorithm that the operators of the plant have adopted to make a schedule manually. The idea of developing a scheduling algorithm is explained by using an example.

Let us suppose the problem of determining the order of processing ten jobs at a batch unit. It is assumed that the changeover cost depends on a pair of successively processed jobs. Even for such a small problem, the number of possible processing orders becomes 10! (= 3.6 million). How do the skilled operators make the schedule of this plant?

They determine the schedule step by step using the characteristics of the jobs and the plant. If there are some jobs with early due dates, they will determine the production order of these jobs first. If there are some similar products, they will try to process these products consecutively, because the changeover costs and set-up time between similar products are usually less than those between different products. By using these heuristic rules, they reduce the number of processing orders to be searched.

The manual scheduling algorithm explained above consists of the following steps:

(1) A set of all jobs (set A in Fig. 14) is divided into two subsets of jobs (set B and set C). Set B consists of jobs with urgent orders.
(2) The processing order of jobs in set B is determined first.
(3) Remaining jobs (jobs in set C) are also classified into two groups (set D and set E). Set D consists of jobs producing similar products.
(4) The processing order of jobs in set D is determined.
(5) Products in set D are regarded as one aggregated job.
(6) The aggregated job (jobs in set D) is combined with jobs in set E. Then, set F is generated.
(7) The processing order of jobs in set F is determined.
(8) The aggregated job in set F is dissolved and its components are again treated as separate jobs.
(9) Finally, by combining set B and set F, a sequence of all jobs can be obtained. In other words, a processing order of ten jobs is determined.

In this case, the problem is divided into nine subproblems. The algorithm is graphically shown in Fig. 14. Each ellipse and circle in the figure corresponds to a set of jobs and a job, respectively. An arrow between the ellipses denotes an operation to solve a subproblem. Here, it should be noted that the same kinds of operations are used several times to solve subproblems. For example, steps (1) and (3) can be regarded as a division of a set of jobs, and steps (2), (4), and (7) are ordering of jobs in a set.

Ideas used here are summarized as follows: First, by taking into account the characteristics of the problem, the scheduling problem is divided into many subproblems. Since the same technique can be used in solving some of the subproblems, these subproblems can be grouped together. In order to solve each group of subproblems, a generalized algorithm is prepared in advance. A scheduling algorithm of the process is generated by combining these generalized algorithms. The production schedule is derived by executing these generalized algorithms sequentially.

One feature of the proposed algorithm is that each subproblem can be regarded as the problem of obtaining one or several new subsets of jobs from a set of jobs. As the problem is divided into many subproblems and the role of each subproblem in the algorithm is clear, we

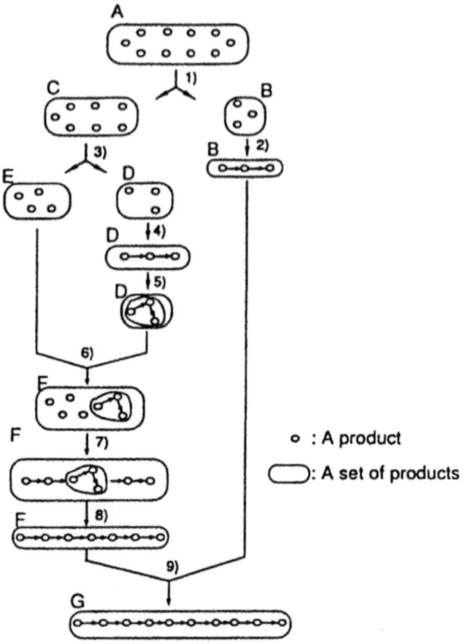

Figure 14. Scheduling Algorithm Using the Characteristics

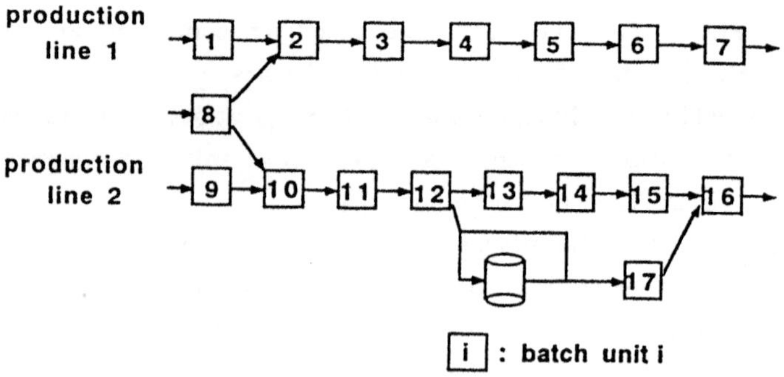

Figure 15. Process Consisting of Parallel Production Lines

can easily identify the part which must be modified in order to adapt the change of restrictions. As the number of jobs treated in each subproblem becomes small, it becomes possible to apply a mathematical programming method to solve each subproblem. Since 1989, a system developed by applying this method has been successfully implemented in a batch resin plant with parallel production lines shown in Fig. 15 [10].

Simulation algorithm

One of the predominant characteristics of batch processes is that the material leaving a batch unit is fluid, and it is sometimes chemically unstable. Therefore, the starting moments of operations must be calculated by taking into account the storage policy between two operations. Furthermore, the operations that can be carried out at night or over the weekend are limited and the simultaneous execution of some operations may be prohibited. So, even if the processing order of jobs at each batch unit is fixed, it is very difficult to determine the optimal starting moments of jobs at each unit that satisfies these constraints. Here we will try to classify the constraints that must be considered in order to determine the starting moments of jobs [6].

Production at a batch unit consists of operations such as filling the unit, processing materials, discharging, and cleaning for the next batch. Each of these operations is hereafter called a "basic operation." In many cases, it is possible to insert a waiting period between two basic operations being successively executed. Therefore, in order to calculate the completion time of each job, the relationship among the starting moments of basic operations must be considered.

A variety of constraints are classified into four groups:

(1) Contraints on Waiting Period

Four types of interstage storage policies have been discussed in the available literature [3],[17], [19],[23]:

(a) An unlimited number of batches can be held in storage between two stages (UIS).

(b) Only a finite number of batches can be held in storage between two stages (FIS).

(c) There is no storage between stages, but a job can be held in a batch unit after processing is completed (NIS).

(d) Material must be transferred to the downstream unit as soon as processing is completed (ZW).

It is possible to express the UIS, NIS, and ZW storage policies by assigning proper values to h_{ij} and h'_{ij} in the following inequalities:

$$t_i + h_{ij} \leq t_j \leq t_i + h_{ij} + h'_{ij} \qquad (1)$$

where

t_i : starting moment of basic operation i,

h_{ij} , h'_{ij} : time determined as a function of basic operations i and j.

Eq. (1) can express not only the UIS, NIS, and ZW storage policies but also some other storage policies, such as the possibility of holding material in a batch unit before the processing operation. When the FIS storage policy is employed between two batch units, it is very difficult to express the relationship between the starting moments of two basic operations by using simple inequality constraints. Therefore, the FIS storage policy should be dealt with separately as a different type of constraint.

(2) Contraint on Working Patterns

The second type of constraint is the restriction with respect to the processing of particular basic operations during a fixed time period. In order to make this type of constraint clearer, we show several examples.

(a) The discharging operation cannot be executed during the night.

(b) No operation can be executed during the night. It is possible to interrupt processing temporarily, and the remaining part of the processing can be executed the next morning.

(c) No operation can be executed during the night. We cannot interrupt processing already in progress as in (b), but it is possible to hold the unprocessed material in a batch unit until the next morning.

(d) A batch unit cannot be used during the night because of an overhaul.

Figure 16 shows schedules for each of the above constraints. In this figure, the possible starting moments for the filling operation are identical, but the scheduling results are completely different.

(3) Utility Constraints

The third type of constraint is the restriction on simultaneous processing of several basic operations. If the maximum level of utilization of any utility or manpower is limited, basic operations that use large amounts of utilities cannot be executed simultaneously. The distinctive feature of this constraint is that the restricted period is not fixed but depends on the starting moments of basic operations which are processed simultaneously.

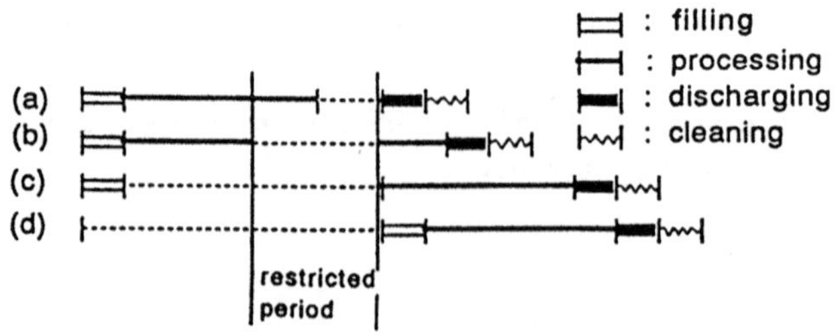

Figure 16. Schedules for Various Types of Working Patterns

(4) Storage Constraint

In an actual batch process, the capacity of each storage tank is finite. If the FIS storage policy is employed between two batch units, we must adjust the starting moments of some basic operations so that the storage tank does not overflow. Holdup at a tank depends not only on the basic operations being executed at that time but also on the operations executed before that time. Therefore, there are many ways to resolve the constraint when the FIS constraint is not satisfied.

By increasing the constraint groups to be considered, calculation time is also increased. It is possible to develop a simulation program that satisfies the constraints on each group independently. Therefore, by selecting the constraints to be considered, schedules can be generated with suitable degrees of speed and precision. Figure 17 shows an example of part of a schedule for the process shown in Fig. 15. In Fig. 17, all operations are prohibited between the period 172 hr to 185 hr, but it is assumed that the holding of material in each batch unit is permitted. The broken line in the lower figure shows the upper bound of a utility, and the hatched area shows the amount of utility used.

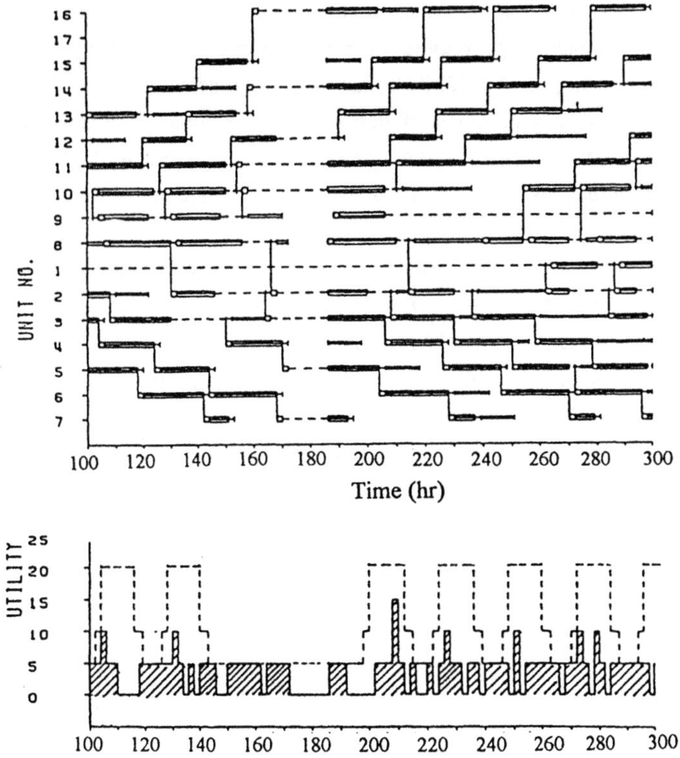

Figure 17. Schedule that Satisfies Every Type of Constraint

5. Multi-Purpose Pipeless Batch Chemical Plant

In a multi-purpose batch plant, many pipes are attached to each batch unit for flexible plant operation. The number of such pipes increases as the number of products increases, and it eventually becomes difficult even for a skilled operator to grasp the operational status of the plant. Meticulous cleaning operations of pipelines as well as of the batch units are required to produce high-quality and high-value-added products. The frequency and the cost of cleaning operations increase when the number of products is increased. Moreover, costs of the peripheral facilities for feeding, discharging, and cleaning operations increase when an automatic operation system is introduced to the plant.

In order to reduce these costs, sharing of these facilities among many batch units and the reduction of the number and the length of pipelines become necessary. With this recognition, much attention has been riveted on a new type of plant called a "pipeless batch plant." In this chapter, characteristics of pipeless batch plants and their present status are explained, and a design method and future problems are discussed.

There are many types of pipeless batch plants proposed by Japanese engineering companies. The most common type involves the replacement of one or more processing stages with a pipeless batch plant [12],[21],[22]. In this type, the pipeless batch plant consists of a number of movable vessels; many types of stations where feeding, processing, discharging, and cleaning operations are executed; and automated guided vehicles (AGV) to carry vessels from one station to another. Waiting stations are sometimes installed in order to use the other stations more efficiently.

A movable vessel on an AGV is transferred from one station to another to execute the appropriate operations as shown in Fig. 18. Figure 19 shows an example of the layout of a pipeless batch plant. This plant consists of six movable vessels, three AGVs, and eight stations for feeding, reacting, distilling, discharging, cleaning, and waiting. About ten commercial plants of this type have been constructed during the last five years. Various kinds of paints, resins, inks, adhesives, and lubrication oils are produced in these plants. It is possible to use movable vessels instead of pipelines. In this case, batch units are fixed, and the processed material is stored in a movable vessel and then fed to the next batch unit.

Figure 20 shows a different type of pipeless plant [5]. In this plant, reactors are rotated around the tower in order to change the coupling between each reactor and the pipes for feeding and discharging.

Characteristics of pipeless batch plants

A pipeless plant has a structure different from that of an ordinary batch plant. Here the characteristics of the pipeless plants are explained from the following three points:

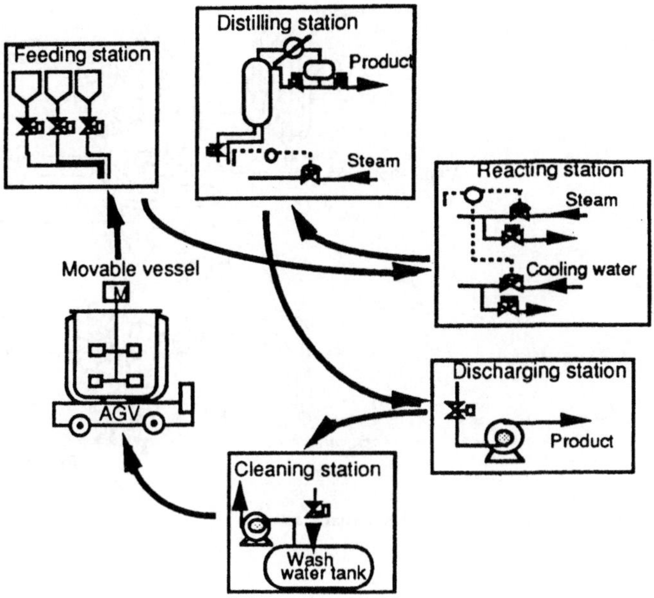

Figure 18. Conceptual Diagram of Pipeless Batch Plant

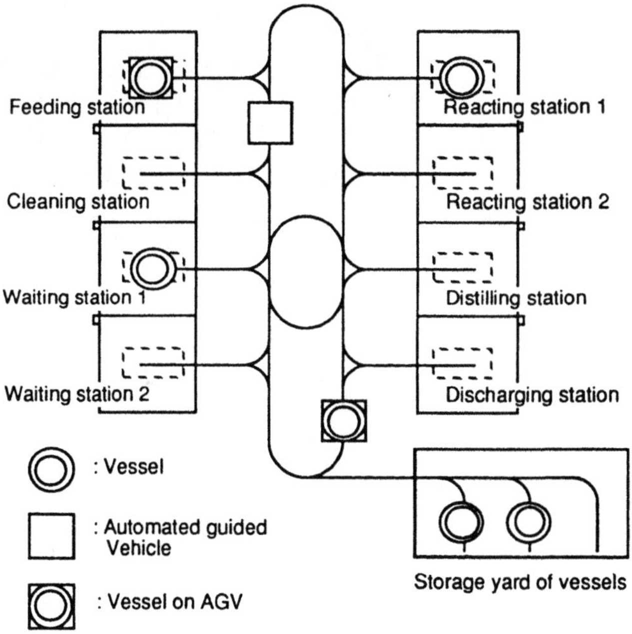

Figure 19. Layout of a Pipeless Batch Plant

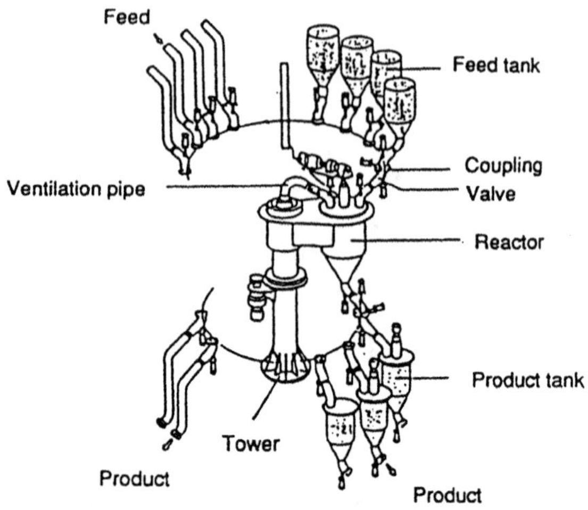

Figure 20. Rotary Type Pipeless Batch Plant

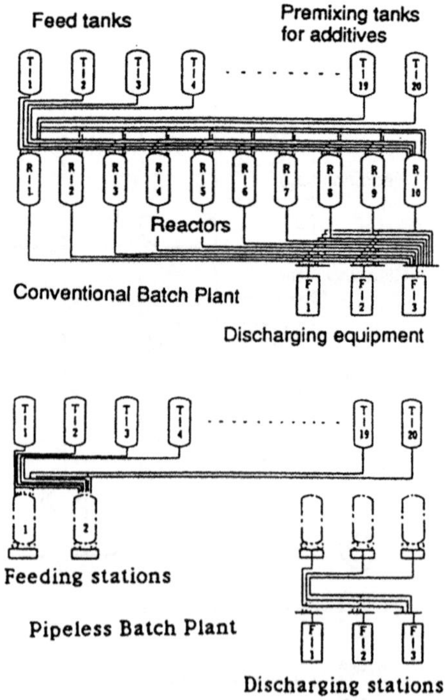

Figure 21. Configuration of Conventional and Pipeless Plants

(1) Reduction of the number of pipelines and control valves

For ordinary batch plants, the number of pipes and control valves increases along with increases of the kinds of raw materials and products. If some raw materials share pipes, it is possible to reduce the number of pipes even for an ordinary batch plant. However, meticulous cleaning of pipelines as well as that of the batch unit is then required. Figure 21 shows examples of plant configurations of a conventional batch plant and a pipeless batch plant [13]. It is clear from this figure that the number of pipelines is drastically decreased, and the plant is greatly simplified by adopting the pipeless scheme.

(2) Effective use of components

The working ratio of a unit is defined as the ratio of the working period of the unit to the whole operating period of the plant. A high working ratio means that the process is operated effectively. Therefore, the working ratio has been used as an index of the suitability of the design and operating policy of a batch plant.

A batch unit consists of many components for feeding, processing, discharging, and cleaning. Not all of them are used at all of the operations to produce a batch of product. The working periods of these components are shown in Table 2. As is clear from Table 2, the vessel is the only component used at every operation to produce a batch of product. In other words, the working ratios of some components in a batch unit are not so high.

The working ratios of the components have not been discussed, because these components have been regarded as being inseparable. The costs of these peripheral components have increased with the introduction of sophisticated control systems for automatic operation of the plant. In the pipeless batch plant, each of these components is assigned to a station or to a movable vessel. Therefore, it is possible to use these facilities efficiently and to reduce the capital costs for these facilities by determining the number of stations and movable vessels appropriately.

Table 2. Working Period of Components

	Feeding	Processing	Discharging	Cleaning	Distilling
Vessel	O	O	O	O	O
Jacket, Heating and Cooling facilities		O			O
Agitator		O			
Measuring tank and Feeding facility	O				
Discharging facility			O		
Cleaning facility				O	
Distilling facility					O

(3) Increase in flexibility

i) Expansibility of the plant

In a pipeless batch plant, the size of each type of station and that of movable vessels can be standardized. Therefore, new stations and/or vessels can be independently added when the production demand is increased.

ii) Flexibility for the production path

As the stations are not connected to each other by pipelines, the production path of each product is not restricted by the pipeline connections. That is, a pipeless plant can produce many types of products with different production paths. The production path of each product can be determined flexibly so that the stations are used effectively.

iii) Flexibility of the production schedule

For many types of stations, the cleaning operation is not required when the product type is changed. Therefore, the production schedule at each station can be determined by taking into account only the production demand of each product. It is possible to develop a JIT system and reduce the inventory drastically.

Design of a Pipeless Batch Plant

The process at which a pipeless plant is introduced must satisfy the mechanical requirement that the vessel be movable by AGV and the safety requirement that the material in the vessel be stable during the transfer from one station to another. The decision as to whether a pipeless plant may be introduced takes into account the above conditions and the cleaning costs of vessels and pipelines. There are many possible combinations for the assignment of the components shown in Table 2 to stations and vessels. For example, the motor of an agitator can be assigned either to the processing station or to the movable vessel. The assignment of the components to the stations and vessels must be decided by taking into account the working ratios of these components and the expansibility of the plant.

When all of the above decisions are made, the design problem of a pipeless batch plant is formulated as follows:

"Determine the number of each type of station, the number and the size of movable vessels, and the number of AGVs so as to satisfy the given production requirement and to optimize the performance index."

When the number of products becomes large and the amount of inventory of each product becomes small, a production schedule must be decided by taking into account the due date of each production requirement. Inventory decreases inhibit the generation of a good schedule. This problem cannot be ignored, because the production capacity of the plant depends on the production schedule as well as on the number and the size of stations and vessels

By taking into account these two factors regarding production capacity, a design algorithm for a pipeless batch plant was proposed [7]. In the proposed algorithm, the upper and the lower bounds of the number of vessels, AGVs,and each type of station are first calculated for each available vessel size. Then, iterative calculations including simulation are used to determine the optimal values of design variables.

Let us try to qualitatively compare the capital costs of the pipeless batch plant and the ordinary batch plant. Here, it is assumed that each batch unit has the functions of feeding, processing, discharging, and cleaning, and that the cost of an ordinary batch unit is equal to the sum of the costs of a vessel and four types of stations.

When the vessel volume is the same for both the ordinary and pipeless plants, the required number of batch units in the ordinary plant is larger than the number of stations of any given type in the pipeless plant. Especially for feeding, discharging, and cleaning operations, the number of each type of stations is very small because the feeding, discharging, and cleaning periods are very short. Feeding, discharging, and cleaning equipment has become expensive in order to automatically execute these functions. Therefore, there is a large possibility that the pipeless plant is more desirable than an ordinary batch plant.

Many mechanical and safety problems must be resolved when a pipeless plant is installed in place of an ordinary batch plant. For example, the material in the vessel must be kept stable during the transfer from one station to another, and vessels and pipes must be coupled and uncoupled without spilling. Therefore, many technical problems must be addressed and resolved in order to increase the range of application of the pipeless plant.

Expansibility and flexibility for the production of different products are very important characteristics of the future multipurpose plant. Methods to measure these characteristics quantitatively must be studied. By assessing these characteristics appropriately, pipeless plants may be widely used as highly sophisticated multipurpose plants.

By installing various sensors and computers to each vessel, it may be possible for each vessel to judge the present condition and then decide autonomously to move from one station to another in order to produce the required product. In such a plant, the information for production management can be distributed to stations and vessels, and the malfunction of one station, vessel, or AGV will not affect the others. It is an autonomous decentralized production system, and is regarded as one of the production systems for the next generation.

6. Conclusion

Due to the increase in product types, inventory costs have become considerably large. In order to cope with this situation, CIM is being promoted actively in batch plants. One way to reduce inventory is to generate a sophisticated schedule and modify it frequently by taking into account the changes in plant condition and demand. In order to execute frequent modification of the schedule, integration of production planning, scheduling, inventory control and production control systems, and the full computerization of scheduling are indispensable. Development of a fully computerized and flexible scheduling system is still an important research area in process systems engineering.

The other way to reduce inventory is to increase the frequency of changeovers. In order to avoid increases of changeover cost and changeover time, improvement of plant hardware and more advanced automation are needed. However, in batch plants, difficulty in handling fine particles and the need for meticulous cleaning obstruct the introduction of automation. In many Japanese companies, introduction of pipeless batch plant is regarded as one of the methods to cope with this dilemma. In pipeless plants, new equipment can be added independently, without having to take other units into consideration, and the production path of each product can be determined flexibly. Expansibility and flexibility for the production of different products are very important characteristics of the future plant. Design methods of multipurpose batch plants considering these characteristics quantitatively must be developed.

References

1. Arima, M.: Multipurpose Chemical Batch Plants, Kagaku Souchi (Plant and Process), vol. 28, no. 11, pp. 43-49 (1986) (in Japanese).
2. Doi, O.: New Production System of Foodstuffs, Seminar on Multi-Product and Small-Quantity Production in Foodstuff Industries, Society of Chemical Engineers, Japan, pp. 25-29 (1988) (in Japanese).
3. Egri, U. M. and D. W. T. Rippin: Short-term Scheduling for Multiproduct Batch Chemical Plants, Comp. & Chem. Eng., 10, pp. 303-325, (1986).
4. Eguchi, K.: New Production Systems in Chemical Industry, MOL, vol. 28, no. 9, pp. 21-28 (1990) (in Japanese).
5. Funamoto, O. : Multipurpose Reaction and Mixing Unit MULTIMIX, Seminar on multi-product and small-quantity production systems , Society of Chemical Engineers, Japan, pp. 21-31 (1989) (in Japanese).
6. Hasebe, S. and I. Hashimoto: A General Simulation Programme for Scheduling of Batch Processes, Preprints of the IFAC Workshop on Production Control in the Process Industry, pp. PS1-7 - PS1-12, Osaka and Kariya, Japan, (1989).
7. Hasebe, S. and I. Hashimoto: Optimal Design a Multi-Purpose Pipeless Batch Chemical Plant, Proceedings of PSE'91, Montebello Canada, vol. I, pp. 11.1-11.12, (1991).
8. Hasebe, S., I. Hashimoto and A. Ishikawa: General Reordering Algorithm for Scheduling of Batch Processes, J. of Chemical Engineering of Japan, 24, pp. 483-489, (1991)
9. Honda, T., H. Koshimizu and T. Watanabe: Intelligent Batch Plants and Crucial Problems in Scheduling, Kagaku Souchi (Plant and Process), vol. 33, no. 9, pp. 52-56 (1991) (in Japanese).
10. Ishikawa, A., S. Hasebe and I. Hashimoto: Module-Based Scheduling Algorithm for a Batch Resin Process, Proceedings of ISA'90, New Orleans Louisiana, pp.827-838, (1990).
11. Kondili, E., C. C. Pantelides and R. W. H. Sargent: A General Algorithm for Scheduling Batch Operations, Proceedings of PSE'88, Sydney, pp. 62-75, (1988).
12. Niwa, T.: Transferable Vessel-Type Multi-Purpose Batch Process, Proceedings of PSE'91, Montebello Canada, vol. IV, pp. 2.1-2.15, (1991).
13. Niwa, T. : Chemical Plants of Next Generation and New Production System, Kagaku Souchi (Plant and Process), vol. 34, no. 1, pp. 40-45, (1992) (in Japanese).
14. Plant Operation Research Group of the Society of Chemical Engineers, Japan: The Current Status of and Future Trend in Batch Plants, Kagaku Kogaku (Chemical Engineering), 45, pp. 775-780 (1981) (in Japanese).

15. Plant Operation Research Group of the Society of Chemical Engineers, Japan: Report on the Present Status of Plant Operation, Kagaku Kogaku symposium series, 19, pp. 57-124 (1988) (in Japanese).
16. Plant Operation Research Group of the Society of Chemical Engineers, Japan: Report on the Present Status of Plant Operation (No.2), unpublished (in Japanese).
17. Rajagopalan, D. and I. A. Karimi: Completion Times in Serial Mixed-Storage Multiproduct Processes with Transfer and Set-up Times, Comp. & Chem. Eng., 13. pp. 175-186, (1989).
18. Sueyoshi, K.: Scheduling System of Batch Plants, Automation, vol. 37, no. 2, pp. 79-84, (1992).
19. Suhami, I. and R. S. H. Mah: An Implicit Enumeration Scheme for the Flowshop Problem with No Intermediate Storage: Comp. & Chem. Eng., 5, pp. 83-91,(1981).
20. Suzuki, K., K. Niida and T. Umeda: Computer-Aided Process Design and Production Scheduling with Knowledge Base, Proceedings of FORCAPD 89, Elsevier, (1990).
21. Takahashi, K. and H. Fujii: New Concept for Batchwise Speciality Chemicals Production Plant, Instrumentation and Control Engineering, vol. 1, no. 2, pp. 19-22, (1991).
22. Takahashi, N.: Moving Tank Type Batch Plant Operation and Evaluation, Instrumentation and Control Engineering, vol. 1, no. 2, pp. 11-13 (1991).
23. Wiede Jr, W. and G. V. Reklaitis: Determination of Completion Times for Serial Multiproduct Processes-3. Mixed Intermediate Storage Systems, Comp. & Chem. Eng., 11, pp. 357-368, (1987).

Batch Processing Systems Engineering in Hungary

Gyula Körtvélyessy

Szeviki, R&D Institute, POB 41, Budapest, H-1428, Hungary

Abstract: The research work in batch processing systems engineering takes place at universities in Hungary. Besides the system purchased from known foreign companies, the Hungarian drug industry has developed their own solution: the Chemiflex reactor in Chinoin Co. Ltd. has been distributed in many places because of its simple programming and low price.

Keywords: Batch processing, pharmaceuticals

Introduction

More than 20 years ago, there were postgraduate courses in the Technical University, Budapest on continuous processing systems. G. A. Almasy, G. Veress and I. M. Pallai [1, 2] were at that time the persons working in this field. Only the mathematical basis of evaluating a computer aided design algorithm from data and the mathematical model of the process could be studied. At that time the main problem in Hungary was that there were not any control devices available which could work in plant conditions. Today, process control engineering can be studied in all of the Hungarian Universities. Some of them can be seen in Table 1.

Table 1. Universities in Hungary

Technical University Budapest	Scientific University of Szeged
University of Veszprem	University of Miskolc
Eotvos Lorand Scientific University Budapest	

General Overview of Batch Processing Systems Engineering in Hungary

The development in Hungary has moved into two directions: Some batch processing systems were purchased completely from abroad together with plants. They originated e.g. from Honeywell, Asea Brown Boveri, Siemens and Eckardt. Naturally, the main user of these systems is the drug industry, therefore the second direction of our development took place in this part of the industry. The office of the author, the Research Institute for the Organic Chemical Industry Ltd. is one of the subsidiary companies of the six Hungarian pharmaceutical firms which can be seen in Table 2. In this review, a survey of independent Hungarian developments made in drug industry is given.

Table 2. Six Hungarian Pharmaceutical Companies That Support the Research Institute for Organic Chemical Industry Ltd.

ALKALOIDA Ltd., **Tiszavasvari**	BIOGAL Ltd., **Debrecen**
CHINOIN Ltd., **Budapest**	EGIS Ltd., **Budapest**
Gedeon Richter Ltd., **Budapest**	REANAL Fine Chemical Works, **Budapest**

Hungarian Research and Developments in Batch Automation

The research work takes place mainly in the universities. In the Cybernetics Faculty of the University of Veszprem, there are projects to develop algorithms for controlling the heating of autoclaves. The other project involves a computer aided simulation based on using PROLOG as a computer language.

Gedeon Richter Pharmaceutical Works Ltd.

Here, they work on fermentation process automation. Figure 1 shows the fermentor and the parameters to be measured and controlled. They are the temperature, the air flow, the pressure, the RPM of the mixer, the pH, oxygen content in solution, the level of foam in the reactor, power consumption in the mixer, the weight of the reaction mass, and the oxygen and CO_2 contents in the effluent air. The volume of the fermentor is 25 liter. The equipment is for development purposes and works quite well. It has been used for optimizing some steroid microbiological oxidation technologies.

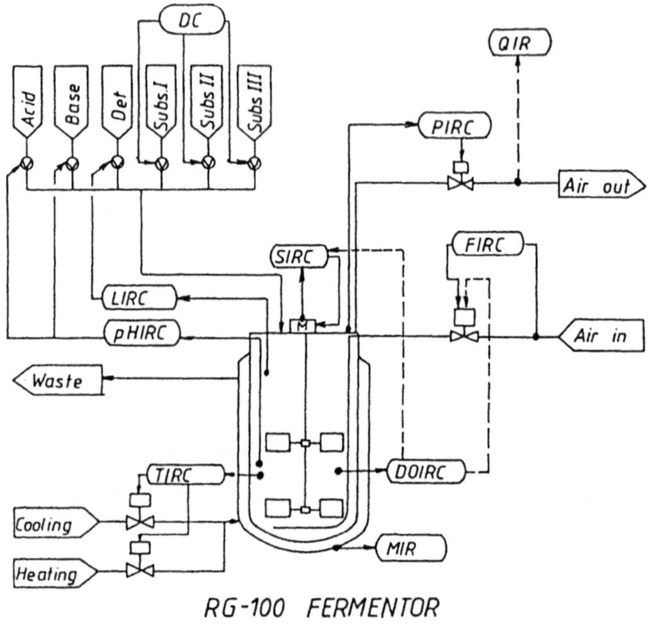

Figure 1: Fermentation Process Automation in Gedeon Richter Ltd.

EGIS Pharmaceuticals Ltd.

The other Hungarian batch automation engineering work was done at the Factory EGIS Pharmaceuticals. They use Programmable Logic Controllers of FESTO for solving problems of specific batch processing system engineering. Some examples of this are: Automatic feeding of aluminum into boiling isopropyl alcohol to produce aluminum isopropylate. The feeding of aluminum is controlled by the temperature and the rate of hydrogen evolution. The problem is that the space, from where aluminum is fed has to be automatically inertized to avoid the mixing of air with hydrogen.

Another operation solved by automatic control at EGIS is a crystallization process of a very corrosive hydrochloride salt, with clarifying. The first washing liquid of the activated carbon has to be used as a solvent in a next crystallization and then the spent carbon has to be backwashed into the waste to empty the filtering device. The main problem here was to find and use the measuring devices which enable a long term work without corrosion.

There is a central control unit room, where these PLC-s are situated and one can follow up the stages of the process. However there is a possibility of manual control in the plant in case of any malfunction.

CHINOIN Pharmaceutical Works Co. Ltd.

In case of CHINOIN, quite a different approach was realized. A few years ago, they developed the CHEMIFLEX Direct system for controlling the temperature of an autoclave for batch production. The short description of this system can be read in a brochure. Now, CHINOIN have developed the CHEMIFLEX Reactor system. The general idea of CHINOIN's approach is that the development of the system has to be made by the specialists at Chinoin, and the client should work only on the actual operational problems. The well-known steps of batch processing control engineering can be seen in Table 3. Yet, CHINOIN uses the so-called multi-phase engineering method. The clients programming work of the system is simple, since there is an in-built default

Table 3. Steps of the Batch Processing Control Engineering

Process	Control
Plant	Plant Management
Production Line	Batch Management
Unit (Reactor)	Recipe
Basic Operation	Phases and Steps
Devices	Regulatory, Sequence (Element) and Discrete Control; Safety Interlocks

control and parameter system. The engineering work can start with the basic operation instead of steps. That is why the time of installation of the system is 2 weeks only and the price is only 10% of the price of the equipment compared with the usual, simple-phase method, where the cost of engineering is the same as the price of the equipment. The measured and controlled values of the

Table 4. Measurements and Controls in the CHEMIFLEX System

Pressure in autoclave	Pressure drop in vapor pipe	Pressure in receivers
Mass of reactor and filling	Rate of circulation in jacket	Liquid level in jacket
Rpm of stirrer	Rate of flow of feed	Level in supply tanks
pH in autoclave	Liquid level in receivers	Conductivity in reactor
Temperature in vapor phase	Temperature in separator	Permittivity in separator
Temperature in autoclave	Jacket temperatures (in/out)	Pressure in jacket

process can be seen in Table 4. The drawing of the whole system is in the Figure 2.

The heating and cooling system in jacket can use water, hot water, cooling media and steam; Chemiflex can change automatically from one system to another. There is a built-in protection in the system to avoid changing e.g. from steam heating to cooling with cooling media. In case it is needed, first filling the jacket with water takes place and then changing to cooling media.

Figure 3 and Table 5. show the whole arrangement opportunities and the operations one can realize with the Chemiflex system, respectively.

Table 5. Operation of the Chemiflex Reactors

Temperature manipulations	Boiling under reflux	Distillation, atmospheric
Distillation, vacuum	Steam-distillation	Evaporation, atmospheric
Distillation, water separation	Evaporation, vacuum	Feeding, time controlled
Feeding, temperature controlled	Inertization	Feeding, pH controlled
Emptying autoclave by pressure	Cleaning autoclave	Filling autoclave by suction

The advantage of the multiphase programming can be realized from the Table 6.

The programming of the Chemiflex system is very simple and the steps to follow can be seen in the Table 7. There are possibilities to the upgraded programmers to use the second programming cycle and change the basic parameters of the system.

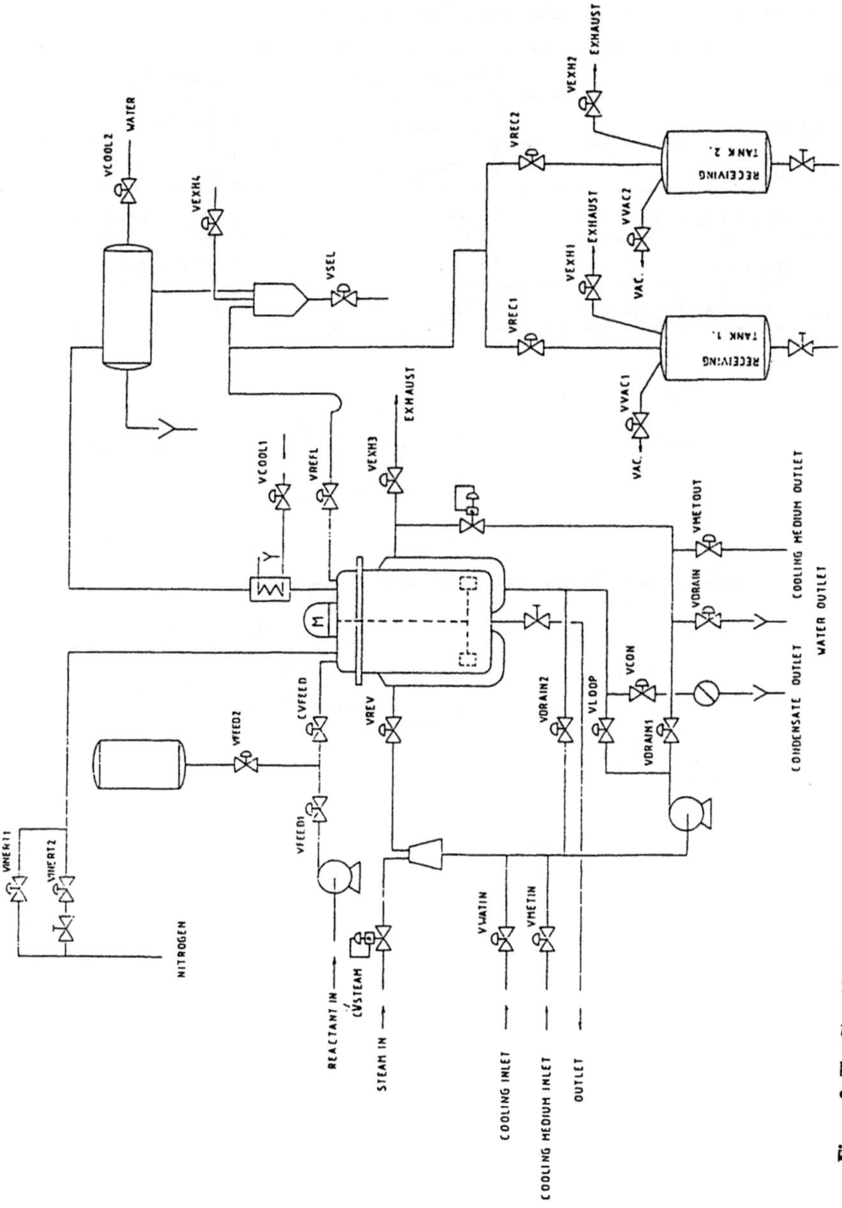

Figure 2: The Chemiflex Reactor

Table 6. Comparison of Simple Phase and Multiphase Batch Engineering

Item	Simple Phase	Multiphase
Programming:	Complicated	Simple
Sol. of Control:	In Development	----
Start of Engineering at:	Step	Basic Operation
Time of Installation:	6-8	0.5 months
Price Compared to Equipment:	100	10 %

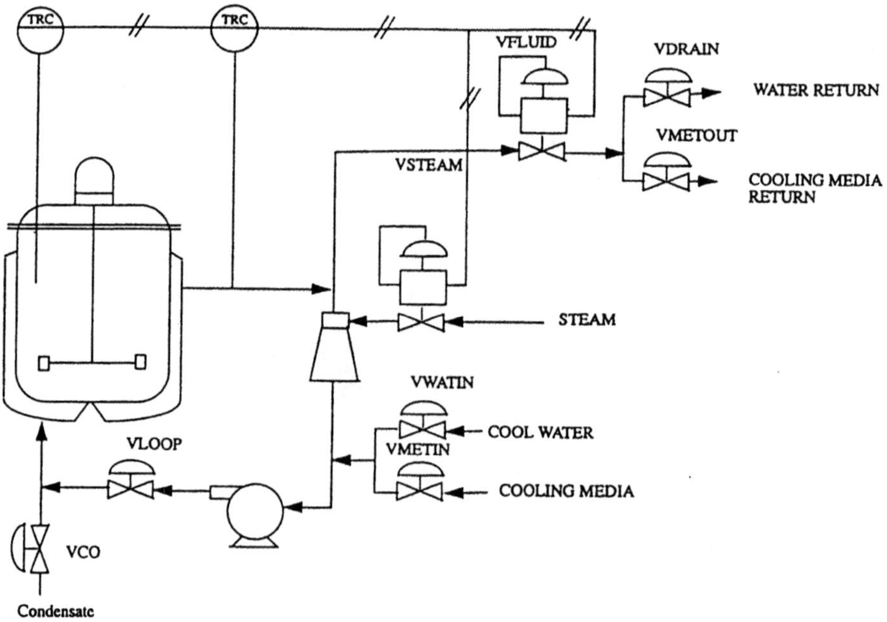

Figure 3: The Different Heating Systems in Chemiflex Reactors

Table 7. Recipe Building

1. Name Filling	2. Name of Phase	3. Select Operation	4. Select Parameters for Operation
5. Fill Basic Parameters	6. Extend Building? Y/N	7. Fill Extended (Added) Parameters	8. Select Extended Parameters

References

1. Almasy, A., Veress, G., Pallai, M.: Optimization of an ammonia plant by means of dynamic programming. Chemical Engineering Science, 24, 1387-1388 (1969)
2. Veress, G., Czulek, A.: Algebraic Investigation of the Connection of Batch and Continuous Operational Units. Hungarian Journal of Industrial Chemistry, Veszprem 4, Supl. 149-154 (1976)

Design of Batch Plants

L. Puigjaner, A. Espuña, G. Santos and M. Graells

Department of Chemical Engineering, Universitat Politecnica de Catalunya
ETSEIB, Avda. Diagonal, 647, E-08028, SPAIN

Abstract: In this paper, a tutorial on batch process design is presented. A brief review of the present status in batch process design is first introduced. The single and multiproduct plant design problems are considered next and alternate methods of solution are compared and discussed. The role of intermediate storage is then analyzed and integrated into the design strategy. Scheduling considerations are also taken into account to avoid equipment oversizing. The complexity of decision making when incorporating production planning at the design stage is brought out through several examples. The paper concludes with a summary of present and future developments in batch plant design.

Keywords: Batch plant, process design, intermediate storage, multiproduct plant, multipurpose plant, flexible manufacturing

Introduction

Chemical Plants are commonly designed for fixed nominal specifications such as capacity of the plant, type and quality of raw materials and products. Also, they are designed with a fixed set of predicted values for the parameters that specify the performance of the system, such as transfer coefficients or efficiencies and physical properties of the materials in the processing. However, chemical plants often operate under conditions quite different from those considered in the design. If a plant has to operate and meet specifications at various levels of capacity, process different feeds, or produce several products, or alternatively when there is significant uncertainty in the parameter values, it is essential to take all these facts into account in the design. That is, a plant must be designed to be flexible enough to meet the specifications even when subjected to various operating conditions. This is even more true for multiproduct and multipurpose plants, where alternate recipes and production routes must be contemplated.

In practice, empirical overdesign factors are widely used to size equipment, in the hope that these factors will compensate for all of the effects of uncertainty in the design. However, this is clearly not a very rational approach to the problem, since there is no quantitative justification for the use of such factors. For instance, with empirical overdesign it is not clear what range of specifications the overdesigned plant can tolerate. Also, it is not likely that the economic performance of the overdesigned plant will be optimum, especially if the design of the plant has been optimized only for the nominal conditions.

In the context of the theory of chemical process design, the need for a rational method of designing flexible chemical plants stems from the fact that there are still substantial gaps between the designs that are obtained with currently available computer- aids and the designs that are actually implemented in practice. One of these gaps is precisely the question of introducing flexibility in the design of a plant. It must be realized that this is a very important stage in the design procedure, since its main concern is to ensure that the plant will economically be able to meet the specifications for a given range of operating conditions.

The design of this kind of process can be viewed at three different levels: selection of the overall structure of the processing network; preliminary sizing of the processing units and intermediate storage tanks, and detailed mechanical design of the individual equipment items. In the present work, we focus on the first two levels because the bulk of the equipment used in this type of processing tends to consist of general purpose, standard items rather than unique, specialized designs. Therefore, synthesis and sizing strategies for single and multiproduct/multipurpose batch plant configurations are reviewed. Present trends and future developments are also dealt with, concluding with a summary of research and development issues in batch plant design.

Literature review: The problem of determining the number of units in parallel and the sizes of the units in each stage such as to minimize the capital cost of the equipment, given the annual product requirements and assuming that all stages operate under ZW, can be posed as a mixed integer nonlinear programming problem [33, 14, 18]. Such optimization problems can be solved using the branch and bound strategy providing that the continuous nonlinear subproblems arising at each node are convex [1]. Because of the severe limitations of this requirement and the generally large computing times needed to solve the MINLP problem, approximate solution procedures have been proposed by Sparrow et al. [33], for the pure batch case and extended to include semi-continuous units by Wiede et al., [40]. Also worth of mention is the work by Flatz [7] who presented a hand calculation procedure to determine a set of equipment sizes and distribute the overall operating time among different products based on a selected unit shared by all the products.

More recently, Yeh and Reklaitis [41] have developed an approximate approach for single product plants which takes into account task merging and splitting. Birewar and Grossmann [3] further demonstrated that task merging may lead to lower total equipment cost. Better performance is achieved by the heuristic design method of multiproduct plants developed by Espuna and Puigjaner [4], which is superior to these earlier methods, typically obtaining designs within at most 3% of the optimal solution in a few seconds of computer time.

Consideration of intermediate storage in the design has been reported by Takamatsu et al., [33] who proposed a combined dynamic programming-direct search procedure which accommodates location and sizing of intermediate storage in the single product case. The results of Karimi and Reklaitis [16, 17] readily show that the storage size is a discontinuous function of the cycle times of the adjacent units and that storage size can be quite sensitive to batch size and cycle time. Thus, the approach of Takamatsu et al. [35], can only yield one of a great number of local minima in the single product case and is computationally impractical because of the "curse of dimensionality" in the multiproduct case. Further work in the design of multiproduct plants with intermediate storage has been reported by Modi and Karimi and Espuna et al. [20, 5].

Simulated annealing has also been used to locate intermediate storage and select operating modes [22]. It has also been demonstrated that a better design can be obtained by incorporating production scheduling considerations at the design stage [3, 4, 6, 10]. Very recently, a comprehensive methodology has been developed that incorporates all the above elements (intermediate storage, prediction scheduling) in an interactive strategy that also allows for non-identical parallel units, in-phase, out-of-phase, mixed operating modes, task merging, task splitting and equipment reuse [12, 28].

The design of multipurpose plants requires detailed prediction planning and scheduling of the individual operations for multiple production routes for each product. Two forms of operation can be considered for these type of plants. In the cyclic multipurpose category, the plant runs in campaign mode, while in the non-cyclic category non-campaign mode is considered, excluding general recipe structure and generating aperiodic schedules [26]. Early works were limited to plants with single prediction route for each product [34, 36, 8]. Faqir and Karimi [9] extended the previous work to allow for multiple prediction routes for each product. More recently, a more general problem formulation was presented [21, 8, 9] which considers flexible unit-to-task allocations and non-identical parallel units. The time horizon is divided into a number of campaigns of varying length within which products may be manufactured in parallel. In these works, semicontinuous units are accommodated but intermediate storage is excluded from design considerations. The only work to present a comprehensive formulation that includes intermediate storage and full consideration of scheduling problem has been recently proposed by Puigjaner and coworkers [23, 12, 29].

Many of the models proposed rely on MINLP formulations, but recently there is some increase on generating MILP models. Voudouris and Grossmann [37] considered the more realistic case of discrete sizes and reformulated the most classical non-linear models on batch plant design as MILP problems. Shah and Pantelides [31] presented a MILP model which also considers the problem of unit-to-task allocation as well as the limited availability of intermediate storage and utilities. Uncertainty in production requirements has been also considered in the design of multipurpose batch plants [32]. The scheduling and prediction planning of multipurpose batch chemical plants has been also addressed [38, 39], considering the formation of single-product and multiple-product campaigns.

To account for the features usually found in real batch processes like batch mixing and splitting, intermediates, raw materials, flexible unit-to-task allocation, Kondili et al [19] introduced the State Task Network (STN) representation that models both the individual batch operations ("tasks") and the feedstocks, intermediate and final products ("states"), which are explicitly included as network nodes. Using the STN representation, Barbosa-Povoa and Macchietto [2] developed an MILP formulation of batch plant design and detailed scheduling, determining the optimal selection of both equipment units and network of connections, considering the horizon time as a set of discrete quanta of time.

Chemical plant layout is also a problem of high interest, since an efficient plant layout can be the biggest cost saver after process design and equipment design. This problem has been recently addressed [15, 27], and can be a growing field in the area of flexible batch chemical processing.

In the following study, we will concentrate on the multiproduct type of prediction networks and indicate the latest developments in this area. The design of multipurpose plants will be enunciated for the cyclic mode of operation and present solution trends will be indicated. Current developments in this field will be discussed in later papers.

The Design Problem

The design problem consists basically in determining the sizing and configuration of equipment items for given production requirements so that capital and/or operating costs are minimized.

Prior to the predesign of this kind of processes, the following information is required:

- list of products and the amount of each to be manufactured and the available production time,
- the individual recipes for each product,
- the size/duty factors for each task,
- the material flow balance for each task of the manufacturing process and flow characterization,
- the equipment available for performing each task, including:
  - the cost/size ratio
  - the processing time of each task to be performed on that unit
- a suitable performance function involving capital and/or operating cost components

to determine:

(a) the number of equipment stages and the task allocations

(b) the intermediate storage requirements

(c) the parallel equipment items in each stage

(d) the size capacities of all equipment items

Thereafter, the objective of the predesign calculation is to optimize the sizing of the processing units by minimizing the selected performance function subjected to specific plant operating conditions.

The following assumptions are made at the predesign stage, and will subsequently be modified when additional information coming from the production planning is obtained:

1. Only single product campaigns are considered. When storage costs are substantial, the demand pattern will determine the proper ordering of production campaigns.
2. Each equipment unit is utilized only once per batch.
3. Parallel equipment units are assigned to the same task and out-of-phase mode is also permitted.
4. Only the overlapping mode of operation is assumed.
5. A continuous range of equipment sizes is assumed to be available.
6. Multiple equipment items of a given type are identical.

7. Instantaneous batch transfer mode (ZW transfer rule).

The first three categories of variable -(a) through (c)-, as indicated above, define the structure of the process network and constitute the synthesis problem, while the last category (d) refers to the sizing problem.

The Single Product Case

It is assumed that the product is to be produced in a process which consists of M different batch equipment types and K types of semicontinuous units. Only parallel batch units operating out-of-phase which reduce the cycle time will be allowed. The size V_j of batch equipment of type j can be calculated -once the size factor S_j for the same equipment and the batch size of the product are known- as follows:

$$V_j = B\,S_j \qquad j = 1, \ldots, M \tag{1}$$

For the overlapping mode of operation, it has been shown [41] that the optimal sizing problem can be formulated as a nonlinear programming problem (NLP):

Minimize $f(V_j, R_k)$

subject to:

$$V_j \geq B\,S_j \qquad j = 1,\ldots,M \tag{2}$$

$$\frac{B\,D_k}{R_k} \leq \sigma_j \qquad k = 1,\ldots,K \tag{3}$$

$$T \geq \frac{\sigma_j + \left[P_j^{(1)} + P_j^{(2)} B^{P_j^{(3)}}\right] + \sigma_j + 1}{m_j^0} \qquad j = 1,\ldots,M \tag{4}$$

$$T \geq \sigma_j \tag{5}$$

those constraints indicate that the sizing of batch units must be done in such a way that processing the product will be met (2), that filling and emptying times are limited by the maximum semicontinuous time involved in semicontinuous equipment k with R_k processing rate and duty factor D_k (3) and, that the cycle time for a given batch size cannot be less than that required by any of the semicontinuous operations which are involved (5). The cycle time is calculated via the general expression (4) with m_{jm}^{o} parallel units operating out-of-phase.

Additionally, the total production time should be less than or equal to the available production time H.

$$\frac{Q}{B}T \leq H \tag{6}$$

where Q is the amount of product required over production time horizon H.

Finally, the available ranges of equipment sizes require that

$$V_j^{min} \leq V_j \leq V_j^{max} \tag{7}$$

$$R_k^{min} \leq R_k \leq R_k^{max} \tag{8}$$

The function to be minimized is of the form

$$f(V_j, R_k) = \sum_{j=1}^{M} m_j \left[c_{j1} + c_{j2} V_j^{c_{j3}}\right] + \sum_{k=1}^{L} m_k \left[c_{k1} + c_{k2} V_k^{c_{k3}}\right] \tag{9}$$

which reduces to

$$\text{minimize} \sum_{j=1}^{M} m_j \left[c_{j1} + c_{j2} V_j^{c_{j3}}\right] \tag{10}$$

in the pure batch case. If m_j is fixed, at the optimum

$$B = QT/H$$

Then, the only variable B is restricted to satisfy

$$\text{Max}\,(V_j^{min}/S_j) \leq B \leq \text{Min}\,(V_j^{max}/S_j) \tag{11}$$

The result is a single-variable optimization for the limiting batch size B.

The Multiproduct Case

The general multiproduct problem can be described in the same terms indicated for the single product case, although the computation time and complexity of the solution increases significantly. Therefore, reasonable simplifications must be introduced. The introduction of appropriate heuristic rules helps to simplify the solution to this problem. In a recent publication [4], the objective function selected only considers the influence of the equipment sizes on the plant investment cost. Thus, the objective function becomes the sum of the individual process equipment costs:

$$f(V_j, R_k) = \sum_{j=1}^{M} [m_{j_0} \cdot m_{j_p}] \cdot [c_{j_1} + c_{j_2} \cdot V_j^{c_{j_3}}] + \sum_{k=1}^{L} m_{k_p} \cdot [c_{k_1} + c_{k_2} R_k^{c_{k_3}}] \tag{12}$$

The plant and process specifications and problem assumptions remain the same as before. The minimization of expression (12) is subject to the reformulated constraints, for all i and j:

$$B_i \leq \left\{ m_j^P \cdot \frac{V_j}{S_{ij}} \right\} \tag{13}$$

$$F_{ij} \geq \left\{ \frac{B_i \cdot \frac{D_{ik}}{R_k}}{m_k^P} \right\} \tag{14}$$

$$E_{ij} \geq \left\{ \frac{B_i \cdot \frac{D_{ik}}{R_k}}{m_k^P} \right\} \tag{15}$$

$$P_{ij} = P_{ij}^{(1)} + P_{ij}^{(2)} \cdot \left[\frac{B_i}{m_j^P} \right]^{P_{ij}^{(3)}} \tag{16}$$

$$t_{ij} = \frac{\left(F_{ij} + P_{ij} + E_{ij}\right)}{m_j^0} \tag{17}$$

$$T_i = \max_j \left\{ F_{ij}, t_{ij}, E_{ij} \right\} \tag{18}$$

$$SP = H - \sum_{i=1}^{N} \frac{Q_i}{B_i} \cdot T_i \geq 0 \tag{19}$$

Again, these constraints ensure that the sizing obtained will meet the intended production requirements (13), that the filling and emptying times are limited by the maximum semicontinuous processing time involved (14, 15), that the limiting cycle time for product i cannot be less than that required for any of the batch operations involved (18), and that the total production time cannot exceed the available time (19). The upper and lower bounds on batch and semicontinuous unit sizes remain as before (7 and 8).

The proposed simplified strategy [4] comprises all of the constraints enumerated above into a single decision parameter which considers that the makespan or completion time required to meet the specified production cannot be greater than the available time (19). Then, the only variables to be adjusted are the equipment sizes. Logically, at the optimum plant production levels the batch size for each product is

$$B_{i_{max}} = \min_j \{m_{j_p} \cdot \frac{V_j}{S_{ij}}\} \tag{20}$$

and the processing time for batch and semicontinuous units can be obtained from (14) and (16) respectively. Then, the limiting cycle time for a given limiting batch size B can be obtained from (17) and (18).

It follows that the total batch processing time for each product will be the maximum of all times calculated above (overlapping mode). Then, the time required to meet the specified production levels can be determined and, consequently, the remaining time for the specific sizing used will be known. Optimum sizing will be obtained by minimizing the objective function keeping the remaining time positive.

The optimization procedure is based on the calculation of partial derivatives of the objective function and associated constraints with respect to the unit sizes. These values will be used to modify the size of unit l (batch or semicontinuous) depending on the step size h_l. The unit l to be modified is selected according to the feasibility of the current sizing point. The unit which most improves the objective function without excessive loss of marginal time will be selected. When a non-feasible point is reached, a unit is selected that gives the highest increase in the marginal time at the lowest penalty cost (objective function). Whenever completion time or boundary restrictions are violated, the step length h_l is decreased accordingly. The optimization procedure ends when h_l values become insignificant. Convergence is accelerated using additional rules which take into account the size reduction for all units not actually involved in processing time calculations and by keeping the step lengths within the same order of magnitude for several computation cycles.

Improving the Design: Solution Strategies

Previous formulations of the design problem may prove to be oversimplified, thus giving rise to an unjustifiable oversizing. Present developments in hardware, together with the use of appropriate optimization techniques, allow introduction of further refinements in the design procedure that will eventually result in improvement of the final design. The following items are also considered at the design stage:

- task allocation of individual equipment and production planning.
- alternative task transfer policies that may incorporate limited storage for intermediates.

Unit-to-task Allocation

Flexible unit-to-task allocation can be achieved by using the binary variable x_{ijm} that identifies all possible assignments of tasks $\mathbf{j} = 1,\dots, \mathbf{J_i}$ to equipment $\mathbf{m} = 1,\dots, \mathbf{M}$ for every product $\mathbf{i} = 1,\dots, \mathbf{I}$

$$x_{ijm} = \begin{cases} 1 & \text{if j-th task is assigned to m-th unit to produce i-th product} \\ 0 & \text{otherwise} \end{cases} \tag{21}$$

subject to

$$\sum_{m=1}^{M} x_{ijm} = 1 \quad ; \quad \sum_{j=1}^{J_i} x_{ijm} = 1 \quad ; \quad \sum_{i=1}^{I} x_{ijm} = 1 \tag{22}$$

Thus, each stage j could have associated with it a set of cycle times given by:

$$t_{ijm} = \frac{x_{ijm}\left[P^{(1)}_{ijm} + P^{(2)}_{ijm}\left(\frac{B_i}{M^p_{jm}}\right)^{P^{(3)}_{ijm}}\right]}{M^o_{jm}} \tag{23}$$

where M^o_{jm} and M^p_{jm} indicate the number of equipment out of phase and in phase at stage j respectively. For simplicity of exposition, only batch equipment has been considered. The extension to include semicontinuous units is straightforward.

The limiting cycle time and the batch size equipment constraints become:

$$t_{ij} = \max_{m} \{ t_{ijm} \} \tag{24}$$

$$T_i = \max_{j} \{ t_{ij} \} \tag{25}$$

$$V_m \geq x_{ijm} S_{ijm} \frac{B_i}{M^p_{jm}} \tag{26}$$

and the objective function is

$$F(V_m) = \sum_{m=1}^{M} M_m^o \quad M_m^P \left\{c_m^{(1)} + c_m^{(2)} V_m^{c_m^{(3)}}\right\} \tag{27}$$

with

$$M_m^o = \sum_{j=1}^{J_i} x_{ijm} M_{jm}^o \tag{28}$$

$$M_m^P = \sum_{j=1}^{J_i} x_{ijm} M_{jm}^P \tag{29}$$

subject to constraints (19, 22, 26), and taking into account (23, 24, 25, 28, 29).

Intermediate Storage

The use of intermediate storage (IS) in batch plants increases production flexibility by de-bottlenecking conflicting process stages, reducing the equipment sizes and alleviating the effects of process parameter variations.

It has also been noted [10] that intermediate storage can serve to decouple upstream and downstream trains. This decoupling can occur in two different forms. If the amount stored is of the order of an entire production campaign, then the trains can operate as two independent processes, but the storage capacity can be selected just to be large enough to decouple the cycle times but not the batch sizes.

In general, the maximum number of available locations, S, is calculated by

$$S = \sum_{m=1}^{M+L} \sum_{\mu=m+1}^{M+L} S_{m\mu} \tag{30}$$

where $S_{m\mu}$ is given by

$$S_{m\mu} = C_{m\mu} A_{m\mu} \left(1 - \prod_{i=1}^{I} (1 - E_{im\mu})\right) \tag{31}$$

where $C_{m\mu}$ indicates the equipment connectivity, $E_{im\mu}$ is a binary variable that represents the stability of intermediates and $A_{m\mu}$, which is also binary, indicates the availability of data for IS.

The problem of sizing and location of intermediate storage and its influence on the overall equipment cost has been studied by Karimi and Reklaitis [161. In multiproduct plants, two possibilities may occur:

- The same storage location is set for all products
- Different storage locations are allowed for each product

Assuming that the storage cost is negligible compared to the equipment cost, as given by equation (12), the insertion of intermediate storage has the following overall consequences [24]:

(a) Minimizing the sum of the individual subtrain costs is equivalent to minimizing the plant cost

(b) The minimum cost with N storage locations $\leq$ the minimum costs with N-1 storage locations

(c) The minimum cost with different storage locations for each product $\leq$ the minimum cost where all locations are the same

But as the intermediate storage costs become relevant, the objective function to be minimized takes the form [29]

$$F\left(V_m, R_l, Z_s\right) = F_{dc}\left(V_m\right) + F_{sc}\left(R_l\right) + F_{is}\left(Z_s\right) \tag{32}$$

The right hand side of expression (32) takes into account the cost of batch F_{dc} (V_m) and semicontinuous equipment F_{sc} (R_l) and the contribution of storage units F_{is} (Z_s) to the plant investment cost. Each term can be calculated as in (27).

As before, the minimization of (32) is subject to the batch size constraint (13), the limiting cycle time requirement (18), upper and lower bounds on equipment sizes (7,8), including intermediate storage, and the "surplus" time constraint (19). Additionally, the minimum productivity constraint for each product in every λ train generated by the allocation of IS also has to be taken into account.

$$\frac{B_i}{T_i} = \min_{\lambda} \left\{ I_{\lambda i} \frac{B_{i\lambda}}{T_{i\lambda}} \right\} \quad ; \quad I_{\lambda i} \in \{0,1\} \tag{33}$$

where the binary variable $I_{\lambda i}$ indicates the presence of IS in train λ.

The sizing of storage tanks depends on the batch size and cycle time in each train associated with it. Calculation of the optimum size for each storage unit, taking into account the requirements for each product, leads to the following general expression, which considers the actual values of the variables (batch size , β, cycle time SC and storage capacity Θ) of the up **u** and down **d** trains associated with it [28].

$$z_{is} = \frac{S_{is}\, \beta_{is}^{u}\, \tau_{is}^{d}}{\min\left\{\max\left(\Theta_{is}^{u}, \tau_{is}^{d}\right), \max\left(\Theta_{is}^{d}, \tau_{is}^{u}\right)\right\}} \qquad (34)$$

Production Scheduling

Typically, in the design of multiproduct plants, the dependency on scheduling considerations is eliminated by the assumption of long production runs for each product and limiting cycle times for a fixed set of similar products.

If the products are dissimilar and/or require the use of different equipment stages, the processing facility may eventually be shared by different products at the same time, giving rise to simultaneous production of several products and, possibly, a lower design cost. This is the situation in multipurpose configurations, where multiple production routes may be allowed for each of the products and even for successive batches of the same product, making it necessary to introduce scheduling considerations at the design stage [11, 13].

When scheduling considerations are incorporated at the design stage, time and demand constraints are needed. These constraints cannot be expressed in terms of cycle time and batch size, as they are not fixed in the general multipurpose case. Instead, time and demand constraints are described as functions of the scheduling variable

$$X_{ijkmn} \in \{0,1\} \quad \forall\, i,j,k,m,n \qquad (35)$$

where

$$X_{ijkmn} = \begin{cases} 1 & \text{if task } \mathbf{j} \text{ of batch } \mathbf{n} \text{ and product } \mathbf{i} \text{ is assigned to the } \mathbf{k}\text{-th use of unit } \mathbf{m} \\ 0 & \text{otherwise} \end{cases}$$

the amount processed in batch **n** for product **i** and its bounds

$$0 \le B_{in} \le B_{in}^{max} \quad ; \quad B_{in}^{max} = \min_{j \le J_i} \left\{ \sum_{m=1}^{M} \sum_{k=1}^{K_m} X_{ijkmn} \frac{V_m}{S_{ijm}} \right\} \qquad (36)$$

the utilization factor η_{km} of unit **m** the **k**-th time it is used (used capacity / nominal capacity)

$$0 \leq \eta_{km} \leq 1 \tag{37}$$

and the overall constraint for batch **n** and product **i**

$$\sum_{m=1}^{M}\sum_{k=1}^{K_m} X_{ijkmn}\,\eta_{km}\frac{V_m}{S_{ijm}} = B_{in} \qquad \forall\, n,i,\, j \leq J_i \tag{38}$$

Hence, operation times and initial and final times (t_{km}, TI_{km}, TF_{km}) can be calculated:

$$t_{km} = \sum_{n=1}^{N}\sum_{i=1}^{I}\sum_{j=1}^{J_i} t_{ijkmn} = X_{ijkmn}\left(P^{(1)}_{ijm} + P^{(2)}_{ijm}\left(\eta_{km}\frac{V_m}{S_{ijm}}\right)^{P^{(3)}_{ijm}}\right) \tag{39}$$

$$TF_{km} = TI_{km} + t_{km} + TW_{km} \tag{40}$$

The waiting time TW eventually required is constrained by the "Finite Wait" time (FW) policy [25, 11, 13].

$$0 \leq TW_{km} \leq \sum_{n=1}^{N}\sum_{i=1}^{I}\sum_{j=1}^{J_i} X_{ijkmn}\, TW^{max}_{ijm} \tag{41}$$

Finally, the global time constraint for the problem is:

$$0 \leq TI_{km} \leq H \qquad \forall\, k,m \tag{42}$$

$$0 \leq TF_{km} \leq H \qquad \forall\, k,m \tag{43}$$

Scheduling variables are also used to determine production for all products:

$$Q_i = \sum_{n=1}^{N}\sum_{m=1}^{M}\sum_{k=1}^{K_m} X_{ijkmn}\left(\eta_{km}\frac{V_m}{S_{ijm}}\right) \qquad \forall\, i,\, j \leq J_i \tag{44}$$

which is subject to the overall demand D_i constraints

$$D_i - Q_i = 0 \qquad \forall\, i \tag{45}$$

It is straightforward to show that this general formulation reduces to the restricted case of the multiproduct plant seen before by assuming on a-priori schedule [6].

$$\sum_{m} x_{ijm} = 1 \; ; \eta_{km} = 1 \tag{46}$$

Overall Solution Strategy for the Design Problem

The solution to the design problem in its full complexity as formulated above can be obtained through a multilevel optimization approach that includes (Fig 1):

1. Predesign without intermediate storage.
2. Optimum sizing resulting after intermediate storage allocation.
3. Campaign preparation (scheduling under constraints) and selection.
4. Predesign of general utilities.

In summary, the solution procedure [10, 28, 30] uses two main computational modules which may be used independently:

- Design Module
- Scheduling and Production Planning Module

The two modules may interact with each other through an optimization algorithm, thus preliminary plant design. Additional design variables are: following an iterative procedure until a valid solution common to both modules is eventually reached.

From start to finish, the optimization procedure always remains under the designer's control in an interactive manner. Thus, specific decisions to be made during the optimization steps, which are difficult to automate, and valuable personal experience with specific processes can be integrated into the optimization procedure in a reasonable way. For instance, the designer may decide to introduce new options which are not included in the set of basic production sequences already generated, or else he may want to eliminate some of the alternatives which seem inadequate or incompatible with personal experience in a specific type of processes.

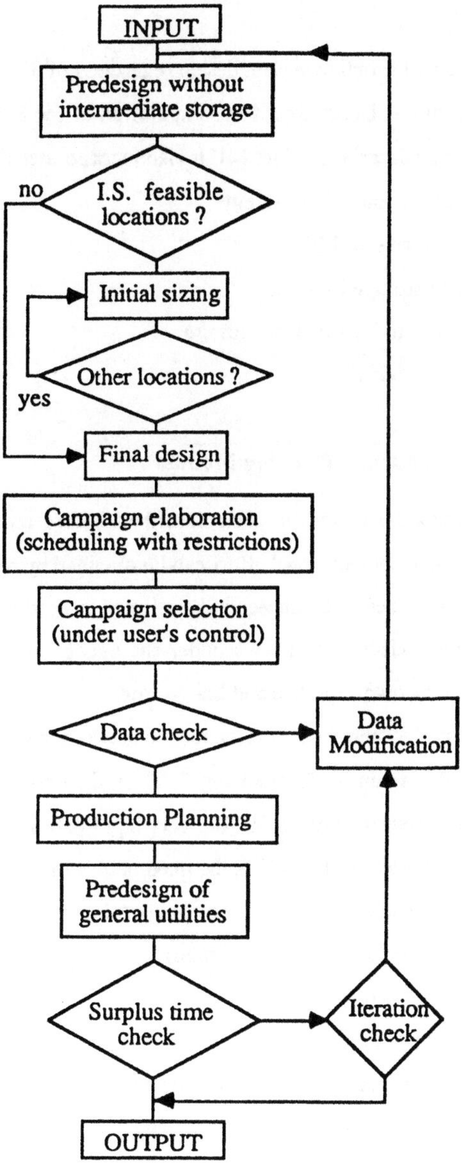

Fig. 1. Simplified flowchart of the optimization algorithm

Design Module

This module produces the optimum plant design regardless of its actual operating conditions. The calculation procedure is based upon the computation of the Surplus Time (SP) in ideal single product campaigns as described before [4]. It also incorporates finite intermediate storage (FIS) analysis which uses the following strategy:

a) Plant Design without FIS
b) intermediate storage location
c) initial sizing with intermediate storage
d) final design with FIS

Scheduling and Production Planning Module

The general formulation of the problem described before requires large MINLP problems. Substantial reduction in computational effort can be obtained by reducing the number of different campaigns to be analyzed and subsequently selected according to an adequate performance criterion. Campaign selection is always under the user's control, who can input appropriate decisions based on his own expertise and know-how.

An optimization algorithm forces the interaction between the two modules through a common optimization variable, again the "surplus time". Thus, the design module will produce optimum equipment sizes and production capacities for every input data set including intermediate storage, assuming that the solution reached offers the maximum profit for all input data. Product recipes and specific requirements of process equipment are taken into consideration to obtain the

1. The overall completion time (makespan)
2. Product market profile
3. Penalty costs for unfilled demand

Then, the production planning module will try to produce the best schedule -for a given plan configuration with specific equipment sizes- in the production campaigns that best suit the market demand profile. Results obtained over long-term periods are described in Table 1.

The alternative solutions given by the production module are subjected to evaluation. The use of appropriate cost-oriented heuristics under the supervision of an experienced user leads to suitable modifications of times and production policies. The design module will incorporate such modifications to obtain a new design which is more suited to actual production requirements, thus increasing the overall production benefits [6].

Table 1. Alternative solutions given by the Production Planning Module

Production Policy	Production Benefits	Final Production	Occupation Time	Maximum required Storage	Maximum Delay
Cover specified demand (NIS/ZW)	Bene f_1	$(P_{1,1},...,P_{1,n})$	T_1	Stock 1	Delay 1
Cover specified demand (UIS/ZW)	Bene f_2	$(P_{2,1},...,P_{2,n})$	T_2	Stock 2	Delay 2
$\Sigma_{Prod.}\Sigma_{Task}$Stock < Value[a]	Bene f_3	$(P_{3,1},...,P_{3,n})$	T_3	Stock 3< Value[a]	Delay 3
.	.	.	.	.	.
.	.	.	.	.	.
Use all available time (NIS/ZW)	Bene f_{m-1}	$(P_{m-1,1},...,P_{m-1,n})$	Time horizon	Stock m-1	Delay m-1
Use all available time (UIS/ZW)	Bene f_m	$(P_{m,1},...,P_{m,n})$	Time horizon	Stock m	Delay m

(a) To be specified by the user

Retrofitting Applications

When the objective is to increase the set of products manufactured in the plant by using existing units and adding new ones for specific treatment of these products, the problem of identification and size selection of the new equipment that makes the best use of the existing ones can be solved by the optimization procedure indicated above, by taking into account the new demand pattern and plant flow-sheet. However, we must remove from the set of available equipment units to be optimized those already existing in the actual plant. Thus, the problem of producing additional products can be solved by an optimal integration of specific "new" units into the existing "general purpose units".

When the objective is to increase the overall production of the plant, the optimal (minimum cost) solution will usually consist in adding new identical parallel equipment units. Consequently, it will only be necessary to identify the places where they should be added and the operating mode of these new units [4].

In the last case, the final solution could result in oversizing if identical in-phase parallel equipment units have been introduced to enlarge batch sizes. To avoid this eventual oversizing, the assumptions made that all parallel equipment units are identical has been relaxed in the design module. The criterion is that all parallel sets of equipment units operating out-of-phase should have the same production capacity, and they should be comprised of M_{jm}^{p} identical units, except for the existing ones plus one (to adjust production capacity). Since this could produce different processing times for parallel in-phase units, the greatest processing time must be selected [29].

A Comparative Test Case Study

A sample problem (example 3 of [3]) is used to illustrate the design strategy described before. Table 2 summarizes the data for this comparative study.

Table 2. Data used for comparative study (from [3], exp. 3)

Production requirements:	$Q_A = 40{,}000\text{kg},\ Q_B = 20{,}000\text{kg}$
Cost coefficients:	$\alpha_j = \$250,\ \beta_j = 0.6$
Horizon time:	$H = 6000$ h.
Size factors (lkg^{-1})	Batch processing times (h)

Product stage	Product A	B	Product stage	Product A	B
1	2	4	1	8	16
2	3	6	2	20	4
3	4	3	3	8	4

Final designs obtained under different test conditions are shown in Table 3.

Table 3. Design and Production Planning results

	A ZW (ref. 2)	B ZW (this work)	C B Including holdups	D C Including NIS	E D Including seasonality	F E Including stock limits
V_1	429	429	501	446	445	534
V_2	643	643	752	668	667	801
V_3	857	857	1003	891	890	1068
B_A	214	215	251	223	222	267
B_B	107	107	125	111	111	133
Cost	35,973.50	35,977.50	39,513.16	36,819.75	36,790.73	41,042.58
Iterations	—	3	5	2	4	10
Surplus time (1)	—	8 + 0	140 + 420	100 + 180	100 + 180	830 + 96
Max. Stock A (2)	—	179	224	213	367	441
Max. Stock B (2)	—	90	112	106	4958	2947
CPU time (3)	3.59 (4)	1.60	5.90	2.43	4.51	10.92

(1) Surplus time is expressed in hours and as the sum of idle time due to lack of demand and useless time remaining until next plant hold-up.
(2) Initial stock of half a batch is assumed. (3) On a SUN workstation SPARC1. (4) On an IBM 3086.

Cases A and B shown similar results for the solution of the design problem when the ZW policy is considered.

In case C, the design with ZW scheduling is faced with a discrete time horizon. A multiproduct campaign AB is considered and the time horizon of 6000h contains short and medium-term periods so that foreseeable shut-downs can be also considered. The time horizon has 10 medium-term periods with an associated demand estimation for each product of one tenth the global demand. Every medium-term period comprises 4 short-term periods of 150h each one including the final shut-down.

Two opposite effects lead to the final design: time savings due to operation overlapping in the multiproduct campaign tend to reduce the plant and its cost. On the other hand, time wasted because of holdups tends to increase it. The final result shows the second effect to be the most important in this case.

Comparison of cases B and C shows that only considering the first effect leads case C to a cheaper plant, but it will not satisfy the demand if a continuous period of 6000h is not available.

In case D, where NIS policy is contemplated, it is shown that production is increased at the same time that wasted time is reduced. As a result we have a lower design cost.

In the following case E, demand seasonality is also considered. Global demand for product B remains the same but demand estimation throughout the whole time horizon is defined for every medium term period as follows:

Medium term	1	2	3	4	5	6	7	8	9	10
Demand estimation	1500	500	500	1500	1000	2000	2000	3000	4000	4000

As no restrictions are imposed, the final design is almost the same as in case D. Also, the production plan is similar because idle time is minimized in the same way. The difference is in the large stock that a constant production plan causes when faced with the given demand profile.

The last case F introduces limitation of stock capacity into the previous case. A maximum of 3000 kg of stock for product B is allowed in the production plan. In this case, a larger plant will be needed and, consequently, a larger surplus time, which is minimized under the stock restriction.

Note that the time horizon provided as input for the design module is chosen as the independent variable. The lowest cost plant satisfying the given overall demand over this continuous time horizon is next sized using the design module. The output provided by this first module is the corresponding capital cost and equipment sizing, the latter being automatically used as the input for the production planning module.

The actual discrete time horizon is used in this second module to accommodate the specified demand and the extra or surplus time obtained after satisfying demand is the resulting output.

The design strategy used is illustrated in Figures 2 to 5. Figure 2 shows the quasi-linear relationship between capital cost and the independent time variable. Obviously, at larger horizon times, smaller and cheaper plants are obtained.

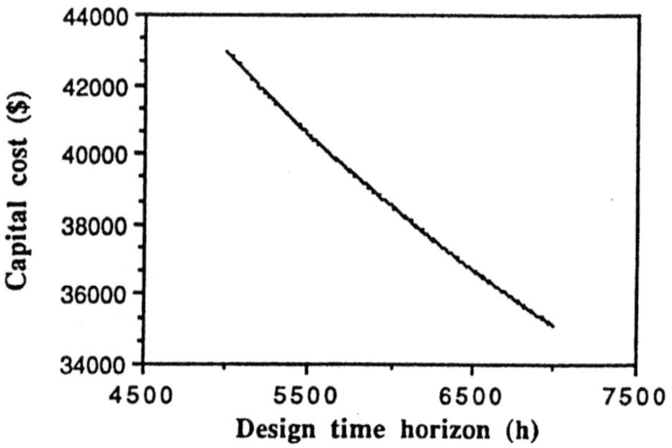

Fig. 2. Capital cost vs. design time horizon

Figures 3, 4 and 5 refer to case D. They reveal the discontinuous behavior of surplus and extra time. Figure 3 shows that extra time is needed if plant design is performed using an input time horizon greater than 6500 h. The resulting plants will not be able to satisfy the demand.

Figure 4 illustrates similar but opposite behavior for surplus time. A design time horizon of less than 6500 h. produces overdesign. Plants sized using larger horizon values cannot satisfy the demand although still remains surplus time, that being the useless time due to shutdowns.

The discontinuous behavior of surplus and extra time observed is due to degeneracy. At a given time horizon, alternate designs with different capital cost are obtained although all they lead to a production plan with the same number of batches and thus the same extra time.

Therefore, the function to be minimized is the sum of extra and surplus time that is shown in Figure 5 at the same time scale as Figure 2. The optimum design time horizon obtained this way will lead to the optimum sizing when introduced in the design module.

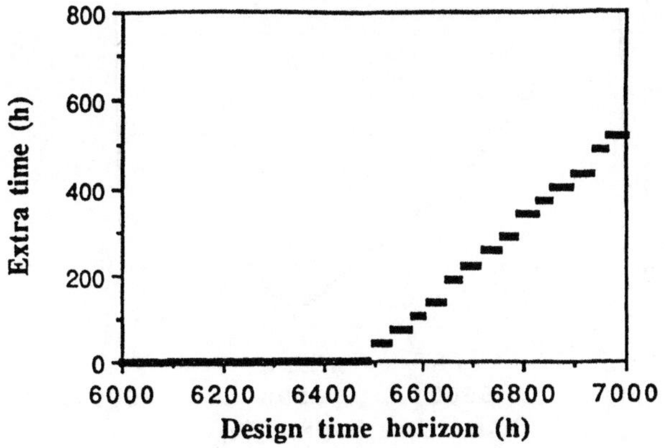

Fig. 3. Extra time vs. design time horizon for case D

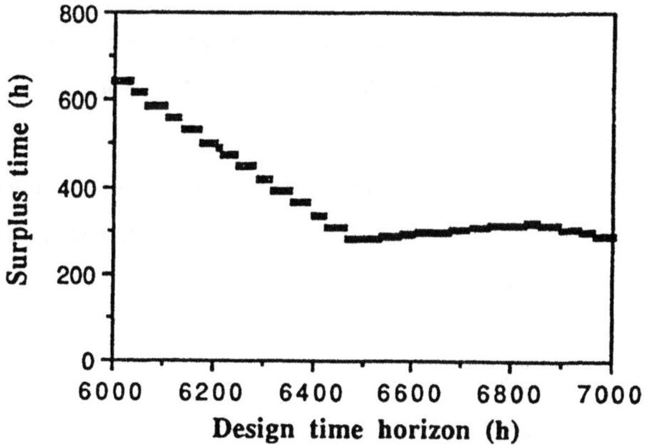

Fig. 4. Surplus time vs. design time horizon for case D

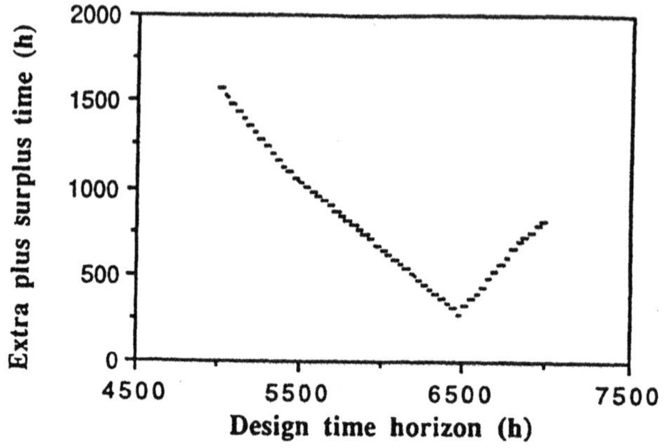

Fig. 5. Extra plus surplus time vs. design time horizon for case D

The Future in Batch Plant Design

In this paper, we have addressed the problems of optimal design of batch plants. Two subproblems ave been identified: equipment sizing and network synthesis. It has been shown how the complexity of problem increases significantly from single to multiproduct and then to multipurpose batch plant design.

Although present formulations of the design problem consider the detailed representation of the batch system constituents even at the subtask level [25], much still remains to be unveiled in these main directions:

- Development of a realistic and more integrated framework for the design and retrofitting of flexible production networks that includes the feedback from planning and scheduling evaluations of train design performance.
- Adequate treatment of more general (i.e. concurrent) recipe structures.
- Design under uncertainty
- Further development of efficient optimization algorithms capable to efficiently solve large-scale industrial problems.
- Energy integration and waste minimization.
- Integrated control strategies
- Development of more intuitive interfaces to overcome the difficulties associated with the use of complex modeling.

Acknowledgments

Support by the European Communities (JOUE-CT90-0043 and JOU2-CT93-0435) and the Comissio Interdepartamental de Recerca i Tecnologia (QFN89-4006 and QFN93-4301) is gratefully appreciated.

Nomenclature

A_{mu}	=	Binary variable that indicates the availability of data for the location of storage between unit **m** and unit μ
B_i	=	Batch production capacity for product **i**
$B_{i\lambda}$	=	Batch production capacity for product **i** in the subplant λ
B_{in}	=	Batch production capacity for product **i** in batch **n**
c_{j1}	=	Independent constant for cost calculation (discontinuous equipment)
c_{j2}	=	Linear constant for cost calculation (discontinuous equipment)
c_{j3}	=	Exponential constant for cost calculation (discontinuous equipment)
c_{k1}	=	Independent constant for cost calculation (semicontinuous equipment)
c_{k2}	=	Linear constant for cost calculation (semicontinuous equipment)
c_{k3}	=	Exponential constant for cost calculation (semicontinuous equipment)
$c_m^{(1)}$	=	Independent constant for cost calculation (discontinuous equipment)
$c_m^{(2)}$	=	Linear constant for cost calculation (discontinuous equipment)
$c_m^{(3)}$	=	Exponential constant for cost calculation (discontinuous equipment)
$C_{m\mu}$	=	Binary variable that indicates if there is physical connection between unit **m** and unit μ
D_i	=	Present market demand of product
D_{ik}	=	Duty factor of semicontinuous equipment **k** for product **i**
E_{ij}	=	Emptying time of the discontinuous stage **j** for product **i**
$E_{im\mu}$	=	Binary variable that indicates the stability of intermediate after unit m and before unit μ for product **i**
F	=	Objective function to be optimized
F_{ij}	=	Filling time of the discontinuous stage **j** for product **i**
h_l	=	fractional change (step size) of unit **1** (batch or semicontinuous) during optimization procedure
H	=	Time horizon
I	=	Total number of products
$I_{\lambda i}$	=	Binary variable that indicates if product **i** is produced in subplant λ
J	=	Total number of tasks in the receipe
K	=	Total number of available semicontinuous equipment
L	=	Total number of available semicontinuous equipment
m_j	=	Number of parallel out-of-phase units for the discontinuous stage **j**
m_j^o	=	Number of parallel out-of-phase units for the discontinuous stage **j**
m_j^p	=	Number of parallel in-phase units for the discontinuous stage **j**
m_k	=	Number of parallel in-phase units for the semicontinuous stage **k**

M	=	Total number of available discontinuous equipment
M_m^o	=	Number of parallel units for the units for the unit **m** operating out-of-phase
M_m^p	=	Number of parallel units for the unit **m** operating in-phase
M_{jm}^o	=	Number of parallel units for the unit **m** in task **j** operating out-of-phase
M_{jm}^p	=	Number of parallel units for the unit **m** operating in-phase
N	=	Total number of batches to be produced in the time horizon
P_{ij}	=	Processing time of the discontinuous stage **j** for product **i**
$P_{ij}^{(1)}$	=	Independent constant for processing time calculation (discontinuous stage)
$P_{ij}^{(2)}$	=	Linear constant for processing time calculation (discontinuous stage)
$P_{ij}^{(3)}$	=	Exponential constant for processing time calculation (discontinuous stage)
$P_{ijm}^{(1)}$	=	Independent constant for processing time calculation (unit **m**)
$P_{ij}^{(2)}$	=	Linear constant for processing time caluculation (unit **m**)
$P_{ijm}^{(3)}$	=	Exponential constant for processing time calculation (unit **m**)
Q_i	=	Production of procuct **i**
R_k	=	Processing rate of the semicontinuous stage **k**
S	=	Total number of possible locations for a intermediate storage unit
SP	=	Surplus time
S_{ij}	=	Size factor of discontinuous stage **j** for product **i**
S_{is}	=	Size factor of storage **s** for product **i**
S_{ijm}	=	Size factor of task **j** for product **i** using equip **m**
$S_{m\mu}$	=	Binary variable that indicates if a storage unit can be located between unit **m** and unit μ
T_i	=	Limiting cycle time for product i
$T_{i\lambda}$	=	Limiting cycle time for product **i** in the subplant λ
TF_{jn}	=	Final time of task **j** and batch **n**
TI_{jn}	=	Initial time of task **j** and batch **n**
TW_{jn}	=	Waiting time of task **j** and batch **n**
t_{ij}	=	Operation time of the discontinuous **j** for product **i**
t_{km}	=	Operation time of the **k**-th use of unit **m**
V_j	=	Sizing of the discontinuous stage **j**

V_m	=	Sizing of the unit **m**
x_{ijm}	=	Binary variable that indicates if task **j** of product i is carried out in equipment **m**
X_{ijkmn}	=	Binary variable that indicates if task **j** of product i in batch **n** is carried out in unit m for the **k**-th time
z_{is}	=	Sizing of the intermediate storage **s** for product **i**
Z_s	=	Sizing of the intermediate storage **s**

Greek Letters

$\beta_{is}{}^{u}$	=	batch size of product i in the upstream subplant of the storage s
$\acute{\eta}_{km}$	=	Utilization factor of unit **m** the **k**-th time it is used
Θ_s^u	=	time needed to fill the storage **s** for product **i**
Θ_s^d	=	time needed to empty the storage **s** for product **i**
σ_j	=	Transfer time of the discontinuous stage **j**
τ_s^u	=	limiting cycle time for the subplant located before the storage **s** for product **i**
τ_s^d	=	limiting cycle time for the subplant located after the storage **s** for product **i**

Subscripts

i	=	Product number
j	=	Task number
k	=	Semicontinuous equipment
m	=	Discontinuous equipment
n	=	Job number
s	=	Storage number

Superscripts

(1)	=	Independent parameter for cost or time calculations
(2)	=	Linear parameter for cost or time calculations
(3)	=	Exponential parameter for cost or time calculations
o	=	out-of-phase
p	=	in-phase
u	=	upstream
d	=	downstream

References

1. Balas, E.: Branch and Bound Implicit Enumeration. Annals of Discrete Mathematics, 5, pp. 185, North-Holland, Amsterdam, 1979.
2. Barbosa-Povoa A.P., Macchietto S.: Optimal Design of Multipurpose Batch Plants. 1. Problem Formulation. Computers & Chemical Engineering, 175, pp. S33-38, 1992.
3. Birewar, D.B., Grossmann, I.E.: Simultaneous Synthesis, Sizing and Scheduling of Multiproduct Batch Plants. Paper presented at AIChE Annual Meeting, San Francisco, 1989.
4. Espuna, A., Puigjaner, L.: Design of Multiproduct Batch Chemical Plants. Computers and Chemical Engng., 13, pp. 163-174, 1989.
5. Espuna, A., Palou, I., Santos, O., Puigjaner, L.: Adding Intermediate Storage to Noncontinuous Processes. Computer Applications in Chem. Eng. (edited by H. Th. Bussmaker and P.D. Iedema), pp. 145-152, Elsevier, Amsterdam, 1990.
6. Espuna, A., Puigjaner, L.: Incorporating Production Planning in to Batch Plant Design. Paper 82f AIChE Annual Meeting, Washington D.C., November, 1988.
7. Flatz, W.: Equipment sizing for Multiproduct Plants. Chemical Engineering, 87, pp. 71-80.
8. Faqir, N.M. Karimi, I.A.: Optimal Design of Batch Plants with Single Production Routes. Ind. & Eng. Chem. Res., 28, pp 1191, 1989a.
9. Faqir, N.M. Karimi, I.A.: Design of Multipurpose Batch Plants with Multiple Production Routes. Conference on the Foundation of Computer Aided Process Design, Snowmass, CO., 1989b.
10. Graells, M., Espuna, A., Santos, G., Puigjaner, L.: Improved Strategy in Optimal Design of Multiproduct Batch Plants, Computer Oriented Process Engineering (edited by L. Puigjaner and A. Espuna), pp. 67-74, 1991.
11. Graells, M., Espuna, A., Santos, G., Puigjaner, L.: Improved Strategy in the Optimal Design of Multiproduct Batch Plants. Computer-Oriented Process Engineering (Ed.: L. Puigjaner and A. Espuna). Elsevier Science Publishers B.V., Amsterdam, pp. 67-73, 1991.
12. Graells, M.: Working Paper. Universitat Politecnica de Catalunya, Barcelona, 1992.
13. Graells, M., Espuna, A., Puigjaner, L.: Modeling Framework for Scheduling and Planning of Batch Operations. The 11th International Congress of Chemical Engineering. Chemical Equipment Design and Automation. CHISA'93, ref. 973, Praha, Czech Republic, 1993.
14. Grossmann, I., Sargent, R.W.H.: Optimum Design of Multipurpose Chemical Plants. Ind. Eng. Chem. Process Des. Dev., 18, pp. 343-348, 1979.
15. Jayakumar S., Reklaitis G.V.: Chemical Plant Layout via Graph Partitioning. I. Single level. Computers & Chemical Engineering, 18, N. 5, pp. 441-458, 1994.
16. Karimi, I.A., Reklaitis, G.V.: Variability Analysis for Intermediate Storage in Noncontinuous Processes: Stochastic Case. I. Chem. Eng. Sym. Series, 92, pp. 79, 1985.
17. Karimi, I.A., Reklaitis, G.V.: Deterministic Variability Analysis for Intermediate Storage in Noncontinuous Processes. AIChE J., 31, pp. 1516, 1985
18. Knopf, F.C., Okos, M.R., Reklaitis, G.V.: Optimal Design of Batch/Semicontinuous Processes. Ind. Eng. Chem. Process Des. Dev., v1, pp. 76-86, 1982.
19. Kondili E., Pantelides C.C., Sargent R.W.H.: A General Algorithm for Short-Term Scheduling of Batch Operations-I. MILP Formulation, Computers & Chemical Engineering, 17, N.2, pp. 211 227, 1993.
20. Modi, A.K., Karimi, I.A.: Design of Multiproduct Batch Processes with Finite Intermediate Storage. Computer Chem. Eng., 13, pp. 127-139, 1989.
21. Papageorgaki, S., Reklaitis, G.V.: 1990.: Optimal Design of Multipurpose Batch Plants. Ind. Eng. Chem. Res., 29, pp. 2054-2062, 1990.
22. Patel, A.N., Mah, R.S.H., Karimi, I.A.: Preliminary Design of Multiproduct Noncontinuous Plants Using Simulated Annealing, AIChE meeting, Chicago, 1990.
23. Puigjaner, L.: Advances in Process Logistics and Design of Flexible Manufacturing. CSChE Conference, Toronto, 1992.
24. Puigjaner, L., Reklaitis, G.V.: Disseny I Operacio de Processos Discontinus. COMMET- CAPE. Course Notes, Vol. I, UPC Barcelona, 1990.
25. Puigjaner, L., Espuna, A. Santos, G., Graells, M.: Batch Processing in Textile and Leather Industry In: Reklaitis, G.V., Sunol, A.K., Rippin, D.W.T., Hortacsu O. (eds.) Batch Process Systems Engineering, NATO ASI Series F, Springer Verlag, Berlin. This volume, p. 808.
26. Reklaitis G.V.: Design of Batch Chemical Plants under Market Uncertainty. Seminar given at Universitat Politecnica de Catalunya, June 1994.

27. Reklaitis G.V.: Chemical Plant Layout via Graph Partitioning. Seminar given at Universitat Politecnica de Catalunya, June 1994.
28. Santos, G., Espuna, A., Graells, M., Puigjaner, L.: Improving the Design Strategy of Multiproduct Batch Plants with Intermediate Storage. AIChE Annual Meeting, Florida, 1992.
29. Santos, G.: Working Paper. Universitat Politecnica de Catalunya, Barcelona, 1993.
30. Santos, G., Espuna, A., Puigjaner, L.: Recent Developments on Batch Plant Design. The 11th International Congress of Chemical Engineering. Chemical Equipment Design and Automation. CHISA'93, ref 973 Praha, Czech Republic, 1993.
31. Shah N., Pantelides C.C.: Optimal Long-term Campaign Planning and Design of Batch Operations. Ind. Eng. Chem. Res., 30, pp. 2308-2321, 1991.
32. Shah N., Pantelides C.C.: Design of Multipurpose Batch Plants with Uncertain Production Requirements. Ind. Eng. Chem. Res., 31, pp. 1325-1337, 1992.
33. Sparrow, R.E., Forder, G.J., Rippin D.W.T.: The Choice of Equipment Sizes for Multiproduct Batch Plants. Ind. Eng. Chem. Process Des. Dev., 14, pp. 197-203, 1975.
34. Suhami, I., Mah, R.S.H.: Optimal Design of Multipurpose Batch Plants. Ind. & Eng. Chem. Proc. Des. Dev., 21(1), pp. 94-100, 1982.
35. Takamatsu, T., Hashimoto, I., Hasebe, S.: Optimal Design and Operation of a Batch Process with Intermediate Storage Tanks. Ind. Eng. Chem. Process Des. Dev. 21, pp. 431, 1982.
36. Vaselenak, 1.A., Grossmann, I.E., Westerberg, A.W.: An Embedding Formulation for Optimal Scheduling and Design of Multipurpose Batch Plants. Ind. Eng. Chem. Res., 26, pp. 139-148, 1987.
37. Voudouris V.T., Grossmann I.E.: Mixed-Integer Linear Programming Reformulations for Batch Process Design with Discrete Equipment Sizes. Ind. Eng. Chem. Res., 31, pp. 1315- 1325, 1992.
38. Wellons M.C., Reklaitis G.V.: Scheduling of Multipurpose Batch Chemical Plants. 1. Formation of Single-Product Campaigns. Ind. Eng. Chem. Res., 30, pp. 671-688, 1991.
39. Wellons M.C., Reklaitis G.V.: Scheduling of Multipurpose Batch Chemical Plants. 2. Multiple Product Campaign Formation and Production Planning. Ind. Eng. Chem. Res., 30, pp. 688-705, 1991.
40. Wiede, W., Jr., Yeh, N.C., Reklaitis, G.V.: Discrete Variable Optimization Strategies for the Design of Multi-product Processes. AIChE Annual Meeting, New Orleans, LA, 1981.
41. Yeh, N.C., Reklaitis, G.V.: Synthesis and Sizing of Batch Semicontinuous Processes. Computers and Chemical Engng., 6, pp. 639-654, 1987.

Predesigning a Multiproduct Batch Plant by Mathematical Programming

D.E. Ravemark and D. W.T. Rippin

Eidgenössische Technische Hochschule, CH-8092 Zürich, Switzerland

Abstract: This paper contains a number of MINLP formulations for the preliminary design of a multiproduct batch plant. The inherent flexibility of a batch plant leads to different formulations depending on which aspects we take into account. The formulations include parallel equipment in different configurations, intermediate storage, variable production requirements, multiplant production, discrete equipment sizes and allowing the processing time to be a function of batch size. A task structure synthesis formulation is also presented. The examples are solved with DICOPT++ and the different formulations are coded in GAMS. The resulting solutions (plants) have different objective functions (Costs) and structure depending on the formulation used. Solution times vary significantly in the different formulations.

Keywords: Batch plant design, MINLP, DICOPT++, Intermediate Storage, Multiplant, Synthesis

1 Introduction

Batch processing is becoming increasingly important in the chemical industry. Trends away from bulk commodity products to higher added-value speciality chemicals, result from both increased customer sophistication (market pull) and high technology chemical engineering (technology push). Batch operations are particularly suitable for speciality chemicals and similar types of complex materials because these operations can be more readily scaled up from benchscale experimental data, designed from relatively modest engineering information and structured to handle multiple products whose individual production requirements are not large enough to justify construction of a dedicated plant. If two or more products require similar processing steps and are to be produced in low-volume, it is economical to use the same set of units to manufacture them all. Noncontinuous plants are very attractive in such situations. Two limiting production configurations are common, multiproduct and multipurpose. In a multipurpose plant or jobshop, the different products follow different routes through the plant. In a multiproduct plant or flowshop, all products follow essentially the same path through the plant and use the same equipment. In this work only the multiproduct plant is modelled.

2 Previous work

The design problem without scheduling considerations using a minimum capital cost design criterion was formulated by Loonkar and Robinson [15] and Robinson and Loonkar

[21]. They solved it as a direct search problem to obtain the minimal capital cost. Semicontinuous equipment was included but they did not include parallel equipment in their formulation, and they used nonoverlapping production of batches (at anytime there is only one batch in the plant).

Sparrow *et. al.* [23] assumed that the cost of semicontinuous equipment was negligible and only considered batch equipment. They developed a heuristic method and a branch and bound method to solve the MINLP problem. They considered the problem with discrete sizes. The heuristic was to size the plant for a hypothetical product (a weighted average of the products) and then sequentially add units in parallel until no improvement was found. The heuristic obtained a continuous solution of the unit sizes and this was rounded up to the nearest discrete size. A comparison between the two methods showed that the branch and bound method produced better solutions (average 1%, max 12%) than the heuristic but the heuristic was order(s) of magnitude faster in computing time.

Grossman and Sargent [9] permitted processing time to be a function of batch size and semicontinuous equipment was not included in their formulation. They relaxed the MINLP problem and solved it as a nonlinear programming problem. Then they rounded the relaxed integer variables to the nearest integer and solved it again. If the gap between the relaxed MINLP and the MINLP with fixed integer variables was small the integer point was assumed optimal. If the gap was large they proposed that one could use a branch and bound method and at every node solve a NLP problem. They also proposed a reformulation of the original relaxed MINLP to a geometric program.

Takamatsu *et.al.* [25] dealt with the optimal design of a single product batch process with intermediate storage tanks. The major limitation of their work was that it dealt with a single product process only.

Suhami and Mah [24] formulated the optimal design problem of a multipurpose batch plant as a MINLP. The production scheduling is done by heuristics and a strategy to generate the nonredundant horizon constraint was presented.

Knopf *et.al.* [12] solved the problem given by Robinson and Loonkar in nonoverlapping and overlapping operation as a NLP. They proposed a logarithmic transformation and showed that this convex NLP reduced the CPU time by 300–400% compared with the original NLP.

Rippin [20] reviewed the general structure of batch processing problems and did a classification of batch processing problems.

Vaselenak *et.al.* [26] formulated the problem of a retrofit design of multiproduct batch plants as an MINLP and solved it with an outer approximation algorithm. In order to circumvent a nonconvex objective function they replaced the function with a piecewise linear underapproximator.

Yeh and Reklatis [28] proposed the partitioning of the design problem into two parts: the synthesis heuristic problem and the equipment sizing subproblem. Their heuristic procedure for sizing yielded near optimal solutions but was applicable only to the single product problem. Their sizing procedure did not include sizing and costing of intermediate storage since they assumed that the cost of storage was negligible. Their synthesis heuristic include splitting and merging of tasks and adding parallel equipment and storage tanks. The synthesis heuristic is partitioned in a train design heuristic and storage heuristic. The train design heuristic is a sequential search which starts from maximum merging with no parallel equipment and tries to improve the train by adding parallel equipment and splitting of tasks. The storage heuristic is to solve the train for no storage and for max storage and add storage to the no storage train until the difference between max storage was small.

Espuna *et.al.* [7] formulated the MINLP problem with parallel units in and out of phase. They solved the NLP sizing subproblem with a gradient search which included heuristic rules for faster convergence. The required production time for a point is evaluated and the difference between this time and the horizon time is called surplus time. Positive surplus time means that the plant is oversized (but feasible) and some unit sizes should be reduced. Negative surplus time means an infeasible plant and the unit sizes have to be increased. The discrete optimization is done by sequentially adding parallel equipment out of phase at the most time limiting stage. If a unit is close to the upper or lower bound, the option exists to add or delete a unit in phase.

Modi and Karimi [16] developed a heuristic procedure for the preliminary design of batch processes with and without intermediate storage. In their procedure a sequence of single variable line searches was carried out, yielding good results with small computational effort. However, in their work the location of storage units was fixed.

Birewar and Grossman, [3] considered synthesis, sizing and scheduling of a multiproduct plant together. Their formulation contained a number of logical constraints to control the selection of units for tasks. They solved their nonconvex MINLP formulation with DICOPT++ but for some of their examples they did not obtain the global optimum.

Patel *et.al.* [18] used simulated annealing to solve the MINLP. They formulated the problem with intermediate storage tanks and parallel equipment in phase, allowing parallel units in phase to be of unequal size. They allow products to be produced in parallel paths that are operated out of phase.

Salomone and Iribarren [22] present a formalized procedure to obtain size factors and processing time.

3 The design problem with parallel units out-of-phase

The design problem has been formulated with a nonlinear objective function involving capital cost of the batch equipment by a number of authors, some also included semicontinuous equipment. We will only consider batch equipment in our formulations.

The problem is to minimize the objective function by choice of M_j parallel units, unit size V_j.

$$Cost = \min \sum_{1}^{J} M_j a_j (V_j)^{\alpha_j} \tag{1}$$

$$V_j \geq B_i S_{i,j} \tag{2}$$

$$LCT_i \geq \frac{T_{i,j}}{M_j} \tag{3}$$

$$H \geq \sum_{i=1}^{N_p} Q_i \frac{LCT_i}{B_i} \tag{4}$$

$$V_j^L \leq V_j \leq V_j^U \tag{5}$$

The goal of predesign of multiple product plants is to optimize the sizing of manufacturing units by minimizing the total capital cost of the units (equation 1). The capital cost of a unit is a simple power function of its size $a_j(V_j)^{\alpha_j}$, where a_j is the cost factor, V_j is the size of unit j and α_j is the cost exponent (less than unity). Equation (2) is the unit size constraint. B_i is the final amount of product i in a batch. $S_{i,j}$ is the size factor, it is the relation between the actual size of a batch of product i in stage j and the final batch size B_i. The unit is sized to accommodate the largest batch processed in that unit. Equation (3) is the limiting cycle time constraint. The processing time for product i in stage j is $T_{i,j}$. The number of parallel equipment items out of phase M_j increases the frequency with which a stage can perform a task and this reduces the stage cycle time. The limiting cycle time LCT_i for product i is the longest of the stage processing times. Equation (4) is the horizon constraint. Q_i is the production requirement of product i. Q_i/B_i is the number of batches of product i and LCT_i is the time between batches. The time to produce all the batches of all the products has to be smaller than or equal to the time horizon H. The size of units in stage j is usually bounded (equation 5) by an upper bound V_j^U and lower bound V_j^L

4 Logarithmic transformation

The design problem formulation in chapter 3 is nonconvex as noted by Kocis and Grossman [13]. The horizon constraint and objective function are nonlinear and nonconvex and using the Outer Approximation/ Equality Relaxation algorithm [6, 14] global optimality cannot be guaranteed. Through logarithmic transformation, the formulation can be modelled as a convex MINLP problem. We define the new (natural) logarithmic transformed variables, $lnV_j = \ln(V_j)$, $lnB_i = \ln(B_i)$, $lnLCT_i = \ln(LCT_i)$ and $lnM_j = \ln(M_j)$, thus "ln" in front of a variable name shows that the variable expresses the (natural) logarithmic value of the variable. Using these new variables the formulation is now:

$$Cost = \min \sum_{j=1}^{J} a_j \exp(lnM_j + \alpha_j lnV_j) \tag{6}$$

$$lnV_j \geq lnB_i + lnS_{i,j} \tag{7}$$

$$lnLCT_i \geq lnT_{i,j} - lnM_j \tag{8}$$

$$H \geq \sum_{i=1}^{N_p} Q_i \exp(lnLCT_i - lnB_i) \tag{9}$$

$$lnV_j^L \leq lnV_j \leq lnV_j^U \tag{10}$$

Now all the nonconvexities have been eliminated and the nonlinearities in this model appear only in the objective function (6) and the horizon time constraint (9), and in both equations the exponential term is convex. Hence when we use the OA/ER algorithm we are guaranteed to find the global optimum.

5 Integer variables by binary expansion

We use DICOPT++ based on the OA/ER described by [27] to solve the problem formulations, and as DICOPT++ cannot accept integer variables, the number of parallel units M_j has to be formulated in terms of binary variables [8].

$$lnM_j = \sum_{k=1}^{M_j^U} \ln(k)\, Y_{k,j} \tag{11}$$

$$1 = \sum_{k=1}^{M_j^U} Y_{k,j} \tag{12}$$

If binary variable $Y_{k,j}$ is equal to 1 then the logarithmic number of parallel equipment lnM_j is equal to $\ln(k)$. Equation (12) ensures that a stage is only assigned one of the possible number (1,2,...,k) of parallel units.

6 MINLP-Formulations

6.1 Parallel units out-of-phase

The problem formulation presented by Kocis and Grossman [13] consists of equations (6–12).

6.2 Parallel units in and out of phase

We add parallel units out of phase if the stage is time limiting and if the stage is capacity limiting we can add parallel units to operate in phase. This increases the largest batch size that can be processed on a stage. The batch from the previous stage is split and assigned to all the units in phase on that stage. Upon completion they are recombined and transferred to the next stage. This does not affect the limiting cycle time but the batch size of a stage is multiplied by the number of in phase units in parallel since we always add equipment of the same size. The formulation is now.

$$Cost = \min \sum_{j=1}^{J} a_j \exp(lnM_j + lnN_j + \alpha_j lnV_j) \tag{13}$$

$$lnV_j + lnN_j \geq lnB_i + lnS_{i,j} \tag{14}$$

$$lnN_j = \sum_{c=1}^{N_j^U} \ln(c) Z_{c,j} \tag{15}$$

$$1 = \sum_{c=1}^{N_j^U} Z_{c,j} \tag{16}$$

and equations (8 – 12)

lnM_j is the number of parallel units operating out of phase and lnN_j is the number of parallel units operating in phase. The binary variable $Y_{k,j}$ is equal to one if stage j

has $k = 1, 2, \ldots, M_j^U$ parallel units operating out of phase. M_j^U is the upper bound on parallel units operating out of phase. The binary variable $Z_{c,j}$ is equal to one if stage j has $c = 1, 2, \ldots, N_j^U$ parallel units operating in phase. N_j^U is the upper bound on parallel units operating in phase. In constraint (14) the number of parallel units in phase is included. This reduces the size of units needed in stage j. The formulation is still convex but it contains twice as many binary variables as formulation 6.1, which will increase the solution time. Many of the possible configurations may not be advantageous and they can be discarded in advance. For example, 4 units in phase and 4 units out of phase means 16 units at that stage. A constraint allowing, for example, only four units in a stage can be included.

$$M_j N_j \leq 4 \tag{17}$$

constraint (17) expressed in logarithmic variables

$$lnM_j + lnN_j \leq \ln(4) \tag{18}$$

Constraint (18) is linear in the logarithmic variables and can be included in the formulation above without increasing the number of nonlinear constraints.

6.3 Unequal sizes of parallel equipment in phase

Parallel equipment is usually assumed to be of equal size. If we only allow equal sizes of parallel equipment operating out of phase we have only one batch size B_i for each product. If equipment out of phase were allowed to be nonidentical we would have several different batch sizes for the same product. This would complicate the formulation and it is not obvious that this would lead to an improvement. Nonidentical equipment in phase may lead to a improvement of the objective function. Batches are split into the two units in phase and then recombined when finished on that stage and we do not have to split the batch 50/50. Due to the economy of scale, it is cheaper to have one large and one small unit than two equal size units with the same total capacity.

The formulation below is for a maximum of two units in phase. It can be expanded to more units but for clarity we restrict it to two units.

$$\begin{aligned}
Cost &= \min \sum_{j=1}^{J} a_j \exp(lnM_j + \alpha_j lnV_j^1) \\
&+ \sum_{j=1}^{J} (f_j)^{\alpha_j} a_j \exp(lnM_j + \alpha_j lnV_j^U) && (19) \\
lnV_j^U + lnN_j &\geq lnB_i + lnS_{i,j} && (20) \\
f_j &\geq \exp(lnN_j) - 1 && (21) \\
lnV_j^1 &\geq lnN_j + lnV_j^U - U(1 - Z_{1,j}) && (22) \\
lnV_j^1 &\geq lnV_j^U - U(1 - Z_{2,j}) && (23) \\
1 &= \sum_{c=1}^{2} Z_{c,j} && (24) \\
(lnV_j^L - lnV_j^U) &\leq lnN_j \leq \ln(2) && (25)
\end{aligned}$$

$$0 \leq f_j \leq Z_{2,j} \tag{26}$$

and equations (8 − 12)

First it must be said that this formulation is only convex for a given vector M_j of parallel units in phase. If the number of units out of phase is included in the optimization the objective function (19) contains bilinear terms. In the size constraint (20) the number of parallel equipment items in phase lnN_j is a continuous variable bounded by (25) and the actual capacity of the parallel units operating in phase on a stage is $lnV_j^U + lnN_j$. lnV_j^1 is the volume of the first unit in parallel. If we only have one unit in phase, equation (22) assigns to the unit the size $lnV_j^U + lnN_j$. Here lnN_j is the logarithmic fraction of the upper bound unit e.g. the volume of the first unit is expressed as a fraction of V_j^U. If we have two units in parallel the equation (23) ensures that the first unit assumes the value lnV_j^U. This is from the simple heuristic (economy of scale) it is cheaper to have one large (at upper bound size) and one smaller unit if we have two units in phase. f_j is the fraction of the size of the second unit in phase compared to the upper bound in size. f_j is equal to zero if we have no parallel units in phase and $0 \leq f_j \leq 1$ if we have two units in parallel (26). In the objective function the cost of the second unit is $(f_j)^{\alpha_j}$ times the cost of a unit at the upper bound. The formulation may not give an optimal solution if we have a lower bound on size on the second unit since we always assume that if the second unit exists the first unit is at an upper bound. The size of the second unit V_j^2 can be calculated by: $V_j^2 = f_j V_j^U$.

6.4 Flexible use of parallel units in and out of phase

We are producing a number of different products in a plant, all with different batch sizes and requirements for size and processing time. A stage can be size limiting for one product and time limiting for another and this leads to the possibility of changing the configuration of equipment for different products. We have equipment working in phase for some products, if the stage is size limiting for these products, and out of phase for others, for which the stage is time limiting. We call this flexible use of parallel equipment. This leads to a formulation with a large number of binary variables.

$$Cost = \min \sum_{j=1}^{J} a_j \exp(lnTotM_j + \alpha_j lnV_j) \tag{27}$$

$$lnV_j + lnN_{i,j} \geq lnB_i + lnS_{i,j} \tag{28}$$

$$lnLCT_i \geq lnT_{i,j} - lnM_{i,j} \tag{29}$$

$$H \geq \sum_{1}^{N_p} Q_i \exp(lnLCT_i - lnB_i) \tag{30}$$

$$lnM_{i,j} = \sum_{k=1}^{M_j^U} \ln(k) Y_{k,i,j} \tag{31}$$

$$lnN_{i,j} = \sum_{c=1}^{N_j^U} \ln(c) Z_{c,i,j} \tag{32}$$

$$1 = \sum_{k=1}^{M_j^U} Y_{k,i,j} \quad (33)$$

$$1 = \sum_{c=1}^{N_j^U} Z_{c,i,j} \quad (34)$$

$$lnTotM_j = lnM_{i,j} + lnN_{i,j} \quad (35)$$

$$lnM_j + lnN_j \leq \ln(4) \quad (36)$$

$$lnV_j^L \leq lnV_j \leq lnV_j^U \quad (37)$$

and equations (8–12) and (14–16)

$M_{i,j}$ is the number of units out of phase at stage j for product i and $N_{i,j}$ is the number of units in phase at stage j for product i. The product of these two is equal to the total number of units at a stage $TotM_j$ and in logarithmic variables it is $lnTotM_j$ (equation 35). Constraint (35) ensures that the total number of parallel equipment at a stage is equal for all products. The binary variable $Y_{k,i,j}$ is equal to one if for product i stage j has k units in parallel out of phase. The binary variable $Y_{c,i,j}$ is equal to one if for product i stage j has c units in parallel in phase. Constraint (36) sets the upper bound on the number of units in and out of phase on a stage. This formulation contains a large number of binary variables.

6.5 Intermediate storage

6.5.1 The role of intermediate storage

The equipment utilization in multiproduct batch plants is in many cases relatively low. By increasing equipment utilization, it is possible to increase the efficiency and profitability of a batch plant. In a perfect plant the stage cycle time of each product for all stages should be equal to its LCT_i. This cannot be achieved in practice due to the difference in processing time at each stage, the best that can be done is to try and make the difference in stage processing time as small as possible. One way of decreasing stage idle time is to add parallel units, another is to insert intermediate storage between stages. The insertion of a storage tank causes the process to be divided into two subprocesses and decouples the operations upstream and downstream of the storage tank. This in turn allows the LCT and the batch sizes on either side to be chosen independently of each other. The installation of a tank would increase the cost of the plant, but due to the decoupling, subprocesses with smaller LCT or larger limiting batch sizes are created, either of which can lead to smaller equipment sizes. It is possible to install storage at appropriate locations in the train which may result in a net reduction of the overall plant cost. Apart from the financial advantage outlined above, the insertion of storage tanks at suitable positions in a batch process yields a number of other benefits as stated by Karimi and Reklaitis [10]. These include an increase in plant availability, a dampening of the effects of process fluctuations and increased flexibility in sequencing and scheduling. They also pointed out that there are a number of drawbacks that are difficult to quantify including the inventory cost of the material stored, maintenance and clean out costs, spare parts costs, labour and supervision costs. The disadvantage of including intermediate storage tanks include the increased likelihood of material contamination, safety hazards, operator errors, processing delay or the requirement for an expensive holding operation such as refrigeration. For a given plant there will be multiple possible locations for an intermediate storage tank. This tank is inserted only to separate the up and down stream batchsizes and limiting cycle times.

6.5.2 Sizing of storage tank

In order to be able to properly assess the cost effects due to storage tanks, we should also include the cost of storage in the objective function. This is made possible by the work of Karimi and Reklaitis [10]. Karimi and Reklaitis developed useful analytical expressions for the calculation of the minimum storage size required when decoupling two stages of operation for a single product process. The major difficulty is that the exact expression for the storage size is a discontinuous function of process parameters. This makes it impossible to use the exact expression in an optimization formulation as functional discontinuities can create problems for most of the optimization algorithms. Karimi and Reklaitis also developed a simple continuous expression which gives a very good approximation to the actual size. They show that for a 2-stage system with identical, parallel units operating out of phase in each stage, as in our case, the following equation gives a very close upper bound for the required storage size for the decoupling of subtrains:

$$VS_s = \max_i \left\{ S_{i,s} \left(B_{\text{up}} \left[1 - \frac{\theta_{i,\text{up}}}{LCT_{i,\text{up}}} \right] + B_{\text{down}} \left[1 - \frac{\theta_{i,\text{down}}}{LCT_{i,\text{down}}} \right] \right) \right\} \tag{38}$$

where the storage size is determined by the requirement of the largest product. The $\theta_{i,\text{up}}$ and $\theta_{i,\text{down}}$ refer to the up and down stream batch transfer times but since it is not part of our model to model the semicontinuous equipment we use the following simplification (39) for evaluating the size of a storage tank. Our alternative equation (39) is linear in normal variables.

$$VS_s \geq S_{i,s}(B_{i,\text{up}} + B_{i,\text{down}}) \tag{39}$$

Modi and Karimi [16] also used these equations (38), (39) for storage size. With logarithmic variables and using the binary variable to show if a storage tank is located in site j (for storage) between stage j and stage $j+1$ we obtain

$$U(1 - X_{j,q}) + U(1 - X_{j+1,q+1}) + VS_j \geq SFS_{i,j}(\exp(lnB_{i,q}) + \exp(lnB_{i,q+1})) \tag{40}$$

The terms $U(1 - X_{j,q})$ and $U(1 - X_{j+1,q+1})$ are to ensure that VS_j (=volume storage between stage j and $j+1$) is given a size only if both $X_{j,q}$ and $X_{j+1,q+1}$ are equal to 1 e.g. that unit j belongs to subtrain q and the next unit $j+1$ belongs to the next subtrain $q+1$. This change of subtrain is caused by a storage tank used to decouple trains. $SFS_{i,j}$ is the size factor for storage, for product i and location j (which is between unit j and unit $j+1$). This equation is nonlinear but convex so we can use it in our model and the resulting optimum is guaranteed to be the global optimum. This equation will add additional nonlinear equations so we can try and simplify the problem. Since the storage tank, when it is inserted, will probably need to be bigger than the smallest possible, we can use twice the larger of the batches instead of the sum of up and down stream batches. The equations are now linear in the logarithmic variables:

$$U(1-X_{j,q})+U(1-X_{j+1,q+1})+lnVS_j \geq \ln(SFS_{i,j})+lnB_{i,q}+\ln(2) \quad (41)$$

$$U(1-X_{j,q})+U(1-X_{j+1,q+1})+lnVS_j \geq \ln(SFS_{i,j})+lnB_{i,q+1}+\ln(2) \quad (42)$$

(41) provides for storage at least double the size of the downstream batch, (42) provides for storage at least double the size of the upstream batch and the storage will of course assume the larger value of these two.

6.5.3 Cost of storage

The cost function for the storage tank is the standard exponential cost function.

$$\sum_{j=1}^{M-1} b_j(VS_j)^{\gamma_j} \quad (43)$$

Where b_j is the cost factor, γ_j is the cost exponent and the costs of the tanks are summed over the M-1 possible locations. If we are using equations (41) and (42) for the sizing of storage we get the logarithmic size of the storage and use equation (44) for costing.

$$\sum_{j=1}^{M-1} b_j \exp(\gamma_j lnVS_j) \quad (44)$$

Alternatively it may be argued that when storage is inserted it will have a size large enough to accommodate all possible batch sizes and may also absorb a little drift in the productivity of the two trains. Then this sizing equation is not necessary at all. We just add a fixed penalty for every storage tank inserted. For example in the form below where $X_{M,q}$ is the binary variable identifying the subtrain to which the last unit ("unit M") belongs, the subtrain index q minus one is equal to the number of storage tanks in the optimal solution. We can just add the total cost of storage to the cost of the equipment items.

$$\text{Total cost of storage} = \text{Cost} * \sum_{q=1}^{M-1}(q-1)X_{M,q} \quad (45)$$

6.5.4 MINLP storage formulation

The formulation uses equation (38) for the sizing of the storage tank and equation (43) is added to the objective function.

$$Cost = \min \sum_{j=1}^{J} a_j \exp(lnM_j + \alpha_j lnV_j) + \sum_{j=1}^{M-1} b_j(VS_j)^{\gamma_j} \quad (46)$$

$$lnV_j \geq lnB_{i,q} + lnS_{i,j} - U(1-X_{j,q}) \quad (47)$$

$$lnLCT_{i,q} \geq lnT_{i,j} - lnM_j - U(1-X_{j,q}) \quad (48)$$

$$lnPRO_i = lnLCT_{i,q} - lnB_{i,q} \quad (49)$$

$$
\begin{aligned}
H &\geq \sum_{i=1}^{N_p} Q_i \exp(lnPRO_i) & (50)\\
U(1-X_{j,q}) + U(1-X_{j+1,q+1}) + VS_s &\geq S_{i,j}(\exp(B_{i,q}) + \exp(B_{i,q+1})) & (51)\\
1 &= \sum_{q=1}^{J} X_{j,q} & (52)\\
X_{j,q} &\geq X_{j+1,q} \quad q = 1 & (53)\\
X_{j,q} &\geq X_{j+1,q+1} \quad q = j & (54)\\
X_{j,q} + X_{j,q+1} &\geq X_{j+1,q+1} & (55)\\
\text{and equations } & (10 \; -12)
\end{aligned}
$$

If a storage tank is inserted the different subtrains can operate with different batch sizes and limiting cycle times but the productivity of both up and downstream trains must be the same. This is ensured by constraint (49). We use this productivity in the horizon constraint (50). Units are sized (47) only to accommodate the batch size of the subtrain that the unit belongs to. The limiting cycle time (48) of a train is the longest stage processing time of the stages that belong to the subtrain. Constraint (52) ensures that every unit is assigned to one and only one subtrain. Constraint (53) ensures that if a unit belongs to the first subtrain, the previous unit belongs to the first subtrain too. Constraint (54) ensures that if unit $j+1$ belongs to subtrain $q+1$ ($j = q$) then the previous unit j belongs to subtrain q ($j = q$). Constraint (55) ensures that if unit $j+1$ belongs to subtrain $q+1$ the previous unit j either belongs to the same subtrain $q+1$ or the previous subtrain q. Constraint (55) ensures that infeasible train structures are prevented.

If in the problem a location for storage is not allowed (for example between unit j' and $j'+1$) we just have to add a constraint

$$X_{j',q} = X_{j'+1,q} \tag{56}$$

This forces unit $j'+1$ to belong to the same subtrain as j' and the location of a storage tank by the solution algorithm is inhibited at this location. We can also force the location of a storage tank between unit j' and $j'+1$ by adding a constraint

$$X_{j',q} = X_{j'+1,q+1} \tag{57}$$

This forces unit $j'+1$ to belong to the next subtrain and the location of a storage tank by the solution algorithm is forced at this location.

6.6 Variable production requirement

In the preliminary design of a plant the production requirement Q_i is likely to be an estimate which might be allowed to move between bounds $Q_i^L \leq Q_i \leq Q_i^U$. Therefore, the production requirement is a possible variable for optimization purposes. It will help us to direct our production facilities towards the more profitable products. We add the following equation in the objective function:

$$\sum_{i=1}^{I} c_i(Q_i^{\text{ref}} - Q_i) \tag{58}$$

(note: large production gives a negative profit but since we are comparing the profit with the capital cost of the plant this is correct. We just have to add the capital cost and the profit from equation (58).)

The change in profit on sales associated with over or under production compared with the nominal value may be offset by changes in capital cost charges incurred for a larger or smaller plant. This equation is linear, for an increase in production the profit is always the same. There are two reasons why this linear function may not be good. First, if the range of product requirements is large the economy of scale leads to the trivial solution that the optimal plant is the largest possible (cost of plant increases exponentially with a factor of ~ 0.6 and profit increases linearly). Second, for the special chemicals that are produced in a multiproduct plant it is probably not true that the marginal value of an extra kg product stays the same. Increasing the production may lead to a decrease in the marginal value of an extra kg product.

Instead we use the function:

$$\sum_{i=1}^{I} c_i \left(\left\{ \frac{Q_i^{\mathrm{ref}}}{Q_i} \right\} - 1 \right) \tag{59}$$

(note: parameter c_i is not the same as in equation (58))

This function is sensitive to large changes in production requirement. It adds a increasing penalty cost if $Q_i < Q_i^{\mathrm{ref}}$ and a decreasing profit as Q_i increases above Q_i^{ref}. The marginal value of an extra kg around Q_i^{ref} is almost constant but if we produce twice as much as Q_i^{ref} the marginal value of an extra kg is is only half of that. Likewise, if we produce only half of Q_i^{ref} the marginal value of an extra kg of product is twice as large. We do not want the plant to produce less than we can sell.

The function is also convex when it is included in a formulation with logarithmic variables. With logarithmic variables, including equation (59) in the objective function

$$Cost = \min \sum_{j=1}^{M} a_j \exp(lnM_j + \alpha_j lnV_j) + \sum_{i=1}^{I} c_i(\exp(lnQ_i^{\mathrm{ref}} - lnQ_i) - 1) \tag{60}$$

We have to change the horizon constraint and include the variable lnQ_i This constraint is also convex.

$$H \geq \sum_{i=1}^{I} \exp(lnQ_i + lnLCT_i - lnB_i) \tag{61}$$

Now we can replace the equations in the formulation for parallel units out of phase (chapter 6.1). We get:

$$\begin{aligned} Cost &= \min \sum_{j=1}^{M} a_j \exp(lnM_j + \alpha_j lnV_j) \\ &+ \sum_{i=1}^{I} c_i(\exp(lnQ_i^{\mathrm{ref}} - lnQ_i) - 1) && (62) \\ H &\geq \sum_{i=1}^{I} \exp(lnQ_i + lnLCT_i - lnB_i) && (63) \\ \text{and equations}(7-8) &\quad \text{and} \quad (10-12) \end{aligned}$$

6.7 Multiplant production

Dividing the multiproduct plant into several multiproduct plants, each producing a fixed subset of products, can be of advantage in increasing equipment utilization, thus decreasing the needed sizes of equipment and also reducing the long term storage costs because we produce each product for a longer period of time.

6.7.1 Cost of long term storage

We assume a constant demand over the time horizon. $Ti_{i,p}$ is the time that we produce product i on plant p. The fraction of the total time that we do not produce product i is $(1 - Ti_{i,p}/H)$. We assume that we have a constant demand over the time horizon and we have to store material to satisfy demand for this time. The average stored amount is a function of the production requirement and the time that we do not produce a product. The cost of long term storage is expressed in equation (64) and is included in the objective function.

$$\sum_{p=1}^{P}\sum_{i=1}^{I}\left(1-\frac{Ti_{i,p}}{H}\right)\frac{Q_i}{2}PC_if_i \tag{64}$$

where

$$Ti_{i,p} = Q_i\frac{LCT_{i,p}}{B_{i,p}} \tag{65}$$

This equation has been proposed by Klossner and Rippin [11]. PC_i is the production cost, taken as the product value for purpose of inventory, and f_i is a weight factor for production cost which may include a discount factor for charges on inventory. Q_i is the production requirement. $Ti_{i,p}$ is the time plant p is dedicated to producing product i.

6.7.2 MINLP multiplant formulation

We define a binary variable $X_{i,p} = 1$ if product i is produced in plant p, otherwise 0. Parallel equipment (binary variable $Y_{k,j,p}$) can be excluded in the interests of a simpler problem. Binary variable $Y_{k,j,p} = 1$ if there are k parallel units at stage j at plant p. The formulation without cost of long term storage:

$$Cost = \min\sum_{p=1}^{P}\sum_{j=1}^{J}a_j\exp(lnM_{j,p}+\alpha_j lnV_{j,p}) \tag{66}$$

$$lnV_{j,p} \geq lnB_{i,p} + lnS_{i,j} - U(1-X_{i,p}) \tag{67}$$

$$lnLCT_{i,p} \geq lnT_{i,j} - lnM_{j,p} - U(1-X_{i,p}) \tag{68}$$

$$Ti_{i,p} = Q_i\exp(lnLCT_{i,p} - lnB_{i,p}) \tag{69}$$

$$PT_{i,p} \geq Ti_{i,p} - H(X_{i,p}) \tag{70}$$

$$H \geq \sum_{i=1}^{N_p}\{PT_{i,p}\} \tag{71}$$

$$lnM_{j,p} = \sum_{k=1}^{M_{j,p}^U}\ln(k)Y_{k,j,p} \tag{72}$$

$$1 = \sum_{k=1}^{M_{j,p}^U} Y_{k,j,p} \tag{73}$$

$$1 = \sum_{p=1}^{P} X_{i,p} \tag{74}$$

$$lnV_j^L \leq lnV_j \leq lnV_j^U \tag{75}$$

Constraint (74) ensures that every product is produced in one and only one plant. $Y_{k,j,p}$ are the binary variables for parallel units out of phase, and $X_{i,p}$ are the binary variables for the plant allocation. $Y_{k,j,p} = 1$ means that unit j in plant p has k number of parallel units. Constraint (70) ensures that if a product is not processed in a plant the total processing time for the product in this plant is equal to the time horizon H in order to make the storage cost equal to zero in this plant. The horizon constraint (71) is that the production time for all units produced in a plant have to be less than the time horizon and if a product is not produced in a plant the $Ti_{i,p}$ is equal to H and this has to be subtracted.

The above formulation is a set partitioning problem. A set covering problem would allow each product to be produced in more than one plant. This can be realized by removing constraint (74) or replacing it by:

$$2 \geq \sum_{p=1}^{P} X_{i,p} \tag{76}$$

This constraint allows any product to be produced in, at most, two plants. Constraint (74) is not sufficient to avoid redundant solutions. A more precise constraint is needed to ensure that the solution of the MILP master problem does not produce redundant solutions. For example if we have two products, the solutions

- product 1 in plant 1 and product 2 in plant 2
- product 1 in plant 2 and product 2 in plant 1

are equal since a plant is defined by the products produced in it. To date, the proper formulation of such a constraint has not been found.

6.8 Discrete equipment sizes - With parallel equipment

In the original problem formulation the optimal sizes of the equipment items in our plant are chosen from a continuous range, subject to upper and lower limits reflecting technical feasibility. Often the sizes of equipment items are not available in a continuous range but rather in a set of known standard sizes available from a manufacturer at a known price. A standard unit larger than required is likely to cost less than a unit specially built to the exact size. Producing equipment in special sizes will probably not be economical. The choice of the most appropriate equipment size is now a discrete decision. This forces us to add more binary decision variables to choose a defined size, but the solution of the problem is more accurate than in the continuous case as we know the cost of the item and we do not have to use some approximate cost function. Allowing parallel units out of phase and several products gives us the formulation

$$Cost = \min \sum_{j=1}^{J} \exp(lnCost_j) \tag{77}$$

$$lnV_j \leq \sum_{g=1}^{G} lnVsize_{j,g} X_{j,g} \tag{78}$$

$$lnCost_j \geq lnM_j + \sum_{g=1}^{G} lnVcost_{j,g} X_{j,g} \tag{79}$$

$$1 = \sum_{g=1}^{G} X_{j,g} \tag{80}$$

and equations (7–12)

$lnCost_j$ is the logarithmic cost of stage j with lnM_j parallel units of size $lnVsize_j$ out of phase. $lnVsize_{j,g}$ is a set of logarithmic standard sizes for units capable of performing tasks in stage j. $lnCost_{j,g}$ is the logarithmic cost of standard units $lnVsize_{j,g}$.

6.9 Discrete equipment sizes - without parallel units

6.9.1 Single product

If we only produce one product the horizon constraint is

$$Q\frac{LCT}{B} = H \text{ or } B = Q\frac{LCT}{H} \tag{81}$$

If the production requirements, the time horizon and limiting cycletime (no or fixed number of parallel units) are known and constant the required batch size can be calculated directly. If we know the batch size we get the required volume in stage j by $V_j = BS_j$ and we just have to round up to the nearest discrete size $Vsize_{j,g}$

6.9.2 Multiproduct plant, long single product campaigns, MINLP model

When no parallel equipment is allowed or the structure of parallel units is fixed, we can formulate the problem in normal (not logarithmic) variables, then the limiting cycle time (LCT_i) is known as it is simply the largest processing times in all stages. With normal variables.

$$Cost = \min \sum_{j=1}^{J} \sum_{g=1}^{G} Vcost_{j,g} X_{j,g} \tag{82}$$

$$V_j \leq \sum_{g=1}^{G} Vsize_{j,g} X_{j,g} \tag{83}$$

$$V_j \geq B_i S_{i,j} \tag{84}$$

$$H \geq \sum_{i=1}^{I} Q_i \frac{LCT_i}{B_i} \tag{85}$$

$$1 = \sum_{g=1}^{G} X_{j,g} \tag{86}$$

Without parallel units the formulation is only nonlinear in the horizon constraint (85), and we can reformulate the MINLP problem as a MILP problem.

6.9.3 Multiproduct plant, long single product campaigns, MILP model

For more than one product, in long single product campaigns with no parallel units (LCT_i is known), we reformulate the problem by transformation of the variables: $1/B_i = invB_i$ to give:

$$Cost = \min \sum_{j=1}^{J}\sum_{g=1}^{G} Vcost_{j,g} X_{j,g} \tag{87}$$

$$\frac{invB_i}{S_{i,j}} \geq \sum_{g=1}^{G} \frac{X_{j,g}}{Vsize_{j,g}} \tag{88}$$

$$H \leq \sum_{i=1}^{I} Q_i LCT_i invB_i \tag{89}$$

$$1 = \sum_{g=1}^{G} X_{j,g} \tag{90}$$

With no parallel equipment or fixed equipment structure and the inverse transformation, the formulation is a MILP.

6.9.4 Multiproduct plant-Multi product campaigns

For all our previous problems we have assumed that products are produced in long single product campaigns, but the incorporation of scheduling in the design can be advantageous as shown by Birewar and Grossman [3].

Birewar and Grossman have in a number of papers [1],[4] and [2] developed linear constraints for the scheduling in design problem. We use their formulation and get:

$$Cost = \min \sum_{j=1}^{J}\sum_{g=1}^{G} Vcost_{j,g} X_{j,g} \tag{91}$$

$$n_i = Q_i invB_i \tag{92}$$

$$n_i = \sum_{k=1}^{N_p} NPRS_{i,k} \tag{93}$$

$$n_k = \sum_{i=1}^{N_p} NPRS_{i,k} \tag{94}$$

$$H \geq \sum_{i=1}^{N_p} \{n_i T_{i,j}\} + \sum_{i=1}^{N_p}\sum_{k=1}^{N_p} NPRS_{i,k} SL_{i,k,j} \tag{95}$$

$$n_i = n_k \tag{96}$$

$$NPRS_{i,k} \geq 0 \tag{97}$$

and equations (88), (90)

$NPRS_{i,k}$ is the number of times that a batch of product i is followed by a batch of product k ($i = 1,2,\ldots,N_p; k = 1,2,\ldots,N_p$). N_p is the number of products. $SL_{i,k,j}$

is the minimum idle time between product i and product k on unit j, in [1] there is a systematic procedure to calculate the slack times. n_i is the number of batches of product i. n_k is the number of of batches of product k. The $NPRS_{i,k}$ should have integer values but since we have the number of batches n_i as a continuous value $NPRS_{i,k}$ is probably not integer. Birewar and Grossman report that this problem has a zero integer gap and therefore a branch and bound search would be effective but, a simple rounding scheme would probably produce sufficiently good solutions.

6.9.5 Continuous sizes - Multiproduct campaigns

If the discrete size constraint is relaxed we get a NLP problem with linear constraints:

$$Cost = \min \sum_{j=1}^{J} a_j (invV_j)^{-\alpha_j} \tag{98}$$

$$invV_j \leq \frac{invB_i}{S_{i,j}} \tag{99}$$

and equations(92) and (93–97)

To expand this convex NLP problem to a convex MINLP problem which includes parallel units is not straight forward.

6.10 Processing time as a function of batchsize

In previous problems the processing time is always assumed constant but this is frequently not true. The processing time is normally dependent on the batch size. We can assume that scaling a batch of product until it becomes twice as large will certainly lengthen the processing time. The change in processing time is dependent on which tasks are executed in the stage. This has previously been noted by other authors, Grossman and Sargent [9] and Yeh and Reklaits [28], and they usually make processing time a function of the batchsize in the form of equation (100) where $R_{i,j}$, $P_{i,j}$ and λ_j are constants.

$$T_{i,j} = P_{i,j} + R_{i,j}(B_i)^{\lambda_j} \tag{100}$$

This will increase the number of nonlinear equations if we model with logarithmic variables but we could take another model that is linear in logarithmic variables.

$$T_{i,j} = R_{i,j}(B_i)^{\lambda_j} \tag{101}$$

$R_{i,j}$ and λ_j are constants. With logarithmic variables this equation is linear and can be added to any formulation, thus allowing for variable processing times without adding nonlinear equations and discrete variables.

$$lnT_{i,j} = lnR_{i,j} + \lambda_j lnB_i \tag{102}$$

6.11 Underutilization of equipment

Equipment is designed for a fixed capacity to handle the largest product batch size. As a result the smallest product batch size using that unit can be far below its design capacity, as noted by Coulman [5]. For example a jacketed stirred tank would have greatly reduced mixing and heat transfer capabilities when it is less than half full. Coulman proposed the constraint

$$\varphi_j V_j \leq B_i S_{i,j} \tag{103}$$

to avoid solutions with large underutilization's for some products. φ_j represents the lowest permitted level of a unit $0 < \varphi_j \leq 1$. $\varphi_j = 1$ means that we do not allow the unit to process batches smaller than the unit size.

This constraint is sufficient if $\varphi_j \leq 0.5$ (e.g. we do not allow units to be less than half full). If $0.5 < \varphi_j \leq 0.75$ and we have two units on a stage a second constraint must be added

$$\varphi_j 2V_j \leq B_i S_{i,j} \tag{104}$$

with the binary variable $Z_{2,j}$ equal to one if on stage j we have two units in parallel in phase. Constraint (104) assumes that we always use both parallel units in phase to process batches. But, we can always use just one unit when we process small batches. Then if we use just one unit the batch size has to be smaller than the unit size.

$$V_j \geq B_i S_{i,j} \tag{105}$$

This constraint is infeasible for the products that use two units, so we define a binary variable $X_{i,j,w}$ that is equal to one if product i in stage j uses w units for processing. The constraints are now with $0.0 < \varphi_j \leq 0.875$ and a minimum of three units in phase on a stage.

$$\begin{aligned}
\varphi_j V_j &\leq B_i S_{i,j} && (106)\\
V_j &\geq B_i S_{i,j} - U(1 - X_{i,j,1}) && (107)\\
\varphi_j 2V_j &\leq B_i S_{i,j} + U(1 - X_{i,j,2}) && (108)\\
2V_j &\geq B_i S_{i,j} - U(1 - X_{i,j,2}) && (109)\\
\varphi_j 3V_j &\leq B_i S_{i,j} + U(1 - X_{i,j,3}) && (110)\\
1 &= \sum_w X_{i,j,w} && (111)\\
Z_{3,j} + Z_{2,j} &\geq X_{i,j,1} && (112)\\
Z_{3,j} + Z_{2,j} &\geq X_{i,j,2} && (113)\\
Z_{3,j} &\geq X_{i,j,3} && (114)
\end{aligned}$$

In figure 1 we will try to explain the underutilization constraint. On the y-axis we have the batch size in a unit and on the x-axis we have the total batch size of all units in phase.

Equation (106) ensures that batches that only use one unit fulfil our underutilization formulation and this equation will always hold irrespective of how many units we actually have on the stage. Only equation (106) is needed when $0.0 < \varphi_j \leq 0.5$ and we do not need any binary variables. This can be seen in the upper graph in figure 1 where the one constraint is shown. Equation (107) ensures that if product i on stage j uses only one unit, the batch size on that stage is smaller than the unit size. Equation (108) ensures

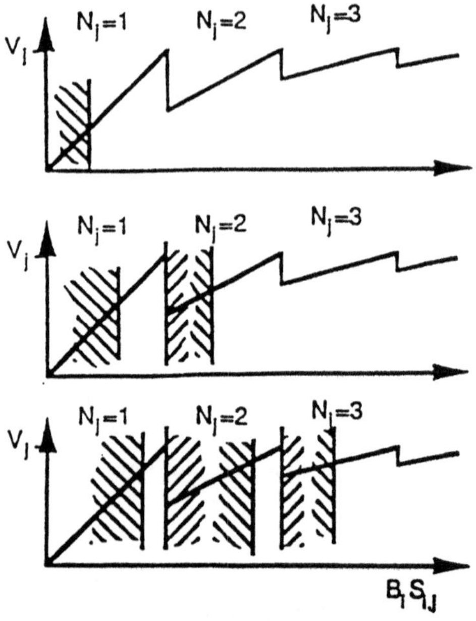

$\varphi_j \leq 0.5$
No discrete variables $X_{i,j,w}$ needed

$0.5 < \varphi_j \leq 0.75$ and $N_j \geq 2$
We have two production "windows" and we need the discrete variables $X_{i,j,1}$ and $X_{i,j,2}$

$0.75 < \varphi_j \leq 0.875$ and $N_j \geq 3$
We have three production "windows" and we need the discrete variables $X_{i,j,1}$, $X_{i,j,2}$ and $X_{i,j,3}$

Figure 1: Underutilization constraints.

that if product i on stage j uses two units, the batch size on that stage is larger than the capacity of two units times the underutilization factor φ_j. These two constraints together with constraint (106) are shown in the middle graph.

Equations (106–108) are needed only for stages j where $0.5 < \varphi_j \leq 0.75$ and we have the possibility to have two units in phase.

Equation (109) ensures that if product i on stage j uses two units, the batch size on that stage is smaller than twice the unit size. Equation (110) ensures that if product i on stage j uses three units, the batch size on that stage is larger than the capacity of three units times the underutilization factor φ_j. All constraints (106–110) are shown in the lower graph.

Logical constraint (111) ensures that a batch at a stage uses only one of the options: one, two or three units for processing. Logical constraints (112), (113) and (114) ensure that only if parallel units exist is it possible to process in them. $Z_{c,j}$ is the binary variable for parallel units out of phase and $Z_{c,j} = 1$ if stage j has c parallel units in phase. The binary variables $X_{i,j,w}$ add a number of discrete variables to any formulation and we should try to minimize the number of discrete variables. For stages j where $\varphi_j \leq 0.5$ all binary variables $X_{i,j,w} = 0$ and for stages j where $0.5 < \varphi_j \leq 0.75$ binary variable $X_{i,j,3} = 0$.

The constraints (106–110) in logarithmic variables (the logical constraints (111–114) stays the same).

$$ln\varphi_j + lnV_j \leq lnB_i + lnS_{i,j} \tag{115}$$

$$lnV_j \geq lnB_i + lnS_{i,j} - U(1 - X_{i,j,1}) \tag{116}$$

$$ln\varphi_j + \ln(2) + lnV_j \leq lnB_i + lnS_{i,j} + U(1 - X_{i,j,2}) \tag{117}$$

$$\ln(2) + lnV_j \geq lnB_i + lnS_{i,j} - U(1 - X_{i,j,2}) \tag{118}$$

$$ln\varphi_j + \ln(3) + lnV_j \leq lnB_i + lnS_{i,j} + U(1 - X_{i,j,3}) \quad (119)$$

If the underutilization constraints differ from product to product on a stage we can replace φ_j with $\varphi_{i,j}$ in the constraint above.

All these constraints are linear in the binary variables.

7 DICOPT++ Example solutions

7.1 Infeasible NLP subproblem - Slack variable

DICOPT++ will first solve a relaxed NLP problem and from this solution relax equality constraints and linearize the nonlinear functions by a first order Taylor approximation to obtain a MILP master problem. The master problem is then solved to obtain a integer point where a NLP subproblem is solved. The relaxed starting point may not give good linear approximations of nonlinear equations for problems with a poor continuous relaxation. The nonlinear equations are linearized far from the integer point that the master problem is trying to predict and the result may be that the master problem predicts integer points that are infeasible.

In a DICOPT++ iteration, no linearizations of nonlinear equations are made if the NLP subproblem is infeasible and only a integer cut is added to the master problem. The master problem has now, from the same bad linearization point, to predict a new integer point and this point is probably also infeasible. These iterations continue until the integer cuts forces the master problem into a feasible area where the problem can be relinearized. In order to speed up the iteration and avoid solving the same master problem several times we add a slack variable. Having an infeasible integer point means that the proposed plant structure cannot fulfil the production requirement even with units at the upper bound. We can make sure that the problem is feasible for all possible integer points by adding a positive slack variable to the horizon constraint and including this slack in the objective function with a large penalty U.

The horizon constraint

$$H + SLACK \geq \sum_{1}^{N_p} Q_i \exp(lnLCT_i - lnB_i) \quad (120)$$

and the objective function

$$Cost = \min \sum_{j=1}^{J} a_j \exp(lnM_j + \alpha_j lnV_j) + U * SLACK \quad (121)$$

Now a feasible integer point will be optimized as usual and an infeasible integer point will optimize the plant to minimize the $SLACK$, which is the extra time over the time horizon needed for fulfilling the production requirements. This is a valid point at which to linearize the nonlinear equations since the problem being convex any point satisfying the constraints will be a valid linearization point.

One disadvantage with this slack variable lies in the termination criterion used in DICOPT++. It terminates the iterations when a subproblem has a larger objective value than the previous one. This means that if from a feasible (but bad) point the master problem projects an infeasible point the iteration is terminated since the objective value of the subproblem will be larger due to the high penalty on the slack variable. The algorithm will present the feasible (but highly sub-optimal) point as the optimal solution.

8 Example data

Processing times, Size factors, Production requirement

$$T_{i,j} = \begin{bmatrix} 7 & 3 & 8 \\ 9 & 2 & 7 \\ 8 & 3 & 2 \end{bmatrix} \quad S_{i,j} = \begin{bmatrix} 3 & 9 & 3 \\ 2 & 10 & 2 \\ 4 & 3 & 9 \end{bmatrix} \quad Q_i = \begin{bmatrix} 260\,000 \\ 260\,000 \\ 260\,000 \end{bmatrix}$$

Cost exponent, Cost factor, Time horizon, Bounds on size

$$\alpha_j = \begin{bmatrix} 0.6 & 0.6 & 0.6 \end{bmatrix} \quad a_j = \begin{bmatrix} 250 & 250 & 250 \end{bmatrix} \quad H = 6000\text{h} \quad 0 \leq V_j \leq 3000$$

Data for intermediate storage

$$SFS_{i,j} = \begin{bmatrix} 9 & 9 & - \\ 10 & 10 & - \\ 4 & 9 & - \end{bmatrix} \quad \gamma_j = \begin{bmatrix} 0.5 \\ 0.5 \\ - \end{bmatrix} b_j = \begin{bmatrix} 350 \\ 350 \\ - \end{bmatrix}$$

Data for flexible production requirement

$$Q_i^{\text{ref}} = \begin{bmatrix} 260\,000 \\ 260\,000 \\ 260\,000 \end{bmatrix} c_i^1 = \begin{bmatrix} 78 * 10^6 \\ 78 * 10^6 \\ 78 * 10^6 \end{bmatrix} c_i^2 = \begin{bmatrix} 96.2 * 10^6 \\ 96.2 * 10^6 \\ 96.2 * 10^6 \end{bmatrix}$$

c_i^1 represents a marginal value of 300 for an extra unit of product around Q_i^{ref}
c_i^2 represents a marginal value of 370 for an extra unit of product around Q_i^{ref}
Discrete equipment sizes

$$Vsize_j = \begin{bmatrix} 500 & 1000 & 2000 & 3000 \end{bmatrix}$$

$$Cost_j = \begin{bmatrix} 10\,407 & 15\,774 & 23\,909 & 30\,494 \end{bmatrix}$$

For problems with no parallel equipment allowed the discrete sizes are:

$Vsize_j$	[400	630	1000	1600	2500	4000	6200	10000	15500]
$Cost_j$	[9103	11955	15774	20913	27334	36239	47138	62787	81684]

9 Results

We have solved the problems presented in chapter 8 with different formulations and we present the results in the next chapter where we give a schematic picture of the structure of the plant, total cost and unit sizes. The computational requirements (on a VAX 9000) and the number of major iterations required by the algorithm are also given. The example problem is only a small one and to show the variations of solution time we give a table with the solution time for randomly generated problems with the different features.

9.1 Summary of results

Formulation 6.1 Parallel units out of phase

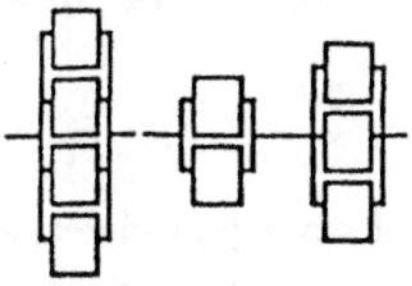

Cost=208 983
V_1 1096
V_2 3000
V_3 2466

CPU=0.98 s
Major iterations=2
Binary variables 12

Formulation 6.2 Parallel units in and out of phase

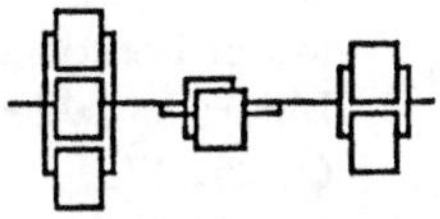

Cost=180 472
V_1 1614
V_2 2633
V_3 3000

CPU=1.73 s
Major iterations=2
Binary variables 24

Formulation 6.3 Parallel units in and out of phase, nonidentical units

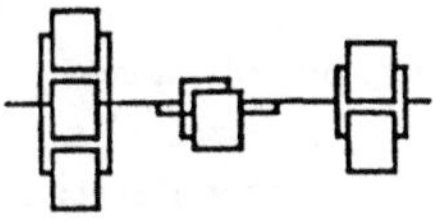

Cost=180 318
V_1 1650
V_2^1 3000
V_2^2 2143
V_3 3000

CPU=2.84 s
Major iterations=3
Binary variables 12

Formulation 6.4 Flexible use of parallel units in and out of phase
This is the plant configuration for products 1 and 2

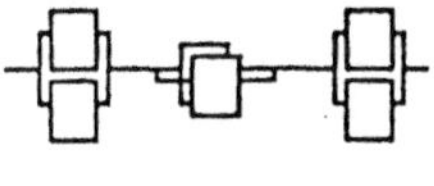

Cost=157 864
V_1 2052
V_2 2692
V_3 2309

CPU=3.26 s
Major iterations=2
Binary variables 72

This is the plant configuration for product 3

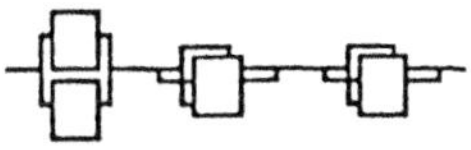

Formulation 6.5 Intermediate storage and parallel units out of phase

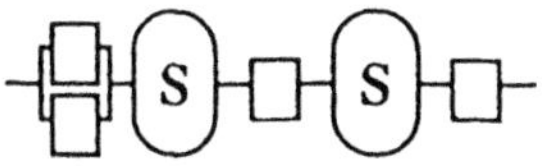

Cost=183 613
V_1 2370
V_2 3000
V_3 2666
VS_j 8871
VS_j 12283
Comments: Size of storage, sum of up and down stream batch size

CPU=20.78 s
Major iterations=3
Binary variables 21

Formulation 6.5 Intermediate storage and parallel units out of phase

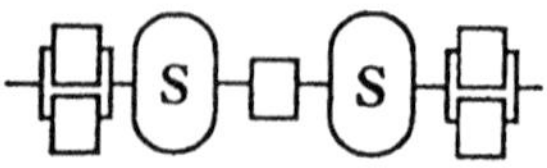

Cost=194 285 CPU=7.01 s
V_1 2370 Major iterations=4
V_2 3000 Binary variables 21
V_3 1333
VS_j 12283
VS_j 9553
Comments: Size of storage, twice the larger of up and down stream batch size

Formulation 6.6 Variable production requirement and units out of phase

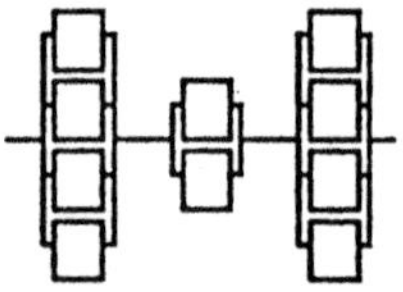

Cost=208 157 CPU=1.58 s
Plant=201 062 Major iterations=2
Margin= 7 096 Binary variables 12
V_1 1000 Q_1 251 171
V_2 3000 Q_2 254 734
V_3 2250 Q_3 251 171

Comments: This is the solution in which the marginal value of an extra kg is 300 "money units" at Q_i^{ref}

Formulation 6.6 Variable production requirement and units out of phase

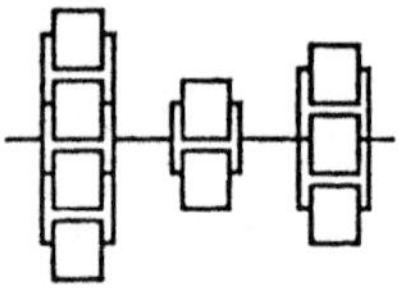

Cost=206 020 CPU=1.11 s
Plant=226 722 Major iterations=2
Margin= -20 702 Binary variables 12
V_1 1000 Q_1 305 555
V_2 3000 Q_2 273 297
V_3 2250 Q_3 264 618

Comments:This is the solution in which the marginal value of an extra kg is 370 "money units" at Q_i^{ref}

Formulation 6.7 Multiplant production

Plant 1 product 1

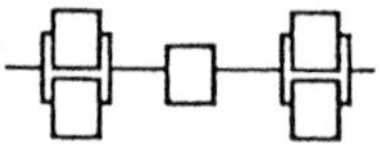

Total Cost=177 731 CPU=37.7 s
Cost=63 216 Major iterations=3
V_1 520 Binary variables 36
V_2 1560
V_3 520

Cost=58 169
V_1 202
V_2 1011
V_3 202

Plant 2 product 2

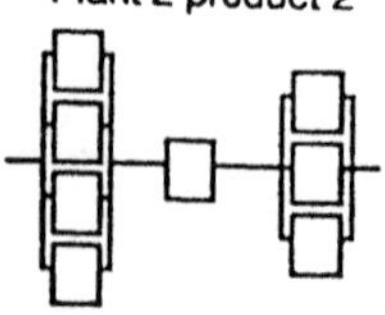

Cost=56 346
V_1 693
V_2 502
V_3 1560

Plant 3 product 3

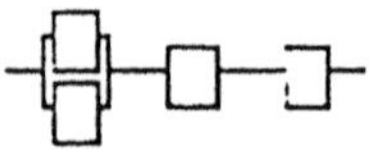

Formulation 6.8 Discrete unit sizes and parallel units out of phase

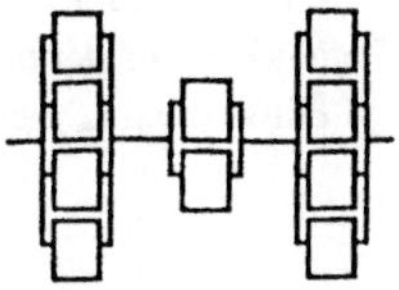

Cost=219 718 CPU=2.3 s
V_1 1000 Major iterations=2
V_2 3000 Binary variables 24
V_3 2000

Formulation 6.9.3 Discrete unit sizes - MILP

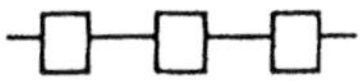

Cost=175 961 CPU=0.25 s
V_1 6200 Binary variables 27
V_2 15500
V_3 6200

Formulation 6.9.4 Discrete unit sizes and multiproduct campaigns

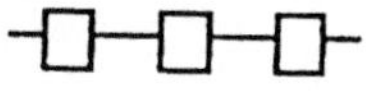

Cost=165 061 CPU=0.31 s
V_1 4000 Binary variables 27
V_2 15500
V_3 6200

Comments: The campaigns used are: 18 batches of product 1, 178 batches of product 1 and 2 alternately, and 378 batches of product 3.

Formulation 6.9.5 Continuous unit sizes and multiproduct campaigns

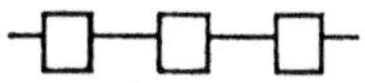

Cost=155 036 CPU=0.13 s
V_1 3597 Binary variables 0
V_2 10790
V_3 8092

Comments: The campaigns used are: 217 batches of product 1 and 2 alternately, 24 batches of product 2 and 3 alternately, and 265 batches of product 3.

Formulation NLP-3 Continuous unit sizes and long single product campaigns

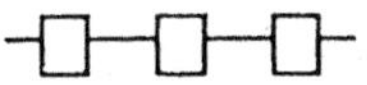

Cost=158 371 CPU=0.15 s
V_1 3726 Binary variables 0
V_2 11180
V_3 8385

Formulation 6.11, Underutilization 50%

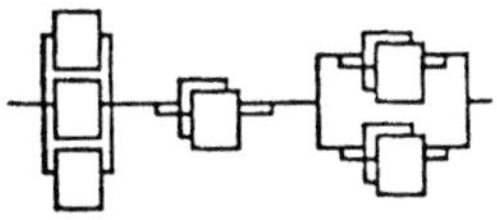

Cost=201 628 CPU=1.72 s
V_1 1545 Major iterations=2
V_2 2318 Binary variables 12
V_3 1738

Formulation 6.11, Underutilization 65%

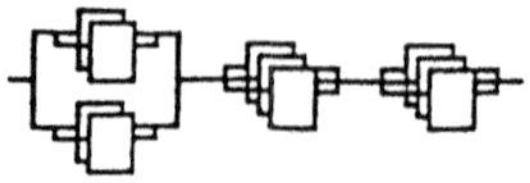

Cost=235 842
V_1 1355
V_2 2937
V_3 1909

CPU=6.56 s
Major iterations=2
Binary variables 36

Comments: Product 1 uses windows {2 2 2} e.g. on all stages two or more units are used. Product 2 uses windows {2 2 1} e.g. on stage three only one of the units are used. Product 3 uses windows {2 1 2}

Formulation 6.11, Underutilization 77%

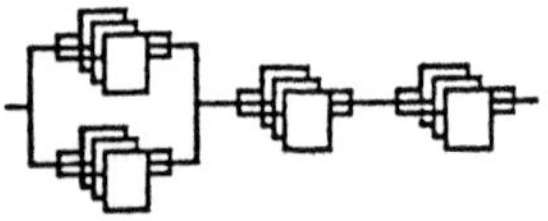

Cost=251 013
V_1 933
V_2 2766
V_3 1909

CPU=7.00 s
Major iterations=2
Binary variables 48

Comments: Product 1 uses windows {3 2 2} e.g. on stage 1 all three units on the other stages only two units are used. Product 2 uses windows {2 3 1} e.g. on stage 1 two units, stage 2 three units and stage3 one unit. Product 3 uses windows {3 1 3}

9.2 Discussion of results

It is difficult to compare all the different formulations with each other but as the superstructure gets larger from one formulation to an other the solution improves (or stays the same). Formulations 6.1 to 6.4 include only parallel units and it can be seen that when the flexibility increases the cost of the plant decreases. The solution time also increases with the flexibility of the formulation. The decrease of cost for the nonidentical units formulation is marginal but for other examples the savings can probably be greater. In the flexible use of parallel units we have to reconfigure the plant, in stage three, between products 1 and 2 and product 3.

The storage formulation (6.5) shows that the cost is reduced by inserting tanks. When we use the linear sizing constraint (storage twice the largest batch size), the third stage will have two units out of phase to reduce the batch size of the last subtrain. The linear storage constraints increase the cost, since the needed storage tank is larger, but reduces the solution time.

The variable production requirement formulation (6.6) is solved for two marginal values of the product. With a marginal value of 300 for all products it is advantageous to produce less than the reference requirement and take the penalty for under production. With a marginal value of 370 for all products we get a larger plant and the increased cost is reduced by the profit of the extra products.

The solution of the multiplant formulation (6.7) is that it is more economical to produce each product in a separate dedicated plant. This problem has the longest solution time of all problem formulations.

When parallel equipment is not allowed the constraint that units are only available in discrete sizes increases the cost. Using multiproduct campaigns reduces the cost for both continuous and discrete sizes.

The underutilization (6.11) constraints increase the cost. The cost increases as the underutilization demand increases. To avoid batches in the forbidden regions the solution has to have many parallel units in phase. The solution time increases as we have to add more binary variables.

9.3 Results on randomly generated problems

In the table below the solution time and the number of binary variables for the different features for different problem sizes are given. BV stands for binary variables. CPUs is the solution time, in seconds, in a VAX 9000. ave. is the average solution time of 5 problems and max is the largest solution time of the 5 problems. Size factors and processing times are randomly generated numbers in the range [1, 10]. All problems have three products (exept 6.7 (multiplant) that has two products) and production requirements $Q_i = 260\ 000$. The time horizon is $H = 6000$ and other data for the features are the same as in Chapter 8.

Problem type		Problem size										
		3 Stages		4 Stages			5 Stages			6 Stages		
		BV	CPUs	BV	CPUs		BV	CPUs		BV	CPUs	
					ave.	max		ave.	max		ave.	max
6.1	MINLP	12	0.98	16	1.71	2.40	20	2.01	2.21	24	3.52	6.23
6.2	MINLP	24	1.73	32	5.53	9.97	40	9.23	16.3	48	18.1	47.0
6.4	MINLP	72	3.26	96	13.1	23.0	120	86.2	166	144	334	789
6.5.4	MINLP	18	7.01	26	7.91	12.0	35	18.9	30.7	45	30.9	40.7
6.6	MINLP	12	1.58	16	2.42	4.42	20	2.96	3.70	24	6.65	8.97
6.7	MINLP	36	37.7	35	9.59	16.9	43	75.17	239	51	109	385
6.8	MINLP	24	2.3	32	9.11	11.6	40	15.7	33.2	48	47.5	101
6.9.3	MILP	27	0.25	36	0.24	0.28	45	0.39	0.46	54	1.31	1.59
6.9.4	MILP	27	0.31	36	0.58	0.71	45	0.76	1.09	54	1.35	1.66
6.9.5	NLP	0	0.13	0	0.15	0.18	0	0.16	0.20	0	0.19	0.24
6.11	MINLP	12	1.72	16	3.92	5.58	20	12.0	14.9	24	44.8	131

Table 1. The solution time on randomly generated test problems.

The number of binary variables is a measure of the size of the problem. But the number of binary variables is not directly proportional to solution time for the different formulations. The binary variables for parallel units have better continuous relaxations than do the binary variables for storage and multiplant. The branch and bound procedure in the master problem can then reduce the search tree by better bounds and this affects the solution time.

Parallel units out of phase (6.1) and parallel units in and out of phase (6.2) have a moderate increase in solution time as the size of the problem increase. Flexible use of parallel equipment (6.4) as a dramatic increase in solution time as the problem gets larger. The storage formulation with linear sizing constraints (6.5.4) has a moderate increase in solution time but the solution found by DICOPT++ is usually not the global optimum due to the termination criterion used. Variable production requirement (6.6) has a slight increase in solution time compared with (6.1). The multplant formulation (6.7) takes much more time to solve than solving three problems (two products) with parallel units.

For the formulation with discrete equipment sizes and parallel units (6.8) the solution time increases more than for (6.1). Without parallel equipment (6.9.3) the formulation reduces to a MILP and this can be solved easily even with additional constraints for multiproduct campaigns (6.9.4). Without discrete equipment sizes the problem reduces to a NLP (6.9.5). With underutilization constraints (6.11) not allowing units to be less that half full the problem becomes harder to solve than when we drop the constraints (6.1).

10 Splitting and merging - Synthesis

In all the previous formulations the task structure is fixed and there is no way to optimize this structure. The search for a good or optimal task structure was previously done in a sequential manner by splitting tasks at the time limiting stage and merging tasks that are not time limiting. Yeh and Reklaitis [28] gave a splitting MINLP formulation with many binary variables for splitting tasks in a single product plant. They formulate the problem with the binary variable $X_{r,k,j} = 1$ if the k th task of stage j is assigned to the r th unit, 0 otherwise. They come to the conclusion that the formulation is too time consuming for a preliminary synthesis procedure. Birewar and Grossman [3] give a very interesting synthesis formulation with a binary variable $Y_{t,j} = 1$ if task t is executed in unit j, 0 otherwise.

Their formulation contains a number of nonconvex functions and it is not surprising that the authors have not found the global optimum to some of their examples. Some of the example solutions given have even an infeasible structure. One drawback of this formulation is the elaborate logical constraints on the binary variables and the "semi binary ($YEX_j, YF_{t,j}$) variables".

Using logarithmic variables to make their problem formulation convex does not work.

10.1 Our synthesis formulation

We presented [19] a similar formulation but with logarithmic variables, This results in a problem with only one non-convex equation (126). We use the binary variable $X_{t,j} = 1$ if task t is executed in unit j, 0 otherwise. If two or more tasks are executed in the same unit we use the largest cost factor of the tasks (equation 128) and the cost factor is expressed in logarithmic form to avoid a non-convex objective function (122). We define a set of units J_t that can perform tasks t and a set of tasks T_j that can be executed in unit j.

The logical constraints are much simpler. There are only two logical constraints in this formulation. Equation (131) ensures that each task is performed in one and only one unit and equation (132) allows a task to be performed in a unit only if the previous task is performed in that unit. The first task in a group of mergeable tasks is always fixed to a unit.

In this formulation we do not have to generate "super units" since the cost (and type) of a unit is a function of the tasks performed in it. We can therefore save some binary variables.

$$Cost = \min \sum_{j=1}^{J} \exp(lna_j + lnM_j + \alpha_j lnV_j) \tag{122}$$

$$H \geq \sum_{i=1}^{N_p} Q_i \exp(lnLCT_i - lnB_i) \tag{123}$$

$$lnV_j \geq lnB_i + lnS_{i,j} - U(1 - X_{t,j}) \tag{124}$$

$$T_{i,j} = \sum_{t=1}^{T} T_{i,t} Y_{t,j} \tag{125}$$

$$\exp(lnT_{i,j}) = T_{i,j} \tag{126}$$

$$lnLCT_i = lnT_{i,j} - lnM_j \tag{127}$$

$$lna_j \geq \ln(a_{t,j}) - U(1 - X_{t,j}) \tag{128}$$

$$lnM_j = \sum_{k=0}^{M_j^U} \ln(k) Y_{k,j} \tag{129}$$

$$1 = \sum_{k=0}^{M_j^U} Y_{k,j} \tag{130}$$

$$1 = \sum_{t=1}^{T} X_{t,j} \tag{131}$$

$$X_{t+1,j} \leq X_{t,j} \tag{132}$$

$$X_{t,j} = 0 \quad t \notin T_j \tag{133}$$

$$lnV_j^L \leq lnV_j \leq lnV_j^U \tag{134}$$

In the binary variable $Y_{k,j}$ for parallel equipment k goes from 0 to M_j^U but k cannot be equal to 0 and instead we use $k = 10^{-4}, 1, \ldots, M_j^U$.

The equation (126) is non-convex. We have to use logarithmic variables in the formulation in order to avoid bilinear horizon constraints. For each product we sum the processing times of the tasks that are executed at a stage to give the stage cycle time (125). Constraint (131) ensures that all tasks are executed in one and only one stage. Constraint (132) ensures that only consecutive tasks are performed at a stage. If any task is performed on a unit it is either the first task on that unit or its immediate predecessor task is also allocated to the same unit. A task can only be performed in units that can execute the task (133).

10.1.1 The noncovex processing time constraint

The equality relaxation in DICOPT++ will relax equation (126) to

$$\exp(lnT_{i,j}) \leq T_{i,j} \tag{135}$$

this equation is non-convex and linearizations obtained by DICOPT++ possibly cut away the global optimum. We can also replace the nonconvex equations with a piecewise linear function as described by Vaselenak et. al. [26] and implement the outer approximation with piecewise linear approximator with APROS [17] in GAMS. This introduces a large number of binary variables.

10.1.2 The linear processing time constraint

We can also replace equation (126) with a number of linear equations. For example when we have three tasks in a group that can be merged we get the linear constraints:

$$lnT_{i,j} \geq \ln(T_{i,t}) - U(1 - X_{t,j}) \tag{136}$$

$$lnT_{i,j} \geq \ln(T_{i,t} + T_{i,t+1}) - U(2 - X_{t,j} - X_{t+1,j}) \tag{137}$$

$$lnT_{i,j} \geq \ln(T_{i,t} + T_{i,t+1} + T_{i,t+2}) - U(3 - X_{t,j} - X_{t+1,j} - X_{t+2,j}) \tag{138}$$

$$lnT_{i,j+1} \geq \ln(T_{i,t+1}) - U(1 - X_{t+1,j+1}) \tag{139}$$

$$lnT_{i,j+1} \geq \ln(T_{i,t+1} + T_{i,t+2}) - U(2 - X_{t+1,j+1} - X_{t+2,j+1}) \tag{140}$$

$$lnT_{i,j+2} \geq \ln(T_{i,t+2}) - U(1 - X_{t+2,j+2}) \tag{141}$$

Task $t, t+1$ and $t+2$ can be merged into one unit j or performed separately in units $j, j+1$ and $j+2$. Four different configurations are possible.

- Unit j performs tasks $t, t+1$ and $t+2$
- Unit j performs tasks t and $t+1$, and unit $j+2$ performs task $t+2$
- Unit j performs task t and unit $j+1$ performs tasks $t+1$ and $t+2$
- Unit j performs task t and unit $j+1$ performs task $t+1$ and unit $j+2$ performs task $t+2$

For the first configuration equation (138) gives the processing time. For the second configuration the equation (137) and equation (141) give the processing time. For the third configuration equation (136) and equation (140) give the processing time. For the fourth configuration the equation (136), (139) and equation (141) give the processing time.

These constraints replace the nonconvex constraint (127) but they have to be formulated depending on the problem. We will show how this is done on our example problem.

10.1.3 Example data

We use the problem 2 stated in Birewar and Grossman [3]. We have to formulate the problem parameters somewhat differently.

Processing time $T_{i,t}$(h)

products	mix1	rxn1	distln	mix2	rxn2	crystl	Production Requirement (kg)
A	2	8	4	1	6	9	600 000
B	2	4	3	1	5	4	600 000
C	1	6	5	3	7	4	700 000
D	3	5	6	2	9	3	700 000
E	3	7	5	2	8	1.5	200 000
F	2.5	4	4	3	4	2	100 000

Size factors $S_{i,t}$

products	mix1	rxn1	distln	mix2	rxn2	crystl
A	3	1	3	4	2	1
B	5	4	4	5	3	4
C	4	2	2	3	2	3
D	3	2	2	3	1	3
E	3	2	2	4	1	4
F	4	4	4	4	4	5

Horizon time (H) = 6 000 h. Cost factor α_j (for all units) = 0.6

$a_{t,j}$ is the cost factor if task t is performed in unit j

Task	Unit					
	1	2	3	4	5	6
mix1	200					
rxn1	300	300				
distln			450			
mix2				200		
rxn2				300	300	
crystl				550	550	250

$\gamma_{t,j}$ is the fixed cost factor if task t is performed in unit j

Task	Unit					
	1	2	3	4	5	6
mix1	45 000					
rxn1	55 000	55 000				
distln			70 000			
mix2				45 000		
rxn2				60 000	60 000	
crystl				95 000	95 000	50 000

The relation between Tasks and unit types

Task	Unit					
	1	2	3	4	5	6
mix1	CI, NJ, A					
rxn1	CI, J, A	CI, J, A				
distln			Des. Col			
mix2				CI, NJ, A		
rxn2				SS, NJ, A	SS, NJ, A	
crystl				SS, J, A	SS, J, A	CI, J

CI cast iron, SS stainless steel, NJ nonjacketed, J jacketed, A agitator and Des. Col. destilation column.

We can see that if task "crystl" is performed in unit 4, this unit is a "super unit" as Birewar and Grossman call it. Tasks "mix2" and "rxn2" have to be performed by the same unit.

10.1.4 Convex MINLP formulation for example

$$
\begin{aligned}
Cost &= \min \sum_{j=1}^{N} \exp(ln\gamma_j + lnM_j) \\
&+ \sum_{j=1}^{N} \exp(lna_j + \alpha_j lnV_j) && (142)\\
H &\geq \sum_{i=1}^{N_p} Q_i \exp(lnLCT_i - lnB_i) && (143)\\
lnV_j &\geq lnB_i + lnS_{i,t} - U(1 - X_{t,j}) && (144)\\
lnLCT_i &\geq lnT_{i,j} - lnM_j && (145)\\
lna_j &= \ln(a_{t,j}) - U(1 - X_{t,j}) && (146)\\
ln\gamma_j &= \ln(\gamma_{t,j}) - ln\gamma_{max}(1 - X_{t,j}) && (147)\\
lnM_j &= \sum_{k=0}^{M_j^U} \ln(k) Y_{k,j} && (148)\\
1 &= \sum_{k=0}^{M_j^U} Y_{k,j} && (149)\\
\sum_{t=1}^{T} X_{t,j} &= 1 && (150)\\
X_{t+1,j} &\leq X_{t,j} \quad t \in T_j && (151)\\
X_{t,j} &= 0 && (152)\\
lnV_j^L &\leq lnV_j \leq lnV_j^U && (153)
\end{aligned}
$$

and the problem specific constraint on the processing time

$$
\begin{aligned}
lnT_{i,1} &\geq \ln(T_{i,1}) - U(1 - X_{1,1}) && (154)\\
lnT_{i,1} &\geq \ln(T_{i,1} + T_{i,2}) - U(2 - X_{1,1} - X_{2,1}) && (155)\\
lnT_{i,2} &\geq \ln(T_{i,2}) - U(1 - X_{2,2}) && (156)\\
lnT_{i,3} &\geq \ln(T_{i,3}) - U(1 - X_{3,3}) && (157)\\
lnT_{i,4} &\geq \ln(T_{i,4}) - U(1 - X_{4,4}) && (158)\\
lnT_{i,4} &\geq \ln(T_{i,4} + T_{i,5}) - U(2 - X_{4,4} - X_{5,4}) && (159)\\
lnT_{i,4} &\geq \ln(T_{i,4} + T_{i,5} + T_{i,6}) - U(3 - X_{4,4} - X_{5,4} - X_{6,4}) && (160)\\
lnT_{i,5} &\geq \ln(T_{i,5}) - U(1 - X_{5,5}) && (161)\\
lnT_{i,5} &\geq \ln(T_{i,5} + T_{i,6}) - U(2 - X_{5,5} - X_{6,5}) && (162)\\
lnT_{i,6} &\geq \ln(T_{i,6}) - U(1 - X_{6,6}) && (163)
\end{aligned}
$$

These constraint have very poor continuous relaxation.

10.1.5 Results

In [3] solutions are presented for four different cases (a) to (d):

- (a) Merging not allowed, Single product campaigns.
- (b) Merging allowed, Single product campaigns, Zero wait scheduling.
- (c) Merging allowed, Multi product campaigns, Zero wait scheduling.
- (d) Merging allowed, Unlimited intermediate Storage.

		Problem		
	(a)	(b)	(c)	(d)
Unit	size	size	size	size
mixl	13 592			
rxnl	9 057	15 250	19 310	18 129
distln	9 091	11 268	12 000	11 250
mix2	12 500	15 000		
rxn2	6 897	8 511	15 000	15 000
crystl	12 500	15 000	15 000	15 000

CPUs for solving the problems (Microvax II)

		2 898 s	1 092 s	577 s

Cost presented in paper

	775 840	713 276	649 146	640 201

We have recalculated the costs from the unit sizes given in [3] with the following results:

	752 685	711 074	649 146	640 201

With the unit sizes given in [3] we have calculated the mimimum total processing time reqired to meet the demand with the following results:

	6805 h	6283 h	6266 h	6266 h

All solutions presented are infeasible (availiable time is 6000 h). The results of (c) and (d) even have a infeasible structures. Even with the four units in (d) (rxn1, distln, rx2, crystl) at the upper bound on size the production requrements cannot be fullfilled within the time horizon. If we cannot solve the problem with Unlimited intermediate Storage the multiproduct campaign is also infeasible.

We have not tried to solve the Multiproduct campaign problem (c) since formulating the campaign constraints will introduce a number of nonconvex (bilinear) constraints and therefore the solution cannot be guaranteed to be globally optimal.

		Proposed solutions	Problem
	(a)	(b)	(d)
Unit	size	size	size
mixl	15 000		
rxnl	10 000	17 800	16 893
distln	10 727	11 867	11 262
mix2	14 298	15 000	15 000
rxn2	7 500	8 900	8 446
crystl	14 298	15 000	15 000

Proposed cost and solution time (seconds on Vax9000)

	786 356	726 328	716 995
CPUs	1.19 s	26.04 s	50.78 s

The large savings in cost reported in the paper (16.3% for MPC and 17.5% for UIS) are mainly due to the fact that the reported solution moves into a region where one unit can be deleted. This structure is in fact infeasible. With a feasible plant the cost savings for UIS is 8.8% and probably less for MPC.

11 Discussion

In this paper we have developed a number of different formulations, which can all be solved to optimality, for the same general problem. The question now arises which of these are the best or how do we get a "total global optimum". There are as we see it two ways. One is to combine all formulations to a super formulation which would then contain all discrete choices but probably be too large to solve even for small problems. The other way is to solve a small part and at the same time generate information about what additional features could with advantage be included. this could provide the basis for an iterative improvement of the solution.

11.1 How the different formulations are related

The model formulations of Section 6, shown in Fig. 2, are also graphically depicted in section 9.1.

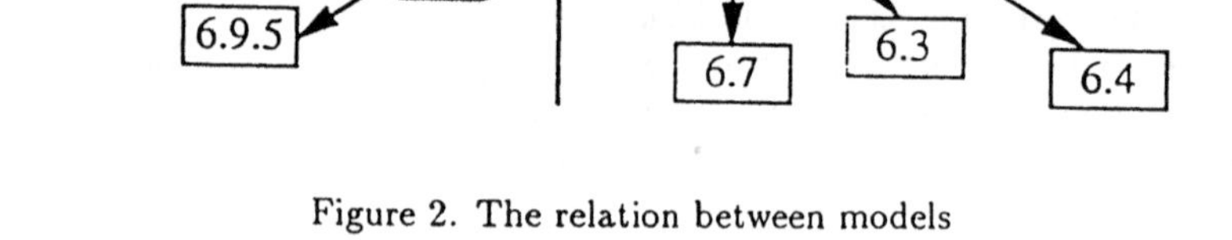

Figure 2. The relation between models

The formulations on the left side of the vertical line in Figure 2 do not include parallel equipment. The arrows pointing upwards indicate that the new formulation will have a larger or equal objective function value but the relative position of a formulation to all others in Figure 2 is meaningless. We start from NLP-3 which is the formulation presented in chapter 3 but without parallel equipment and without bounds on unit size and assuming that the products are produced in long single product campaigns. From this formulation we can:

- impose an upper bound on size and allow parallel units out of phase which leads to formulation 6.1
- add a constraint that units are only available in discrete sizes which leads to formulation 6.9.3
- allow multiproduct campaigns which leads to formulation 6.9.5

From formulation 6.9.3 we get formulation 6.9.4 by allowing multiproduct campaigns. Formulation 6.1 is the MINLP sizing problem with parallel units out of phase and from this we can get a number of other formulations:

- allowing parallel units to operate in phase leads to formulation 6.2
- allowing intermediate storage leads to formulation 6.5
- allowing the products to be produced in different plants leads to formulation 6.7
- allowing variable production requirement leads to formulation 6.6
- imposing the constraint that units are only available in discrete sizes leads to formulation 6.8

from formulation 6.2, with parallel units in and out of phase, we can:

- allow parallel equipment with flexible operation, different operation for different products, which leads to formulation 6.4
- allowing units in phase to have unequal sizes leads to formulation 6.3
- add the underutilization constraints in chapter 6.11

The storage formulation can be formulated as in chapter 6.5.4 or with the linear storage sizing constraints in chapter 6.5.2.

12 Conclusions

More than a dozen examples have been presented of the basic problem of equipment sizing for a multiproduct batch plant. Almost all of the examples require integer choices. They are presented in a common format and results are given for the solution of a basic problem in which the various extensions are incorporated in turn. Some of the extensions or formulations for particular extensions have been presented by previous authors, but several are new. These include the flexible use of parallel items, both in and out of phase, with different configurations for different products, the choice of intermediate storage location, the trade off between marginal product value and increased equipment cost, constraints preventing underutilization of equipment, transformation of the discrete sizing problem to an integer linear program, the multiplant problem with parallel equipment items and a new MINLP formulation of the process synthesis problem of task splitting and merging.

The simple demonstrative examples are used to show the result of the different formulations. The problem size is given in terms of the number of binary variables in the result summary (chapter 9.3) and the problem difficulty is indicated by the solution times.

The MINLP formulation seems to be quite well suited for some problems i.e. parallel equipment items, variable production requirements and underutilization. For others including storage location and multiplant selection, unless better MINLP formulations can be found it seems preferably to use enumeration, or for larger problems a stochastic method such as simulated annealing (Patel et.al. (1991)) which effectively makes a partial enumeration of promising alternatives.

13 Nomenclature

a_j	Cost factors
$a_{t,j}, a_{max}$	Cost factors
b_j	Cost factor, storage
$B_i, B_{i,q}, B_{i,j}$	Batch size
$B_{\text{up}}, B_{\text{down}}$	Batch size
c_i	Marginal value of product
$Cost_j$	Cost of stage
f_i	Production cost factor
J	Number of stages
$LCT_i, LCT_{i,j}$	Limiting cycle time
n_k, n_i	Number of batches
N_p	Number of products
$M_j, M_{i,j}, M^U$	Parallel units out of phase
$N_j, N_{i,j}, N^U$	Parallel units in phase
$NPRS_{i,k}$	Multi product campaigns
PC_i	Production cost
PRO_i	Inverse productivity
Q_i, Q^{ref}	Production requirements
$R_{i,j}$	Time factor
$S_{i,j}, S_{i,t}$	Size factor
$SFS_{i,j}$	Size factor storage
$SL_{i,k,j}$	Slack time
$T_{i,j}, T_{i,t}$	Processing time
$Ti_{i,p}$	Total processing time
$TotM_j$	Total number of parallel units
$V_j, V_{j,p}$	Unit size (Volume)
VS_j, VS_s	Storage volume
$Vsize_{j,g}$	Set of discrete sizes
$Vcost_{j,g}$	Set of discrete costs
$X_{j,q}, X_{i,p}$	Binary variables
$X_{i,j,w}, X_{j,g}$	Binary variables
$Y_{k,j}, Y_{k,i,j}, Y_{t,j}$	Binary variables
$Z_{c,j}, Z_{c,i,j}$	Binary variables
$P_{i,j}$	Time constant
$PT_{i,p}$	Production time
U	Large skalar

Greek letters

α_j,	Cost exponent
φ_j,	Underutilization
λ_j,	Time exponent
γ_j,	Cost exponent for storage

Subscript

i	Product
j	Stage
k, c	For parallel units
q	Subtrain
g	Discrete equipment
p	Plant
t	Task

Superscript

max	Maximum
U	Upper (bound)
L	Lower (bound)

Transformation of variables

lnX	Variable expressing the log. value of variable X
$invX$	Variable expressing the inverse of variable X

Logarithms and exponential's

$\ln(X)$	Taking the logarithm of parameter X
$\exp(X)$	The exponential of parameter or variable X

REFERENCES

1. D.B. Birewar and I.E. Grossmann. Efficient optimization Algorithms for Zero-Wait Scheduling of Multiproduct Batch Plants. *Ind. Eng. Chem. Res.*, 28: 1333-1345, 1989.
2. D.B. Birewar and I.E. Grossmann. Simultaneous Production Planning and Scheduling in Multiproduct Batch Plants. *Ind. Eng. Chem. Res.*, 29: 570-580, 1990.
3. D.B. Birewar and I.E. Grossmann. Simultaneous Synthesis, Sizing and Scheduling of Multiproduct Batch Plants. *Ind. Eng. Chem. Res.*, 29: 2242-2251, 1990.
4. D.B. Birewar and I.E. Grossmann. Incorporating Scheduling in the Optimal Design of Multiproduct Batch Plants. *Computers and Chem. Eng.*, 13(1/2): 141-161, 1989.
5. G.A. Coulman. Algorithm for Optimal Scheduling and Revised Formulation of Batch Plant Design. Ind. Eng. Chem. Res., 28: 553, 1989.
6. M.A. Duran and I.E. Grossmann. A Mixed-Integer Nonlinear Programming Algorithm for Process System Synthesis. *AIChE J.*, 32(4); 592-606, 1986.
7. A. Espuna, M. Lazaro, J.M. Martinez, and L. Puigjaner. Efficient and Simplified Solution to the Predesign Problem of Multiproduct Plants. Computers and Chem. Eng., 13: 163-174, 1989.
8. R.S. Garfinkel and G.L. Nemhauser. *Integer Programming*. Wiley: New York, 1972.
9. I.E. Grossmann and R.W.H. Sargent. Optimal Design of Multipurpose Chemical Plants. *Ind. Eng. Chem. Process. Des. Dev.*, 18(2), 1979.
10. I.A. Karimi and G.V. Reklaitis. Intermediate Storage in Noncontinuous Processes Involving Stages of Parallel Units. *AIChE J.*, 31: 44, 1985.
11. J. Klossner and D.W.T. Rippin. Combinatorial Problems in the Design of Multiproduct Batch Plant - Extension to Multiplant and Partly Parallel Operation. *Presented at the AIChE Annual Meeting, San Francisco*, Nov. 1984.
12. F.C. Knopf, M.R. Okos, and G.V. Reklaitis. Optimal Design of Batch/Semicontinuous Processes. *Ind. Eng. Chem. Process. Des. Dev.*, 21: 79-86, 1982.
13. G.R. Kocis and I.E. Grossmann. Global Optimization of Nonconvex Mixed-Integer Nonlinear Programming (MINLP) Problems in Process Synthesis. *Ind. Eng. Chem. Res.*, 27: 1407-1421, 1988.
14. G.R. Kocis and I.E. Grossmann. Relaxation Strategy for the Structural Optimization of Process Flowsheets. *Ind. Eng. Chem. Res.*, 26: 1869-1880, 1987.
15. Y.R. Loonkar and J.D. Robinson. Minimization of Capital Investments for Batch Processes. *Ind. Eng. Chem. Process. Des. Dev.*, 9(4), 1970.
16. A.K. Modi and I.A. Karimi. Design of Multiproduct Batch Processes with Finite Intermediate Storage. *Computers and Chem. Eng.*, 13(1/2): 127-139, 1989.
17. G.E. Pales and C.A. Floudas. APROS: Algorithmic Development Methodology for Discrete-Continuous Optimization Problems. *Operations Research*, 37(6): 902-915, 1989.
18. A.N. Patel, R.S.H. Mah, and I.A. Karimi. Preliminary Design of Multiproduct Non-Continuous Plants Using Simulated Annealing. *Computers and Chem. Eng.*, 1991.
19. D.E. Ravemark and D.W. T. Rippin. Structure and Equipment for Multiproduct Batch Production. *Presented at AIChE 1991 Annual Meeting*, Nov. 1991.
20. D.W. T. Rippin. Design and Operation of Multiproduct and Multipurpose Batch Chemical Plants - An Analysis of Problem Structure. *Computers and Chem. Eng.*, 7(4): 463-481, 1983.
21. J.D. Robinson and Y.R. Loonkar. Minimizing Capital Investments for Multiproduct Batchplants. *Process Technol. Int.*, 17(11), 1972.
22. H.E. Salomone and O.A. Iribarren. Posynomial Modeling of Batch Plants: A Procedure to Include Process Decision Variables. *Computers and Chem. Eng.*, 16(3): 173-184, 1992.
23. R.E. Sparrow, G.I. Forder, and D.W.T. Rippin. The Choise of Equipment Sizes for Multiproduct Batch Plants. Heuristics vs. Branch and Bound. *Ind. Eng. Chem. Process. Des. Dev.*, 14(3), 1975.
24. I. Suhami and R.S.H. Mah. Optimal Design of Multipurpose Batch Plants. *Ind. Eng. Chem. Process. Des. Dev.*, 21:94-100, 1982.
25. T. Takamatsu, I. Hashimoto, and S. Hasebe. Optimal Design and Operation of a Batch Process with Intermediate Storage Tanks. *Ind. Eng. Chem. Process. Des. Dev.*, 21: 431-440, 1982.
26. J. Vaselenak, I.E. Grossmann, and A.W. Westerberg. Optimal Retrofit Design of Multiproduct Batch Plants. *Ind. Eng. Chem. Res.*, 26: 718-726, 1987.
27. J. Wiswanathan and I.E. Grossmann. A Combined Penalty Function and Outer-Approximation Method for MAINLY Optimization. *Computers and Chem. Eng.*, 14(7): 769-782, 1990.
28. N.C. Yeh and G.V. Reklaitis. Synthesis and Sizing of Batch/Semicontinuous Processes: Single Product Plants. *Computers and Chem. Eng.*, 11(6): 639-654, 1987.

The Influence of Resource Constraints on the Retrofit Design of Multipurpose Batch Chemical Plants

Savoula Papageorgaki[1], Athanasios G. Tsirukis[2], and Gintaras V. Reklaitis[1]

1. School of Chemical Engineering, Purdue University, W. Lafayette, IN 47907, USA
2. Department of Chemical Engineering, California Institute of Technology, Pasadena, CA 91125, USA

Abstract: The objective of this paper is to study the effects of resource availability on the retrofit design of multipurpose batch chemical plants. A mixed integer nonlinear model will be developed to address retrofit design arising from changes in the product demands and prices and revisions in the product slate (addition of new products, removal of old products, modifications in the product recipes). Resource constraints on the availability of utilities such as electricity, water and steam, manpower, etc. will be incorporated into the formulation. In addition, the option of resource expansion to accommodate the needs of the new plant will be explored.

A decomposition solution strategy will be presented to allow solution of the proposed MINLP optimization model in reasonable computation time. The effectiveness of the proposed model and solution strategy will be illustrated with a number of test examples.

Keywords: Batch design, retrofit, resource, mathematical programming.

Introduction

Retrofit design is defined as the redesign of an existing facility to accommodate revisions in the product slate and/or readjustments in product demands and feedstock availability, as well as to improve the operability of the process by means of increases in the process flexibility and reduction of the operating costs and the energy consumption. This problem is an important one to process operations because of the need to respond to variations in the availability and prices of feedstocks and energy, the short life cycles of many specialty chemicals and the continuing pressure to develop and accommodate new products. The availability of resources can significantly influence the feasibility and quality of the design during retrofit as resource limits may impose constraints on the extent of retrofit modifications.

The objective of this paper is to study the effects of resource availability on the retrofit design of multipurpose batch chemical plants. Retrofit in batch processes has been only sparingly investigated with most of the attention directed at the retrofit of multiproduct batch plants. A complete survey on the existing approaches can be found in [6]. The authors of this paper developed a MINLP optimization model and a subsequent solution strategy for the retrofit design of multipurpose batch plants in view of changes in the product demands/prices and/or revisions in the product slate. The issue of resource availability during retrofit, however, has not been addressed to date, although resource restrictions may impose the need for extensive modifications during retrofit.

Problem Statement

The deterministic retrofit design problem for a general multipurpose batch facility can be defined as follows [6]:

Given:

1. A set of products, the current production requirements for each product and its selling price and an initial configuration of equipment items used for the manufacture of these products.

2. A set of changes in the demands and prices of the existing products over a prespecified time period (horizon) and/or a set of modifications in the product slate in the form of addition of new products, elimination of old products or modification of the product recipe of existing products.

3. A set of available equipment items classified according to their function into *equipment families*, the items of a particular family differing in size or processing rate. Items that are members of the same equipment family and have the same size belong to the same *equipment type*.

4. Recipe information for each product (new and old) which includes the task precedence relationship, a set of processing times / rates and a corresponding set of size / duty factors, both associated with every feasible task - equipment pair. In general, the processing time may be specified as a function of the equipment capacity.

5. The set of feasible equipment items for each product task.

6. The status (stable or unstable) and the transfer rules for the intermediates produced between tasks.

7. Resource utilization levels or rates and change-over times between products with their associated costs.

8. Inventory availability and costs.

9. A suitable performance function involving capital and/or operating costs, sales revenue and inventory costs.

Determine :

(a) A feasible equipment configuration which will be used for the manufacture of each product in the plant (new and old),

(b) The sizes of new processing units and intermediate storage vessels and the number of units required for each equipment type (new and old),

so as to optimize the given performance function.

Model Formulation

The mixed integer nonlinear programming (MINLP) formulation developed by [6], to address the optimal retrofit design of general multipurpose batch chemical plants with no resource considerations, can be extended to incorporate resource restrictions. The key structural choice variable required by this formulation is the binary variable X_{imegk} which is defined as follows:

$$X_{imegk} = \begin{cases} 1 & \text{if } \textit{task m of product i is performed in unit type e in equipment group g} \\ & \text{and } \textit{in campaign k} \\ 0 & \textit{otherwise} \end{cases}$$

The variable set also includes variables denoting number of units, number of groups, batch sizes, cycle times, number of batches, campaign lengths and production amounts. The constraint set consists of seven principal subsets:

1. Assignment and Connectivity constraints
2. Production Demand constraints
3. Cycle Time and Horizon constraints
4. Equipment Size and Equipment Number constraints
5. Batch Size and Batch Number constraints
6. Direct and Derived Variable Bounds
7. Degeneracy Reduction Constraints

Finally, the objective function is a maximization of net profit (or minimization of -net profit) including the capital cost of the new equipment, the operating cost associated with the new and old equipment and the sales revenue resulting from the increased production of the old products and the production levels of the new products over a given time period.

Since resource availability will be considered in this work, an additional constraint set will be introduced into the formulation, namely

8. Resource Utilization Constraints

along with the appropriate set of variables [7]. If resource expansion is also considered, then an additional linear term will be introduced in the objective function as will be shown below. The set of resources includes all the plant utilities that supplement the operation of equipment items in a batch plant. For example, the reaction task of some product has specific temperature requirements which imply the use of heating or cooling fluids, electric filters consume electricity and almost every processing task requires the attendance of a human operator. The present paper deals with the class of renewable resources whose availability levels are replenished after their usage. Examples of renewable resources are manpower, electricity, heating and cooling flowrates, water and steam, etc. Simple extensions to the proposed formulation can accommodate the existence of consumable resources such as raw materials, capital, etc.

Let us now define the following sets:

$RES = \{ s \mid \text{resource } s \text{ is available to the plant} \}$, $S1_s = \{ i \mid \text{product } i \text{ uses resource } s \}$ and,

$S2_s = \{ m \mid \text{task } m \text{ uses resource } s \}$.

Furthermore, let r_{simegk} denote the utilization level of resource s by task m of product i performed in unit type e and in group g during campaign k. Then, by assuming nonlinear dependence of the resource utilization level on the split batch size BS_{imegk}, we get the following equation [7]

$$r_{simegk} = \eta_{sime} NU_{imegk} + \theta_{sime} NU_{imegk} BS_{imegk}^{\mu_{sime}}$$

where $s \in RES$; $i \in S1_s$; $m \in TA_i \cap S2_s$; $e \in P_{im}$; $g=1,...,NG^{max}_{imk}$; $k=1,...,K$. Furthermore, NU_{imegk} denotes the number of units of type e contained in group g that is assigned to task m of product i during campaign k, and, η_{sime}, θ_{sime} and μ_{sime} are given constants.

In addition, let RS_s denote the utilization level of resource s. Then, the total amount of resource usage is constrained by RS_s as follows [7]:

$$\sum_{i \in S1_s} \sum_{m \in TA_i \cap S2_s} \sum_{e \in P_{im}} \sum_{g=1}^{NG^{max}_{imk}} r_{simegk} \leq RS_s \qquad s \in RES\ ;\ k=1,...,K$$

Finally, let prs_s denote the cost coefficient associated with the utilization of resource s. Then, the following term will be added to the objective function to account for resource expansion:

$$\sum_{s \in RES} prs_s \, (RS_s - RS_s^{min})$$

The complete optimization model is shown in Appendix I. The extended formulation exhibits the same characteristics as the original retrofit model, namely, nonconvexity of the objective function and of the set of feasible solutions which leads to existence of many local optima, and, combinatorial complexity which results in prohibitively high computation times for problems of practical scope. Consequently, rather than attempting to solve the proposed model directly, a formulation specific decomposition of the original MINLP will be developed.

Solution Procedure

The solution procedure builts on our earlier developments for the retrofit design of multipurpose plants with no resource restrictions [6]. Specifically, the original MINLP problem (posed in minimization form) will be decomposed into two subproblems, an upper and a lower bound (master) subproblem, which are relaxations of the original model and which will be solved alternately until one of the termination criteria is met. The form of the relaxed subproblems, however, will be different to accommodate the incorporation of the resource constraints. Two versions of the decomposition scheme were proposed by [6] with flowcharts depicted in Figures 1 and 2. The two relaxed subproblems are described in the following sections.

Master Problem

The master problem is a broad relaxation of the original MINLP problem. The corresponding formulation is shown in Appendix II. Details on the derivation of constraints (II.3), (II.4), (II.5), (II.6)-(II.13) can be found in [6], whereas the derivation of (II.2) follows easily from the definition of the variable $VC_{imegk} = NU_{imegk} V_e$ and constraint (I.19). Integer cuts corresponding to infeasible assignments identified by the upper bound subproblem (described below) or cuts corresponding to previously identified assignments may also be included in the formulation. The following proposition describes the sufficient condition under which the master problem provides a valid lower bound on the optimal solution of the original MINLP:

Proposition 1. The master problem is a relaxation of the original MINLP model and provides a valid lower bound on the optimal solution of the original MINLP, if $b_e \leq 1$ for all e.

The proof appears in [6].

The master problem is nonconvex due to the nonlinear terms involved in constraints (I.12), (I.14), (I.15), (I.18), (I.27), (I.29), (I.30), (II.2) and (II.3). This form of the problem cannot be convexified, since most of the variables involved in the formulation have zero as their natural lower bound. Following the procedure developed in [6], we assume that the lower bounds on selected variables are equal to ε instead of 0, where ε is a very small positive number. Then the following proposition is true:

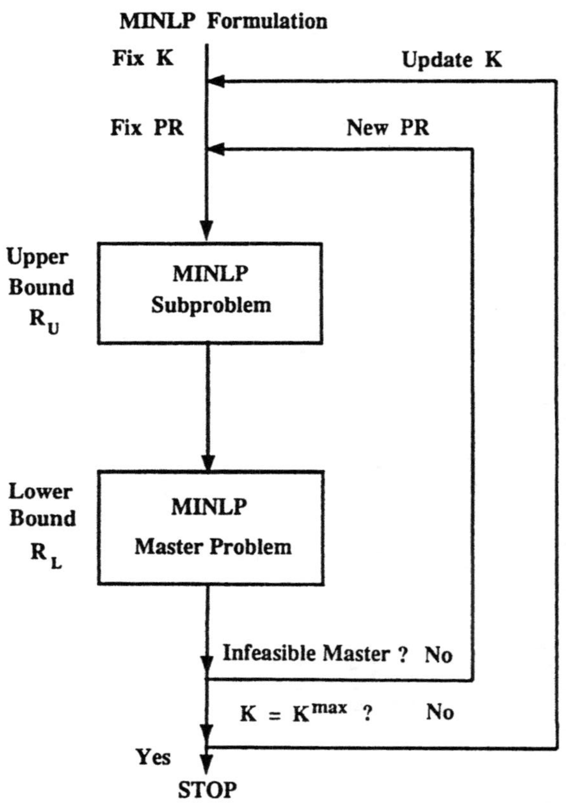

Figure 1. First version of proposed Decomposition Algorithm

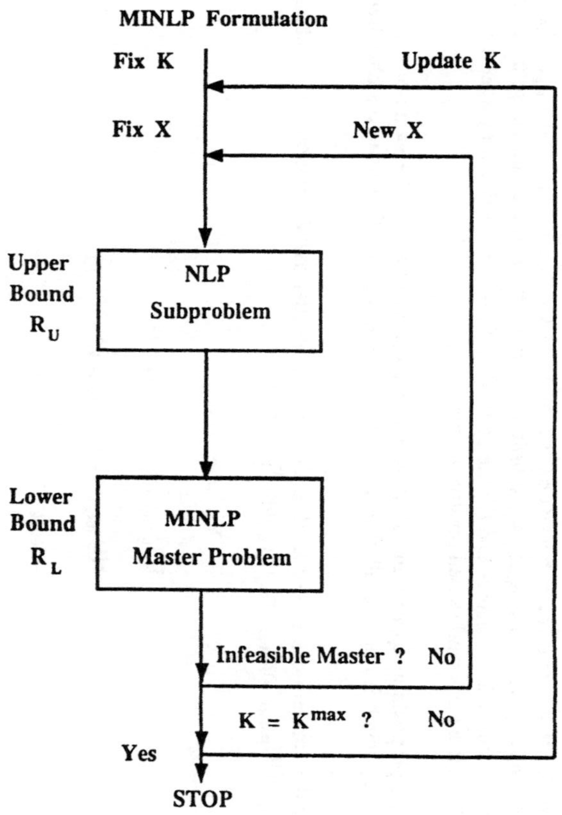

Figure 2. Second version of Proposed Decomposition Algorithm.

Proposition 2. If the zero lower bounds on the variables CAP_e, VC_{imegk}, NU_{imegk}, Q_{ime}, T_k, $T_{L_{ik}}$, Y_{ik}, n_{ik}, np_{imgk} and BS_{imegk} are substituted with ε, then (i) the optimal solution to the master problem will not change and (ii) the optimal profit will be modified by a term $o(\varepsilon)$, provided that ε is a sufficiently small positive number.

Proof of this proposition appears in [6]. A practical guide to selecting the value of ε is to choose a value which is much smaller than the values of the cost coefficients in the objective function.

Now the problem can be convexified through exponential transformations of variables Y_{ik}, BS_{imegk}, NU_{imegk}, np_{imgk}, n_{ik}, $T_{L_{ik}}$ and T_k, namely

$$Y_{ik} = \exp(sy_{ik})$$

$$BS_{imegk} = \exp(bs_{imegk})$$

$$NU_{imegk} = \exp(v_{imegk})$$

$$np_{imgk} = \exp(snp_{imgk})$$

$$n_{ik} = \exp(sn_{ik})$$

$$T_k = \exp(\tau_k)$$

$$T_{L_{ik}} = \exp(tl_{ik})$$

After this substitution, constraints (I.11), (I.12), (I.15), (I.17), (I.18), (I.20), (I.27)-(I.30), (II.2), (II.3) and (II.4) take the following form

$$\sum_{k=1}^{K} \exp(sy_{ik}) = P_i \tag{III.2}$$

$$\sum_{k=1}^{K} \sum_{g} \exp(v_{imegk} + bs_{imegk} + snp_{imgk}) \leq Q_{ime} \tag{III.3}$$

$$t_{imgk} \geq t^0_{ime} X_{imegk} + \alpha_{ime} \exp(\beta_{ime}\, bs_{imegk}) \tag{III.4}$$

$$\sum_{k} \exp(\tau_k) \leq H \tag{III.5}$$

$$\tau_k \geq sn_{ik} + tl_{ik} \tag{III.6}$$

$$N_e \geq \sum_{i} \sum_{m} \sum_{g} \exp(v_{imegk}) \tag{III.7}$$

$$\sum_{e} \exp(\nu_{imegk} + bs_{imegk}) \le B_i^{\max} \tag{III.8}$$

$$\sum_{g} \exp(snp_{imgk}) \exp(-sn_{ik}) \le 1 \tag{III.9}$$

$$\sum_{i}\sum_{m}\sum_{e}\sum_{g} \eta_{sime} \exp(\nu_{imegk}) + \theta_{sime} \exp(\mu_{sime}\, bs_{imegk} + \nu_{imegk}) \le RS_s \tag{III.10}$$

$$VC_{imegk} \ge S_{ime} \exp(bs_{imegk} + \nu_{imegk}). \tag{III.11}$$

$$\sum_{g}\sum_{e} \frac{VC_{imegk}}{S_{ime}} \ge \exp(sy_{ik} - sn_{ik}) \tag{III.12}$$

$$sn_{ik} \ge sy_{ik} - \ln(B_i^{\max}) \tag{III.13}$$

of which only the first is nonconvex (nonlinear equality). To remedy the situation, this constraint will be replaced by two equivalent constraints

$$\sum_{k=1}^{K} \exp(sy_{ik}) \le P_i \tag{III.2a}$$

$$\sum_{k=1}^{K} \exp(sy_{ik}) \ge P_i \tag{III.2b}$$

the first of which is convex and will be retained in the formulation and the second is nonconvex and it will be linearly approximated as follows

$$\sum_{k=1}^{K} (\,\delta_{ik}\, sy_{ik} + \theta_{ik}\,) \ge P_i \tag{III.2c}$$

where $\delta_{ik} = \dfrac{\exp(sy_{ik}^{\min}) - \exp(sy_{ik}^{\max})}{sy_{ik}^{\max} - sy_{ik}^{\min}}$

and $\theta_{ik} = \dfrac{\exp(sy_{ik}^{\max}) - \exp(sy_{ik}^{\min})}{sy_{ik}^{\max} - sy_{ik}^{\min}}\, sy_{ik}^{\min} - \exp(sy_{ik}^{\min})$

Notice that a piecewise linear overestimation [2] to more closely approximate this constraint can also be constructed at the cost of introducing additional binary variables in the model.

Constraint (I.14), however, remains nonconvex. By considering the logical equivalent of this constraint,

$$T_{L_{ik}} \geq \frac{t^{*}_{imgk}}{NG_{imk}} X_{imegk}$$

where

$$t^{*}_{imgk} \geq t^{0}_{ime} + \alpha_{ime} \exp\left(\beta_{ime}\, bs_{imegk}\right)$$

and by rearranging the terms as follows

$$NG_{imk} \geq \frac{t^{*}_{imgk}}{TL_{ik}} X_{imegk} \qquad \text{(III.14)}$$

we introduce a form which can be linearized according to the procedure introduced by [3] for bilinear constraints involving a continuous variable (in this case $\frac{t^{*}_{imgk}}{TL_{ik}}$) multiplied by an integer variable (in this case X_{imegk}). For this purpose, the following variables are introduced and substituted in the model

$$SRG_{imgk} = \frac{t^{*}_{imgk}}{T_{L_{ik}}} \qquad \text{(III.15)}$$

$$RG_{imegk} = SRG_{imgk}\, X_{imegk} \qquad \text{(III.16)}$$

After this substitution, constraint (III.14) takes the following equivalent form

$$NG_{imk} \geq RG_{imegk} \qquad \text{(III.14*)}$$

and, the bilinear equality (III.16) can be substituted by the following equivalent set of linear inequalities

$$SRG_{imgk} - SRG^{\max}_{imgk}\,(1 - X_{imegk}) \leq RG_{imegk} \qquad \text{(III.16a)}$$

$$SRG^{\min}_{imgk}\,(1 - X_{imegk}) \leq RG_{imegk} \qquad \text{(III.16b)}$$

$$SRG_{imgk} - SRG^{\min}_{imgk}\,(1 - X_{imegk}) \geq RG_{imegk} \qquad \text{(III.16c)}$$

$$SRG^{\max}_{imgk}\,(1 - X_{imegk}) \geq RG_{imegk} \qquad \text{(III.16d)}$$

Finally, the nonlinear equality (III.15) takes the following convex form after substituting for t^{*}_{imgk}

$$SRG_{imgk} \geq t^{0}_{ime} \exp\left(-tl_{ik}\right) + \alpha_{ime} \exp\left(\beta_{ime}\, bs_{imegk} - tl_{ik}\right) \qquad \text{(III.15*)}$$

Finally, the bounding constraints (I.16), (I.21), (I.22), (I.31)-(I.33), (II.6), (II.8) and (II.9) take a slightly different but equivalent form to account for the non-zero lower bound on the variables and their exponential transformation:

$$tl_{ik} \leq (\ln(T_{ik}^{\max}) - \ln\varepsilon) \sum_{m \in TA_i} \sum_{g=1}^{NG_{imk}^{\max}} \sum_{e \in P_{im}} X_{imegk} + \ln\varepsilon \qquad \forall\, i, k \tag{I.16*}$$

$$v_{imegk} \geq \ln\varepsilon\,(1 - X_{imegk}) \qquad \forall\, i, m, e, g, k \tag{I.21*}$$

$$v_{imegk} \leq (\ln(N_e^{\max}) - \ln\varepsilon)\, X_{imegk} + \ln\varepsilon \qquad \forall\, i, m, g, e, k \tag{I.22*}$$

$$bs_{imegk} \geq \ln(B_i^*)\, X_{imegk} + \ln\varepsilon\,(1 - X_{imegk}) \qquad \forall\, i, m, e, g, k \tag{I.31*}$$

$$snp_{imgk} \geq \ln(np_{imgk}^*)\, Z_{imgk} + \ln\varepsilon\,(1 - Z_{imgk}) \qquad \forall\, i, m, g, k \tag{I.32*}$$

$$snp_{imgk} \leq (\ln(np_{imgk}^{\max}) - \ln\varepsilon)\, Z_{imgk} + \ln\varepsilon \qquad \forall\, i, m, g, k \tag{I.33*}$$

$$Q_{ime} \leq (\ln P_i^{\max} - \ln\varepsilon) \sum_{k=1}^{K} \sum_{g=1}^{NG_{imk}^{\max}} X_{imegk} + \ln\varepsilon \qquad \forall i, m, e \tag{II.6*}$$

$$VC_{imegk} \leq (V_e^{\max} N_e^{\max} - \varepsilon)\, X_{imegk} + \varepsilon \qquad \forall\, i, m, e, g, k \tag{II.8*}$$

$$sy_{ik} \leq (\ln(P_i^{\max}) - \ln\varepsilon)\, PR_{ik} + \ln\varepsilon \qquad \forall\, i, k \tag{II.9*}$$

The new formulation (III) consists of minimizing equation (II.1) subject to constraints (III.2a), (III.2c), (III.3)-(III.13), (III.14*), (III.15*), (III.16a)-(III.16d), (II.5), (II.6*), (II.7), (II.8*), (II.9*), (II.10)-(II.13), (I.2)-(I.10), (I.16*), (I.23)-(I.26), (I.31*)-(I.33*), (I.35)-(I.39), (I.45)-(I.49), (I.51). Clearly, formulation (III) is a relaxation of formulation (II) due to the linear underestimation of the exponential terms in constraint (III.2c). As already mentioned, the new formulation constitutes a convex MINLP model which can be solved for its corresponding globally optimal solution. DICOPT (software implementation of the OA/ER algorithm [4]) can be used to solve the convex master problem utilizing MPSX to solve the MILP subproblems and MINOS 5.0 [5] for the NLP subproblems of the corresponding MINLP. Note that the OA/ER algorithm guarantees the global optimal solution for convex MINLP's.

Upper Bound Subproblem

In the first version of the decomposition algorithm, the upper bound subproblem corresponds to the original MINLP with the values of the integer variables PR_{ik} fixed. Consequently, the upper bound subproblem subproblem remains a MINLP model, but contains less binary variables than the original MINLP, since the product-campaign assignment is fixed and thus, several sets of binary variables can be eliminated from the model. In the second version of the decomposition scheme, the upper bound subproblem is a NLP model, since it corresponds to the original MINLP with the values of the binary variables X_{imegk} fixed. In both cases, the value of the objective function provides an upper bound on the optimal solution of the original MINLP. However, the problem formulation is nonconvex and cannot be convexified through variable transformations.

DICOPT++ (software implementation of the AP/OA/ER algorithm [9]) can be used for the solution of the nonconvex MINLP upper bound subproblems in the first version of the decomposition procedure and MINOS 5.0 can be used to solve the NLP upper bound subproblems in the second version of the algorithm.

Example

A multiproduct plant involving 4 products and 4 stages [8] is considered in this example. Since it is assumed that there is a one to one correspondence between stages and equipment families, four different equipment families are available in the plant. An initial equipment configuration involving one unit in each of the stages 1,2 and 4 and 2 out-of-phase units in stage 3 is given. For each of the existing units, the addition of a single unit in- and out-of-phase is considered. Consequently, the resulting maximum number of single-unit equipment groups that are allowed in each stage is 2 for stages 1,2 and 4, and 3 for stage 3. In addition, there are 14 equipment types available in the plant that are detailed in Table I. Note that, since no upper bounds on the equipment sizes have been explicitly given by [8], the sizes of the existing equipment will be used as upper bounds. In addition, since the proposed model requires non-zero lower bounds on the equipment capacities, a minimum capacity of 500 has been assumed for each equipment type. The unit processing times (assumed to be constant in this example) and size factors and the upper bounds on the annual production requirements and selling prices are given in Tables II and III. The authors approximated the capital cost of equipment by a fixed-charge model which is incorporated into our formulation in the following equivalent form

$$\sum_{e=1}^{NEQ} (\gamma_e V_e N_e + \lambda_e N_e)$$

The cost coefficients γ_e and λ_e are given in Table IV. Notice that since $CAP = V_e N_e$ in the master problem, the values of coefficients γ_e and λ_e will be used for coefficients c_e and d_e. Also notice that the value of coefficient λ_{G_i} has been corrected from the value of 10180 to the value of 44573 to agree with the reported results.

Additional assumptions that further simplify the model are that all products must use all units in the plant and thus, there is no product dependence of the structural variables, no operating costs are considered and the old units must be retained in the plant. As a consequence, a two-subscript binary variable suffices for the representation of the structural decisions that must be made at the design stage

Table I. Available Equipment Items

equipment item	capacity range (L)	N_e^{max}
R1 *,R2,R3	4000 *, 500-4000	1 *,1,1
L1 *,L2,L3	4000 *, 500-4000	1 *,1,1
F1 *,F2 *,F3,F4,F5	3000 *, 3000 *, 500-3000	1 *,1 *,1,1,1
G1 *,G2,G3	3000 *, 500-3000	1 *,1,1

* : existing units

Table II. Size Factors (L/kg/batch) and Processing Times (h/batch)

( ) = Processing Time

product / eq. type	R1,R2,R3	L1,L2,L3	F1-F5	G1,G2,G3
A	7.9130(6.3822)	2.0815(4.7393)	5.2268(8.3353)	4.9523(3.9443)
B	0.7891(6.7938)	0.2871(6.4175)	0.2744(6.4750)	3.3951(4.4382)
D	0.7122(1.0135)	2.5889(6.2699)	1.6425(5.3713)	3.5903(11.9213)
E	4.6730(3.1977)	2.3586(3.0415)	1.6087(3.4609)	2.7879(3.3047)

Table III. Upper Bounds on Demands and Selling Prices

product	projected demand (kg/yr)	price ($/kg)
A	268,200	1.114
B	156,000	0.535
D	189,700	0.774
E	166,100	0.224

$$X_{eg} = \begin{cases} 1 & \textit{if unit type e is assigned in equipment group g} \\ 0 & \textit{otherwise} \end{cases}$$

The rest of the variables and the constraints in the model are simplified accordingly. The results obtained after the solution of the model with no resource considerations [6] are shown in Table V. The corresponding design configuration depicted in Figure 3 yields a profit of $516,100 and it shows that the optimal policy is to purchase one unit that must operate in-phase in stage 4. Note that the second version of the proposed decomposition scheme has been used to solve this problem, since the values of the integer variables PR_{ik} are fixed due to the multiproduct nature of the plant in consideration. In addition, note that the modeling language GAMS [1] on an IBM 3090 was used for the execution of the program.

Let us now assume that resource restrictions have been imposed on the process. The set of resources includes steam (denoted by *ST*), cooling flowrates (*CL*), electricity (*EL*) and manpower (*MP*). The resource utilization rates are assumed to be posynomial functions of the split batch size, described by (I.29). The corresponding values of the constants η_{se}, θ_{se} and μ_{se} are presented in Table VI. Assume that no resource expansion is considered at this point, rather there is a maximum availability for each resource (RS_s^{max}) that is shown in Table VII. Therefore, the variable RS_s will assume a constant value equal to RS_s^{max} and the resource expansion term in the objective function will be deleted from the formulation. Details on the problem size and the computational requirements for the master problem and the NLP subproblem during the decomposition procedure are given in Table VIII.

The results obtained are shown in Table IX. Note that no expansion of the plant is suggested due to the imposed resource constraints which are rather restrictive in this case. Consequently, the profit of $461,400 that can be made during the operation of the plant is lower compared to the profit of $516,100 that can be made after the expansion of the plant with the addition of one unit in stage 4. This is due to the fact that, in the former case, the production level of product D is considerably lower than its upper bound value and, although no new equipment purchase is made, the revenue due to the production levels is lower than the increased revenue due to the new production targets in the latter case minus the investment cost for the additional unit.

Let us now solve a different version of this problem by retaining the same maximum resource availability RS_s^{max} and by slightly changing selected values of the constants η_{se} and μ_{se} (Table X). The proposed decomposition algorithm required three major iterations to attain the optimal solution which now suggests the purchase of two units that must operate in- and out-of-phase in stage 1. The results obtained are shown in Table XI and the corresponding design configuration is depicted in Figure 3. Note that the profit of $504,900 made in this case is again greater than the profit made with no equipment purchase ($461,400) because the new production target for product D is increased due to the addition of the in-phase unit in stage 1.

Finally, we solve a third version of the problem in which we consider the possibility of resource expansion. The upper (RS_s^{max}) and lower (RS_s^{min}) bounds on the utilization level of resource s, and the resource cost coefficients prs_s are given in Table XII. The values of the constants η_{se}, θ_{se} and μ_{se} presented in Table VI are considered in this case. The proposed

Table IV. Capital Cost Coefficients

unit type / cost coefficient	γ_e	λ_e
R1,R2,R3	0.1627	15280
L1,L2,L3	0.4068	38200
F1,F2,F3,F4,F5	0.4881	45840
G1,G2,G3	0.1084	44573

Table V. Solution with no resource restrictions (only nonzero values)

unit type	V_e	N_e
R1	4000	1
L1	4000	1
F1,F2	3000,3000	1,1
G1,G2	3000,3000	1,1

product	n_i	T_{L_i},h	P_i,kg
A	530.6	6.382	268,200
B	88.3	6.794	156,000
D	122.8	11.921	189,700
E	166.6	3.305	142,600

- Net Profit : $516,100

Table VI. Resource Coefficients η_{se}, θ_{se} , μ_{se}

resource / eq. type	R1,R2,R3	L1,L2,L3	F1-F5	G1,G2,G3
ST	4,1.36e-2,1	3,3e-2,1	-	3,3e-2,1.3
CL	4,2e-3,1.2	1,3e-2,0.75	-	1,3e-2,0.75
EL	3,1e-3,1.1	2,5e-4,1	2,3.4e-3,1.1	2,5e-4,1
MP	3,2e-2,0.75	1,1.55e-1,0.3	2,1.83e-5,0.5	1,1.55e-1,0.3

Table VII. Maximum Resource Availability R_s^{max}

resource	R_s^{max}
ST	100
CL	50
EL	70
MP	50

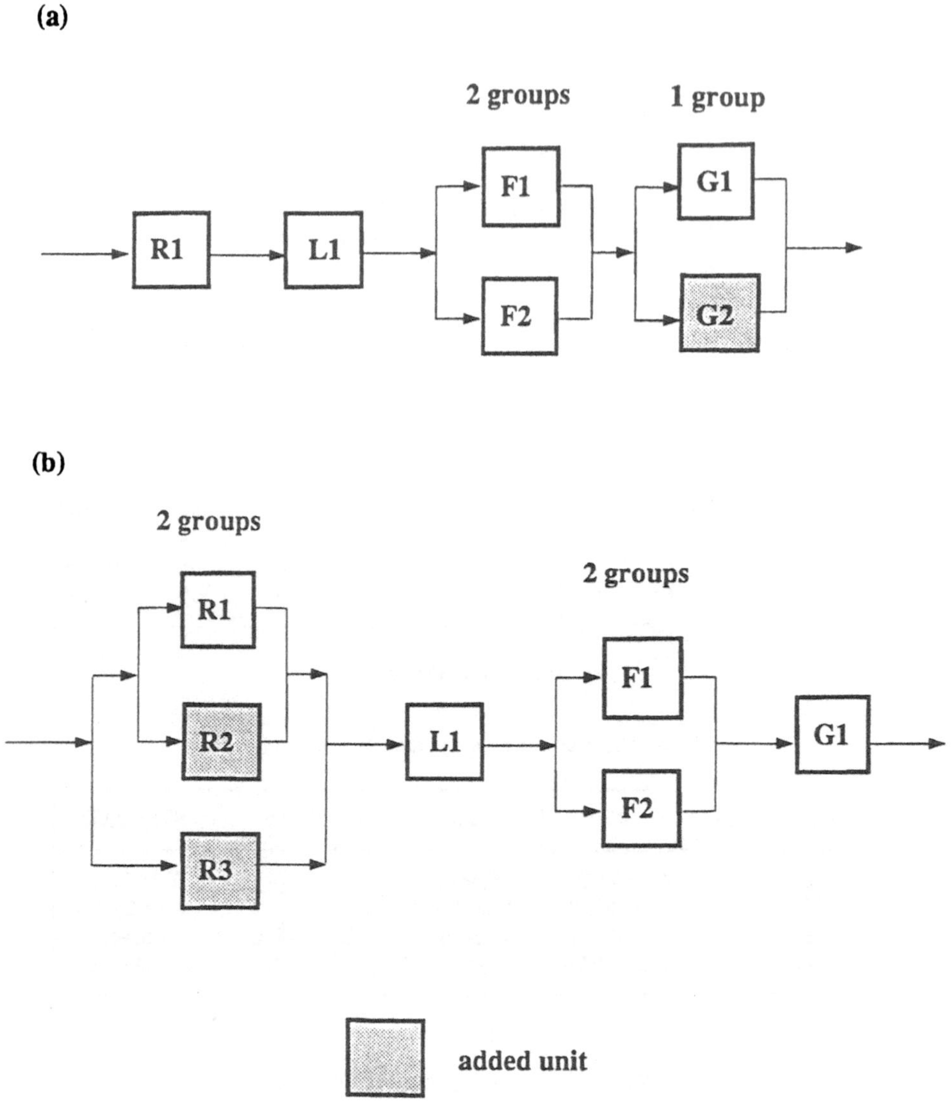

Figure 3. Design configuration for case with (a) no resource restrictions and (b) resource restrictions (second version of example).

Table VIII. Decomposition Algorithm Performance

Version	Iteration	Subproblem	Obj. function	No of Eqns/Vars/0-1 Vars	CPU time*(sec)
1	1	MINLP1	-487,000	361/147/9	18.8
		NLP1	-461,400	108/90	0.3
	2	MINLP2	infeasible	363/147/9	5.
					Total = 24.1
2	1	MINLP1	-536,200	361/147/9	6.2
		NLP1	-496,900	116/99	0.5
	2	MINLP2	-520,900	362/147/9	14.3
		NLP2	-504,900	124/104	0.5
	3	MINLP3	infeasible	363/147/9	29.8
					Total = 51.3
3	1	MINLP1	-479,300	361/151/9	8.1
		NLP1	-440,000	116/103	0.4
	2	MINLP2	-462,000	362/151/9	17.4
		NLP2	-422,400	116/103	0.4
	3	MINLP3	-458,400	363/151/9	18.3
		NLP3	-456,500	116/99	0.5
	4	MINLP4	infeasible	364/151/9	13.8
					Total = 58.9

* IBM 3090

Table IX. Solution with resource restrictions (only nonzero values)

unit type	V_e	N_e
R1	4000	1
L1	4000	1
F1,F2	3000,3000	1,1
G1	3000	1

product	n_i	T_{L_i},h	P_i,kg
A	530.6	6.382	268,200
B	176.5	6.794	156,000
D	64.9	11.921	54,200
E	194.	3.305	166,100

- Net Profit : $461,400

Table X. Modified Resource Coefficients η_{se}, θ_{se}, μ_{se}

resource / eq. type	R1,R2,R3	L1,L2,L3	F1-F5	G1,G2,G3
ST	3,1.36e-2,0.63	3,3e-2,1	-	3,3e-2,1.3
CL	3,2e-3,1.2	1,3e-2,0.75	-	1,3e-2,0.75
EL	2,1e-3,1.1	2,5e-4,1	2,3.4e-3,1.1	2,5e-4,1
MP	3,2e-2,0.75	1,1.55e-1,0.3	2,1.83e-5,0.5	1,1.55e-1,0.3

Table XI. Solution with resource restrictions (case with modified resource coefficients)

unit type	V_e	N_e
R1,R2,R3	4000,1029,500	1,1,1
L1	4000	1
F1,F2	3000,3000	1,1
G1	3000	1

product	n_i	T_{L_i},h	P_i,kg
A	467.3	4.739	268,200
B	176.5	6.417	156,000
D	179.7	11.921	150,200
E	154.4	3.305	166,100

- Net Profit : $504,900

Table XII. Bounds on Resource Utilization and Cost Coefficients

resource	RS_s^{min}	RS_s^{max}	prs_s
ST	10	150	500
CL	10	100	200
EL	10	140	140
MP	10	100	200

Table XIII. Resource Utilization Levels (case with resource expansion)

resource	RS_s^1	RS_s^2
ST	90.62	113.4
CL	19.96	22.2
EL	39.4	41.5
MP	13.7	15.4

1 Case without equipment expansion (initial equipment configuration)
2 Case with equipment expansion (addition one in-phase unit at stage 4)

algorithm required four major iterations to obtain the optimal solution (Table VIII) that suggests the addition of an in-phase unit at stage 4, similarly to the case without resource restrictions. A lower profit ($456,500) can be made in this case, however, due to the resource expansion term added in the objective function. Table XIII shows that the resource utilization level had to increase in order to accommodate the addition of the new equipment unit in stage 4.

Note that, in all cases, the resource availability led to different plant equipment inventory expansions. This example shows that the incorporation of resource restrictions into the retrofit design formulation is necessary to more accurately predict the extent of process modifications during retrofit.

Conclusions

The design problem for the retrofit of a general multipurpose plant with resource restrictions is posed as a nonconvex mixed integer nonlinear program (MINLP) which accommodates changes in the product demands, revisions in the product slate, addition and/or elimination of equipment units, batch size dependent processing times and resource utilization rates and resource expansion. The proposed model is developed as an extension of the corresponding model for the retrofit design of a general multipurpose plant with no resource considerations [6].

The complexity of the proposed model makes the problem computationally intractable for direct solution using existing MINLP solution techniques. Consequently, a formulation specific decomposition strategy is developed, which builds on our earlier developments for the retrofit design problem with no resource restrictions. The proposed solution strategy cannot guarantee the global optimal solution due to the nonconvexity of the upper bound subproblems.

The solution of a test example clearly showed that incorporation of resource restrictions into the retrofit design formulation has a great influence on the feasibility and quality of process modifications during retrofit.

Nomenclature

N	number of products
E	number of batch equipment types
NEQ	number of new equipment types
F	number of equipment families
K	maximum number of campaigns
H	total available production time

i	index on products
m	index on tasks
e	index on equipment types
g	index on equipment groups
k	index on campaigns
TA_i	set of tasks for product i
P_{im}	set of feasible equipment types for task m of product i
U_e	set of tasks that can be executed by equipment type e
L_f	set of equipment types that belong to equipment family f
$S1_s$	set of products using resource s
$S2_s$	set of tasks using resource s
Y_{ik}	amount of product i produced during campaign k
Q_i	yearly production requirement for product i
S_{ime}	size factor of task m of product i in equipment type e
t_{imgk}	group processing time of task m of product i in group g during campaign k
t^0_{ime}, α_{ime}, β_{ime}	processing time coefficients
a_e , b_e	cost coefficients for equipment type e
c_e , d_e	cost coefficients for eq. type e used in master problem
p_i	unit profit for product i
ω_{ime}	operating cost coefficient for task m of prod. i in eq. type e
V_e	size of units of equipment type e
N_e	number of units of equipment type e
P_i	production demand for product i
Q_{ime}	amount of product i produced during task m in eq. type e
X_{imegk}	0-1 assignment variable for task m of product i in equipment type e in group g during campaign k
n_{ik}	number of batches of product i produced during campaign k
np_{imgk}	number of batches of product i processed by group g during task m in campaign k
BS_{imegk}	split batch size produced during task m of product i in eq. type e in group g during campaign k
NU_{imegk}	number of units of type e that are contained in group g assigned to task m of product i during campaign k
NG_{imk}	number of equipment groups assigned to task m of product i during campaign k
$T_{L_{ik}}$	limiting cycle time of product i during campaign k
T_k	length of campaign k
CAP_e	total capacity of equipment type e ($=V_e N_e$)
PR_{ik}	priority index denoting assignment of product i to campaign k
η_{sime}, θ_{sime}, μ_{sime}	resource coefficients
r_{simegk}	utilization level of resource s by task m of product i

	in unit e in group g during campaign k
RS_s	utilization level of resource s

Appendix I: Original MINLP Formulation (I)

$$\min \sum_{e=1}^{NEQ} a_e N_e (V_e)^{b_e} + \sum_{i=1}^{N} \sum_{m \in TA_i} \sum_{e \in P_{im}} \omega_{ime} Q_{ime} - \sum_{i=1}^{N} p_i P_i \sum_{s \in RES} prs_s (RS_s - RS_s^{\min}) \quad (I.1)$$

s.t.

$$\sum_{k=1}^{K} \sum_{g=1}^{NG_{imk}^{\max}} \sum_{e \in P_{im}} X_{imegk} \geq 1 \qquad i=1,\dots,N\ ;\ m \in TA_i \quad (I.2)$$

$$\sum_{(i,m) \in U_e} \sum_{g=1}^{NG_{imk}^{\max}} X_{imegk} \leq N_e \qquad e=1,\dots,E\ ;\ k=1,\dots,K \quad (I.3)$$

$$X_{imegk} + X_{imqjk} \leq 1 \qquad \forall\ i,\ k\ ;\ m \in TA_i\ ;\ e,q \in P_{im}\ ;\ e \in L_f\ ;\ q \in L_h\ ;\ f \neq h\ ;\ g,j=1,\dots,NG_{imk}^{\max} \quad (I.4)$$

$$X_{imegk} + X_{im+1ljk} + X_{im-plqk} \leq 2 \qquad \forall\ i,\ k\ ;\ m=2,\dots,|TA_i|-1\ ;\ p=1,\dots,m-1\ ;\ e \in P_{im}\ ;\ e \in L_f\ ;\ l \in \{P_{m+1i} \cap P_{m-qi}\}\ ;\ l \in L_h\ ;\ f \neq h\ ;\ g=1,\dots,NG_{imk}^{\max}\ ;\ j=1,\dots,NG_{im+1k}^{\max}\ ;\ q=1,\dots,NG_{im-pk}^{\max} \quad (I.5)$$

$$X_{imegk} \leq \sum_{g=1}^{NG_{iqk}^{\max}} \sum_{e \in P_{iq}} X_{iqegk} \qquad \forall\ i,\ k\ ;\ m,q \in TA_i\ ;\ e \in P_{im} \quad (I.6)$$

$$Z_{imgk} \geq Z_{img+1k} \qquad \forall\ i,\ k\ ;\ m \in TA_i\ ;\ g=1,\dots,NG_{imk}^{\max}-1 \quad (I.7)$$

$$Z_{imgk} \geq X_{imegk} \qquad \forall\ i,k\ ;\ m \in TA_i\ ;\ g=1,\dots,NG_{imk}^{\max}-1\ ;\ e \in P_{im} \quad (I.8)$$

$$Z_{imgk} \leq \sum_{e \in P_{im}} X_{imegk} \qquad \forall\ i,\ k\ ;\ m \in TA_i\ ;\ g=1,\dots,NG_{imk}^{\max} \quad (I.9)$$

$$\sum_{e \in P_{im}} Q_{ime} \geq P_i \qquad \forall\ i\ ;\ m \in TA_i \quad (I.10)$$

$$\sum_{k=1}^{K} Y_{ik} = P_i \qquad \forall\ i,\ k \quad (I.11)$$

$$\sum_{k=1}^{K} \sum_{g=1}^{NG_{imk}^{\max}} NU_{imegk}\, BS_{imegk}\, np_{imgk} \leq Q_{ime} \qquad \forall\ i\ ;\ m \in TA_i\ ;\ e \in P_{im} \quad (I.12)$$

$$\sum_{g=1}^{NG_{imk}^{max}} \sum_{e \in P_{im}} NU_{imegk}\, BS_{imegk}\, np_{imgk} \geq Y_{ik} \qquad \forall\ i,\ k\ ;\ m \in TA_i \qquad (I.13)$$

$$T_{L_{ik}} \geq \frac{t_{imgk}}{NG_{imk}} \qquad \forall\ i,\ k\ ;\ m \in TA_i\ ;\ e \in P_{im}\ ;\ g=1,...,NG_{imk}^{max} \qquad (I.14)$$

$$t_{imgk} \geq t_{ime}^{0}\, X_{imegk} + \alpha_{ime}\, BS_{imegk}^{\beta_{ime}} \qquad \forall\ i,\ k\ ;\ m \in TA_i\ ;\ e \in P_{im}\ ;\ g=1,...,NG_{imk}^{max} \qquad (I.15)$$

$$T_{L_{ik}} \leq T_{L_{ik}}^{max} \sum_{m \in TA_i} \sum_{g=1}^{NG_{imk}^{max}} \sum_{e \in P_{im}} X_{imegk} \qquad \forall\ i,\ k \qquad (I.16)$$

$$\sum_{k=1}^{K} T_k \leq H \qquad (I.17)$$

$$T_k \geq n_{ik}\, T_{L_{ik}} \qquad \forall\ i,\ k \qquad (I.18)$$

$$V_e \geq S_{ime}\, BS_{imegk} \qquad \forall\ i,\ k\ ;\ m \in TA_i\ ;\ e \in P_{im}\ ;\ g=1,...,NG_{imk}^{max} \qquad (I.19)$$

$$N_e \geq \sum_{(i,m) \in U_e} \sum_{g=1}^{NG_{imk}^{max}} NU_{imegk} \qquad \forall\ i\ ;\ m \in TA_i\ ;\ e \in P_{im} \qquad (I.20)$$

$$NU_{imegk} \geq X_{imegk} \qquad \forall\ i,\ k\ ;\ m \in TA_i\ ;\ e \in P_{im}\ ;\ g=1,...,NG_{imk}^{max} \qquad (I.21)$$

$$NU_{imegk} \leq N_e^{max}\, X_{imegk} \qquad \forall\ i,\ k\ ;\ m \in TA_i\ ;\ e \in P_{im}\ ;\ g=1,...,NG_{imk}^{max} \qquad (I.22)$$

$$NG_{imk} = \sum_{g=1}^{NG_{imk}^{max}} g\, W_{imgk} \qquad \forall\ i,\ k\ ;\ m \in TA_i \qquad (I.23)$$

$$W_{imgk} = Z_{imgk} - Z_{img+1k} \qquad \forall\ i,\ k\ ;\ m \in TA_i\ ;\ g=1,...,NG_{imk}^{max}-1 \qquad (I.24)$$

$$W_{imgk} = Z_{imgk} \qquad \forall\ i,\ k\ ;\ m \in TA_i\ ;\ g=NG_{imk}^{max} \qquad (I.25)$$

$$\sum_{g=1}^{NG_{imk}^{max}} W_{imgk} \leq 1 \qquad \forall\ i,\ k\ ;\ m \in TA_i \qquad (I.26)$$

$$\sum_{e \in P_{im}} NU_{imegk}\, BS_{imegk} \leq B_i^{max} \qquad \forall\ i,\ k\ ;\ m \in TA_i\ ;\ g=1,...,NG_{imk}^{max} \qquad (I.27)$$

$$n_{ik} \geq \sum_{g=1}^{NG_{imk}^{max}} np_{imgk} \qquad \forall\ i,\ k\ ;\ m \in TA_i \qquad (I.28)$$

$$r_{simegk} = NU_{imegk}\, (\eta_{sime} + \theta_{sime}\, BS_{imegk}{}^{\mu_{sime}}) \qquad s \in RES\ ;\ i \in S1_s\ ;\ m \in TA_i \cap S2_s$$

$$e \in P_{im}\ ;\ g=1,...,NG_{imk}^{max}\ ;\ k=1,...,K \qquad (I.29)$$

$$\sum_{i \in S1_s} \sum_{m \in TA_i \cap S2_s} \sum_{e \in P_{im}} \sum_{g=1}^{NG_{imk}^{max}} r_{simegk} \leq R_s \qquad s \in RES\ ;\ k=1,...,K \qquad (I.30)$$

$$BS_{imegk} \geq B_i^{*}\, X_{imegk} \qquad \forall\ i,\ k\ ;\ m \in TA_i\ ;\ e \in P_{im}\ ;\ g=1,...,NG_{imk}^{max} \qquad (I.31)$$

$$np_{imgk} \geq n^*_{imgk}\, Z_{imgk} \qquad \forall\ i,\ k\ ;\ m \in TA_i\ ;\ g=1,\ldots,NG^{\max}_{imk} \qquad (I.32)$$

$$np_{imgk} \leq n^{\max}_{ik}\, Z_{imgk} \qquad \forall\ i,\ k\ ;\ m \in TA_i\ ;\ g=1,\ldots,NG^{\max}_{imk} \qquad (I.33)$$

$$V^{\min}_e \leq V_e \leq V^{\max}_e \qquad \forall\ e \qquad (I.34)$$

$$0 \leq N_e \leq N^{\max}_e \qquad \forall\ e \qquad (I.35)$$

$$P^{\min}_i \leq P_i \leq P^{\max}_i \qquad \forall\ i \qquad (I.36)$$

$$0 \leq Q_{ime}\ ,\ Y_{ik} \leq P^{\max}_i \qquad \forall\ i\ ;\ m \in TA_i\ ;\ e \in P_{im} \qquad (I.37)$$

$$0 \leq NG_{imk} \leq \max_{e \in P_{im}} \{N^{\max}_e\} \qquad \forall\ i,\ k\ ;\ m \in TA_i \qquad (I.38)$$

$$0 \leq NU_{imegk} \leq N^{\max}_e \qquad \forall\ i,\ k\ ;\ m \in TA_i\ ;\ e \in P_{im}\ ;\ g=1,\ldots,NG^{\max}_{imk} \qquad (I.39)$$

$$0 \leq T_k \leq H \qquad \forall\ k \qquad (I.40)$$

$$0 \leq T_{L_{ik}} \leq \max_{m \in TA_i} \max_{e \in P_{im}} \{t^{\max}_{ime}\} \qquad \forall\ i,\ k \qquad (I.41)$$

$$t^0_{ime} \leq t_{imgk} \leq t^0_{ime} + \alpha_{ime} \frac{V^{\max}_e}{S_{ime}}^{\beta_{ime}} \qquad \forall\ i,\ k\ ;\ m \in TA_i\ ;\ e \in P_{im}\ ;\ g=1,\ldots,NG^{\max}_{imk} \qquad (I.42)$$

$$0 \leq n_{ik}\ ,\ np_{imgk} \leq \frac{H}{T^*_{L_{ik}}} \qquad \forall\ i,\ k\ ;\ m \in TA_i\ ;\ g=1,\ldots,NG^{\max}_{imk} \qquad (I.43)$$

$$0 \leq BS_{imegk} \leq \frac{V^{\max}_e}{S_{ime}} \qquad \forall\ i,\ k\ ;\ m \in TA_i\ ;\ e \in P_{im}\ ;\ g=1,\ldots,NG^{\max}_{imk} \qquad (I.44)$$

$$0 \leq r_{simegk} \leq R^{\max}_s \qquad s \in RES\ ;\ i \in S1_s\ ;\ m \in TA_i \cap S2_s$$
$$e \in P_{im}\ ;\ g=1,\ldots,NG^{\max}_{imk}\ ;\ k=1,\ldots,K \qquad (I.45)$$

$$RS^{\min}_s \leq RS_s \leq RS^{\max}_s \qquad s \in RES \qquad (I.46)$$

$$\sum_i PR_{ik} \geq \sum_i PR_{ik+1} \qquad \forall\ ;\ i\ \ k=1,\ldots,K-1 \qquad (I.47)$$

$$PR_{ik} \geq X_{i1e1k} \qquad \forall\ i,\ k\ ;\ e \in P_{i1} \qquad (I.48)$$

$$PR_{ik} \leq \sum_{e \in P_{i1}} X_{i1e1k} \qquad \forall\ i,\ k \qquad (I.49)$$

$$V_e \geq V_{e+1} \qquad \forall\ e\ ;\ e,e+1 \in L_f \qquad (I.50)$$

$$N_e \geq N_{e+1} \qquad \forall\ e\ ;\ e,e+1 \in L_f \qquad (I.51)$$

$$T^*_{L_{ik}} = \max_{m \in TA_i} \min_{e \in P_{im}} \{\frac{t^0_{ime}}{NG^{\max}_{imk}}\} \qquad \forall\ i,\ k$$

$$B^*_i = \min_{m \in TA_i} \min_{e \in P_{im}} \frac{V^{\min}_e}{S_{ime}} \qquad \forall\ i$$

$$B_i^{\max} = \min_{m \in TA_i} \{ \sum_{e \in P_{im} \cap L_f} N_e^{\max} \frac{V_e^{\max}}{S_{ime}} \} \qquad \forall\ i,\ k$$

Appendix II: Master Problem Formulation (II)

$$\min \sum_{e=1}^{NEQ} (c_e\, CAP_e + d_e\, N_e) + \sum_{i=1}^{N} \sum_{m \in TA_{im}} \sum_{e \in P_{im}} \omega_{ime}\, Q_{ime} - \sum_{i=1}^{N} p_i\, P_i + \sum_{s \in RES} prs_s\, (RS_s - RS_s^{\min}) \text{(II.}$$

$$c_e = a_e \frac{(V_e^{\max})^{b_e} - (V_e^{\min})^{b_e}}{V_e^{\max} - V_e^{\min}}$$

$$d_e = a_e\, (V_e^{\min})^{b_e} - a_e\, V_e^{\min} \frac{(V_e^{\max})^{b_e} - (V_e^{\min})^{b_e}}{V_e^{\max} - V_e^{\min}}$$

s.t.

$$VC_{imegk} \geq S_{ime}\, NU_{imegk}\, BS_{imegk} \qquad \forall\ i\ ; m \in TA_i\ ; e \in P_{im}\ ; g = 1,\ldots,NG_{imk}^{\max} \qquad \text{(II.2)}$$

$$\sum_{g=1}^{NG_{imk}^{\max}} \sum_{e \in P_{im}} \frac{VC_{imegk}}{S_{ime}} \geq \frac{Y_{ik}}{n_{ik}} \qquad \forall\ i,k\ ; m \in TA_i \qquad \text{(II.3)}$$

$$n_{ik} \geq \frac{Y_{ik}}{B_i^{\max}} \qquad \forall\ i,\ k \qquad \text{(II.4)}$$

$$CAP_e \geq \sum_{i=1}^{N} \sum_{m \in TA_i} \sum_{g=1}^{NG_{imk}^{\max}} VC_{imegk} \qquad \forall\ k\ ; e \in P_{im} \qquad \text{(II.5)}$$

$$Q_{ime} \leq \frac{V_e^{\max}\, N_e^{\max}\, H}{S_{ime}} \sum_{k=1}^{K} \sum_{g=1}^{NG_{imk}^{\max}} \frac{1}{T_{L_{ik}}^{*}} X_{imegk} \qquad \forall\ i\ ; m \in TA_{im}\ ; e \in P_{im} \qquad \text{(II.6)}$$

$$VC_{imegk} \geq V_e^{\min}\, X_{imegk} \qquad \forall\ i,k\ ; m \in TA_i\ ; e \in P_{im}\ ; g = 1,\ldots,NG_{imk}^{\max} \qquad \text{(II.7)}$$

$$VC_{imegk} \leq V_e^{\max}\, N_e^{\max}\, X_{imegk} \qquad \forall\ i,k\ ; m \in TA_i\ ; e \in P_{im}\ ; g = 1,\ldots,NG_{imk}^{\max} \qquad \text{(II.8)}$$

$$Y_{ik} \leq P_i^{\max}\, PR_{ik} \qquad \forall\ i,\ k \qquad \text{(II.9)}$$

$$CAP_e \geq CAP_{e+1} \qquad \forall\ e \quad e,e+1 \in L_f \qquad \text{(II.10)}$$

$$V_e^{\min}\, N_e^{\min} \leq CAP_e \leq V_e^{\max}\, N_e^{\max} \qquad \forall\ e \qquad \text{(II.11)}$$

$$0 \leq VC_{imegk} \leq V_e^{\max}\, N_e^{\max} \qquad \forall\ i,m,e,g,k \qquad \text{(II.12)}$$

$$R_L \leq R_U \qquad \text{(II.13)}$$

\+ (I.2)-(I.12), (I.14)-(I.18), (I.20)-(I.33), (I.35)-(I.49), (I.51)

REFERENCES

1. Brook, A.; Kendrick, D.; Meeraus, A.: GAMS, A User's Guide. Redwood City, CA: Scientific Press 1988.
2. Garfinkel, R.S.; Nemhauser, G.L.: Integer Programming. New York: Wiley 1972.
3. Glover, F.: Improved Linear Integer Programming Formulations of Nonlinear Integer Problems. In Management Science, 22(4), 455-459 (1975).
4. Kocis, G.R.; Grossmann, I.E.: Global Optimization of Nonconvex MINLP problems in Process Synthesis. In Ind. Eng. Chem. Res., 27, 1407-1421 (1988).
5. Murtagh, B.A.; Saunders, M.A.: MINOS 5.0 User's Guide. Technical Report SOL 83-20. Stanford University Systems Optimization Laboratory, 1983.
6. Papageorgaki, S.; Reklaitis, G.V.: Retrofitting a General Multipurpose Batch Chemical Plant. Industrial and Engineering Chemistry Research, Vol. 32, 345-363 (1993).
7. Tsirukis, A.G.: Scheduling of Multipurpose Batch Chemical Plants. PhD Dissertation, Purdue University, W. Lafayette, IN (1991).
8. Vaselenak, J.A.; Grossmann, I.E.; Westerberg, A.W.: Optimal Retrofit Design of Multiproduct Batch Plants. In Ind. Eng. Chem. Res., 26, 718-726 (1987).
9. Viswanathan, J.; Grossmann, I.E.: A Combined Penalty Function and Outer Approximation Method for MINLP Optimization. In Comp. Chem. Engng., 14, 769-782 (1990).

Design of Operation Policies for Batch Distillation

S. Macchietto and I. M. Mujtaba

Centre for Process Systems Engineering, Imperial College, London SW7 2BY, UK

Abstract: The batch distillation process is briefly reviewed. Control variables, operating decisions and objectives are identified. Modeling aspects are discussed and a suitable representation for operations is introduced. Techniques for the dynamic simulation and optimization of the operation are reviewed, in particular the control vector parameterization method. Optimization formulations and results are presented for typical problems: optimization of a single distillation step, distillation with recycle of off-cuts, multiperiod optimization, reactive batch distillation and the online optimization of a sequence of batches in a campaign. Outstanding research issues are identified.

Keywords: Batch Distillation, Modeling, Operation, Dynamic Simulation, Optimization, Optimal Control.

Introduction

Batch distillation is perhaps one of the oldest unit operations. It was discovered by many ancient cultures as a way to produce alcoholic beverages, essential oils and perfume, and its basic operation had been perfected long before the advent of phase equilibrium thermodynamics, let alone of computer technology. Today, batch distillation is widely used in the production of fine chemicals and for specialized productions and is the most frequent separation method in batch processes [49]. Its main advantages are the ability of separating several fractions of a feed mixture in a single column and of processing several mixtures in the same column. When coupled with reaction, batch distillation of one or more products permits achieving much higher conversions than otherwise possible.

Although distillation is one of the most intensely studied and better understood processes in the chemical industry, its batch version still represents an interesting field for academic and industrial research, for a variety of reasons: i) even for a simple binary mixture there are many alternative operations possible, with complex trade-offs as a result of the many degrees of freedom available, hence there is ample scope for optimization ii) the process is intrinsically dynamic, hence its optimization naturally results in an optimal control problem, for which problem formulations and numerical solution techniques are not yet well established. However, advances made both in dynamic optimization techniques and computing speeds make it possible to consider rather more complex operation policies iii) advances in plant control make it now feasible to implement much more sophisticated control policies that was possible with earlier control technology and hence to achieve in practice any potential benefits predicted iv) finally, batch distillation is also of interest as merely an excellent representative example of a whole class of complex dynamic optimization problems.

The purposes of this paper are i) to summarize some recent advances in the development of optimal operation policies for a variety of batch distillation applications and ii) to highlight some research issues which are still outstanding. Although, there are obvious interactions between a batch column design and its operation [52], in the following coverage it will be assumed that the column design is given a priori and that an adequate dynamic model (including thermodynamic and physical properties) has been developed. Attention will be focused on the problem of establishing a priori the optimal values and time profiles of the variables controlling the operation for a given feed mixture. It is assumed that a suitable control system can be separately designed later for accurately tracking the optimal profiles predicted. It is in this sense that we talk of "design of operation policies". The control approach, of establishing the optimal policies on-line in conjunction with a state estimator, to take into account model mismatch and disturbances, will not be considered here. Finally, we will concentrate mainly on the optimal operation of a single batch, rather than of an entire campaign.

The paper is structured as follows: first, a brief reminder is given of the batch distillation process and of the main control and operation decision choices available. Some representation and modeling aspects are considered next, followed by simulation issues and optimization issues. Finally, a set of illustrative examples are given for the optimization of typical batch distillation problems involving single and multiple distillation steps, the optimization of off-cut recycles, of a whole batch and of reactive batch distillation. An example is also given of the use of the above techniques in conjunction with an on-line control system, for the automation of a batch campaign. Many of the examples summarized here have been presented in detail elsewhere. Suitable references are given in the main body of the paper.

The Process - A Brief Review

The Batch Distillation Process

The basic *operation* for processing of a charge in a batch column (a *batch*) is illustrated with reference to the *equipment* in Figure 1 (a general introduction is given in [79]). A quantity of fresh feed is *charged* (typically cold) into a still pot and *heated* to its boiling point. The column is brought to the right pressure and temperature during an initial *startup period*, often carried out at total reflux, during which a liquid holdup builds up in the top condensate receiver and internally in the column. Initial pressure, flow, temperature and composition profiles are established. A *production period* follows when distillate is withdrawn and collected in one or more fractions or "cuts". The order of appearance of species in the distillate is determined by the phase equilibria characteristics of the mixture to be separated (for simple distillation, the composition profiles will follow well defined distillation curves [84, 9]. A typical instant distillate composition profile is given in Figure 2. The distillate is initially rich in the lowest boiling component (or azeotropic mixture), which is then progressively depleted. It then becomes richer in the next lowest boiling

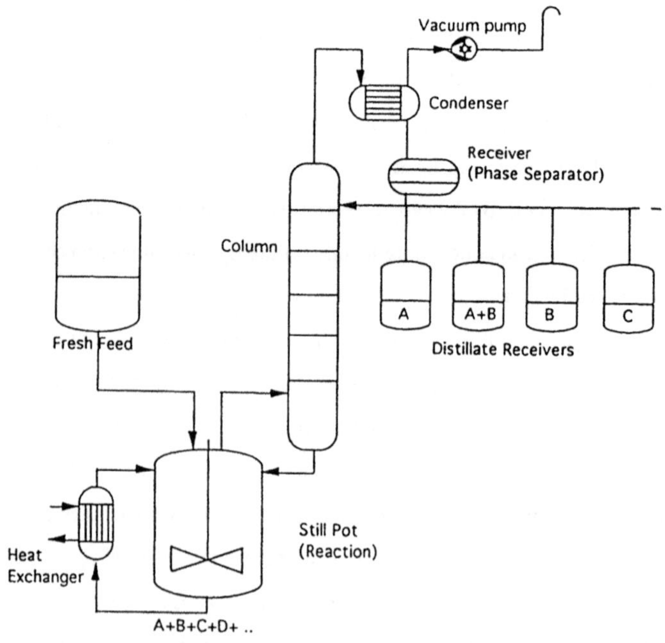

Figure 1. Typical Configuration of a Conventional Batch Distillation Column

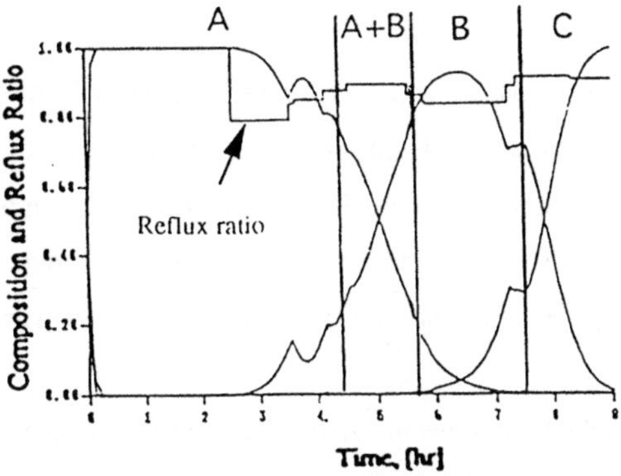

Figure 2. Typical Distillate Composition (mole fraction) Profiles vs. Time, with Fractions Collected

component, etc. Diverting the distillate flow to different receivers permits collecting distillate product cuts meeting desired purity specifications. Intermediate cuts ("off cuts" or "slop cuts") may also be collected which will typically contain material not meeting purity specifications. The operation is completed by a brief ***shut down*** period, when the heat supply to the reboiler is terminated and the liquid holdups in the column collapse to the bottom. The heavy fraction remaining in the pot may be one of the desired products. Valuable constituents in the offcuts may be recovered by reprocessing the offcut fractions in variety of ways. The column is then prepared for the next batch. Several variations of the basic process are possible. Additional material may be charged to the pot during the batch. Reaction may occur in the pot, or sometimes in the entire column. Esterification reactions are often conducted this way [21, 20, 6]. Vacuum may be applied to facilitate separation and keep boiling temperatures low, so as to avoid thermal degradation problems. Two liquid phases may be present in the distillate, in which case the condensate receiver has the function of a two phase separator. In some cases, the fresh feed is charged to an enlarged condenser reflux drum which thus acts as the pot and the column is used in a stripping mode (inverted column), with high boiling products withdrawn from the bottom, as described by Robinson and Gilliland [77]. Alternative configurations involving feeding the fresh feed in the middle of the column (which therefore has both stripping and a rectifying sections) were also mentioned by [11, 2, 41]. In this paper, attention will be concentrated on the conventional batch distillation system (Figure 1) since the techniques discussed are broadly applicable to those alternative configurations with only minor extensions.

Operation Objectives

The purpose of batch distillation is typically to produce selected product fractions having desired specifications. For each product, these specifications are expressed in terms of the mole fraction of a key component meeting or exceeding a specified value. Additionally, the mole fraction of one or more other species (individually or cumulatively) in some fractions should often not exceed specified values. These quality specifications are therefore naturally expressed as (hard) inequality constraints.

Additionally, it is typically of interest to maximize the recovery of (the most) valuable species, to minimize energy requirements and to minimize the time required for all operations (not just for fresh feed batches, but also for reprocessing off-cuts, if any). Each of these quantities (recoveries, energy requirements, time) may also have defined upper and/or lower limits. Clearly, there may be conflicting requirements. We may observe that rather than posing a multiobjective optimization problem, it is much easier to select just one of the desired quantities as the objective function and define the others as constraints (e.g. maximize recovery subject to purity, time and energy limits). In fact, the easiest way to combine multiple objectives is to formulate an overall economic objective function which properly weighs all factors of interest in common, in well understood monetary terms.

Operation Variables and Trade-offs

The maximum vapor boilup and condensate rate that can be produced for a given feed mixture and the liquid holdups in the column are essentially established by the column design characteristics, fixed a priori (column diameter, number of equilibrium stages and column internals, pot, reboiler and condenser type and geometry). For a given charge, the main operation variables available for control are the reflux ratio, the heating medium flow rate (or energy input to the reboiler or vapor boilup rate, varied by means of the energy input to the reboiler), the column top pressure and the times during which distillate is collected in each of the different distillate receivers. Specifying all these variables determines the amount and composition of each of the fraction collected (hence recoveries) and other performance measures (e.g. total time, energy used, etc.). A number of trade-offs must be considered.

Increasing *reflux ratio* will increase the instant distillate purity, giving a smaller distillate flow rate and thus requiring longer time to produce a given amount of distillate, with higher energy requirements for reboiler and condenser. On the other hand, a larger amount of distillate meeting given purity specifications may be collected this way. Productivity (in terms of the amount of distillate produced over the batch time) may go down as well as up with increasing reflux ratio, presenting an interesting optimization problem. A useful upper bound for the top distillate composition achievable at any one time is given by the total reflux operation.

Traditionally, *constant reflux ratio* (on grounds of simplicity of application) and *constant distillate purity* (requiring progressively increasing reflux ratio) have been considered. In order to achieve a given final composition in a specific accumulated distillate fraction, initially richer distillate may be mixed with lower quality distillate near the end of the production cut. This is obtained with the constant reflux policy. Thermodynamic considerations suggest that any mixing is a source of irreversibility and hence will have to be paid for somehow, providing a justification for the constant distillate composition operation. This argument however ignores the relative costs of product, column time and energy. In general, some intermediate reflux ratio policy will be optimal. Other reflux ratio strategies have been used in practice, for example one characterized by alternating total reflux (no distillate) and distillate product withdrawal (no reflux) [5].

With regards to the *vapor boilup rate*, it is usually optimal to operate at the highest possible energy input, except when hydraulic considerations become limiting (entrainment). When hydraulics is not considered (at one's own risk), the optimal policy can thus be established a priori.

A constant (top) *pressure* level is often selected once for the entire batch distillation, or different but constant pressure levels, if necessary, may be used during each production cut. Pressure may be decreased as the batch progresses as a way to maintain the boiling point in the pot below or at a given limit for heat labile mixtures or to increase relative volatility for difficult separations.

The choice of the *timing for each production cut* is important. With reference to Figure 2, two rather obvious points may be noted. First, it is possible to achieve a high purity specification on one component in an individual fraction by choosing its beginning and end time so as to remove the lower purity front and tail, respectively, in the previous and in the subsequent cut. This, however, may make achieving any specifications on the earlier and subsequent cuts much harder and even impossible. Thus, the operations of all cuts are interacting rather strongly. Second, as already observed, it is possible to eliminate fronts and tails of undesired components as off-cut fractions. This however will affect recoveries and will make offcut reprocessing important.

With respect to the *offcuts*, several choices are available. A fraction not meeting product specifications may be simply a waste (with associated loss of material and possibly a disposal cost) or it may be a lower quality by-product, with some residual value. Offcuts may also be reprocessed. Here, several alternatives are possible. The simplest strategy is to collect all off-cuts produced during one batch in a single distillate receiver and then add the entire amount to the pot, together with fresh feed, to make up the next charge. This will increase recovery of valuable species, at the expense of a reduction in the amount of fresh feed that can be processed in each batch, leading again to an interesting trade-off [51]. For each offcut, the reflux ratio profile used and the duration of tile off-cut production step (hence, the amount and composition of the accumulated offcut) must be defined. The addition of the offcuts from one batch to the pot may be done either at the beginning of or during the subsequent batch, the time of addition representing one further degree of operational freedom. The second principle of thermodynamics gives us again useful indications. Since mixing is accompanied by irreversibility, the addition of an offcut to the pot should be done when the compositions of the two are closest. This also suggest an altogether different reprocessing policy, whereby distinct off-cut fractions from a batch charge are not mixed in the same distillate receiver, but rather collected in separate receivers (assuming sufficient storage capacity is available). Each off-cut fraction can then be individually recycled to the pot at different times during the next batch. In fact, re-mixing of fractions already separated can also be reduced if the same off-cut material produced in successive batches is collected (in a much larger distillate storage) and then reprocessed later as one or more batches, with the full charge made up by the stored off-cut [59, 73]. This strategy is motivated by the fact that each off-cut is typically rich in just a couple of components, and hence their separation can be much easier this way. In practice, the choice of a reprocessing policy will depend on the number and capacity of storage vessels.

It may also be noted that unlike its continuous counterpart, a batch distillation column can give a very high separation even with few separation stages. If a desired purity cannot be achieved in one distillation pass, an intermediate distillate off-cut can be collected and distilled a second (and third, etc.) time.

In summary, we may identify two types of operation choices to be made. Some choices define the *structure* of the operation (the sequence of products to be collected, whether to produce intermediate cuts or not, whether to reprocess off-cut fractions immediately or store them for subsequent distillation, whether to collect off-cuts in a single vessel, thereby re-mixing material already separated, or to store them individually in distinct vessels. These are *discrete* (yes/no) decisions. For a specific instance of these decisions (which we will call an *operation strategy*), there are continuous decision variables, the main ones being the reflux ratio profiles and the duration of each processing step, with possibly pressure and vapor boilup rate as additional control variables.

Even for simple binary mixtures, there are clearly many possible operation strategies. With multicomponent mixtures, the number of such structural options increases dramatically. For each strategy, the (time dependent and time independent) continuous control variables are highly interrelated. Selecting the best operation thus presents a rather formidable problem, the solution of which clearly requires a systematic approach. Formally, the overall problem could be posed as a mixed integer nonlinear dynamic process optimization problem (MINDPOP), with a general economic objective function and all structural and continuous operation variables to be optimized simultaneously. So far, however, only very much simpler subsets of this problem have been reported. The operation strategy is typically fixed a priori, and just (some) continuous operation variables are optimized. This approach will be followed in this paper as well. Fixed structure operations for which optimization solutions have been proposed include the already mentioned minimum time problem (P1), maximum distillate problem (P2), and maximum profit problem (P3) (Table 1).

Table 1. Selected References on a priori Control Profile Optimization-Conventional Batch Distillation Columns

Reference	Column Model	Mixture	Phase Equilibria	Optimisation Problem
Converse & Gross (1963)	Simple	Binary	CRV	P2
Converse & Huber (1965)	"	"	"	"
Coward (1967)	"	"	"	P1
Robinson (1969)	"	"	"	"
Robinson (1970)	"	Multicomponent	"	"
Mayur and Jackson (1971)	"	"	"	"
Kerkhof and Vissers (1978)	"	Binary	"	P3
Murty et al. (1980)	"	"	"	P2
Hansen and Jorgensen (1986)	"	"	"	P1
Diwekar et al. (1987)	Simple[1]	Multicomponent	"	P2
Mujtaba (1989)	Rigorous	"	Rigorous	P1/P2/P3
Fahrat et al. (1990)	Simple[3]	"	Simple	P2-multiperiod
Mujtaba and Macchietto (1991)	"	Binary	Rigorous	P1
Diwekar and Madhavan (1991)	"	"	"	P3
Logsdon & Biegler (1992)	Simple[1]	"	CRV	P2
Jang (1992)	Rigorous	Multicomponent	Rigorous	P1/P2
Diwekar (1992)	Simple[2]	"	CRV	P1/P2/P3
Mujtaba and Macchietto (1992b)	Rigorous	"	Rigorous	P3-multiperiod

CRV = Constant Relative Volatility. 1 - short-cut model of continuous distillation. 2 - same as 1 but modified for column holdup and tuned for nonideality. 3.- short-cut model, no holdups.

Benefits

That improved performance batch distillation should be achieved by clever manipulation of the available operation parameters (relatively to simpler, constant control) is intuitively appealing. However, reviews of previous optimization studies have indicated that benefits are often small [78]. This argument has been used to dismiss the need for sophisticated operation policies. On the other hand, these results were often obtained using highly simplified models and optimization techniques, a limited subset of alternative policies (constant distillate composition vs. constant reflux ratio) and objective functions only indirectly related to economic performance. Different benefits are calculated with different objective functions. Kerchief and Vassar [46], for example, showed that small increases in distillate *yield* (order 5%) can translate into 20-40% higher profit. This clearly calls for more comprehensive consideration of operating choices, dynamic models, objective functions, and constraints.

Representation and Modeling Issues

Representation

There is a need to formalize the description of the operating procedures. It may be convenient to consider a batch distillation operation as composed by a series of steps, each terminated by a significant event (e.g. completion of a production cut and switch to a different distillate receiver). Following [64] the main structural aspects of a batch distillation operation are schematically represented here as a State Task Network (STN) where a state (denoted by a circle) represents a specified material, and a task (rectangular box) represents the operation (task) which transforms

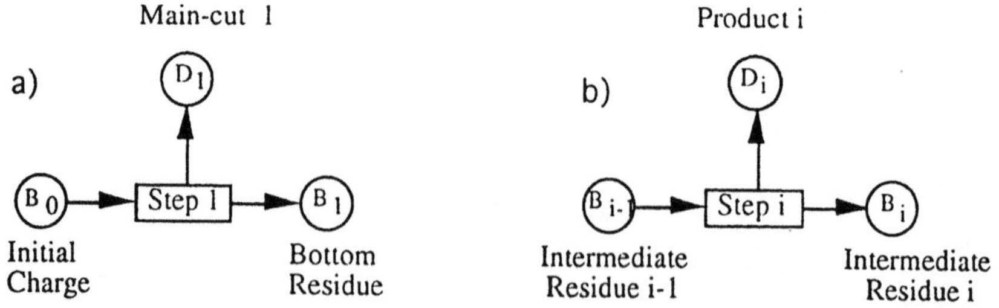

Figure 3. a) State Task network for a Simple Batch Distillation Operation Producing Two Fractions
b) Basic Operation Module for Separation into One Distillate and One Residue Fractions

the input state(s) into the output state(s). A task may consist of one or more steps. For example, Figure 3 shows a simple batch distillation operation with one task (Step 1) producing the two fractions Main-cut 1 and Bottom Residue from the state Initial Charge.

States are characterized by a name an amount and a composition vector for the mixture in that state. Tasks are characterized by an associated dynamic model and operational attributes such as duration the time profiles of reflux ratio and other control variables used during the task etc. Additional attributes of a distillation task are the values of all variables in the dynamic model at the beginning and at the end of the step. The states in the STN representation should not be confused with any state variables present in the dynamic model. For example, in Figure 1 the overall amount and composition (B_0 and x_{B0} respectively) of the Initial Charge STN state are distinct from the initial pot holdup and composition. The latter may be assigned the same numerical value as the former (a specific model initialization procedure). It is also possible for simulation and optimization purposes to neglect the startup period altogether and initialize the distillation task Step 1 by assuming that some portion of the charge is initially distributed along the column or that the initial column profiles (holdups composition etc.) are those obtained at total reflux (two other distinct initialization procedures). Of course, whatever initialization procedure is used the initial column profiles must be consistent (i.e. mass balance) with the amount and composition of the Initial Charge state in the STN. The initialization procedure is therefore a mapping between the STN states and the initial states in the dynamic model for that task. Similarly the amount and composition B_1 and x_{B1} of the STN state Bottom Residue are not the holdup and composition of the reboiler at the end of Step 1 but those which are obtained if all holdups in the column at the end of Step 1 are collected as the Bottom Residue STN state. The STN representation originally proposed in. [45] for scheduling is extended by the use of a dynamic model for a task in the place of a simple fixed time split fraction model.

The advantages of the above representation are: makes the structure of the operation quite explicit ii) it enables writing overall mass balances around an individual task (or a sequence of tasks) iii) suitable definition of selected subsets of states and task attributes make it possible to easily define a mix of dynamic simulation and optimization problems for a given operation strategy (STN structure) iv) different initialization procedures or even model equations may be defined for different tasks v) it enables the easy definition of alternative operation strategies by adjoining a small set of basic task modules used as building blocks. As an example the batch distillation of a multicomponent mixture is represented in Figure 4 as the combination of 2 main distillate and 1

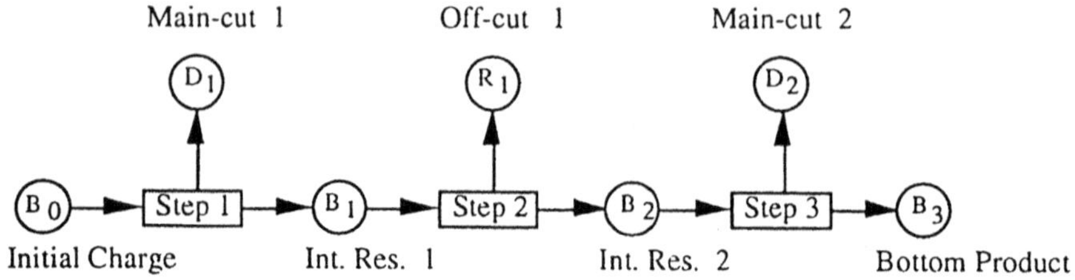

Figure 4. STN for Batch Distillation Operation with Two Main Distillate Cuts and One Intermediate Off-cut

off-cut production steps each of these steps being represented by essentially the same generic "module" given in Figure 3b. Similarly, an operation is shown in Figure 5 consisting of a distillate product cut (Step 1) followed by an off-cut production step (Step 2). The off-cut produced in a batch (state Off-cut Recycle of amount R and composition X_R) is recycled by mixing it in the pot residue with the next batch immediately before performing the main cut step to give an overall mixed charge amount B_c (of composition x_{BC})- This operation (with states and tasks suitably indexed as in Figure 3b.) can be used as a building block to define for example the cyclic batch operation strategy for a multicomponent mixture defined in Figure 6 (states omitted) [63].

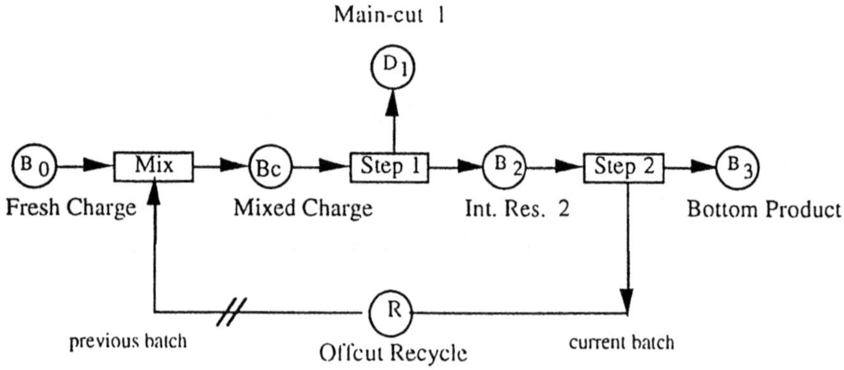

Figure 5. Basic Module for Separation Operation into One Distillate and One Residue Fractions, with Recycle of an Intermediate Off-cut to the next Batch

Modeling

Distillation modeling issues have been extensively discussed both for continuous and batch applications (e.g. [27, 44, 69, 4]) and will not be reviewed here in detail. The models required for

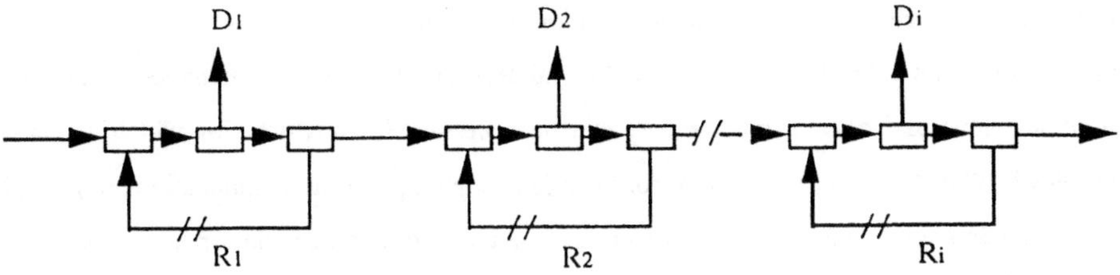

Figure 6. Operation Strategy for Multicomponent Batch Distillation with Separate Storage of Off-cuts and Sequential Off-cut Recycle to the Next Batch

batch distillation are in principle no different than those required for continuous distillation dynamics. What is of interest is the ability to predict the time responses of temperatures compositions and amounts collected for generic mixtures, columns and operations. It may be argued that models for batch distillation must cover a wider operations range, since startup and shutdown are part of the normal operation. For optimization purposes, the models must often be integrated many times and, hence, speed of execution is important. Some issues particularly relevant to batch columns are:

Modeling detail - Short-Cut vs. "Rigorous". As with any modeling exercise, a balance must be struck between the accuracy of the predicted responses availability of data, and speed of execution. Therefore, the "right" level of "rigorousness" depends on advances in thermophysical property predictions, numerical algorithms and computer hardware, and on purpose of use. In the past, most work on batch distillation used fairly simple short-cut models which relied on a number of assumptions such as constant relative volatility equimolar overflow no holdup and no hydraulic models etc. When used, the dynamic mass and energy balances have also been simplified in many ways. In some cases, accumulation terms have been neglected selectively or even altogether with dynamics approximated by a sequence of steady state calculations (e.g. [67, 35] for recent examples). These simplifications were dictated by the available technology and lead to useful results (e.g. [28]). At the other end of the spectrum, full dynamic distillation models have been proposed with very detailed phenomena described (e.g. [80]).

Without entering into a lengthy discussion, it appears that at present fairly "rigorous" dynamic models with full dynamic mass and energy balances for all plates and thermophysical properties predicted from generic comprehensive thermophysical packages can be readily integrated for use in batch simulation and optimization. The main justifications for shortcut models (simplicity , speed of solution, and ability to tailor special solution algorithms) appear to be less and less valid particularly in the light of the effort needed to validate the short cut approximations. Simplified thermodynamic and column models may still be necessary for mixtures for which a relative volatility is all that can be estimated, or for very demanding optimization applications (e.g. [48]). The desirable approach to follow, however, is no doubt to develop simulation and optimization tools suitable for as generic a model form as possible, but to leave the user the possibility to adopt simplifying assumptions which may be appropriate for specific instances.

Production period. Standard continuous dynamic distillation models can be used, represented by the general form:

$$f(x, x', t, u, v) = 0 \qquad t_0 \le t \le t_f \qquad \text{(eq. 1)}$$

with initial conditions $x_0 = x(t_0)$ and $x'_0 = x'(t_0)$. In eq. 1, f is a vector of differential and algebraic equations (DAEs), (mass and energy balances, vapor liquid equilibrium relations, thermophysical property defining equations, reaction kinetic equations, etc.), x and x' are vectors of (differential and algebraic) state variables and their time derivatives (differential variables only), t is the time, u is a vector of time varying control variables (e.g the reflux ratio) and v is a vector of time independent control variables (e.g. pressure).

A modeling issue particularly important for batch distillation regards the treatment of liquid holdups. Holdups may have quite significant effects on the dynamic response of the column (e.g. [19, 62]), and therefore should be considered whenever possible (zero liquid holdup means that a composition change in the liquid reflux reaches the still infinitely fast, clearly not the case). Assumptions of constant (mass, molar or volume) holdup are often used, and these already account for the major dynamic effects (delay in composition responses). Where necessary, (i.e. low pressure drop per plate) more detailed hydraulic models for plates and downcomers should be used. The extra information added to the model should be considered in light of the low accuracy often attached to generic hydraulic models and other estimated (guessed?) quantities (e.g. Murphee efficiencies). High quality relations, regressed from experimental data, are however often available for proprietary applications and their use is then justified. Vapor-liquid equilibrium has been invariably assumed so far for batch distillation, but there is no reason in principle why rate models cannot be used.

Similar arguments apply to the modeling of heat transfer aspects (reboiler, heat losses) and control loops. If the heat exchange area decreases as the pot holdup decreases, or the heat transfer coefficients vary significantly during a batch, for example, the assumptions of constant heat supply or constant vapor boilup rate may no longer be valid and it may be necessary to use a more detailed heat transfer model, accounting for the reboiler geometry [3]. If perfect control action (say, on the reflux ratio) is not a good assumption, equations describing the control loop dynamics may be included in eq. 1. In this case, the controller set-points will become the manipulated quantities and possible optimization variables.

Startup period. This may be defined as the period up to the time when the first distillate is drawn. It may be divided into two steps. In the first step, the column fills up and some initial profiles are established. In the second step total reflux is used until the top distillate is suitably enriched. For the first step, a model must be able to describe the establishment of initial liquid and vapor profiles, holdups and compositions along the column and in the condenser from an empty, cold column. Flows may be intermittent and mixing behavior on the plates and in the downcomers will be initially very different from the fully established situation, with vapor channeling weeping of liquid, sealing of downcomers, etc. This would demand accurate representation of hydraulic and phase behavior in rather extreme conditions. The use of vapor-liquid models based on perfect mixing and equilibrium under these circumstances is clearly suspect. Thermal effects due to the column being initially cold may be as important as other effects and the exact geometry of the column is clearly important [57]. Some of these aspects have been modeled in detail in [80], based on work for continuous distillation [36].

What is more usually done is to assume some far simpler mechanism for the establishment of the initial vapor and liquid profiles which does not need a detailed mechanistic model. For example: i) finite initial plate and condenser liquid holdups may be assigned at time zero at the same conditions as the pot (boiling point) (e.g. [19, 48]): ii) the column is considered initially (as a single theoretical stage, with the vapor accumulating in the condenser receiver at zero reflux. When sufficient liquid has been accumulated, this is redistributed as the internal plate holdups, filling the plates from the top down (e.g. [50, 38]) or from the bottom up [3]).

For the second startup step the same dynamic model may be used for the production steps. From an operation point of view, the practical question is whether a total reflux operation should be used at all, and if so, for how long. The duration of the total reflux operation (if at all necessary) can be optimized [19, 59].

An even cruder approximation is to ignore the startup period altogether, consider it as an instantaneous event and initialize the column model using the assumption of total reflux, steady-state (or as in [4], at the steady-state corresponding to a finite reflux ratio with no distillate production, obtained by returning the distillate to the pot). Of course, this procedure does not permit calculating the startup time.

Clearly, different startup models will provide different starting values for the column dynamic profiles in the first production step. Whether this matters was considered [1], comparing simulation results with four different models of increasing complexity for the first startup a stage. In all four cases, the procedure was followed by a second startup step at total reflux until stable profiles were achieved (steady state). The main conclusions were that the (simulated) startup) time can be significant (especially for difficult separations), is essentially due to the second step, is roughly proportional to the major holdups (in the condenser drum and, for difficult separations, those in the column) and that all four models gave approximately the same results (startup time and column conditions).

In the absence of experimental data to confirm or reject either approach, the use of the simpler procedures (i and ii above) to model the first step of the startup period would appear to be a reasonable compromise.

Shut down period. This period is typically not modeled in detail, since draining of the column and condenser following interruption of heating is usually very rapid compared to all other periods and separation is no longer affected. However, the final condenser receiver holdup may be mixed with the last distillate fraction, collected separately or mixed with the bottoms fraction, thus the exact procedure followed should be clearly defined (this is clearly irrelevant when holdups are ignored).

Transfers, additions, etc. Fresh feed or recycle addition are often modeled as instantaneous events. Adding side streams with finite flow rates to a generic column model, if required, is however a simple matter. These additional feeds (if any) may then be chosen as additional control variables.

In principle, different models of the type defined by eq. 1 may be defined for distinct distillation tasks. For example, referring to the operation in Figure 4, the number of species present in the final separation task (Step 3) may involve only a small subset of the species present in the initial charge. For numerical reasons, it may well be better to eliminate the equations related to the depleted species, giving a different set of equations for Step 1 and Step 3. Other modeling formulation details affect the numerical solution. For example, we found it is better to use an internal reflux ratio definition (L/V, with range 0-1) rather than the usual external one (L/D, ranged 0-infinity).

Simulation Issues

The simulation problem may be defined as follows:

Given: A DAE batch distillation model	$f(x, x', t, u, v)=0$	(eq. 1)
Values for all control variables	$u(t), v$	
Initial conditions	t_0, x_0 and x'_0	
Termination conditions	based on $t_f, x(t_f), x'(t_f), u(t_f)$	
Calculate: Time profiles of all state variables	$x(t), x'(t)$	

Dynamic models used in batch distillation are usually stiff. The main techniques used to integrate the above DAE system are BDF (Gear's) methods [37, 43] and orthogonal collocation methods [85]. Discontinuities will typically be present due to *discrete* events such as the instantaneous change in a control variable (e.g. a reflux ratio being changed between two values) and the switching of receivers at the end of a step. The presence of algebraic equations adds constraints between the state variables and their derivatives (and a number of complications). With respect to the initial conditions, only a subset of all x_0 and x'_0 variables may then be fixed independently, the remaining ones having to satisfy eq. 1 at the initial time (consistent initialization). A similar situation occurs with the discrete changes. If an algebraic variable can only be calculated from the right hand side of a differential equation (for example, the vapor leaving the stage from the energy balance), then some algebraic equation may have to be differentiated before the algebraic variable can be calculated (the number of differentiations required being called the index of the DAE system). A number of ad-hoc solutions had been worked out in the past to overcome these problems. For example, the derivative term in the energy balance could be approximated (e.g by a backwards finite difference, yielding an algebraic equation). To avoid discontinuities when changing reflux ratios, a continuous connecting function may be used [25]. These aspects are now far better understood, and even if it is still not possible to always solve higher index systems, it is usually possible to reformulate the equations so as to produce index 1 systems in the first place and to address implicit and explicit discontinuities. Requirements related to the consistent definition of the initial conditions, solvability of the system, DAE index, model formulation so as to avoid index higher than one and (re)initialization procedures with discontinuities were discussed in particular by [69, 4].

General purpose simulation packages such as BATCHES [17] and gPROMS [8] allow the process engineer to build combined discrete-event/differential algebraic models for simulation studies and have built in integration algorithms dealing with the above issues. In particular, they are able to deal effectively with model changes between different stages and discrete events. They are therefore suitable for general batch distillation simulations. General purpose dynamic simulators for essentially continuous systems such as SPEEDUP [68] can also be used, although the ability to handle generic events and model changes is more limited. A number of codes have also been developed specifically for batch distillation simulation (e.g. [4, 31, 35]) or adapted from continuous distillation [80]. A batch simulation "module" is available in several commercial steady state process simulators (e.g. the BATCHFRAC program [12], Chemcad III, ProsimBatch [71]). These are tailored to handle the specific events involved (switching of receivers, intermediate addition of materials, etc.) and in general use a single equation model for the entire batch operation and generic thermophysical property packages.

Optimization Issues

As noted, there are both discrete and continuous operation decisions to be optimized. At present dynamic optimization solutions have only been presented dealing with problems with fixed operation structure and with the same model equations for all separation stages. In the following we will therefore consider the optimization of the continuous operating decisions for an operation strategy selected a priori. Some work on the optimization of the operations structure is briefly discussed in the last section.

Equations and variables may be easily introduced into eq. 1 to define integral performance measures, for example, the total energy used over a distillation step. This permits calculating additional functions of all the states, controls, etc. and to define additional inequality constraints and an objective function, in a general form:

Inequality constraints	$g = g\,(t_f, x(t_f), x'(t_f), u, v)$	(eq. 2)
Objective function	$J = J\,(t_f, x(t_f), x'(t_f), u, v)$	(eq. 3)

The optimization problem may then be defined as follows:

Given: Initial conditions	t_o, x_o and x'_o	
Find: Values of all control variables	v, $u(t)$	
So as to: Minimize the objective function	Min J	(eq. 3)
Subject to: Equality constraints (DAE model)	$f(x, x', t, u, v) = 0$	(eq. 1)
Inequality constraints	$g\,(t_f, x(t_f), x'(t_f), u, v) \leq 0$	(eq. 2)

Upper and lower bounds may be defined on the control variables, u(t) and v, and on the final time. Termination conditions may be implicitly or explicitly defined as constraints. Additional inequality constraints may be defined for state variables not just at the end, but also at interior points (*path constraints*). For example, a bottom temperature may be bounded at all times. Some initial conditions may also have to be optimized.

The main problem in this formulation is the need to find optimum *functions* u(t) i.e. an infinite set of values of the controls over time. The main numerical techniques used to solve the above optimal control problem are the control vector parameterization (CVP) method (e.g. [58, 34]) and the collocation method (e.g. [10, 26, 74, 48]. Both transform the control functions into discrete forms approximated by a finite number of parameters.

The CVP method (used in this paper) discretizes each continuous control function over a finite number, defined a priori, of *control intervals* using a simple basis function (parametric in a small number of parameters) to approximate the control profile in each interval. For example, Figure 7 shows a piecewise constant approximation to a control profile with 5 intervals. With the initial time given, two parameters are sufficient to describe the control profile in each interval a control level and the final time of the interval. Thus the entire control profile in Figure 7 is defined by 10 parameters. These can then be added to any other decision variable in the optimization problem to form a finite set of decision variables. The optimal control problem is then solved using a nested procedure: the decision variables are set by an optimizer in an outer level and, for a given instance of these variables, a dynamic simulation is carried out to calculate objective function and constraints (eqs. 1-3). The outer problem is a standard (small scale) nonlinear programming problem (NLP), solvable using a suitable method such as Sequential Quadratic Programming (SQP). Parameterizations with linear, exponential, etc. basis functions may be used [34]. Since the

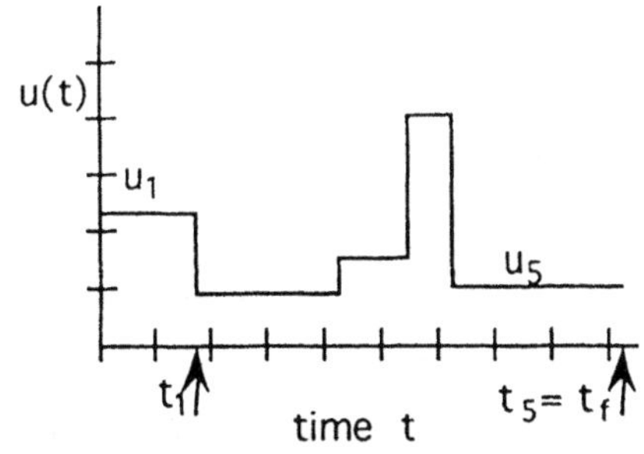

Figure 7. Piecewise Constant Discretization of Continuous Control Function

DAEs are solved for each function evaluation, this has been called a feasible path approach. Its main advantage is that the approach is very flexible. General purpose optimizers and integrators may be used, and the set of equations and constraints may be very easily changed.

The collocation methods discretize both the control functions and the ordinary differential equations in the original DAE model, using collocation over finite elements. The profiles of all variables are approximated using a set of basis functions, with coefficients calculated to match any specified initial conditions, final conditions and interior conditions (additional conditions being provided by the continuity of profiles across the finite elements). The end result is a large system of algebraic equations, which together with constraints and objective function form a large NLP problem. For the same problem, however, the degree of freedom for this large scale NLP is the same as for the small scale NLP in the CVP approach. The optimization may be solved using suitable NLP algorithms, such as a SQP with Hessian decomposition [48]. Since the DAEs are solved at the same time as the optimization problem, this has been called an infeasible path approach. The main advantage claimed for this approach is that it avoids the repeated integrations of the two level CVP method, hence it should be faster. However, comprehensive comparisons between the collocation and the CVP methods have not been published, so it is not possible to make definite statements about the relative merit.

The *path constraints* require particular treatment. Gritsis [39] and Logsdon and Biegler [48] have shown that they can be handled in practice both by the CVP and orthogonal collocation methods (in particular, by requiring the constraints to hold at a finite number of points, coincident with control interval end points or collocation points). A general treatment of the optimal control of constrained DAE systems is presented in [70].

Computer codes for the *optimization* of generic batch distillation operations have been developed by [59, 31]. To my knowledge, no commercial code is presently available.

Application Examples

In this section, some examples are given of the application of the ideas and methods outlined above, drawn from our own work over the last few years. The examples are used to illustrate the current progress and capabilities, in particular with respect to the generality of operation strategies, column and thermodynamic models and objective functions and constraints that can be handled.

Solution Models and Techniques

All examples were produced using a program developed specifically for batch distillation simulation and optimization by [68] and successively extended to handle more features. For a particular problem, a user supplies the definition of a batch *column configuration* (number of theoretical stages, pot capacity, etc.), defines the fresh charge mixture to be processed and selects a *distillation model* and *thermodynamic options*. Two column models have been used, mainly to show that the solution techniques are not tied to a specific form of the model. The simpler column model (MC1) is based on constant relative volatility and equimolar overflow assumptions. A more rigorous model (MC2) includes dynamic mass balances and general thermodynamics, with the usual assumptions of negligible vapor holdup, adiabatic plates, perfect mixing for all liquid holdups, fixed pressure and equilibrium between vapor-liquid. A total condenser is used with no sub-cooling and finite, constant molar holdups are used on the plates and in the condenser receiver. To maintain solution time reasonably low, the energy balances are modeled as algebraic equations (i.e. energy dynamics is assumed to be much faster than composition dynamics). Thermodynamic models, vapor liquid equilibria calculations and kinetic reaction models are supplied as subroutines (with analytical derivatives, if available). Rigorous, general purpose thermodynamic models, including equations of state and activity coefficient models may therefore be used for nonideal mixtures. A simple constant relative volatility model (MT1) may still be used, if desired, in conjunction with the more rigorous dynamic model MC2.

A desired *operation strategy* (number and sequence of product cuts, off-cuts, etc) is defined *a priori*. A simple initialization strategy is used for the first distillation task. The fresh charge is assumed to be at its boiling point and a fraction of the initial charge is distributed on the plates and in the condenser (according to the specified molar holdups). For successive tasks, the initial column profiles are initialized to be the same as the final column profiles in the preceding task. Adjustments may be made for secondary charges and to drop depleted component equations from the model. For each STN task, the *control variables* are identified and the number of discretization *control intervals* is selected (a CVP method with piecewise constant parameterization is used) and initial values are supplied for all control levels and control switching times. Reflux ratio, vapor boilup rate and times are available as possible control variables (additional quantities, if present, may also be used for control, such as the rate of a continuous feed lo an intermediate stage). A variety of *specifications* may be set for individual distillation steps, including constraints on selected end purities and amounts collected (other specifications are described in the examples). Finally, an *objective function* is selected, which may include those traditionally used (max distillate, min time), one based on productivity (amounts over time) or on economics (max profit). In the latter case, suitable cost coefficients must also be supplied for products, off-cuts, feed materials and utilities. From these problem definitions, programs are generated for computing all required equation residuals and the analytical Jacobians, together with a driver for the specific simulation/optimization case.

A robust general purpose SQP code [15] is used to solve the nonlinear programming optimization. A full matrix version is used here, although a sparse version for large scale problems (with decomposition) is also available. The DAEs system is integrated by a robust general purpose code, DAEINT, based on Gear's BDF method. The code includes procedures for the consistent initialization of the DAE variables and efficient handling of discontinuities. The gradients needed by the SQP method for optimization are efficiently calculated using adjoint variables [58], requiring the equivalent of approximately only two integrations of eq. 1 to calculate all gradients (an alternative, method for generating the gradients would be to integrate the dynamic sensitivity equations alongside the model equations, [13]. Outputs are in tabular and simple graphical form.

Simulation and Sequential Optimization of Individual Distillation Steps

The first example deals with a four component mixture (propane, butane, pentane and hexane) to be separated in a 10 equilibrium stage column and was initially presented in [12]. The operation consists of 5 distillation steps with two desired products. A fraction with high propane purity is collected first, followed by a step where the remaining propane is removed. A high purity butane fraction is then collected, with removal of pentane in an off-cut in the fourth step. The final step removes the remaining pentane in the distillate and leaves a high purity hexane fraction as the bottom residue. A secondary charge is made to the pot after the second step (Figure 8).

Boston et al. presented [12] simulation results for an operation with constant reflux ratio during each step and different values in distinct steps. Simulations of their operation (same reflux ratio and duration for each step) were carried out in [59] with two thermodynamic models, MT2 (an ideal mixture model with VLE equilibrium calculated using Raoult's law and Antoine's model for vapor pressure, ideal gas heat capacities for vapor enthalpies from which liquid enthalpies were obtained by subtracting the heat of vaporization) and MT3 (SRK equation of state). Results with the former thermodynamic model and with the more rigorous dynamic column model MC2 are

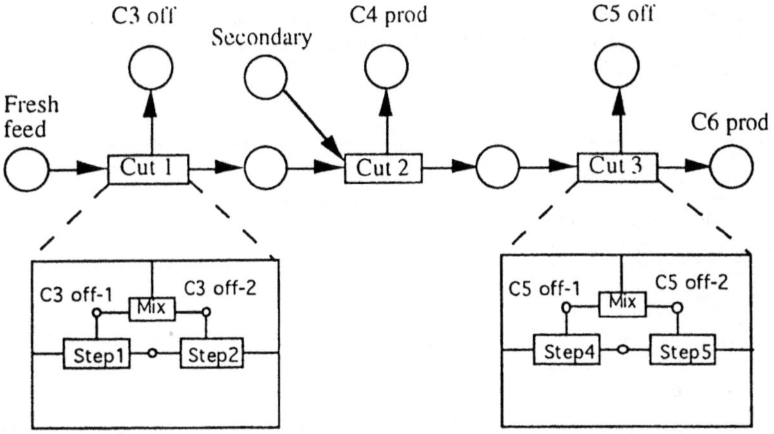

Figure 8. Batch Distillation of Quaternary Mixture [12]-Operating Strategy. The two distillation tasks Cut1 and Cut3 are composed of two off-cut production steps each

reproduced in Table 2. They are very similar to those reported in the original reference, with inevitable minor differences that are due to difference in thermodynamic models, integration tolerances, etc. This operation was taken as a base case.

Mujtaba [12] considered the optimization of the same process, using the above operating policy as a base case, as follows. In the absence of an economic objective function in the original reference, it was assumed that the purities obtained in each of the cuts (in terms of the molar fraction of the key component in that cut) were the desired ones and that the reflux ratio policy could be varied. A variety of objective functions were defined. Table 3 reproduces the results obtained for the *first* distillation step with three objective functions: minimum time, maximum distillate and maximum productivity, with an end constraint on the mole fraction of propane in the accumulated distillate ($x_{propane} = 0.981$). All results are reported in terms of a common measure of productivity, the amount of C3 off-1 distillate produced over the time for Step 1 and for two ways of discretizing the control variable, into a single interval and five intervals, respectively. As expected, the productivity of this Step varies depending on the objective function used, increases when more control intervals are used and is maximum when productivity itself is used as the objective function. As a comparison, the base case operation had a productivity of 2.0 lbmol/hr for Step 1 (8.139 lbmole produced in 4.07 hr). With a single reflux ratio level, the minimum time problem and the maximum distillate problem resulted (within tolerances) in the same operation as

Table 2. Batch Distillation of Quaternary Mixture [12]. Simulation Results with MC2 Column Model (dynamic, more rigorous) and MT2 Thermo Model(ideal).Total Time Tf=30.20 hr; Overall Productivity (A+B)/Tf=3.17 lbmol/hr

Column					
No. of Internal Stages (ideal)	8				
Condenser	total no	subcooling			
Liquid Holdup - stage (lbmol)	4.93 10^{-3}				
Liquid Holdup - cond. (lbmol)	4.93 10^{-2}				
Charges	Fresh feed	Secondary			
(1) Propane	0.1	0.0			
(2) Butane	0.3	0.4			
(3) Pentane	0.1	0.0			
(4) Hexane	0.5	0.6			
Amount (lbmol)	100	20			
at time	initial	after Step2			
Operation - Specified					
Task	Step1	Step2	Cut2	Step4	Step5
Products	C3 off-1	C3 off-2	C4 prod	C5 off-1	C5 off-2 C6 prod
Specified:					
Reflux ratio (external)	5	20	25	15	25
Time (hr)	4.07	1.81	18.27	4.31	1.78
Distillate rate (lbmol/hr)	2	2	2	2	2
Pressure (bar)	1.03	1.03	1.03	1.03	1.03
Operation - Results					
Top Vapour rate (lbmol/hr)	12	42	52	32	52
Instant Distillate (mole fraction)					
Propane	0.754	0.031	....	....	....
Butane	0.246	0.969	0.254	....	....
Pentane	....	....	0.745	0.613	0.091
Hexane	....	....	....	0.387	0.909
Accum. Distillate (mole fraction)					
Propane	0.981	0.850	....	....	....
Butane	0.019	0.150	0.988	0.017	0.012
Pentane	....	....	0.012	0.940	0.778
Hexane	....	....	....	0.043	0.210
Amount (lbmol)	8.139	11.760	36.548=A	8.619	12.180
Still Pot (mole fraction)					
Propane	0.021	....	...	....	....
Butane	0.325	0.319	0.001	....	....
Pentane	0.109	0.113	0.133	0.023	0.002
Hexane	0.545	0.567	0.866	0.977	0.998
Amount (lbmol)	91.860	88.240	71.680	63.061	59.380=B

the base case (for the required purity, the system behaves as a binary and the optimization essentially solves a two-point boundary value problem). Taking both the amount produced and the time required into account (i.e. the productivity objective function), however, permits improving this step by over 50%. Further improvements are achieved with more control intervals. Two of the optimum control policies calculated for the maximum productivity problem are also reported in Table 3. The desired propane purity is achieved in all cases.

Table 3. Batch Distillation of Quaternary Mixture [12]-Optimization of Step 1 with MC2 Column Model (dynamic, more rigorous) and MT2 Thermo Model (ideal)

Problem - Optimise Step 1	Min Time	Max Distillate	Max Productivity
Specified:			
Top Vapour rate (lbmol/hr)	12	12	12
Pressure (bar)	1.03	1.03	1.03
Product state C3 off 1 - mole fraction C3	0.981	0.981	0.981
Amount(lbmol)	8.139		
Time (hr)		4.07	
Controls:			
Reflux ratio (external) (r); End time (t1)	r, t1	r	r, t1
Optimal Operation			
1 control interval			a)
End Time: t1 (hr)	4.01		1.75
Amount of C3 off-1: C (lbmol)		8.15	5.67
Productivity = A/t1 (lbmol/hr)	2.02	2.00	3.24
5 control intervals			b)
End Time: t1 (hr)	2.82		1.64
Amount of C3 off-1: C (lbmol)		9.26	5.88
Productivity = C/t1 (lbmol/hr)	2.881	2.275	3.59
Optimal reflux ratio policies (control level/end time, level/time, ...)			
a) (r, t) = (0.73/1.75)			
b) (r, t) = (0.30/0.636, 0.664/0.59, 0.695/0.91, 0.727/1.22, 0.758/1.64)			

Optimization of the first step, as described, provided not only the optimal values of the control parameters for the step, but also values of all state variables in the column model at the end of the step. These are used to calculate the initial values of all the state variables in the model for the next step. If the same equations model is used in the two steps, the simplest initialization policy for step 2 is simply to set the initial values for step 2 to the final values of step 1 (used in all examples. unless noted). However, it is also possible to change the model (for example, to use a different thermodynamic option. to eliminate equations for components no longer present) or to carry out calculations for impulsive events defined at the transition between the two steps (for example, to determine the new pot amount and composition upon instantaneous addition of a secondary charge). The next step may then be either simulated (if all controls are specified) or optimized (if there are some degrees of freedom). In principle, a different objective function could be used for each step.

Results for such a sequential optimization of the five distillation steps in the operation, using the same distillation model throughout, minimum time as the objective function for each step and the same purity specifications as in the base case gave the results summarized in Table 4, in terms of overall productivity (total amount of the two desired products/ total time). Significant increases

Table 4. Batch Distillation of Quaternary Mixture [12] - Optimization of 5 Steps in Sequence (min. time for each step) with MC2 Column Model (dynamic, more rigorous) and MT2 Thermo Model (ideal)

Problem: Sequential Optimisation (min time for each step)	Step1	Step2	Step3	Step4	Step5	Overall
Specified:						
Top Vapour rate (lbmol/hr)	12	42	52	32	52	
Pressure (bar)	1.03	1.03	1.03	1.03	1.03	
Product state	C3 off1	C3 off-2	C4 prod	C5 off-1	C6 prod	
(Key component) mole fraction	(1)0.981	(1)0.850	(2)0.988	(3)0.940	(4)0.998	
Amount (lbmol)	8.139	11.760	36.548	8.619	59.380	
Controls:						
Reflux ratio (external), r; times	r, t1	r, t2	r, t3	r, t4	r, t5	
Optimal Operation						
1 control interval per step:						
Time (hr)	4.01	1.56	9.20	3.49	1.55	19.31
Amount (lbmol) A=C4 prod, B=C6 prod			36.566=A		59.44=B	
Productivity = (A+B)/Tf (lbmol/hr)						**4.84**
5 control intervals per step:						
Time (hr)	2.82	1.37	2.57	2.83	1.54	11.13
Amount (lbmol) A=C4 prod, B=C6 prod			36.567=A		59.44=B	
Productivity = (A+B)/Tf (lbmol/hr)						**8.62**

in performance (over 50%) relative to the base case are achieved even with a single (but optimal) profiles are given in Figure 9 for the whole operation. The dynamic model used does not include hydraulic equations, therefore, the optimal operation should be checked for weeping, entrainment, flooding, etc.

Recycle Optimization

Policies for recycles were discussed in general terms in the initial sections. Off-cut recycle optimization methods have been discussed for the simpler, binary case in [56, 16, 60], as well [51, 72, 63] as different special cases of multicomponent mixtures.

With reference to the specific operation strategy in Figure 5, if the off-cut produced in Step 2 in one batch is characterized by the same amount and composition as that charged to the pot in the *previous* batch, then, for constant fresh charge and with the same control policies applied in subsequent batches, the operation described in Figure 5 will be a quasi-steady state one, that is subsequent batches will follow the same trajectories, produce the same product and intermediate states, etc. This cyclic condition may be stated in several ways, for example by writing the composition balance for the charge mixing step (Mix) as:

Instant Distillate Composition Minimum Time Problem

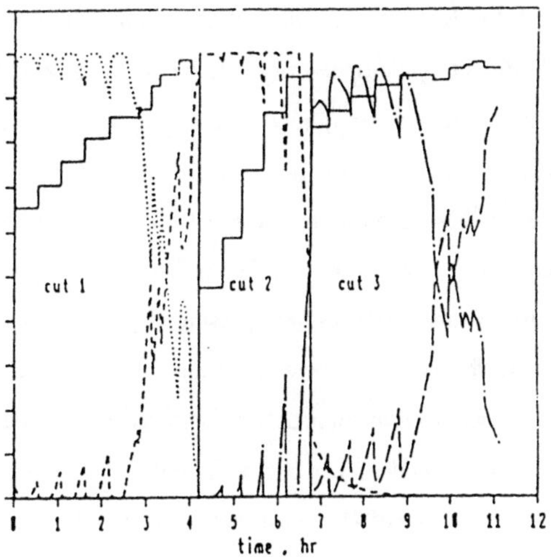

Accumulated Distillate Composition Minimum Time Problem

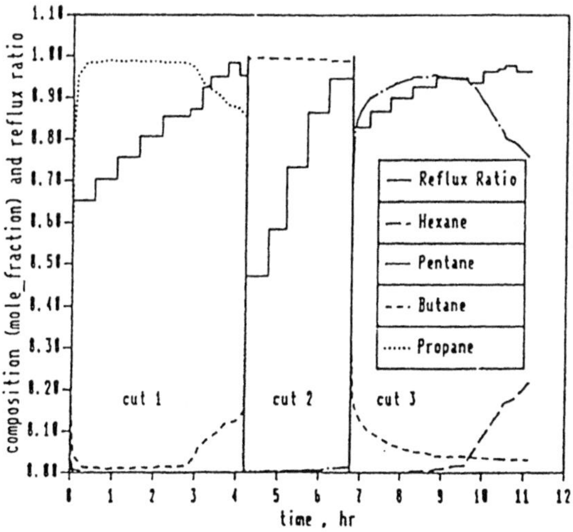

Figure 9. Batch Distillation of Quaternary Mixture[12]-Sequential Optimization of all 5 Steps.Optimum Reflux ratio Profile and Distillate Composition (mole fractions)

$$B_0\, x_{B0} + R\, x_R = B_c\, x_{Bc} \qquad \text{(eq. 5)}$$

where B_0, x_{B0} and B_c, x_{Bc} the fresh feed and mixed charge amounts and composition, supply the *initial* conditions for Step 1 and R, x_R are the amount and composition of the collected off-cut at the *final* time of Step 2. Different cyclic operations may be obtained corresponding to different off-cut fraction amount and composition. An optimal recycle policy can be found by manipulating simultaneously the available controls (e.g. reflux ratio profile and duration) of all distillation steps within the loop (Step 1 and Step 2), so as to optimize a selected objective function, while satisfying constraints 5 (or equivalent ones) in addition to other constraints (on product purity, etc.). A solution of this problem for binary mixtures was presented [60] using a two level problem formulation, with minimum time for the cyclic operation as the objective function. For given values of R and x_R, a sequence of two optimal control problems is solved in an inner loop. The minimum time operation of Step 1 is calculated as described above, followed by the minimum time operation for Step 2, with purity and amount of the off-cut constraints to match the given values. In an outer loop, the total time is minimized by manipulating R and x_R as the (bounded) decision variables.

An alternative, single level formulation for this problem was also developed by [59] where the total time is minimized directly using the mixed charge quantities (B_c and x_{Bc}) as decision variables, together with reflux ratio as control variable, as follows:

$$\min_{B_c,\, x_{Bc},\, r(t)} \quad J = t1 + t2 = tf$$

subject to: DAE model

bounds on all control variables

interior point constraints, e.g.: $x_{D1}(t1) \geq x_{D1}^*$

$D_1(t1) \geq D_1^*$

end point constraints, e.g.: $x_{B3}(tf) \geq x_{B3}^*$

cyclic conditions (eq. 5)

Here tl and t2 are the duration of production Steps 1 and 2, respectively and tf is the total time of the cyclic operation. The starred values are specifications (only two of which are independent, the other two being obtained from a material balance over the STN states Fresh Charge, Maincut 1 and Bottom Product in Figure 5). The control vector is discretized, as usual, into a number of control intervals. However, the end time of Step 1, tl, is defined to correspond to one of the control interval boundaries. The constraints on the main distillate fraction are now treated as interior point constraints and the cyclic conditions are treated as end point constraints. This

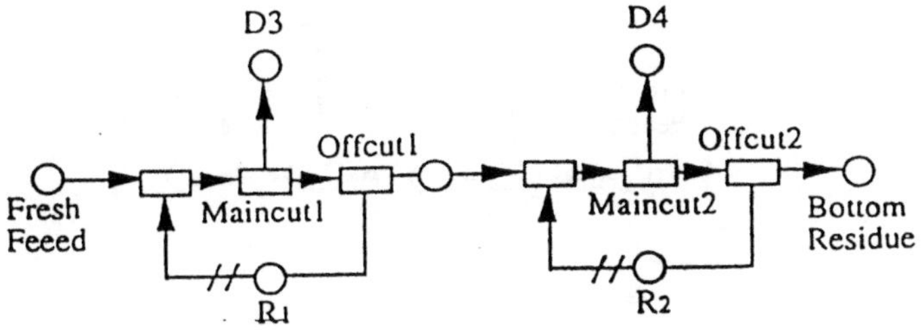

Mixture: butane(1), propane(2), n-hexane (3)

State	Amount (kmol)	(component) mole fraction
Fresh feed	6.0	<0.15, 0.35, 0.50>
Specifications		
D3	0.9	(1) 0.935
D4	2.0	(2) 0.82

Top Vapour rate=3.0kmol/hr

<u>Optimal Operation</u>

R1	0.64	(1) 0.276
R2	0.41	(2) 0.40

Min. time for maincut1 + Offcut 1: 2.05 hr

Min. time for maincut2 + Offcut 2: 1.69 hr

Total batch time 3.74 hr

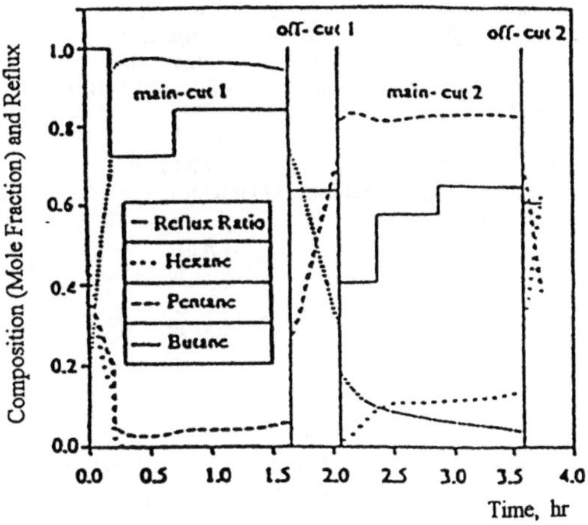

Figure 10. Batch Distillation of Ternary Mixture with production and recycle of intermediate off-cut. Thermodynamic model MT2 (ideal), Column Model MC2 (dynamic, more rigorous)

formulation was found to result in rather faster solution that the two level formulation, while giving the same results.

This approach may also be used for multicomponent mixtures and more complex operation strategies, such as that shown in Figure 6. An example for a ternary mixture (butane, pentane and hexane) with specifications on two main products was reported in [63]. The production strategy involves four steps (two main product fractions, each followed by an off-cut production step, with each off-cut recycled independently), as shown in Figure 10. Using the single level formulation, the minimum distillation time for the first recycle loop was calculated first, providing initial column conditions for the second loop, for which the minimum time was then also calculated. The results for this *sequential* optimization of individual recycle loops are summarized in Figure 10. Comparison of the total time required with that for the optimal operation involving just two product steps and no off-cuts (Figure 11) shows a significant reduction of 32% in the batch time. The same products (amounts and compositions) are obtained with both operations.

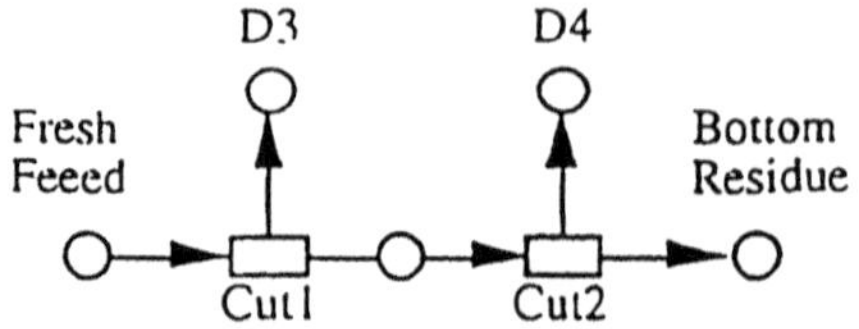

Mixture: butane(1), propane(2), n-hexane (3)

State	Amount (kmol)	(component) mole fraction
Fresh feed	6.0	<0.15, 0.35, 0.50>

Specifications

D3	0.9	(1) 0.935
D4	2.0	(2) 0.82

Top Vapour rate=3.0kmol/hr

Optimal Operation

Min. time for Cut1:	3.8 hr
Min. time for Cut2:	1.71 hr
Total batch time	5.51 hr

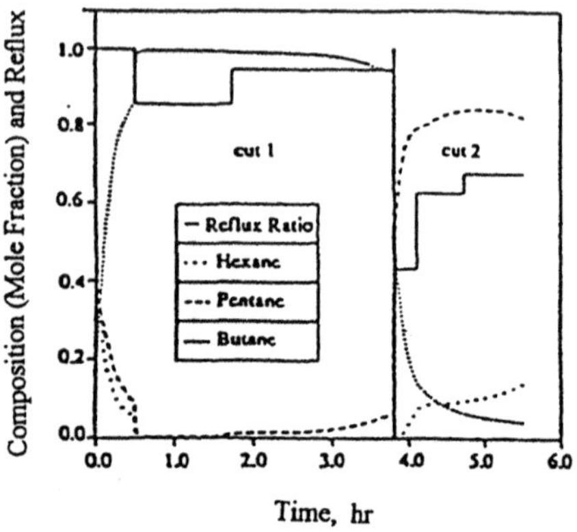

Figure 11. Batch Distillation of Ternary Mixture with no Off-Cuts. Thermodynamic Model MT2 (ideal), Column Model MC2 (dynamic, more rigorous)

Multiperiod Optimization

In the above examples, optimization was performed of distillation steps individually and sequentially. Clearly, this is not the same as optimizing an overall objective function (say, minimum total time), nor are overall constraints taken into account (say, a bound on overall energy consumption). An *overall* optimization was however discussed for all steps in a single recycle loop and the same two approaches in that section may also be used to optimize several adjacent steps or even all the production periods in an operation. We refer to this as multiperiod optimization. This has been discussed by Fahrat et al. [34], however with shortcut models and simple thermodynamics and in [64] with more general models.

Only a small set of decision variables is required to define a well posed optimal control problem for individual steps. With the fresh feed charge given, specification of a key component purity and an extensive quantity (amount of distillate or residue, or a recovery) for each distillation step permits calculating the minimum time operation for that step, and hence for all steps in sequence. Typical quantities of interest for overall economic evaluation of the operation (amounts produced, energy and total time used, recoveries, etc.) are then explicitly available. A general overall economic objective function may be calculated utilizing unit value/cost ($/ kmol) of all products, off-cut materials, primary and (if any) secondary feeds and unit costs of utilities (steam, cooling water). For example an overall profit ($/batch) may be defined as:

J1=Sum (product/off-cut values)-Sum (feed costs)-Sum (utility, setup, changeover costs) (eq. 6)

or in terms of profit per unit time ($/batch):

J2 = J1 / batch time (eq. 7)

with significant changeover and setup times also included in the batch time. Overall equality and inequality constraints may be similarly written.

The decision variables may then be selected in an outer optimization step to maximize an overall objective function subject to the overall constraints. Many purities are typically specified as part of the problem definition and recoveries represent sensible engineering quantities for which reasonably good initial guesses can often be provided. The outer level problem is therefore a small nonlinear programming problem solvable using conventional techniques. The inner problem has a block diagonal structure and can be solved as a sequence of small scale optimal control problems one for each STN task.

This solution method was proposed in [64] who presented results for batch distillation of a ternary mixture in a 10 stage column with two distillate product fractions (with purity specification) and one intermediate off-cut distillate (with a recovery specified). Here, the more rigorous dynamic model was used with thermodynamic described by the SRK equation of state (Thermo model MT3). The amounts of all fractions the composition of the recycle off-cut and the reflux ratio profiles were optimized so as to maximize the *hourly profit*, taking into account energy costs. The optimal operation is summarized in Figure 12 with details given in the original reference. The required gradients of objective functions and constraints for the outer problem were obtained by finite difference (however exploiting the block diagonal structure of the problem) with small effort. Analytic sensitivity information could be used if available.

For the quaternary example [12] previously discussed the operation in Figure 8 is considered again this time with Step 1 and Step 2 merged into the singe task Cut 1 (propane elimination) and Step 4 and 5 merged into the single task Cut 3 (pentane elimination). A problem is formulated whereby the overall productivity (amounts of states C4 prod and C6 produced over the total time)

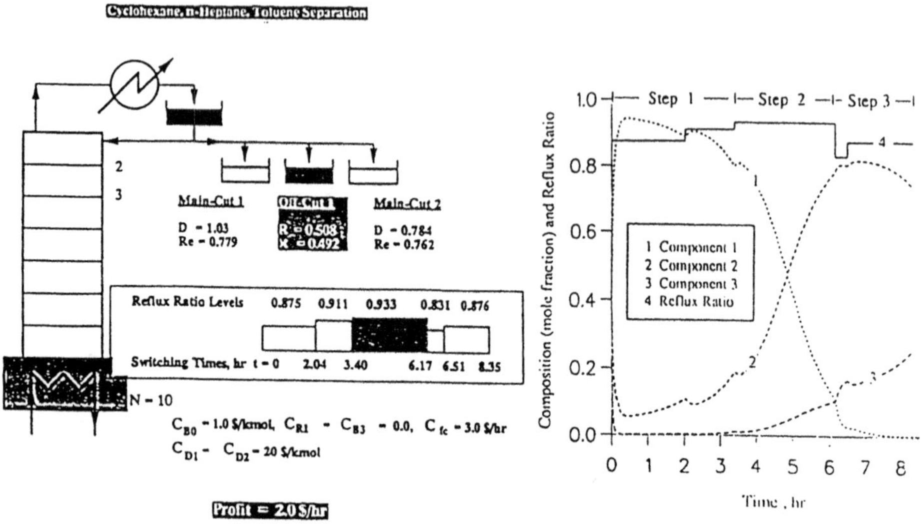

Figure 12. Multiperiod Optimization with Ternary Mixture - Maximum hourly profit, SRK EOS, dynamic column model MC2, Specifications on maincut 1 and maincut 2 product purity and Cyclohexane recovery in Off-cut. Optimal operation and instant distillate composition profiles are shown

is maximized subject to the same product purity constraints considered in the sequential optimization (Table 4). The pentane purity in the C3 off-fraction butane recovery in Cut 2 and hexane recovery in Cut 3 are now considered as variables to be optimized in addition to the reflux ratio profiles and times. A two level multiperiod solution leads to the optimal operation detailed in Table 5. Smaller amounts of product are produced, however in far less time, leading to an overall productivity almost twice as high as for the optimal sequential solution and four times higher than the base case. This is no doubt also due to the use here for all times of the largest vapor rate (52 lbmol/hr) which was utilized in the previous cases (only in Steps 3 and 5). This optimization required 4 functions and 3 gradient evaluations for the outer loop optimization and less than 5 hrs CPU time on a SUN SPARC10 machine.

Table 5. Multiperiod Optimization of Quaternary Mixture [12] with 3 Separation tasks. Maximum overall productivity, MC2 column model (dynamic, more rigorous) and MT2 thermo model (ideal)

Multiperiod Optimisation max Productivity = (A+B)/Tf	Cut1	Cut2	Cut3	Overall
Specified:				
Top Vapour rate (lbmol/hr)	52	52	52	
Pressure (bar)	1.03	1.03	1.03	
State (Key comp.) mole fraction		C4 prod (2) 0.988	C6 prod (4) 0.998	
State (Key comp.) task recovery	C3 off (1) 0.996			
Optimised:	initial <bounds>	initial <bounds>	initial <bounds>	
State	C3 off	C4 prod	C6 prod	
(Key comp.) mole fraction	(1)0.981<0.8-0.95>			
(Key comp.) recovery for task		(2)0.90<0.85-0.95>	(4)0.95<0.90-0.98>	
Controls				
Variables (No. contr. intervals)	r, t (3)	r, t (3)	r, t (3)	
Initial guess control level	0.8, 0.8, 0.8	0.8, 0.8, 0.8	0.8, 0.8, 0.8	
end time	0.5, 1.0, 1.5	1.0, 2.0, 3.0	1.0, 2.0, 3.0	
Optimal Operation				
Time (hr)	1.66	1.28	3.05	5.99=Tf
Productivity =(A+B)/Tf (lbmol/hr)				**14.55**
Product State (mole fraction)	C3 off	C4 prod	C5 off C6 prod	
Propane	0.869	0.001	... ...	
Butane	0.131	0.988	0.258 ...	
Pentane	...	0.011	0.450.....0.002	
Hexane	...	...	0.292 0.998	
Amount (lbmol)	11.433	31.361=A	21.186 55.814=B	
Optimal control level	0.711,0.987,0.969	0.469,0.599,0.723	0.648,0.877,0.945	
end time	0.37, 0.90, 1.66	0.57, 0.96, 1.28	0.59, 1.44, 3.05	

As with recycles, it is possible to utilize the single level problem formulation for the general multiperiod case.

Reactive Batch Distillation

Reactive batch distillation is used to eliminate some of the products of a reaction by distillation as they are produced rather than in a downstream operation. This permits staying away from (reaction) equilibrium and achieving larger conversions of the reactants than possible in a reactor followed by separation. With reference to Figure 1, reaction may occur just in the pot/reactor (for example, when a solid catalyst is used) or in the column and condenser as well (as with liquid phase reactions). Suitable reactions systems are available in literature [6].

With the batch distillation model (eq. 1), written in general form, extensions to model the reactive case amount to just small changes in the mass and energy balances to include a generation term and to the addition of reaction kinetic or equilibrium equations for all stages where reaction occurs (pot, or all stages). Energy balances do not even need changing if written in terms of total enthalpy. Modeling details and simulation aspects are given in many references (i.e. [25, 4]).

From the operation point of view, there are some interesting new aspects with respect to ordinary batch distillation. Reaction and separation are tightly coupled (reaction will affect temperatures, distillate and residue purities, batch time, etc. while the separation will affect reaction rates and equilibrium, reactant losses, etc.). There is one more objective, essentially the extent of reaction, but no new variables to manipulate (unless some reactants are fed semi-continuously, in which case the addition of time and rate may he new controls), making this an interesting dynamic optimization problem. Optimization aspects were first discussed in [33] and more recently in [86, 40, 65] who also review recent literature.

With the optimal control problem formulated as above, no change is needed to handle the reactive case, other than to supply a slightly modified model and constraints. Optimal operating policies were presented in [65] for several reaction systems and column configurations. Typical results are summarized in Figure 13 for the esterification of ethanol and acetic acid to ethyl acetate and water, with *maximum conversion* of the limiting reactant used as the objective function, reflux ratio as the control variable, given batch time and a constraint on the final purity of the main reaction product (80% molar ethyl acetate, separated in a single distillate fraction). This formulation is the equivalent of the maximum distillate problem in ordinary batch distillation. Similar results were also calculated for maximization of a general economic objective function, *hourly profit* = (value of products - cost of raw materials - cost of utilities)/ batch time, with the batch time also used as a control variable. Again, profit improvements in excess of 40% were achieved by the optimal operation with respect to quite reasonable base cases, obtained manually by repeated simulations.

On-Line Application - Optimization of All Batches in a Distillation Campaign

The above examples involved the a priori definition of the optimal control profiles for a batch, assuming that all batches would be carried out in the same way. Here, we wish to show how such techniques may be applied in an automated, flexible batch production environment.

The application was developed for demonstration purposes and involves a simple binary mixture of Benzene and Toluene, to be separated into a distillate fraction and a bottom residue, each with specified minimum purity (benzene mole fraction = 0.9 in the distillate, toluene mole fraction = 0.9 in the residue). A batch column is available, of given configuration. A quantity of

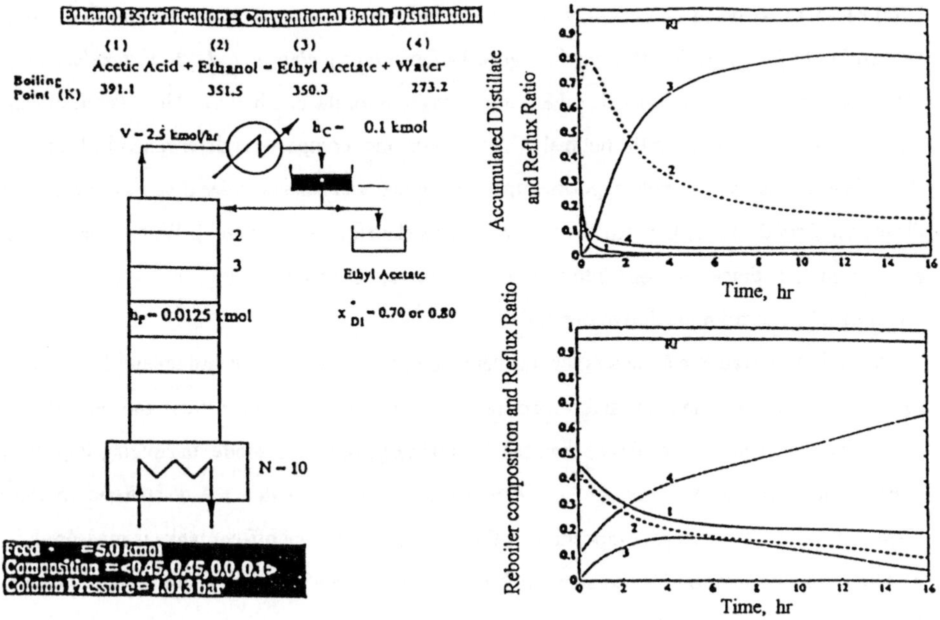

Figure 13. Reactive Batch Distillation - Maximum Conversion of Acetic Acid to Ethyl Acetate, with purity specification on the Ethyl Acetate (Distillate) product. Column model MC2, correlated K-values and kinetic rates

feed material becomes available, its exact composition being measured from the feed tank. Other tanks (one for each product plus one for an intermediate) are assumed initially empty. The quantity of feed is such that a large number of batches are required. The following procedure is adopted:

1. An operation structure is selected for the typical batch, in this case involving an off-cut production and recycle to the next batch (Figure 5). Given the measured fresh feed charge composition (mole fraction benzene = 0.6), pot capacity and product purity specifications, the *minimum total time*, *cyclic* operation (reflux ratios profiles, times, etc.) is calculated as 0.6 kmol of distillate, optimal off-cut of 0.197 kmol at mole fraction benzene = 0.63, reflux ratio r = 0.631 for 2.64 hr (benzene production) then r= 0.385 for 0.51 hr (off-cut production), leaving 0.6 kmol of bottom product. A single control level was chosen for each step for simplicity, with a multiperiod, single level formulation for the optimization problem.

2. First batch - Since the off-cut recycle tank is initially empty, the cyclic policy cannot be implemented at the beginning and some way must be established to carry out one or more initial

batches so as to reach it. One way (not necessarily the best one), is to run the first batch (according to the operation strategy in Figure 14 (*secondary charge and off-cut production.*). The off-cut from the previous batch, OFF-0 is known (zero for the first batch). The desired off-cut OFF-1, is specified as having the optimal cyclic amount and composition determined in 1 above. With distillate and residue product purifies specified, a material balance gives the target values for distillate and product amounts (e.g. 0.4744 kmol of distillate for the first batch). With these targets, a minimum total time operation for this batch is calculated. In an ideal world, the cyclic operation could then be run from batch two onward.

3. The optimal operation parameters for this batch are passed to a control system and the batch is carried out. The actual amounts and compositions produced (distillate cut, off-cut and residue cut) will no doubt be slightly different than the optimal targets, due to model mismatch, imperfect control, disturbances on utilities, etc. Product cuts from the batch are discharged to their respective tanks. The actual composition in the product and in the off-cut tanks is measured.

4. The optimal operation for the next batch is calculated again using the operation structure in Figure 14 and the calculation procedure outlined for the first batch but using the measured amounts and composition of the off-cut produced in the previous batch. Because of this, the optimal policy for the batch will be slightly different from that calculated in step 1. Any variations due to disturbances control problems missed targets etc. in previous batches will also be compensated by the control policy in the current batch. Steps 3 and 4 are repeated until all the fresh feed is processed (a special operation could be calculated for the last batch so as to leave the off-cut tank

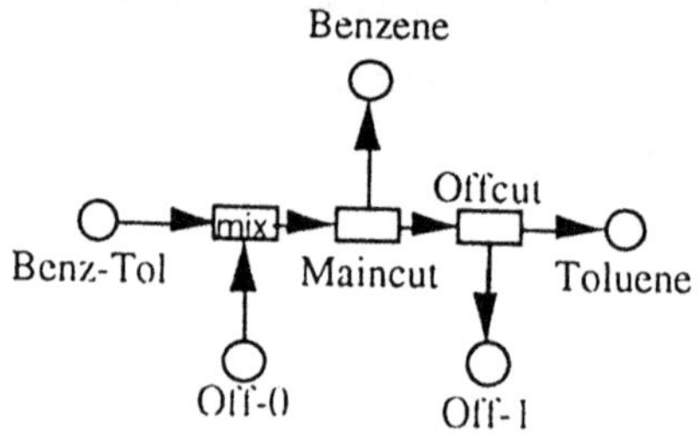

Figure 14. Batch Distillation of Binary Mixture (Benzene, Toluene)-Operating strategy with addition of secondary charge (off-cut from previous batch) and production of intermediate off-cut. Column model MC2 (dynamic, more rigorous), thermo model MT2 (ideal)

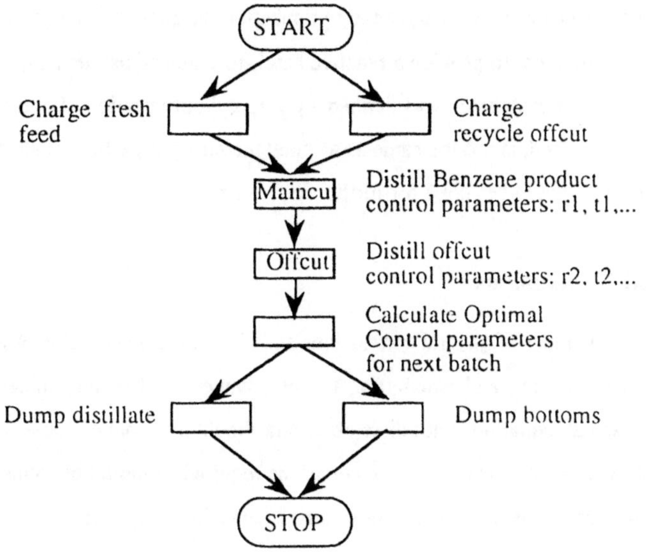

Figure 15. Control Procedure for the Automatic Execution of one Batch by the Control System

empty).

The procedure controlling the actual batch execution is shown schematically in Figure 15. The main control phases correspond to the distillation tasks in the STN definition of the operation with additional control phases for all transfers and details with regards to the operation of individual valves control loops etc. After the main distillation steps (and quality measurements on the fractions produced) the sequence automatically executes an optimal control phase kicking off the program for the numerical calculation of the optimal policy for the next batch. The resulting parameters are stored in the control system database and used in the control phases for the next batch.

A complete implementation of this strategy within an industrial real time environment (the RTPMS real time plant management and control system by IBM) was presented in [53] with the optimal control policies calculated as previously discussed batch management carried out by the SUPERBATCH system [22, 23, 54] and standard control configurations (ratio control on reflux rate and level control on the condenser receiver all PID type). Actual plant behavior was simulated by a detailed dynamic model implemented in a version of the Speedup general purpose simulator directly interfaced to the control system [68].

This application indicates that the dynamic optimization techniques discussed can indeed be used in an on-line environment to provide a reactive batch-to-batch adjustment capability. Again, the operation strategy for each batch was defined a priori and optimization of individual batches in sequence, as just described, is not the same as a simultaneous optimization of all the batches in the campaign, so there should be scope for further improvements.

Discussion and Conclusions

We may now draw a number of general conclusions and highlight some outstanding problems.

1) With regards to modeling and simulation, recent advances in handling mixed systems of algebraic and differential equations with discrete events should now make it possible to develop and solve without (too many) problems the equations required to model in some detail all the operations present in batch distillation. The question remains, in my opinion, of how detailed the models need be in particular to represent the startup period, heat transfer effects in the reboiler, and hydraulic behavior in *sufficient* detail.

2) It is presently possible to formulate and solve a rather general class of dynamic optimization problems with equations modeled as DAEs, for the optimization of individual batch distillation steps. Both the control vector parameterization method utilized in this paper and collocation over finite elements appear to work well. These solution techniques are reaching a certain maturity, that is, are usefully robust and fast (a few minutes to a few hours for large problems). Algorithmic improvements are still needed to handle more efficiently very large scale problems and for some special cases (e.g. high index problems). Some attention is also required to formulation of constraints and specifications so that the dynamic optimization problem is well posed. A variety of problems for which specialized solution methods were previously developed can now be all effectively solved in a way which is largely independent from the specific column and thermodynamic models objective function and specifications used. These advances should shift the focus of batch distillation studies towards the use of more detailed dynamic column models and towards optimization of more difficult processes (reactive azeotropic extractive with two liquid phases etc.) which did not easily match the assumptions (simple thermodynamics etc.) of the short cut methods.

3) With regards to the optimization of several batch distillation steps in a specified sequence (multiperiod problem), two approaches were presented here, one based on a two level

decomposition, taking advantage of the natural structure of batch distillation, the other on a single level formulation. In this area, further work is needed to establish whether one approach is better than the other or indeed for altogether new approaches.

4) The problem of choosing the optimal sequence of steps for the processing of one batch (an operation strategy), as well as the control variables for each step, has not been given much attention, no doubt because of the difficulty of the mathematical problem involved. Systematic methods are needed to select in particular the best strategy for reprocessing off-cuts and more in general, for processing multiple mixtures. Some initial results were presented in literature [82, 83] where a nonlinear programming (NLP) formulation was proposed. The use MINLP for simulation and solution was suggested, but not developed. In this area, there is clearly scope for novel problem formulations and solutions. Similarly, the optimization of each batch in a campaign so as to maximize the performance of the whole campaign does not appear to have been considered other than in the context of scheduling, with extremely simplified "split fraction" and fixed time models, e.g. in [45].

5) One of the assumptions made initially was that of perfect control response. The integration of open loop operation design and closed loop optimization is clearly relevant, as are the sensitivity, controllability and robustness properties of any optimal operation policies. These issues are beyond the scope of this paper, but some initial results are discussed in [73, 81], while an interesting method for model based control of a column startup was presented in [7].

6) Current developments in hardware speed, optimal control algorithms and control and supervisory batch management systems are such that sophisticated optimal operations can be calculated and implemented on-line, not only with respect to the optimal control policies for a batch, but also with regards to batch-to-batch variations, as demonstrated by the last application example. "Keeping it constant", which used to be a practical advantage, is no longer a constraint.

7) Finally, while a number of earlier studies indicated that performance improvements obtained by optimizing the reflux ratio policies were often small, if not marginal, more recent work appears to point to double digit benefits in many cases. As discussed above in one of the examples, this is possibly due to the consideration of a wider range of operating choices, more realistic models and objective functions. Whether the benefits predicted using the more advanced operation policies are indeed achieved in practice is an interesting question which awaits confirmation by the presentation of more experimental, as well as simulated results.

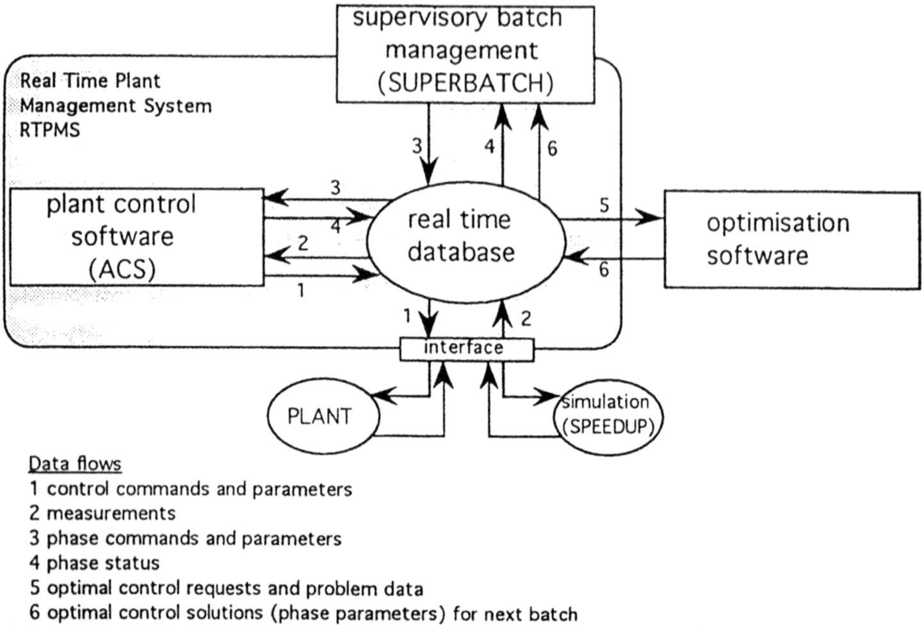

Figure 16. Schematic Structure of Control Software

Acknowledgments

This work was supported by SERC/AFRC, whose contributions are gratefully acknowledged.

References

1. Abdul Aziz, B. B., S. Hasebe and I. Hashimoto, Comparison of several startup models for binary and ternary batch distillation with holdups. in Interactions Between Process Design and Process Control, IFAC Workshop, J.D. Perkins ed., Pergamon Press, pp 197-202, 1992
2. Abram, H.J., M. M. Miladi and T. F. Attarwala. Preferable alternatives to conventional batch distillation. IChemE. Symp Series No. 104. IChemE, Rugby, UK, 1987
3. Albet, J.,. Simulation Rigoureuse de Colonnes de Distillation Discontinue a Sequences Operatoires Multiples. PhD Thesis, ENSIGC, Toulouse, 1992
4. Albet, J., J. M. Le Lann, X. Joulia and B. Koehret. Rigorous simulation of multi component multi sequence batch reactive distillation. Proceedings Computer Oriented Process Engineering, Elsevier Science Publishers B.V., Amsterdam, p. 75. 1991
5. Barb, D. K. and C. D. Holland, Batch distillation. Proc. of the 7th World Petroleum Con., 4, 31,.1967
6. Barbosa, D. and M. F. Doherty, The influence of chemical reactions on vapor-liquid phase diagrams. Chem Eng. Sci., 43 (3), 529, 1988
7. Barolo, M., G.B. Guarise, S. Rienzi and A. Trotta, On-line startup of a distillation column using generic model control. Comput. Chem. Engng., 17S, pp 349-354,(1992
8. Barton, P. and C. C. Pantelides, The modeling and simulation of combined discrete/continuous processes. Proc., PSE'91, Vol. 1, pp.20., Montebello, Canada, 1991
9. Bernot, C., M.F. Doherty, and M. F. Malone. Patterns of composition change in multicomponent batch

distillation. Chem. Eng. Sci., 45 (5), 1207, 1990
10. Biegler, L.T., Solution of dynamic optimization problems by successive quadratic programming and orthogonal collocation. Comput. Chem. Engng., 8, pp 243-248, 1984
11. Bortolini and Guarise (1970). Un nuovo metodo di distillazione discontinua. Ing. Chim. Ital., Vol. 6, pp. 1-9, 1970
12. Boston J.F., H.J. Britt, S. Jirapongphan and V.B. Shah, An advanced system for the simulation of batch distillation operation. FOCAPD, 2, p. 203, 1981
13. Caracotsios, M. and W.E. Stewart, Sensitivity analysis of initial value problems with mixed ODEs and algebraic equations. Comput. Chem. Engng., 9, pp. 359-365, 1985
14. Chang, Y.A. and J.D. Seider, Simulation of continuous reactive distillation by a homotopy-continuation method. Comput. Chem. Engng., 12. 12, p. 1243, 1988
15. Chen, C.L., A class of successive quadratic programming methods for flowsheet optimization. PhD Thesis. Imperial College. University of London, 1988
16. Christensen, F.M. and S.B. Jorgensen, Optimal control of binary batch distillation with recycled waste cut. Chem. Eng. J., 34, 57, 1987
17. Clark, S. M. and G. S. Joglekar, General and special purpose simulation software for batch process engineering. This volume, p 376
18. Converse, A.O. and G.D. Gross, Optimal distillate-rate policy in batch distillation. IEC Fund, 2(3), p. 217., 1963
19. Converse, A.O. and C.I. Huber, Effect of holdup on batch distillation optimization. IEC Fund., 4 (4), 475, 1965.
20. Corrigan, T.E. and W.R. Ferris, A development study of methanol-acetic acid esterification. Can. J. Chem. Eng., 47(6), 334, 1969
21. Corrigan, T.E. and J.H. Miller, Effect of distillation on a chemical reaction. IEC PDD, 7(3), 383, 1968
22. Cott, B. J., An Integrated Management System for the Operation of Multipurpose Batch Plants. PhD Thesis, Imperial College, University of London, 1989
23. Cott, B.J. and S. Macchietto, An integrated approach to computer-aided operation of batch chemical plants. Comput. Chem. Engng., 13, 11/12, pp. 1263-1271, 1989
24. Coward, I.,. The time optimal problem in binary batch distillation. Chem. Eng. Sci., 22, 503, 1967
25. Cuille, P.E. and G.V. Reklaitis, Dynamic simulation of multicomponent batch rectification with chemical reaction. Comput. Chem. Engng., 10, 4, 389, 1986
26. Cuthrell, J.E. and L.T. Biegler, On the optimization of differential-algebraic process systems, AIChE J., 33, 1257, 1987
27. Distefano, G.P., Mathematical modeling and numerical integration of multicomponent batch distillation equation. AIChE J., 14,1, p. 176, 1968
28. Diwekar, U.M., Unified approach to solving optimal design-control problems in batch distillation. AIChE J., 38(10), 1571, 1992
29. Diwekar, U.M. and J.R. Kalagnanam, An application of qualitative analysis of ordinary differential equations to azeotropic batch distillation, AIChE Spring National Meeting, New Orleans, March 29 - April 2, 1992
30. Diwekar, U.M. and K.P. Madhavan, Optimal design of multicomponent batch distillation column. Proceedings of World Congress III of Chemical Engng., Sept., Tokyo, 4, 719, 1992
31. Diwekar, U.M. and K.P. Madhavan, BATCHDIST- A comprehensive package for simulation, design, optimization and optimal control of multicomponent, multifraction batch distillation columns design. Comput. Chem. Engng., 15 (12), 833, 1991
32. Diwekar, U.M., K.P. Madhavan and R.K. Malik, Optimal reflux rate policy determination for multicomponent batch distillation columns. Comput. Chem. Engng., 11, 629, 1987
33. Egly, H., V. Ruby and B. Seid, Optimum design and operation of batch rectification accompanied by chemical reaction. Comput. Chem. Engng., 3, 169, 1987
34. Farhat, S., M. Czernicki, M., L. Pibouleau, L. and S. Domenech, Optimization of multiple-fraction batch distillation by nonlinear programming. AIChE J., 36(9), 1349, 1990
35. Galindez, H. and A. Fredenslund, Simulation of multicomponent batch distillation processes. Comput.

Chem Engng., 12(4), 281, 1988
36. Gani, R., C.A. Ruiz and I. T. Cameron, A generalized model for distillation columns - I. Model description and applications. Comput. Chem. Engng., 10(3), 181, 1986
37. Gear, C. W., Simultaneous numerical solution of differential-algebraic equations. IEEE Trans. Circuit Theory, CT- 18, 89, 1971
38. Gonzales-Velasco, J. R., M.A. (Gutierrez-Ortiz, J. M. Castresana-Pelayo and J A. Gonzales-Marcos. Improvements in batch distillation startup. IEC Res., 26, pp. 745, 1987
39. Gritsis, D., The Dynamic Simulation and Optimal Control of Systems described by Index Two Differential-Algebraic Equations. PhD. Thesis, Imperial College, University of London, 1990
40. Gu, D. and A.R. Ciric, Optimization and dynamic operation of an ethylene glycol reactive distillation column. Presented at the AIChE Annual Meeting, Nov. 1-6, Miami Beach, USA, 1992
41. Hasebe, S., B.B. Abdul Aziz, I. Hashimoto and T. Watanabe, Optimal design and operation of complex batch distillation column. in Interactions Between Process Design and Process Control, IFAC Workshop, J.D. Perkins ed., Pergamon Press, p 177, 1992
42. Hansen, T. T. and S.B. Jorgensen, Optimal control of binary batch distillation in tray or packed columns. Chem. Eng. 1., 33, 151, 1986
43. Hindmarsh, A.C., LSODE and LSODI, two new initial value ordinary differential equation solvers. Ass. Comput. Mach., Signam Newsl., 15(4), 10, 1980
44. Holland C.D. and A. I. Liapis, Computer methods for solving dynamic separation problems. McGraw-Hill Book Company, New York, 1983
45. Kondili, E., C.C. Pantelides, R.W.H. Sargent, A general algorithm for scheduling of batch operations. Proceedings 3rd Intl. Symp. on Process Systems Engineering, pp. 62-75. Sydney, Australia, 1988
46. Kerkhof, L.H. and H.J.M. Vissers, On the profit of optimum control in batch distillation. Chem. Eng. Sci., 2, 961, 1978
47. Logsdon, J.S. and L.T. Biegler, Accurate solution of differential-algebraic optimization problems. Ind. Eng. Chem. Res., 28, 1, pp. 1628-1630, 1989
48. Logsdon, J.S. and L.T. Biegler, Accurate determination of optimal reflux policies for the maximum distillate problem in batch distillation. AIChE National Mt., New Orleans, March 29 - April 2, 1992
49. Lucet, M., A. Charamel, A. Chapuis, G. Guido and J. Loreau, Role of batch processing in the chemical process industry. This volume, p. 43
50. Luyben, W.L., Some practical aspects of optimal batch distillation design. IEC PDD, 10, 54, 1971
51. Luyben, W.L., Multicomponent batch distillation. 1. Ternary systems with slop recycle. IEC Res., 27. 642, 1988
52. Macchietto, S., Interactions between design and operation of batch plants. in interactions between process design and process control, IFAC Workshop, J.D. Perkins ed.,Pergamon Press, pp. 113-126, 1992
53. Macchietto, S., B. J. Cott, I. M. Mujtaba an J C. A. Crooks, Optimal control and on line operation of batch distillation. AIChE Annual Meeting, San Francisco, Nov. 5-10, 1989
54. Macchietto, S., C. A. Crooks and K. Kuriyan, An integrated system for batch processing. This volume, p. 750
55. Mayur, D.N. and R. Jackson, Time optimal problems in batch distillation for multicomponent mixtures and for columns with holdup. Chem. Eng. J., 2, 150, 1971
56. Mayur, D.N., R.A. May and R. Jackson, The time-optimal problem in binary batch distillation with a recycled waste-cut. Chem. Eng. J. ,1, 15, 1970
57. McGreavy, C. and G.H. Tan, Effects of process and mechanical design on the dynamics of distillation column. Proc. IFAC Symposium Series, Bournemouth, England, Dec. 8-10, p. 181, 1986
58. Morison, K.R., Optimal control of processes described by systems of differential and algebraic equations. PhD. Thesis, University of London, 1984
59. Mujtaba, I.M., Optimal operational policies in batch distillation. PhD Thesis, Imperial College, London, 1989
60. Mujtaba, I.M. and S. Macchietto. Optimal recycle policies in batch distillation - binary mixtures. Recents progres en Genie des Proceeses, (S. Domenech, X. Joulia and B. Koehret Eds.) Vol. 2, No. 6,

pp. 191-197, Lavoisier Technique et Documentation, Paris, 1988

61. Mujtaba, I.M. and S. Macchietto, Optimal control of batch distillation. in IMACS Annals on Computing and Applied Mathematics-Vol 4: Computing and Computers for Control Systems, (P. Borne, S.G. Tzafestas, P.Breedveld and G. Daophin Tanguy, eds.), I.C. Baltzer AG, Scientific Publishing Co., Basel, Switzerland, pp.55-58, 1989
62. Mujtaba, I.M. and Macchietto, S, The role of holdup on the performance of binary batch distillation. Proc., 4th International Symp. on PSE, Vol. 1, 1.19.1, Montebello, Quebec, Canada, 1991
63. Mujtaba, I.M. and S. Macchietto, An optimal recycle policy for multicomponent batch distillation. Computers Chem. Engng., 16S, pp. 273-280, 1992
64. Mujtaba, I.M. and S. Macchietto, Optimal operation of multicomponent batch distillation. AIChE National Mt., New Orleans, USA, March 29-April 2, 1992
65. Mujtaba, I.M. and S. Macchietto, Optimal operation of reactive batch distillation", AIChE Annual Meeting, Nov. 1-6, Miami Beach, USA, 1992
66. Murty, B.S.N., K. Gangiah and A. Husain, Performance of various methods in computing optimal control policies. Chem Eng. J.,19, p. 201, 1980
67. Nad, M. and L. Spiegel, Simulation of batch distillation by computer and comparison with experiment. Proceedings CEF'87, 737, Taormina, Italy, 1987
68. Pantelides, C. C. Speedup - Recent Advances in Process Simulation. Computers Chem. Engng., 12 (7), 745, 1988
69. Pantelides, C.C, D. Gritsis, K.R. Morison and R.W.H. Sargent, The mathematical modeling of transient system using differential-algebraic equations. Comput. Chem. Engng., 12(5), 449, 1988
70. Pantelides, C. C. Sargent, R. W. H. and V. S. Vassiliadis, Optimal control of multistage systems described by differential-algebraic equations. AIChE Annual Meeting, Miami Beach, Fla, USA, 1988
71. ProsimBatch, Manuel Utilisateur. Prosim S. A., Toulouse, 1992
72. Quintero-Marmol, E. and W.L. Luyben, Multicomponent batch distillation. 2. Comparison of alternative slop handling and operating strategies. IEC Res., 29, 1915, 1990
73. Quintero-Marmol, E. and W.L. Luyben, Inferential model-based control of multicomponent batch distillation. Comput. Chem. Engng., 47(4), 887, 1992
74. Renfro, J.G., A. M. Morshedi and O.A. Asbjornsen, Simultaneous optimization and solution of systems described by differential/algebraic equations. Comput. Chem. Engng., 11(5), 503, 1987
75. Robinson, E. R., The optimization of batch distillation operation. Chem. Eng. Sci., 24, 1661, 1969
76. Robinson, E.R. , The optimal control of an industrial batch distillation column, Chem Eng. Sci., 25, 921, 1970
77. Robinson, C.S. and E.R. Gilliland, Elements of fractional distillation. 4th ed., McGraw-Hill, 1950
78. Rippin, D. W. T., Simulation of single and multiproduct batch chemical plants for optimal design and operation. Comput. Chem. Engng., 7, pp. 137-156, 1983
79. Rose, L.M., Distillation design in practice, Elsevier, New York ,1985
80. Ruiz, C.A., A generalized dynamic model applied to multicomponent batch distillation. Proceedings CHEMDATA 88, 13-15 June, Sweden, p. 330.I, 1988
81. Sorensen, E. and S. Skogestad control strategies for a combined batch reactor/batch distillation process. This volume, p. 274
82. Sundaram, S. and L.B. Evans, Batch distillation synthesis. AIChE Annual Meeting, Chicago, 1990
83. Sundaram, S. and L.B. Evans, Synthesis of separations by batch distillation. Submitted for publication, 1992
84. VanDongen, D.B. and M.F. Doherty, On the dynamics of distillation processes: Batch distillation. Chem. Eng. Sci., 40, 2087, 1985
85. Villadsen, J. and M. I. Michelsen, Solution of differential equation models by polynomial approximation. Prentice Hall, Englewood Cliffs, NJ, 1978
86. Wilson, J.A. Dynamic model based optimization in the design of batch processes involving simultaneous reaction and distillation. IChemE Symp. Series No. 100, p. 163, 1987

Sorption Processes

Alirio E. Rodrigues[1] and Zuping Lu

Laboratory of Separation and Reaction Engineering, School of Engineering, University of Porto
4099 Porto Codex, Portugal

Abstract: The scientific basis for the design and operation of sorption processes is reviewed. Examples of adsorptive processes, such as liquid phase adsorption, parametric pumping and chromatography, are discussed. A new concept arising from the use of large-pore materials is presented and applied to the modeling of HPLC and pressure swing adsorption processes. Numerical tools to solve the model equations are addressed.

Keywords: Sorption processes, chromatography, intraparticle convection, modeling, parametric pumping, pressure swing adsorption, high pressure liquid chromatography.

Chapter Outline

In the first part, we will present a definition and the objectives sorption processes. Then, the fundamentals of design and operation of such processes will be reviewed, examples of liquid phase adsorption and parametric pumping presented, and chromatographic processes will be discussed showing various modes of operation. The concept of Simulated Moving Bed (SMB) will be introduced. The second part will be devoted to a new concept in High Pressure Liquid Chromatography (HPLC) and Pressure Swing Adsorption (PSA) processes will be shown. Finally, numerical tools used for solving model equations will be briefly reported and ideas for future work will be given.

---

[1] Lecturer

1. Sorption Processes

1.1. DEFINITION AND OBJECTIVES

Sorption processes are a sub-set of percolation processes ,i.e., processes in which a fluid flows through a bed of particles, fibers or membranes exchanging heat/mass or reacting with the support [1]. Examples of percolation processes are ion exchange, adsorption, chromatography, parametric pumping, pressure-swing adsorption (PSA). Sorption processes can be carried out with different objectives: i) separation of components from a mixture, ii) purification of diluents, iii) recovery of solutes.

1.2. BASIS FOR THE DESIGN OF SORPTION PROCESSES

The basis for the design of sorption processes, as for any other chemical engineering operation, are:

i) conservation equations (mass, energy, momentum ,electric charge)
ii) kinetic laws (heat transfer, mass transfer, reaction)
iii) equilibrium laws at the interfaces
iv) boundary and initial conditions
v) optimization criteria

The factors governing the behavior of sorption processes are: equilibrium isotherms, hydrodynamics, kinetics of film mass/heat transfer, kinetics of intraparticle mass/heat transfer (diffusion and convection) [2]. Sorption in fixed-bed is really a wave propagation process. Figure 1a shows the evolution of the concentration front at different times (*concentration profiles* in the bed), starting with a clean bed until complete saturation; Figure 1b shows the *concentration history* at the bed outlet, i.e., solute concentration as a function of time. Several quantities can be defined at this point. The *breakthrough time* t_{bp} is the time at which the outlet solute concentration is, say, 0.01 c_{in}; the *stoichiometric time* t_{st} is the time corresponding to complete saturation if the concentration history was a discontinuity. The *useful capacity* (storage) of the bed S_u is the amount of solute retained in the column until t_{bp}; the *total capacity* S_∞ is the amount of solute retained in the bed until complete saturation. The *length of the mass transfer zone* (MTZ) is $u_{shock}\, t_{MTZ}$ and the *Length of Unused Bed* (LUB) is $L(1- t_{bp}/t_{st})$. The stoichiometric time is the first quantity to be calculated in designing a fixed-bed adsorption process; it is obtained from an overall mass balance leading to:

$$t_{st} = \frac{\varepsilon V}{U}\left(1+\frac{1-\varepsilon}{\varepsilon} \frac{q_{in}}{c_{in}}\right) \qquad [1]$$

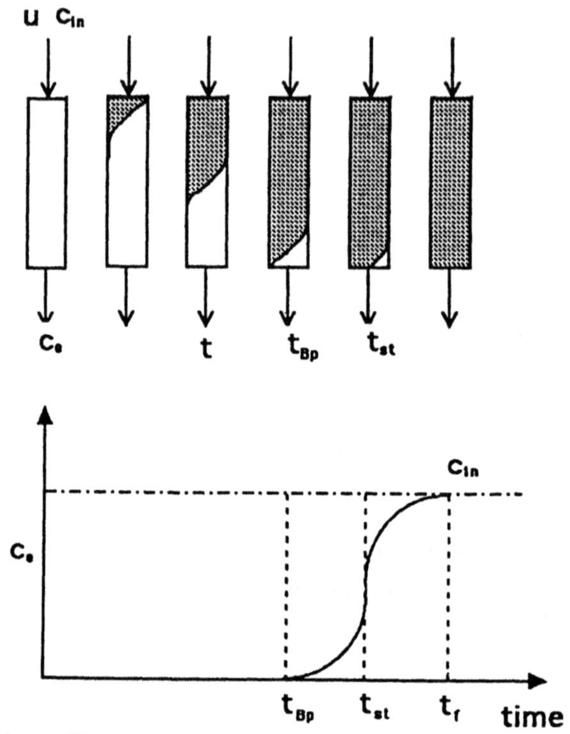

Figure 1. Concentration profiles and concentration history

where V is the bed volume, U is the flowrate, q_{in} is the amount sorbed in equilibrium with the inlet concentration of the sorbed species c_{in} and ε is the interparticle porosity. Introducing the space time $\tau = \frac{\varepsilon V}{U}$ and the capacity factor $\xi = \frac{1-\varepsilon}{\varepsilon}\frac{q_{in}}{c_{in}}$, we get $t_{st} = \tau\,(1+\xi)$.

1.3. EQUILIBRIUM MODEL. COMPRESSIVE AND DISPERSIVE FRONTS.

The main features of fixed-bed sorption behavior can be captured by using an *equilibrium model* based on the assumptions of instantaneous equilibrium at the fluid/solid interface at any point in the column, isothermal operation, plug flow for the fluid phase and negligible pressure drop.

For a single solute, tracer system, the model equations for the equilibrium model are:

mass balance for species i :

$$v\frac{\partial c_i}{\partial z} + \frac{\partial c_i}{\partial t} + \frac{1-\varepsilon}{\varepsilon}\frac{\partial q_i}{\partial t} = 0 \qquad [2]$$

Sorption equilibrium isotherm

$$q_i = f(c_i) \qquad [3]$$

The velocity of propagation of a concentration c_i is:

$$u_{c_i} = \frac{v}{1+\frac{1-\varepsilon}{\varepsilon} f'(c_i)} \qquad [4]$$

It appears that the nature of the sorption equilibrium isotherm is the main factor governing the behavior of the fixed-bed process. For unfavorable isotherms, $f''(c_i) > 0$; therefore, u_{c_i} decreases when c_i increases and the front is dispersive [3]. For favorable isotherms, $f''(c_i) < 0$ and so u_{c_i} increases with c_i; the front is compressive leading to a shock which propagates with the velocity u_s given by:

$$u_s = \frac{v}{1+\frac{1-\varepsilon}{\varepsilon}\frac{\Delta q}{\Delta c}} \qquad [5]$$

where Δq and Δc are calculated between the feed state (c_{in}, q_{in}) and the presaturation state of the bed; for a clean bed the presaturation state is (0, 0).

1.4. KINETIC MODELS

Sorption kinetics has been included in the analysis of sorption processes in two different ways. The first one is through a kinetic law similar to a reaction rate as in the Thomas model [4]:

$$\frac{\partial \bar{q}_i}{\partial t} = k_l [\ c_i (q_0 - \bar{q}_i) - r\,\bar{q}_i (c_0 - c_i)\ ] \qquad [6]$$

where r is the reciprocal of the constant separation factor. The Thomas model can be simplified to the Bohart model [5] in the case of irreversible isotherms (r=0), to the Walter model [6] in the case of linear isotherms, etc. as shown elsewhere [7]. Thomas model has been applied recently in the area of affinity chromatography [8, 9]. This class of models is what we call "chemical kinetic type models". The second way of treating sorption kinetics is by describing intraparticle mass transport. This class of models is called "physical kinetic type models". A typical model in this group is Rosen's model [10]. It considers homogeneous diffusion inside particles, film diffusion and linear isotherms. However, the general particle equation is:

$$\frac{\partial \varphi}{\partial r} + \frac{2}{r}\frac{\partial c_{ve}}{\partial t} = 0 \qquad [7]$$

where φ is the flux of species i through the sphere at radial position r and c_{ve} is the solute concentration in the volume element. According to the particle structure and the model used to describe diffusion inside particles we may have:

i) homogeneous diffusion : $\varphi = -D_h \frac{\partial q_i}{\partial r}$; $c_{ve} = q_i$

ii) pore diffusion: $\varphi = -\varepsilon_p D_p \frac{\partial c_{pi}}{\partial r}$; $c_{ve} = \varepsilon_p c_{pi} + q_i$

iii) pore + surface diffusion in parallel

$$\varphi = -\varepsilon_p D_p \frac{\partial c_{pi}}{\partial r} - D_h \frac{\partial q_i}{\partial r}; \quad c_{ve} = \varepsilon_p c_{pi} + q_i$$

iv) pore + surface diffusion in series

$$\varphi = -\varepsilon_p D_p \frac{\partial c_{pi}}{\partial r}; \quad c_{ve} = \varepsilon_p c_{pi} + \bar{q}_i$$

where $\bar{q}_i$ is the average adsorbent concentration in the microspheres calculated with the homogeneous diffusion model.

1.5. METHODOLOGY FOR THE DESIGN OF SORPTION OPERATIONS

The methodology for the design of sorption processes is based on the measurement of sorption equilibrium isotherms (batch equilibration), film mass transfer (shallow-bed technique), intraparticle diffusivity (Carberry type adsorber operating in batch or as a CSTR) and axial dispersion (fixed-bed) by simple, independent experiments. Model parameters are then introduced in the adsorber model in order to predict its dynamic behavior under conditions different from those used at the laboratory scale [11, 12].

1.5.1. *Adsorption from liquid phase*

This methodology was tested for a single component system : adsorption from liquid phase of phenol onto a polymeric adsorbent (Duolite ES861; Rohm and Haas). Adsorption equilibrium isotherms measured at 20 °C and 60 °C are shown in Figure 2.

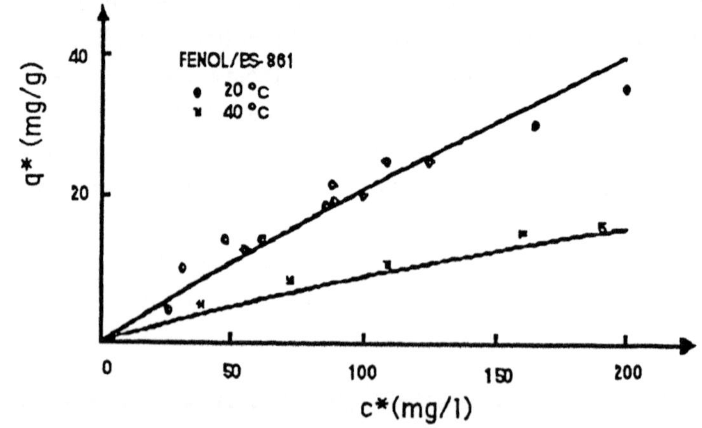

Figure 2. Adsorption isotherms at 20 °C and 60 °C for the system: phenol/duolite ES-861

Intraparticle effective diffusivities were measured in a stirred adsorber of Carberry type shown in Figure 3. The response of the batch adsorber was measured for two different particle diameters d_p=0.077 cm d_p=0.06 cm and d_p=0.034 cm (Figure 4).

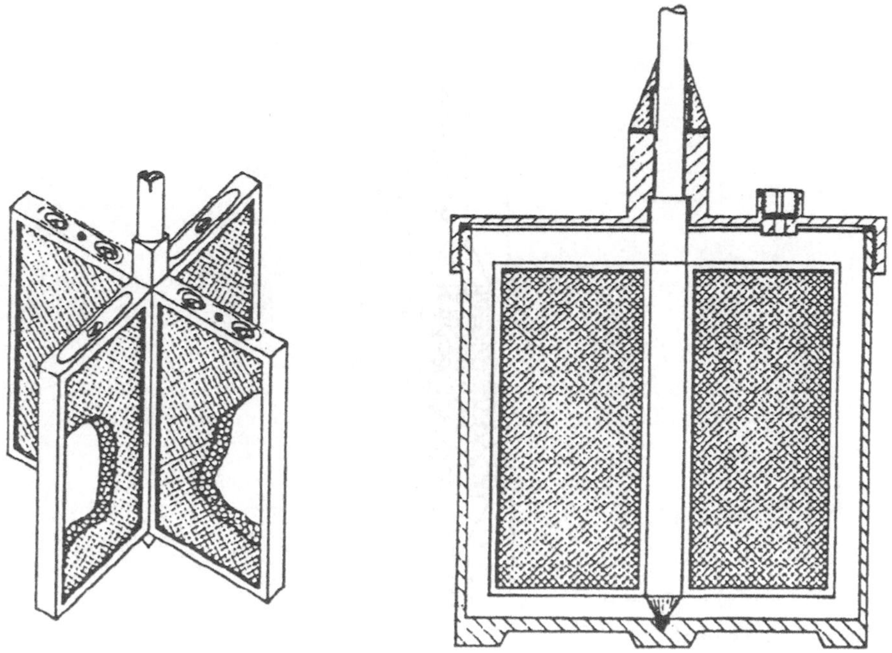

Figure 3. Sketch of the Carberry type adsorber

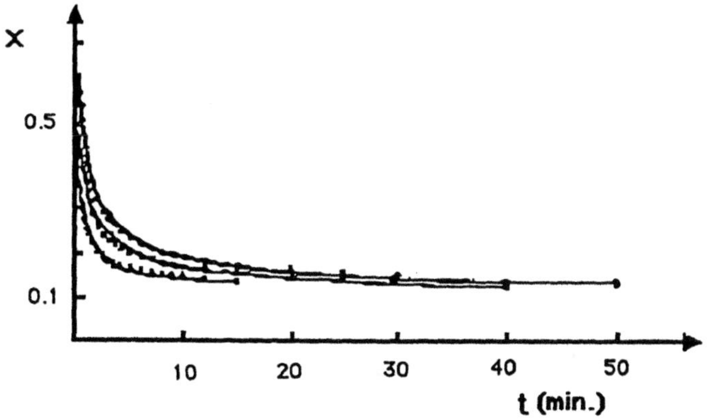

Figure 4. Batch adsorption phenol on to Duolite ES-861 for different particle diameters

Film mass transfer coefficients were measured by using the shallow bed technique shown in Figure 5. The response of the system to a step in phenol concentration at the inlet is shown in Figure 6. Modeling studies showed that the initial part of the response curve does not depend on the value of the intraparticle diffusivity and can be used to estimate the number of film mass transfer units N_f.

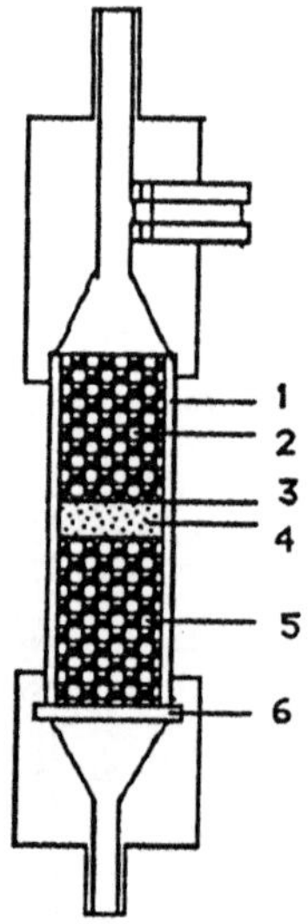

Figure 5. Shallow-bed

(1 - Column; 2&5 - glass beads; 3 - inox gauze; 4 - shallow bed; 6 - porous glass support)

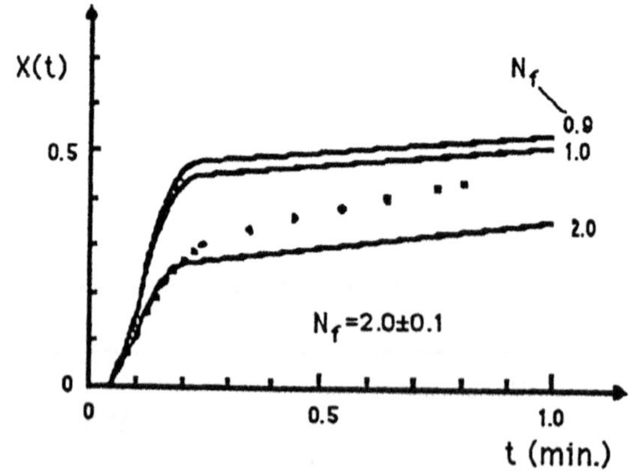

Figure 6. Response of a shallow-bed: outlet concentration versus time

Finally, breakthrough curves were obtained in fixed bed runs. Experimental and model predicted results are shown in Figure 7. We should stress that model parameters were measured by independent experiments; no fitting of breakthrough curves was involved. Table I summarizes operating conditions and model parameters.

Table I. Operating conditions and model parameters for fixed-bed runs

Run	U(ml/min)	c_{in}(mg/l)	ξ	N_f	N_d	Pe
1	158.7	82	93.6	36.3	0.261	18.7
2	115.2	91.6	79.7	35.8	0.395	24.0
3	54.4	91.6	82.7	57.0	0.824	36.0
4	16.8	82.4	95.1	132.6	2.468	68.0

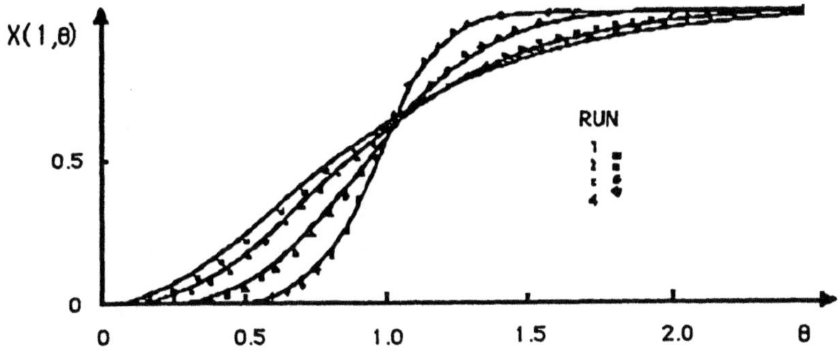

Figure 7. Breakthrough curves of phenol in a fixed-bed of Duolite ES-861

Scaling for large beds is not a major problem. In fact only the parameter related with hydrodynamics will be affected.

Sorption processes are cyclic in nature in the sense that saturation, adsorption or load is followed by regeneration or desorption. In the case of phenol/polymeric adsorbent system the regeneration is made with sodium hydroxide and therefore a problem of desorption with chemical reaction arises. The basis for the regeneration is that the amount of phenol adsorbed as a function of the total concentration (phenol + phenate) changes with pH as shown in Figure 8.

When both steps (saturation and regeneration) are understood it is quite easy to predict the cyclic behavior. Model and experimental results are shown in Figure 9; the model used for the regeneration step is a shrinking core model in which the inner core contains phenol and the outer core contains phenate from the reaction between NaOH and phenol.[13,14].

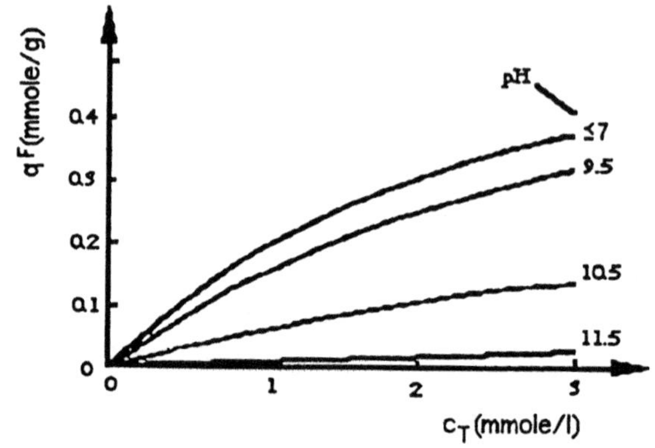

Figure 8. Effect of the pH on the adsorption equilibrium of phenol on Duolite ES-861

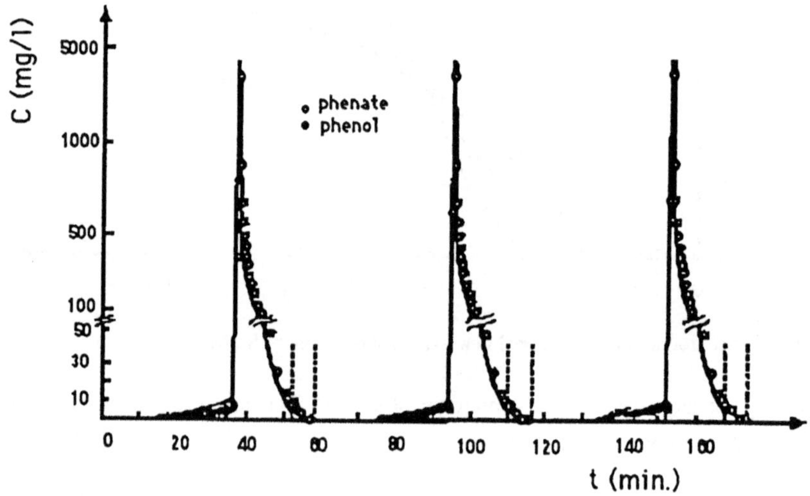

Figure 9. Cyclic behavior

1.5.2. *Parametric pumping*

Parametric pumping was invented by Wilhelm *et al.* in 1966 [15]; it is a cyclic separation process based upon the effect of an intensive thermodynamic variable (temperature, pressure or pH) on the adsorption equilibrium isotherm together with a periodic change of the flow direction. There are two modes of operation of thermal parametric pumping : a) direct mode, in which the cyclic temperature is imposed through the column wall and b) the recuperative mode in which the temperature change is imposed at the bed extremities being carried by the fluid.

A linear equilibrium model enables us to understand the behavior of a thermal parapump. Model equations are:

mass balance for species i :

$$v\frac{\partial c_i}{\partial z} + \frac{\partial c_i}{\partial t} + \frac{1-\varepsilon}{\varepsilon}\frac{\partial q_i}{\partial t} = 0 \qquad [8]$$

sorption equilibrium isotherm

$$q_i = K(T)\, c_i \qquad [9]$$

The key parameter in the analysis is the separation factor

$$b = \frac{a}{1+m_o} \qquad [10]$$

where $a = \frac{m(T_1)-m(T_2)}{2}$, $m_o = \frac{m(T_1)+m(T_2)}{2}$ and $m(T) = \frac{1-\varepsilon}{\varepsilon} K(T)$.

The concentration velocity in each half-cycle is

$$u_j = \frac{v_j}{1+m_o} \qquad \text{(j=1 cold downward flow; j=2 hot upward flow)} \qquad [11]$$

It has been shown [16] that $b = \tanh [(\frac{\gamma}{2}\frac{T^*}{1-T^*/4}]$ where $T^* = \Delta T/\bar{T}$ and $\gamma = (-\Delta H)/RT$. For the system phenol/Duolite ES861 b=0.32 to 0.36 .

Laboratory work in narrow columns (1.5 cm in diameter) was carried out many years ago [17,18] with good results. For a semicontinuous parametric pump the reduced concentration in the bottom reservoir after n cycles is given by:

$$\frac{\langle y_{bp}\rangle_n}{y_o} = (\frac{1-b}{1+b})^n \qquad [12]$$

and the top product concentration at steady-state is :

$$\frac{\langle y_{tp}\rangle_\infty}{y_o} = 1+\frac{\phi_b}{\phi_t} \qquad [13]$$

where ϕ_b and ϕ_t are the reflux ratios to the bottom and top of the column, respectively.

Recent work [19] was carried out in wide columns (9 cm in diameter). The completely automated unit is shown in Figure 10. The system was characterized from the hydrodynamic,

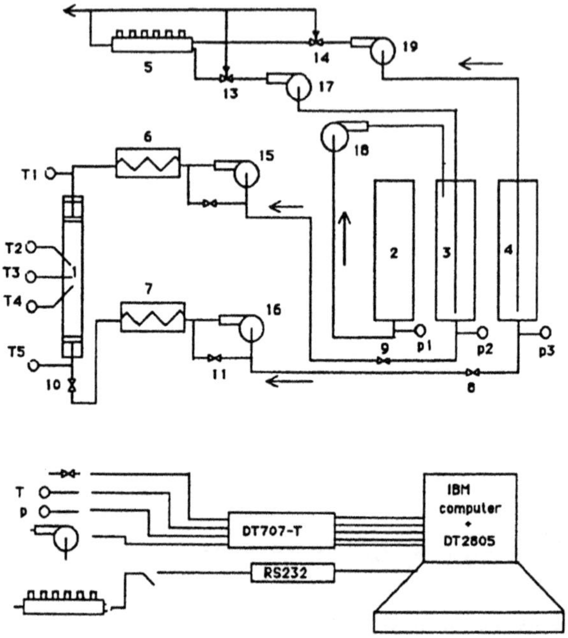

Figure 10. Experimental set-up

1.- glass column G90-Amicon. 2- Feed reservoir. 3.- Top reservoir. 4.- Bottom reservoir. 5.- Fraction collector. 6-7.- Heat exchange. 8-12.- two-ways solenoid valves. 13-14. - three-ways solenoid valves. 15-19.- Peristhaltic pumps. T1-T5. - Thermocouples type K. P1-P3.- Pressure transducers

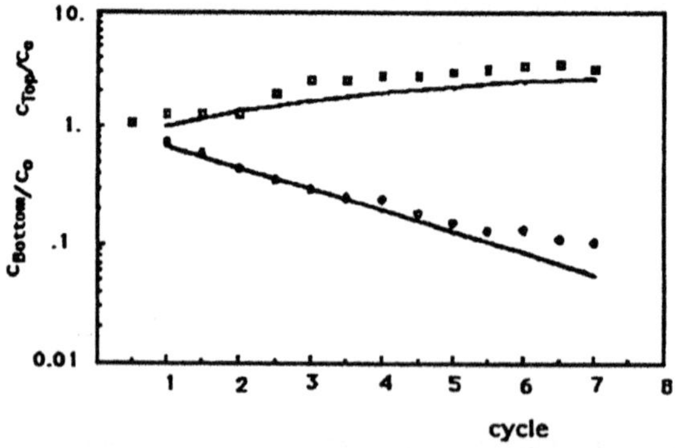

Figure 11. Model and experimental results for semicontinuous parametric pumping

heat /mass transfer point of view. A typical result is shown in Figure 11. Conditions were: feed concentration = 98mg/l; bed initially equilibrated with feed composition at 60 °C; average flowrate = 290 ml/min; top product flowrate = 12 ml/min; bottom product flowrate = 30 ml/min; average cycle time=156 min; ϕ_b=0.1; ϕ_t=0.04; V_U=20300 ml; V_D=24900 ml; $U\pi/w$=22400 ml.

A complete model for parametric pumping was developed including all relevant mechanisms; results are in Figure 11 for comparison with experimental results.

1.6. CHROMATOGRAPHY

1.6.1. *Operating modes*

Chromatography is an old operation discovered by M.Tswett [20]. The definition of chromatography given by IUPAC is a wordy and lengthy one; however, the main characteristic of chromatography, i.e, separation occurs as a result of different species velocities, is missing. Several modes of operation are listed below [21]:

a) Elution chromatography

The sample to be separated is injected in a continuous stream of eluent; the main problem is the dilution of components to be separated.

b) Gradient elution

The elution of the samples is carried out with different eluents.

c) Frontal chromatography

The column is continuously fed with the mixture to be separated until complete saturation of the sorbent which is then regenerated.

d) Displacement chromatography

After loading with the sample, a displacer with higher affinity is fed to the column.

Figure 12 shows the above mentioned operating modes of chromatography.

Elution chromatography has several drawbacks: a)dilution of the species as they travel down the bed and b) only a fraction of the bed is effectively used.

Several operating modes can be envisaged to improve process performance such as recycle chromatography, mixed recycle chromatography, segmented chromatography and two-way chromatography [22, 23]. Figure 13 shows schematically these operating modes. Figure 14 compares results obtained in the separation of a binary mixture by elution, mixed recycle and two-way chromatography. Process performance can be improved using these new operating modes, namely in terms of eluent consumption.

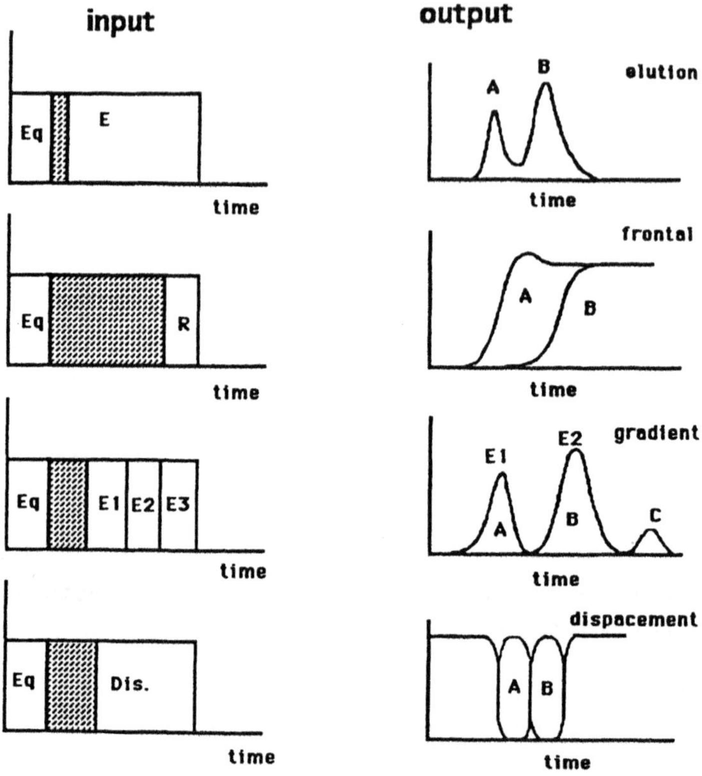

Figure 12. Operating modes of chromatography

1.6.2. *Simulated moving bed (SMB)*

Along the line of improving the use of the adsorbent the idea of SMB was developed; it is in my opinion one of the more interesting ideas in chemical engineering [24].

In a moving bed sketched in Figure 15 both solid and fluid phases flow in counter-current. However, there is a problem of attrition. The idea was then to simulate the behavior of a moving bed by keeping the particles fixed in a column and moving the position of feed and withdrawal streams. Schematically, the SMB can be represented as in Figure 16. Many applications of this technology are currently being used in industry such as the Parex process for the recovery of p-xylene from a mixture of C_8 isomers, Sarex process, etc. [25, 26]. Modeling and experimental studies in SMB have been worked out by Morbidelli *et al.* [27]. Figure 17 shows a typical diagram for the separation of p-xylene from a mixture containing m-xylene, o-xylene and ethylbenzene.

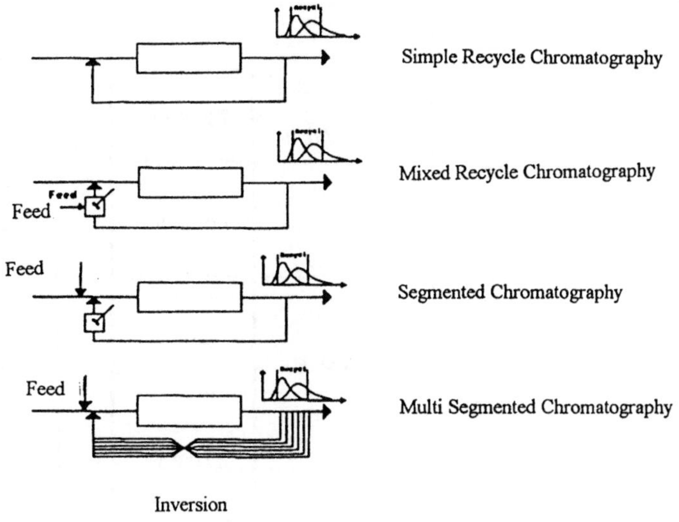

Figure 13. Enhanced operating modes of chromatography

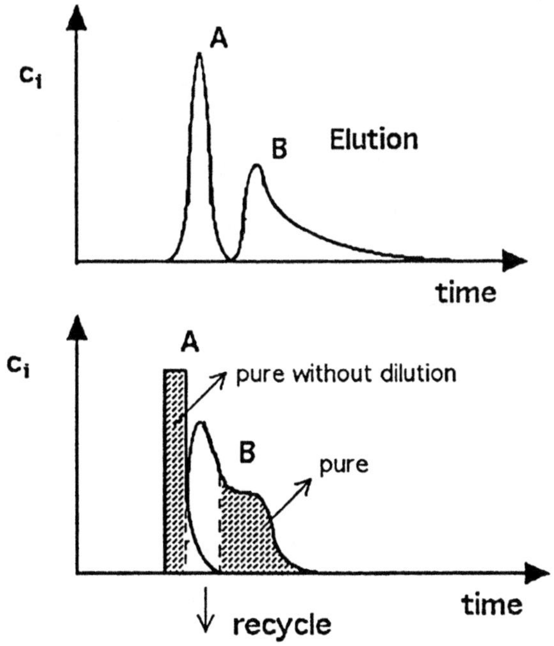

Figure 14. Comparison between elution and recycle chromatography for separation of binary mixture

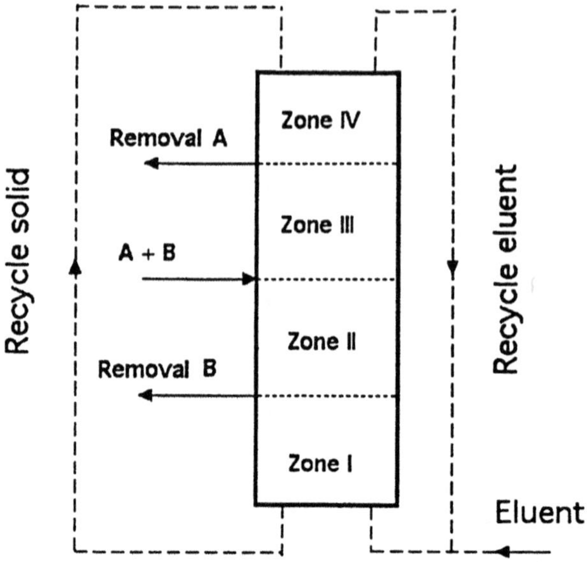

Figure 15. Moving-bed

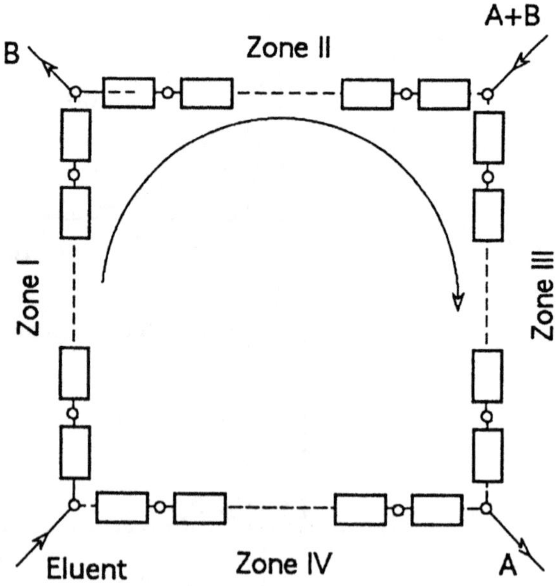

Figure 16. Simulated moving-bed

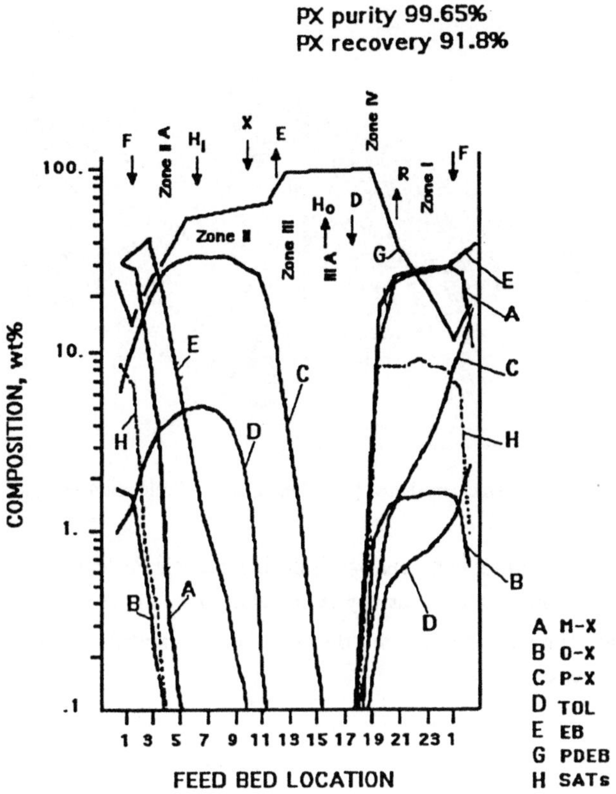

Figure 17. A typical profile in a Parex process

2. A NEW CONCEPT: DIFFUSIVITY ENHANCEMENT BY INTRAPARTICLE CONVECTION. APPLICATION TO HPLC AND PSA PROCESSES.

The need for large pore transport pores in catalyst preparation was recognized by Nielsen *et al.* [28] and others. The importance of flow through the pores was addressed by Wheeler in his remarkable paper [29]. He concluded that viscous flow due to pressure drop would only be important for high pressure gas phase reactions. He also wrote the correct steady state model equation for the particle taking into account diffusion, convection and reaction. Unfortunately, he did not solve the model equation and so he missed the point recognized later by Nir and Pismen [30] : in the intermediate region of Thiele modulus the catalyst effectiveness factor is enhanced by intraparticle convection (Figure 18).

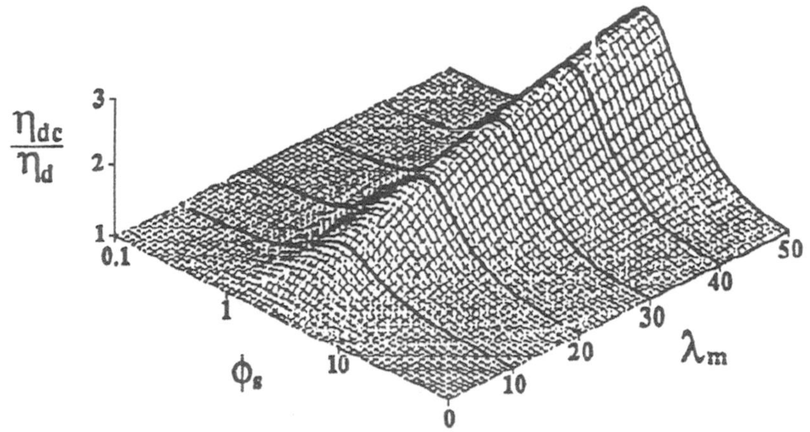

Figure 18. η_{dc}/η_d versus Thiele modulus ϕ_s and intraparticle Peclet number λ_m

The important parameter in the analysis is the intraparticle Peclet number $\lambda = v_0 \ell / D_e$ relating the intraparticle velocity v_0 with diffusivity D_e in a slab with half thickness ℓ. In 1982 Rodrigues et al. [31] showed that the reason was that diffusivity is augmented by convection; therefore the augmented diffusivity is $\tilde{D}_e$ is:

$$\tilde{D}_e = D_e \frac{1}{f(\lambda)} \qquad [14]$$

where

$$f(\lambda) = \frac{3}{\lambda}\left(\frac{1}{\tanh\lambda} - \frac{1}{\lambda}\right) \qquad [15]$$

The enhancement factor of diffusivity is thus $1/f(\lambda)$, shown in Figure 19.

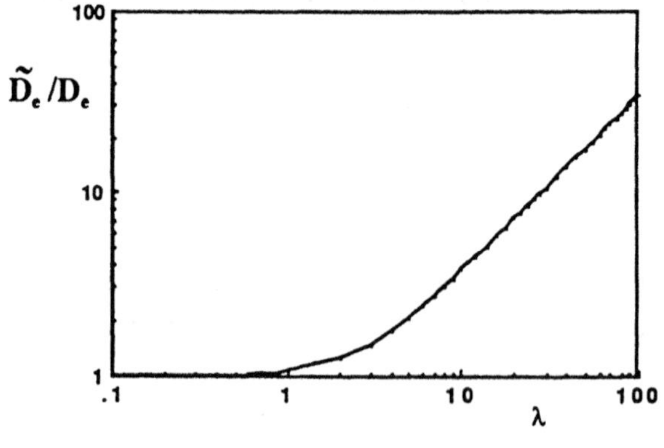

Figure 19. Enhancement factor versus intraparticle Peclet number λ

This is the reason why perfusion chromatography using large-pore packings has improved performance compared with conventional supports, although the original patent [32, 33] failed to mention the key result in Equation 14. In fact, in HPLC using large-pore, permeable packings the classic Van Deemter equation for HETP (height equivalent to a theoretical plate) has to be modified and replaced by Rodrigues equation [34, 35, 36] to take into account the contribution of intraparticle flow.

2.1. High Pressure Liquid Chromatography (HPLC)

A model for linear HPLC includes the following dimensionless equations in terms of axial coordinate $x=z/L$, particle coordinate $\rho=z'/l$ and $\theta=t/t$:

species mass balance in the outer fluid phase:

$$\frac{1}{Pe}\frac{\partial^2 c_i}{\partial x^2}-\frac{\partial c_i}{\partial x}-\{\varepsilon_b\frac{\partial c_i}{\partial \theta}+b\varepsilon_p(1-\varepsilon_b)\frac{\partial \langle c_i'\rangle}{\partial \theta}\}=0 \qquad [16]$$

where $b=1+\frac{1-\varepsilon_p}{\varepsilon_p}m$ and $Pe=\frac{u\,L}{\varepsilon_b D_{ax}}$.

mass balance inside particle for species i:

$$\frac{\partial^2 c_i}{\partial \rho^2}-\lambda\frac{\partial c_i}{\partial \rho}=b\alpha\frac{\partial c_i'}{\partial \theta} \qquad [17]$$

where $\alpha=\tau_d/\tau$ and $\lambda=\tau_d/\tau_c$ ($\tau_c=\varepsilon_p\ell/v_o$, $\tau_d=\varepsilon_p\ell^2/D_e$, $\tau=L/u$)

equilibrium adsorption isotherm

$q_i'=mc_i'$

Boundary and initial conditions

$x=0$ $\quad c_i'=Md(\theta)$

x $\quad c_i$ limited

$\theta=0$ $\quad c_i=c_i'=0$ $\quad$ x and ρ

$\rho=0$ and $\rho=2$, $c_i'=c_{is}'$

The HETP is defined as σ^2L/μ_1^2 and is given by:

$$HETP=\frac{2L}{Pe}+\frac{2}{3}\frac{\varepsilon_p(1-\varepsilon_b)b^2\alpha L}{[\varepsilon_b+\varepsilon_p(1-\varepsilon_b)b]^2}f(\lambda) \qquad [18]$$

or

$$HETP = A+\frac{B}{u}+C\,f(\lambda)\,u \qquad \text{Rodrigues equation} \qquad [19]$$

where $C=\frac{2}{3}\frac{\varepsilon_p(1-\varepsilon_b)b^2\tau_d}{[\varepsilon_b+\varepsilon_p(1-\varepsilon_b)b]^2}$. Equation 18 was derived by Rodrigues *et al* [34]. In the classic Van Deemter equation [37] $f(\lambda)=1$ (no intraparticle convection). At high superficial velocity or high λ, $f(\lambda)=\frac{3}{\lambda}$ and the HETP reaches a plateau at:

$$\text{HETP}=A+2\frac{\varepsilon_p{}^2(1-\varepsilon_b)b^2\ell}{a_1[\varepsilon_b+\varepsilon_p(1-\varepsilon_b)b]^2} \qquad [20]$$

since $v_o=a_1u$.

It appears that intraparticle convection contributes to a better efficiency of the chromatographic column since we obtain lower HETP than with conventional adsorbents and also the speed of separation can be increased without loosing efficiency.

Figure 20 shows Van Deemter equation for HETP of columns packed with conventional supports (dashed line) and Rodrigues equation for large-pore packings (full line). Figures 22a and b show the response of a chromatographic column to an impulse and step input of concentration, respectively. Dashed lines refer to conventional supports and full lines to large-pore supports.

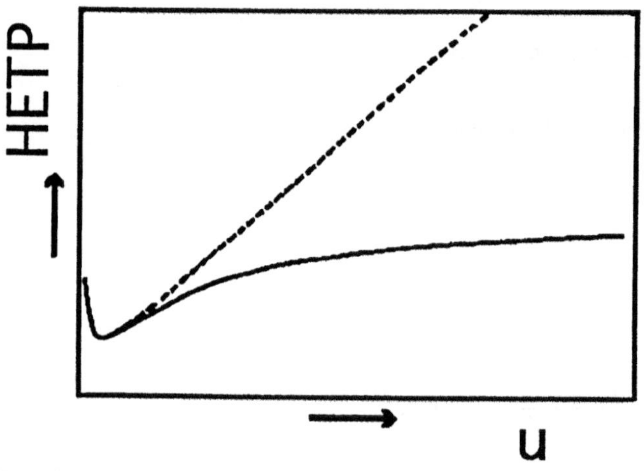

Figure 20. HETP as function of the superficial velocity

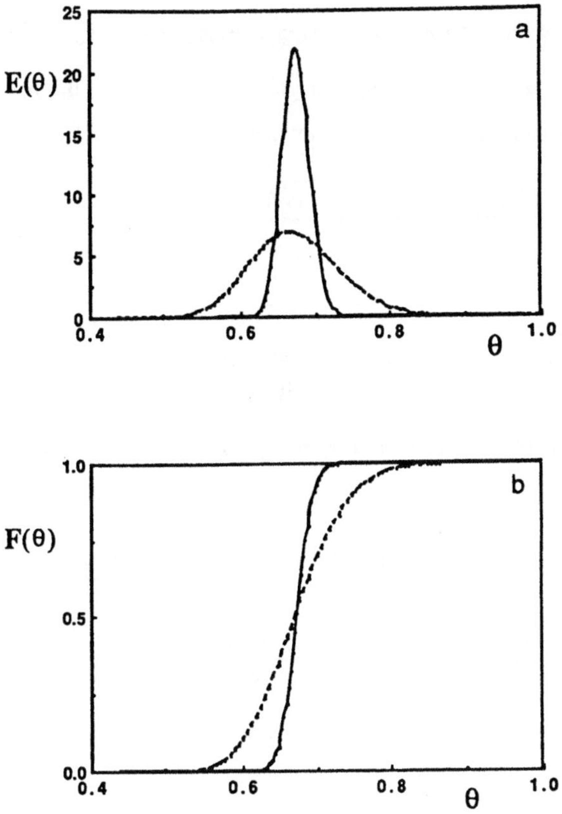

Figure 21. Response of a HPLC column to impulse and step functions

(Dashed lines : conventional supports; full lines: large-pore packings)

2.2. Pressure Swing Adsorption (PSA)

In gas phase adsorption processes intraparticle convective flow should also be considered if there are important large pressure gradients as it is the case of pressurization and blowdown steps in PSA. We have been involved in modeling PSA and details can be found in a recent series of papers [38-42]. Model equations in dimensionless form are given below using the following variables:

$$x = \frac{z}{L},\ \rho = \frac{z'}{\ell},\ f = \frac{c}{c_o} = \frac{P}{P_o},\ f' = \frac{c'}{c_o} = \frac{P'}{P_o},\ u^* = \frac{u}{u_o},\ v^* = \frac{v}{v_o},\ \theta = \frac{t}{\tau_o}$$

where $c_o = \frac{P_o}{RT}$ is the total concentration at atmospheric pressure, $v_o = \frac{B_p}{\mu}\frac{\Delta P_o}{L}$ is a reference intraparticle velocity, $\tau_o = \frac{\varepsilon L}{u_o}$ is the reference space time and u_o is the bulk fluid superficial velocity at the bed inlet in steady-state at a given pressure drop $\Delta P_o = P_h - P_\ell$ (here $P_\ell = P_o$), i.e.:

$$u_o = \frac{-La_1/P_h + \sqrt{(La_1/P_h)^2 + 2La_2(1 - (P_\ell/P_h)^2)}}{2La_2} \qquad [21]$$

with

$$a_1 = \frac{150\mu(1-\varepsilon)^2}{d_p^2\varepsilon^3}; \qquad a_2 = \frac{1.75\rho_o(1-\varepsilon)}{P_o d_p \varepsilon^3} \qquad [21a]$$

Intraparticle diffusion + convection model

Mass balances inside the adsorbent particle:

$$\frac{\partial}{\partial\rho}\left(\frac{f}{b_4+f}\frac{\partial y_A'}{\partial\rho} + \frac{y_A'}{b_4}\frac{\partial f}{\partial\rho}\right) - \lambda_o\frac{\partial(v^* f' y_A')}{\partial\rho} = \alpha_o(1+\xi_p)\frac{\partial(f' y_A')}{\partial\theta} \qquad [22]$$

$$\frac{\partial}{\partial\rho}\left(\frac{1}{b_4}\frac{\partial f}{\partial\rho}\right) - \lambda_o\frac{\partial(v^* f')}{\partial\rho} = \alpha_o\left[\frac{\partial f'}{\partial\theta} + \xi_p\frac{\partial(f' y_A')}{\partial\theta}\right] \qquad [23]$$

where $\xi_p = \frac{1-\varepsilon_p}{\varepsilon_p}\, m$ is the adsorbent capacity parameter. The boundary conditions are:

$$\rho = 0, \qquad y_A' = y_A - \frac{\partial y_A}{\partial x}\beta_R; \quad f' = f - \frac{\partial f}{\partial x}\beta_R \qquad [24]$$

$$\rho = 1, \qquad y_A' = y_A + \frac{\partial y_A}{\partial x}\beta_R; \quad f' = f + \frac{\partial f}{\partial x}\beta_R \qquad [25]$$

and initial condition:

$$\theta = 0, \qquad y_A' = y_o; \quad f' = f_o; \qquad \forall x,\rho \qquad [26]$$

where $f_o = f_\ell$ for pressurization and $f_o = f_h$ for blowdown.

Momentum equation for the fluid inside particle:

$$v^* = \frac{v}{v_o} = -\beta_\ell\frac{\partial f}{\partial\rho} \qquad [27]$$

Mass balances for the bulk fluid phase in a bed volume element:

Species A

$$\frac{\partial}{\partial x}\left(\frac{u^* f}{Pe}\frac{\partial y_A}{\partial x}\right) - \frac{\partial(u^* f y_A)}{\partial x} = \frac{\partial(f y_A)}{\partial \theta} + \frac{1-\varepsilon}{\varepsilon} N_A \qquad [28]$$

Overall

$$\frac{\partial(u^* f)}{\partial x} + \frac{\partial f}{\partial \theta} + \frac{1-\varepsilon}{\varepsilon} N = 0 \qquad [29]$$

where dimensionless fluxes of species A and overall are given by:

$$N_A = \frac{\varepsilon_p}{\alpha_o}\left[\left(-\frac{f}{b_4 + f'}\frac{\partial y_A'}{\partial \rho} - \frac{y_A'}{b_4}\frac{\partial f'}{\partial \rho} + \lambda_o v^* f' y_A'\right)\Big|_{\rho=0} - \left(-\frac{f}{b_4 + f'}\frac{\partial y_A'}{\partial \rho} - \frac{y_A'}{b_4}\frac{\partial f'}{\partial \rho} + \lambda_o v^* f' y_A'\right)\Big|_{\rho=1}\right] \qquad [30]$$

$$N = \frac{\varepsilon_p}{\alpha_o}\left[\left(-\frac{1}{b_4}\frac{\partial f'}{\partial \rho} + \lambda_o v^* f'\right)\Big|_{\rho=0} - \left(-\frac{1}{b_4}\frac{\partial f'}{\partial \rho} + \lambda_o v^* f'\right)\Big|_{\rho=1}\right] \qquad [31]$$

Momentum equation for the bulk fluid:

$$-\frac{\partial f}{\partial x} = b_5 u^* + b_6 f (u^*)^2 \qquad [32]$$

Boundary conditions associated with eqs.28 and 29 are :

Pressurization:

$$x = 0, \qquad y_A - \frac{1}{Pe}\frac{\partial y_A}{\partial x} = y_f; \quad f = f_h \qquad [33]$$

$$x = 1, \qquad \frac{\partial y_A}{\partial x} = 0; \quad \frac{\partial f}{\partial x} = 0 \qquad [34]$$

The initial condition is:

$$\theta = 0, \qquad y_A = y_o; \quad f = f_o; \qquad \forall x, 0 < x \le 1 \qquad [35]$$

where $f_o = f_\ell$ for pressurization .

The definitions of the parameters, β_R, β_ℓ, b_1, b_2, b_3, b_4, b_5, b_6, Pe, λ, λ_o, α, α_o, Φ can be found in Table II.

Intraparticle diffusion model

In the absence of intraparticle convection, $\lambda_o = 0$; therefore, the intraparticle diffusion + convection model reduces to the intraparticle diffusion model.

Equilibrium model:

If there are no mass transfer resistances inside the particle, the intraparticle diffusion model reduces to the equilibrium model.

Model equations are :

In the bulk fluid phase:

$$\frac{\partial}{\partial x}\left(\frac{u^*f}{Pe}\frac{\partial y_A}{\partial x}\right) - \frac{\partial(u^*fy_A)}{\partial x} = [1 + \varepsilon_p \frac{1-\varepsilon}{\varepsilon}(1+\xi_p)]\frac{\partial(fy_A)}{\partial\theta} \quad [36]$$

$$\frac{\partial(u^*f)}{\partial x} = \frac{\partial f}{\partial\theta} + \varepsilon_p \frac{1-\varepsilon}{\varepsilon}[\frac{\partial f}{\partial\theta} + \xi_p \frac{\partial(fy_A)}{\partial\theta}] \quad [37]$$

Table II. The definition of parameters

$b_1 = \frac{\varepsilon\gamma_1 D_{mo}}{Lu_o}$	$\frac{1}{Pe} = \frac{\varepsilon D_{ax}}{Lu} = \frac{b_1}{u^*f} + b_2$
$b_2 = \frac{\varepsilon\gamma_2 d_p}{L}$	$\lambda = \lambda_o v^*(b_4 + f)$
$b_3 = \frac{1}{\lambda_o}$	$\lambda_o = \frac{\tau_f \ell^2}{D_{mo}} \frac{v_o}{\varepsilon_p \ell}$
$b_4 = \frac{D_{mo}}{D_{kA}}$	$\alpha = \alpha_o u^*(b_4 + f)$
$b_5 = \frac{a_1 L}{P_o} u_o$	$\alpha_o = \frac{\tau_f \ell^2}{D_{mo}} \frac{u_o}{\varepsilon L}$
$b_6 = a_2 L u_o^2$	$\beta_R = \frac{\ell}{2L}$
$\Phi = \frac{P_\ell}{P_h}$	$\beta_\ell = \frac{L}{\ell}\frac{P_o}{\Delta P_o}$
$\xi_p = \frac{1-\varepsilon_p}{\varepsilon_p} m$	

Simulations show that the final profile of the mole fraction of the adsorbable species calculated with the intraparticle diffusion/convection model is between profiles predicted by the diffusion and the equilibrium model (Figure 22).

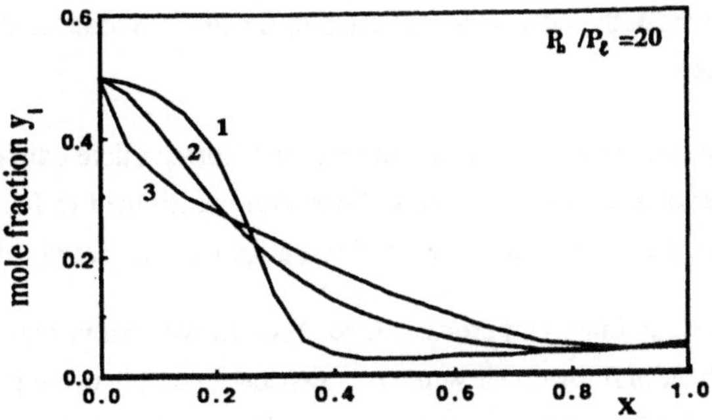

Figure 22. Final axial mole fraction profiles in pressurization with equilibrium (1), diffusion (3) and diffusion/convection model

3. CONCLUSIONS

Sorption processes are becoming more important in chemical process industries as a result of biotechnology developments, energy and environment constraints.

The first message we tried to give is on the methodology used to study sorption processes based on the measurement of parameters governing the behavior of fixed beds by simple independent experiments ;those parameters are then used in the fixed-bed adsorber model.

The second message is related with modeling. It is first of all a learning tool before it becomes a design, operation and process optimization aid. Numerical methods used for the solution of various model equations include: collocation methods, finite differences, moving finite element methods, Lax-Wendroff with flux correction.

Available packages are PDECOL, COLNEW, FORSIM together with our own software [43-44].

The third message is to encourage development of new processes and concepts. New processes arise sometimes by coupling separation and reaction, or by using flow reversal or by simulating moving bed as Broughton did. The new concept of augmented diffusivity by convection is a rich one. In separation processes large-pore packings contribute to the improvement of sorption efficiency; in fact, column responses are driven from diffusion-controlled to equilibrium-controlled limits by intraparticle convective flow. The result is equivalent to that observed with reactions in large-pore catalysts; in such case conversion at the reactor outlet moves from the diffusion-controlled to the kinetic-controlled limit as a result of intraparticle convection.

HPLC is a separation process in which the effect of intraparticle convection is important when large-pore particles are used. Proteins separation contributed to focus the attention of researchers ; therefore the interest on perfusion chromatography increased recently.

PSA is an industrially important process. It has been shown that intraparticle convection is important in pressurization/blowdown steps. It is expected that the cycle performance will be influenced by intraparticle convective flow when cycles are short as in rapid pressure swing adsorption.

The last message is related with multicomponent systems. A crucial step is the prediction of multicomponent sorption equilibria. Most of the calculations are based in multicomponent equilibrium isotherms which are easy to implement in fixed-bed models. However ,even if a model based on Ideal Adsorption Solution (IAS) is used fixed-bed calculations become time consuming since at each step one has to solve the iterative algorithm for IAS leading to equilibrium composition. We believe that we have to focus on the representation of multicomponent equilibrium first; then describe it by some working relationships to be finally used in the fixed-bed model avoiding the time-consuming iterations.

References

1. Rodrigues, A.: Modeling of percolation process. In Percolation Processes: Theory and Applications, ed. A. Rodrigues and D. Tondeur, pp 31-81, Nijtholf and Noordhoff, 1981
2. Rodrigues, A.: Mass transfer and reaction in fixed-bed heterogeneous systems: application to sorption operations and catalytic reactors. In Disorder and Mixing, ed. E. Guyon, et. al, pp 215-236, Kluwer Acad. Pubs., 1988
3. De Vault, D.: The theory of chromatography. J. Am. Chem. Soc., 65, 532 (1943)
4. Thomas, H. C.: Heterogeneous ion exchange in a flow system. J. Am. Chem. Soc., 66, 1664 (1944)
5. Bohart, G. and Adams, E.: Some aspects of the behavior of charcoal with respect to diclorine. J. Am. Chem. Soc., 42, 523(1920)
6. Walter, J.: Rate dependent chromatography adsorption. J. Chem. Phy., 13, 332(1945)
7. Rodrigues, A.: Theory of linear and nonlinear chromatography. In Chromatographic and Membrane Processes in Biotechnology, ed. C. Costa and J. Cabral, Kluwer Academic. Pubs., 1991
8. Arnold, F., Schofield, S. and Blanch, H.: Analytical affinity chromatography. J. Chromat. 355, 1-12 (1986)
9. Chase, F.: Prediction of performance of preparative affinity chromatography. J. Chromat. 297, 179-202 (1984)
10. Rosen, J.: Kinetics of fixed-bed system for solute diffusion into spherical particles. Ind. Eng. Chem. 20(3). 387 (1952)
11. Rodrigues, A.: Percolation theory I - Basic principles. In Stagewise and Mass Transfer Operations, ed. J. Calo and E. Henley, AIChEMI., 1984
12. Rodrigues, A. and Costa, C.: Fixed-bed processes: a strategy for modeling. In: Ion Exchange: Science and Technology, pp 272-287, M.Nijthoff, 1986
13. Costa, C. and Rodrigues, A.: Design of cyclic fixed-bed adsorption processes I. AIChE J. 31, 1645-1654 (1985)
14. Costa, C. and Rodrigues, A.: Intraparticle diffusion of phenol in macroreticular adsorbents. Chem. Eng. Sci., 40, 983-993 (1985)

15. Wilhelm, R. Rice, A. and Bendelius, A.: Parametric pumping: A dynamic principle for separating fluid mixtures. Ind. Eng. Chem. Fund.,5 , 141-144 (1966)
16. Ramalho, E., Costa, C., Grevillot, G. and Rodrigues, A.: Adsorptive parametric pumping for the purification of phenolic effluents. Separations Technology, 1, 99-107 (1991)
17. Costa, C., Grevillot, G., Rodrigues, A. and Tondeur, D.: Purification of phenolic waste water by parametric pumping. AIChE J., 28, 73 (1982)
18. Almeida, F. Costa, C., Grevillot, G. and Rodrigues, A.: Removal of phenol from waste water by recuperative mode parametric pumping. In: Physicochemical Methods for Water and Wastewater Treatment. ed. L. Pawlowski, pp.169-178, Elsevier, 1982
19. Ferreira, L. Costa, C. and Rodrigues, A.: Scale up of adsorptive parametric pumping. Annual AIChE mt., LA, 1991
20. Tsewtt, M.: On a novel class of adsorption phenomena and their use in biochemical analysis. Trudi Varshavskogo obstchestva estestvoispitatelei, 40, 20-39 (1903)
21. Nicoud, R. and Bailly, M.: Choice and optimization of operating mode in industrial chromatography. In: PREP-92, ed. M. Perrut, 1992
22. Bailly, M. and Tondeur, D.: Reversibility and performance in productivity chromatography. Chem. Eng. Process., 18, 293-302 (1984)
23. Bailly, M. and Tondeur, D.: Two way chromatography. Chem. Eng. Sci., 36, 455-469 (1981)
24. Broughton, D.: Adsorptive separations-liquids. In: Kirk-Othmer Enc. of Chem. Techn., Vol.1, J. Wiley, 1978
25. De Rosset, A.: Neuzil, R. and Broughton, D.: Industrial application of preparative chromatography. In: Percolation Processes: Theory and Applications, ed. A. Rodrigues and D. Tondeur, pp 249, Nijtholf and Noordhoff, 1981
26. Johnson, J.: Sorbex: continuing innovation in liquid phase adsorption. In: Adsorption: Science and Technology, pp383-395, M.Nijthoff, 1989
27. Stord, G., Masi, M. and Morbidelli, M.: On counter current adsorption separation processes. In Adsorption: Science and Technology, pp 357-382, M. Nijhoff, 1989
28. Nielsen, A., Bergd, S. and Troberg, B.: Catalyst for the synthesis of ammonia and method of producing it. US Patent 3,243,386, (1966)
29. Wheeler, A.: Reaction rates and selectivity in catalyst pores. Adv. in Catalysis, 3, 250- 337 (1951)
30. Nir, A. and Pismen, L.: Simultaneous intraparticle convection, diffusion and reaction in a porous catalyst. Chem. Eng. Sci., 32, 35-41 (1977)
31. Rodrigues, A, Ahn, B. and Zoulalian, A.: Intraparticle forced convection effect in catalyst diffusivity measurement and reactor design. AIChE J., 28, 541-546 (1982)
32. Afeyan, A. Regnier, F. and Dean, R.: Perfusive chromatograph. US Pat. 5,019,270, (1990)
33. Afeyan, A. Gordon, N., Mazsareff, I.,Varady, L.,Fulton, S., Yang, Y. and Regnier, F.: Flow through particles for the HPLC separation of biomolecules: Perfusion chromatography. J. Chromat., 519, 1-29 (1990)
34. Rodrigues, A.E., Lu, Z.P. and Loureiro, J.M.: Residence Time Distribution of Inert and Linearly Adsorbed Species in Fixed-Bed Containing 'Large-Pore' Supports: App. in Sep. Eng. Chem. Eng. Sci., 46, 2765-2773 (1991)
35. Rodrigues, A.E., Lopes, J.C., Lu, Z.P., Loureiro J. M. and Dias, M.M.: Importance of Intraparticle Convection on the Performance of Chromatographic Processes. 8th. Intern. Symp. Prep. Chromat. "PREP-91", Arlington, VA; J. Chromatography, 590. 93-100 (1992)
36. Rodrigues, A.: An extended Van Deemter equation (Rodrigues equation) for the performance of chromatographic processes using large-pore, permeable packings. submitted to LC-GC, 1992
37. Van Deemter, J. Zuiderwveg, F. and Klinkenberg, A.: Longitudinal diffusion and resistance to mass transfer as causes of nonideality in chromatography. Chem. Eng. Sci., 5, 271-289 (1956)
38. Lu, Z.P., Loureiro, J.M., LeVan, M.D. and Rodngues, A.E.: Effect of Intraparticle Forced Convection on Gas Desorption from Fixed-Bed Containing 'Large-Pore' Adsorbents. Ind. Eng. Chem. Res., 35, 1530 (1992)
39. Lu, Z.P., Loureiro, J.M., LeVan, M.D. and Rodrigues, A.E.: Intraparticle Convection Effect on Pressurization and Blowdown of Absorbers. AIChE J.,38, 857-867 (1992)
40. Rodrigues, A.E., Loureiro, J.M. and LeVan, M.D.: Simulated Pressurization of Adsorption Beds. Gas Separation and Purification, 5, 115 (1991)
41. Lu, Z.P., Loureiro, J.M., LeVan, M.D. and Rodrigues, A.E.:Intraparticle Diffusion/ Convection Models in Pressurization and Blowdown: Nonlinear Equilibrium", Sep. Sci. Technol., 271(4), 1857-1874 (1992)
42. Lu, Z.P., Loureiro, J.M., LeVan, M.D. and Rodrigues, A.E.: Pressurization and Blowdown of an Adiabadc Adsorption Bed: IV Diffusion/Convection Model. Gas Sep & Purif. 6 (2), 89-100, (1992)
43. Sereno, C. Rodrigues, A. and Villadsen, J.: Solution of partial differential equations systems by the moving finite element method. Computers Chem Engng., 16, 583-592 (1992)
44. Loureiro, J. and Rodrigues, A.: Two solution methods for hyperbolic systems of partial differential equations in chemical engineering. Chem Eng. Sci.,46, 3259-3267 (1991)

Monitoring Batch Processes

John F. MacGregor and Paul Nomikos

Department of Chemical Engineering, McMaster University, Hamilton, Ontario, Canada L8S 4L7

Abstract: Two approaches to monitoring the progress of batch processes are considered. The first approach based on nonlinear state estimation is reviewed and the problems in implementing it are discussed. A second approach based on multi-way principal components analysis can be developed directly from historical operating data. This approach leads to multivariate statistical process control plots which are very powerful in detecting subtle changes in process variable trajectories throughout the batch. This method is evaluated on a simulation of the semi-batch emulsion polymerization of styrene/butadiene.

Keywords: Batch monitoring, state estimation, statistical process control, fault detection

Introduction

Monitoring batch reactors is very important in order to ensure their safe operation, and to assure that they produce a consistent and high quality product. Some of the difficulties limiting our ability to provide adequate monitoring are the lack of on-line sensors for product quality variables, the highly nonlinear nature and finite duration of batch reactors, and the difficulties in developing accurate mechanistic models that characterize all the chemistry, mixing, and heat transfer in these reactors.

Current approaches to achieving consistent, reproducible results from batch reactors are based on the precise sequencing and automation of all the stages in the batch sequence. Monitoring is usually confined to checking that these sequences are followed, and that certain reactor variables such as temperature are following acceptable trajectories. In some cases, on-line energy balances are used to keep track of the instantaneous reaction rate, and the conversion or the residual reactant concentrations in the reactor [2, 11, 19].

In this paper, we consider two advanced model-based approaches to batch monitoring. The first approach is based on state estimation, and combines fundamental nonlinear dynamic models of these batch processes with on-line measurements in order to provide on-line, recursive estimates of the fundamental states of the process. Since this approach is well known and has an abundant literature, we only provide an overview of its main ideas and some of the key references.

The second approach is based on empirical multivariate statistical models which are easily developed directly from past production data on the large number of process variables such as temperatures, pressures and flows measured throughout the batch. This new approach is based on Multi-way Principal Component Analysis (MPCA) and Projection to Latent Structure (PLS) methods. It will be discussed in greater detail and some examples will be presented.

Both of the above approaches rely upon the observability of the states or events of interest. If the data contain no information or very little information on certain states or events, then no effective monitoring scheme for them is possible.

State Estimation

Theoretical or mechanistic models for batch or semi-batch reactors usually take the form of a set of ordinary nonlinear differential equations in the form

$$\frac{dx^d}{dt} = f^d(x, u, t) \tag{1}$$

$$y = h(x, u, t) \tag{2}$$

$$x_0 = x(t = 0) \tag{3}$$

where x represents the complete vector of internal states, x^d is the deterministic subset of the differential state vector x described through eq. (1), y is the vector of measured outputs that is related to x through eq. (2), and u is a time-varying vector of known manipulated inputs. The complete state vector x is assumed to be made up of a deterministic component x^d and a stochastic component x^s. Included in x^s are model parameter and disturbance states that may vary with time in some stochastic manner and may be unknown initially.

The objective of the state estimation problem is to predict the internal states $x(t)$ from a limited set of sampled,and noise-corrupted measurements $y(t_k)$. It is assumed that some elements of x_0 will be initially unknown and that some states will be time-varying disturbances and/or fixed parameters that must be estimated.

These are several requirements needed for the successful application of state estimation. One must have a good mechanistic model that captures the main physical and chemical phenomena occurring in the reactor. A set of measurements (y) is necessary that not only makes the states of interest observable, but also ensures that the state estimation errors will be small enough to detect the state deviations of interest. As pointed out by MacGregor et al [11] and Kozub and MacGregor [8], a common error in formulating filters is neglecting to incorporate adequate disturbance and/or parameter states (x^s). These are needed to eliminate biases in the state estimates and to yield a robust estimator when there are modelling errors or unknown disturbances in the system. Failing to incorporate such nonstationary stochastic states leads to a proportional type of state estimator without the integral behaviour necessary to eliminate the biases.

The most common form of nonlinear state estimator is Extended Kalman Filter (EKF) but various other forms of second order filters, reiterative Kalman Filters, and nonlinear optimization approaches have been suggested. Kozub and MacGregor [8, 9] investigated the use of the EKF, the reiterative EKF and a nonlinear optimization approach in monitoring semi-batch emulsion polymerization reactors, and concluded that the EKF with reiteration and a second filter to estimate the initial states was the preferred approach. In other situations linear Kalman Filters based on semi-empirical models are adequate. Stephanopoulos and San [17] used such filters with non-stationary growth parameter states to monitor fed-batch fermentation reactors.

The state estimator provides on-line recursive estimates of important process states $\mathbf{x}^d(t_k|t_k)$ and stochastic disturbance or parameter states $x^s(t_k \mid t_k)$ thereby enabling one to monitor the progress of the batch reactor. The number of non-stationary stochastic states or parameters estimated cannot be greater than the number of measured output variables (y). One can also monitor the performance of the filter itself by plotting the innovations $(y(t_k) - y(t_k \mid t_{k-1}))$. The variance of the innovations is sometimes used to adapt the Kalman Filter gain.

On-line reactor energy balances are very effectively implemented using Kalman Filters [11, 2, 14]. The objective is to combine the energy balance equations with simple flowrate, and temperature measurements taken around the reactor and its cooling system to track stochastic states such as the instantaneous heat release due to reaction (q_R), and the overall heat transfer coefficient (UA). Unsafe or undesirable batch reactor conditions such as the beginning of a run-away reaction, or excessive fouling of the reactor heat transfer surfaces can then be detected. In semi-batch reactors where reactants are being fed continuously these state estimates can be used to detect the unsafe accumulation of reactants in the reactor that may occur if there is a temporary reaction break-down due to poisoning, etc. [15].

Kalman Filters based on models which include more detailed kinetic phenomena and more informative sensors can be used to monitor species concentrations and molecular properties. Kozub and MacGregor [8] illustrate this in monitoring a semi-batch styrene-butadiene emulsion polymerization reactor. Particle size and concentration, and polymer composition and structure were monitored together with stochastic states such as impurity concentrations. Such detailed models allow for detection of more specific problems such as particle coagulation, impurity contamination, or feedrate errors. These state estimators can also be used to implement nonlinear control over polymer property development [9].

An alternative approach aimed at providing more specific fault detection and diagnosis is to run several parallel filters, each based on a different set of plausible events or faults. Based on the innovations ($x(t_k) - x(t_k \mid t_{k-1})$) from each filter the posterior probability of each model being valid can be evaluated at each sampling interval. A high probability for any model which characterizes an undesirable event would lead to an alarm and a response. Such an approach was used by King [7] to monitor batch reactors for the onset of undesirable side reactions.

A major difficulty in practice with monitoring batch reactors by state estimators is the need for detailed mechanistic models, and some specific on-line sensors related to the variables of interest. Even when these models are developed some parameters in both the models and the filters must be adjusted to ensure that the resulting filters can track the actual processes. The advantage of such an approach is that it is "directional" in nature since the incorporation of mechanistic understanding into the choice of the state vector $x = (x^d, x^s)^T$ allows one to make inferences about the nature of any faults as well as their magnitudes.

Empirical Monitoring Approaches

Although good theoretical models of batch processes and on-line sensors for fundamental quality properties are often unavailable, nearly every batch process has available frequent observations on many easily measured process variables such as temperatures, pressures, and flowrates. One may have up to 50 measurements or more every few seconds throughout the entire history of a batch. Furthermore, there is usually a history of many past successful (and some unsuccessful) batches. From this data it should be possible to build an empirical model to characterize the operation of successful batch runs. The major difficulties are how to handle the highly correlated process variables, and the large number of multivariate observations taken throughout the batch history.

In this section, we develop some multivariate statistical process control methods for monitoring and diagnosing problems with batch reactors which make use of such data. The approach used in based more on the statistical process control (SPC) philosophy of Shewhart [16] than that of feedback control. In SPC one usually assumes that under normal operation, with only common cause variations present, the system will operate in some stable state of statistical control, and will deviate from this behaviour only due to the appearance of special causes. The approach is therefore to develop some statistical monitoring procedures to detect any special event as quickly as possible, and then look for an assignable cause for the event. Through such a procedure one can gradually make continuous improvements to the process. Traditionally univariate SPC charts such as the Shewhart chart have been used to monitor single variables. However, these approaches are inappropriate when dealing with large multivariate problems such as the one being treated here. Statistical process control charts based on multivariate PCA ad PLS methods have been developed for steady-state continuous processes [10], but for batch processes where the data consists of finite duration, time varying trajectories little has been done. Therefore, in this paper we develop monitoring procedures based on multi-way principal components analysis (MPCA). This method extracts the essential information out of the large number of highly correlated variables and compresses it into low-dimensional spaces that summarize both the variable and time histories of successful batches. It then allows one to monitor the progress of new batches by comparing their progress in these spaces against that of the past reference distribution.

Multivariate factor analysis methods (closely related to principal components) have recently been used by Bonvin and Rippin [3] to identify stoichiometry in batch reactors. This represents an approach which combines some fundamental knowledge with a multivariate statistical approach to monitor more specific features in batch reactors, but does not provide a general framework for on-line monitoring of the progress of batch reactors.

Multi-Way Principal Components Analysis (MPCA)

The type of historical data one would usually have available on a batch process is illustrated in Figure 1. For each batch ($i = 1, ..., I$), one would measure J variables at K time periods throughout the batch. Thus one has a three-dimensional array of data $\underline{\mathbf{X}}(i, j, k)$; $i = 1, ..., I$; $j = 1, ..., J$ and $k = 1, ..., K$. The top plane in the array represents the data on the time trajectories for all J variables in the first batch. Similarly, the front plane represents the initial measurements on all J variables for each of the I batches.

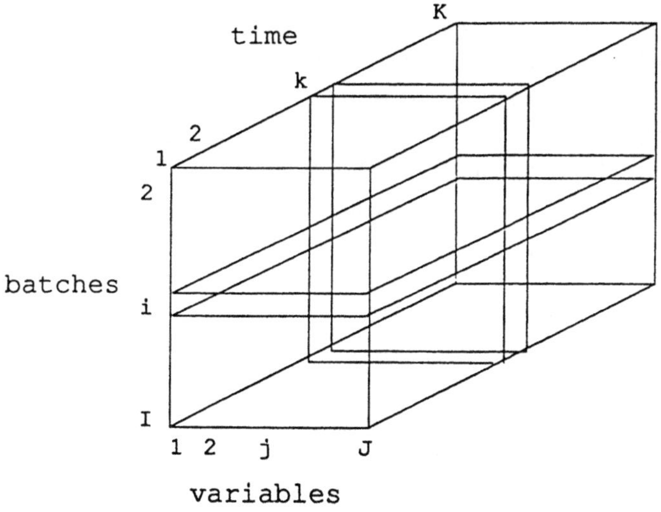

Figure 1: Data array $\underline{\mathbf{X}}$ for a typical batch process.

If we only had available a two-dimensional matrix (**X**) such as the matrix of variables versus batch at a given time k, then ordinary principal components analysis (PCA) could be used to decompose the variation in it into a number of principal components. After mean centering (ie subtracting the mean of each variable), the first principal component is given by that linear combination of variables exhibiting the greatest amount of variation in the data set ($\mathbf{t}_1 = \mathbf{X}\,\mathbf{p}_1$). The second principal component ($\mathbf{t}_2$) is that linear combination, orthogonal to the first one, which exhibits the next greatest amount of variation, and so forth. With highly correlated variables, one usually finds that only a few principal components ($\mathbf{t}_1, \mathbf{t}_2, ..., \mathbf{t}_A$) are needed to explain most of the significant variation in the data. The (I×J) data matrix can then be approximated as sum of A rank one matrices

$$\mathbf{X} = \mathbf{T}\,\mathbf{P}^T = \sum_{a=1}^{A} \mathbf{t}_a\,\mathbf{p}_a^T$$

where the "score" vectors $\mathbf{t}_a$ are mutually orthogonal and represent the values of the principal components for each object (i). The loading vectors $\mathbf{p}_a$ show the contribution of each variable to the corresponding principal component. Principal components analysis is described in most multivariate statistics texts [1, 6]. However, the projection aspects of PCA and the NIPALS algorithm for computing principal components sequentially that are used in this paper are best described in Wold et al [18].

Since in the batch data array of Figure 1, we are interested in analyzing the variation with respect to variables, batches and time a three-dimensional PCA is needed. Such methods have been developed for the analysis of multivariate images [5], and we use a variant of this approach in this paper.

Multi-way PCA decomposes the $\underline{\mathbf{X}}$ array into score vectors ($\mathbf{t}_a$; a = 1, 2, ..., A) and loading matrices $\mathbf{P}_a$ such that

$$\underline{\mathbf{X}} = \sum_{a=1}^{A} \mathbf{t}_a \otimes \mathbf{P}_a + \mathbf{E}$$

There are three basic ways in which the array can be decomposed but the most meaningful in the context of batch monitoring is to mean center the data by subtracting the means of the variables for each time over the I batches. In this way, the variation being studied in MPCA is the variation about the average trajectories of the variables for the I batches. In this way, the major nonlinear behaviour of the batch process is removed through subtracting the mean trajectories, and linear MPCA will be used to analyze variations about these mean trajectories.

The loading matrices $\mathbf{P}_a$ (a = 1, 2, . . ., A) will then summarize the contributions of the variables at different times to the orthogonal score vectors $\mathbf{t}_a$. These new variables $t_a = \underline{\mathbf{X}} \circ \mathbf{P}_a$ are those exhibiting the greatest variation in the variables over the time of the batch.

The NIPALS algorithm for MPCA follows directly from that for ordinary PCA, and the steps are given below.

0. scale the array $\underline{\mathbf{X}}$, set $\underline{\mathbf{E}} = \underline{\mathbf{X}}$
1. take randomly a column from $\underline{\mathbf{E}}$ and put it as t

Start of minor iterations

2. $\mathbf{P} = \underline{\mathbf{E}}' \bullet \mathbf{t}$
3. $\mathbf{P} = \mathbf{P} / \| \mathbf{P} \|$
4. $\mathbf{t} = \underline{\mathbf{E}} \circ \mathbf{P}$
5. if the **t** has converged then go to step 6 else go to step 2
6. $\underline{\mathbf{E}} = \underline{\mathbf{E}} - \mathbf{t} \otimes \mathbf{P}$
7. go to step 2 for the calculation of the second principal component

Post-Analysis and On-line Monitoring of Batch Processes Using MPCA Plots

The information extraction and data compression ideas of MPCA can be used to perform post-analysis of batch runs to discriminate between similar and dissimilar runs, and to develop on-line methods for monitoring the progress of new batches. We shall concentrate here on the development of on-line monitoring methods. The approach follows closely that of Kresta et al [10] for monitoring the operating performance of continuous processes.

From the historical data base, a representative sample of successful batch runs can be selected. These would normally comprise all those batches that resulted in good product quality. The variable trajectory data array ($\underline{\mathbf{X}}$) as shown in Fig. 1 can be assembled using the data from these runs, and an MPCA analysis performed on this array.

The progress of these "good" batches can be summarized by their behaviour in the reduced principal components space $\mathbf{T} = (\mathbf{t}_1, \mathbf{t}_2, \ldots, \mathbf{t}_A)$. The behaviour of these batches with time will be confined to a region in this space. This region will therefore define a reference distribution against which we can assess the performance of other past batches or new batches. The principal components calculated for other good batches should fall close to the hyperplane defined by **T**, and they should fall in the region of this plane defined by the previous good batches. The acceptable regions in the T-space can be defined using multivariate Normal distribution contours with variances calculated from the estimated score vectors ($\mathbf{t_a}$), or if the reference sample contains a sufficient number of batches an approximate 99% contour can be defined directly as the contours enclosing approximately 99% of the scores from these batches. Post-analysis, that is the analysis past batches for which the complete history is known, can be summarized by plotting the final value of the t-scores for any given batch and comparing it against this reference distribution of t-scores from other batches.

A problem arises in the on-line monitoring of new batches because measurements on the variables are not available over the complete batch as they were with past batch runs. Instead, measurements are only available up to time interval k. There are several approaches to handling this missing data, and we shall use here the rather conservative approach of setting all the values of the scaled, mean-centered variables beyond the current time k to zero. This means that we are giving the new batch the benefit of the doubt by implying that the remaining portion of the batch history will have no deviation from the mean trajectory. Therefore, in monitoring a new batch the following procedure is used:

1. Take the new vector of measurements at time k
2. Mean center and scale them as with the reference set.
3. Add this new observation as the k-th row in $\mathbf{X_{new}}$ and set the rows from (k+1) onward equal to zero.
4. Calculate the new scores $\mathbf{t_a} = \mathbf{X_{new}}\,\mathbf{P_a}$; $\mathbf{E} = \mathbf{X_{new}} - \sum_{a=1}^{A} \mathbf{t_a} \otimes \mathbf{P_a}$
5. Return to step 1.

In monitoring the progress of a new batch there are several ways in which an excursion from normal operation can show up. If the process is still operating in the same way as the batches in the reference data base, but simply exhibits some larger than normal variations, this behaviour should show up as the scores ($\mathbf{t_a}$'s) for the new batch moving outside the control region in the T-space. However, if a totally new fault not represented in the reference data base were to occur, at least one new principal component vector would be needed to describe it. In this case the computed score values of the new batch would not be predicted well by the MPCA model since they would fall off the reduced space of the reference T-plane. To detect all such new events that would show up in this way we plot the squared prediction error (SPE) or equivalently the squared perpendicular distance from the reference T-plane for each new observation from the new batch. To assess the significance of any increase in SPE we place an upper control limit on the SPE above which there is only approximately a 1% probability of occurring if the new batch is on target. This control limit can be calculated in various ways from the variations of the calculated SPE's in the reference set [13].

Example: Semi-Batch Styrene-Butadiene Emulsion Polymerization

Styrene-butadiene rubber (SBR) is made by semi-batch emulsion polymerization for use in adhesives, coatings, footwear, etc. A detailed modelling study on the SBR processes was performed by Broadhead et al [4]. A modification of this model was used in a simulation study to evaluate these MPCA monitoring methods. Using typical variations in the initial charge of materials and impurities, and in the process operations a number of batches were simulated. Fifty batches which gave final latex and molecular properties within an acceptable region were selected to provide a reference data array. On-line measurements were assumed to be available on nine variables: the feedrates of styrene and butadiene monomers, the temperatures of the feed, the reactor contents and jacket contents, the latex density, the total conversion, and the instantaneous heat release from an energy balance. Using 200 time increments over the duration of the batch the reference data set $\underline{\mathbf{X}}$ was a $(50 \times 9 \times 200)$ array.

To evaluate the ability of MPCA to discriminate between "good" and "bad" batches a post-analysis was performed using the 50 good batches plus one bad batch. In one case the "bad" batch had a 33% higher level of organic impurities in the butadiene monomer feed to the reactor right from the beginning its operation. The other "bad" batch had a 50% higher level of organic impurities, but this time the contamination started half-way through its cycle (at time = 100). The score plots for the first two principal components ($\mathbf{t}_1$, $\mathbf{t}_2$) are shown in Figure 2. The two "bad" batches, denoted as point "51", clearly do not belong to the family of normal batches. Therefore in this case the MPCA plots were easily able to detect abnormal operation of the batch.

In order to implement an on-line monitoring scheme for new batches a MPCA model was developed from the historical records of 50 good batches. Four significant principal components were needed to capture the predictable variation about the average trajectories of the variables in the batches. Plots of the first two principal components (t_1, t_2) and the SPE versus time are shown in Figure 3 for the evolution of a new "good" batch run. The principal components remain

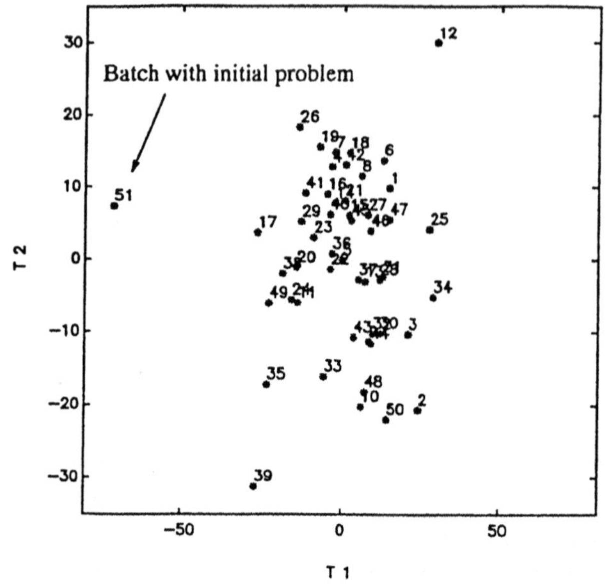

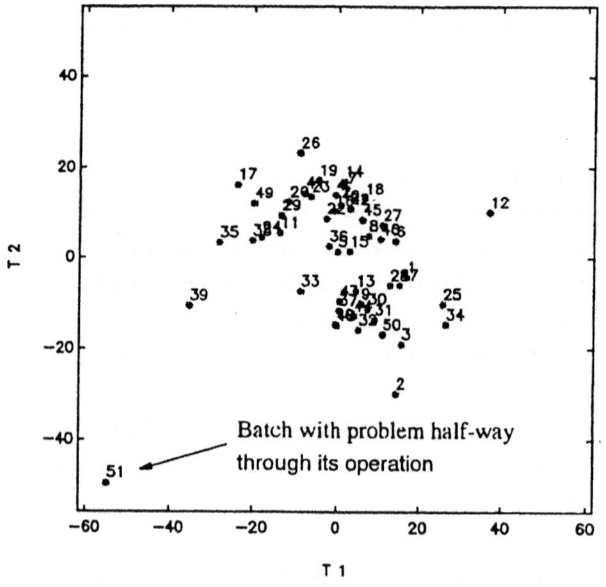

Figure 2: Post analysis of batch data in the score space.

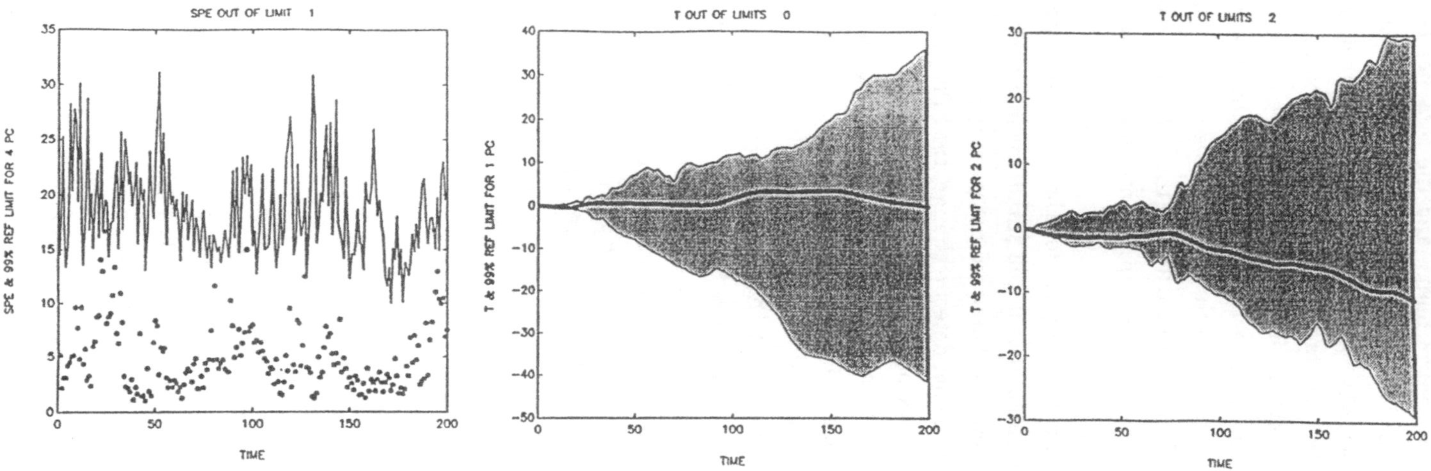

Figure 3: Monitoring of a new "good" SBR batch.

between their upper or lower control limits, and the SPE lies below its control limit throughout the duration of the batch indicating that the progress of this new batch was well within the acceptable range of variation defined by the reference distribution. The SPE control limits shown here are approximately 99% limits based on the reference distribution with only 50 samples, and are therefore quite erratic. Improved estimates of this upper control limit can be obtained.

Figure 4 shows the same monitoring plots for the new SBR batch in which there is a 33% higher level of organic impurities in the butadiene monomer feed to the reactor starting at time zero. The principal component plots and the SPE plot detect this event very quickly.

Figure 5 shows the monitoring plots for the new SBR batch in which at time 100 the level of organic impurities in the butadiene monomer feed increased by 50%. The final product from this batch did not meet specifications. Although the principal component plots do not detect a change, the SPE plot rapidly detects the occurrence of this new event around $t = 100$.

To further verify that these multivariate monitoring methods are very powerful for detecting problems which occur in batch processes, the trajectories of the individual variable measurements for the three batch runs just considered are plotted in Fig. 6. It can be seen that there is not much observable difference among the three runs. If all the trajectories from the good batches in the reference set were plotted on this Figure any differences between these and the bad batches would be almost undetectable through visual inspection. The power of the multivariate MPCA method results from using the joint covariance matrix of all the variable trajectories. By doing this, it utilizes not just the magnitude of the deviation of each trajectory from its mean, but the correlations among them in order to detect abnormal operation.

Summary

Methods for on-line monitoring of the process of batch processes have been presented. Theoretical model-based approaches based on state estimation were briefly reviewed, and new empirical methods based on multi-way principal components analysis were presented. The latter statistical monitoring approach was evaluated on simulations of an SBR semi-batch emulsion polymerization reactor, and was shown to provide rapid detection of operating problems.

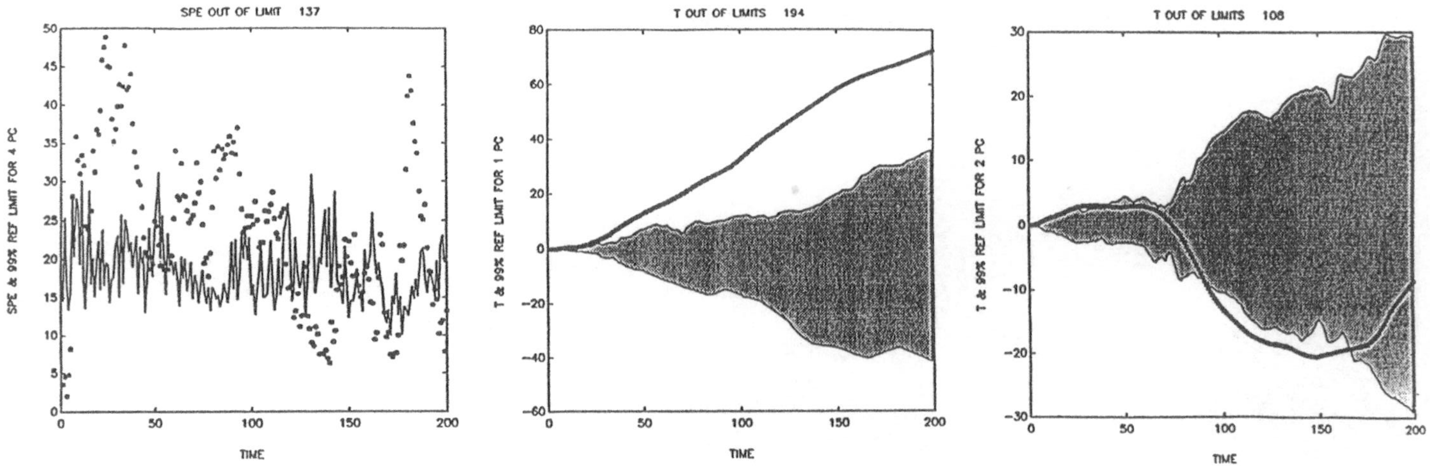

Figure 4: Monitoring of a new SBR batch with impurity contamination in the butadiene feedrate starting at time zero.

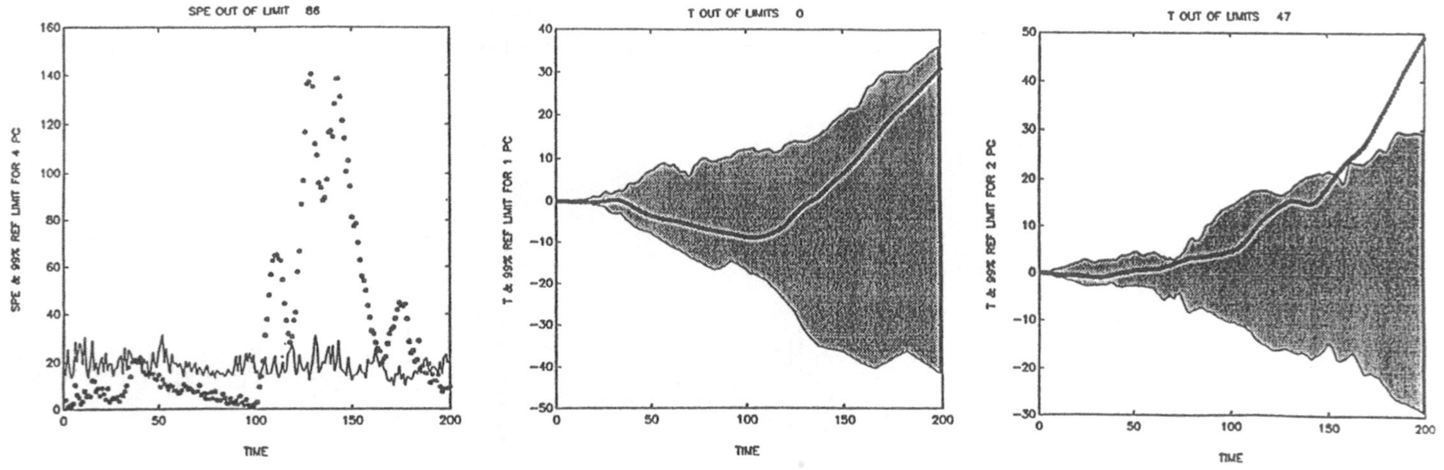

Figure 5: Monitoring of a new SBR batch with impurity contamination in the butadiene feedrate starting at time 100.

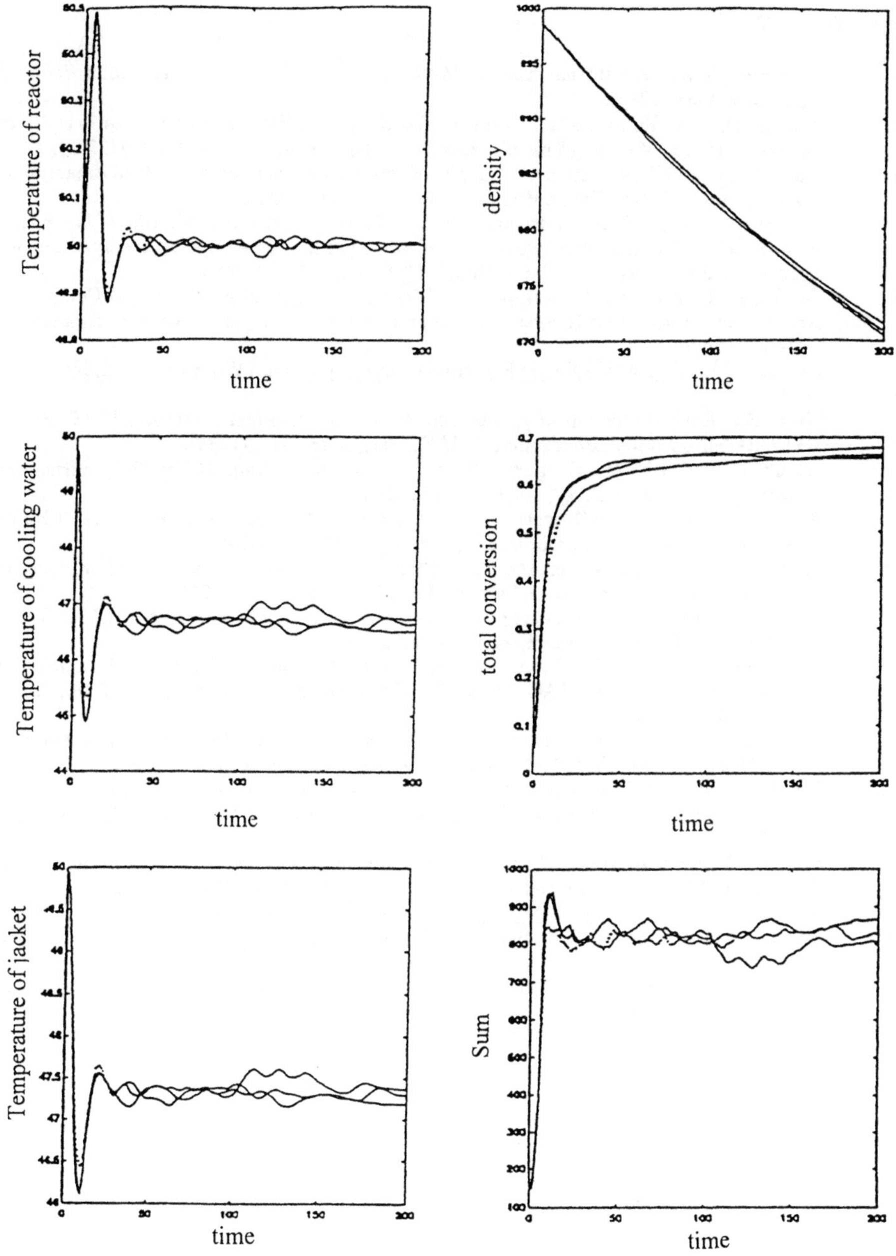

Figure 6: Trajectories of some of the measurements variables during one good batch (solid) and the two batches with early (dotted) and late (dashed) impurity contamination.

Literature

1. Anderson, T.W.: An Introduction to Multivariate Statistical Analysis, John Wiley & Sons, New York (1984).
2. Bonvin, D., P. de Valliere and D. Rippin: Application of Estimation Techniques to Batch Reactors. Part I: Modeling Thermal Effects. Comp. Chem. Eng., 13, pp. 1-9 (1989).
3. Bonvin, D. and D.W. Rippin: Target Factor Analysis for the Identification of Stoichiometric Models, Chem. Eng. Sci., 45, 3417-3426 (1990).
4. Broadhead, T.O., A.E. Hamielec and J.F. MacGregor: Dynamic Modelling of the Batch, Semi-Batch and Continuous Production of Styrene/Butadiene Copolymers by Emulsion Polymerization, Makromol. Chem., Suppl. 10/11, pp. 105-128 (1985).
5. Geladi, P., H. Isaksson, L. Linkqvist, S. Wold and K. Esbensen: Principal Component Analysis of Multivariate Images, Chemometrics and Intelligent laboratory Systems, 5, 209-220 (1989).
6. Jackson, J.E.: A User's Guide to Principal Components, John Wiley and Sons, New York (1991).
7. King, R.: Early Detection of Hazardous States in Chemical Reactors, IFAC Symp. DYCORD'86, pp. 93-98, Bournemouth, U.K., Pergamon Press (1986).
8. Kozub, D. and J.F. MacGregor: State Estimation for Semi-Batch Polymerization Reactors, Chem. Eng. Sci., 47, 1047-1062 (1992a).
9. Kozub, D., and J.F. MacGregor: Feedback Control of Polymer Quality in Semi-Batch Copolymerization Reactors, Chem. Eng. Sci., 47, 929-942 (1992b).
10. Kresta, J.V., J.F. MacGregor and T.E. Marlin: Multivariate Statistical Monitoring of Process Operating Performance, Can. J. Chem. Eng., 69, 35-47 (1991).
11. MacGregor, J.F.: On-line Energy Balances via Kalman Filtering, Proc. IFAC Symp. PRP-6, pp. 35-39, Akron, Ohio, Pergamon Press (1986).
12. MacGregor, J.F., D. Kozub, A. Penlidis, A.E. Hamielec: State Estimation for Polymerization Reactors, IFAC Symp. DYCORD '86, Bournemouth, pp. 147-152, U.K., Pergamon Press (1986).
13. Nomikos, P.: Multivariate Statistical Process Control of Batch Processes, Ph.D. transfer report, Dept. of Chem. Eng., McMaster University, Hamilton, Canada (1992).
14. Schuler, H. and C.U. Schmidt: Calorimetric State Estimators for Chemical Reactor Diagnosis and Control: Review of Methods and Applications, Chem. Eng. Sci., 47, 899-915 (1992).
15. Schuler, H. and K. de Haas: Semi-batch Reactor Dynamics and State Estimation, IFAC Symp. DYCORD'86, Bournemouth, UK, pp. 135-140, Pergamon Press (1986).
16. Shewhart, W.: Economic Control of Quality, Van Nostrand (1931).
17. Stephanopoulos, G. And K.Y. San: Studies on On-line Bioreactor Identification, I: Theory, Biotechnol. Bioengng., 26, 1176 (1984).
18. Wold, S., K. Esbensen and P. Geladi: Principal Component Analysis, Chemometrics and Intelligent Laboratory Systems, 2, 37-52 (1987).
19. Wu, R.S.: Dynamic Thermal Analyser for Monitoring Batch Processes, Chem. Eng. Progress, Sept. 1985, pp. 57-61 (1985).

Tendency Models for Estimation, Optimization and Control of Batch Processes

Christos Georgakis

Chemical Process Modeling and Control Research Center and Department of Chemical Engineering. Lehigh University, Bethlehem, PA 19015, USA

Abstract: This paper summarizes recent progress in the area of estimation and control of batch processes. The task of designing effective strategies for the estimation of unmeasured variables and for the control of the important outputs of the process is linked to our need to optimize the process and its success is depended upon the availability of a process model. For this reason we will provide a substantial focus on the modeling issues that relate to batch processes. In particular we will focus attention on the approach developed in our group and referred to as ***"tendency modeling"*** that can be used for the estimation, optimization and control of batch processes. Several batch reactor example processes will be detailed to illustrate the applicability of the general approach. These relate to organic synthesis reactors and bioreactors. The point that distinguishes tendency modeling from other modeling approaches is that the developed Tendency Models are multivariable, nonlinear, and aim to incorporate all the available fundamental information about the process through the use of material and energy balances. These models are not frozen in time as they are allowed to evolve. Because they are not perfectly accurate they are used in the optimization, estimation and control of the process on a tentative basis as they are updated either between batches or more frequently. This iterative or adaptive modeling strategy also influences the controller design. The controller performance requirements and thus the need of a more accurate model increase as successive optimization steps guide the process operation near its constraints.

Keywords: Tendency modeling, batch reactors, process control, state estimation, process optimization

Introduction

Batch processing is an important segment of the chemical process industries. A growing proportion of the world's chemical production by volume and a larger proportion by value is made in batch plants. In contrast to continuous processes, batch processes related to the production of fine and specialty chemicals, pharmaceuticals, polymers, and biotechnology are characterized by the largest present and future economic growth among all sections of the chemical industry [1]. This trend is expected to continue as industry pursues the manufacture of low volume, high value added chemicals, particularly in developed countries with few indigenous raw materials.

In comparison to continuous processes, batch processes are characterized by a greater flexibility of operation and a rapid response to changing market conditions. Typically, a single piece of process equipment may be used for manufacturing a large variety of products utilizing several different unit operations such as reactors, distillation columns, extraction units, etc. As a result, batch plants have to be cycled frequently and monitored carefully, thereby, requiring higher labor costs per unit volume of product throughput. At the same time most of the batch processes, particularly those related to the production of fine and specialty chemicals, are characterized by significant price differences between the reactants and the products.

Unlike continuous processes, batch processes seldom operate at steady state. This results in a lack of reproducibility. Most batch processes suffer from batch-to-batch variation in the quality of the product due to imprecise measurement and control of operating conditions. In the case of continuous processes, the off-specification material produced during the start-up and transient operation of the plant can be blended with the good products during normal operation. Batch processes do not enjoy this luxury. The batch process is in transition most of the time, and the limited time cycle of a batch does not allow for many corrective actions. If the product resulting from a batch is not of desired quality, it usually has to be discarded. Because the added value of the product in relationship to the reactants is so high, the economics of process improvements are dependent more on whether the batch made an acceptable product rather than whether the amount of energy or reactants used was the minimum possible. There are significant economic benefits which can be realized from the optimization of a great variety of batch and semi-batch processes. The challenge, however, is different from that of the traditional and continuously operated chemical industry.

Many batch processes are characterized by small annual volumes of production. Frequently, the annual requirement for a particular product can be manufactured in a few weeks. The plant is then adapted, and if necessary, re-configured to produce the next product. This makes the development of detailed models for each process or product economically unattractive. The frequent process changes that characterize this technology seldom provide enough time for the development of any model at all. In the absence of such a systematic organization of our knowledge of the process through a process model, the operation is quite sub-optimal. On-line information is limited to the available process measurements and the most often used control action is temperature and/or pressure control. In some rare cases, the controller might utilize the energy balance of the unit [20, 21]. In almost all cases, one lacks any information as to how much the process operation can be improved. Any improvement attempts have to be based on a trial and error approach without the benefit of the quantitative suggestions of a model.

In the chemical and pharmaceutical industries, emphasis is presently placed on the quality control of the product. For example, in emulsion polymers the demand for special properties and improved performance has recently led to the increased interest in the more detailed understanding of the inner workings of the process. In the area of bioreactors, the need for substantial on-line information is presently needed to ensure that each batch meets regulatory guidelines and provides a reproducible product quality. Such time dependent knowledge can induce improvements in the quality of the product produced and can also help increase the process productivity. The more readily available time dependent information are those provided through the on-line measurements of the process. These measurements might not be directly related to product quality, necessitating the off-line measurements at the end of the batch. At that time, one can find if the product quality is appropriate but can do nothing to correct an undesirable situation. For this reason, an on-line estimation of the quality related parameters is needed through the help of a process model. Since the development of a detailed and accurate model might be uneconomical, one important issue to consider is how accurate the model needs to be to achieve the appropriate estimation of the unmeasured quality variables.

To account for some of these difficulties the investigation of a general purpose modeling, estimation, optimization, and control strategy has been initiated by some researchers [13, 14, 15, 25, 26, 28, 34] which is directly applicable to several batch processes. This strategy aims to properly account for our lack of detailed process knowledge and the great variety of batch or semi-batch reactors. At the same time it takes advantage of new plant data collected during the process operation to update the model as many times as necessary. The iterative modeling and optimization methodology proposed initially by Filippi et al. [14, 15], as well as the alternative approach proposed by Hamer [17] have been modified and extended to develop a more complete methodology as detailed by Rastogi et al. [28, 31]. For example, a systematic procedure was proposed, based on statistical design of experiment techniques for determining preliminary experimental runs that will provide the data for initialization of the modeling and optimization cycle.

In the following sections, we will detail a comprehensive framework needed for the modeling, optimization and control of batch processes based on the use of a tendency model of the process. Since such a model is not always very accurate, emphasis will be placed on the model updating strategy. For this reason, we have called the proposed strategy "Tendency Modeling, Optimization and Control". In defining such strategy, we have tried not to be constrained by present practice and hardware limitations. On the other hand, we feel that the strategy defined is a realistic and achievable one. Several successful industrial applications of the approach have provided considerable positive feedback.

Even though all the technical details concerning such a strategy are not presently resolved, substantial progress has been achieved. The following sections will provide a brief overview of the most recent progress.

Generic Characteristics of TeMOC

In this section, we will examine generic research issues related to the proposed strategy of tendency modeling, optimization and control of batch processes. These include modeling, state estimation, control, and optimization. The overall Tendency Modeling, Optimization and Control (TeMOC) strategy can be summarized in the diagram of Figure 1. This diagram describes the comprehensive structure of the proposed approach for the modeling, optimization, and control of chemical reactors or other units. Activities involved in each of the schematic boxes will be summarized here. The central box in the middle of the diagram depicts the process. Both on-line and off-line measurements are assumed available from the process. Because measurements are not always correct and reliable, two additional boxes could have been added to denote that some data reconciliation and gross error detection algorithm should be employed right after the measurements become available. Since off-line measurements are not as frequent and are characterized by substantial time delays between the time the measurement is made and the

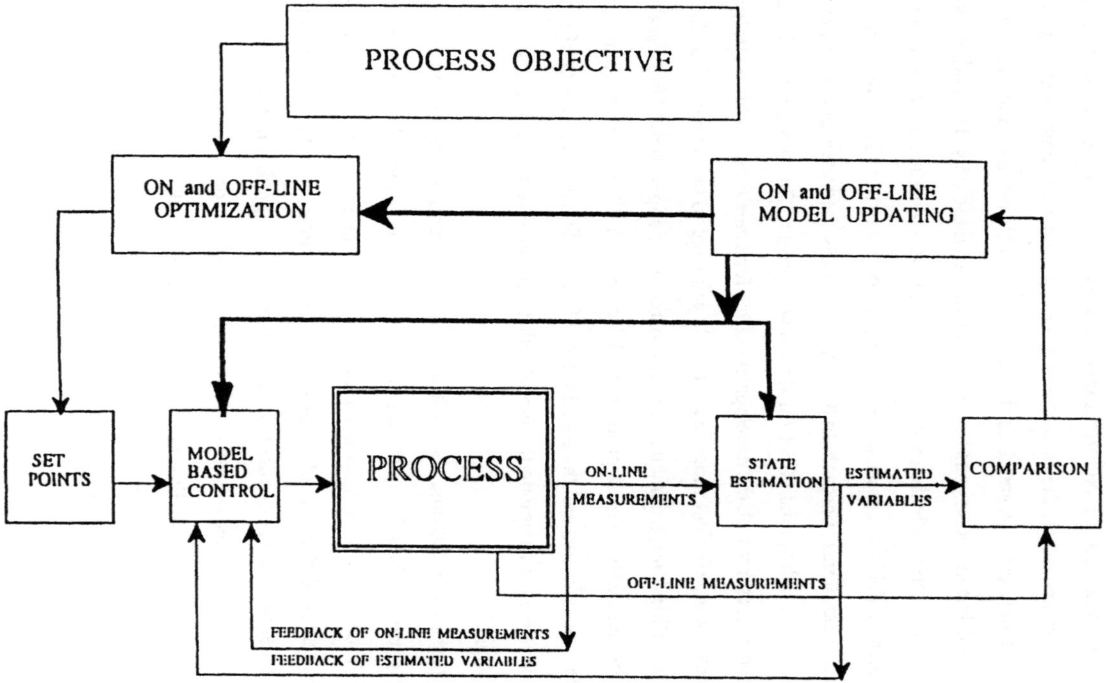

Figure 1: Comprehensive Schematic Diagram of the TeMOC Methodology

laboratory results are available, an important task, denoted by the state estimation box, has been introduced. This algorithm utilizes the on-line measurements and a dynamic or steady state model of the process to estimate process variables that are not available on-line. Such variables can and should include the off-line measurements but are not limited to these only. Depending on how this algorithm is designed, it can handle the existence of noise in the on-line measurements and can also cope with a dynamic model of the process that is not perfectly accurate. The existence of a model is critical for the success of this approach. In the past the models used were accurate first principle ones and thus required substantial effort for their development. This has limited the use of this technique to applications for which the development of the fundamental model is a straightforward task. In the future, more efficient modeling methods will need to be developed to facilitate the application of state estimation techniques. In most batch plant applications, one usually expects that there will be substantial initial process-model mismatch. This can be quantified by the difference between the estimated and the actual values of the off-line measurements of the process. This difference can be used to update the model either off-line or on-line. We will return to this point in a short while.

On the top part of the diagram, the indicated task represents the definition of the process objectives that are used in the subsequent task of process optimization. The optimization task is also achieved with the use of an appropriate process model. The results of the optimization task are expressed in terms of desired set point or set point profiles with time. The model based controller ensures that these set points are indeed met during the operation of the process. For this purpose the model base controller will utilize, in general, both direct on-line measurements as well as those that are estimated on-line through the state estimation task.

It is quite obvious that there are three important tasks where some model of the process is utilized: State Estimation, Control and Optimization. Since the model development and updating activities are time consuming and technical manpower intensive, there is a very substantial incentive to initially develop and subsequently update the same model for the control, estimation and optimization purposes.

In the following sections, we will first describe the progress made to date on the methodological issues of the TeMOC approach. We will also comment on the additional challenges that need to be met in order to further the comprehensive character of the strategy described above. After this has been achieved we direct attention to some references where application examples demonstrate the successful use of the proposed approach.

Modeling

Despite the central role that individual batch unit operations, and in particular batch reactors, play in the overall performance of an industrial batch process, their description in the form of a mathematical model is either nonexistent or not very accurate. While steady state models for continuous units are widely used for the design and operation of such processes, the development and use of models for batch processes is not as widely practiced. Since batch process units are never in steady state, the necessary models need be dynamic ones quantitatively describing the time evolution of the unit or process. Dynamic models have presently started to become widely used in studies of continuous processes. Small efficiency improvements in such processes can result in economical benefits that more than compensate for the development costs of the process model. Many other chemical and specialty chemical processes have not widely benefited from such dynamic models because of the apprehension that the development cost of a specific process model might outweigh the potential process benefits. This apprehension is also present in batch processes.

To make models more widely available, one needs to address and resolve two issues. The first issue relates to the cost of model development: It should be reduced! The second issue relates to the effective use of the developed model: It should be used for more than one purpose, i.e., optimization, estimation of unmeasured variables, as well as control. Reduction of model development costs can be achieved by development of a systematic modeling approach that we have called "Tendency Modeling". Past modeling strategies, usually for continuous processes, implied that the model was developed once and remained valid for a substantial part of its process life. This was justified because processes did not change their operational characteristics very often and the amount of available on-line data were not substantial. These assumptions are no longer valid. Nowadays, the operation of many continuous processes varies week by week and even day by day. This was and is true with batch processes, but it is quickly becoming true for continuous processes as well. As feed-stock characteristics and product specifications change, the model structure and its parameters also need to change. A model must be flexible enough to adapt to these changes. Meanwhile, the wide use of digital control computers in plants has availed a much larger number of experimental data that can be used to update and improve the model.

The proposed "Tendency Modeling" approach that has been developed over the past few years can serve as the primary vehicle for answering the modeling needs for batch and other processes. It is based on the following main principles:

* i) The initial and subsequent versions of the "Tendency Model" should involve a set of material and energy balances of the unit. For this purpose utilize as much initial information as is available on the reaction stoichiometry, chemical kinetics, heat and mass transfer, thermodynamics, and mixing characteristics of the unit. If the initial information is not sufficient then design and perform additional experiments to collect the necessary information.
* ii) Estimate the accuracy of the model and take this information into account in all possible uses of the model, as for the design of estimation, optimization, and/or control algorithms.
* iii) As new experimental data become available from the operation of the unit, update the model parameters as well as its structure by the effective use of all relevant on-line and off-line process data.

In contrast to previous ones, the proposed "Tendency Modeling" approach is an evolutionary one and aims to systematize the ad hoc model updating that is widely used in industry today. A simpler version of such an evolutionary modeling approach has been used in the past in a different and restrictive manner - adaptive control. However, the types of models used in adaptive control are linear input/output dynamic models that do not provide any fundamental knowledge about the internal workings of the process. Because of their restrictive character, these models have not been used for purposes other than control. They are only linear and cannot be used for optimization; their input/output nature make them inappropriate for estimation of unmeasured parameters. There is also some similarity between the proposed "Tendency Modeling" approach and evolutionary optimization (EO) in that the model is used for the optimization of the process and that it evolves with time. On the other hand, evolutionary optimization uses input-output statistical and often static models compared to the dynamic and nonlinear models that are used in the Tendency Modeling approach. Further more, evolutionary optimization models do not utilize any of our knowledge of the inner workings of the process. This means that EO utilizes identical statistical input-output models for the optimization of a batch reactor as for the optimization of a batch distillation column. Models used in evolutionary optimization cannot be used for the estimation of unmeasured variables or for the control of the unit.

The proposed "Tendency Modeling" methodology aims to develop nonlinear models based on material and energy balances around the unit operation of interest, such as a chemical reactor. The nonlinear character of these models enables their use in optimizing the reactor or process. Since these models are developed by writing material and energy balances, they provide the possibility of estimating process variables that are not measured on-line. Periodic updating of tendency models, either on-line or off-line, will combine and enhance the advantages of their fundamental character by continuously increasing their accuracy. There are generic research challenges that the tendency modeling approach must still resolve to become a useful and successful tool in all process examples. They are summarized in the following open research questions:

* i) When is parameter updating of the nonlinear tendency model sufficient and how does one select the parameters that should be updated?
* ii) When is it necessary to update the structure of the model, i.e., change the reaction rate expression, consider partial mixing, or introduce heat or mass transfer limitations?

While methods for parameter updating of linear models have been considered in the literature, work needs to be done for updating the structure of the process model using nonlinear models. Here updating of the model structure implies a discrete change of one model to another in order to take more detailed account of, for example, imperfect mixing or mass (and/or heat) transfer limitations.

State Estimation

State estimation techniques for linear models were first introduced by Kalman [22, 23] about thirty years ago. They were extended to nonlinear systems and found their most extensive use in aerospace control applications. It is not totally clear why such technologies have not yet found a more extensive use in chemical process applications. One can argue that the model developmental cost in the aerospace industry is distributed over a number of identical products (e.g. airplanes), while chemical plants are usually one of a kind. One can further speculate that the nonlinear character of chemical processes has been an inhibiting factor. Unlike aerospace applications, the available model of a chemical process is not always as accurate. Recent research on the control of emulsion copolymerization [7, 8, 9, 10, 11] has demonstrated that the use of a less-than-perfect model can lead to successful estimation of unmeasured parameters-with substantial economic benefit resulting from the simultaneous increase in the product quality and process productivity. The ever-increasing emphasis nowadays on controlling product quality variables that cannot be measured on-line further increases the need for successful application of state estimation techniques.

The following issues need to be addressed to achieve significant progress in the use of state estimation techniques:

* i) Practicing research engineers need to be more effectively informed about the power and limitations of the method. Tutorial papers need to be written, simple processes examined as examples, and demonstration software provided for application engineers to develop an intuitive rather than mathematical understanding of the power of the method.
* ii) The success of existing state estimation techniques depends on the accuracy of the model, which needs to be examined in a more systematic and generic fashion. Motivated by the specific application results in the reactor control area, this activity shows promise for a significant technical contribution. Understanding what model accuracy is needed to make state estimation successful will further enhance the applicability of this technique to chemical processes.
* iii) Methods need to be developed that utilize the observed process model mismatch, usually called the innovation residual, to help in the on-line updating of the process model and further enhance the usefulness of state estimation techniques. Model updating can mean parameter updating or updating the form of the model. Parameter adaptive state estimation techniques can be used in updating the model when we are certain which major parameters are to be updated. However, new techniques are needed for updating the model structure. One such technique is to perform two state estimations in parallel with two different models in order to see which one provides a more confident estimation of the unmeasured variables.
* iv) In most of the past research activities one has assumed that a dynamic model is needed for the estimation of the unmeasured variables from the ones that are measured on-line. Because a steady state model is easier to develop than a dynamic model one needs to examine when this simplification can be made without a substantial sacrifice in the accuracy of the estimate. Furthermore one needs to explore the development of input-output models between measured and estimated variables so that the dependence of the estimation task on a fundamental model of the process is not as critical.

Control

In the design of control strategies for batch processes, one needs to first focus on the selection of the proper variables to be controlled. As in most other processes, the objective of the controller is to ensure that the desired quality of the product is achieved in spite of disturbances that might enter the process. Since the final product quality is affected by the operation of several units, many of which are downstream from the one under consideration, it is not easy to define and select the quality related variables that need to be controlled in each unit. Furthermore, even if these variables are identified it might not be possible to measure then on line. For example, one knows very well that the quality of emulsion polymers is often dependent on the molecular weight, which is impossible to measure on-line. This then necessitates the development of an estimation algorithm discussed in the previous section. If the expected process and products improvements are substantial, then the recommended approach is to design and implement an estimation algorithm and then control the estimated variables. In many cases, the possible process or product improvements have not been considered and estimated. The easy way out then is to accept that we can only control the variables that we can directly measure. It is often hoped that this will indirectly help in the control of the product quality and in some cases it does. However, it is quite possible that the maximum process benefits might not be achieved through this approach.

Once the selection of the controlled and manipulated variables is made, the remaining task is to design the controller strategy. The challenges here are that the relationship between manipulated and controlled variables is almost always nonlinear and often quite multivariable. The controller design becomes more challenging when more than one variable related to product quality, such as polymer composition, particle size, and molecular weight, is controlled simultaneously. Often the challenge becomes even greater if temperature runaway, due to the exothermicity of the reaction, can lead to an explosion. In this case, the possibly more urgent task of temperature control must be coordinated with the economically more important product quality control strategy. This is not an easy task in many applications, such as bulk polymerization or certain organic synthesis reactions, because of the substantial interactions between temperature and compositions. Many of the model based control strategies that have been developed for the control of continuous processes could be utilized in the control of batch processes. Their use of the available model of the process in the prediction of the future differences between the measured variables and their appropriate set points could be an effective way to design the controller algorithm. The major limitation that can be cited is that most, but not all, model predictive control strategies utilize a linear model of the process. Because batch processes are nonlinear in their dynamics, substantial room exists for the application of nonlinear model predictive control strategies. One can mention the use of the Reference System Control (RSC) strategy that has been proposed by Bartusiak et al. [3, 4], and further examined by Bartee et al. [2]. This is an almost identical strategy to the one proposed by Lee and Sullivan [24] and often referred to as Generic Model Control.

Optimization

Batch process optimization is the final and most important task that needs to be undertaken in order to improve performance of the process. Such optimization tasks can be divided into two broad and overlapping categories. The first one deals with the optimization of the operation of the process unit. The second category of optimization challenges relates to the optimal scheduling of the different unit tasks to perform the overall process objective. While the second class of problems is very important as well, we will focus attention here onto the first class. The issues examined here with respect to the optimization of the operation of each batch units, are also of relevance to the optimal scheduling of the overall process.

Improvements in the process will either reduce operating costs, equivalently increase the productivity, or increase the quality of the product produced, or both. A substantial number of mathematical optimization techniques are available in the literature [5, 18, 19, 33, 12, 6] and can be readily utilized if an accurate model is available. One needs to also mention the substantial progress made recently by the application of Sequential Quadratic Programming on the operation of batch processes. In this case only a tendency model will be available, and process optimization will be performed concurrently with efforts to increase the accuracy of the model. Our interests then should be to develop algorithms that will ensure simultaneous convergence of model updating and process optimization tasks. We need to develop algorithms which determine the global, rather than local, process optimum and result in the largest process improvement. To achieve this, one needs to develop a strategy that decides whether the next batch run should be used to either improve the process or increase the model's accuracy. As Rippin, Rose, and Schifferli [32] have demonstrated, the optimization of the process through an approximate model can be trapped into a local optimum. These authors have also proposed an extended design procedure which continues model parameter improvement with performance verification in the neighborhood of the predicted optimum. While achieving the global rather than the local optimum is an important issue, one should keep in mind that by guiding three or four process units to some local optimum might be more beneficial rather than guiding one process unit to its global optimum. Nevertheless, the impact of the accuracy of the model on the process optimum needs to be studied further and we might consider denoting such a challenge as "Tendency Optimization".

Example Process Applications

To properly elucidate the arguments provided above, one needs to also refer to some specific example applications of the proposed Tendency Modeling approach. We will provide here some comments about the application to organic synthesis reactions done as part of the doctoral thesis by Rastogi [28, 30, 31]. The reaction considered was the epoxidation of oleic acid to epoxide and the evolution of the Tendency Model and the optimization of the process are summarized in Figure 2. Here the value of an economic performance index ($/day/liter processed) is plotted against the number of experimental runs. The initial eight runs were designed by a factorial design of experiments procedure [28] to provide the initial data needed to start the Tendency Modeling approach. All experiments indicate that operation of the process was not economical resulting in a loss rather than a profit. It is worth also mentioning that all experiments were operated in a batch mode, all reactants were fed in the reactor at initial time. Utilizing these data, the first version of the tendency model (M0) of three reactions and power-law kinetics was identified. Because a negative reaction order was calculated on the oleic acid reaction, it was decided to feed the oleic acid in semibatch mode [30]. The optimization of the next batch through this model predicted a possible profit of 73.4 $/day/liter but the corresponding experiment only achieved a profit of 6.9 $/day/liter. One needs to remark that inaccuracies of the M0 model are to blame for the substantial difference between the predicted and the experimentally achieved value of the profit function. At the same time one can easily observe that this tendency model, however inaccurate, guided the process to profit making operation (Run 9) as compared to the previous eight runs. With this additional data and a closer look at the process model mis-match, the structure of the kinetic equations was changed [29, 30] and Model M1 was obtained. Optimization of the next batch run through this model led to the prediction that a profit of 74.2 $/day/liter can be achieved. Experimental run 10 implemented the optimal profile of model M1 and resulted in a profit of 53.2 $/day/liter. With these additional experimental data, Tendency Model M2 was obtained by refitting the parameter of model M1. With experimental Run 11, it was shown that both model predictions and experimental data converged very close to each other successfully ending the cycle of tendency model updating and process optimization.

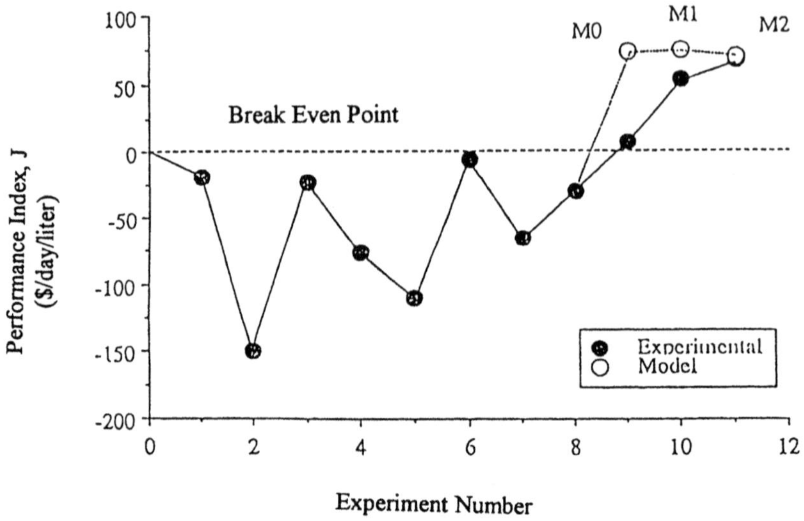

Figure 2. Summary of the Evolution of the Performance Index with Experimental Number

An additional and successful comparison of the overall approach to experimental data of a process of industrial interest was also presented by Marchal [27]. The applicability of the Tendency Modeling approach to the case of bioreactors has been addressed by Tsobanakis et al. [35]. Rastogi [29] also reports of a very successful application to an industrial process of Air Products and Chemicals, Inc. Productivity was increased by 20.

One might end by making the comment that while additional industrial applications are expected to be undertaken and completed in the future, the challenge of further extending the methodology to more quantitatively account for the accuracy of the model is a real and worthy one.

References

1. E. Anderson. Specialty chemicals in a mixed bag of growth. Chem. Eng. News, 20,1984
2. J.F. Bartee, K.F. Bloss, and C. Georgakis. Design of nonlinear reference system control structures. Paper presented at AIChE National Meeting, San Francisco, 1989
3. R.D. Bartusiak, M.J. Reilly, and C. Georgakis. Designing nonlinear control structures by reference system synthesis. Proceedings of the 1988 American Control Conference, Atlanta, Georgia, June 1988,
4. R.D. Bartusiak, M.J. Reilly, and C. Georgakis. Nonlinear feedforward/feedback control structures designed by reference system synthesis. Chem. Eng. Sci., 25,1989
5. Denbigh. Optimal temperature sequence in chemical reactors. Chem. Eng. Sci., 8:125-132, 1958
6. M.M. Denn. Optimization by Variational Methods. Robert E. Krieger Publishing Company, 1978
7. J. Dimitratos, M. El-Aasser, A. Klein, and C. Georgakis. Composition control and kalman filtering in emulsion

copolymerization. Proceedings at the 1988 American Control Conference, Atlanta, Georgia, 1988
8. J. Dimitratos, M. El-Aasser, A. Klein, and C. Georgakis. Control of product composition in emulsion copolymerization. Proceedings, 3rd International Workshop, Polym. React. Eng., Berlin, Sep. 27-29, 1989
9. J. Dimitratos, M. El-Aasser, A. Klein, and C. Georgakis. Digital monitoring, estimation and control of emulsion copolymerization. Proceedings of the 1989 American Control Conference, Pittsburgh, PA, 1989
10. J. Dimitratos, M. El-Aasser, A. Klein, and C. Georgakis. Dynamic modeling and state estimation for an emulsion copolymerization reactor. Comp. Chem. Engng., 13:21-33, 1989
11. J. Dimitratos, M. El-Aasser, A. Klein, and C. Georgakis. An experimental study of adaptive kalman filtering in emulsion copolymerization. Chem. Eng. Sci., 46:3203-3218, 1991
12. T.F. Edgar and D.M. Himmelblau. Optimization of Chemical Processes. McGraw-Hill, 1988
13. C. Filippi, J. Bordet, J. Villermaux, S. Marchal-Brassely, and C. Georgakis. Batch reactor optimization by use of tendency models. Comp. and Chem. Engng., 13:35-47, 1989
14. C. Filippi, J.L. Graffe, J. Bordet, J. Villermaux, J.L. Barney, P. Bonte, and C. Georgakis. Tendency modeling of semibatch reactors for optimization and control. Chem. Eng. Sci., 41:913, 1986
15. C. Filippi-Boissy. PhD thesis, L' Institute National Polytechnique de Loraine, Nancy, France, 1987
16. V. Grassi. Communication at pmc's industrial advisory committee meeting, October. 1992
17. J.W. Hamer. Stoichiometric interpretation of multireaction data; application to fed-batch fermentation data. Chem. Eng. Sci., 44:2363-2374, 1989
18. J. Horak, F. Jiracek, and L. Jezova. Adaptive temperature control in chemical reactors. a simplified method maximizing productivity of a batch reactor. Czech. Chem. Comm., pages 251-261, 1982
19. H. Horn. Feasibility study of the application of self-tuning controllers to chemical batch reactors. Oxford Univ. Lab Report, 1978
20. M.R. Juba and J.W. Hamer. Process and challenges in batch process control. Paper presented at the third International Conference on Process Control, 1986
21. A. Jutan and A. Uppal. Combined feedforward-feedback servo control scheme for an exothermic batch reactor. Proc. Des. Dev., 23:597-602, 1984
22 R.E Kalman. A new approach to linear filtering and prediction problems. J. basic Eng, March:35-46, 1960
23. R.E. Kalman and R.S. Bucy. New results in linear filtering and prediction theory. J. Basic Eng., March:95-108, 1961
24. P.L. Lee and G.R. Sullivan. Generic model control- theory and applications. Paper presented at IFAC Workshop, June 1988, Atlanta, 1988
25. S. Marchal-Brassely. PhD thesis, L' Institute National Polytechnique de Loraine, Nancy, France, 1990
26. S. Marchal-Brassely, J. Villermaux, J.L. Houzelot, J.L Barnay, and C. Georgakis. Une methode interative efficace d'optimisation des profils de temperature et de debit d'alimentation pour la conduite optimale des reacteurs discontinus. Proc.of 2eme Congres Frcncais De Genie Des Procedes, Toulouse, France, pages 441-446, 1989
27. S. Marchal-Brassely, J. Villermaux, J.L. Houzelot, and J.L. Barney. Optimal operation of a semi-batch reactor by self-adaptive models for temperature and feed-rate profiles. Chem. Eng. Sci., 47:2445-2450, 1992
28. A. Rastogi. Evolutionary optimization of batch processes wing tendency models. PhD thesis, Lehigh U., 1991
29. A. Rastogi. Personal communications, September. 1992
30. A. Rastogi, J. Fotopoulos, C. Georgakis, and H.G. Stenger. The identification of kinetic expressions and the evolutionary optimization of specialty chemical batch reactors using tendency models. Paper presented at 12th Int. Symposium for Chemical Reaction Engineering, Torino, Italy. Also in Chem. Eng. Sci., 47:2487-2492., 1992
31. A. Rastogi, A. Vega, C. Georgakis, and H.G. Stenger. Optimization of catalyzed epoxidation of unsaturated fatty acids using tendency models. Chem. Eng. Sci., 45:2067-2074, 1990
32. D.W.T. Rippin, L.M. Rose, and C. Schifferli. Non-linear experimental design with approximate models in reactor studies for process development. Chem. Eng. Sci., 35:356-363, 1980
33. C.D. Siebenthal and R. Aris. Studies in optimization vii. the application of Pontryagin's method to the control of a batch and tubular reactor. Chem. Eng. Sci., 19:747-746, 1964
34. P. Tsobanakis, S. Lee, J. Phillips, and C. Georgakis. Adaptive stoichiometric modeling and state estimation of batch and fed-batch fermentation processes. Presented at the 1989 Annual AIChE Meeting, San Francisco, 1989
35. P. Tsobanakis, S. Lee, J. Phillips, and C. Georgakis. Issues in the optimization, estimation, and control of fed-batch bioreactors using tendency models. 5th Int. Con. on Computer Applications in Fermentation Tech. and 2nd IFAC Sym. on Modeling and Control of Biotechnology Processes, Keystone, Colorado, March 29 - April 2, 1992

Control Strategies for a Combined Batch Reactor/Batch Distillation Process

Eva Sørenson and Sigurd Skogestad

Department of Chemical Engineering, University of Trondheim-NTH, N7034 Trondheim

Abstract: A batch reactor may be combined directly with a distillation column by distilling off the light component product in order to increase the reactor temperature or to improve the product yield of an equilibrium reaction. The controllability of such a system is found to depend strongly on the operating conditions, such as reactor temperature and composition of distillate, and on the time during the run. In general, controlling the reactor temperature (one point bottom control), is difficult since the set point has to be specified below a maximum value in order to avoid break-through of heavy component in the distillate. This maximum value may be difficult to know a priori. For the example considered in this study control of both reactor temperature and distillate composition (two-point control) is found to be difficult. As with one point bottom control, the reactor temperature has to be specified below a maximum value. However, energy can be saved since the vapor flow, and thereby the heat input to the reactor, can be decreased with time. Controlling the temperature on a tray in the column (one point column control) is found to give the best performance for the given process with no loss of reactant and a high reactor temperature although no direct control of the reactor temperature is obtained.

Keywords: Reactive batch distillation, controllability, control strategies

1 Introduction

Batch distillation is used in the chemical industry for the production of small amounts of products with high added value and for processes where flexibility is needed, for example, when there are large variations in the feed composition or when production demand is varying. Batch reactors are combined with distillation columns to increase the reaction temperature and to improve the product yield of equilibrium reactions in the reactor by distilling off one or more of the products, thereby driving the equilibrium towards the products.

Most often the control objective when considering batch processes is either i) to minimize the batch time or ii) to maximize the product quality or yield. Most of the papers published on batch distillation focus on finding optimal reflux ratio policies. However, sometimes the control objective is simply to obtain the same conditions in each batch. This was the case for the specific industrial application which was the starting point for our interest in this problem and which is to be presented later.

Few authors have considered the operation of batch distillation with chemical reaction although these processes are inherently difficult to control. The analysis of such systems in terms of controllability has so far only been considered by Sørensen and Skogestad [11]. Roat *et al.* [8] have developed a methodology for designing control schemes for continuous reactive distillation columns based on interaction measures together with rigorous dynamic simulation. However, no details about their model were given.

Modelling and simulation of reactive batch distillation has been investigated by Cuille and Reklaitis [2], Reuter *et al.* [7] and Albet *et al.* [1]. Cuille and Reklaitis [2] developed a model and solution strategies for the simulation of a staged batch distillation column with chemical reaction in the liquid phase. Reuter *et al.* [7] incorporated the simulation of PI-controllers in their model of a batch column with reaction only in the reboiler. They stated that their model could be used for the investigation of control structure with the aid of Relative Gain Array analysis (RGA) but no details were given. Albet *et al.* [1] presented a method for the development of operational policies based on simulation strategies for multicomponent batch distillation applied to reactive and non-reactive systems.

Egly *et al.* [3], [4] considered optimization and operation of a batch distillation column accompanied by chemical reaction in the reboiler. Egly *et al.* [3] presented a method for the optimization of batch distillation based upon models which included the non-ideal behavior of multicomponent mixtures and the kinetics of chemical reactions. The column operation was optimized by using the reflux ratio as a control variable. Feeding one of the reactants during the reaction was also considered. In a later paper [4], they also considered control of the column based upon temperature measurements from different parts of the column. The optimal reflux ratio policy was achieved by adjusting the distillate flow using a non-linear control system. However, no details where given about neither the column/reactor nor the control system.

The purpose of this paper is to investigate the possible difficulties in controlling a coupled system of a reactor and a distillation column, and also to give some alternative control strategies based on an industrial example. First, a model of the industrial process, consisting of a batch reactor with a rectifying column on top, is developed. Based on a linearized version of this model, we compare different operating points to show how the model differs, that is, whether the same controller settings can be used for different reactor conditions or reactor temperatures. In the various operating points we also consider the stability of the system and the response to step changes in flows. We consider *two-point control*, when both the top and the bottom part are controlled, as well as *one point control*, when only one part of the column/reactor is controlled. A Relative Gain Array (RGA) analysis is used for the investigation of control structures in two point control. Finally, the similarities and differences between our process and a conventional continuous distillation column is considered.

The reaction is reported to be of 0'th order and due to limited data, we also assume the rate to be independent of temperature. However, interesting observations can still be made concerning the coupling between the formation of product in the reboiler and the separation in the column above. Indeed, later work [6], has confirmed that this simplification does not affect the conclusions. The influence of disturbances on the system, e.g. in reaction rate or in temperature measurements, has not been considered in this study.

Column:	6 trays + condenser
Reaction:	$0.5\ R_1 + 0.36\ R_2 + 0.14\ R_3 \rightarrow P(s) + W$
Volatile components:	W ($T_b = 100\ ^0C$) and R_2 ($T_b = 188\ ^0C$)
Non-volatile components:	R_1 ($T_b = 767\ ^0C$), R_3 ($T_b = 243\ ^0C$) and P (solid)
Vapor pressure:	R_1: $\ln P^s_{R_1} = -4009.3 + 176750.0/T_i + 6300.0 \log T_i - 0.51168 T_i$ (Pa)
	R_2: $\ln P^s_{R_2} = 25.4254 - 6091.95/(-22.46 + T_i)$ (Pa)
	R_3: $\ln P^s_{R_3} = 231.86 - 18015.0/T_i - 31.753 \log T_i + 0.025 T_i$ (Pa)
	W: $\ln P^s_W = 23.1966 - 3816.44/(-46.13 + T_i)$ (Pa)
Relative volatility (W and R_2):	8-32
Startup time:	30 min
Total reaction time:	15 hr
Pressure in column/reactor:	1 atm/1.2 atm
Reaction rate, r:	1.25 kmol/hr
Initial vapor flow, V:	16.8 kmol/hr
Hydraulic time constant, τ:	0.0018 hr =6.5 s
Initial holdups:	reactor: 24 kmol
	condenser: 1.6 kmol
	trays: 0.09 kmol
Initial amounts in reactor:	R_1: 10.4 kmol (Acid)
	R_2: 7.5 kmol (Alcohol) (20 % excess)
	R_3: 3.2 kmol (Alcohol)
	P: 0.0 kmol (Ester)
	W: 2.5 kmol (Water)

Table 1: Process data for simulation.

2 Process example

The motivation for this study was an industrial equilibrium esterification reaction of the type

$$\xi_1 R_1 + \xi_2 R_2 + \xi_3 R_3 \rightleftharpoons P(s) + W$$

where R_1 is a dibasic aromatic acid, R_2 and R_3 are glycols, P is the solid polymer product and W is the bi-product water. The reaction takes place in a reactor heated by a heating jacket with heat oil. The equilibrium is pushed towards the product side by distilling off the low boiling by-product W from the reactor. Only reactant R_2 and the by-product W are assumed to be volatile, and the binary separation between these two components takes place in the column. The reaction rate was reported to be of zero order; independent of compositions. Due to lack of data we also assume the rate is independent of temperature. A summary of the process data is given in Table 1. In the industrial unit the amount of reactant R_2 in the feed was 20 % higher than necessary to yield complete conversion of the reaction, and this was also assumed in most of our simulations. This was done to account for the possible loss of the reactant in the distillate.

The existing operating practice was to use *one-point top control*; the temperature at the top of the column T_T was kept constant at about 103 ^0C which gave a distillate composition of 0.004 (about 2 weight%) of the heavy component R_2 and thereby a loss of this component. The vapor flow was kept constant by using maximum heating of the

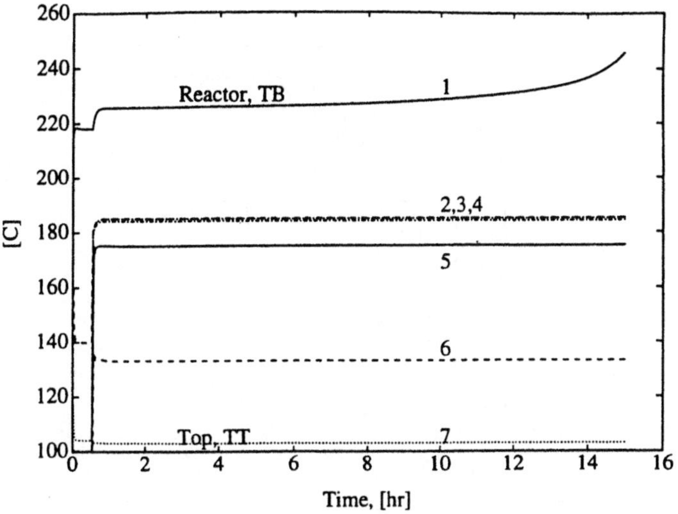

Figure 1: The existing temperature profile in column/reactor.

reactor and the condenser level was controlled by the distillate flow D. The temperature profile at different locations in the column as a function of time is given in Fig. 1. The reactor temperature T_B is almost constant at the beginning but increases as reaction proceeded. The conditions on tray 2, 3 and 4 are practically equal because the column has more stages than needed for the desired separation. With the existing control scheme there is no direct control of the reactor temperature, T_B and more severely, it gives a varying loss of the heavy component, reactant R_2. This leads to a varying quality of the product P between batches.

3 Mathematical model

In this section we consider the mathematical description of the batch distillation column and reactor shown in Fig. 2 and described in the previous section. The equations for the individual stages consist of the total mass balance, the mass balance for each component, tray hydraulics and phase equilibrium and are valid under the following assumptions:

A1 A staged model is used for the distillation column.

A2 A multicomponent mixture in the reactor, but a binary mixture in the distillation column is considered.

A3 Perfect mixing and equilibrium between vapor and liquid on all stages is assumed.

A4 The vapor phase holdup is negligible compared to the liquid phase holdup.

A5 The stage pressures and the plate efficiencies are constant.

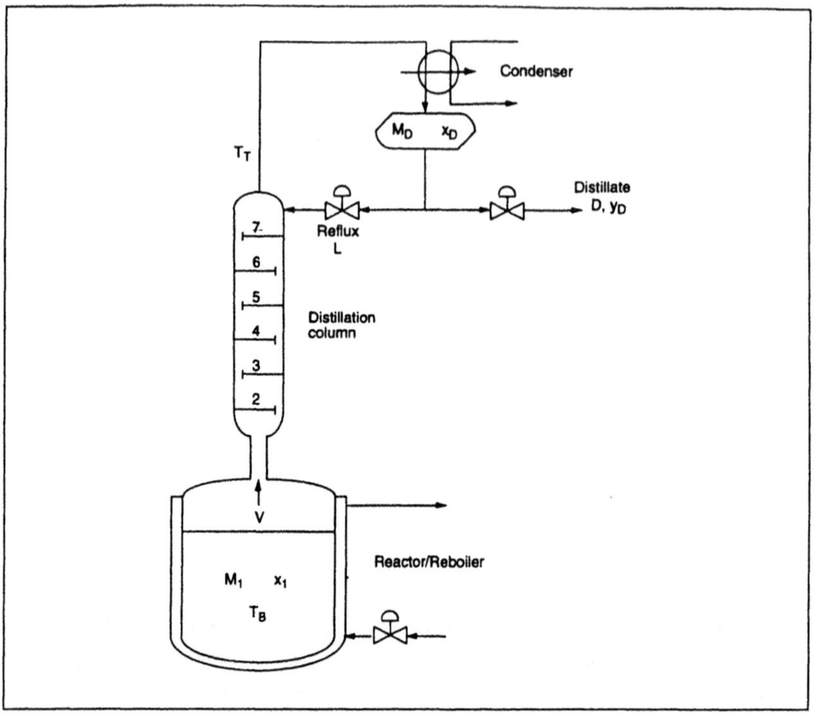

Figure 2: Batch distillation column/reactor.

A6 Constant molar flows are assumed (no energy balance).

A7 Linear tray hydraulics is considered.

A8 Total condensation with no subcooling in the condenser is assumed.

A9 The chemical reaction is limited to the reactor.

A10 Raoult's law for the vapor-liquid equilibrium holds.

Below i denotes the stage number and j the component number ($j = 1, 2$ are the volatile components W and R_2). The following differential and algebraic equations result.

reactor/reboiler, i=1 :

$$dM_1/dt = L_2 - V + \sum_{j=1}^{4} \xi_j r \tag{1}$$

$$d(M_1 x_{1,j})/dt = L_2 x_{2,j} - V y_{1,j} + \xi_j r \quad , j = 1 \tag{2}$$

reaction components not distilled :

$$d(M_1 x_{1,j})/dt = \xi_j r \quad , j = 3, 4 \tag{3}$$

column tray, i=2,N :

$$dM_i/dt = L_{i+1} - L_i \tag{4}$$

$$d(M_i x_{i,j})/dt = L_{i+1}x_{i+1,j} + Vy_{i-1,j} - L_i x_{i,j} - Vy_{i,j} \quad , j = 1 \tag{5}$$

condenser, i=N+1 :

$$dM_{N+1}/dt = V - L_{N+1} - D \tag{6}$$

$$d(M_{N+1}x_{N+1,j})/dt = Vy_{N,j} - L_{N+1}y_{D,j} - Dy_{D,j} \quad , j = 1 \tag{7}$$

linearized tray hydraulics :

$$L_i = L_{oi} + \frac{M_i - M_{oi}}{\tau} \tag{8}$$

liquid-vapor equilibrium :

$$y_{i,j} = \frac{\alpha_i x_{i,j}}{1 + (\alpha_i - 1)\, x_{i,j}} \tag{9}$$

relative volatility :

$$\alpha_i = f\,(T_i) = \frac{P_1^s\,(T_i)}{P_2^s\,(T_i)} \tag{10}$$

temperatures :

$$P_i = \sum_{j=1}^{2(4)} x_j P_j^s\,(T_i) \tag{11}$$

On each stage the composition of component $j = 2$ (R_2) is obtained from $\sum x_j = 1$. Note that all four components were used to calculate the reactor temperature using eq. 11, but only the two lightest components were considered in the column. The model is highly non-linear in vapor composition y and in temperature T. On vector form the differential equation system to be solved can be written

$$d\mathbf{x}/dt = \mathbf{f}[\mathbf{x}(t), \mathbf{u}(t)] \tag{12}$$

In addition there is a set of algebraic equations, equations (8)-(11)

$$0 = \mathbf{g}[\mathbf{x}(t), \mathbf{u}(t)] \tag{13}$$

Eq. (12-13) constitute a set of differential-algebraic equations (DAE). The equations are solved using the equation solver LSODE [5]. The startup conditions are total reflux and no reaction.

3.1 Linear model

In order to investigate the controllability of a process using available tools, a linear model is needed. Based on the non-linear model described by eq. (12) and (13) a linear model can be developed by linearizing the equation system at a given operating point. For continuous processes normally only one operating point considered; that of the steady state conditions. The linear model is then found by linearizing around this operating point and will be valid for small deviations from the steady state. When considering batch processes there is no such steady state; the conditions in the reactor or column are changing with time and the model is linearized along a trajectory. A linearized model of

Controlled variables ($\mathbf{y}$):	Manipulated variables ($\mathbf{u}$):
condenser holdup M_D	distillate flow D
distillate composition y_D	reflux flow L
reactor temperature T_B	vapor flow V

Table 2: Controlled and manipulated variables.

the process, representing deviations from the "natural drift" along the trajectory with D, L and V constant, can be described by the following equations:

$$\begin{aligned} d\mathbf{x}/dt &= A\mathbf{x} + B\mathbf{u} \\ \mathbf{y} &= C\mathbf{x} \end{aligned} \tag{14}$$

Where

$$\begin{aligned} \mathbf{x} &= [\Delta x_1, \Delta M_1.., \Delta n_3, ..]^T \\ \mathbf{y} &= [\Delta M_D, \Delta y_D, \Delta T_B]^T \\ \mathbf{u} &= [\Delta D, \Delta L, \Delta V]^T \end{aligned}$$

Laplace transformation yields:

$$\mathbf{y}(s) = G(s)\mathbf{u}(s) \tag{15}$$

The control problem will thus have the controlled and manipulated variables as given in Table 2. It is then assumed that the vapor flow V can be controlled directly, even though the real manipulated variable is the heat input to the reactor.

4 Analysis of linear model

4.1 Operating procedures

The linear model depends on the operating point. To study these variations we initially considered four different operating procedures:

I The existing operating practice, $T_T = 103^0$ C (one point top control, V constant)

II $T_B = 200^0$ C (one-point bottom control, V constant)

III $T_B = 222^0$ C (one-point bottom control, V constant)

IV $T_B = 228^0$ C (one-point bottom control, V constant)

Temperature profiles for the four operating procedures are given in Fig. 3. For operating procedure I, II and III the conditions are more or less constant with time, whereas procedure IV has a changing temperature profile with large variations at the beginning of the batch but more or less stabilizing midway through the batch. For operating point I ($T_T = 103\ ^oC$), the front between light and heavy component is kept high in the column giving a loss of the heavy component R_2. For procedure II ($T_B = 200\ ^oC$), the front is low and the composition of heavy component R_2 is almost negligible from tray 3 and up, giving a very pure distillate. However, the reactor temperature is low and it is unlikely

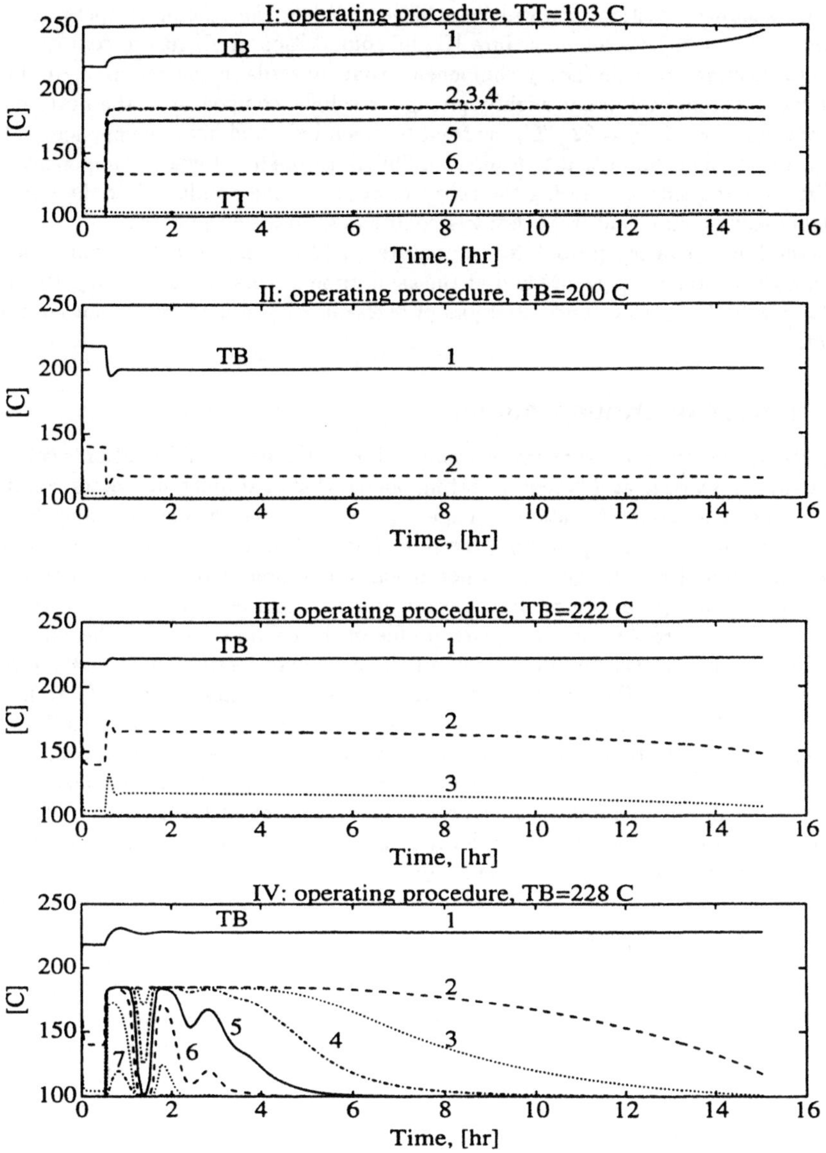

Figure 3: Temperature profiles in column/reactor for different operating procedures.

that the assumed reaction rate will be achieved. When the reactor temperature is increased to $T_B = 222\ ^0C$ for procedure III the composition of R_2 in the column section increases pushing the light/heavy component front upwards in the column. At the end of the batch the front detracts slightly giving more light component in the bottom part. For procedure IV, at $T_B = 228\ ^0C$, the front between light and heavy component is lifted so high up in the column that it leads to a "break-through" of heavy component R_2 in the distillate and thereby causing the large variations in the profile. After the loss of R_2 the light/heavy component front detracts continuously during the batch.

Of the four operating procedures, procedure III ($T_B = 222\ ^0C$) is the only one with both a high reactor temperature and at the same time no loss of reactant R_2. Procedure IV ($T_B = 228\ ^0C$) gives a substantial loss of reactant R_2 and is therefore not considered further.

4.2 Linear open-loop model

To illustrate how the process behavior changes during the batch the equation system (eq. 12 and 13) is linearized at different operating points; that is at different reactor conditions or times during a batch. (Notation: An operating point is specified as *procedure-time*, eg. I-8 is the conditions with operating procedure I after 8 hr reaction time.) These linear models were found by first running a non-linear simulation of the process with control loops implemented (level control in the condenser and temperature control of tray 1 or of the reboiler) in order to obtain a given profile in the column/reactor. The simulations were then stopped at the specified time, all the controller loops opened and the model linearized numerically. (We would get the same responses, in Δy_{DH} and ΔT_B from steps in ΔL and ΔV if the condenser level loop was to remain closed with the distillate flow D during the linearization. This is because L and V have a direct effect on compositions and the effect of the level loop is a second order effect which vanishes in the linear model.) The resulting linear model is thus an *open loop description* of the process at the given time and conditions; it describes how the system responds to changes when no controllers (or only the level controller) are implemented.

4.3 Step responses

To illustrate how the process behavior changes with conditions in the reactor we consider step changes to the linearized models. The effect of a step in the vapor flow V on y_{DH} and T_B (deviation from nominal value) for three different operating procedures after 8 hr is given in Figure 4. The variation of the linear model within batch III is illustrated by Fig. 5. The responses in y_{DH} for different reactor conditions (top part of Fig. 4) are similar but differ in magnitude. This is because in operating point II-8, where we have a very low reactor temperature, we have a very pure distillate. The increase in reflux will only increase the purity marginally. Whereas in operating point I-8, we have a distillate which is less pure so the increase will be larger. We note from Fig. 5 that the variations within the batch are large for the response in reactor temperature. The main reason for the changes in the responses for this temperature (lower part of Fig. 4 and 5) is the concentration of water in the reactor. That is, a higher water concentration gives a larger effect.

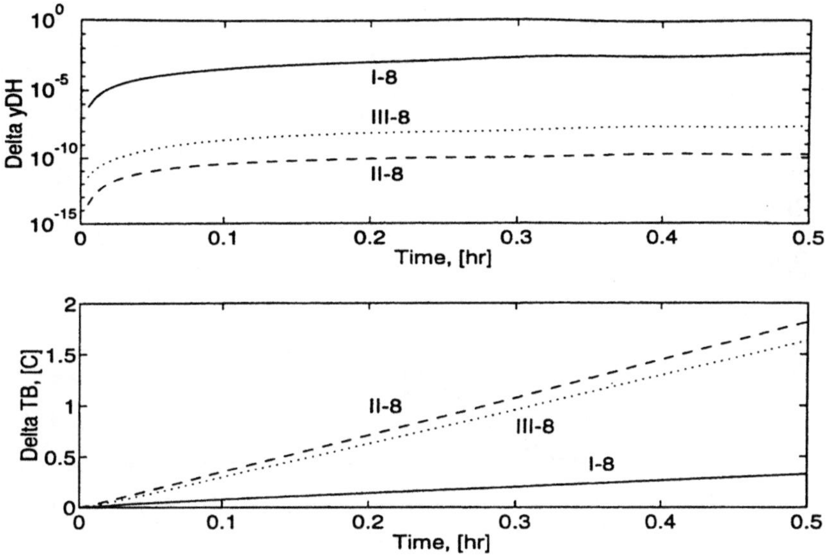

Figure 4: Step in vapor flow ($\Delta V = 0.1$) for linear model: effect on Δy_{DH} and ΔT_B for procedure I-8 ($T_T = 103\ ^0C$), II-8 ($T_B = 200\ ^0C$) and III-8 ($T_B = 222\ ^0C$).

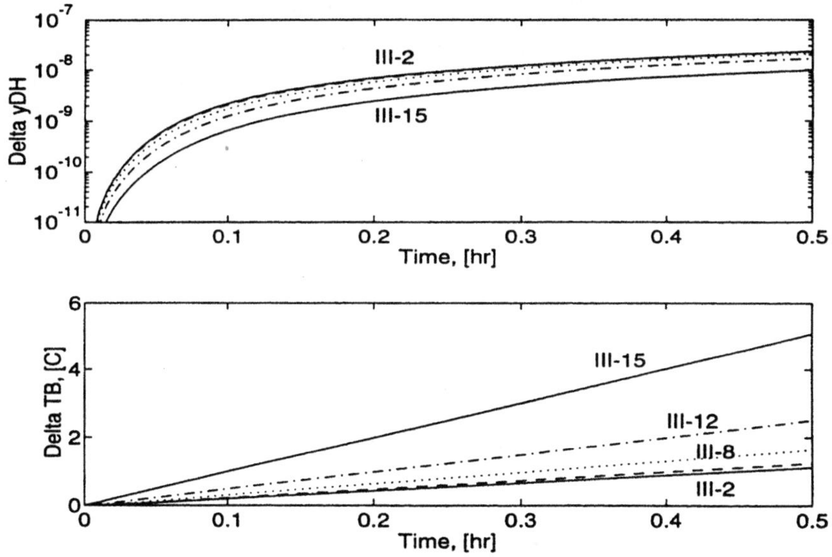

Figure 5: Step in vapor flow ($\Delta V = 0.1$) for linear model: effect on Δy_{DH} and ΔT_B for procedure III-2 to III-15 ($T_B = 222\ ^0C$).

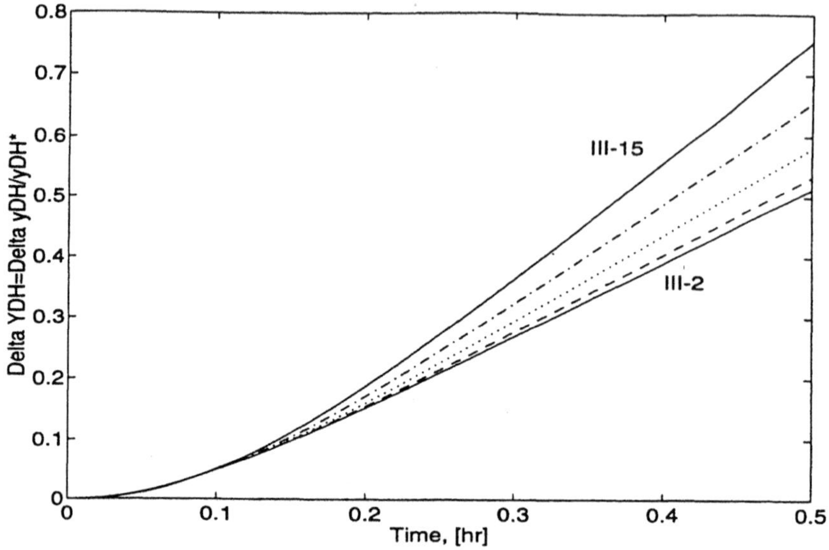

Figure 6: Logarithmic transformation for linear model: Different times during batch for procedure III ($T_B = 222\ ^0C$).

4.4 Reducing the non-linearity for top composition

An interesting feature in Fig. 4 and 5 is that the responses in y_{DH} to step changes have a similar initial shape on a log-scale. This is actually a general property for distillation [9]. The inherent nonlinearities in this variable can therefore be reduced by using a log transformation on the distillate composition y_D:

$$Y_D = -\ln(1 - y_D) \tag{16}$$

which in deviation variables becomes

$$\Delta Y_D = \frac{\Delta y_D}{1 - y_D^*} \quad , \quad \Delta Y_{DH} = \frac{\Delta y_{DH}}{y_{DH}^*} \tag{17}$$

The responses in mole fraction of heavy component R_2 in the distillate after the transformation, Y_{DH}, is given in Fig. 6 for operating points III-2 to III-15. These responses are of the same order of magnitude and the non-linearity is thereby reduced. From Fig. 4 and 5 there is no obvious transformation that can be suggested to deal with the non-linear effect for the reactor temperature.

5 Control strategies

The varying loss of reactant R_2 in the distillate and the lack of direct control of the reactor temperature were the major problems with the existing operating practice. In the control part of this study the following control strategies are compared:

- one-point bottom control (controlling the reactor temperature directly)
- two-point control (controlling both the distillate composition and the reactor temperature)
- one-point column control (controlling the temperature on a tray in the column)

The control parameters for the PI-controllers used in the simulations are given in Table 3. Note that an integral time of $\tau_i = 0.1hr = 6min$ was used in all the simulations and that the transformed variable Y_D was used instead of y_D for two-point control.

level control:	$K_p = -500$ and $\tau_i = 0.1$ ($M_D \rightarrow D, L$)
bottom control:	$K_p = 1.0$ and $\tau_i = 0.1$ ($T_B \rightarrow L$)
two-point control:	$K_p = 0.456$ and $\tau_i = 0.1$ ($Y_D \rightarrow L$)
	$K_p = -4.0$ and $\tau_i = 0.1$ ($T_B \rightarrow V$)
column control:	$K_p = 1.0$ and $\tau_i = 0.1$ ($T_5 \rightarrow L$)

Table 3: Control parameters used in the simulations.

5.1 One-point bottom control

The objective with one-point bottom control is to keep the reactor temperature constant at the highest possible temperature as this will maximize the rate of reaction for a temperature dependent reaction. The reflux flow is used as manipulated variable and the vapor flow is kept at its maximum value ($V = V_{max} = 16.8$ kmol/hr). However, it is very difficult to achieve a high value of T_B and at the same time avoid "break-through" of the heavy component R_2, in the distillate. This is illustrated in Fig. 7 which shows how the mole fraction of R_2, y_{DH}, changes when the set point for the temperature controller in the reactor increases from $T_{B,set} = 224.5^0C$ to $T_{B,set} = 225^0C$. An increase of 0.5^0C causes the mole fraction of reactant R_2 to increase by a factor of 25. The loss of reactant is only temporary, and y_{DH} is reduced to ≈ 0 after about 1 hr. The break-through is caused by the fact that when the specified temperature is above a certain maximum value where most of the light component W is removed, then a further increase is only possible by removing the heavy component, reactant R_2. If the set point temperature is specified below the maximum value, in this case $\approx 224.0^0C$, good control of the system ($T_B \approx T_{B,set}$ and $y_{DH} \approx 0$) is achieved. The system can, however, become unstable at the end of the batch depending on the choice of control parameters in the PI-controller. This due to the non-linearity in the model causing the system to respond differently to changes at different times during the batch as illustrated in Fig. 5.

Another alternative for raising the reaction temperature, and thereby the reaction rate for a temperature dependent reaction, is to let the set point follow a given trajectory, e.g. a linear increase with time. Again, the maximum reactor temperature to avoid break-through will limit the possible increase and break-through is inevitable if it is specified too high. Fig. 8 illustrates a run when the set point follows a linear trajectory from 220^0C at $t = 0.5$ hr to 245^0C at $t = 15$ hr. The loss of reactant R_2 is substantial, almost 10 % of the feed of this component. By lowering the endpoint temperature to 230 ^0C, loss of reactant is avoided (not shown).

5.2 Two-point control

By using *two-point control* it may be possible to control both the top and the bottom part of the distillation column by implementing two single control loops in the system. In this way energy consumption can be reduced since it will no longer be necessary to keep the vapor flow V, and thereby the temperature or amount of heating oil, at its maximum value. In the case of the esterification process, it is desirable to control not only the reactor temperature T_B but also the composition of the distillate y_D, i.e. the loss of reactant R_2.

Two different control configurations are considered for the batch column:

LV-configuration Controlling the condenser level using the distillate flow D leaving the reflux flow L and the vapor flow V to control the distillate composition Y_D and the reactor temperature T_B:

$$\begin{array}{ccc} M_D & \longleftrightarrow & D \\ Y_D, T_B & \longleftrightarrow & L, V \end{array}$$

DV-configuration Controlling the condenser level using the reflux flow L leaving the distillate flow D and the vapor flow V to control the distillate composition Y_D and the reactor temperature T_B:

$$\begin{array}{ccc} M_D & \longleftrightarrow & L \\ Y_D, T_B & \longleftrightarrow & D, V \end{array}$$

5.2.1 Controllability analysis of two point model

Open-loop step responses for both configurations are given in Fig. 9 and 10 for operating point III-8 ($T_B = 222\ ^0C$ at $t = 8hr$). The term "open-loop" should here be put in quotes because we are not talking about an uncontrolled column, but assume that the condenser level is perfectly controlled ($M_D \longleftrightarrow D$ or $M_D \longleftrightarrow L$) and we consider the effect of the remaining independent variables on the composition and reactor temperature.

From Fig. 9 it can be seen that for the LV-configuration the responses to steps in L and V are similar but in opposite direction. For the DV-configuration the responses by a step in D are similar as for the step in V for the LV-configuration. However, the responses to a step in V is very small. This is a general property for distillation.

In a distillation column there are large interactions between the top and the bottom part of the column, a change in the conditions in one end will lead to a change in the other end as well. Because of these interactions a distillation column can be difficult or almost impossible to control. The interactions in a system can be analyzed by various tools (see e.g. Wolff [12]), amongst them the RGA, or Relative Gain Array. Systems with no interactions will have an RGA-value of 1. The larger the deviation from 1, the larger the interaction and the more difficult the process is to control. Pairing control loops on steady-state RGA-values less than 0 should be avoided.

The magnitude of the 1,1-element of the RGA for both the LV- and DV-configuration is given as a function of frequency in Fig. 11 for operating procedure III-8 ($T_B = 222^0C$). From the figure it can be seen that for the LV-configuration the RGA is very high at low frequencies (when the system is approaching a steady state). This shows that the interaction reduce the effect of the control input (L, V) and make control more difficult.

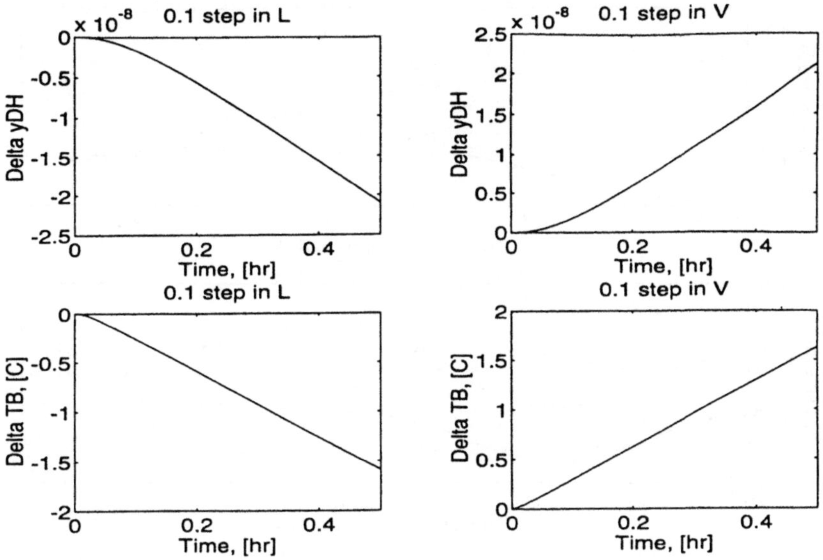

Figure 9: Linear open-loop step responses for LV-configuration for operating point III-8.

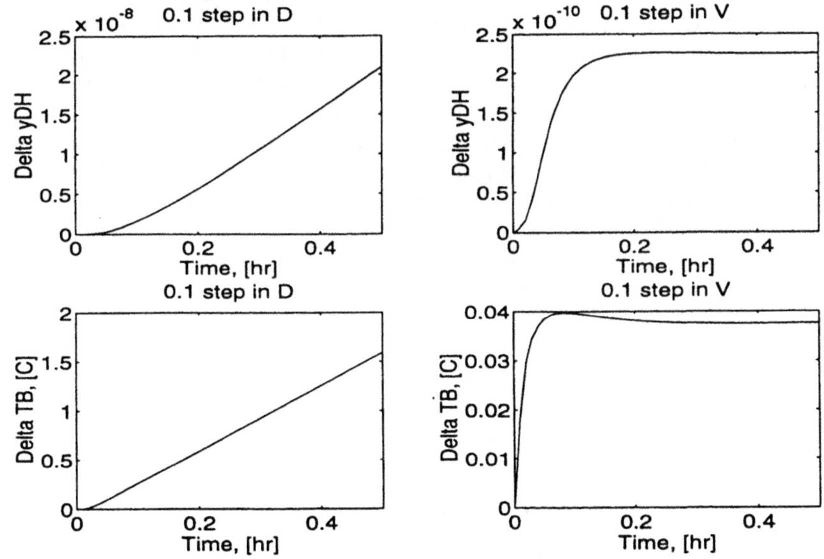

Figure 10: Linear open-loop step responses for DV-configuration for operating point III-8. (Note that the y-axis scaling is 100 times smaller for changes in V).

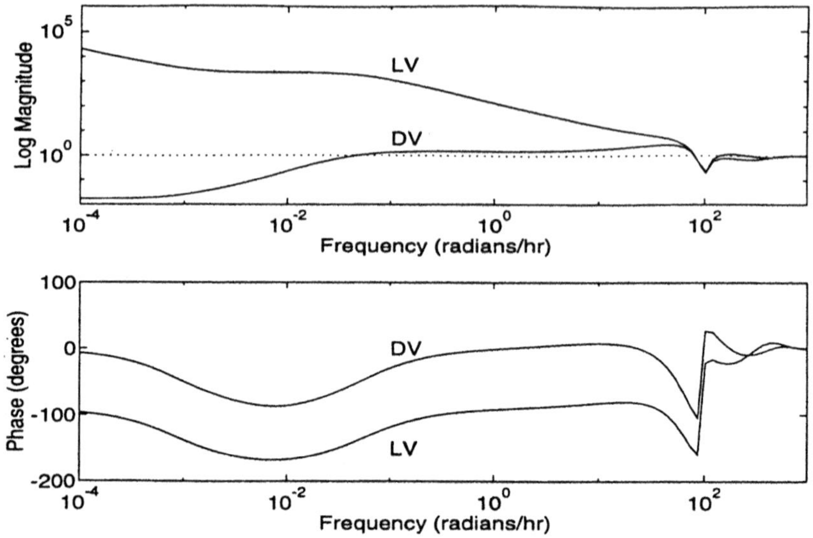

Figure 11: RGA for LV- and DV-configuration for linear model in operating point III-8.

RGA for DV is generally lower at all frequencies. This difference between configurations is the same as one would observe in a continuous distillation column.

However, the control characteristics from the RGA-plot for the LV-configuration are not quite as bad as it may seem. For control the steady-state values are generally of little interest (particularly in a batch process since the process will never reach such a state), and the region of interest is around the system's closed-loop bandwidth (response to changes), which is in the frequency range around 10 rad/hr (response time about 6 min). We note that the RGA is closer to 1 here and that the difference between the two configurations is much less. From the high-frequency RGA, which is close to 1, we find that for decentralized control, the loop pairing should always be to use the vapor flow V to control the reactor temperature T_B and either the reflux flow L or the distillate flow D to control the distillate composition or the loss of reactant R_2, Y_D. This is in agreement with physical intuition.

$$\begin{aligned} T_B &\longleftrightarrow V \\ Y_D &\longleftrightarrow L, D \\ M_D &\longleftrightarrow L, D \end{aligned}$$

5.2.2 Non-linear simulation of two-point model

Closed-loop simulations confirm that two-point control may be used if fast feedback control is possible. However, as in the case for one-point bottom control, we still have the problem of specifying a reasonable set-point for the bottom temperature to avoid breakthrough of reactant R_2 in the distillate. An example of two-point control of the process using the LV-configuration is given in Figure 12 with the following set point for the controllers: $T_{B,set} = 225\ ^0C$ and $y_{DH,set} = 0.0038$. (Note that we control the transformed distillate composition Y_D instead of y_D in order to reduce the non-linearity in the model.) It can be seen that only a minor break-through of reactant occurs during the run. The reactor temperature T_B is kept at its set point while the distillate composition y_{DH} is slightly lower than its set point showing that it is difficult to achieve tight control of both ends of the column at the same time. It should also be noticed how the vapor flow decreases with time which shows that energy can be saved using two-point control. Control using DV-configuration give similar results (not shown).

5.3 One-point column control

In the existing operating practice the temperature at the top of the column was controlled. The set point was 103 ^0C which gave a composition of 0.4 % of reactant R_2 in the distillate. By lowering the set point to e.g 100.1 ^0C the distillate would be purer, but the column would become very sensitive to measurement noise, and this system would not work in practice.

One alternative is to measure the composition y_D and use this for feedback. However, implementing an analyzer (or possibly an estimator based on the temperature profile) is costly and often unreliable. A simpler alternative is to place the temperature measurement further down in the column, e.g. a few of trays below the top tray, since this measurement will be less sensitive to noise. In this investigation the temperature on tray 5 is chosen as the new measurement to be used instead of the one on the top tray. The vapor flow is kept fixed at its maximum value ($V = V_{max} = 16.8$ kmol/hr). With this control configuration ($T_5 \leftrightarrow L$) there is no direct control of the reactor temperature. However, with an appropriate choice of set point, $T_{5,set}$, loss of reactant R_2 could easily be avoided and one of the main causes of the operability problems thereby eliminated.

The temperature profile for one-point column control with set point $T_{5,set} = 130\ ^0C$ is shown in Fig. 13. The conditions are "stable" (i.e. no break-through of reactant R_2) throughout the batch. The reactor temperature increases towards the end and the mole fraction of heavy component in the distillate y_{DH} is less than or equal to 0.0001 at all times. Also note that this control procedure with V fixed at its maximum will yield the highest possible reactor temperature. This may be important in some cases when the reaction is slow.

6 Reducing amount of reactant

The proposed operating procedure with one-point column control gives a lower reactor temperature than the existing one-point top control procedure with $T_T = 103\ ^0C$. In the existing procedure the amount of reactant R_2 in the feed is about 20 % higher than needed for the reaction and all the above simulations were based on this. This is done to account for the loss of the reactant during the run. By using one-point column control with $T_5 = 130\ ^0C$, loss of reactant can be avoided and the surplus of R_2 is therefore not needed. By removing the excess 20 % of the reactant from the feed (such that the initial charge of R_2 is 6.25 kmol) the obtainable reactor temperature increases by about 2^0C at the beginning of the batch to about 40^0C towards the end as illustrated in Fig.

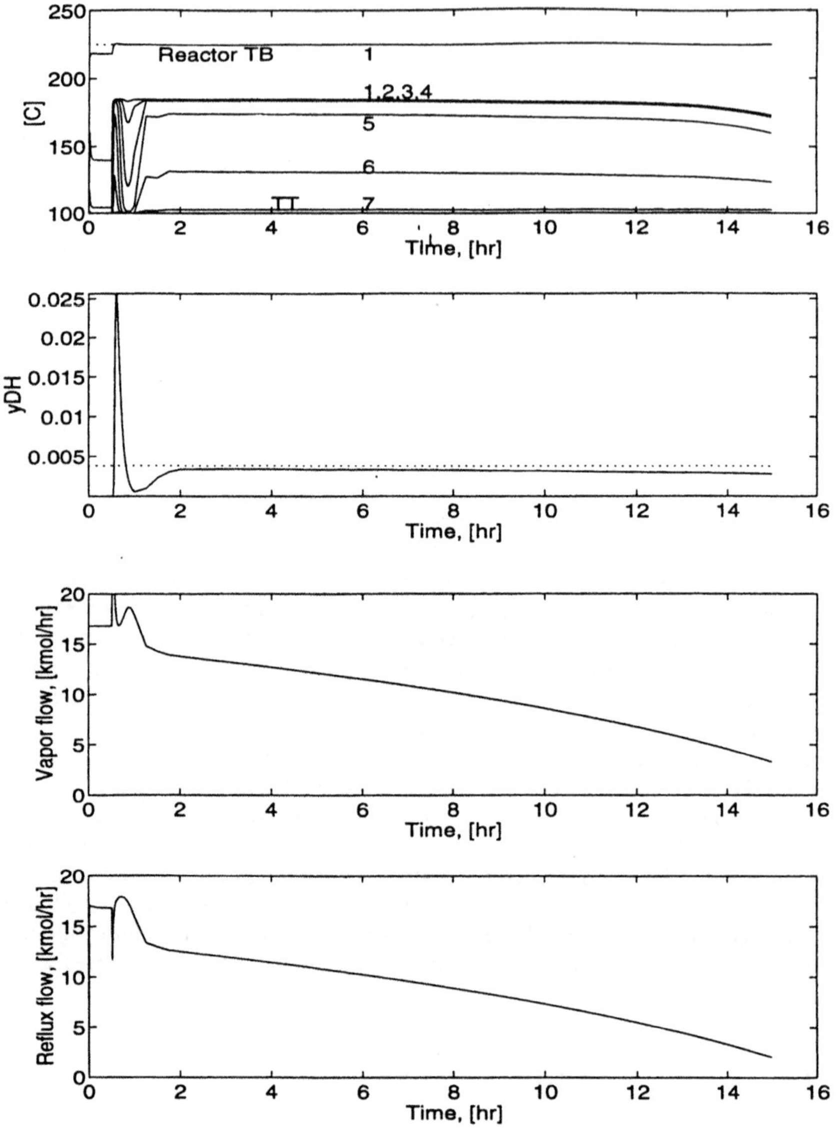

Figure 12: Two-point control. Temperature profile, distillate composition, vapor flow and reflux flow for LV-configuration with set points $T_{B,set} = 225\ ^0C$ and $y_{DH,set} = 0.0038$.

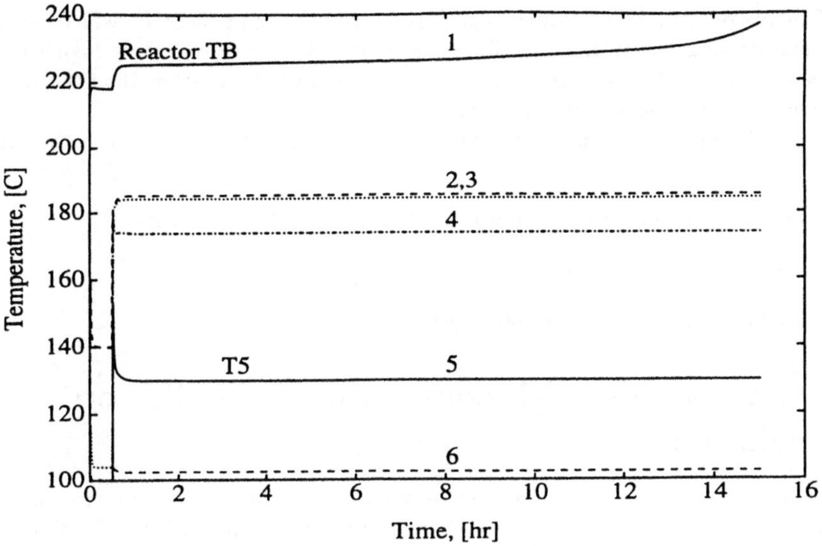

Figure 13: One-point column control. Temperature profile with $T_{B,set} = 130\ ^0C$.

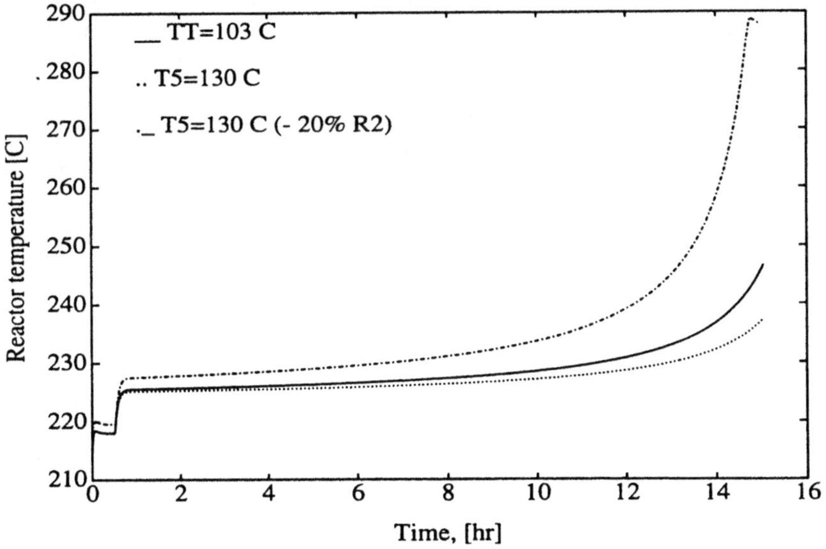

Figure 14: Effect of reducing the amount of reactant R_2 in the feed.

14. The reason for this is the high vapor pressure of the component R_2 which lowers the temperature as given by Eq. 11. Since the temperature is considerably higher towards the end of the batch when the excess R_2 is removed, the total batch time can therefore be reduced for a temperature dependent reaction.

In conclusion, by moving the location of the temperature lower down in the column, we

1. Increase the reactor temperature and thus reduce the batch time
2. Avoid loss of reactant R_2
3. Maintain more constant reactor conditions.

7 Comparison with conventional distillation columns

A comparison of our column with a conventional batch distillation column, shows significant differences in terms of control. For example, the common "open-loop" policy of keeping a fixed product rate (D) or reflux ratio (L/D) does not work for our column because of the chemical reaction (see also [6]). If the distillate flow D is larger than the amount of light component W formed by the reaction, the difference must be provided for by loss of the intermediate boiling reactant R_2. For optimal performance we want to remove exactly the amount of bi-product W formed. Therefore *feedback from the top is needed.* In fact, our column is very similar to a conventional continuous distillation column, but with the feed replaced by a reaction and with no stripping section.

By comparing our reaction batch column with a conventional continuous column we find that most conclusions from conventional columns carry over. As for a continuous column RGA(1,1) ≈ 0 at steady state (low frequency) for the DV-configuration for a pure top product column (see Fig. 11) implying that the reflux flow should be used to control the reactor temperature [10]. However, for control the pairing must be selected based on the RGA(1,1)-values around the bandwidth (10 rad/hr) implying that the vapor flow should always be used to control the reactor temperature for two-point control as was done in the simulations.

8 Conclusion

In this paper a dynamic model of a combined batch reactor/distillation process has been developed. Based on a linearized version of the model the controllability of the process depending on different reactor conditions and different times during a batch has been analyzed. The responses of the industrial example has been found to change considerably with operating point.

Controlling the reactor temperature directly using *one-point bottom control*, will give a more consistent product quality. However, since the response changes with time (gain between T_B and V), a non-linear controller might be needed to avoid instability. Moreover, because of the moving light/heavy component front in the column it is difficult to find the right set point temperature that does not give a break-through of heavy component in the distillate. This set point temperature will therefore in practice have to be specified low enough to ensure an acceptable performance.

Two-point control allows both the reactor temperature and the distillate composition to be controlled. By using two-point control energy will be saved compared with one-point control as the vapor flow can be reduced. However, one encounters the same problems of specifying the set point for the reactor temperature as for one-point bottom control.

The existing operating practice, controlling the temperature at the top of the column, is poor, sensitive to noise and leads to a varying loss of reactant R_2 and thereby varying product quality. The measuring point should therefore be moved from the top tray and further down in the column. The proposed new procedure of *one-point column control*, where the temperature on tray 5 is controlled, has several advantages:

- No loss of reactant R_2 (compared to controlling the top temperature)
- Need not worry about maximum attainable reactor temperature (compared to controlling the reactor temperature directly by one-point bottom control)
- No interactions with other control loops (compared to two point control)

With this new operating policy addition of excess reactant R_2 to the initial batch can be avoided. Thus, the batch temperature can be increased and the batch time thereby reduced.

NOTATION

A	system matrix	T_i	temperature, K
B	system matrix	T_b	boiling point, C
C	system matrix	T_B	reactor temperature, K
D	distillate flow, $kmol/hr$	T_T	temperature at top of column, K
$G(s)$	transfer function	$\mathbf{u}$	control vector
L	reflux flow, $kmol/hr$	V	vapor flow, $kmol/hr$
L_i	internal liquid flow, $kmol/hr$	$\mathbf{x}$	state vector
L_{0i}	initial liquid flow, $kmol/hr$	$x_{i,j}$	molfraction of light comp. (W) in liquid
M_i	liquid holdup, $kmol$	y_D	molfraction of light comp. (W) in distillate
M_B	liquid holdup in reactor, $kmol$	y_{DH}	molfraction of heavy comp.(R_2) in distillate $= 1 - y_D$
M_D	liquid holdup in cond., $kmol$		
M_{0i}	initial liquid holdup, $kmol$	Y_D	logarithmic molfraction of light comp. (W) in distillate $= -\ln(1 - y_D)$
P_i	pressure on tray i, Pa		
P_j^s	vapor pressure, Pa	$y_{i,j}$	molfraction of light comp. (W) in vapor
r	reaction rate, $kmol/hr$	$\mathbf{y}$	measurement vector

Greek letters		*scripts*	
α_i	relative volatility	i	tray number
Δ	deviation from operating point	j	component number
τ	hydraulic time constant, h^{-1}	*set*	set point
ξ_j	stoichiometric coefficient	*	nominal value

References:

1. Albet, J., J.M. Le Lann, X. Joulia and B. Koehret: "Rigorous Simulation of Multicomponent Multisequence Batch Reactive Distillation", Proc. COPE'91, Barcelona, Spain, 75-80 (1991).
2. Cuille, P.E. and G.V. Reklaitis: "Dynamic Simulation of Multicomponent Batch Rectification with Chemical Reactions", Comp. Chem. Engng., 10(4), 389-398 (1986).
3. Egly, H., V. Ruby and B. Seid: "Optimum design and operation of batch rectification accompanied by chemical reaction", Comp. Chem. Engng., 3, 169-174 (1979).
4. Egly, H., V. Ruby and B. Seid: "Optimization and Control of Batch Rectification Accompanied by Chemical Reaction", Ger. Chem. Eng., 6, 220-227 (1983).
5. Hindmarsh, A.C.: "LSODE and LSODI, two new initial value ordinary differential equation solvers", SIGNUM Newsletter, 15(4), 10-11 (1980).
6. Leversund, E.S. S. Macchietto, G. Stuart and S. Skogestad: "Optimal control and on-line operation of reactive batch distillation", Comp. Chem. Eng., Vol. 18, Suppl., S391-395 (1994) (supplement from ESCAPE'3, Graz, July 1993).
7. Reuter, E., G. Wozny and L. Jeromin: "Modeling of Multicomponent Batch Distillation Processes with Chemical Reaction and their Control Systems", Proc. CHEMDATA'88, Gothenburg, 322-329 (1988).
8. Roat, S.D., J.J. Downs, E.F. Vogel and J.E. Doss: " The integration of rigorous dynamic modeling and control system synthesis for distillation columns: An industrial Approach", Presented at CPC'3 (1986).
9. Skogestad, S. and M. Morari: "Understanding the dynamic behavior of distillation columnsn, Ind. & Eng. Chem. Research, 27(10), 1848-1862 (1988).
10. Shinskey, F.G: "Distillation Control", 2. ed., McGraw-Hill Inc., 1984.
11. Sorensen, E. and S. Skogestad: "Controllability analysis of a combined batch reactor/distillation process", AIChE 1991 Annual Meeting, Los Angeles, paper 140e (1991).
12. Wolff, E.A., S. Skogestad, M. Hovd and K.W. Mathisen: "A Procedure for controllability analysis", presented at the IFAC Workshop on interactions between process design and control, Imperial College, London, Sept. 6-8, (1992)

A Perspective on Estimation and Prediction for Batch Reactors

Mukul Agarwal

TCL, Eidgenossische Technische Hochschule, Zurich, 8092 CH

Abstract: Estimation of states and prediction of outputs for poorly known processes in general, and batch reactors in particular, have conventionally been approached using empiricism and experience, with apparently inadequate regard for the underlying reasons and structure. In this work, a consistent perspective is presented that clarifies some important issues, explains the causes behind some of the intuition-based tactics, and offers concrete guidelines for a logical approach to the estimation problem.

Keywords: Estimation, Identification, Prediction, Model-mismatch, Extended State

1 Introduction

The *science* of estimation using observers and filters originated from the needs of electrical and aeronautical applications [4,7]. In subsequent chemical-engineering applications, this science steadily metamorphosed into an art of estimation using such techniques as the Extended Kalman Filter and the Luenberger Observer. The more the chemical process differed in crucial aspects from the processes that originated the science, the more the engineer became an artist using whim, fancy, creativity, and trial-and-error to achieve a pleasing, or at least an acceptable, final result. As in any art, success came to be validated by the final result itself, regardless of how many unsatisfactory versions were discarded in the process or what kind of creative license was used to deviate from the established norm. Indeed, the creative twists and deviations took on a validity of their own and sneaked a place in the norm. Little surprise then that the art of estimation has its lore, which is inundated with *effect*-oriented prescriptions and claims that do not always have roots in any *cause* [1-3,5-8]. Some of these prescriptions and claims have evolved from empirical observation on numerous real and simulated applications; others have been borrowed directly from the science in spite of invalidity in process applications of the theoretical assumptions that they are based upon. For example:

- The distinction between parameters and states lies in their -dynamics.
- The innovation is an indicator of filter performance. Whether estimating parameters or states, strive to get the output residual to be a white, or at least a zero-mean, sequence
- The covariance of the state errors is an indicator of the accuracy of the obtained estimates.
- Inaccurately known parameters can simply be coestimated by inclusion in the extended state vector.
- Given the process and the model, the best tuning is fixed and only needs to be divined.
- Filters require a stochastic model, observers a deterministic model.

- The tuning of a state-space filter depends solely, or mainly, on the noises in the input and output signals.
- Estimators are tuned a priori without needing any measured data, except to deduce the noise level.
- When the covariance of the estimated states indicates filter divergence, increase the noise in the dynamic model equations.
- If the estimator does not work, try to somehow model the uncertainty.

Batch processes are characterized by strongly nonlinear and time-varying behavior. Batch reactors, in particular, are moreover difficult to model and do not permit easy on-line measurement of crucial properties such as concentration. Together these characteristics render estimation and prediction for batch reactors specially intractable, and use of much of the lore specially treacherous.

This presentation attempts to regard the popular art of estimation from a scientific perspective, in the hope of reinstating some science to guide the development of useful estimators and predictors, and to discourage the propagation of artistic ones.

2 Process and Model

A simple semi-batch-reactor model assuming a single second-order exothermic reaction and a cooling jacket that allows removal and measurement of the generated heat of reaction may take the form:

$$\frac{dc}{dt} = -kc^2 + F \tag{1}$$

$$q = (-\Delta H)kc^2 \tag{2}$$

where t is time, c unknown concentration of the limiting reagent, k known kinetic rate constant, F measured input flow rate of the limiting reagent, $(-\Delta H)$ known heat of reaction, and q measured rate of generated heat of reaction. This **model** corresponds to the general state-space form:

$$\frac{dx}{dt} = f(x, p, u) \tag{3}$$

$$y = h(x, p, u) \tag{4}$$

where c is the *conventional*, possibly extended, state that comprises all the estimated variables with both zero and non-zero dynamics, **p** is the *conventional* parameter that is known a priori, **u** the measured process input, **y** the measured process output, and **f** and **h** known general nonlinear functions.

The inevitable model-mismatch causes deviation between the above model and the true process. Regardless of the nature of the mismatch, the **true process** can be described as:

$$\frac{dx^*}{dt} = f(x^*, p^*, u) + e_x^* \qquad \textbf{(5)}$$

$$y^* = h(x^*, p^*, u) + e_x^* \qquad \textbf{(6)}$$

where $\mathbf{x}^*$, $\mathbf{p}^*$, and $\mathbf{y}^*$ are the true state, the true parameter, and the true output, respectively; $\mathbf{e}_x^*$ and $\mathbf{e}_y^*$ are the errors in the dynamic and measurement equations, respectively; and the input u and the functions **f** and **h** are identical to those used in the model.

3 Errors e_x* and e_y*

The errors $\mathbf{e}_x^*$ and $\mathbf{e}_y^*$, appearing in the process equations 5 and 6 describe the entire deviation due to model-mismatch, and are time-dependent, in general. The mismatch could be due to structural incompleteness in the model equations 3 and 4, due to discrepancy between the model parameters **p** and the true parameters **p***, due to inaccuracy of the initial-state value available to the model, due to noise in the measurements of **u** and **y**, due to unmeasured disturbances, and due to possible approximations such as linearization that might be involved in any application of the model equations. All these sources of mismatch are incorporated in the instantaneous deviation represented by $\mathbf{e}_x^*$ and $\mathbf{e}_y^*$ which are unknown by definition.

At any given time, the estimation of **x** using the model equations requires a prescription of the relative instantaneous values of the unknown errors $\{\mathbf{e}_x^*, \mathbf{e}_y^*\}$. The prescribed values, $\{\mathbf{e}_x, \mathbf{e}_y\}$, are not necessarily sought to lie closest possible to the true $\{\mathbf{e}_x^*, \mathbf{e}_y^*\}$. The end-effect of this prescription[1] is essentially to single out, at any given time, one particular estimate from an infinite number of possible, equally valid estimates that comprise the space bounded by all cases of extreme prescriptions (i.e., each element of $\{\mathbf{e}_x, \mathbf{e}_y\}$ being zero or very large).

Since the algorithms proposed in the literature for estimation cannot directly use prescribed values of $\{\mathbf{e}_x, \mathbf{e}_y\}$, this prescription is invariably made in an indirect manner, and has taken a variety of forms. The most popular or well-known form is utilized by the Kalman Filter algorithm, which specifies $\{\mathbf{e}_x, \mathbf{e}_y\}$ to be zero-mean white noises with possibly time-varying covariances. Prescription of $\{\mathbf{e}_x, \mathbf{e}_y\}$ is then realized through prescription of the covariances. In theory, these covariances could be, and indeed for poorly modeled batch processes must be, different at each different time. But, in practice, there is rarely enough prior reason or knowledge to prescribe each covariance as more time-variant than one or a few step-wise constant values over the entire run.

[1] What is commonly called "tuning" of an estimator is consistently referred to in this work as "prescription of $\{\mathbf{e}_x, \mathbf{e}_y\}$" in order to emphasize its origin and to distinguish it from the true errors $\{\mathbf{e}_x^*, \mathbf{e}_y^*\}$ in the dynamic and measurement equations.

In the Kalman Filter algorithm, the prescribed covariances affect the resultant instantaneous estimate via a series of steps involving an array of intermediate matrices and vectors, each of which has a qualitatively foreseeable effect on the resultant outcome. Prescription of $\{\mathbf{e_x}, \mathbf{e_y}\}$ could therefore, instead of being made through prescription of the covariances, be delegated just as easily to prescription of any of the intermediate matrices and vectors [5]. Numerous ad-hoc methods have been devised to do just that, e.g., covariance resetting, high-gain filters, or gain bounding.

Another popular device effects the $\{\mathbf{e_x}, \mathbf{e_y}\}$ prescription by restricting the time duration over which the model equations are deemed to be valid [1,2]. By limiting the data memory using exponential forgetting or a moving window, these methods collapse the multidimensional $\mathbf{e_x}$ space to a single, albeit in general time-varying, dimension. The attendant simplification in prescribing $\{\mathbf{e_x}, \mathbf{e_y}\}$ comes at the cost of a restricted space of possible outcomes attainable by using the reduced-dimension prescription. An even more indirect form of prescribing $\{\mathbf{e_x}, \mathbf{e_y}\}$ is commonly used by observers, through the convergence speed of the estimates.

Regardless of which indirect form of prescribing $\{\mathbf{e_x}, \mathbf{e_y}\}$ is used, it serves, in essence, to "tune" the attained instantaneous estimate to any location within the space comprised by all possible prescriptions. Not surprising then is abundance in the literature of excellent estimates or predictions obtained both in simulations and in experiments, where the employed prescription is either not reported or simply stated without justification. In these cases, the excellent results could readily be achieved by accordingly tuning the prescription off-line on the application run or on past runs. In other cases, the process is so well known that either the dynamic model in equation 3, or the measurement model in equation 4, or both, are nearly perfect so that $\mathbf{e_x}^*$, or $\mathbf{e_y}^*$, or both, are negligibly small, and excellent results are expected with $\mathbf{e_x} << \mathbf{e_y}$, or $\mathbf{e_y} << \mathbf{e_x}$, or all possible prescriptions, respectively [7].

In yet other cases, the prescription is made truly on-line and a-priori using, instead of a result-oriented tuning, some criterion that is not always justified. For example, $\{\mathbf{e_x}, \mathbf{e_y}\}$ has been prescribed to simply match the known covariances of the noises in the input and the measured output, disregarding the possibly large contributions to $\{\mathbf{e_x}^*, \mathbf{e_y}^*\}$ from model-mismatch, parameter discrepancy, and approximations [3,5]. Or the elements of $\{\mathbf{e_x}, \mathbf{e_y}\}$ have been prescribed corresponding to the nominal values of the respective outputs and state changes, which amounts to setting without justification as equal all elements of the $\{\mathbf{e_x}, \mathbf{e_y}\}$ corresponding to a properly scaled model. The ordinary least-squares estimator tacitly neglects $\mathbf{e_x}^*$. In the limited-memory algorithms, the one-dimensional prescription of forgetting factor is limited to a few favorite values, with the lone justification of them being "typical" or "experience-based". The observers go even a step further by completely ignoring the accuracy of the estimates or predictions while prescribing $\{\mathbf{e_x}, \mathbf{e_y}\}$, and basing the prescription solely on the convergence dynamics desired by the end-use, e.g., control [8].

4 Source of Prescription

The confusion, and the silence, prevalent in the literature about the means and justification for the prescription of $\{\mathbf{e_x}, \mathbf{e_y}\}$ points perhaps to a reluctance to meet the problem head-on. A candid perspective in this respect therefore seems in order.

Since $\{e_x^*, e_y^*\}$ is unknown by definition, and there is no way to a priori divine it, prescription of $\{e_x, e_y\}$ *has* to be made a posteriori based on data from past runs or past data from the current ran. Clearly, the $\{e_x, e_y\}$-prescription should be congruent with the goal of the exercise, at least a posteriori for these past data. The proper source of $\{e_x, e_y\}$-prescription therefore derives directly from the goal of estimation itself. Given measured data up to the current time, there are two main, in general mutually irreconcilable, goals of estimation:

1. Predict the process output at a certain horizon in the future, or
2. Estimate the real process state at the current time,

with best possible accuracy. That these goals are mutually irreconcilable[2] is evident from the fact that, so long as e_x^*, or e_y^* is not negligible, the model equations 3 and 4 and the process equations 5 and 6 process the same state value differently to give different future output values. When the second goal is met perfectly, the current estimate of the state equals $\mathbf{x}^*(t)$. Setting $\mathbf{x}(t) = \mathbf{x}^*(t)$ and processing the equations 3 and 4 results in $y(t_+) \neq y^*(t_+)$ at any future time t+, unless both $e_x^*$and e_y^* are negligible. The output then cannot be predicted accurately, and the first goal cannot be met simultaneously. By similar reasoning, state estimate obtained to fulfil the first goal will have to differ from the true current state $\mathbf{x}^*(t)$, and consequently will not meet the second goal.

The complementary cases of estimating the current output or predicting the future state are of secondary interest. Estimating the current process output would merely mean filtering out the noise in its available current measurement. This is not a goal of the exercise, specially for poorly known batch processes where contribution to e_y^*, of the measurement noise is overwhelmed by the other contributions, such as due to model-mismatch.[3] In this situation, the measured values can be refined relatively little, and only through model-independent signal-processing techniques.[4] The other complementary case, predicting the future process state, is of interest only once the prerequisite second goal can be, and has been, met. Issues relating to prediction of the future state, given satisfactory estimate of the current state under the second goal, is deferred to a later section.

The first goal, prediction of the future process output, is chosen whenever the output is the sole property of interest and the state itself is of no direct import. For example, predictive control of output or output-based supervision constitutes exactly this situation. The other goal of estimating the current real state holds whenever the state itself is paramount for optimization of a run or for fault diagnosis, or is to be controlled directly using, say, optimal control or PID control. The process output is, in this case, of little interest, except to help deduce the state.

It is also possible to have subsets or combinations of the two goals. The first goal could refer

---

[2]The concept of reconcilability is important throughout the later discussion. A goal is considered irreconcilable if it comprises sub-goals that cannot be met simultaneously for the given system due to mutual contradiction.

[3]In case the other contributions are negligible compared to the measurement noise, e.g., when an output simply equals a state, then estimate of the current output is readily obtained from estimate of the current state. In case the other contributions are comparable to the measurement noise, then the output is also included as an extension to the state. Both these cases fall under the second goal.

[4]Indeed, in the following discussion it will be assumed that for batch processes the noises in all measurements of process inputs and outputs are negligible compared to other errors in $\{e_x^*, e_y^*\}$, so that for all practical purposes the measured values (or values filtered through signal-processing techniques) can be considered to be the true values.

to only some of the process outputs, or the second goal to only some of the process states. A combined goal could simultaneously include some of the outputs and some of the states.

5 States and Parameters

Despite some controversy in the literature, perhaps the most widely believed distinction between states and parameters is based on dynamics. In this view, a boundary is drawn at some low, arbitrary level of dynamics. Variables with faster dynamics are called states and those with slower dynamics are deemed parameters. The boundary does not lie at zero dynamics due to slow drifts that parameters often exhibit [5,7]. This distinction is useful for control purposes, and may even be useful for observer-based estimation where the emphasis is on the response speeds instead of the accuracy of estimation. In the general field of estimation, this distinction still holds for known parameters and states. Extrapolation of the dynamics-based distinction to include the unknown, to-be-estimated variables, as is common in the literature, is -misleading.

For estimated variables, the *essential distinction* between a state and a parameter stems from the two goals of estimation discussed above. Variables that are estimated so as to serve the goal of predicting the process output accurately cannot strive to maintain identity with a real, physical variable, and are therefore to be thought of as "parameters" that give a best-fit output prediction. This is in analogy to the well-known subset of our first goal, namely identification, where all the estimated variables have zero dynamics. In identification, the emphasis is on the output; the identified parameters can turn out to be whatever they might please, and have no true physical value. Similarly for estimation with the first goal, regardless of whether a particular element of a in equation 3 has significant or zero dynamics, all the estimated $\mathbf{x}$ variables have to sacrifice identity with any real, physical variable in order to serve the output prediction. All $\mathbf{x}$ variables, in this case, are essentially best-fit "parameters" that have no physical meaning and merely give best output predictions.

In the complementary case of the second goal, all $\mathbf{x}$ variables in equation 3, regardless of whether they have significant or zero dynamics, not only have real physical meaning but must be measurable at least once, as discussed above. All variables therefore qualify as "states" that have physical meaning and can be compared with the true physical value. Thus, even a zero-dynamics variable such as the-heat of reaction, if included in the second goal, must be independently measurable and qualifies as a state. Of course, if some of the $\mathbf{x}$ variables, typically the ones with zero dynamics, are exempt from the second goal, and might serve some reconcilable outputs in the first goal, then these variables are regarded as parameters.

Since the conventional dynamics-based terminology of states and parameters had to be used in the previous sections, it is consistently maintained in the remaining sections in order to avoid ambiguity.

6 Prediction of Future State

Predictive control of the state variables or, as is customary in batch reactors, optimization of a criterion dependent on the future process states, requires accurate values of the state during or at the end of a time horizon. In such cases, prediction of the process state at a certain horizon in the future is an important goal. The trivial solution of simply integrating equation 3 starting with the current-state estimate would work only when $\mathbf{e}_x{}^*$ is negligible.

For the more realistic case of significantly large $\mathbf{x}$, a promising option is to use this goal itself to deduce the $\{\mathbf{e}_x, \mathbf{e}_y\}$-prescription, analogous to the case of the first goal in the previous sections. The prescription is consequently chosen so as to minimize the cumulative deviation between the prediction obtained by processing each state estimate through equation 3 and the corresponding measured value of the state available in the past data. The current-state estimate itself, in this case, would have to be some bestfit value different from the true current value, and the estimated variable would be regarded as a parameter in our terminology. A disadvantage of this option is that it promises reliable prediction only at the end of the horizon and not at other sampling instants within the horizon. If prediction is needed successively during the horizon, then an independent estimator would have to be used for each time at which a prediction is needed.

An ad-hoc alternative is to first estimate the current state using the second goal as in the previous sections, and then integrate equation 3 starting with this estimate while correcting the integrated value at each future sampling time using some correction measure obtained from the recent estimation results. In its simplest form, this correction measure would simply equal the difference between the current-state estimate and its prediction based on the previous-state estimate and on equation 3. In addition to the difference at the current time, some more most recent differences could be used to refine the correction measure as well as to indicate its reliability. Although ad-hoc, this alternative has the advantage of delivering all predictions during the horizon.

7 Goal Reconciliation

The first goal mentioned above can, alone or in partial combination with the second goal, lead to an irreconcilable objective. In general, the combined goal is irreconcilable whenever the number of states and outputs included in the goal exceeds the total number of states that are estimated. Thus, an irreconcilable goal can be rendered reconcilable by increasing the number of estimated states. This involves extending the state vector through inclusion of existing or new variables appearing in the measurement equations. To illustrate this point, consider the reactor model in equations 1 and 2 with fixed values of the parameters k and $(-\Delta H)$. For this model, either the first goal alone or the second goal alone is reconcilable, but both together are not. If the goal is to predict the future q, then the current real c cannot be estimated, and vice versa. Two variables q and c cannot together be sought, while only one state c gets estimated. One of the goals must be abandoned, or another state must be estimated.

One way to add a state is to include the output q directly as a state, so that the model becomes:

$$\frac{dc}{dt} = -kc^2 + F \tag{7}$$

$$\frac{dq}{dt} = (-\Delta H)k(2c)(-kc^2+F), \quad q(0) = (-\Delta H)kc^2(0) \tag{8}$$

$$q_m = q \tag{9}$$

where the measured output has been renamed as q_m for distinction. In this extended model, the combined goal of predicting the future q. and estimating the current real c is reconcilable as two variables c and q get estimated. In keeping with our terminology introduced in the previous section, c is here a state, and q is a parameter that merely serves to enable accurate prediction of q_m.

Another way to get the additional degree of freedom is to extend the state to include a process parameter appearing in the measurement equation 2. Letting k be free in equations 1 and 2 leads to the model:

$$\frac{dc}{dt} = -kc^2 + F \tag{10}$$

$$\frac{dk}{dt} = 0 \tag{11}$$

$$q = (-\Delta H) kc^2 \tag{12}$$

Again the extended model allows the combined goal to be reconcilable. In this case, it just happens that our terminology for state and parameter coincides -with the conventional one, since we are not interested in the estimated value of k so long as it leads to the best possible prediction of q.

Yet another way to add a state is to create an artificial variable, such as a bias b added to the measurement model, to give the model:

$$\frac{dc}{dt} = -kc^2 + F \tag{13}$$

$$\frac{db}{dt} = 0, \qquad b(0) = 0 \tag{14}$$

$$q = (-\Delta H) kc^2 + b \tag{15}$$

This renders the combined goal reconcilable. Notice though that reconcilability would not result, if the bias b were added to the dynamic model instead, giving the model:

$$\frac{dc}{dt} = -kc^2 + F + b \tag{16}$$

$$\frac{db}{dt} = 0, \qquad b(0) = 0 \tag{17}$$

$$q = (-\Delta H)kc^2 \tag{18}$$

In this case, the added state b does not affect the output q independently of the effect of the desired state c on the output. It therefore provides no additional degree of freedom for q and does not lead to goal reconciliation.

8 Deduction of the Prescription

Once the goal and/or the model has been modified so that the goal is reconcilable, the prescription of $\{\mathbf{e_x}, \mathbf{e_y}\}$ can be deduced from past measurements. The strategy for prescribing $\{\mathbf{e_x}, \mathbf{e_y}\}$ for each of the above two goals is conceptually similar. The best possible setting of the indirect form of $\{\mathbf{e_x}, \mathbf{e_y}\}$-prescription (e.g., covariances, or forgetting factor) must be sought by trial-and-error (or equivalently using a higher-level optimization). For any assumed setting of the prescription, first the algorithm is used over the available past data to deliver corresponding state estimates.

Then, for case of the first goal, the state estimate corresponding to each data point is translated using the model equations 3 and 4 into a prediction of the process output as far into the future as dictated by the goal. Each prediction is then compared with its corresponding measured value from the data set,[5] and the cumulative deviation is regarded as a measure of badness of the assumed setting of the prescription. The setting that minimizes, or reduces to a satisfactory level, this cumulative deviation is then regarded as the $\{\mathbf{e_x}, \mathbf{e_y}\}$-prescription to be used in the next implementation of the algorithm.

In case of the second goal, availability of some state measurements in the past data is a *prerequisite* for a meaningful $\{\mathbf{e_x}, \mathbf{e_y}\}$-prescription. In other words, if the past data does not include any measurement of the state, then $\{\mathbf{e_x}, \mathbf{e_y}\}$ cannot be prescribed for the purpose of estimating the real process state. That is, there is no way to estimate the real process state in an application run, if the state cannot be measured, at least once, in a "tuning" run. Given such measurement, the trial-and-error procedure follows similarly as in the other case. Those state estimates for which a corresponding state measurement is available in the data set are compared with that measured value, and the cumulative deviation is again regarded as a measure of the badness of the assumed setting of the $\{\mathbf{e_x}, \mathbf{e_y}\}$-prescription.

In the case of a partial or a combined goal, the cumulative deviation minimized by the trial-and-error procedure stems from only those outputs and/or states that are included in the goal.

---

[5]Except for some predictions at the end for which no corresponding measured value is available.

Whenever the goal includes more than one variable (output or state), the user must assign relative weights to these variables for determining their contribution to the cumulative deviation that dictates the $\{e_x, e_y\}$-prescription. This requirement is obvious when the different variables are mutually irreconcilable, but holds even when the goals are reconcilable. In the latter case, only a $\{e_x, e_y\}$-prescription that changes with each sampling time could fit all the past data perfectly, obviating the need for userspecification of relative weights. The perfect fit, however, is undesirable, as discussed in the next section. An imperfect fit inevitably entails left-over error that must be distributed among the different goal variables by design.

The design decision that the user must make about the desired relative accuracy of each of the goal variables is independent of the true or suspected nature of the corresponding components of $\{e_x^*, e_y^*\}$. Still, having to specify weighting factors is undesirable in practice, because it forces an arbitrary compromise in meeting the different goals. The *sensitivity* of the obtained estimates and the achieved goals with respect to these weights depends on the left-over error. For irreconcilable goals, unavoidably large leftover errors can render the weights so sensitive that the user effectively ends up having to practically specify one particular outcome from a wide range of possible, equally valid estimates. For reconcilable goals, the weight specification can be made relatively insensitive by parameterizing the prescription so as to avoid excessive left-over error due to underfit, as described in the next section.

9 Parameterization of the Prescription

The $\{e_x, e_y\}$-prescription deduced by trial-and-error can be used successfully in the further application of the algorithm only if the subsequent process data has similar characteristics as the past data that was used for the trial-and-error prescription. If not, then still this is the best that can be done, and the goal cannot be met any better. In case the past data includes the current run, the prescription could periodically be updated on line, and the most recent setting used for the next implementation of the algorithm. Such on-line adaption of the prescription is specially useful when prolonged unmeasured disturbances constitute a significant contribution to $\{e_x^*, e_y^*\}$.

In theory, the $\{e_x, e_y\}$-prescription can be different at each sampling time,[6] but that would make it *overfit* the past data and render it unreliable for application to new data. In practice, the other extreme is preferred where each element of $\{e_x, e_y\}$ is restricted to a single constant value for the entire past data, with the associated inevitable *underfit*. The superior middle ground of allowing each element to take a few step-wise constant[7] values is forgone for practical reasons. There is no way to decide how many constant values ought to be allowed, how each constant value ought to be ordered within a run, and how the values obtained for one run are to be applied to another run that might not have a similar course of state changes with time.

Being forced to use a single constant value for each element of $\{e_x, e_y\}$ involves considerable

---

[6]For the second goal of true-state estimation, the prescription can be different only at the points of successive state measurements.

[7]Instead of a step-wise constant curve, a linear or higher-order curve is obviously possible, but rarely justified as an a-priori choice.

loss of freedom in fulfilling the goal. Due to the underfit caused by underparamctcrization of the prescription, the goal might get fulfilled quite accurately in some parts of the run, but relatively larger inaccuracies may remain in other parts of the run. A reduction in the degree of underfit is possible through two kinds of modification of the model. The first kind involves making the model inherently richer in the sense of enhancing the knowledge it embodies. This could be done, for example, by including an additional measured variable or by modeling more accurately a parameter that was previously set to a constant value. Process considerations and modeling limitations, however, often rule out this option. The other kind of modification, which can be realized more readily, entails endowing the model with further degrees of freedom for the $\{\mathbf{e_x}, \mathbf{e_y}\}$-prescription. This reduces the degree of under-parameterization of the prescription by increasing the number of elements in $\{e_x, e_y\}$, while each element is still restricted to a single constant value.

The extra degree of freedom, or the higher-order parameterization, for $\{\mathbf{e_x}, \mathbf{e_y}\}$-prescription is achieved by extending the model to include certain estimated variables with fixed initial values. The sole purpose of these estimated variables is to increase the number of elements in $\{\mathbf{e_x}, \mathbf{e_y}\}$. No true value is sought for them; in fact, they might not even have any physical meaning. These extension states are strictly parameters in our terminology from a previous ' section. Theoretical observability with respect to these states is implied by restricting their initial values to be fixed and constant for all runs, past and future.

One way to modify the extended model of equations 10, 11, and 12 for this purpose would be to let another physical parameter $(-\Delta H)$ be free, to give the model:

$$\frac{dc}{dt} = -kc^2 + F \tag{19}$$

$$\frac{dk}{dt} = 0 \tag{20}$$

$$\frac{d(-\Delta H)}{dt} = 0, \qquad (-\Delta H)(0) = (-\Delta H)_0 \tag{21}$$

$$q = (-\Delta H)kc^2 \tag{22}$$

where $(-\Delta H)_0$ is fixed a priori and is constant for all runs. A somewhat equivalent effect can be achieved by using an additive bias b in the measurement model, to give:

$$\frac{dc}{dt} = -kc^2 + F \tag{23}$$

$$\frac{dk}{dt} = 0 \tag{24}$$

$$\frac{db}{dt} = 0, \qquad b(0) = 0 \tag{25}$$

$$q = (-\Delta H)kc^2 + b \tag{26}$$

Both these extension options may be inferior to the original model of equations 10, 11, and 12, since the numerical observability of the modified model may be low.

A better way that does not deteriorate numerical observability is to modify the extended model of equations 10, 11, and 12 as:

$$\frac{dc}{dt} = -kc^2 + F \tag{27}$$

$$\frac{dk}{dt} = a \tag{28}$$

$$\frac{da}{dt} = 0, \qquad a(0) = 0 \tag{29}$$

$$q = (-\Delta H)kc^2 \tag{30}$$

Conceptually, this gives the effect of allowing in the original model the prescription element corresponding to k to have two values instead of one, without having to worry about how the values are to be ordered within and between runs. A similar effect can again be sought using instead an additive bias b as:

$$\frac{dc}{dt} = -kc^2 + F + b \tag{31}$$

$$\frac{dk}{dt} = 0 \tag{32}$$

$$\frac{db}{dt} = 0, \qquad b(0) = 0 \tag{33}$$

$$q = (-\Delta H)kc^2 \tag{34}$$

The above explains to some extent why the applications in the literature commonly resort to large extended-state vectors [3,7,8]. A minimal extension of the state vector leads to goal reconciliation, allowing them to attain good estimates of the original states as well as good single-step predictions of the measured output. Further extension of the state vector allows the use of single-value prescription of the $\{\mathbf{e}_x, \mathbf{e}_y\}$-elements without serious deterioration of quality due to underfit. The extension of state should, however, be made so as to not only retain theoretical observability (by using fixed initial values, if necessary) but also numerical observability. Including too many variables in the extended state may compromise numerical observability and jeopardize, due to likely overfit on the past data, the validity of the deduced prescription for application on future data. The extension states should be regarded strictly as parameters and no connection with true physical values should be sought for them.

10 A Look at Some Common Practices

The above perspective affords some insight into certain common practices in the estimation and identification literature.

In identification, the state consists of several parameters, the output consists of relatively few, or one, measured variables, and the goal is only to predict the future output. Since the number of estimated states exceeds the number of outputs to be predicted, the goal is reconcilable. However, many identification algorithms collapse the e.' space to a single forgetting factor, thereby losing all those degrees of freedom, and rendering the goal irreconcilable for more than one output. The price is paid in that the left-over errors are larger, the predictions are sensitive to the e weight specification for the outputs, and the overall quality of prediction is poorer [6].

Most estimation algorithms yield some covariance matrix for the estimated states. Many users tend to trust this covariance as a measure of the accuracy of the obtained estimates [7]. It is clear from the above discussion that if the goal of estimation is to predict future outputs, then the estimated variables are to be seen as parameters for which no true value exists. In this case, it is meaningless to talk of a measure of the accuracy of these estimates. If the goal is to estimate the true states, then an indication of the accuracy of obtained estimates can be taken only from the degree of fit to past data that was achieved during the trial-and-error determination of the prescription used. There is no independent way to check how well the obtained prescription extrapolates to new data, except to have some state measurements in the new data. The covariance matrix again is of little use as a measure of accuracy of the estimates in the new data. As a corollary, it is not an indicator of filter divergence either.

Perhaps the most striking aspect of much of the literature on estimation applications is the apparent confusion with regard to the goal of the estimation exercise and the procedure used to prescribe $\{\mathbf{e}_x, \mathbf{e}_y\}$. The goal is often not stated explicitly, but in most cases the tacit goal is estimation of the current true states. The associated prescription, whether obtained by trial-and-error or by adaption, is nonetheless based on striving invariably some property of the residuals, such as zero-meanness, whiteness, or match with theoretical covariance [3,5,7]. This may be a result of carrying over the experience from identification problems. Having thus inadvertently borrowed the additional goal of predicting the output correctly, the applications often slip into a two-goal situation that is irreconcilable. Failing to achieve the goals by using any prescription, the user is then forced to extend the model to make ends meet, as outlined in a previous section. Even having gained reconcilability and enough degrees of freedom in this way, it is perplexing how most applications end up showing good state estimates purportedly without ever using state measurements to determine the prescription. The reason for this might lie in the belief that the states must be taken as unmeasurable, except for verification of shown results, and in not realizing that it is fair, indeed indispensable, to use some state measurements to get the prescription. In fact many applications do not even concede having used an off-line "tuning" at all.

11 Conclusion

The perspective presented above clarifies some issues that have not been clear in the estimation literature. In any estimation exercise, it is paramount to first clearly define its goal. The next step is to check whether the goal is reconcilable and leaves some extra degrees of freedom to allow use of constant tuning parameters. If not, the model should be appropriately enriched or extended. Then data must be collected to enable determination of the tuning. These data must include measured values of all variables that appear in the stated goal, also when some of these variables happen to be states. The data may originate from the past values gathered in the application run itself, or in independently made runs. The tuning is then obtained by trial-and-error, higher-level optimization, or adaption. The objective thereby is to find that tuning which, for the collected past data, gives states estimates that best reach the stated goal on the same data. The same tuning is then used in application hoping that the underlying characteristics of the process-model-algorithm-operation combination remain comparable to those of the past-data runs. The accuracy of the obtained state estimates or output predictions can be verified only by comparison with corresponding additional measurements, and cannot be deduced from the covariance matrix that the algorithm might deliver. If the accuracy is not satisfactory, one or more of the above steps will have to be changed. The steps are therefore intermingled, in practice, and do not necessarily have to follow the order in which they are stated above.

References

1. Biegler, L.T., Damiano, J.J., and Blau, G.E. Nonlinear Parameter Estimation: A Case Study Comparison. *AICHE J.*, **32(I)**, 29-45, (1986).
2. Eykhoff, P. A Bird's Eye View on Parameter Estimation and System Identification. *Automatisierungstechnik*, **36(11)**, 413-479, (1988).
3. Goldmann, S.F. and Sargent, P.W.H. Applications of Linear Estimation Theory to Chemical Processes: A Feasibility Study. *Chem. Eng. Sci.*, **26**, 1535-1553, (1971).
4. Jazwinski, A.H. *Stochastic Processes and Filtering Theory*, Academic Press, New York, (1970)
5. Liang, D.F. Exact and Approximate State Estimation Techniques for Nonlinear Dynamic Systems. *Control Dynamic Systems*, **19**, 1-80, (1983).
6. Ljung, L., and Gunnarsson, S. Adaptation and Tracking in System Identification - A Survey, *Automatica*, **26**, 7-21, (1990).
7. Sorenson, H.W., editor, *Kalman Filtering. Theory and Application*, IEEE Press, New York, (1985).
8 Zeitz, M., Nonlinear Observers. Regelungstechnik, **27 (8)**. 241-272, (1979), (in German)

A Comparative Study of Neural Networks and Nonlinear Time Series Techniques for Dynamic Modeling of Chemical Processes

A. Raich, X. Wu, H.-F. Lin, and Ali Cınar

Department of Chemical Engineering, Illinois Institute of Technology, Chicago, IL 60616, USA

Abstract: Neural networks and nonlinear time series models provide two paradigms for developing input-output models for nonlinear systems. Methodology for developing neural networks with radial basis functions (RBF) and nonlinear auto-regressive (NAR) models are described. Dynamic input-output models for a MIMO chemical reactor system are developed by using standard back-propagation neural networks with sigmoid functions, neural networks with RBF and time series NAR models. The NAR models are more parsimonious and more accurate in predictions.

Keywords: Nonlinear dynamic models, input-output models, nonlinear autoregressive models, CSTR model, neural networks, radial basis functions.

1. Introduction

Although most chemical processes are nonlinear systems, traditionally linear input-output models have been used in describing their dynamic behavior. The existence of a well developed linear control theory enhances the use of linear models. Linear models may provide enough accuracy in the vicinity of the linearization point, but they have limited predictive capability when the system has been subjected to large disturbances. Recently, the interest in describing chemical processes by nonlinear input-output models has increased significantly. This is partly due to the shortcomings of linear models in representing nonlinear processes. Advanced process monitoring and model-based control techniques are expected to give better results when process models that are more accurate over a wider range of operating conditions are used. Model-predictive control approaches which are becoming popular in chemical process industries permit direct use of nonlinear models in control algorithms. Another important reason for the increase in nonlinear model development activities is the availability and the popularity of new tools such as neural networks which provide automated tools for constructing nonlinear models. Yet, several other paradigms for nonlinear model development have been available for over two decades [4]. Consequently, it would be useful to model some test processes using various approaches, compare their prediction accuracies, and assess their strengths and shortcomings. Neural networks (NN) and nonlinear auto-regressive (NAR) models are the two paradigms utilized in this study. The input-output functions of the feedforward NN utilized are radial basis functions (RBF) and sigmoid functions. RBFs [14, 15], also called local receptive fields, necessitate only one "hidden" layer and yield an identification problem that is linear in the parameters. The NAR modeling approach provides a linear in the parameters identification as well. This has a significant impact on the computational effort needed in finding the optimal values of the parameters.

In this study, various dynamic models are developed for modeling a multivariable ethylene oxidation reactor system. The "process data" are generated with a detailed model of the reactor developed by material and energy balances, and kinetic data from experimental studies. The reactor equations have both multiplicative and exponential type nonlinearities. The models developed are used for making one-step-ahead and 5-steps-ahead predictions of reactor outputs.

The paper is structured as follows. In Section 2, NNs with Gaussian RBF are outlined. NAR modeling methodology is presented in Section 3. The reactor system and the generation of input-output data is described in Section 4. Predictions of both types of models for various cases are discussed in Section 5.

2. Neural Networks with Radial Basis Functions

Neural networks have been utilized to create suitable nonlinear models, especially for use in pattern recognition, sensor data processing, forecasting, process control and optimization [1, 5, 6, 8, 9, 11, 17, 19, 21, 22, 23, 24, 25, 26]. Sigmoid functions have been most popular as input-output functions in NN nodes. RBFs provide alternative nonlinear input-output functions which are "locally tuned". A single-layered NN using RBFs can approximate any nonlinear function to a desired accuracy [3]. By overlapping the local receptive fields, signal-to-noise ratios can be increased to provide improved fault tolerance [14]. Control with Gaussian RBF networks has been demonstrated to be stable, with tracking errors converging towards zero [20]. First order-lag-plus deadtime transfer functions in the network nodes has also been discussed [24].

RBF approximation is a traditional technique for interpolation in multidimensional space. A RBF expansion with n inputs and a scalar output generates a mapping $f_r: \Re^n \rightarrow \Re$ according to

$$f(\mathbf{x}) = w_0 + \sum_{i=1}^{n_c} w_i \cdot g(\|\mathbf{x} - \mathbf{c}_i\|) \tag{1}$$

where $\mathbf{x} \in \Re^n$, $g(\cdot)$ is a function from $\Re^+$ to $\Re$, n_c is the number of RBF centers, w_i, $0 \leq i \leq n_c$ are the weights or parameters, $\mathbf{c}_i \in \Re^n$, $1 \leq i \leq n_c$ are the RBF centers, and $\| \cdot \|$ denotes the Euclidian norm. The equation can be implemented in a multilayered network (Figure 1) where the first layer is the inputs, the second layer performs the nonlinear transformation and the top layer carries out the weighted summation only. Notice that the second layer is equivalent to all hidden layers and the nonlinear operation (with sigmoid functions) at the output layer nodes of the NN with sigmoid functions. A frequent choice for the RBF, $g(\| \mathbf{x}-\mathbf{c}_i \|)$, is the Gaussian function

$$g(\mathbf{x}) = exp(-\|\mathbf{x} - \mathbf{c}_i\|/\beta_i^2) \tag{2}$$

where, β_i is a scalar width such as the standard deviation. Given the numerical values of the centers c_i and of the width, β_i, determination of best values of the weights, w_i, to fit the data is a standard model identification problem which is linear in the parameters. If the centers and/or the width are not predetermined and are adjustable parameters whose values are to be determined along with weights, then the RBF network becomes equivalent to a multi-layered feedforward NN, and an identification which is nonlinear in the parameters must be carried out.

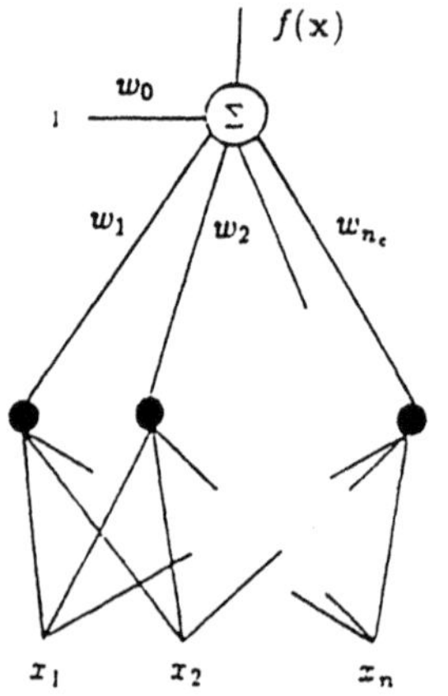

Figure 1: Neural network structure with radial basis functions as nonlinear functions in nodes.

A popular algorithm for choosing the centers and the widths is *k-means clustering*, which partitions the data set $\mathbf{X} = [\mathbf{x}_j]$ into *k* clusters, finds their centers so as to minimize the total distance $\| \cdot \|$ of the x vectors from their nearest center. Widths β_i can then be calculated to provide sufficient overlap of the Gaussian functions around these centers and ensure a smooth, continuous interpolation over **X** while keeping the RBFs local enough so that only a small portion of the network contributes for relating an input vector $\mathbf{x}$ of **X** to the respective output. This localization makes the function with the closest center $\mathbf{c}_i$ to the input $\mathbf{x}$ the strongest voice in predicting the corresponding output. The number of nodes in the nonlinear transformation layer is set equal to the number of clusters, k. For each of the k clusters or nodes, a Gaussian width, β_i can be found to minimize the objective function

$$E = \frac{1}{2}\sum_{i=1}^{k}[\sum_{j=1}^{n} \mathrm{g}_i(\mathbf{x}_j)[(\mathbf{x}_j - \mathbf{c}_i)/\beta_i]^2 - p] \tag{3}$$

where *n* is the number of columns in **X** (the number of input vectors used in training) and p is an overlap parameter which assures that each cluster partially overlaps with its neighboring clusters. With the RBF relations in each node fixed, the weights can be chosen to map the inputs to a variety of outputs according to

$$f_l(\mathbf{x}_j) = w_{l0} + \sum_{i=1}^{k} w_{li} \cdot \mathrm{g}(\|\mathbf{x}_j - \mathbf{c}_i\|) \tag{4}$$

where l is the number of outputs. Hence, for each output q only the weights w_{qi} are specific to that output, while the centers and widths are common to all outputs. With the k-means clustering algorithm to select the clusters and compute their centers, optimization of the Gaussian widths will fix the form of $g(x_j)$ and node weights can be easily optimized to model a multivariate mapping of X to $f_l(X)$. The selection of the cluster members, computation of cluster means and widths are done first, as *unsupervised* learning. The computation of the weights is carried out as the training of the neural net, i.e. *supervised* learning.

3. Nonlinear Time Series Models

Classical model structures used in non-linear system identification have been the functional series expansions of Volterra or Wiener which map past inputs into the present output. This moving average type approach results in a large number of coefficients in order to characterize the process. Input/output descriptions which expand the current output in terms of past inputs and outputs provide parsimonious models. The non-linear autoregressive moving average with exogenous inputs (NARMAX) model [12], the bilinear model, the threshold model and the Hammerstein model belong to this class. In general, NARMAX models consist of polynomials which include various linear and nonlinear terms combining the inputs, outputs and past errors. Once the model structure, the monomials to be included in the model, has been selected, the identification of the parameters can be formulated as a standard least squares problem which can be solved using various well-developed numerical techniques. The number of all candidate monomials to be included in a NARMAX model ranges from about a hundred to several thousands for moderately nonlinear systems. Determination of the model structure stepwise regression type of techniques become inefficient. Instead, methods of model structure determination must be developed and included as a vital part of the identification procedure. An orthogonal algorithm which efficiently combines structure selection and parameter estimation for stochastic systems has been proposed by Korenberg [10] and later extended to MIMO nonlinear stochastic systems [12].

In this paper, a special case of NARMAX model, the nonlinear autoregressive (NAR) model, and the classical Gram-Schmidt (CGS) orthogonal decomposition algorithm using Akaike Information Criterion (AIC) are presented.

Nonlinear Model Representation and CGS Orthogonal Decomposition Algorithm

A discrete time multivariable nonlinear stochastic system with m outputs and r inputs can be described by the NARMAX model [12]

$$y(t)=f(y(t-1),...,y(t-n_y),u(t-1),...,u(t-n_u),e(t-1),...,e(t-n_e))+e(t) \quad (5)$$

where

$$y(t)=\begin{pmatrix} y_1(t)\\ \vdots \\ y_m(t)\end{pmatrix},\quad u(t)=\begin{pmatrix} u_1(t-1)\\ \vdots \\ u_r(t-1)\end{pmatrix},\quad e(t)=\begin{pmatrix} e_1(t)\\ \vdots \\ e_m(t)\end{pmatrix} \quad (6)$$

are the system output, input and noise respectively; n_y, n_u, n_e are the maximum lags in the output, input and noise; $\{e(t)\}$ is a zero mean independent sequence; and f (·) is some vector valued nonlinear function.

A special case of the NARMAX model is the NAR model

$$y(t)=f(y(t-1),...,y(t-n_y))+e(t) \quad (7)$$

which can be expanded as

$$y_q(t)=f_q(y_1(t-1),...,y_1(t-n_y),...,y_m(t-1),...,y_m(t-n_y))+e_q(t),\quad q=1,...,m \quad (8)$$

Writing $f_q(\cdot)$ as a polynomial of degree l yields

$$y_q(t)=\theta_0^{(q)}+\sum_{i_1=1}^{n}\theta_{i_1}^{(q)}x_{i_1}(t)+\sum_{i_1=1}^{n}\sum_{i_2=i_1}^{n}\theta_{i_1i_2}^{(q)}x_{i_1}(t)x_{i_2}(t)$$
$$+\dots+\sum_{i_1=1}^{n}\cdots\sum_{i_l=i_{l-1}}^{n}\theta_{i_1\dots i_l}^{(q)}x_{i_1}(t)\dots x_{i_l}(t)+e_q(t),\quad q=1,...,m \quad (9)$$

where

$$n=m\times n_y \quad (10)$$

$$x_1(t)=y_1(t-1),\quad x_2(t)=y_1(t-2),\cdots,\quad x_{mn_y}(t)=y_m(t-n_y) \quad (11)$$

All terms $x_{i_1}(t)...x_{i_l}(t)$ in Eq. (9) are given. Hence, for each q, $1\le q\le m$, Eq. (9) describes a linear regression model of the form

$$y_q(t)=\sum_{i=1}^{M}p_i(t)\theta_i+\xi(t),\quad t=1,\cdots,N \quad (12)$$

where $M = \sum_{i=1}^{l} m_i$ with $m_i = m_{i-1} \cdot (n_y \cdot m + i - 1)/i$, N is the time series data length, $p_i(t)$ are the monomials of degree upto l which consist of various combinations of x_1(t) to x_n(t) (n defined in Equation (10)), $\xi(t)$ is the residual, and θ_i are the unknown parameters that will be estimated. $p_1(t) = 1$ is the constant term. In matrix form Equation (12) becomes

$$Y_q = P\Theta + \Xi \tag{13}$$

where

$$Y_q = \begin{pmatrix} y_q(1) \\ \vdots \\ y_q(N) \end{pmatrix}, \quad P = (p_1 \;\; \dots \;\; p_M), \;\; p_i = \begin{pmatrix} p_i(1) \\ \vdots \\ p_i(N) \end{pmatrix}, \quad \Theta = \begin{pmatrix} \theta_1 \\ \vdots \\ \theta_M \end{pmatrix}, \quad \Xi = \begin{pmatrix} \xi(1) \\ \vdots \\ \xi(N) \end{pmatrix} \tag{14}$$

Usually a model consisting of a few monomials describe the dynamic behavior of most real processes to desired accuracy. This is a small subset of the monomials in the full model. Consequently, the development of an algorithm that can select the significant monomial terms efficiently is of vital importance. A method has been developed for the combined problem of structure selection and parameter estimation [2]. The task is to select a subset P_s of the full model set P and to estimate the corresponding parameter set θ_s. Several least squares solution methods can be utilized, but methods based on orthogonal decomposition of P offer a good compromise between computation accuracy and computation "cost".

Gram-Schmidt Orthogonal Decomposition Algorithm The classical Gram-Schmidt (CGS) orthogonal decomposition of P yields

$$P = WA\,, \quad A = \begin{pmatrix} 1 & \alpha_{12} & \alpha_{13} & \dots & \alpha_{1M} \\ 0 & 1 & \alpha_{23} & \dots & \alpha_{2M} \\ \vdots & \vdots & \vdots & \ddots & \vdots \\ 0 & 0 & 0 & \dots & \alpha_{M-1,M} \\ 0 & 0 & 0 & \dots & 1 \end{pmatrix}, \quad W = (w_1 \dots w_M) \tag{15}$$

where A is an M x M unit upper triangular matrix and W is an N x M matrix with orthogonal columns that satisfy $W^T W = D$ with the positive diagonal matrix D.

The selection of the specific monomials that will make up the model structure and the estimation of their coefficients can be combined by extending the orthogonal decomposition techniques. Let P_S be a subset of P with M_s columns such that $M_s < M$ and $M_s \leq N$. Factorize P_S, into $W_s A_s$ where W_s is an N x M_s matrix with M_s, orthogonal columns and A_s is an M_s x M_s unit upper triangular matrix. The residuals can be expressed as

$$\hat{\Xi} = \begin{pmatrix} \hat{\xi}(1) \\ \vdots \\ \hat{\xi}(N) \end{pmatrix} = Y_q - P_s\hat{\Theta}_s = Y_q - W_s g_s \quad \text{with} \quad g_s = D^{-1} W_s^T Y_q. \tag{16}$$

Denoting the inner product as $<\cdot,\cdot>$, and rearranging Equation (16) as $Y_q = W_s g_s + \hat{\Xi}$, the sum of squares of the dependent variable Y_q is

$$< Y_q, Y_q > = \sum_{i=1}^{M_s} g_i^2 < w_i, w_i > + < \hat{\Xi}, \hat{\Xi} > \tag{17}$$

The reduction in the residual due to the inclusion of w_i in the regression can be measured by an error reduction ratio Ω_i defined as the proportion of the dependent variable variance explained by w_i,

$$\Omega_i = \frac{g_i^2 < w_i, w_i >}{< Y_q, Y_q >} \tag{18}$$

The error reduction ratio can be used for extracting W_s from W and consequently P_s from P by utilizing the CGS decomposition procedure. Let k denote the stages of the iterative procedure. At the first stage (k = 1), set $w_1^{(i} = p_i$ for $i = 1, \cdots, M$ and compute

$$g_{k=1}^{(i} = \frac{< w_1^{(i}, Y_q >}{< w_1^{(i}, w_1^{(i} >}, \quad \Omega_1^{(i} = \frac{(g_1^{(i})^2 < w_1^{(i}, w_1^{(i} >}{< Y_q, Y_q >} \tag{19}$$

Select as w_1 the $w_1^{(i}$ that causes the maximum error reduction ratio: $w_1 = w_1^{(i}$ such that $\Omega_1^{(i} = \max$ $\Omega_1^{(i}, i = 1, \cdots, M)$. Similarly, the first element of g, is $g_1 = g_1^{(i}$.

At the kth stage, excluding the previously selected j's compute for $i = 1, \cdots, M$ $(i \neq j)$

$$\alpha_{1,k}^{(i} = \frac{< w_1, p_i >}{< w_1, w_1 >}, \quad \cdots, \quad \alpha_{k-1,k}^{(i} = \frac{< w_{k-1}, p_i >}{< w_{k-1}, w_{k-1} >} \tag{20}$$

$$w_k^{(i} = p_i - \sum_{l=1}^{k-1} \alpha_{lk} w_l, \quad g_k^{(i} = \frac{< w_k^{(i}, Y_q >}{< w_k^{(i}, w_k^{(i} >} \tag{21}$$

Let $\Omega_k^{(j} = \max(\Omega_k^{(i}, 1 \le i \le M$ (i $\ne$ *all previous j*)). Then $w_k = w_k^{(j}$ is selected as the *k*th column of W_s, together with the *k*th column of A_s, $\alpha_{lk} = \alpha_{lk}^{(j}$ ($l = 1, \cdots, k$-1) the *k*th element of g_s, $g_k = g_k^{(j}$, and $\Omega_k = \Omega_k^{(j}$.

The selection procedure can continue until a prespecified residual threshold is reached. However, since the goal is to develop a model that will be used for prediction, it is better to balance the reduction in residual and increase in model complexity. This would reduce the influence of noise in the data on the development of the process model. The Akaike Information Criteria (AIC) is used to guide the termination of the modeling effort.

$$AIC(k) = N log(\frac{1}{N} < \hat{\Xi}, \hat{\Xi} >) + 2k \tag{22}$$

Addition of new monomials to the model is ended when AIC is minimized. The subset model parameter estimate θ_s can be computed from $A_s\theta_s = g_s$ by backward substitution.

4. Multivariable Chemical Reactor System

A reactor model based on mass and energy balances is available for simulating the behavior of ethylene oxidation in a nonadiabatic internal recycle reactor [18, 16]. Ethylene reacts with oxygen to produce ethylene oxide. A competing total oxidation reaction generates CO_2 and water vapor. Furthermore, the ethylene oxide produced can dissociate to yield CO_2 and H_2O. All three reactions are highly exothermic. The reactor input variables that are perturbed are inlet flowrate, inlet ethylene concentration, and inlet temperature. The output variables are outlet ethylene and ethylene oxide concentrations and outlet temperature. Data is generated by perturbing some or all inputs by pseudo-random binary sequences (PRBS). The two main cases studied are based on PRBS forcing of (a) inlet ethylene concentration and total inlet flow rate, and (b) inlet ethylene concentration, total flow rate and inlet temperature. The first case has multiplicative interactions while the second case provides exponential interactions as well. Results for the second case is reported in this communication. Data was collected at three different PRBS pulse duration: 5, 10, or 25 s. In all input-output models, only reactor output data are used. The length of the time series is 1000. For NN training all 1000 values are used while for NAR models 300 data points are utilized. A second data set has been developed for each case, in order to assess the predictive capabilities of the models developed.

5. Nonlinear Input-output Models of the Reactor System

Dynamic Models Based on Neural Network

Dynamic models are constructed to predict the one-step-ahead and 5-steps-ahead values of reactor outputs based on past values of outputs. The current and the most recent four previous values of all reactor output variables are fed to the NN as the inputs and either the next or the five time steps ahead value of the same variable is used as the output. Consequently, for a NN with generic input(s) z(t) and generic output y(t)

$$y(t) = z(t+1) = f([z(t)\ z(t-1)\ z(t-2)\ z(t-3)\ z(t-4)]) \tag{23}$$

for the one-step-ahead prediction, while y(t) = z(t +5) for the 5-steps-ahead prediction.

Neural Network with Sigmoid Functions: A commercial package has been utilized for developing the standard neural networks with sigmoid functions. Various sizes of hidden nodes have been tested and a single hidden layer with 12 hidden nodes has provided the best prediction accuracy [13].

Neural Network with Radial Basis Functions: Training the network is done by both supervised and unsupervised methods. Clustering using the K-means algorithm and optimization of widths is unsupervised, dependent only on the network inputs, not the specific output to be modelled. Once the centers and widths are determined, supervised learning of the weights for each output variable is done. The K-means algorithm is implemented in FORTRAN to choose centers for the Gaussian functions. Davison-Fletcher-Powell method with unidimensional Coggins search [7] is used for the optimization of Gaussian function widths and for the selection of weights for the various reactor outputs. The network parameters (centers, widths, and weights) were saved for each of the training cases for use with additional reactor data sets to test the predictive accuracy of networks developed.

Networks with various numbers of hidden nodes were trained and tested to find the number of nodes yielding the smallest predictive sum of squared error and Akaike Information Criterion.

Dynamic NAR Models

The pool of monomials to be used in the NAR models included all combinations of the current and immediate past 3 values of all three reactor output variables, resulting in 82 candidates. Models with past 4 values made little improvement. NAR modeling was conducted with a 300 line C program, having typical execution times of 3 to 5 minutes on the Vax station for time series lengths of 300 for all 3 variables.

Discussion of Results

Neural Network Models. There were no clear trends in the optimal number of nodes, sum of squared errors or AIC for the best networks for 1-ahead or 5-ahead predictions, with or without noise in data (Table I). Generally, as PRBS pulse duration increased, the number of nodes increased. For one at a time variable prediction, while outlet ethylene concentration was nearly always the most difficult variable to learn (minimum error was reached with higher numbers of nodes than the other two variables), trends across time horizon or absence/presence of noise were not apparent.

The prediction plots (Figures 2-5), show that much of the prediction error is at level changes, when a variable alternates from a fairly steady low value to a fairly steady high value. In all plots, actual data is shown in solid lines, NN-RBF predictions with dotted lines and NAR predictions with broken lines. Ethylene, ethylene oxide concentrations and temperature are denoted by y2, y3, y4, respectively. At such abrupt changes, the NN often predicts a sharper change than actually occurs, and then retreats to a steady value not as extreme as the real value. This tendency could be due to the size of the window of past values used for prediction: with a smaller window than the 5 past values

1-Ahead Predictions

	NO NOISE Training	NO NOISE Prediction	.5 % NOISE Training	.5 % NOISE Prediction
PRBS Forcing Period of 5 s				
best K	3	3	3	3
Ce	0.234	0.246	0.086	0.095
Ceo	0.001	0.002	0.001	0.002
T	0.004	0.003	0.046	0.048
SSE	0.239	0.251	0.133	0.144
n-SSE	0.053	0.075	0.028	0.047
AIC	0.596		0.544	
PRBS Forcing Period of 10 s				
best K	3	3	3	3
Ce	0.122	0.193	0.122	0.193
Ceo	0.001	0.003	0.001	0.003
T	0.001	0.007	0.001	0.007
SSE	0.124	0.203	0.124	0.203
n-SSE	0.034	0.091	0.034	0.091
AIC	0.557		0.557	
PRBS Forcing Period of 25 s				
best K	10	10	3	3
Ce	0.044	0.047	0.064	0.076
Ceo	0.001	0.001	0.004	0.004
T	0.001	0.002	0.009	0.008
SSE	0.046	0.049	0.076	0.089
n-SSE	0.014	0.017	0.059	0.077
AIC	1.036		0.608	

5-Ahead Predictions

	NO NOISE Training	NO NOISE Prediction	.5 % NOISE Training	.5 % NOISE Prediction
PRBS Forcing Period of 5 s				
best K	3	3	3	3
Ce	0.418	0.439	0.421	0.441
Ceo	0.001	0.002	0.001	0.002
T	0.005	0.004	0.006	0.004
SSE	0.425	0.445	0.428	0.448
n-SSE	0.089	0.112	0.088	0.111
AIC	0.662		0.662	
PRBS Forcing Period of 10 s				
best K	10	10	3	3
Ce	0.238	0.381	0.325	0.478
Ceo	0.0023	0.0033	0.0030	0.0026
T	0.0101	0.0309	0.0078	0.0059
SSE	0.250	0.416	0.335	0.487
n-SSE	0.081	0.123	0.108	0.127
AIC	1.172		0.695	
PRBS Forcing Period of 25 s				
best K	3	3	5	5
Ce	0.213	0.238	0.216	0.248
Ceo	0.0035	0.0034	0.0034	0.0042
T	0.0084	0.0085	0.0051	0.0077
SSE	0.225	0.250	0.225	0.260
n-SSE	0.081	0.093	0.080	0.108
AIC	0.649		0.870	

n-sse = Weighted average of the squared error where the weight is equal to 1/(average of the real value for that variable)

Table 1. Sum of squared errors and AIC for best neural networks

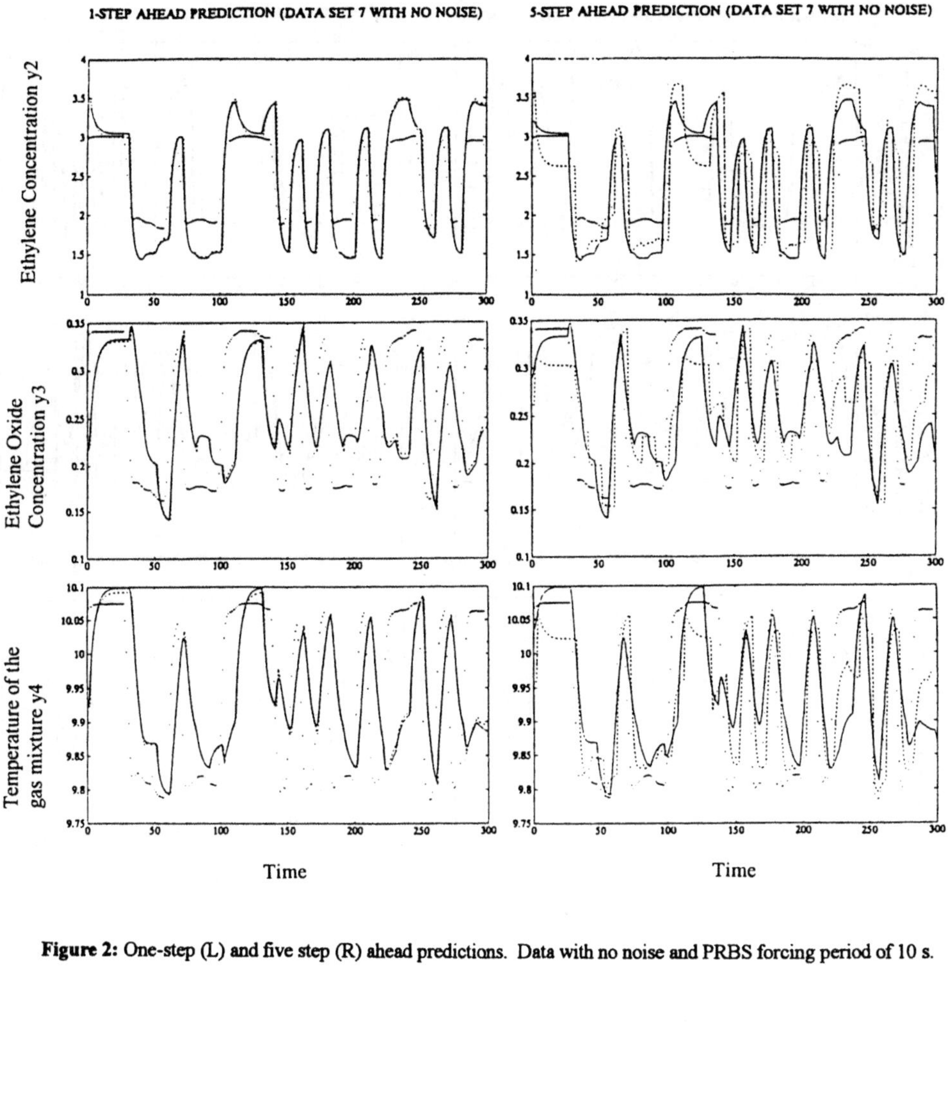

Figure 2: One-step (L) and five step (R) ahead predictions. Data with no noise and PRBS forcing period of 10 s.

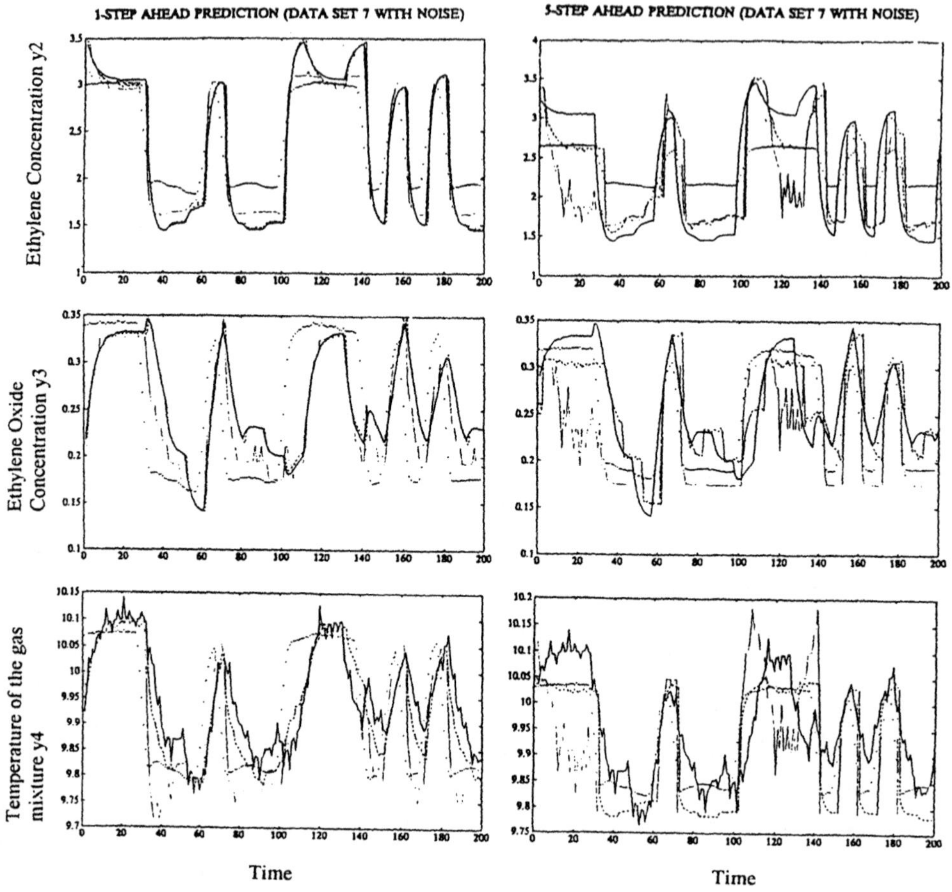

Figure 3: One-step (L) and five step (R) ahead predictions. Data with no noise and PRBS forcing period of 10 s. Lighter solid lines show prediction with neural networks with sigmoid functions.

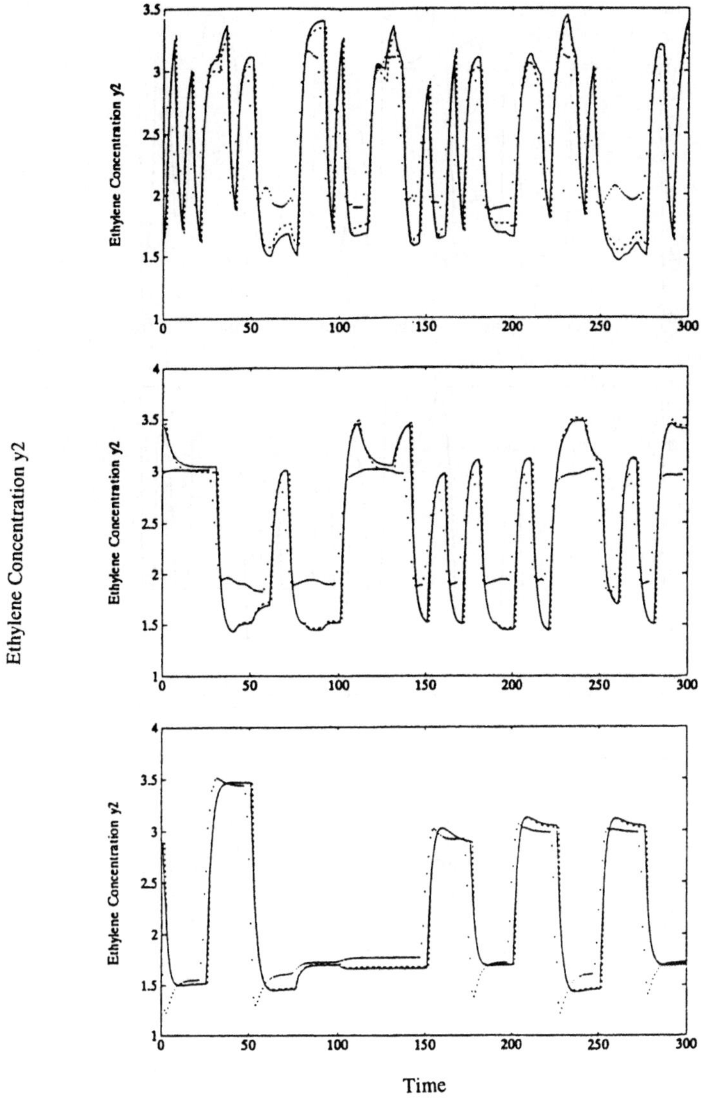

Figure 4: One-step ahead predictions of ethylene concentration. Data with no noise and PRBS forcing period of 5, 10, 25s.

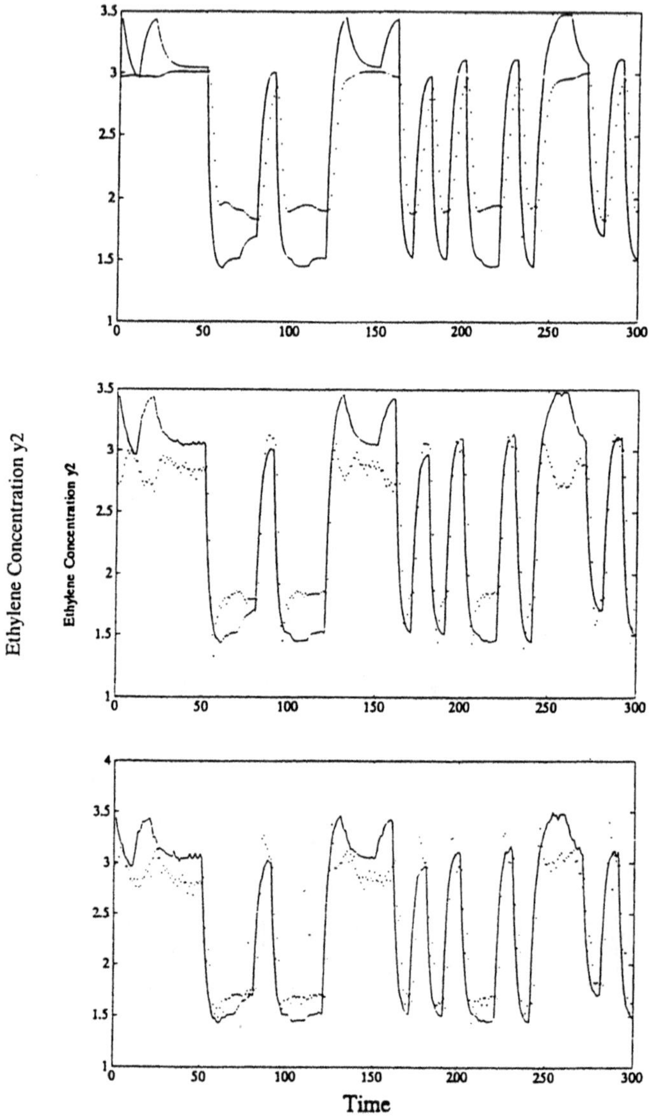

Figure 5: One-step ahead predictions of ethylene concentration. Data with no noise, 1% and 2% noise and PRBS forcing period of 10 s.

used fairly equally in clustering, more recent values would have more impact on prediction, perhaps enabling better "steady" predictions. The predictions based on neural networks with sigmoid functions (NN-S) are also shown in Figure 3 (thin solid lines). The predictions with NN-S are better for 1-ahead predictions of ethylene and ethylene oxide concentrations, but worse for 5-ahead predictions compared with NNRBF predictions. The period of the forcing functions of reactor inputs have a large effect on the accuracy of NN prediction (Figure 4). As the PRBS period increased, the fit and the predictions improved. As expected, in general excessive noise in data degraded the prediction accuracy (Figure 5). However, with a 'small" amount of noise, NNs trained with noisy data not only yielded errors of the same magnitude as those trained without noise, but the NNs with noise made predictions, in some cases, with smaller total errors than the corresponding NNs without noise, sometimes with only half as much error (Table II). There is no clear trend in improvement of prediction accuracy as a function of the number of nodes used. For the case considered in Figure 6, three nodes yielded the smallest SSE, the next best was 15 nodes. A reason for the poor performance of the NN with respect to NAR may be due to clustering data from multivariable systems The clusters may be "nonrepresentative". A feedforward NN with sigmoid functions was developed by utilizing a commercial package. After various adjustments of the data, and long training times (over 2 days with a PC/286 with a math coprocessor) much better fit was obtained. This may indicate the limitations of clustering when a multivariable NN is being developed. Additional work is currently being conducted on this issue. Normalization of the data used in training the NN with RBF did not yield any appreciable improvement.

1-Ahead Predictions

	TRAINING				PREDICTION			
NOISE %	0	0.5	1	2	0	0.5	1	2
# Nodes	3	3	15	15	3	3	15	15
Ce	0.1223	0.1224	0.0446	0.0621	0.1931	0.1927	0.1347	0.1101
Ceo	0.0006	0.0007	0.0027	0.0018	0.0033	0.0033	0.0020	0.0021
T	0.0010	0.0013	0.0021	0.0079	0.0067	0.0069	0.0055	0.0394
SSE	0.1240	0.1245	0.0496	0.0719	0.2031	0.2030	0.1423	0.1517
n-SSE	0.0336	0.0337	0.0547	0.0416	0.0907	0.0908	0.0588	0.0561
AIC	0.5569	0.5572	1.2995	1.2729				

5-Ahead Predictions

	TRAINING				PREDICTION			
NOISE %	0	0.5	1	2	0	0.5	1	2
# Nodes	10	3	3	10	10	3	3	10
Ce	0.2380	0.3247	0.3253	0.2517	0.3813	0.478	0.4779	0.3999
Ceo	0.0023	0.003	0.0030	0.0041	0.0033	0.0026	0.0026	0.0025
T	0.0101	0.0078	0.0084	0.0172	0.0309	0.0059	0.0066	0.0565
SSE	0.2505	0.3355	0.3368	0.2731	0.4157	0.4865	0.4873	0.4589
n-SSE	0.0827	0.1083	0.1089	0.1151	0.1234	0.1266	0.1281	0.1133
AIC	1.1750	0.6946	0.6955	1.2280				

Table 2. Effect of noise on NN selection. SSE for data with PRBS period of 10 s.

1-Ahead Prediction

Variables	without Noise: Models Terms	without Noise: Prediction Errors	with 0.5% Noise: Model Terms	with 0.5% Noise: Prediction Errors
PRBS Input forcing period: 5 s.				
Ce	3	2.67E-2	3	2.67E-2
CeO	5	6.87E-5	6	7.53E-5
T	6	1.34E-4	9	4.28E-3
SSE		2.69		3.12
AIC		-1.39E+3		-1.33E+3
PRBS Input forcing period: 10 s.				
Ce	2	1.67E-2	2	1.67E-2
CeO	4	3.07E-5	4	3.27E-5
T	4	6.38E-5	7	2.34E-3
SSE		1.68		1.91
AIC		-1.54E+3		-1.49E+3
PRBS Input forcing period: 25 s.				
Ce	3	8.88E-3	3	8.91E-3
CeO	4	2.51E-5	4	2.62E-5
T	5	3.49E-5	9	4.35E-4
SSE		8.94E-1		9.36E-1
AIC		-1.72E+3		-1.70E+3

5-Ahead Prediction

Variables	without Noise: Models Terms	without Noise: Prediction Errors	with 0.5% Noise: Model Terms	with 0.5% Noise: Prediction Errors
PRBS Input forcing period: 5 s.				
Ce	3	3.67E-5	3	3.90E-1
CeO	5	1.48E-3	6	1.60E-3
T	6	4.12E-3	9	8.10E-3
SSE		5.61E-1		3.99E+1
AIC		-1.86E+3		-5.69E+2
PRBS Input forcing period: 10 s.				
Ce	2	3.09E-1	2	2.82E-1
CeO	4	1.11E-3	4	9.98E-4
T	4	2.87E-3	7	7.67E-3
SSE		3.12E+1		3.00E+1
AIC		-6.59E+2		-6.65E+2
PRBS Input forcing period: 25 s.				
Ce	3	1.61E-1	3	1.59E-1
CeO	4	7.24E-4	4	7.23E-4
T	5	1.46E-3	9	1.89E-3
SSE		1.63E+1		1.62E+1
AIC		-8.49E+2		-8.44E+2

Table 3. Sum of squared errors in prediction with best NAR models

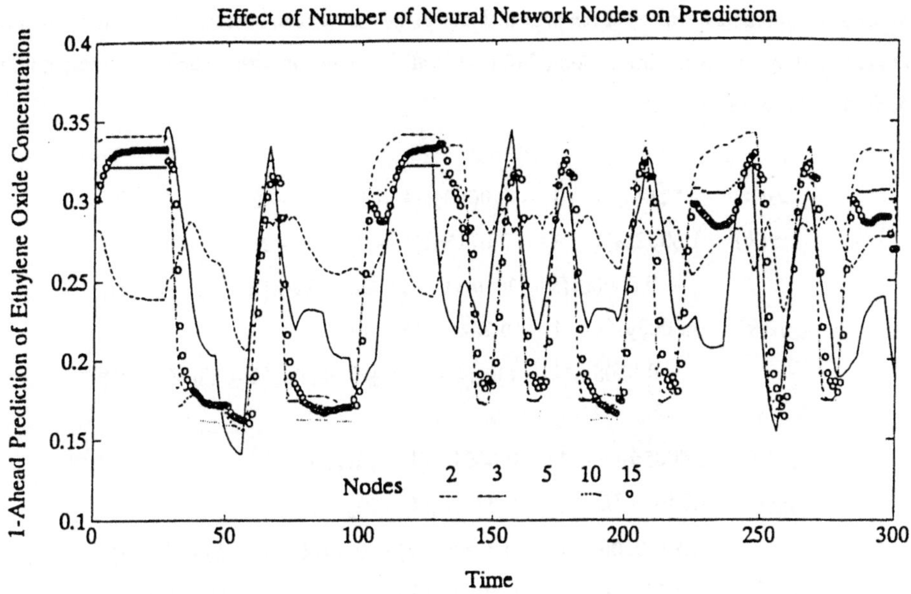

Figure 6. One-step ahead ethylene concentration predictions. Data with no noise and PRBS forcing period of 10 s.

Nonlinear Auto-Regressive Model: For all cases considered, NAR models provided more accurate predictions than NN models (Table III). Since predicted values are used in predicting multi-step ahead values,,the accuracy is affected by the prediction horizon (Figures 2-3). Noise in data necessitates more monomials, and 3 to 9 monomials are enough to minimize AIC. The effect of each monomial added to the NAR model is tabulated for one case in Table IV. The last terms in each column show the effect of the

Monomial Number	Terms Ce	AIC Ce	SSE Ce	AIC CeO	SSE CeO	Terms CeO
1	$y_2(k-1)$	-1247.6	7.22	-2876.9	0.0503	$y_3(k-1)$
2	$y_2(k-2)y_4^2(k-1)$	-1458.9	3.75	-3374.1	0.0109	$y_3(k-2)y_4(k-1)$
3	$y_3(k-1)$	-1456.3	3.73	-3378.1	0.0106	$y_2^2(k-2)y_4(k-2)$
4	...	...	...	-3384.3	0.0103	$y_2(k-10y_2(k-2)$
5				-3380.6	0.0102	$y_2^2(k-1)y_4(k-1)$
6				...	...	...

Table 4. SSE after adding each monomial-PRBS input period 10 s.

next candidate monomial which has not been included in the model. The best NAR models for reactor data generated by PRBS forcing period of 10 s and with no noise or with white measurement noise superimposed are, respectively

$$\begin{aligned} y_2(k) &= 1.6113y_2(k-1) - 0.0062y_4(k-1)^2y_2(k-2) \\ y_3(k) &= 1.6952y_3(k-1) - 0.0715y_3(k-2)y_4(k-2) \\ &\quad + 0.0005y_2(k-1)^2y_4(k-1) - 0.0039y_2(k-1)y_2(k-2); \\ y_4(k) &= 1.8003y_4(k-1) - 0.7548y_4(k-2) \\ &\quad - 0.0005y_4(k-1)y_4(k-2)^2 + 0.0260y_2(k-1)y_3(k-2) \end{aligned} \tag{24}$$

$$\begin{aligned} y_2(k) &= 1.6085y_2(k-1) - 0.0061y_2(k-2)y_4(k-1)^2 \\ y_3(k) &= 1.7402y_3(k-1) - 0.0736y_3(k-2)y_4(k-1) \\ &\quad + 0.0005y_2(k-2)^2y_4(k-2) - 0.0044y_2(k-1)y_2(k-2); \\ y_4(k) &= 1.7456y_4(k-1) - 0.0070y_2(k-1)^3 - 0.1053y_4(k-1)^2 \\ &\quad + 0.0821y_2(k-1)^2y_3(k-2) + 0.0029y_4(k-1)^2y_4(k-2) \\ &\quad - 0.1508y_2(k-2)y_3(k-2) + 0.0030y_2(k-1)^2y_4(k-2)\ . \end{aligned} \tag{25}$$

The best models for reactor data with no noise and PRBS forcing periods of 5 and 25 s are, respectively

$$\begin{aligned} y_2(k) &= 1.5557y_2(k-1) - 0.0064y_2(k-2)y_4(k-1)^2 + 0.0002y_4(k-1)^3 \\ y_3(k) &= 1.7402y_3(k-1) - 0.0718y_3(k-2)y_4(k-2) + 0.0355y_2(k-1) \\ &\quad - 0.0139y_2(k-2) + 0.0003y_2(k-2)^3 - 0.0003y_2(k-1)y_4(k-2)^2 \\ y_4(k) &= 1.2795y_4(k-1) - 4.9542y_3(k-2)^3 + 414.9254y_3(k-1)^2 \\ &\quad - 0.0031y_4(k-1)^3 - 0.0410y_2(k-2)y_3(k-2)y_4(k-1) \\ &\quad + 35.6044y_3(k-1)y_3(k-2)y_4(k-1) - 40.1064y_3(k-1)^2y_4(k-1) \\ &\quad + 0.0565yy_2(k-2)^2y_3(k-2) - 362.9179y_3(k-1)y_3(k-2) \end{aligned} \tag{26}$$

$$
\begin{aligned}
y_2(k) &= 1.6514y_2(k-1) - 0.6306y_2(k-2) - 0.0008y_2(k-2)^2 y_4(k-1) \\
y_3(k) &= 1.7662y_3(k-1) - 0.7761y_3(k-2) + 0.0111y_2(k-1) - 0.0001y_2(k-2)y_4(k-2)^2 \\
y_4(k) &= 0.6039y_4(k-1) + 0.3873y_4(k-2) + 0.1299y_2(k-1) \qquad (27) \\
&\quad - 0.0059y_2(k-1)y_2(k-2)y_4(k-2) + 0.0047y_2(k-2)^3 \\
&\quad + 0.1053y_2(k-1)^2 y_3(k-1) - 0.2363y_2(k-1)y_3(k-2)y_4(k-1) \\
&\quad + 2.0335y_2(k-1)y_3(k-2) + 0.1319y_2(k-2)y_3(k-1)
\end{aligned}
$$

6. Conclusions

Nonlinear time series modeling techniques and neural network techniques offer methods that can be implemented easily for modeling processes with severe nonlinearities. In this study, the NAR models have provided more accurate predictions than NN models with radial basis functions. These results are certainly not conclusive evidence to draw general conclusions. However, they indicate that other nonlinear modeling paradigms can be used as easily and may provide as good models as the NN approach. Both approaches have strong points: NN models may be trained directly for multistep ahead predictions, while NAR models are more parsimonious and the functional relationships can provide physical insight. The availability of general purpose software and the capability to capture nonlinear relations have made NN a popular paradigm. This popularity has fueled the interest in other nonlinear modeling paradigms. We are hopeful that future studies will provide powerful nonlinear modeling methods as well as guidelines and heuristics in selecting the most appropriate paradigm for specific types of modeling problems.

References

1. Bhat, N. and T. J McAvoy (1990): Use of Neural Networks for Dynamic Modeling and Control of Chemical Process Systems, Computers Chem. Engng 14 (4/5) 573
2. Chen S., S. A. Billings and W. Lou (1989): Orthogonal least squares methods and its application to non-linear system identification, Int. J. Ctrl., 50 (5), 1873-1896
3. Cybenko, G. (1989): Approximations by Superpositions of a Sigmoidal Function, Math. Cont. Signal & Systems,
2 303-314
4. Haber R. and H. Unbehauen (1990): Structure Identification of Nonlinear Dynamic Systems - A survey on Input/Output Approaches, Automatica, 26, 651-677
5. Haesloop, D. and B. Holt (1990): A Neural Network Structure for System Identification, Proc. Ameri. Cntrl Conf. 2460
6. Hernandez, E. and Y. Arkun (1990): Neural Network Modelling and an Extended DMC Algorithm to Control Nonlinear Systems, Proc. Ameri. Cntrl Conf. 2454
7. Himmelblau, D. M. (1972): Applied Nonlinear Programming, McGraw-Hill, New York,
8. Holcomb, T. and M. Morari (1990): Analysis of Neural Controllers, AIChE Annual Meeting. Paper No. 16a.
9. Hoskins, J. C. and D. M. Himmelblau (1988): Artificial Neural Network Models of Knowledge Representation in Chemical Engineering Comput. Chem. Engng 12, 881
10. Korenberg M. J. (1985): Orthogonal Identification of Nonlinear Difference Equation Models, Models, midwest Symp. on Circuits and Systems, Louisville, KY
11. Leonard, J. A. and M. Kramer (1990): Classifying Process Behavior with Neural Networks: Strategies for Improved Training and Generalization, Proc. Ameri. Cntrl Conf. 2478
12. Leontaritis I. J. and S. A. Billings (1985): Input-output parametric models for nonlinear systems, Int. J. Ctrl., 41, 303-344
13. Lin, Han-Fei (1992): Approximate Dynamic Models with Back-Propagation Neural Networks, Project Report, Illinois Institute of Technology
14. Moody and Darken (1988): learning with Localized Receptive Fields, Research Report YALEU/DCS/RR-649, Yale Computer Science Department, New Haven, Connecticut
15. Niranjan M. and F. Fallside (1988): Neural Networks and Radial Basis Functions in Classifying Static Speech Patterns, Report No. CUED/F-INFENG/TR 22, University Engineering Department, Cambridge, England
16. Ozgulsen F., R. A. Adomaitis and A. Cinar (1991): Chem. Eng. Sci., in press
17. Pollard, J. F., D. B. Garrison, M. R. Broussard and K. Y. San (1990): Process Identification using Neural Networks, AIChE Annual Meeting. Paper No. 96a
18. Rigopoulos, K. (1990): Selectivity and Yield Improvement by Forced Periodic Oscillations: Ethylene Oxidation Reaction, Ph D. Thesis, Illinois Institute of Technology, Chicago,IL
19. Roat, S. and C. F. Moore (1990): Application of Neural Networks and Statistical Process Control to Model Predictive Control Schemes for Chemical Process Industry, AIChE Annual Meeting. Paper No. 16b.
20. Sanner and Slotine (1991): Gaussian Neural Networks for Direct Adaptive Control, Proc. Amer. Control Conf., 2153
21. Ungar, L. H., B. A. Powell and S. N. Kamens (1990): Adaptive Networks for Fault Diagnosis and Process Control, Computers Chem. Engng 14 (4/5) 561
22. Venkatasubramanian, V., R. Vaidyanathan and Y. Yamamato (1990): Process Fault Detection and Diagnosis Using Neural Networks: I. Steady State Processes, Comput. Chem. Engng 14, 699
23. Whiteley, J. R. and J. F. Davis (1990): Backpropagation Neural Networks for Qualitative Interpretation of Process Data, AIChE Annual Meeting. Paper No. 96d
24. Willis, M. J., G. A. Montague, A. J. Morris, and M. T. Tham (1991): Artificial Neural Networks:- A Panacea to Modelling Problems?, Proc. Ameri. Cntrl Conf. 2337
25. Yao, S. C. and E. Zafiriou (1990): Control System Sensor Failure Detection via Network of Local Receptive Fields, Proc. Ameri. Cntrl Conf. 2472
26. Ydstie, B. E. (1990): Forecasting and Control Using Adaptive Connectionist Networks, Computers Chem. Engng 14 (4/5) 583

Systems of Differential-Algebraic Equations

R. W. H. Sargent

Imperial College of Science, Technology and Medicine, Centre for Process Systems
Engineering, London, SW7 2BY, UK

Abstract: The paper gives a definition and a general description of the properties of systems of differential-algebraic equations. It includes a discussion of index and regularity, giving simple illustrative examples. It goes on to describe methods for index-reduction and numerical methods for solution of high-index problems.

Keywords: Differential-algebraic equations, high-index problems, ordinary differential equations, regularity, index-reduction, numerical solution, applications

1. Introduction - What is the Problem?

Most chemical engineering degree courses give a good grounding in the theory of differential equations, and in numerical methods for solving them. However, dynamic models for most process systems consist of mixed systems of differential and algebraic equations and these sometimes have unexpected properties. It is the purpose of this talk to describe the properties of such systems, and methods for their numerical solution.

1.1 Systems of Ordinary Differential Equations (ODEs)

Let us start by recalling the general approach to the numerical solution of ordinary differential equations (ODEs) of the form:

$$\dot{x}(t) = f(t, x(t)), \text{ where } t \in R, \; x(t) \in R^n \text{ and } f: R \times R^n \to R^n \tag{1.1}$$

For numerical solution we seek to generate a sequence $\{x_n\}$, n = 0, 1, 2, ... which approximates the true solution, $x_n \approx x(t_n)$ at a sequence of times $\{x_n\}$, n = 0, 1, 2... using equation (1.1) in the form:

$$\dot{x}_k = f(t_k, x_k), \tag{1.1a}$$

where $\dot{x}_k \approx x(t_k)$.

Linear multistep methods for the numerical solution of ordinary differential equations make use of formulae of the form:

$$x_k = \gamma_k h_k \dot{x}_k + \varphi_{k-1}, \tag{1.2}$$

where $h_k = t_k - t_{k-1}$, γ_k is a scalar, and φ_{k-1} a function of past values $x_{k'}$, $\dot{x}_k$, $k' = k-1, k-2, \ldots$. Of course, the value of γ_k and the form of the function φ_{k-1} depend on the particular formula.

The method is called explicit if $\gamma_k=0$, yielding explicitly $x_k = \varphi_{k-1}$, and implicit if $\gamma_k \neq 0$, in which case (1.1 a) and (1,2) have to be solved to obtain x_k. An explicit method involve less computation per step, but the step-length is limited by stability considerations, and if these dictate a very small step-length it is worth using an implicit method, allowing a larger step. In this case, an explicit method is often used to give an initial prediction of x_k, followed by iterative solution of (1.1a) and (1.2) to yield a corrected value. Usually Newton's method is used for the iteration, or rather its simplified version in which the Jacobian matrix $f_x(t_k, x_k)$ is held fixed and only re-evaluated every few steps.

For initial-value problems, we require the solution for $t \geq t_0$ given initial values $x(t_0) = x_0$. Of course, in applying (1.2) the past consists only of x_0, $\dot{x}_0$, so (1.2) can only contain two parameters, and hence approximates $x(t_1)$ only to first order. As we generate further points in the sequence, we have more information at our disposal and can use a formula giving a higher order of approximation. Multistep methods in use today have reached a high degree of refinement, incorporating automatic error estimation and approximate optimization to choose the order of approximation of the formula, the step-length, and the frequency of re-evaluation of the Jacobian matrix.

Runge-Kutta methods make use of several evaluations of $\dot{x}(t)$ from (1.1) per step, but the same general principles apply. For fuller details of both kinds of method, the reader is referred to the excellent treatise by Butcher [3].

1.2 Systems of Differential-Algebraic Equations (DAEs)

To illustrate the use of such standard methods, let us consider the following simple example:

Example 1 - Stirred tank batch reactor.

We consider a perfectly stirred tank in which the simple endothermic reactions:

$$A \underset{1}{\rightarrow} B \underset{2}{\rightarrow} C$$

take place. The reactor is heated by a coil in which steam condenses (and the condensate is removed as it is formed via a steam-trap). The instantaneous steam flow-rate is F_s and the condensation temperature T_s. The contents of the reactor are at temperature T, with molar concentrations a, b, c of components A, B, C respectively, and total volume V. Both reactions are first order, with rate expressions

$$k_1 = k_{10} \exp(-E_1/RT), \qquad k_2 = k_{20} \exp(-E_2/RT) \qquad (1.3)$$

where R is the gas constant and k_{10}, E_1, k_{20}, and E_2 are given constants.

If A, B, C represent the total amounts of the corresponding components present in the reactor, we have immediately:

$$A = Va, \qquad B = Vb, \qquad C = Vc. \qquad (1.4)$$

The dynamic molar balances on each component are then given by:

$$\dot{A} = -k_1 A, \qquad \dot{B} = k_1 A - k_2 B, \qquad \dot{C} = k_2 B, \qquad (1.5)$$

while the energy balance yields

$$\dot{H} = F_s . \Delta H_s (T_s) = U S (T_s - T) \qquad (1.6)$$

where $\Delta H_s (T_s)$ is the latent heat of condensation of the steam at T_s, S is the heat-transfer surface area of the coil, U the overall heat-transfer coefficient (assumed constant), and H is the total heat content of the material in the reactor, given by:

$$H = h_A A + h_B B + h_C C \qquad (1.7)$$

where h_A, h_B, and h_C are partial molar enthalpies.
We also have

$$= v_A\ A + v_B\ B + v_C\ C \tag{1.8}$$

where v_A, v_B, v_C, are partial molar volumes, and these partial molar quantities are given by

$$v_A = v_A(T, a, b, c), \qquad h_A = h_A(T, a, b, c) \ldots \tag{1.9}$$

This example shows how mixed differential-algebraic systems naturally arise in modelling the dynamics of chemical processes. Material and energy balances usually give rise to first-order differential equations in time, while volume relations, chemical kinetic expressions and physical property relations give rise to algebraic relations between the instantaneous values of the variables.

To use a standard ODE integration method to solve these equations, we need to convert the equations to the form of (1.1), which implies eliminating the variables whose derivatives do not appear in the equations (henceforth called the "algebraic variables"), leaving relations between the others (called the "differential variables").

We could carry out this elimination symbolically, but it is equivalent to do it numerically at each step, using the following algorithm:

Given A, B, C, H:

1. Solve (1.4), (1.7), (1.8), (1.9) for T, V, a, b, c.
2. Compute k_1, k_2 from (1.3) and F_s from (1.6)
3. Compute $\dot{A}$, $\dot{B}$, $\dot{C}$, $\dot{H}$ from (1.5) and (1.6).

For the initial conditions, we would be given the initial contents of the reactor, hence T(0), V(0), a (0), b (0), c(0), from which A, B, C, H are easily calculated using (1.4), (1.7) and (1.9), and remaining variables as in steps 2 and 3 above.

The approach used in this example can be generalized to deal with any differential-algebraic system in the semi-explicit form:

$$\dot{x}(t) = f(t, x(t), y(t)), \tag{1.10}$$

$$0 = g(t, x(t), y(t)), \tag{1.11}$$

where $x(t) \in R^n$ represents the differential variables, $y(t) \in R^m$ the algebraic variables, $f: R \times R^n \times R^m \to R^n$, $g: R \times R^n \times R^m \to R^m$, and (1.11) can be solved for y (t) in terms of t and x (t).

It will not in general be possible to solve (1.11) analytically, but (as in the above example) it can be solved numerically. If g (t, x, y) is nonlinear in y we shall need an iterative solution, and if Newton's method is used, we shall need to evaluate the Jacobian matrix g_y (t, x, y).

Alternatively, we note that since (1.11) holds at every time t, then its total derivative with respect to time must also be zero, yielding

$$0 = g_t + g_x \cdot \dot{x} + g_y \cdot \dot{y}, \tag{1.12}$$

where for simplicity, we have dropped the arguments, which are the values of t, x (t), y (t) on the solution trajectory. If the Jacobian matrix g_y of $\dot{y}$ in (1.12) is nonsingular, the equation can be solved for $\dot{y}$ (t), and we can eliminate $\dot{x}$ (t) using (1.10) to yield symbolically:

$$\dot{y} = -[g_y]^{-1} [g_t + g_x f], \tag{1.13}$$

though of course we would normally carry out these operations numerically. Equations (1.10) and (1.13) now give an ODE system in (x, y), to which standard integration methods can be applied.

The most general form of differential-algebraic system is:

$$f(t, \dot{x}(t), x(t), y(t)) = 0, \tag{1.14}$$

where again $t \in R$, $x(t) \in R^n$, $y(t) \in R^m$,

but now $f: R \times R^n \times R^n \times R^m \to R^{n+m}$,

and in general f (·) can be nonlinear in all its arguments.

However, if (1.14) can be solved for $\dot{x}$ (t), y (t), given t and x (t), we can again use a standard ODE integration method, with the y (t) variables being obtained as a by-product of the solution for $\dot{x}$ (t).

In all cases considered above, we have assumed that the equations can be solved for the variables in question, but unfortunately this is often not the case, as illustrated by next example:

Example 2 - Continuous Stirred-Tank Reactor.

We again consider the system described in Example 1, but this time there is a continuous feed, with volumetric flow-rate F^0, temperature T^0 and composition a^0, b^0, c^0, and continuous product withdrawal with flow-rate P, with of course temperature and composition identical to those of the tank contents. The describing equations are now:

$$k_1 = k_{10} \exp(-E_1/RT), \qquad k_2 = k_{20} \exp(-E_2/RT) \tag{1.15}$$

$$A = Va, \qquad B = Vb, \qquad C = Vc, \qquad H = Vh \tag{1.16}$$

$$\dot{A} = F^0 a^0 - Pa - Vk_1 a \tag{1.17}$$

$$\dot{B} = F^0 b^0 - Pb + Vk_1 a + Vk_2 b \tag{1.18}$$

$$\dot{C} = F^0 c^0 - Pc + Vk_2 \tag{1.19}$$

$$\dot{H} = F^0 h^0 - Ph + US(T_s - T) \tag{1.20}$$

$$F_s \Delta H_s(T_s) = US(T_s - T) \tag{1.21}$$

$$v_A A + v_B B + v_C C = V \tag{1.22}$$

$$h_A a + h_B b + h_C c = h \tag{1.23}$$

$$v_A = v_A(T, a, b, c), \qquad h_A = h_A(T, a, b, c), \ldots \tag{1.24}$$

If the feed (F^0, T^0, a^0, b^0, c^0) and product flow (P) are given as functions of time, there is no difficulty in evaluating all the remaining variables in terms of A, B, C, H, using the scheme given in Example 1.

However, if P is adjusted to maintain the hold-up V constant at a given value, it is impossible to determine P or the derivatives $\dot{A}$, $\dot{B}$, $\dot{C}$, $\dot{H}$ from the above equations. This is evident from the fact that P, an algebraic variable, appears only in the equations which determine the derivatives.

Nevertheless, it is clear on physical grounds that the system is well defined!

Closer examination reveals that A, B, C, H are related through (1.22), so their derivatives must also be related, and the required equation is obtained by differentiating (1.22). Substitution of $\dot{A}$, $\dot{B}$, $\dot{C}$ from (1.17), (1.18), (1.19) will then yield on algebraic equation, which can be solved for P. The values of the remaining variables can then be obtained as before.

This example shows that locally unique solutions can exist, even if the Jacobian matrix for the system is singular, and also that further information is implicit in the requirement that the algebraic equations must be satisfied at **every** instant of time - hence differentiating them provides further relevant equations.

The next example shows that a single differentiation may not be enough.

Example 3 - The Pendulum

We consider a pendulum consisting of a small bob suspended by a string of negligible weight. We shall use Cartesian coordinates, as indicated in Figure 1, and assume that the string is of unit length and the bob of unit weight (i.e. mg = 1).

The describing equations are:

$$\dot{x} = u, \qquad \dot{y} = v \tag{1.25}$$

$$\dot{u} = zx, \qquad \dot{v} = 1 - zy \tag{1.26}$$

$$x^2 + y^2 = 1 \tag{1.27}$$

Here, we have 4 differential variables (x, y, u, v) and one algebraic variable (z), which again appears only in the equations defining the derivatives.

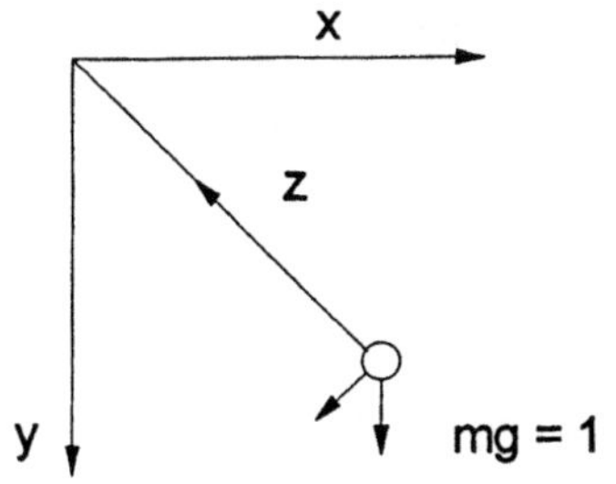

x = horizontal distance from fulcrum   u = horizontal velocity

y = vertical distance from the fulcrum   v = vertical velocity

z = tension of the string

Figure 1. The Pendulum

Again, it is clear that u and v are related, and by differentiating (1.27) we obtain:

$$xu + yv = 0 \tag{1.28}$$

This still does not allow us to determine z, but again $\dot{u}$ and $\dot{v}$ are related, as shown by differentiating (1.28): $x\,\dot{u} + \dot{x}\,u + y\,\dot{v} + \dot{y}v = 0$,

and substitution from (1.25) and (1.26) yields:

$$z = (y + u^2 + v^2)/(y^2 - x^2) \tag{1.29}$$

Thus, two successive differentiations of (1.27) were required to obtain all the necessary information for solution.

2. Properties of DAE Systems

Let us examine the structure and properties of DAE systems a little more closely.

From now on, it will be more convenient not to identify separately the algebraic variables, and we shall consider the general form:

$$f(t, x(t), \dot{x}(t)) = 0 \tag{2.1}$$

with $t \in R$, $x(t) \in R^n$, $f\colon R \times R^n \times R^n \to R^n$

We shall also restrict ourselves to the case where $f(\cdot)$ is continuous in $x(t)$, $\dot{x}(t)$, though not necessarily in t, since this covers most cases of practical interest.

We shall need the following formal definitions:

Definition 1 A solution of (2.1) is a function x(t) defined and continuous on $[t_0, t_f]$, and satisfying [2.1] almost everywhere on $[t_0, t_f\ ]$,.

Definition 2 A system (2.1) is regular if:

a) It has at least one solution.

b) A solution through any point (t, x) is unique.

Definition 3 The **index** of system (2.1) is the smallest non-negative integer m such that the system:

$$\left.\begin{aligned} & f(t, x(t), \dot{x}(t)) = 0, \\ & \frac{d^r}{dt^r} f(t, x(t), \dot{x}(t)) = 0, \qquad r = 1, 2, \ldots m, \end{aligned}\right\} \tag{2.2}$$

defines $\dot{x}(t)$ as a **locally unique** function of t, x (t),

Obviously, m will in general vary as t, x (t) vary, so the index is a local property.

The existence of a finite index at a point (t, x) is a necessary and suffcient condition for the existence of a unique solution trajectory through (t, x), and this solution can be extended so long as the index remains bounded. Thus, a system is regular on any domain of (t, x) on which a bounded index exists.

Of course, the index may fail to exist at a given point because derivatives of $f(\cdot)$ of sufficienty high order do not exist at this point. However, such points will in general form a set of measure zero, defining submanifolds in the full space of variables. Pathological functions for which derivatives of a certain order are undefined over a full-dimensional region are unlikely to arise from modeling physical systems.

It also follows from the definition that the index is invariant under nonlinear algebraic one-to-one transformations of the variables or equations.

The following examples further illustrate some of the implications of the above:

Example 4

Consider the system:

$$\dot{x} - 2y^2 + 1 = 0 \tag{2.3}$$

$$\dot{y} - 2z^2 + 1 = 0 \tag{2.4}$$

$$x^2 + y^2 - 1 = 0 \tag{2.5}$$

We note that, z occurs only in the equation defining $\dot{y}$ so the index must be greater than one. Differentiating (2.5) and substituting from (2.3) and (2.4) we obtain:

$$x(2y^2-1)+y(2z^2-1)=0 \tag{2.6}$$

This determines z in terms of x and y (though not uniquely), but we must differentiate again to obtain an equation for $\dot{z}$:

$$(2y^2-1)^2+(2z^2-1)^2+4xy(2z^2-1)\,4yz\,\dot{z}=0 \tag{2.7}$$

Now, it can be shown that neither y=0 or z = 0 is consistent with (2.5), (2.6) and (2.7), so (2.7) uniquely defines $\dot{z}$, and the index is two for all x, y, z. It follows that the system is regular.

However, if we set $x=\pm 1/\sqrt{2}$, it follows from (2.5) that $y=\pm 1/\sqrt{2}$ and from (2.6) that $z=\pm 1/\sqrt{2}$, and these are consistent with (2.3), (2.4) and (2.7) almost everywhere. Hence, there is an infinity of solutions with x, y and z switching independently between $+1/\sqrt{2}$ and $-1/\sqrt{2}$ arbitrarily often!

Nevertheless, there is no contradiction; none of these "solutions" are a solution in the sense of Definition 1, since they are not continuous.

Of course, there may be situations in which we are interested in such discontinuous solutions, but we must realize that they fall outside the scope of the standard theory.

Examle 5

Consider the system:

$$\dot{x}_1=x_1+x_2-y \tag{2.8}$$

$$\dot{x}_2=x_1+x_2-y \tag{2.9}$$

$$x_1=2x_2+u=0 \tag{2.10}$$

where u(t) is a piece-wise constant control function (i.e. $\dot{u}(t)=0$ almost everywhere). Differentiating (2.10) and substituting from (2.8) and (2.9):

$$3x_1-x_2-y=0, \qquad \text{a.e.} \tag{2.11}$$

Now suppose that $u(t)=1 \quad , \quad 0\le t<t_1$

$u(t)=2 \quad , \quad t_1\le t<t_2$

and $x_1=1$ at $t=t_1^-$

Then, at $t=t_1^-$ we have $x_2=-1$ from (2.10) and $y=4$ from (2.11).

But, what are the values of x_1, x_2, y at t_1 ? From (2.10), it is clear that x_1 or x or both must have a jump at t_1, but nothing more can be deduced from the above, and the solution for $t_2\ge t_1$ is not uniquely defined.

Nevertheless, the system satisfies all the conditions for (2.1), and clearly has an index of two everywhere, so it is regular. Again, however, there is no contradiction, because a solution in the sense of Definition 1 cannot exist at times where *u* jumps.

However, we are often interested in solving optimal control problems for which jumps in the control are allowed, and hence in solutions which may be discontinuous at such points.

In a real physical situation, we would in fact have information on the behaviour of physical quantities in the presence of discontinuities in the controls. For example, if we have a tank of liquid, and the control *u* represents the **rate** of addition of hot feed, there cannot be a jump in the temperature of the tank contents. On the other hand, if *u* represents an instantaneous addition of a finite quantity of feed (an "impulse" of feed), there will be a corresponding jump in the temperature of the contents, (obtained by energy balance).

The most important point to note is that specification of behaviour at points of discontinuity is a part of the model, and is not implicit in the DAE system itself. We obtain the complete solution by piecing together segments on which the DAE system does have a solution in the sense of Definition 1, using these "junction conditions" for the purpose.

Example 6

Consider the system:

$$\dot{x} + x^2 + y\,z - 1 = 0 \tag{2.12}$$

$$\dot{y} + x\,y - x\,z = 0 \tag{2.13}$$

$$x^2 + y^2 - 1 = 0 \tag{2.14}$$

Again, we must differentiate (2.14) and substitute from (2.12) and (2.13) to obtain an equation for z:

$$x\,(1 - x^2 - yz) + y\,(xz - xy) = 0.$$

Thus, using (2.14):

$$x\,y\,(y - z) + y\,x\,(z - y) = 0,$$

and the equation is satisfied identically!

Obviously, this will remain true for further differentiations, so we can obtain no more information not already implicit in the original formulation (2.12) - (2.14). However this is not enough to determine $\dot{z}(t)$, or even z (t), so the index is not defined.

In fact, we are free to choose any arbitrary function z (t), and any initial values for x and y consistent with (2.14), and x (t), y (t) will then be uniquely defined.

This example shows the dangers of **assuming** that a system has a finite index, and hence inferring that it is regular.

Having given warnings of possible pitfalls, it may now be helpful to discuss the structure and properties of linear, constant-coefficient DAE systems, which have the general form:

$$A.\dot{x}(t) + B.x(t) = c(t), \tag{2.15}$$

where A and B are n x n matrices and x(t), $\dot{x}(t)$, c(t) are n-vectors.

The analysis of such systems dates back to Kronecker, and an excellent detailed analysis of the general case will be found in Gantmacher [5]. Here, we will treat only the regular case, where regularity of the system in the sense of Definition 2 coincides with regularity of the matrix pencil $[A + \lambda B]$, where λ is a scalar.

Such a pencil is said to be regular if det $|A + \lambda B|$ is not identically zero for all λ; otherwise it is said to be singular. It then follows that for a regular pencil, $|A + \lambda B|$ is nonsingular except when λ coincides with one of the n roots of the equation:

$$\det |A + \lambda B| = 0. \tag{2.16}$$

For a regular pencil $|A + \lambda B|$, there exist nonsingular n x n matrices P and Q such that

$$P[A + \lambda B]Q = \begin{vmatrix} I_0 & \cdot & \cdots & \cdot \\ \cdot & N_1 & \vdots & \cdot \\ \vdots & \vdots & \ddots & \vdots \\ \cdot & \cdot & \cdots & N_m \end{vmatrix} + \lambda \begin{vmatrix} B_0 & \cdot & \cdots & \cdot \\ \cdot & I_1 & \cdots & \cdot \\ \vdots & \vdots & \ddots & \vdots \\ \cdot & \cdot & \cdots & I_m \end{vmatrix} \tag{2.17}$$

where

$$I_r = \begin{vmatrix} 1 & \cdot & \cdots & \cdot \\ \cdot & 1 & \cdots & \cdot \\ \vdots & \vdots & \ddots & \vdots \\ \cdot & \cdot & \cdots & 1 \end{vmatrix}, \quad N_r = \begin{vmatrix} \cdot & 1 & \cdots & \cdot \\ \cdot & \cdot & \cdots & \cdot \\ \vdots & \vdots & \ddots & 1 \\ \cdot & \cdot & \vdots & \cdot \end{vmatrix} \tag{2.18}$$

and I_r, N_r, $r = 0, 1, \ldots m$, are n_r x n_r matrices.

The **index of nilpotency** of the pencil is $\max\limits_r \; n_r$

Now, to relate this to the system (2.15), let us define:

$$x(t) = Q.z(t) \qquad , \qquad \bar{c}(t) = P.c(t) \tag{2.19}$$

Then from (2.15), (2.17), (2.18) and (2.19):

$$\left.\begin{aligned} \dot{z}_0 + B_0 z_0 &= \overline{c_0} \\ N_r \, \dot{z} + z_r = \overline{c_r}, \quad r &= 1, 2, \ldots m \end{aligned}\right\} \tag{2.20}$$

and for a typical N_r:

$$\left.\begin{aligned} \dot{z}^{(i+1)} + z_r^{(i)} &= \overline{c_r}^{(i)}, \quad i = 1, \ \ldots (n_r - 1) \\ z_r^{(n_r)} &= \bar{c}^{(n_r)}, \end{aligned}\right\} \tag{2.21}$$

from which we deduce:

$$z_r^{(n_r - i)} = \frac{d^t}{dt^r} \bar{c}^{(n_r)} + \frac{d^{r-1}}{dt^{r-1}} \bar{c}^{(n_r - 1)} + \ldots + \frac{d}{dt} \bar{c}^{(n_r - i + 1)} + \bar{c}^{(n_r - i)} \tag{2.22}$$

From (2.20) and (2.22) we see that the system completely decomposes into (m+l) non-interacting subsystems. The first of these (r = 0) is a standard linear constant-coefficient ODE, explicit in $\dot{z}_0$, whose general solution involves n_0 arbitrary constants. The other m systems contain no arbitrary constants, and the z_r are obtained from the RHS vector and successive differentiations of it.

Thus, to obtain $\dot{z}_r$ we need n_r differentiations of $\overline{c_r}$ and it is clear that the index of the DAE system (as given in Definition 3) is equal to the index of nilpotency of the matrix pencil $[A + \lambda B]$.

This analysis clearly shows that the system index is a property of the system equations, and independent of the specification of the boundary conditions, but caution is necessary here, as the next example shows:

Example 7. Flow through Tanks.

We consider the flow of a binary mixture through a sequence of n perfectly stirred tanks, as

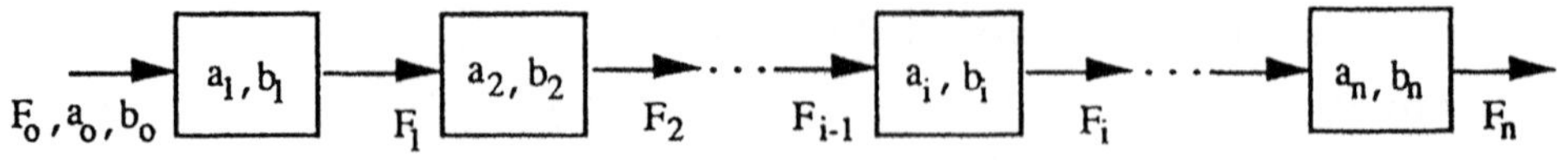

Figure 2

illustrated in Figure 2. The molar concentrations of the components leaving tank i are a_i, b_i, with volumetric flow-rate F_i while the feed to the first tank has constant flow-rate F_0 and time-varying composition a_0, b_0. For simplicity, we assume that all the tanks have the same volume, V.

The describing equations are:

$$V\frac{da_i}{dt} = F_{i-1}a_{i-1} - F_i a_i \quad , \quad i = 1, 2, \dots n, \tag{2.23}$$

$$V\frac{db_i}{dt} = F_{i-1}b_{i-1} - F_i b_i \quad , \quad i = 1, 2, \dots n, \tag{2.24}$$

$$V_a a_i + V_b b_i = 1 \quad , \quad i = 1, 2, \dots n, \tag{2.25}$$

where v_a, v_b are partial molar volumes of the two components, here assumed constant for simplicity.

As in Example 2, equation (2.25) relates the two differential variables, so we differentiate it to yield:

$$v_a\frac{da_i}{dt} + v_b\frac{db_i}{dt} = 0 \, ,$$

Then substituting from (2.23) and (2.24) and using (2.25) yields:

$$F_{i-1} = F_i \quad , \quad i = 1, 2, \dots n.$$

Since all flows are equal to F_0, which is constant, we can transform the time variable:

$$\tau = F_0\, t / V, \tag{2.26}$$

whereupon (2.23) and (2.24) can be written:

$$\left.\begin{aligned} \dot{a}_i &= a_{i-1} - a_i \\ \dot{b}_i &= b_{i-1} - b_i \end{aligned}\right\} \quad i = 1,2, \dots n \tag{2.27}$$

were $\dot{a}_i$ denotes $da_i/d\tau$ etc., and we see that the system decomposes into two independent sub-systems.

Now, if we specified the feed composition a_0 (t), b_0 (t), $t \geq t_0$, and the initial concentrations in the tanks: a_i (t_0), b_i (t_0), i = 1, 2, ... n, (2.27) represents a constant-coefficient ODE system (hence of index zero), which can be integrated from $t = t_0$.

On the other hand, if we require the feed composition to be controlled so that the outlet concentrations a_n (t), b_n (t) follow a specified trajectory for $t \geq t_0$ then $\dot{a}_n$ (t), $\dot{b}_n$ (t), are thereby specified, and we must use (2.27) to compute a_{n-1}, b_{n-1} .The same then applies to tank (n -1), and so on recursively, until $a_0(t)$, $b_0(t)$ are computed.

In this second case, we were not free to choose the initial concentrations in the tanks, and the solution was obtained by successive differentiations instead of integration, showing that the index is (n+1).

Thus, the index of the system is crucially dependent on the type of boundary condition specified!

The paradox is resolved by noting that the "system" just referred to is the physical system, and the corresponding DAE system is different for the two cases. In the first case the, differential variables are $a_i(t)$, $b_i(t)$, i = 1, 2, ... n, $a_0(t)$, $b_0(t)$ and are given driving functions, while in the second case the variables are and the specifications $a_n^*(t)$, $b_n^*(t)$ are given driving functions.

From the above properties of the linear constant-coefficient system, in which regularity and index of the DAE system coincide with regularity and index of the matrix pencil [A + λ B], one is led to suspect that the same relations might hold in the nonlinear case for the local linearization of the DAE system. The matrix pencil in question would then be $[f_{\dot{x}} + \lambda \, . f_x]$, but unfortunately no such correspondence exists, even in the linear time-varying case (except if the system has index one or zero), as illustrated by the following examples:

Example 8. - Regularity and Index

a) Consider the system:

$$t^2\dot{x} + t\dot{y} - y = t \tag{2.28}$$

$$t\dot{x} + \dot{y} + x = 0 \tag{2.29}$$

We have: $\det |f_{\dot{x}} + \lambda f_x| = \det \begin{vmatrix} t^2 & , & t-\lambda \\ t+\lambda & , & 1 \end{vmatrix} = \lambda^2$

Hence, the matrix pencil is regular everywhere. However, multiplying (2.29) by t and subtracting (2.28), we obtain:

$t_x + y = -t$

Differentiating and substituting for $\dot{y}$ in the LHS of (2.29) then yields:

$t_{\dot{x}} + \dot{y} + x = -1$

This contradicts (2.29), so the system is inconsistent and hence has no solution.

b) For the system of example 6 we have

$$\det |f_{\dot{x}} + \lambda f_x| + \det \begin{vmatrix} 1 = 2\lambda x & \lambda z & \lambda y \\ \lambda(y-z) & 1+\lambda x & -\lambda x \\ 2\lambda x & 2\lambda y & 0 \end{vmatrix} = 2\,(y-z)$$

Thus, the pencil is regular, except along the line $y = z$,

Moreover, we have the factorization:

$$\begin{vmatrix} 2\varphi_1 x, & 0, & -\varphi_1 xz \\ 2x, & 2y, & -x(1+z/y) \\ 0, & 0, & 1 \end{vmatrix} \left[f_{\dot{x}} + \lambda f_x \right] \begin{vmatrix} 0, & 1/2\varphi_1, & 0 \\ 0, & 1/2\varphi_1 y, & 1/2y \\ 1/2\varphi_1 xy, & \varphi_2, & 0 \end{vmatrix} = x$$

$$= \begin{vmatrix} \cdot & 1 & \cdot \\ \cdot & \cdot & 1 \\ \cdot & \cdot & \cdot \end{vmatrix} + \lambda \begin{vmatrix} 1 & \cdot & \cdot \\ \cdot & 1 & \cdot \\ \cdot & \cdot & 1 \end{vmatrix}$$

where $\varphi_1 = (y - z)\,(y/x + x/y^2)$, $\qquad \varphi_2 = (z - zy) / z\,\varphi_1\, y^2$

This factorization is well defined, and the factors are nonsingular, if $y \neq z$ and x, y are non-zero, so that the index of nilpotency is three under these conditions.

However, we saw in Example 6 that the DAE index is not defined, and there is an infinity of solutions through each point, so the system is not regular in the sense of Definition 2.

c) Consider the system:

$$\dot{x} - t\,\dot{y} = t \tag{2.30}$$

$$x - ty = 0 \tag{2.31}$$

$$\text{We have:} \det \left| f_{\dot{x}} + \lambda f_x \right| = \det \begin{vmatrix} 1 & -t \\ \lambda & -\lambda t \end{vmatrix} = 0 \text{ all } \lambda.$$

Hence, the the pencil is singular everywhere, and there is no factorization of the form of (2.17), so the index of nilpotency is not defined.

On the other hand, differentiating (2.31) and substituting for $\dot{x}$ from (2.30) yields

$$\dot{x} = t\dot{y} + y = t\dot{y} + t.$$

Whence $y = t$ and $x = t^2$. Thus, the DAE index is two, and we have a unique solution.

3. Reformulation of High-index Problems

The complicated properties of DAE systems with index greater than one (commonly referred to as "high index" systems) and the difficulties associated with their solution has led some engineers [9] to assert that high index models are in some sense ill-posed, and hence avoidable by "proper" process modelling.

Certainly systems for which the index is undefined over a full-dimensional region (such as that in Example 6) are usually functionally singular, indicating redundant equations and an under-determined system. However, we have seen that a system is regular and well-behaved so long as a bounded index exists, and as we shall see later, such a system is reducible to an index-one system, so there is nothing "improper" about such systems.

The natural way of modelling the system may well be the high-index form, and the mathematical reduction procedure may destroy structure and lead to a system which is not easily interpreted in physical terms.

It is however always useful to investigate how the high index arises. In general, it is because something is assumed to respond instantaneously, so this focusses on underlying assumptions, for which the relative advantages and disadvantages can then be assessed.

For example, high index in Example 2 arises because we assumed that the product flow-rate can be instantaneously adjusted to maintain the hold-up V exacty constant. In practice, this may be achieved by overflow, or more often by adjusting a control valve in the product pipe to maintain the level constant.

We could instead model this controller as a proportional-integral-derivative (PID) controller

$$\left.\begin{aligned} P - P^* &= K_P (V - V^*) + K_I I + K_D \frac{d}{dt} (V - V^*) \\ \frac{dI}{dt} &= V - V^* \end{aligned}\right\} \qquad (3.1)$$

where V* is the desired (set-point) valve of V, and P* arises from setting up the controller - for example:

$$\text{At } t = t_0: \quad P(t_0) = P^*, \quad V(t_0) = V^*, \quad I(t_0) = 0 \qquad (3.2)$$

Then, if we use no derivative action ($K_D = 0$) we find that the expanded model now has index one.

This is certainly a more realistic model, not significantly more complicated than the original, though we do have to choose appropriate values for the parameters K_P, K_I. Note, however that if we add derivative action we would again have an index two model!

Example 9

A further example is afforded by modelling the dynamics of the steam heating-coil of the above reactor, replacing the instantaneous steady-state model of equation (1.21). The equations are (see [10]):

$$V_s \, d\frac{\rho}{dt} = F_s - L \qquad (3.3)$$

$$V_s \, c_s \frac{dT_s}{dt} = F_s \, c_s \, (T_s^0 - T_s) + L \, . \, \Delta H_s - U \, S \, (T_s - T) \qquad (3.4)$$

$$P_s = R \, T_s \, P_s = \overline{P} \exp(\overline{T}/T_s). \qquad (3.5)$$

where the inlet steam is superheated vapour at temperature T_s^0, assumed to obey the ideal gas law and a simple vapour-pressure equation, with constant specific heat c_s and latent heat ΔH_s. The volume of the coil is V_s and the condensate, removed as it is formed via a steam-trap, has a flow-rate L.

L appears only in the differential equations, so (3.5) must be differentiated, showing that this sub-system has index two.

Again, the high index arises from an idealized control system, for the steam-trap is in reality a level controller. However, to model it more realistically we should have to introduce a condensate phase with its own balance equations, as well as the controller relations, and one might then feel that this is more complicated than the differentiation of (3.5).

Alternatively, we could look for further simplification, for example by neglecting the variation of vapour hold-up (not the hold-up itself), setting $dp_s/dt = 0$. Then, $L = F_s$ and the system has index one.

Thus, we have two contrasting approaches. The first is to identify instantaneous responses and then model their dynamics more accurately. The second is to identify an algebraic variable which appears only in differential equations, and to convert one of these to an algebraic equation by neglecting the derivative term.

Another technique for obtaining lower index models, particularly useful in considering mechanical systems, is to look for invariants of the motion (e.g. energy, momentum or angular momentum), then transform the variables to include these invariants as independent variables. The pendulum considered in Example 3 nicely illustrates this:

Example 10 - The Pendulum revisited.

If we use radial rather than Cartesian coordinates, we can make direct use of the fact that the pendulum is of constant length, hence avoiding the need for the algebraic relation (1.27).

Using the same assumptions as before, the system can now be modelled in three terms of only variables: the angle to the vertical (θ), the velocity of the bob (V), and again the tension in the string (z), yielding:

$$\dot{\theta} = V \quad , \tag{3.6}$$

$$\dot{\theta} = -\sin\theta \quad , \tag{3.7}$$

$$z = v^2 + \cos\theta \quad , \tag{3.8}$$

This is clearly an index one system!

It would in fact seem that we have a counter-example to our assertion in Section 2 that the index is invariant under a one-to-one algebraic transformation, since the transformation from Cartesian to radial coordinates is just such a transformation:

$$\left.\begin{aligned} x = r\sin\theta,\quad y = \cos\theta \\ u = V\sin\varphi,\quad v = V\cos\varphi \end{aligned}\right\} \tag{3.9}$$

$$\left.\begin{aligned} \dot{r} = V\cos(\varphi-\theta),\quad r\theta = V\sin(\varphi-\theta) \\ \dot{V} = \cos\varphi - z_T\cos(\varphi-\theta),\quad V\dot{\varphi} = z_T\sin(\varphi-\theta) - \sin\varphi, \\ r = \pm 1 \end{aligned}\right\} \tag{3.10}$$

However, direct application of (3.9) to (1.25), (1.28), (1.27) yields the system:

This still has an index of three, and two differentiations are required to reduce it to (3.6) - (3.8).

Whilst such reformulations are useful, and occasionally instructive, it is obvious that we need more systematic methods of solving high-index systems, and we turn to this problem in the next section.

4. The Solution of DAE Systems

Gear and Petzold [7] were the first to propose a systematic method of dealing with high-index systems, and we have in fact used their method in solving the problems in Examples 2-6.

The method involves successive stages of algebraic manipulation and differentiation, starting from the general system (2.1), and at each stage the system is reduced to the form:

$$\left.\begin{aligned} \dot{x}_r &= X_r\,(t,x,\dot{y}_r) \\ 0 &= f_r\,(t,x,\dot{y}_r) \end{aligned}\right\} \qquad (4.1)$$

where is ($\dot{x}_r$, $\dot{y}_r$) a partition of $\dot{x}$.

The algorithm can be formally stated as follows:

Gear-Petzold Index-Reduction Algorithm

0. Set: $r = 0$, $\dot{y}_0 = x$, $\dot{x} = \emptyset$,
   $f_0(\cdot) = f(\cdot)$, $X_0(\cdot) = \emptyset$.
1. Solve $f_r(t, x, \dot{y}_r)$ for as many derivatives as possible, and eliminate these equations to yield:
   $$\left.\begin{aligned} \dot{x}'_{r+1} &= X_{r+1}(t, x, \dot{y}_{r+1}) \,, \\ 0 &= \eta_{r+1}(t,x) \,, \end{aligned}\right\} \qquad (4.2)$$

   where $(\dot{x}'_{r+1}, \dot{y}_{r+1})$ is a partition of $\dot{y}_r$
2. Form $\dot{x}_{r+1} = (\dot{x}_{r+1}, \dot{x}'_{r+1},)$
   Substitute $\dot{x}_{r+1} = \chi_{r+1}(t, x, y_{r+1})$ into $X_r(t, x, \dot{y}_r)$ and append $\chi_{r+1}(t, x, \dot{y}_{r+1})$ to form $X_{r+1}(t, x, \dot{y}_{r+1})$
3. If $\dot{y}_{r+1} = \emptyset$, ST O P.
4. Differentiate $\eta_{r+1}(t,x)$ with respect to t, then substitute $\dot{x}_{r+1} = X_{r+1}(t, x, \dot{y}_{r+1})$ to form $f_{r+1}(t, x, \dot{y}_{r+1})$
5. Set r: = r+1 and return to step 1.

On termination, we have the explicit O.D.E. system:

$$\dot{x} = X(t, x). \tag{4.3}$$

Of course, we need appropriate initial conditions for this system, and these need to satisfy the algebraic equations $\eta_{r+1}(t,x)$, $r = 0, 1, 2, \ldots$ generated by the algorithm.

We illustrate this algorithm by applying it to the system of Example 7:

Example 11. Gear-Petzold Index Reduction.

Given:

$$\left.\begin{array}{ll} \dot{a}_i = a_{i-1} - a_i , & i = 1,2,\ldots n, \\ a_n = a_n^*(t) & \end{array}\right\} \tag{4.4}$$

r = 0 The system is already in solved from for all the derivatives in the equations, so we have $\dot{y}_1 = \dot{a}_0$, $\dot{x}_1 = \dot{x}'_1 = (\dot{a}_1, \dot{a}_2, \ldots, \dot{a}_n)$.

Differentiating the algebraic equation and substituting for derivatives yields

$$\dot{a}_n = a_{n-1} - a_n = \dot{a}_n^*$$

r = 1 The form of the system is unchanged, with $\dot{y}_z = \dot{y}_1$, $x_2 = x_1$. Differentiation and substitution yield:

$$a_{n-2} - 2a_{n-1} + a_n = \dot{a}_n^*$$

r = n Differentiation and substitution yields

$$a_0 = na_0 + \sum_{i=1}^{n} (-1)^i \binom{n+1}{i+1} a_i + (a_n^*)^{(n+1)} \tag{4.5}$$

r = n + 1 Equation (4.5) is in solved form, so we append it to the ODEs in (4.4) to yield the final system, and since $\dot{y}_{r+1} = \varnothing$ we stop.

In this case, the algebraic equations are of the general form:

$$\sum_{i=0}^{r} (-1)^i \binom{r}{i} a_{n-1} = \left(a_n^*\right)^{(r)}, \quad r = 0, 1, \ldots n, \tag{4.6}$$

and are sufficient to determine uniquely all $a_i(t_0)$, $i = 0, 1, \ldots n$.

We note that the system of ODEs contains only the (n+1) - derivative $(a_n^*)^{(n+1)}$, so that changing a_n^* by an arbitrary polynomial of order n:

$$\Delta a_n^* = c_0 + c_1 t + c_2 t^2 + \ldots + c_n t^n$$

would not change the ODEs. Of course, choosing initial conditions consistent with (4.6) would then ensure that the coefficients $c_0, c_1, \ldots c_n$ were all zero, but in practical computation rounding and truncation errors may cause non-zero c_i, and an error in c_n would be multiplied by t_n, causing rapid error growth.

Brenan et al [2] discuss this problem of "drift" in satisfying the algebraic equations $\eta_{r+1}(t,x)$, $r = 0, 1, 2, \ldots$ and report on various proposals in the literature for stabilizing system (4.3). The simplest and most effective proposal was made by Bachmann et al. [1], who noted that the algebraic equations in (4.2) should be retained instead of differential equations. Since by construction these have maximum rank, they determine a corresponding subset y of the elements of x in terms of the remainder, z.

The equations defining the derivatives of these y-variables in (4.3) can then be dropped, leaving a semi-explicit index-one system of the form:

$$\left.\begin{aligned} \dot{z}(t) &= X(t,x) \\ 0 &= G(t,x) \end{aligned}\right\} \qquad (4.7)$$

where G (t, x) represents the union of the algebraic equations $\eta_{r+1}(t,\ x)$, $r = 0, 1, \ldots.$ Consistent initial conditions are obtained simply by assigning values for $z\ (t_0)$.

In fact, Bachmann et al. described their algorithm only for linear systems, simply stating that "the algorithm can also be used for nonlinear systems" without giving details. Chung and Westerberg [4] describe a very similar conceptual algorithm for index reduction, though they then propose solution of the reduced index-one problem, rather than retaining algebraic equations.

Although the proposal of Bachmann et al deals satisfactorily with the stability problem, and the problem of identifying the means of specifying consistent initial conditions, there is still the fundamental weakness that step 1 of the algorithm requires symbolic algebraic manipulation.

One consequence of this is that only a limited class of systems can be treated. Since one differentiation occurs (in step 4) on each iteration, and an ODE system is finally generated, it is clear that the index of the system is equal to the number of iterations.

However, since (4.3) was generated symbolically, the reduction is true over the whole domain of definition of $f(t,\ x,\ \dot{x})$, implying that the index must be constant over this domain. We cannot therefore treat systems whose index varies over the domain.

More generally, the explicit solution of the form (4.2) cannot always be generated symbolically, and in any case symbolic manipulation is a significant task which is not easily automated.

These problems are circumvented by the numerical algorithm put forward very recently by Pantelides et al. [11]. Here, the index-reduction procedure is carried out numerically during the integration, each time that the Jacobian matrix is re-evaluated.

Thus, given $f_{\dot{x}}(t, x, \dot{x})$ we start by obtaining the factorization:

$$Q_0 f_{\dot{x}} P_0 = \begin{bmatrix} f_{\dot{x}}^1 \\ 0 \end{bmatrix} \tag{4.8}$$

where $f_{\dot{x}}^1$ is a truncated unit upper-triangular matrix, P_0 is a permutation matrix and Q_0 is either orthogonal or generated by Gaussian elimination using row and column interchanges.
We then define:

$$\begin{bmatrix} f^1 \\ g^1 \end{bmatrix} = Q_0 f, \begin{bmatrix} \dot{f}^1 \\ \dot{g}^1 \end{bmatrix} = Q_0 \dot{f} \text{ etc.} \tag{4.9}$$

where it should be noted that $\dot{f}^1$ should ***not*** be interpreted as df^1/dt since Q_0 may itself be a nonlinear function of $(t, x, \dot{x})$.

Now, $f_{\dot{x}}^1 \in R^{r_1}$, where r_1 is the rank of $f_{\dot{x}}$, and if this remains constant over a sub-domain of, $(t, x, \dot{x})$ then (4.9) defines a function $g^1(\cdot) \in R^{n-r_1}$ with $g_{\dot{x}}^1 = 0$ over the sub-domain, and hence $g^1(\cdot)$ may be written $g^1(t, x)$, independent of $\dot{x}$. In general, the rank r_1 and hence the dimension of $g^1(\cdot)$, may vary from point to point but from the continuous differentiability assumption this will happen only on a set of measure zero.
If $f_{\dot{x}}$ is nonsingular, there are no null rows in (4.8), and we have immediately:

$$f_{\dot{x}}^1 \delta\dot{x} = -f^1,$$

which can be used in a Newton iteration to solve (2.1) for $\dot{x}(t)$, recalling that $f_{\dot{x}}^1$ is now unit upper-triangular.
Otherwise, we form and factorize:

$$Q_1 \begin{bmatrix} f_{\dot{x}}^1 \\ \dot{g}_{\dot{x}}^1 \end{bmatrix} P_1 = \begin{bmatrix} f_{\dot{x}}^2 \\ 0 \end{bmatrix}, \text{ defining } \begin{bmatrix} f^2 \\ \dot{g}^2 \end{bmatrix} = Q_1 \begin{bmatrix} f^1 \\ \dot{g}^1 \end{bmatrix}$$

Again $\dot{g}_{\dot{x}}^2 = 0$, and with the assumption of continuous differentiability $\dot{g}^2(t, x)$ is independent of $\dot{x}$ almost everywhere.

Again, if $f_{\dot{x}}^2$ is nonsingular we have

$$f_{\dot{x}}^2 \delta x = -f^2$$

and we can use this to solve (2.1).

The recursion can be continued with the general form:

$$Q_r \begin{bmatrix} f_{\dot{x}}^r \\ \left(g_{\dot{x}}^{(r)}\right)^r \end{bmatrix} P_r = \begin{bmatrix} f_{\dot{x}}^{r+1} \\ 0 \end{bmatrix}; \begin{bmatrix} f^{r+1} \\ \left(g^{(r)}\right)^{r+1} \end{bmatrix} = Q_r \begin{bmatrix} f^r \\ \left(g^{(r)}\right)^r \end{bmatrix} \tag{4.10}$$

with $\left(g^{(r)}\right)^{r+1}$ independent of $\dot{x}$.

Eventually, when r =m (the index) we shall have $f_{\dot{x}}^m$ nonsingular and:

$$f_{\dot{x}}^m . \delta\dot{x} = -f^m. \tag{4.11}$$

As before, the algebraic equations $\left(g^{(r)}\right)^{r+1} = 0$, $r = 0,1, ..m$, can be collected to form the system:

$$G(t,x) = 0 \tag{4.12}$$

The Jacobian matrix G_x of this system can be factorized to yield:

$$QG_xP = \left[\overline{G_y}\ \overline{G_z}\right] \tag{4.13}$$

where $\overline{G_y}$ is unit upper-triangular and (y, z) is a partition of x. Given z, we can then use:

$$\overline{G}y \cdot \delta y = -Q.G. \tag{4.14}$$

As in the Bachmann algorithm, we use the integration formula (cf.(1.2):

$$z = \gamma h.\dot{z} + \varphi \tag{4.15}$$

to determine new values of z.

The algorithm is then:

Pantelides - Sargent Algorithm

Given an estimate of x and x:

1. Set r =0, $f_{\dot{x}}^{0} = f_{\dot{x}}$, $(g_{\dot{x}}^{(0)})^{0} = \emptyset$
2. Compute Q_r, P_r, $f_{\dot{x}}^{r+1}$, $[f^{r+1}, (g^{(r)})^{r+1}]$ using (4-10)
3. If $(g^{(r)})^{r+1} = \emptyset$, go to step 6.
4. Add $(g^{(r)})^{r+1}$ to G and $(g_x^{(r)})^{r+1}$ to G_x
5. Set r:=r+1 and return to step 2.
6. Compute $\dot{x} := \dot{x} + \delta\dot{x}$ using (4.11)
7. Factorize G_x using (4.13)
8. Compute $y := y + \delta y$ using (4.14)
9. Compute z from (4.15)
10. Repeat from step 1 to convergence.

Again, as in the Bachmann case, consistent initial conditions are obtained by specifying only $z(t_0)$.

In fact, this can be generalized to specifying the same number of algebraic initial conditions:

$$J(t_0, x(t_0), \dot{x}(t_0)) = 0, \tag{4.16}$$

such that the Jacobian matrix:

$$\begin{vmatrix} f_{\dot{x}}^{m} & f_{x}^{m} \\ \cdot & G_x \\ J_{\dot{x}} & J_x \end{vmatrix} \tag{4.17}$$

is non-singular. Equations (4.16) then replace equations (4.15) in step 9 when the algorithm is used to evaluate initial conditions.

As in standard ODE practice, the Jacobians $f_{\dot{x}}^{m}$, G_x need not be recomputed at each iteration, nor even at each integration step, and we note that the integration formula can be explicit or implicit.

As described above, it seems as if a numerical rank determination is implied in the factorization in (4.10). However, in practice the factorization can be discontinued as soon as the remaining elements fall below a suitable threshold value. Additional differentiations caused by a finite threshold do not invalidate the algorithm, and are even beneficial in dealing with the ill-conditioning. For the same reason, changes in the rank or index cause no difficulties.

The algorithm can be implemented using automatic differentiation for the requisite differentiations of f (·), and otherwise only standard numerical linear algebra is involved. Of course, sparse matrix techniques can be used to deal with large-scale problems, and in principle the algorithm will deal with any systems with bounded index. In practice, the index may be limited by the limitations of the automatic differentiation package in generating successively higher-order derivatives.

As shown by Pantelides et al. [10], high index problems are common in chemical engineering, and multistage countercurrent systems can in some circumstances give rise to very high index systems. It is not therefore surprising that there has been a search for direct solution methods which do not involve the successive differentiations required in index-reduction algorithms.

The earliest such technique was proposed by Gear [6]. He noted that the general implicit linear multistep integration formula (1.2) can be solved for $\dot{x}_k$ in terms of x_k and past data, and the result substituted into (2.1) to yield a set of n equations in the n unknowns x_k. The solution could then be used in conjunction with standard ODE techniques for optimal choice of order and step-length for the formula in question.

If the corrector formula is solved by Newton's method, we require nonsingularity of the corresponding Jacobian matrix $[f_{\dot{x}} + \gamma h_k . f_x]$, and this also guarantees uniqueness of the solution generated. However we note that there will be such a solution for any initial condition, whereas we have seen that true solutions must satisfy a set of algebraic equations. Clearly, we must start with consistent initial values, but the question then arises of whether the generated solution remains consistent. It has in fact been shown (see [6]) that for systems of index zero or one, and semi-explicit systems of index two, the order of accuracy of the solution is maintained so long as the initial conditions are consistent to the order of accuracy of the integrator formula, and at each step the equations are solved to the same order. However, there are counter-examples for higher index systems, and even for general index-two systems.

It will have been noted that the underlying Jacobian matrix is of the same form as the matrix pencil arising from the local linearization of the equations, so that regularity of this pencil ensures non singularity of the Jacobian for all but a finite set of values of h_k. Unfortunately, as we have seen, regularity of the pencil has no connection with regularity of the DAE system, so we cannot conclude that the Jacobian will be nonsingular if the DAE system is regular. Moreover, since by definition the matrix $f_{\dot{x}}$ is singular for all DAE systems with index greater than zero, the condition number for $[f_{\dot{x}} + \gamma h_k . f_x]$ tends to infinity as $h_k \to 0$. In fact, it can be shown that the condition number is $O[h^{-m}]$, where m is the index. This is unfortunate, since we rely on reducing h_k to achieve the specified integration accuracy, but the solution then becomes increasingly sensitive to rounding errors.

This problem was studied by Bachmann et al.[l], and the table below is taken from their numerical results for the solution of the system (cf. Example 7)

:

$$\left.\begin{array}{ll} \dot{a}_i = a_{i-1} - a_i, & i=1,\ 2,\ \ldots\ n \\ a_n = 1-\exp(-t/2), & 0 \le t \le 1. \end{array}\right\} \tag{4.18}$$

h / n	1.0	0.1	0.01	0.001	0.0001
4	0.958	0.371 E - 2	0.378 E - 3	0.164 E - 4	0.438
8	0.154 E +2	0.441 E - 3	0.412 E + 1	0.368 E + 9	0.326 E+17
12	0.246 E +3	0.141 E + 11	0.168 E +10	0.990 E + 22	*
16	0.394 E +4	0.423 E + 17	0.170 E +19	*	*

* indicates failure with a "singular' matrix.

The table gives the error in a_0 at t = 1 for the integration of this system using the implicit Euler method, starting with the analytically derived consistent initial values at t = 0.
Even for n = 4, the accuracy attainable is limited by the above ill-conditioning, and the system is violently unstable for higher values of n. Of course, the results could be improved by use of a higher-order formula, but it is clear that the process is intrinsically unstable.

It seems that the higher derivative information is essential to provide reliable predictions, but usually we are interested in behaviour over an extended period and this prediction problem would be avoided by a global approximation over the whole period of interest, as is the case for two-point boundary value problems. Thus, we might expect satisfactory results from use of collocation or finite-element techniques. However, there seems to be no published work on an analysis of these techniques as applied to high-index DAEs.

An approach in the same spirit has recently been proposed by Jarvis and Pantelides [8], in which the integration over a fixed time interval is converted into an optimal control problem.

Here, the system is taken as described by (1.14):

$$f(t, \dot{x}(t), x(t), y(t)) = 0 \tag{4.19}$$

and at each step we linearize the system and carry out a factorization of the matrix $[f_{\dot{x}}, f_y]$ as in (4.8):

$$Q_0 \left[f_{\dot{x}}\ f_y\right] P = \begin{vmatrix} U_{11} & U_{12} \\ \cdot & \cdot \end{vmatrix} \tag{4.20}$$

where again U_{ll} is unit upper-triangular, and the vector $(\dot{x}, y)$ has been partitioned into $[z_l, z_2]$. The linearized equation is then:

$$\begin{vmatrix} U_{11} & U_{12} \\ \cdot & \cdot \end{vmatrix} \begin{bmatrix} \delta_{z1} \\ \delta_{z2} \end{bmatrix} = -Qf = -\begin{bmatrix} \overline{f_1} \\ \overline{f_2} \end{bmatrix} \tag{4.21}$$

Of course, for consistency we must have $\overline{f_2} \approx 0$ within rounding errors, and then the iteration:

$$U_{11}\ \delta z_1 = -\overline{f_1} \tag{4.22}$$

until $\| \overline{f_1} \| \leq \varepsilon$ effectively solves a subset of (4.19) for z_1, given z_2:

$$z_1(t) = \varphi(t, x(t), z_2(t)) \tag{4.23}$$

Thus, if $z_2(t)$ is a known function, we can use (4.23) to integrate over the time interval of interest, say $[t_0, t_f]$. Hence, we treat $z_2(t)$ as a control function, and choose it to minimize:

$$P = \int_{t_0}^{t_f} \left\|\overline{f_2}\right\| dt \tag{4.24}$$

subject to satisfying (4.23) at each t.

This problem can be solved by choosing a suitable parametrization of $z_2(t)$:

$$z_2(t) = \psi(t, p), \tag{4.25}$$

where p is the set of parameters, which converts the optimal control problem into a nonlinear programming problem in p.

Again of course the factorization (4.20) need not be performed at every step, as in standard ODE practice. In fact, we still have the problem of consistent initialization, since as we have seen, the initial $x(t_0)$, $y(t_0)$ must satisfy a set of algebraic equations, but these can be correctly established by using the Pantelides-Sargent algorithm described earlier. The advantage of the optimal control approach is that this need only be done at the initial point, or at subsequent points at which discontinuities occur.

A number of high-index problems have been successfully solved using this approach, including the pendulum problem (Example 3) and versions of the canonical linear problem described below with index up to 12.

The analysis of this last problem is instructive in indicating the requirements and limitations of these various methods:

Example 12

The canonical linear system (2.21) yields the system:

$$\left.\begin{aligned} \dot{x}_i &= x_{i+1} \quad , \quad i = 1, 2, \ldots, (n-1), \\ x_1(t) &= \cos t \end{aligned}\right\} \qquad (4.26)$$

The advantage of using cos t as the driving function is that the solution is bounded by ± 1 and the analytical solution is immediately available:

For n=6:

$$x_1 = \cos t = 1 - t^2/2! + t^4/4! + O(t^6)$$
$$x_2 = \sin t = t + t^3/3! - t^5/5! + O(t^7)$$
$$x_3 = -\cos t = -1 + t^2/2! - t^4/4! + O(t^6)$$
$$x_4 = \sin t = t - t^3/3! + t^5/5! + O(t^7)$$
$$x_5 = \cos t = 1 - t^2/2! + t^4/4! + O(t^6)$$
$$x_6 = -\sin t = -t + t^3/3! - t^5/5! + O(t^7)$$

The partition given by the factorization is

$$z_1 = [x_1, x_2, x_3, x_4, x_5] \quad , \quad z_2 = x_6$$

As an illustration, we use the simple parametrization:

$$x_6(t) = \bar{x}_6, \text{ all } t.$$

In this case, there are no degrees of freedom in the initial conditions, so all initial values other than x_1 are unknown parameters:

$$x_2(0) = \overline{x_2}, \quad x_3(0) = \overline{x_3}, \quad x_4(0) = \overline{x_4}, \quad x_5(0) = \overline{x_5}$$

Thus, the optimal control problem is:

$$\min_{\bar{x}} \int_0^{t_f} \|x_1(\tau) - \cos \tau\| \, dt$$

subject to the differential equations in (4.26).

Again, for given parameters we can solve the system analytically:

$$x_1 = 1 + \bar{x}_2 t + \bar{x}_3 t^2/2! + \bar{x}_4 t^3/3! + \bar{x}_5 t^4/4! + \bar{x}_6 t^5/5!$$
$$x_2 = \bar{x}_2 + \bar{x}_3 t + \bar{x}_4 t^2/2! + \bar{x}_5 t^3/3! + \bar{x}_6 t^4/4!$$
$$x_3 = \bar{x}_3 + \bar{x}_4 t + \bar{x}_5 t^2/2! + \bar{x}_6 t^3/3!$$
$$x_4 = \bar{x}_4 + \bar{x}_5 t + \bar{x}_6 t^2/2!$$
$$x_5 = \bar{x}_5 + \bar{x}_6 t$$
$$x_6 = \bar{x}_6$$

Comparing these expressions with the analytical solution, we ought to have $\bar{x}_6 = 0$.

To determine this value from evaluation of $f(\cdot)$ in a local integration formula would require the determination of $\{x_1(t) - \cos t\}$ to an accuracy of at least $O[h^5]$. However this error propagates, and to obtain $\bar{x}_6 = 0$ from the optimal control problem requires evaluation of the integral only to $O[t_f^{\,5}]$.

Of course, in either case we rapidly lose accuracy in the variables representing higher derivatives, and only methods which directly use higher derivative information, like the index-reduction methods, can obtain the requisite accuracy in all the variables.

5. Conclusions

Lumped-parameter dynamic models for chemical process systems usually give rise to a system of DAEs, and the same is true of distributed-parameter systems solved by discretization of the space variables. These systems often have an index of two or three, and higher index systems can arise, particularly in modelling behaviour under abnormal conditions (eg. during start-up or under potentially hazardous conditions).

Although physical insights and reformulation can often be useful in reducing the index, this is likely to be the preserve of expensive specialists, and as the size of systems being studied grows, this approach becomes more and more time-consuming, and less and less effective. It is therefore important to have reliable and accurate general-purpose numerical methods for solving such systems, which can be very large-scale, involving possibly hundreds of thousands of variables.

This paper has attempted to describe the essential nature of the problem and some of the difficulties which arise, and to review the current state of the art in techniques for solving these problems.

In order to capture all facets of the behaviour, as represented by the model, there seems to be no alternative but to use a method which makes use of the full system defining the index (2.2), such as the index-reduction methods, and we are just beginning to see the emergence of effective numerical algorithms in this area, such as the Pantelides-Sargent algorithm.

Particularly in large-scale systems there is often only a small proportion of the variables which are of detailed interest, and if their behaviour does not depend on higher derivative information it may be acceptable to use methods which use only the DAE system itself (2.1). Some analysis is required to establish the possibilities, but this is provided by using a full algorithm, at the initial point, which is in any case necessary for consistent initialization.

In this area, we pointed out the intrinsic instability of Gear's original method for higher index systems, and argued that no method based on a purely local approximation is likely to be effective. This leaves the field to global approximation methods based on collocation, finite elements, or the hybrid-optimal control approach of Jarvis and Pantelides. We can look forward to more results in this area, particularly in optimal control applications and other applications giving rise to distributed boundary value problems.

References

1. Bachmann, R., L. Brull, T. Mrziglod and U. Pallaske, "On Methods for Reducing the Index of Differential-Algebraic Equations", Comput. Chem. Engng., 14, pp 1271-1273 (1990)
2. Brenan., K.E., S.L. Campbell and L.R. Petzold, "Numerical Solution of Initial-Value Problems in Differential-Algebraic Equations", North Holland, New York (1989)
3. Butcher, J.C., "The Numerical Analysis of Ordinary Differential Equations", John Wiley & Sons, (Chichester, 1987)
4. Chung, Y., and A.W. Westerberg, "A Proposed Numerical Algorithm for Solving Nonlinear Index Problems", Ind. Eng. Chem. Res., 29, pp 1234-1239, (1990)
5. Gantmacher, F.R., "Applications of the Theory of Matrices", Interscience, (New York, 1959)
6. Gear, C.W., " The Simultaneous Numerical Solution of Differential-Algebraic Equations", IEEE Trans. Circuit Theory, CT-18, pp 89-95 (1971)
7. Gear, C.W., and L.R. Petzold, "ODE Methods for the Solution of Differential-Algebraic Systems", SIAM J. Numer. Anal. 21, pp 716-728 (1984)
8. Jarvis, R.B., and C.C. Pantelides, "A Differentiation-free Algorithm for Solving High-Index DAE Systems", Paper 146g, AIChE Annual Meeting, Miami Beach, 1-6 November, 1992
9. Marquardt, W., "Dynamic Process Simulation - Recent Progress and Future Challenges", in Y. Arkun and W.H. Ray (Eds), " Chemical Process Control-CPC IV", pp 131-180, AIChE (New York 1991)
10. Pantelides, C.C., D.M. Gritsis, K.R. Morison and R.W.H. Sargent, "The Mathematical Modelling of Transient Systems Using Differential-Algebraic Equations", Comput. Chem. Engng., 12, pp 449-454 (1988)
11. Pantelides, C.C., R.W.H. Sargent and V.S. Vassiliadis, "Optimal Control of Multistage Systems Described by Differential- Algebraic Equations", Paper 146h, AIChE Annual Meeting, Miami Beach, 1-6 November, 1992

Features of Discrete Event Simulation

Steven M. Clark, Girish S. Joglekar

Batch Process Technologies, Inc., P. O. Box 2001, W. Lafayette, IN 47906, USA

Abstract: The two most important characteristics of batch and semicontinuous processes which demand special methodology from the simulation standpoint are the continuous overall changes with time as well as the discrete changes in the state of the process at specific points in time. This paper discusses general purpose combined discrete/continuous simulation methodology with focus on its application to batch processes.

The modeling of continuous overall change involves the solution of simultaneous differential/algebraic equations, while the modeling of discrete changes in the process state requires algorithms to detect the discrete changes and implement the actions associated with each discrete change. The workings of the time advance mechanism which marches the process in time are discussed with the help of a simple batch process.

Keywords: Simulation, batch/semicontinuous

1. Introduction

A major portion of the research and development activity over the past 25 years in the area of the simulation of chemical processes has been targeted to steady state processes. As a result, steady state process engineering has benefited significantly from the use of simulation based decision support tools to achieve reduced capital costs and improved control systems. Employing simulation for steady state process engineering has reached such a level of confidence and maturity that today any quantitative process decision without its use would be inconceivable.

The batch/semicontinuous processes significantly lag behind the steady state processes in the availability of versatile simulation based decision support tools. The complex nature of this mode of operation poses challenging problems from the perspective of efficient solution methodology and information handling. The key features of batch/semicontinuous processes are as follows.

Batch/semicontinuous processes are inherently dynamic in nature, that is the state of the process changes with time. The state of a process is defined as a set of variables which are adequate to provide the necessary description of the process at a given time. Some variables, which describe the state, change continuously with time. For example, the concentration of species in process vessels may change constantly due to reactions, the level of material in vessels changes as material is withdrawn or added, the flowrate and conditions of a vapor stream (temperature, pressure and composition) may change due to the changes in the bulk conditions. Alternatively, the values of some variables may change instantaneously. These discrete changes may be introduced on the start and finish of operations and by specific operating decisions. For example, a processing vessel may be assigned to a new operation after completing an operation, a reaction step may be terminated when the concentration of a particular species reaches the desired value, or an operator may be released after completing an assignment.

In a typical batch/semicontinuous process several products are made concurrently over a period of time, each according to a unique recipe. A recipe describes the details of the operations associated with the manufacture of that product, for example which ingredients to mix, the conditions for ending a particular process step, the pieces of equipment suitable for an operation.

The key operating decisions which influence the overall performance of a process are concerned with the assignment of equipment to an operation and the assignment of materials and shared resources as and when needed during the course of the operation. The materials, resources and equipment are shared by the entire process and typically have limited availability. For example, only one operator may be available to manage 10 reactors in a process, or an intermediate storage tank may be allowed to feed material to a maximum of two downstream mixing tanks at a time.

Therefore, in addition to the ability to model the dynamics of physical/chemical transformations, the design of a simulator for batch processes must include the ability to implement operating decisions.

Several simulators are available for modeling the dynamics of chemical processes [1]. These range from programs written in a procedural languages, such as FORTRAN, for performing specific simulations to general purpose continuous simulation languages. Also, various solution techniques have been employed to solve the underlying differential equations, such as the sequential modular approach using Runge-Kutta, or the equation-oriented approach using implicit integrators. However, none of these simulators have been designed to handle discrete changes in the process state, and in some cases to handle simultaneous algebraic equations. These shortcomings make them unsuitable for applications in batch process simulation.

Several discrete simulators are now available, such as SLAM [9], GPSS [10], for applications predominantly in the discrete manufacturing sector. Some of these tools have been successfully used in simulating batch processes [5]. These simulators also provide an executive for the combined discrete/continuous simulation. However, since these simulators are merely general purpose simulation languages, they require that the user program the necessary process dynamics models and the logic associated with the processing of discrete changes. Since these simulators were designed mainly for discrete manufacturing systems, the algorithms for solving differential equations are not only very inefficient, but also unable to solve algebraic equations. Therefore, even though they provide the basic solution methodology, the discrete simulators have enjoyed very limited success in modeling batch/semicontinuous processes.

The design of a general purpose process modeling system for batch/semicontinuous processes has received considerable attention over the past few years, and has resulted in the development of simulators such as DISCO [6], UNIBATCH [4], BATCHES [3] and the gPROMS [1].

This paper presents the methodology which is central to a combined discrete/continuous simulator. The discussion of the methodology is based on its implementation in the

BATCHES simulator. However, the concepts are applicable to any combined discrete/continuous simulator. The special requirements of a general purpose batch process simulator have been discussed in another paper [2].

2. The Time Advance Mechanism

Central to a combined discrete/continuous simulator is an algorithm, the **time advance mechanism**, which marches the process being modeled in time. The four key components of the time advance mechanism are:

1. Manipulation of the event calendar
2. Solution of the differential/algebraic system of equations
3. Detection and processing of discrete changes
4. Implementation of operating decisions

The role played by each component in the time advance mechanism will be discussed in this section.

2.1 The Event Calendar

The events during a simulation represent discrete changes to the state of the process. The events are of two types: **time events** or **state events.**

A time event is that discrete change whose time of occurrence is known a priori. For example, if the recipe for a particular step requires the contents of a vessel to be mixed for 1.0 hour, then the end-of-mixing event can be scheduled 1.0 hour after the start of mixing. Of course, the state of the process will usually change when an operation ends.

A state event occurs when a linear combination of state variables crosses a value, the **threshold**, in a certain direction. An example of a state event is given in Figure 1. The exact time of occurrence of a state event is not known a priori. For example, suppose the recipe for a particular step requires the contents of a vessel to be heated to 400 K.

Therefore, the time at which the contents of the vessel performing that step reach 400 K cannot be predicted when the heating is initiated. As a result, during simulation the time advance mechanism must constantly check for the occurrence of state events.

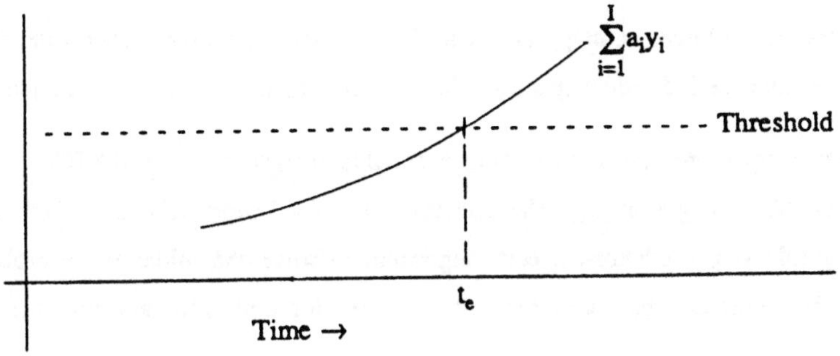

Figure 1. An example of a state event

The **event calendar** is a list of time events ordered on their scheduled time of occurrence. Associated with each event is an **event code** and additional descriptors which determine the set of actions, called the **event logic**, which are implemented when that event occurs. In the example given above, the end-of-mixing event may initiate a heating step in the vessel, or may initiate the addition of another ingredient to the vessel.

The event calendar is always in a state of transition. The events are removed from the calendar when they occur, and new events are added to the calendar as they are scheduled by the event logic and other components of the time advance mechanism. Several time events on the event calendar may have the same time of occurrence.

2.2 Solution of Differential/Algebraic Equations

The continuous overall change in the state of the system can be represented by a system of non-linear differential/algebraic equations, the **state equations**, of the following form:

$$F(y, \dot{y}, t) = 0$$

where, **y** is the vector of dependent variables, the **state variables**, and $\dot{\mathbf{y}} = \mathbf{dy/dt}$. The initial conditions, $\mathbf{y}(0)$ and $\dot{\mathbf{y}}(0)$ are given. In the BATCHES simulator, the state vector comprises only those variables which could potentially change with time. For example, if the material is heated using a fluid in a jacketed vessel and there are no material inputs and outputs, and no phase change, the composition and the total amount do not change. Only the temperature, volume, enthalpy and heat duty change with time. Therefore, there is no need to integrate the individual species balance equations nor the total mass equation.

The state equations are solved using a suitable integrator. The BATCHES simulator uses the DASSL integrator [8]. The integrator uses backward difference formulas (BDF) with a variable step predictor-corrector algorithm to solve the initial value problem. The integrator is robust and has been extensively tested for both stiff and non-stiff system of equations.

2.3 Detection and Processing of Discrete Changes

As described earlier, the exact time of occurrence of a state event is not known a priori. Therefore, after each integration step the time advance mechanism invokes an algorithm to detect whether a state event occurred during that integration step. To detect the occurrence of a state event the values of the desired linear combination of state variables before and after the integration step are compared with the threshold. For example, suppose a heating step is ended when the temperature of a vessel becomes higher than 400 K. If the temperature before an integration step is less than 400 K and after that step it is greater than or equal to 400 K, then a state event is detected during that step. The direction for crossing in this case is 'crossing the threshold of 400 K from below'. Thus, if the temperature before an integration step is higher than 400 K and after that step it is lower than 400 K then a state event as described above is not detected.

The next step after detecting a state event is to determine its exact time of occurrence, the **state event time**. The state event time is the root of the equation:

$$y_a - thr_a = 0$$

where, y_a is the linear combination of variables for which the state event was detected, and thr_a is the threshold. The upper and lower bounds on the root are well defined, namely, the time before and after the current integration step. The DASSL integrator maintains a polynomial for each state variable as part of its predictor-corrector algorithm. As a result, a Newton interpolation which has a second order convergence can be used for determining the state event time [7]. During a given integration step more than one state event may be detected. As shown in Figure 2, variables y_1 and y_2 crossed thresholds thr_1 and thr_2, respectively, during the same integration step. When multiple state events are detected, the state event time for each event is determined and the event(s) with the smallest state event time is(are) selected, t_{e_2} in this case. The simulation time is reset to that value and the state variables are interpolated to match the new simulation time.

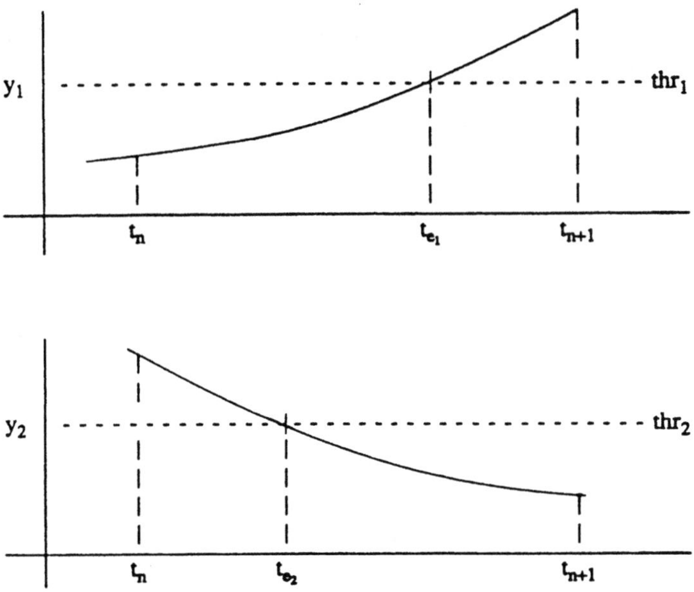

Figure 2. Example of multiple state events in one time step

The processing of an event consists of implementing the actions associated with the specified event logic. For example, ending a filling step may result in releasing the transfer line used during filling and advancing the vessel to process the next step. In BATCHES a library of elementary actions is provided, such as shutting off a transfer line, opening a valve, releasing a utility. An event logic consists of a combination of these elementary actions. Also, customized user written logic can be incorporated into BATCHES to implement complex heuristics which cannot be conveniently modeled with the existing event logic library.

2.4 Implementation of Operating Decisions

The operations associated with a recipe perform the necessary physical and chemical transformations to produce the desired end products from raw materials. The operations can be broadly divided into two categories: those operations which are initiated to pull the upstream material, and those operations which are initiated because the upstream material is pushed downstream. The operations in the first category must be independently initiated while those in the second category are initiated by the operations upstream when they become ready to send the material downstream. In BATCHES you can specify processing sequences which define the order in which the 'pull' type operations are independently initiated. The time advance mechanism reviews the processing sequences at each event to check whether a piece of equipment could be assigned to initiate an operation. Also, BATCHES uses a prioritized first-in-first-out (FIFO) queue discipline to process the requests for assigning equipment generated by the operations which push material downstream. The queues also are reviewed at each event by the time advance mechanism to determine whether any requests could be fulfilled. The processing sequences and priorities used in the queue management are specified by the user. Hence, by suitably manipulating the processing sequences and the priorities the user can influence the assignment of equipment to operations and the movement of material in the process. The priorities are also used for resolving the competition for shared resources.

3. Process Variability

The operating conditions in a typical batch process are characterized by random fluctuations. For example, the duration of an operation may not be the same for every instance,. but instead vary within a range, or a piece of equipment may breakdown randomly forcing delays due to repairs. The random fluctuations significantly affect the overall performance of a process and their effects should be included in the decision making process.

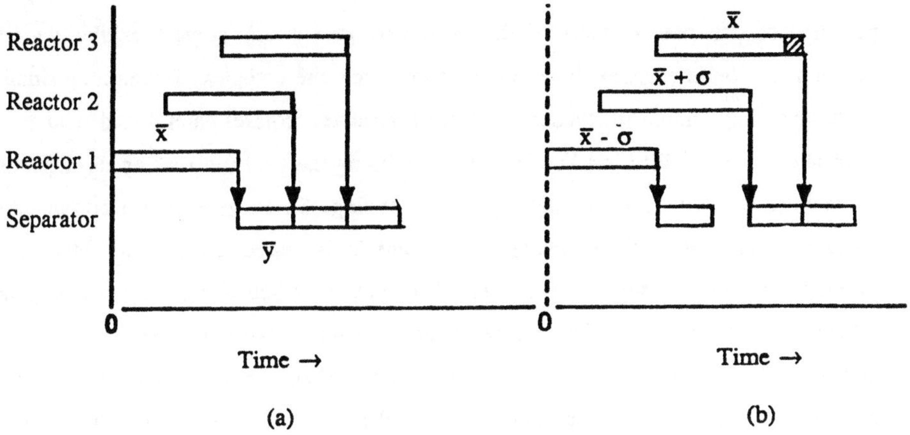

Figure 3. Effect of fluctuations in reactor cycle time on the makespan

Suppose a process consists of three reactors and a separator. Each reactor batch is processed in the separator which produces the final product. Also, suppose that the successive batches initiated on the reactors are offset by a fixed duration. If one assumes no process variability, with the reactor batch cycle time of $\bar{x}$ and the separator batch cycle time of $\bar{y}$, the Gantt chart for processing three reactor batches is shown in Figure 3a. With no process variability, all the operations are well synchronized resulting in the total elapsed time of $(\bar{x} + 3\bar{y})$ between the initiation of the first reactor batch and the completion of the last separator batch. However, in reality the reactor batch cycle time may not be constant for each batch. Suppose the reactor batch cycle times are normally distributed with the mean of $\bar{x}$ and the standard deviation of σ, and suppose the cycle times of the three reactor batches are $(\bar{x} - \sigma)$, $(\bar{x} + \sigma)$, and $\bar{x}$. As shown in Figure 3b, the reactors and the separator are no more well synchronized due to the variability. The shorter first batch results in some separator idle time, while the longer second batch introduces delays in the third reactor batch. Therefore, due to the interactions through the separator the effective time to complete the third reactor batch is $(\bar{x} + \sigma)$ instead of $\bar{x}$. The total elapsed time between the initiation of the first reactor batch and the completion of the last separator batch is $(\bar{x} + 3\bar{y} + \sigma)$.

The simple example discussed above illustrates the effect of process variability on its performance. In reality, most of the operations in a batch process exhibit variability. Furthermore, the interactions between operations are quite complex. Therefore, a simulation study involving random fluctuations in process variables requires careful statistical analysis. Typically, the variability is taken into account during the decision making through the use of confidence intervals or through hypothesis testing [5]. In general, the computation of confidence intervals or the testing of hypothesis is based on results obtained from experiments with the simulation model. The most common technique used to generate different conditions for simulation experiments is changing the initial seeds of the random-number streams used for sampling the parameters. Also, the simulation run length, data truncation to reduce the bias introduced by the initial transients and the number of experiments are some of the important factors which must be considered for reliable statistical analysis.

To represent process variability, the design of a simulator for batch processes must provide the capability to sample the appropriate model descriptors and to collect the necessary data for statistical analysis. The average, minimum, maximum and standard deviation for a variables are commonly required for statistical analysis.

4. Combined Discrete/Continuous Simulation Methodology

In this section the combined discrete/continuous simulation methodology is illustrated using a simple batch process.

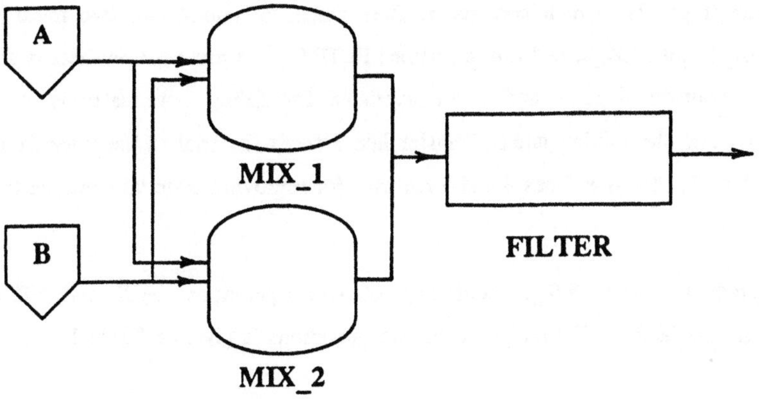

Figure 4. Process flow diagram of a batch process

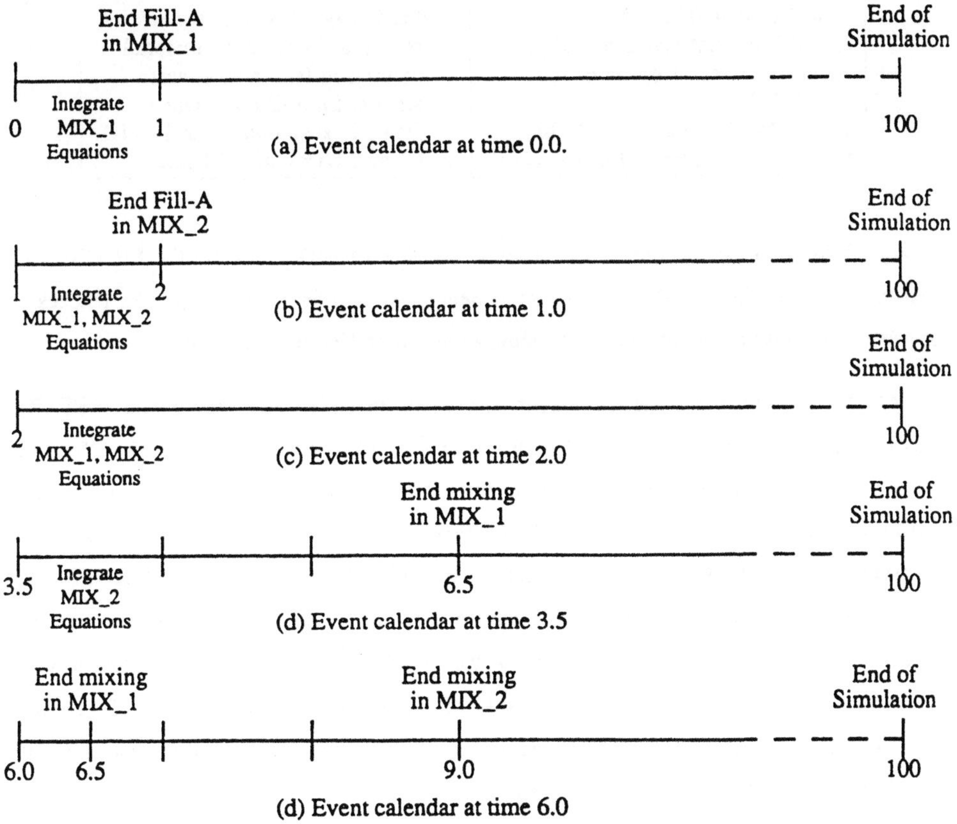

Figure 5. Changes in the event calendar with the progression of time

Consider a process which consists of three pieces of equipment; two mixing tanks, named MIX_1 and MIX_2, and a filter, named FLTR1. The process flowsheet is shown in Figure 4. Transfer lines 1 and 2 are available for filling raw materials A and B, respectively, into the mixing tanks. Transfer line 3 feeds material to the filter from either mix tank A or B. Transfer lines 4 and 5 are used for removing material from the two filter outputs.

One product, named PR_1, consisting of two operations, MIX and FILTER is processed in this facility. The recipe of the two operations is given in Table 1.

Table 1. Recipe of MIX and FILTER steps

MIX	FILTER
1. Fill 20 kg Raw Material A in 1.0 hour (FILL-A) 2. Fill Raw material B at 40 kg/hr until the vessel becomes full (FILL-B) 3. Mix for 3 ± 0.2 hours (STIR) 4. Filter the contents (FEED-FLTR)	1. Filter the contents of either MIX_1 or MIX_2 at the rate of 60 kg/hr. 90% of the material coming in leaves as a waste stream. Stop filtering when 20 kg is accumulated in FLTR1. 2. Clean FLTR1 in 0.5 hour.

The MIX step can be performed on either MIX_1 or MIX_2, while the FILTER step can be performed on FLTR1. A processing sequence to initiate 2 mixing tanks batches is specified. Suppose an event to stop the simulation run at 100 hr is specified.

At the beginning of the simulation the event calendar has two events, 'START SIMULATION' at time 0.0, and 'STOP SIMULATION' at time 100.0. The 'START SIMULATION' event forces the review of the processing sequence which in turn start the FILL-A elementary step in MIX_1. Since there is only one transfer line available to transfer raw material A, the MIX step cannot be initiated MIX_2. Also, the FILTER step cannot be initiated because none of the mixers is ready yet to send material downstream. Since the FILL-A elementary step is completed in 1 hr, a time event is scheduled at time 1.0. The filling step in MIX_A entails solving the differential equations for the species mass balance and the total amount equations. Figure 5a shows the event calendar after all the events at 0.0 are processed. Since FILL-A requires the solution of differential equations, the simulator starts marching in time by integrating the state equations. The goal of the simulator is to advance the process up to the next time event time. Since there are no state event conditions active during Fill-A step, the integration halts at time 1.0 because of the time event.

At time 1.0, the Fill-A elementary step is completed and FILL-B is started in MIX_1. Also, since transfer line 1 is released after the completion of FILL-A in MIX_1, the MIX step is initiated in MIX_2. A new event to end Fill-A in MIX_2 is scheduled at time 2.0. Figure 5b shows the event calendar after all the events at 1.0 are processed. Fill-B ends based on a state event (MIX_1 becoming full), hence there is no event to end FILL-B in MIX_1 on the event calendar. The FILTER still cannot be initiated. The new set of equations to be integrated consist of the species mass balance and total mass equations for both MIX_1 and MIX_2. Since, there is one active state event condition to end Fill-B step in MIX_1, the integrator checks for a state event after each integration step. The next known event is at time 2.0 when Fill-A is completed in MIX_2.

Suppose no state event is detected and the integration halts at time 2.0. After completing Fill-A, MIX_2 has to wait because Fill-B is still in progress in MIX_1 and there is only one transfer line available for transferring raw material B. Therefore, after processing the events at time 2.0, there is only one event on the calendar as shown in Figure 5c. The state vector consists of the species balance and total mass equations for FILL-B in MIX_1.

Suppose a state event is detected at time 3.5 because MIX_1 became full. Transfer line 2 is released and MIX_1 is advanced to the STIR elementary step which has no state equations. The duration of STIR varies randomly between 2.8 and 3.2 hr. Suppose for the first batch the duration is 2.9 hr. An event is scheduled at time 6.4 to mark the end of the STIR elementary step in MIX_1. Also, Fill-B is initiated in MIX_2. Figure 5d shows the event calendar after all the events at 3.5 are processed. The state vector consists of the species balance and total mass equations for Fill-B in MIX_2. One state event condition is active, namely MIX_2 becoming full. The goal of the simulator is to advance the process to 6.4. However, a state event is detected at 6.0, and the integration is halted.

At 6.0, MIX_2 is advanced to the STIR elementary step which has no state equations. Suppose the duration of the STIR elementary step for the second batch is 3.1 hr. As a result, a time event is scheduled at 9.1 to mark the end of the STIR elementary step in MIX_2. Figure 5e shows the event calendar after all the events at 6.0 are processed. Since both MIX_1 and MIX_2 are processing the STIR elementary step there are no state equations to be integrated and the time is advanced to 6.4 hr.

At 6.4, the FEED-FLTR elementary step is started in MIX_1, and FILTER step in FLTR1. The state vector consists of the species mass balance and total mass equations for MIX_1, and species balance and total mass equations for FLTR1. Two state events active when integration is resumed, one to mark the end of the FEED-FLTR elementary step when MIX_1 becomes empty, and one to stop filtering when the total mass in FLTR1 reaches 300.0 kg. No new time events are added to the event calendar.

Suppose MIX_1 becomes empty at 8.4 hr. As a result the flow into FLTR1 is stopped, and MIX_1 is released. Since there are no more batches to be made, MIX_1 remains idle for the rest of the simulation. 12 kg is accumulated in the FLTR1. After processing all events at 8.4, the time is advanced to 9.1 since there are state equations to be integrated.

At 9.1, the FEED-FLTR elementary step is started in MIX_2, and the FILTER step is resumed in FLTR1. The state vector consists of the species mass balance and total mass equations for MIX_2, and species balance and total mass equations for FLTR1. Two state events are active when integration is resumed, and there is only event on the event calendar, namely, end simulation at 100.0.

At 10.433, the FILTER step is ended because 20 kg is accumulated in FLTR1. Therefore, the flow is stopped with 40 kg still left in MIX_2. Cleaning is initiated in FLTR1, and an event is scheduled at 10.933 hr to mark the end of cleaning of FLTR1. Since there are no equations to be integrated the time is advanced to 10.933 hr.

At 10.933, FLTR1 is reassigned to the FILTER step and the filtration of the rest of the material in MIX_2 is resumed. At 11.6 hr MIX_2 becomes empty and the filtration is halted with 4.0 kg left in FLTR1. Since there are no more batches to be made and no more material to be processed the time is advanced to 100.0 hr and the simulation is ended.

The example given above illustrates how the time advance mechanism works in a combined discrete/continuous simulator.

5. Conclusions

The combined discrete/continuous simulation methodology discussed in this paper is used in several simulators for modeling discrete manufacturing systems as well as batch and semicontinuous processes. The time advance mechanism uses the event calendar to coordinate the execution of the following key blocks: Solution of the differential/algebraic system of equations, Detection and processing of discrete changes, and Initiating operations in equipment.

6. References

1. Barton, P.I., and Pantelides, C.C.: The Modelling and Simulation of Combined Discrete/Continuous Processes. International Symposium of Process Systems Engineering, Montebello, Canada, August 1991
2. Clark, S., and Joglekar, G.S.: Simulation Software for Batch Process Engineering. NATO Advanced Study Institute. This volume, p. 376
3. Clark, S., and Kuriyan, K.: BATCHES - Simulation Software for Managing Semicontinuous and Batch Processes. AIChE National Meeting. Houston, April 1989
4. Czulek, A.J.: An Experimental Simulator for Batch Chemical Processes. Comp. Chem. Eng. 12, 253 - 259 (1988)
5. Felder, R., Mcleod, G., and Modlin, R.: Simulation for the Capacity Planning of Specialty Chemicals Production. Chem. Eng. Prog. 6, 41 - 61 (1985)
6. Helsgaun, K.: DISCO - a SIMULA Based Language for Continuous Combined and Discrete Simulation. Simulation. 1, July 1980
7. Joglekar, G.S., Reklaitis, G.V.: A Simulator for Batch and Semicontinuous Processes. Comp. Chem. Eng. 8, 315 - 327 (1984)
8. Petzold, L.R.: A description of DASSL : A differential/Algebraic system solver. IMACS World Congress. Montreal, Canada, August 1982
9. Pritsker, A.A.B.: Introduction to Simulation and SLAM II. Systems Publishing Corporation 1986
10. Schreiber, T.: Simulation Using GPSS. John Wiley 1974

Simulation Software for Batch Process Engineering

Steven M. Clark, Girish S. Joglekar

Batch Process Technologies, Inc., P. O. Box 2001, W. Lafayette, IN 47906, USA

Abstract: Simulation is ideal for understanding the complex interactions in a batch/semicontinuous process. Typically, multiple products and intermediates are made in a batch process, each according to its given recipe. Heuristics are often employed for sequencing of operations in equipment, assignment of resources and control of inventories. A decision support tool which can integrate both design and operating features is necessary to accurately model batch processes.

The details of the BATCHES simulator, designed specifically for the needs of batch processes, are discussed. The simulator provides a unique approach for representing a process in a modular fashion which is data driven. The library of process dynamics models can be used to control the level of detail in a simulation model. Its integrated graphical and database user interface facilitates model building and analysis of results.

Keywords: Simulation, batch/semicontinuous

1. Introduction

A general purpose simulation tool for batch/semicontinuous processes must be able to meet their special requirements from the computational standpoint as well as from the model representation and analysis standpoint. The combined discrete/continuous simulation methodology necessary to model the process dynamics, accompanied with discrete changes in the process state is discussed by Clark [1]. Apart from the special computational requirements, the large amount of information necessary to build a simulation model of batch processes requires innovative modeling constructs to ease the data input and make process representation flexible, modular and intuitively clear. Also, the simulation results must be presented in a way so as to allow the analysis of the time dependent behavior of the process, and provide information about the overall performance measures.

Over the past few years, several modeling systems have been reported in the literature, gPROMS [1], UNIBATCH [4], BOSS [5], which incorporate a combined discrete/continuous simulation methodology necessary for a general purpose simulator for batch/semicontinuous process engineering. However, none of these systems adequately address the special requirements of batch processes from the process representation, data management and analysis standpoint. These needs must be fulfilled for a wider acceptance and sustained use of a tool for batch process engineering.

In this paper, the modeling constructs and the data input and analysis capabilities provided by the BATCHES simulator are discussed.

2. Process Equipment and Recipes

In a typical batch process, several products are manufactured in the given set of equipment items. Each product is made according to a specific recipe. A recipe describes the series of operations which are performed in the manufacture of a product. Each operation, in turn, consists of a series of elementary processing steps performed in the piece of equipment assigned to that operation. A process equipment network is given in Figure 1, while the description of a recipe is given in Figure 2.

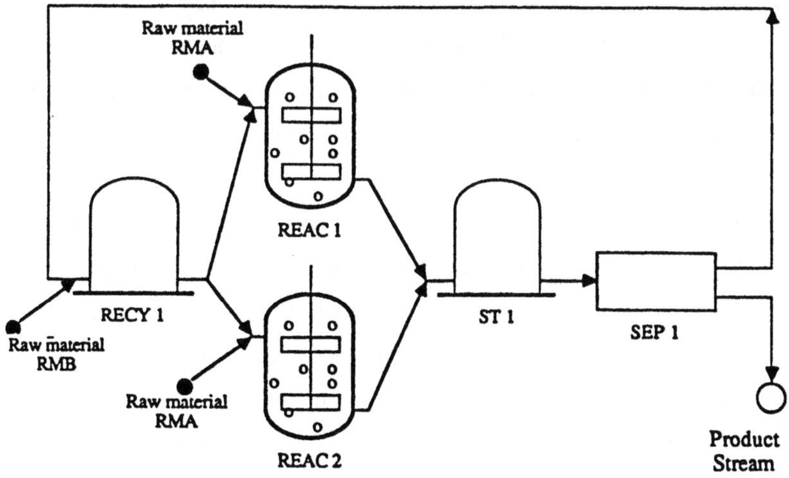

Figure 1. Example of a process equipment network

- **Recycle storage operation:**
  1. Fill raw material 'RMB' at 1000.0 kg/hr until the tank becomes full
  2. Allow recycle of material from the separator and withdrawal of material by downstream reactors

- **Reactor operation:**
  1. Transfer 100.0 kg of raw material 'RMA' in one hour. Use operator 'REACTOR-OP' during filling.
  2. Preheat the contents of the reactor for one hour using 'STEAM'. Use operator 'REACTOR-OP' during heating.
  3. Allow the contents to react for one hour.
  4. Transfer 100.0 kg of material from the recycle storage tank in one hour. Use operator 'REACTOR-OP' during filling.
  5. Let the reaction continue for one more hour.
  6. Cool the contents and let the material age for one hour.
  7. Transfer the contents into the intermediate storage tank in one hour.

- **Intermediate Storage Operation:**
  1. Allow transfer of material into the tank from the reactors and withdrawal of material by the downstream separator

- **Separator operation:**
  1. Separate continuously the contents of the storage tank into a product and a recycle stream. Send the recycle stream to the recycle storage tank.

Figure 2. Description of operations in a recipe

In a batch process, operations and equipment items have a many to many relationship, that is, several operations may be performed in a given piece of equipment, and several pieces of equipment may be suitable to perform a given operation. This represents a significant departure from the steady state processes where an operation and a piece of equipment have a one to one relationship.

The recipe in Figure 2 further shows that, during the course of an operation, a series of physical and chemical changes may take place in the assigned piece of equipment. For example, during the RXN operation, the first two elementary steps represent physical changes, followed by a chemical change, and so on. As a result, the mathematical equations which describe the process dynamics are different for each elementary step. This also is markedly different from a steady state operation in which a unit involves one physical/chemical change. An operation in a batch process represents a series of 'unit operations' which are performed in the assigned piece of equipment.

The BATCHES simulator provides two constructs, the equipment network and the recipe network, to model the many to many relationship between equipment items and operations, and to model the recipes of various products

2.1 Process Equipment Network

A process equipment network represents the physical layout and connectivity of the equipment items in the given process. The equipment parameters describe the physical characteristics of each equipment item such as volume, or heat transfer area. Also, any physical connectivity constraints are specified in the equipment network through the use of transfer lines. For example, some pieces of equipment from a stage may be connectible to only a few pieces of equipment from another stage. Similarly, to transfer material between two stages, a manifold may be available which allows only one active transfer at a time, or from a storage tank only one active transfers may be allowed at any given time.

2.2 Recipe Network

Each product in a batch process is represented by a recipe network. In a recipe network, an operation is represented by the BATCHES construct task, while an elementary processing step is represented by the construct subtask. Figure 3 shows the recipe network for the recipe described in Figure 2. The appropriate choice of task and subtask parameters allows the user to represent the recipe details. Thus, the basic building block for recipe networks is a subtask.

Subtask Models: The most important subtask descriptor is the model used to represent the corresponding elementary processing step. A library of 31 subtask models is provided with BATCHES. Each subtask model in the library represents specific physical/chemical transformations. For example, filling, emptying, reaction under adiabatic conditions, continuous separation of material and so forth. The models range from a simple time delay model to a complex batch reactor model. The pertinent material and energy balances associated with the underlying transformations are described as a set of simultaneous differential/algebraic equations (DAE), the model equations. The formulation of model equations is based on certain assumptions. For example, the following assumptions are made in one of the models, named '=FILLING': there is only one active phase in the bulk, the subtask has no material outputs, each material input is at constant conditions, namely, constant temperature, pressure and composition of species. Based

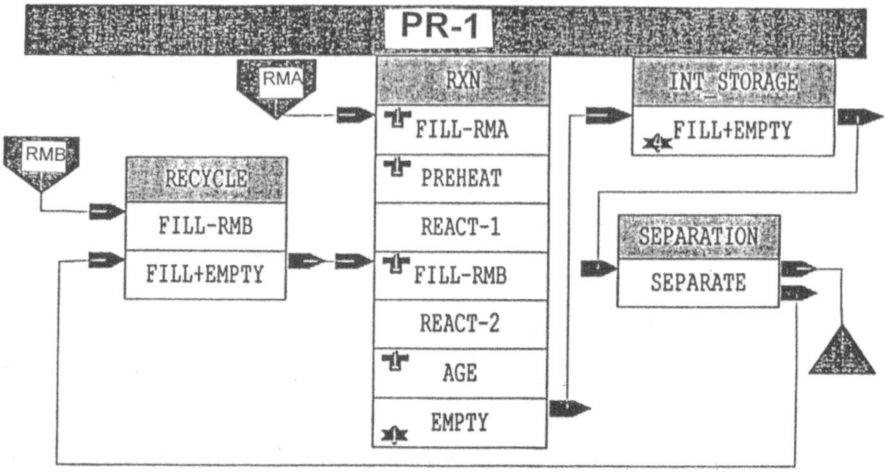

Figure 3. Example of a recipe network

on these assumptions the '=FILLING' model is described by the following equations:

$$M \frac{dx_j}{dt} + x_j \frac{dM}{dt} = \sum_{i=1}^{I} \sum_{p=1}^{\pi_i} x_{ipj} F_{ip} \qquad j = 1,n$$

$$\frac{dM}{dt} = \sum_{i=1}^{I} \sum_{p=1}^{\pi_i} F_{ip}$$

$$\frac{dE}{dt} = \sum_{i=1}^{I} \sum_{p=1}^{\pi_i} E_{ip}$$

$$E = MH_l(T,P,x_j)$$

$$\frac{dP}{dt} = 0$$

$$\rho = \rho_l(T,P,x_j)$$

$$\theta \rho V = M$$

where,

n	number of components	V	effective volume of equipment
x	mass fraction	π	number of phases
E	total enthalpy	ρ	density
F	mass flowrate	θ	volume fraction
H	enthalpy per unit mass	I	subtask input index
I	number of subtask inputs	j	component index
M	total mass	l	liquid phase
P	Pressure (Pa)	p	phase index
T	Temperature (K)		

Thus, when a piece of equipment executes a subtask which uses this model, the generalized set of equations given above is solved for the specific conditions in that piece of equipment at that time. Since the number of species associated with each recipe could be different, the number of equations with each instance of a model could be different. Furthermore, since the species associated with each recipe will be different, the mass fraction variables with each instance of a model could be associated with different species. The subtask models in the library provide the modularity necessary for a general purpose simulator. If the models in the library do not meet the requirements of a particular elementary processing step, new models can be added to the library.

The other important subtask parameters are: subtask duration, state event description, operator and utility requirements. Typically, the available resources are shared by the entire process, and the instantaneous rate of consumption of a resource cannot exceed a specified value at any time. For example, the maximum steam consumption rate may be constrained by the design capacity of the boiler, or only one operator of a particular type may be available per shift. Individual operations compete for the resources required for successfully executing the elementary processing steps in that operation.

The other building blocks for a recipe network are flow lines, raw materials, and infinite sinks.

Flow Lines: Material is transferred from one piece of equipment to another during specific elementary processing steps. The transfer of material is represented by a flow line connecting the appropriate subtasks. The flow parameters describe the amount of material being transferred and the flow characteristics such as flowrate, and flow profiles.

Raw Materials: A raw material is an ingredient which is available whenever a subtask needs it, and is characterized by its temperature, pressure, and composition. A raw material is represented by a pentagon shown in Figure 3 and is identified by a unique name.

Infinite Sinks: An infinite sink represents the material leaving a process. An infinite sink is represented by a triangle shown in Figure 3, and is identified by the subtask and the output index connected to it. The simulator continuously monitors the cumulative amount and the average composition of the material withdrawn from the process.

2.3 Link Between Equipment and Recipe Networks

For every task in a recipe an ordered list of equipment items suitable to process that task is specified. At a given time, if there is a choice, the piece of equipment best suited to perform the task is selected. A particular piece of equipment suitable for several tasks appears on the lists associated with the corresponding tasks. The suitable equipment lists provide the link between the equipment and recipe networks.

2.4 Executing an Operation in Equipment

After a piece of equipment is assigned to perform an operation, all subtasks in that task are implemented according to the recipe.

Typically, before the execution of an elementary step certain conditions are checked. For example, a filling step may start only when sufficient amount of upstream material, a transfer line, and an operator are available. A heating step may entail checking the availability of a particular resource. The execution of a subtask begins only when the following conditions are satisfied:

- The required amount of input material is available upstream
- Equipment items are available downstream for every output
- A transfer line is available for each material transfer
- Operators and utilities are available in the specified amounts

2.5 Advantages of the Two-Network Representation

Representing a batch process as a set of two network types provides several advantages. First of all, it is a natural representation of how a batch process operates, because there is a natural dichotomy between the physical equipment which merely are 'sites', and the standard operating procedure for manufacturing each product using the sites. As a result, a two-network model becomes easy to understand at a conceptual level as well as at the implementation level.

The two-network representations provide an efficient mechanism for building simulation models of multiple manufacturing facilities. For example, very often the same products are made in different physical locations which may have different equipment specifications and configurations. To model such facilities one equipment network can be constructed for each facility, with one common set of recipe networks for the products made in them. To model a specific facility the appropriate combination of equipment network and recipe networks can be selected. Similarly, in a given facility several products can be made over an extended period of time, such as one year. However, for a simulation study, the time horizon may be much shorter, for example one day or one week, during which the user may want to consider only a few recipes. In such cases, to create a process model the appropriate subset of recipe networks and the appropriate equipment network can be selected.

3. Decision and Control Logic

During the operation of a batch process decisions are constantly made which affect its overall performance, for example, the sequence of making products, assignment of equipment to operations, assignment of resources to elementary steps, and transfer of material between equipment items. Also, actions which help in synchronizing operations are implemented based on the state of the process at that time. In a recipe network, various task and subtask parameters allow the user to select the suitable decision and control logic options.

Processing Sequences: For a simulation run, sequences for initiating the operations are specified. The identity of the operation to be initiated and a stopping criteria are specified for each entry in the sequence. For example, make 5 reactor batches of recipe A, followed by sufficient reactor batches of recipe B to produce 100 kg of reaction mixture and so on. Typically, those operations which mark the beginning of the processing of material associated with a product are independently initiated through processing sequences. The processing sequences merely specify the desired sequence. The actual start and end times of operations, determined by the simulator as it marches the process in time, are some of the key performance indicators.

Equipment and Resource Assignment: The requests for assigning equipment items to initiate tasks are handled by First In First Out (FIFO) queues which are ordered by user specified priorities. Similarly, the requests for assigning operators and utilities are handled by FIFO queues.

The decision to assign a piece of equipment to initiate a task is governed by the logic option associated with that task. For example, one of the options may assign a suitable piece of equipment as soon as one becomes available and then searches for upstream material, downstream equipment to send material to and so on. Another option may not assign a piece of equipment if upstream material is not available which prevents unnecessary reservation of equipment. Thus, by selecting an appropriate option one can accurately represent how assignments are done in the actual process.

Selection of Upstream Equipment: The selection of upstream equipment items which provide the input material is governed by a logic flag. The following options are available to select upstream equipment and start the material flow:

- Wait until the required amount becomes available in a single piece of equipment upstream
- Wait until the required amount becomes available cumulatively in one or more pieces of equipment upstream
- Start the transfer of material as soon as some material becomes available upstream and keep searching for additional material until the amount requirement is satisfied

Conditional Branching: Often, during an operation a different set of elementary steps is executed based on quality control considerations or the state of the material at that particular time. For example, in the RXN recipe shown in Figure 2, 80% of the batches may require additional processing of 1 hour after the aging step to bring the material within allowed specifications. This is modeled by specifying conditional branching information with the AGING subtask.

User Defined Logic: Certain complex decision logic implemented in a process may be outside the scope of the existing logic options available through various parameter choices. In that case, the user can incorporate customized decision logic into BATCHES. The simulator provides a set of utility subroutines which can access the process status. By retrieving the appropriate information, the user can formulate and implement any desired set of actions triggered by the process status.

4. Shortcomings of Discrete Simulators

The discrete simulators such as SLAM, GPSS, in principle, provide the basic discrete/continuous simulation methodology necessary for modeling batch processes. However, these simulators are merely general purpose simulation languages, and therefore require considerable expertise and effort for simulating batch processes. In spite of their availability for more than 15 years, only relatively few chemical process applications have been reported in the literature. In general, the discrete simulators have not been accepted as effective tools for simulating batch processes because of several limitations. However, it must be noted that because these simulators have been written in procedural languages like FORTRAN and provide a mechanism to incorporate user written code there is no inherent limit on adapting them to satisfy a particular need, provided the user has the luxury of time and the expertise required for developing the customized code.

Process Representation: In discrete simulators, the manufacturing 'activities' are performed on 'entities' by 'servers'. The servers and activities would be equivalent to process equipment and elementary processing steps in a recipe, respectively. An entity is a widget which can be very loosely compared to a batch of material. A widget keeps its identity and its movement can be tracked in the process. Also, when a server completes an activity, modeled as a time delay, the entity is released and the server can be assigned to process another entity. In a batch process, during a task several elementary steps are performed in the assigned piece of equipment. Also, not all elementary steps can be modeled as pure time delays. For example, a step can end based on a state event, or it can end based on interactions with other pieces of equipment. During a task, material from several upstream steps may be transferred into a piece of equipment and several downstream steps may withdraw material from it. Thus, a batch of material may constantly change its identity. Also, whenever there is a transfer of material between elementary steps, pieces of equipment must be available simultaneously from several processing stages. For example, in order to filter material from a feed tank and store the mother liquor in a storage tank, three pieces of equipment ('servers'), one from each stage, must be available simultaneously.

All parallel servers in a discrete process are assumed to be identical and suitable for an activity. The parallel pieces of equipment in a batch process are seldom identical, thus resulting in equipment dependent cycle times for processing steps and also equipment dependent batch sizes. Also, some pieces of equipment in a stage are often not suitable for certain tasks because of safety and corrosivity considerations. Additionally, there may be constraints due to connectivity on transfer of material between certain pairs of equipment items in parallel stages.

Push and Pull: The key mechanism for the movement of entities assumed in the discrete simulators is 'push entities downstream', that is, whenever a server finishes an activity it releases the entity and the entity waits in a queue for the assignment of a server to perform the next activity. However, in batch processes, two mechanisms for the movement of material are prevalent, namely 'push' and 'pull'. In the 'pull' mechanism, an elementary processing step searches upstream for the required material and transfers ('pulls') the required amount which may be a fraction of an upstream batch. For example, a large batch of a catalyst may be prepared which is consumed by the downstream reactors in small amounts.

Process Dynamics: The discrete simulators use an explicit integration algorithm such as Runge-Kutta for solving the system of differential equations describing the processing steps. The explicit methods cannot solve a system of DAEs very effectively and therefore the differential equations must be explicit as given below:

$$\dot{y} = F(y, t)$$

Also, the explicit methods are not recommended for solving stiff equations. Since even a simple dynamic process model such as '=FILLING' illustrated in Section 2.1 requires the solution of DAEs, the discrete simulators have a severe limitation. The problem is compounded by the fact that the more complex models such as reactor and evaporation have implicit DAEs and are generally stiff.

In discrete simulators, the differential equations and the state events must be defined prior to a simulation run. Applied to the simulation of multiproduct batch processes that would translate into defining the equations for all feasible combinations of subtasks and equipment items, and the associated state event conditions, along with the logic to set the values and derivatives of the variables which would be inactive at a given time, prior to a simulation run. While not impossible, this is an overwhelming requirement for the team generating and maintaining the model. Also, defining all of the possible combinations in the state vector and checking the state event conditions at the end of every integration step would result in tremendous computational overhead considering that at any time at the most one combination could be 'active' per equipment item. The modularity of BATCHES eliminates all the programming as well as computational overheads mentioned above.

Hard-wired Model: In discrete simulators, since the event logic, dynamic process models and operating characteristics are implemented through user written code, the models tend to be very problem specific. Therefore, any 'what if...' which is outside the scope of the existing code can be studied only after suitably changing the code. Since the main objective of a simulation study is to evaluate various alternatives, the prospect of having to change the software code in order to evaluate the impact of a change restricts its use to experts. Since BATCHES is data driven and modular, the changes to the process can be made and evaluated easily.

Simulation Output: The output report from discrete simulators is restricted to utilization and queue statistics about the entities processed. In addition to this information, the BATCHES simulator provides mass balances and a detailed breakdown of cycle times. The former is necessary for the determining the production rates of useful as well as side or waste stream, while the latter is crucial in pinpointing the bottlenecks. The BATCHES simulation output is discussed in the next section.

5. Simulation Input and Output

A simulation project involves many complex activities, and requires an understanding of systems and information flow. Activities on a project include collecting data, building models, executing simulations, generating alternatives, analyzing outputs, and presenting results. Making and implementing recommendations based on the results need to be a part of a simulation project. The BAtches SHell, BASH, is a software package which creates an environment that supports all of these activities. To provide this support, BASH has been integrated with BATCHES and contains database and graphic capabilities.

5.1 BASH Overview

The BATCHES simulation models are data driven. To build a simulation model, one must specify appropriate values of various parameters associated with the modeling constructs. BASH provides interactive graphical network builders and forms to build and alter simulation models. Models built in this fashion are cataloged and easily recalled.

BASH provides the capability to store, retrieve, and organize simulation generated outputs. The use of a database makes possible the flexible presentation of project results for comparison and validation purposes.

BASH has a complete system for providing both graphical and tabular outputs. It segregates formats from the data to be displayed and allows general or stylized formats to be built independently. Over the years, formats used on various projects become available for reuse.

An animation is a dynamic presentation of the changes in the state of a system or model over time. A BASH animation is presented on a facility model. Icons are used to represent the elements of the system. During a simulation run, traces are collected for user specified events. The animation rules translate an event trace into actions on the screen.

The architecture of BASH is shown in Figure 4. The outer ring in Figure 4 indicates the user interfaces to BASH. The BASH language gives the user access to the BASH modules for network building, schedule building, building simulation specifications, control building, data entry, format preparation, and facility, rule and icon building. In addition, the BASH language allows the user to specify directly operations such as data analysis, report generation, graphics generation, and animation.

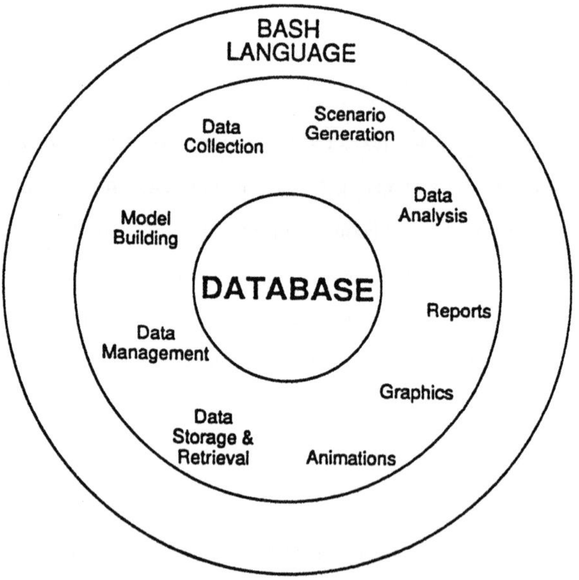

Figure 4. BASH architecture

5.2 Simulation Output

The BATCHES simulator generates textual summary reports and time series data for analyzing the performance of a simulation run.

The summary report consists of mass balance summary, and equipment and resource utilization statistics. Additionally, the cycle time summary for each piece of equipment, along with a breakdown of the time lost due to waiting for material and resources, is reported. An example of the cycle time and wait time summary for a piece of equipment, which is suitable to perform task {PR 1, RXN}, is given below.

```
***********************************************
* BATCH CYCLE TIME AND WAITING TIME STATISTICS *
*               TIME IN (hr)                   *
***********************************************

  **** EQUIPMENT NAME : REAC 1

    (P,T)             BATCHES TOT-PROC-TM AV-CYCLE-TM TOT-ACTIV-TM TOT-WAIT-TM
    PR 1
     RXN                    8    183.0        22.88        112.0         71.00

  DURING SUBTASK              TOTAL TIME SPENT WAITING FOR              ACTIVE
                  UPSTR-EQP   DOWNSTR-EQP  OPS+UTILS    S/C-CHAIN       TIME
  FILL-RMA           0.           0.         54.00         0.          8.000
  PREHEAT            0.           0.         6.000         0.          8.000
  REACT-1            0.           0.           0.          0.          32.00
  FILL-RMB         3.000          0.           0.          0.          16.00
  REACT-2            0.           0.           0.          0.          32.00
  AGE                0.           0.         8.000         0.          8.000
  EMPTY              0.           0.           0.          0.          8.000
  ------------------------------------------------------------------------------
     ((TOTAL))     3.000          0.         68.00         0.          112.0

                              BATCHES TOT-PROC-TM TOT-ACTIV-TM TOT-WAIT-TM
    TOTAL FOR REAC 1                8    183.0        112.0        71.00
```

First, the name of the piece of equipment is printed. Next, the name of the task and the number of batches completed, the total processing time, average cycle time (Total processing time/number of completed batches), the total active time and the total waiting time for the corresponding task are printed. The total processing time is the sum of the total active time and the total waiting time. This is followed by a detailed breakdown of the waiting and active times for each subtask. The time spent waiting for upstream material and/or transfer line (UPSTR-EQP), the downstream equipment (DOWNSTR-EQP), operators and/or utilities (OPS+UTIL). Column S/C-CHAIN denotes the time spent waiting for either an upstream subtask to send information to proceed or a downstream subtask to initiate the flow of material. The last column denotes the time spent in actively processing the subtask.

The cycle time statistics are critical in identifying bottlenecks. For example, if the waiting time for operators/utilities is significant, then increasing their availability may resolve the problem, or if waiting time for downstream equipment is significant, adding a piece of equipment in the appropriate stage may resolve the problem.

5.1 Time Series Output

During a simulation run, one can collect time series data for the selected variables, for example, the amount of material in a specific piece of equipment, the process inventory, the cumulative production, utility usage and so on. The time series data are useful in characterizing the dynamic behavior of a process. For example, when do the peaks occur, how saturated a particular variable is, is there any periodicity etc. Also, by comparing the time series data for the desired variables one can measure the amount of interaction or establish correlations between them. For example, are equipment items demanding a particular utility at the same time, are the product inventories related to the unavailability of operators and so on. Most importantly, by comparing time series data from various simulation runs one can study the effects of changes on the dynamic behavior of a process.

Each simulation run is one 'what if...' defined by a particular combination of values of model descriptors, a 'scenario'. The time series data for a simulation run are stored in the BASH database under the scenario name.

Typically, the first step in the analysis of time series data is to present the data in a suitable graphical form such as plots, Gantt charts, pie charts and so on. BASH provides a wide variety of presentation graphics options including the ones mentioned above [2]. Examples of a plot and a Gantt chart are shown in Figures 5 and 6. For detailed quantitative analysis, BASH has functionalities to compute summaries for characterizing the time series data such as minimum, maximum, average, standard deviation, frequency distributions and so on. Also, the database provides additional data manipulation capabilities such as extracting data within a specific time window, or extracting data occurrences within a specific range of values. After extracting data which satisfy certain characteristics one can either present them graphically or compute summaries. Thus, very detailed information required for analysis can be easily derived. For example, the total

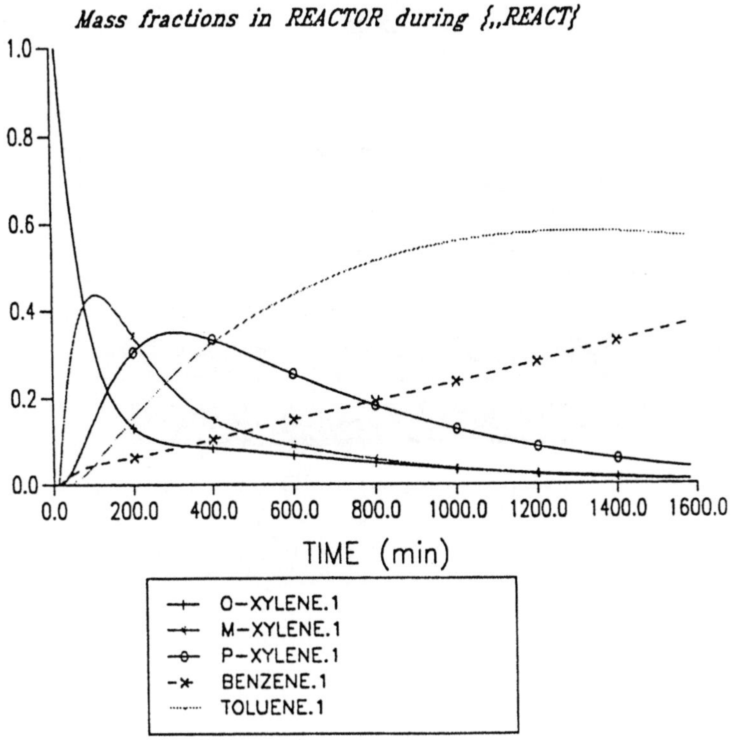

Figure 5. Example of an X-Y plot showing concentration profiles in a piece of equipment during a subtask

time when the process inventory was higher than a specific value, or the time spent by a piece of equipment in processing a particular task and so on.

The BASH database facilitates comparison of results from various simulations. The data from multiple simulation runs can be presented in multiple windows, or alternatively the desired variables from multiple simulation runs can be simultaneously displayed on one graphical output.

6. Conclusions

The BATCHES simulator, designed for the special requirements of batch/semicontinuous processes, provides a unique approach for representing a process in a modular fashion which is

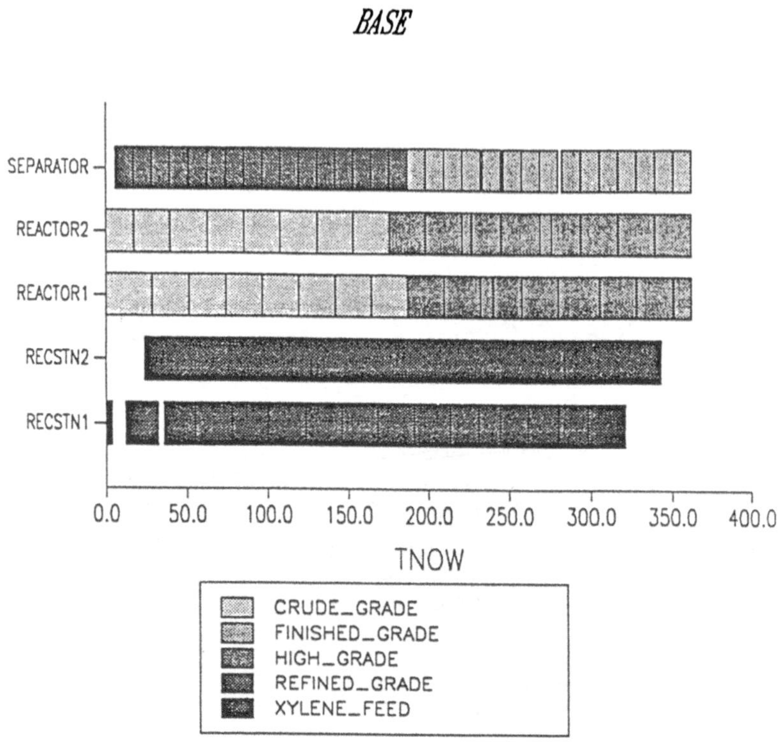

Figure 6. Example of a Gantt chart based on recipe in equipment. (Equipment names are shown in the left hand column)

data driven. Its integrated graphical and database user interface facilitates model building and analysis of results.

7. References

1. Barton, P.I., and Pantelides, C.C.: The Modeling and Simulation of Combined Discrete/Continuous Processes. International Symposium of Process Systems Engineering, Montebello, Canada, August 1991
2. BASH User's Manual. Batch Process Technologies, Inc., West Lafayette, IN, (1992)
3. Clark S., and Joglekar, G.S.- Features of Discrete Event Simulation. NATO Advanced Study Institute, This volume, p. 361
4. Czulek AJ.: An Experimental Simulator for Batch Chemical Processes. Comp. Chem. Eng. 12, 253- 259 (1988)
5. Joglekar, G.S., Reklaitis, G.V.: A Simulator for Batch and Semicontinuous Processes. Comp. Chem. Eng. 8, 315 -327 (1984)

The Role of Parallel and Distributed Computing Methods in Process Systems Engineering*

Joseph F. Pekny

School of Chemical Engineering, Purdue University, West Lafayette IN 47907, USA

Abstract: Parallel computers are becoming available that offer revolutionary increases in capability. These capability increases promise to shatter notions of intractability and make a substantial difference in how scientists and engineers formulate and solve problems. This tutorial will explore some process systems areas where a parallel computing approach and advances in computer technology promise a substantial impact. The tutorial will also include a summary of trends in underlying technologies, current industrial applications, and practical difficulties. An important underlying theme of the tutorial is that researchers must begin to devise flexible algorithms and algorithm design environments to help close the widening gulf between software and hardware capability. In particular, since parallel and distributed computers complicate algorithm design and implementation, advanced software engineering methods must be used to reduce the burden.

Keywords: parallel computing, distributed computing, special purpose algorithms, process engineering, algorithm design

1. Introduction

Computers are a critical enabling technology for the Chemical Process Industries. Indeed, no modern chemical plant could function or could have been designed without the benefit of the computer. The last thirty years have seen a 10,000 fold increase in the performance per unit cost of computers [41]. Each order of magnitude increase in capability has brought about a qualitative and quantitative improvement in process efficiency. This improvement manifests itself in better energy, equipment, and manpower utilization as well as in improved safety, inventory management, and flexibility. Commensurate with the important role of computers, the study of process systems engineering has intensified. Advances in methodology, algorithms, and software as well as ever expanding applications has been made possible by the rapid advances in computer technology. In this tutorial, we will explore the impact

---

* excepted in part from J. P. Pekny, "Parallel Computing Methods in Chemical Engineering", course notes, ChE 697P, School of Chemical Engineering, Purdue University, West Lafayette, IN 47907

that high performance computing will have on the direction of process systems engineering in the near future. The tutorial is divided into three parts: (i) advances in computer technology and basic concepts, (ii) designing parallel algorithms, and (iii) using optimal computer architectures through distributed computing.

2. Basic Concepts and Advances in Computer Technology

Parallel computing is a dominant theme of high performance computing in this decade. One operational definition of parallel computing is the application of spatially distributed data transformation elements (processors) to cooperatively execute a computational task. The underlying goal of all parallel computing research is to solve currently intractable problems. In science and engineering at large, there are a number of applications that make nearly unlimited demands on computer capability. For example,

- protein, polymer conformations and design
- global climate prediction
- artificial intelligence
- image, voice recognition
- computer generated speech
- quantum chromodynamics - calculate mass of elementary particles
- computational chemistry - quantum mechanics
- astronomical calculations (e.g. galaxy formation)
- turbulence - prediction and control
- seismology and oil exploration
- fractals and chaos
- wind tunnel simulations
- gene sequencing

Within Chemical Engineering a number of areas demand dramatic increases in computer capability. For example,

- Simulations for Molecular Description of ChE Processes
- Process Design, Control, Scheduling, and Management
- Transport Phenomena, Kinetics (Fluid flow simulation, Combustion, Chemical vapor deposition, Reactor profiles)
- Boundary Element Methods (Suspension simulations, Protein dynamics)
- Molecular Dynamics (Protein solvation, Macromolecular chain conformations)
- Monte Carlo Simulation (Thermodynamics, Reaction pathways, Polymerization/pyrolysis reactions)

- Expert Systems and AI (Fault diagnosis, Operator assistance)

In the 1960s, computational research was sometimes considered the weakest contributor of the theoretical, experimental, and computational triad underlying science and engineering. However continued dramatic improvements in computer capability will result in an increasing reliance on computational methods. Computational research presents exciting opportunities since rarely in the history of science and engineering have the underlying tools improved so dramatically in such a short period of time.

The hardware performance increases to be realized over the next decade are fundamentally different than the hardware performance increases realized over the last twenty years which largely came about without any effort on the part of applications researchers. In particular, the boundaries separating applications researchers, computer scientists, and computer architects is blurring since the design of high performance computers and support tools is largely dependent on specific applications and the choice of algorithms is dictated by hardware considerations. The users of high performance computing can still largely be shielded from the complexities of parallel computing by suitably packaging the algorithms and applications software.

2.1. Trends in Computer Technology

Based on projected improvements in circuit packing densities, manufacturing technology, and architecture, there is every reason to expect computer capability gains at the same or an accelerated rate for the foreseeable future [7]. In fact, [7] projects that high performance computers could achieve 10^{12} operations per second by 1995 and 10^{14} operations per second by 2000 or roughly a factor of 100 and 10,000 times faster than the highest performance computers today, respectively. Supporting these processing rates will be main memories of 10 to 100 gigawords and disk storage up to a terabyte [42]. Mirroring the advances in parallel supercomputing, desktop, laptop, and palm-top computers should increase in capability to the point that these classes of computers will be as or more powerful than current supercomputers with differentiation occurring as to the types of peripherals that are supported, i.e. input devices, graphics, disk storage, and I/O capability [42].

Because computer circuitry is approaching fundamental limits in speed, the high performance computers of the future will certainly possess hundreds to many thousands of processors. As flexible computer manufacturing techniques are perfected, cheap and high performance special purpose architectures will proliferate for common calculations. Already manufacturers cheaply produce special purpose control hardware and high performance special purpose machines exist for digital signal processing [7]. Within the process industries we can expect to see further special purpose hardware arise to perform such computations as expensive thermodynamic property calculations, sophisticated control algorithms, and sparse matrix computations for simulation and design calculations. There is every reason to expect that the most useful and widely needed algorithms will be compiled into silicon especially since the process industries are expected to be among the top five consumers of high performance computing [7].

The fundamental improvements in digital electronics technology will impact each of the components of computer architecture and hence application capabilities. Below, a short projection is given for capabilities that can be expected in the near future for each of the major architecture components.

2.1.1. Processing Elements

Speed improvements in computer processing elements are limited to a constant factor, however costs for a given unit of capability will continue to fall dramatically for the foreseeable future. Fabricators hope to make 1,000 Million Instruction Per Second (MIPS) processors available within the next few years (Sun Sparc, Digital Alpha, IBM Power Series, HP Precision Architecture) for use within engineering workstations [21]. Supercomputer vendors have entered into agreements to build machines based on hundreds or thousands of these processors which have substantial support for parallel processing built within them. Indeed, these processors will use a substantial amount of parallelism internally to achieve the projected speeds (instruction pipelining, multiple instruction launches, 64-bit words, etc.). Special purpose processing elements have been and are being developed for important operations from linear algebra (matrix inversion, multiplication), multimedia presentation (ray tracing, animation with high resolution graphics/audio), and data transformation (compression, encryption, Fast Fourier Transformation, convolution). With the advent of effective silicon compilers, computational researchers will have the option of implementing part or all of an algorithm in hardware for maximum speed. There are technologies on the horizon such as optical and quantum well devices that promise to dramatically improve (one or more orders of magnitude over conventional technology) the sequential speed of processing elements but they are probably at least ten years away from having a practical impact. Thus the proliferation of parallel processing is virtually assured into the next century. When new implementation technologies are practical, they will not usher in a new era of sequential computing since they can also be combined into parallel architectures for maximum performance. The experience gained with parallel algorithms and hardware will ultimately guarantee this result.

2.1.2. Volatile Memory

The memory hierarchy within computers will continue to be stratified with the introduction of additional caching layers between fast processors and the most abundant volatile memory [15]. The management of this hierarchy in a parallel environment continues to be the focus of intense research interest. There is also a trend to make some memory elements active in the sense that they perform simple processing activities in addition to passively storing data. This is just another manifestation of parallel processing. Passive memory capacity will continue order of magnitude improvements in capacity but not access speed over this decade. Techniques for guaranteeing the reliability and integrity of large amounts of passive memory will become increasingly important as computing systems utilize gigabytes and terabytes of memory capacity.

2.1.3. Long Distance Interconnection

Bandwidth and latency are two fundamental measurements of interconnection technology. The latency of a computer network is defined to be the amount of time required for a small message to be sent between two computers. Network bandwidth is defined as the rate at which a large message can be transmitted between two computers. Actual values for latency and bandwidth depend on the geographic location of the computers, network hardware and software technology, and the amount of network traffic. Latency is limited by the speed of light but there is no fundamental limit to bandwidth. The next five years will see dramatic improvements in bandwidth (as much as a factor of 10,000) as fiber optic technology matures. Advanced interconnection technology will allow cooperative parallel computing over regional, national, and international distances. Quite likely specialized supercomputing centers will arise with dedicated hardware for performing certain tasks. The availability of high bandwidth interconnection technology will allow automatic utilization of these centers during the course of routine calculations.

2.1.4. Permanent Storage

The next few years will be characterized by a proliferation of alternative permanent storage strategies: CD-ROM, Read-Write Optical Disk, etc. In addition, basic magnetic storage technology will continue to make dramatic improvements in cost per bit of storage, access times to data, and bandwidth through new combinations of conventional disk technology such as Redundant Arrays of Inexpensive Disks (RAID) [36]. Bandwidth and reliability increases will primarily be a result of data parallelism as researchers and vendors perfect the use of interleaved disk drives. By the end of the decade, disk drive capacity per unit cost should increase several orders of magnitude (factor of 1,000). The decreasing cost of integrated circuits will make disk controllers and interface electronics ever more sophisticated which is just another example of parallelism. The general trend is order of magnitude improvements in storage capacity at declining cost and increased accessibility of data.

2.1.5. Sensor Technology

Dramatic cost reductions in all forms of digital electronics will continue to promote the introduction of lower cost sensors within all industries. As a result this decade will see an unprecedented increase in the volume of data made available concerning the manufacture, distribution, and consumption of products. Thus there will be a great demand to transform this data into useful knowledge. Parallel computers will play a critical role in transforming this data into knowledge via increasingly sophisticated models. High speed computer interconnection networks will dramatically decrease the cost of instantaneously transporting large amounts of data.

2.1.6. Interface Technology

Interface technology serves the dual purpose of facilitating user convenience and inducing intuition in applications. Most improvements in interface technology will take place in the area of audio-graphics. High resolution, three dimensional, color graphics depictions of complicated simulations and calculations will soon become commonplace. During the latter part of this decade voice and handwriting recognition as a means of user input should become commonplace.

2.1.7. Overall Component Technology Conclusion

The only barrier to continued dramatic advances in computer technology lie in the ability to sustain performance increases for processing elements. Parallel computing should circumvent this limitation. Thus the overall trend in computing through the 1990's should be sustained order of magnitude improvement in all hardware aspects. A continuing barrier to a commensurate increase in the usefulness of computers is the inability to improve algorithm design and software implementation advances at the same rate. A significant amount of research effort will have to be devoted to parallel algorithm design and software implementation. The interaction of algorithms with architecture will force applications researchers, such as chemical engineers, to become more familiar with all aspects of parallel computing, software engineering, and algorithm design.

2.2. Trends in Software Engineering

The development of special purpose computer hardware mitigates the difficulty of developing parallel computing software since manufacturers will hardcode algorithms in silicon. However, the single greatest barrier to fully exploiting incipient computer technology lies in the development of reliable and effective software. The process industries will benefit from computer science research in the area of automatic parallelizing compilers, parallel algorithm libraries, and high level parallel languages but current research indicates that effective parallel algorithms require applications expertise as well as an understanding of parallel computing issues [17, 33]. Thus the process industries will have to sponsor research into the development of parallel algorithms for industry specific problems that are computationally intensive [26, 34]. Even putting parallel processing issues aside, the development of high quality software will remain a hurdle to taking full advantage of this capability. Indeed, several studies have shown that software development productivity gains lag far behind gains in hardware since the 1960s [12]. Computer scientists have introduced notions such as object-oriented programming which promotes large scale reuse of code, ease of debugging, and integration of applications and Computer Aided Software Engineering (CASE) tools that reduce development, debugging, and maintenance effort but much research remains before computers achieve the potential offered by the hardware alone.

2.3. Progress in Computer Networking

So far we have only outlined the expected advances in stand-alone computing capability. However, perhaps the greatest potential for revolutionizing computer-aided process operations over the remainder of the decade lies in combining this promised capability with pervasiveness and high speed networks. Pervasiveness of computing technology will be made possible by the fact that 100 million operations per second computers with megabytes of memory will become available for only a few dollars and they will only occupy a single chip involving a few square inches [42]. Such digital technology will allow manufacturers to imbue even passive process equipment with a certain level of intelligence. Thus we will have pipes that not only always sense flowrates, temperatures, and pressures but also warn when they might rupture, tanks that always keep track of their contents, and equipment that suggests how more economic use may be obtained or suggest to field technicians how it might be serviced. At the very least, pervasive digital technology will make an extraordinary amount of information available about processes and the component equipment. Some researchers even suggest that hundreds of embedded computers will exist in an area the size of an average room [42]. Wireless networks and high speed fiber optic networks will enable the enormous amount of information to be instantaneously transported to any location in the world at very little cost [41].

By 1995, limited deployment of one gigabit/second fiber optic networks will occur and by early next century gigabit/second fiber networks and 0.25 to 10 million bit/second wireless networks promise to be as common as copper wire based technology [41]. In fact, limited gigabit/second fiber networks are now becoming available in experimental testbeds sponsored by the U.S. government [19, 41]. Global deployment of such high speed networks will offer a multitude of opportunities and challenges to the process industries. For example, the intimate linking of geographically distributed plant information systems with high speed networks will make close coordination of plants a competitive necessity. An order originating at one point on the globe will be scheduled at the most appropriate plant based on transportation costs, plant loadings, raw material inventories, quality specifications, etc. A single order for a specialty chemical could instantaneously spawn an entire web of orders for intermediates and raw materials not just within a single company but across many suppliers and vendors. Processes and inventories will be controlled with great efficiency if scheduling and management systems can be constructed to take advantage of this information. A field technician servicing a piece of process equipment could call up schematics, illustrations, a complete service record, and operating history as well as consult with experts or expert systems all from a lap-top or palm-top computer. High bandwidth networks will allow expert consultants to view relevant process information, consult with operators and engineers, view process equipment, and exchange graphics all without leaving the office. The abundance of information will offer greatly increased opportunities for improvements in on-line process monitoring, diagnosis, and control. The research challenge is clearly to develop algorithms that can profitably exploit this information. Certainly a large amount of information is already gathered about processes, although most engineers will attest that it is put to little use. Digital pervasiveness will ensure that at least an order of magnitude more of information will become available about processes but greatly increased computer capabilities and high speed networks promise that this information can be put to productive use.

From a modeling and simulation perspective, high speed networks offer the opportunity for large scale distributed computing whereby demanding calculations are allocated among computers located around the globe. In a distributed environment, heterogeneous computing becomes possible. Different portions of a calculation can be transported to the most appropriate special purpose architecture and then reassembled. In such an environment, aggregate computer power is the sum of the machines available on the network. The research challenges are clear: Which problems are amenable to distributed solution? How should problems be partitioned? How will distributed algorithms be designed and implemented in a cost efficient manner? The key challenge is to learn to deal with network latency which arises due to the finite speed of light. Latencies prevent computational processes from communicating with arbitrarily high frequencies, however, large bandwidths allow them to exchange enormous quantities of information when they do communicate. Distributed computing algorithms must be designed to accommodate this reality.

Having offered an overview of emerging computer technology, we will now discuss basic concepts necessary for designing and assessing parallel algorithms.

2.4. Terminology and Basic Concepts

Parallel computing can be studied from many different perspectives including fabrication of hardware, architecture, system software design and implementation, and algorithm design. This section will be confined to those concepts that are of immediate importance to designing effective algorithms for process systems applications. We begin with simple metrics for measuring parallel algorithm performance and then discuss those architectural attributes that control the nature of parallel algorithms.

2.4.1. Granularity

The concept of granularity is used qualitatively to describe the amount of time between interprocessor communication. A computational task is said to be of coarse granularity when this time is large and of fine granularity when this time is small. Granularity provides a broad guide as to the type of parallel computer architecture with may be effectively utilized. As a rule of thumb, coarse granularity tasks are very easy to implement on any parallel computer. Some practical problems result in coarse granularity tasks, e.g. decision problems that arise from design, optimization, and control applications.

2.4.2. Speedup and Efficiency

Speedup and efficiency are the most common metrics used to rate parallel algorithm performance. For the simplest of algorithms, speedup and efficiency can be computed theoretically, but for most practical applications speedup is measured experimentally. The most stringent definition of speedup is

$$Speedup = \frac{Time\ for\ 1\ processor,\ most\ efficient\ algorithm}{Time\ for\ n\ processor\ algorithm}$$

In practice, speedup is usually reported as

$$Speedup = \frac{Time\ for\ 1\ processor,\ using\ n\ processor\ algorithm}{Time\ for\ n\ processor\ algorithm}$$

The definition for efficiency uses the notion of speedup

$$Efficiency = \frac{Speedup}{Number\ of\ Processors} \times 100\%$$

The goal of the parallel algorithm designer is to achieve 100% efficiency and a speedup equal to the number of processors. This is usually an unrealistic goal. A more achievable goal is to develop an algorithm with efficiency bounded away from zero with an increasing number of processors. Many papers use the second definition of speedup which often gives a faulty impression of an algorithm's quality. The first definition of speedup is consistent with the goal of using parallel computing to solve intractable problems, while the "practical" definition is not necessarily consistent. To see that this is so, consider the speedup that is possible with a sorting algorithm that generates all possible permutations of n items and saves a permutation that is in sorted order. Obviously, such an algorithm is inefficient but since speedup and efficiency are relative measures, a parallel algorithm judged using the second definition of speedup could look attractive. Problem size is often a central issue. Consider matrix addition. If the number of processors is larger than the number of elements in the matrix then efficiency will suffer. As long as the number of elements exceeds the number of processors then processor efficiency may be acceptable (depending on the computer architecture). Early research in parallel computing was centered on determining the limits to speedup and efficiency.

2.4.3. Amdahl's Law

In the late 1960s, Gene Amdahl argued that parallel computing performance was fundamentally limited. His arguments can be concisely stated using the following:

$$S(n) \leq \frac{1}{f + \frac{(1-f)}{p}} \leq \frac{1}{f}$$

where f is the fraction on an algorithm that is inherently sequential and $S(n)$ is the speedup achievable with n processors. This expression of Amdahl's law simply says that, even if the execution time of the parallel portion of an algorithm is made to vanish, speedup is limited by the inherently sequential portion of the algorithm. Amdahl's law is often cited as the reason why parallel computing will not be effective. In fact there is some controversy over how to measure f, but by any measure, f is very small for many engineering and scientific applications. Later, we will discuss the concept of hierarchical parallelism as a way to counter Amdahl's law. The essential premise behind hierarchical parallelism is that there is no truly sequentially calculation in the sense that parallelism is possible at some level. Indeed, the fraction of an algorithm that is inherently sequential largely depends on the type of computer architecture which is available.

2.4.4. Computer Architecture

A central issue in parallel computing is the organization of the multiple functional units and their mode of interaction and communication. Such architectural issues are of importance to algorithm designers since they impact the type of algorithms which can be effectively implemented. Parallel computing architectures may be classified along a number of different attributes. Probably the simplest classification scheme is due to Flynn [10] who designated computers as SISD (Single Instruction, Single Data), SIMD (Single Instruction, Multiple Data), or MIMD (Multiple Instruction, Multiple Data). SISD computers were the conventional design from the start of the computer era to the 1990s in which one processor performs all computational tasks on a single stream of data. SIMD computers are embodied as an array of processors which all perform the same instruction on a different streams of data, e.g. for a parallel matrix addition, each processor adds different elements together. MIMD computers represent the most flexible parallel computer architecture in which multiple processors act independently on different data streams. Unfortunately, MIMD machine performance is hampered by communication and coordination issues which can impact algorithm efficiency.

2.4.4.1. Topology of Parallel Computing Systems

Because placement of data and the ability to communicate it in a time critical fashion is crucial to the success of parallel algorithms, application researchers must be aware of the relationship among hardware components. Parallel computing vendors employ three basic strategies in the construction of systems: (1) use large numbers of cheap and relatively weak processors, e.g. pre-CM5 Connection Machine Computers, (2) use dozens of powerful off-the-shelf microprocessors, e.g. Intel Hypercube which consists of 16 to 512 Intel i860 processors, and (3) use small number of expensive and very powerful processors, e.g. Cray YMP/832 which possesses eight 64-bit vector processors. Architectures relying on large numbers of weak processors for high performance demand high quality parallel algorithms with a very small sequential character. To date, only very regular computations have been suitable for massively parallel computers such as the Connection Machine. Vendors have had trouble making money with strategy two in recent years since by the time a parallel computer was designed and built microprocessor speed improved resulting in the need to redesign the architecture or market a machine barely competitive with the latest generation of microprocessor. As technological limits are reached with microprocessors this should become less of a problem, although these limits may be somewhat circumvented by increasing the amount of parallelism available in microprocessors. Supercomputer vendors such as Cray have been very profitable using strategy three, however in the Fall of 1990 they committed their research efforts to developing a next generation of supercomputers using hundreds and perhaps thousands of powerful processors.

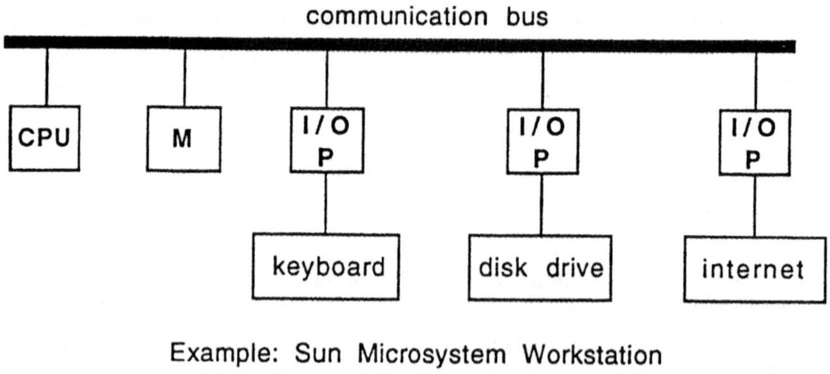

Figure 1. Traditional (Von Neumann) sequential computer architecture.

For purposes of algorithm design, the most important architectural features are the relationship of processors to memory and each other. Figure 1 illustrates the traditional sequential computer architecture. A defining feature of this architecture is the communication bus over which instructions and data are moved among the processor, memory, and input/output devices. The rate at which the bus transmits data is a controlling factor in determining overall computer performance. Figure 2 shows a parallel extension of the traditional architecture that places multiple processors and memory modules along the communication bus.

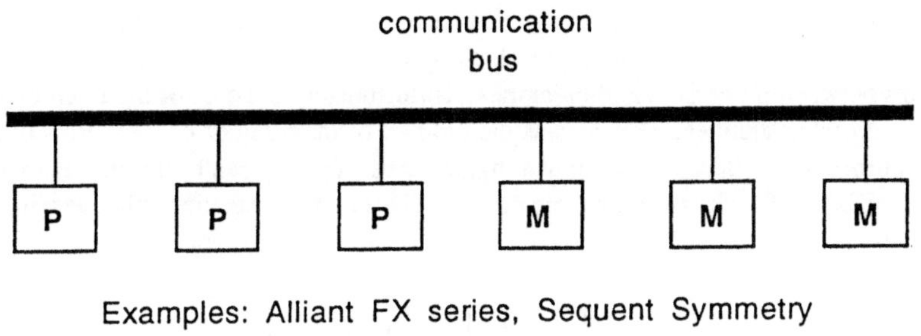

Figure 2. Extension of traditional architecture to multiple processors.

The bus architecture can support O(10) processors until bus contention degrades system efficiency. Cache memory located on a processor can be used to reduce bus contention by reducing bandwidth requirements. However, the coherency of data in various processor caches is a complicating factor in computer and algorithm design.

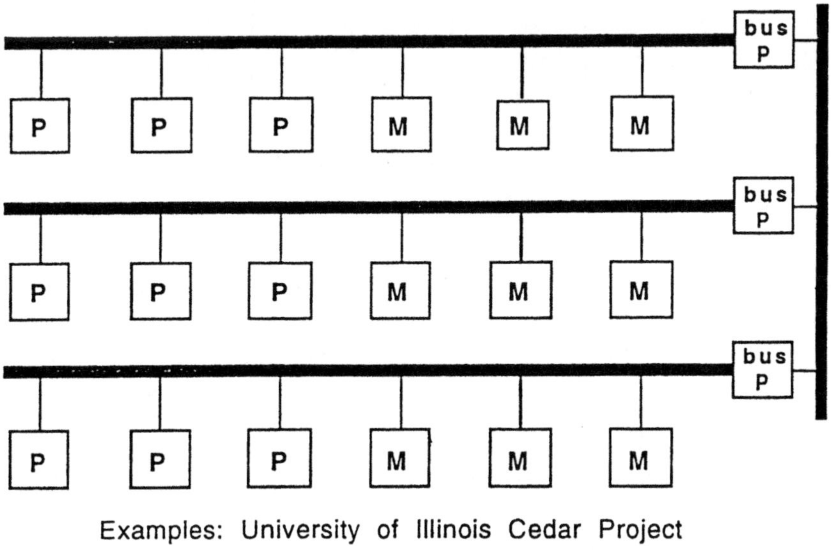

Figure 3. Hybrid architecture avoids bus contention.

Even though the bus architecture can support only a small number of processors, it can typically be incorporated into hybrid architectures. Processors communicate through the common memories. Some mechanism must exist for arbitrating collisions among processors trying to access a given memory. A memory hot spot is said to exist when several processors attempt to access a single memory. Algorithms should be designed to avoid these hot spots by controlling the way in which data is accessed and by distributing data across different memory modules. Figure 3 illustrates how the bus architecture may be combined in a hybrid pattern to support a larger number of processors. Data frequently used by a processor must be stored in a local memory while infrequently used data may be stored in a remote memory. In general, the placement of data in a parallel computer controls algorithm effectiveness. Figure 4 depicts a crossbar architecture which tends to minimize contention by providing many paths between processors and memories. Unfortunately the cost of the interconnection matrix in a crossbar architecture scales as the product of the number of memories times the number of processors. Because of this scaling, the crossbar becomes prohibitively expensive for large numbers of processors and memories. The crossbar can be embedded in hybrid architectures.

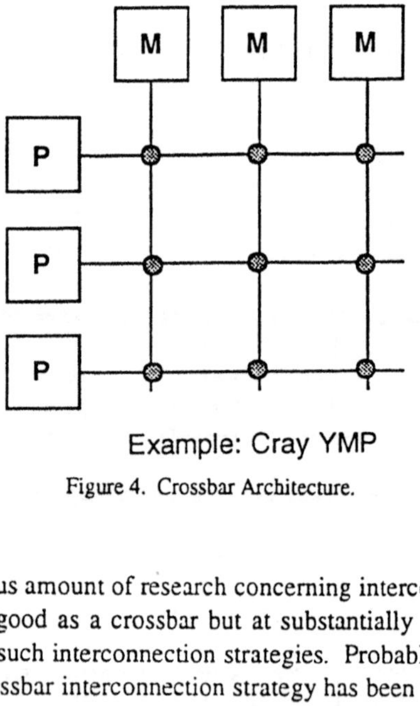

Figure 4. Crossbar Architecture.

There has been an enormous amount of research concerning interconnection strategies whose performance is nearly as good as a crossbar but at substantially less cost. We will briefly examine a few of the best such interconnection strategies. Probably the most widely investigated alternative to the crossbar interconnection strategy has been the hypercube architecture shown in Figure 5 for three dimensions. In general consider the set of all points in a d-dimensional space with each coordinate equal to zero or one. These points may be considered as the corners of a d-dimensional cube. If each point is thought of as a processor/memory pair and a communication link is established for every pair of processors that differ in a single coordinate, the resulting architecture is called a hypercube. A number of architectures can be mapped to the hypercube in the sense that the hypercube may be configured to simulate the behavior of the architecture. The hypercube architecture possesses a number of useful properties and many symmetric multiprocessors implemented in the late 1980s were based on the hypercube architecture.

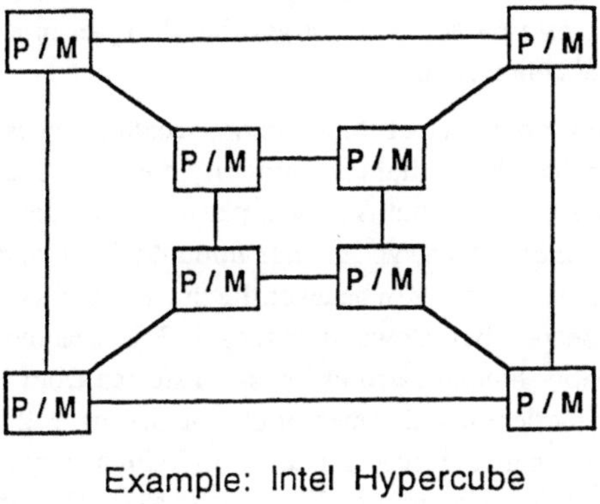

Figure 5. Hypercube architecture in three dimensions.

The cost of the hypercube interconnection strategy scales as O(p log(p)) where p is the number of processors. Figure 6 illustrates the pipeline architecture which is widely used.

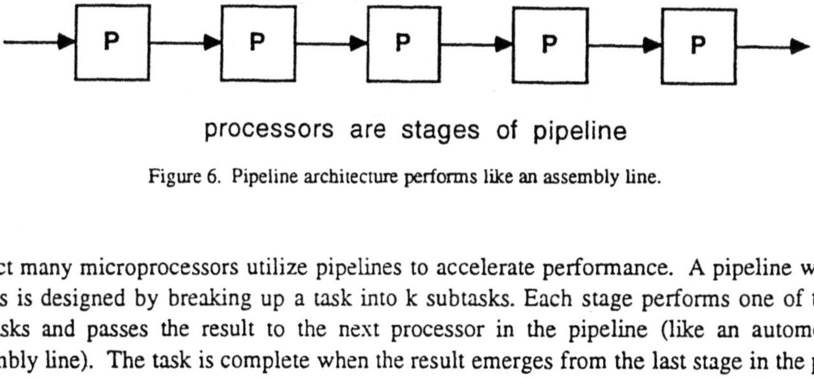

Figure 6. Pipeline architecture performs like an assembly line.

In fact many microprocessors utilize pipelines to accelerate performance. A pipeline with k stages is designed by breaking up a task into k subtasks. Each stage performs one of the k subtasks and passes the result to the next processor in the pipeline (like an automobile assembly line). The task is complete when the result emerges from the last stage in the pipeline. The pipeline architecture is only effective if several identical tasks need to be processed. If each subtask takes unit time and there are a large number of tasks (say n) to be processed, the speedup of a pipeline architecture with k stages is kn/(k-1 + n) or approximately k. In practice, pipeline architectures are limited by how fast data can be fed into them. The feed rate is controlled by the memory subsystem design.

2.4.4.2. Memory Subsystems

Memory subsystem performance is usually enhanced by organizing memory modules to function in parallel. Assuming that a memory module can supply data at a unit rate, k interleaved memory modules can supply data at a rate of k units. In order for this to work, data has to be organized in parallel in such a fashion that the k pieces of simultaneously available data are useful. This places a burden on the algorithm designer or a compiler to situate data so that it may be accessed concurrently.

Figure 7 illustrates a memory caching system which is necessary because processors are typically much faster than the available commodity memory. Memory sufficiently fast to keep up with processors is available but is very expensive. In order to counter this situation, computer architects use a memory cache. A small amount of fast memory (M1) is used at the top of the cache to feed the processor. Whenever the processor asks for an address that is not available in M1, the cheaper but slower memory M2 is searched. If M2 contains the requested memory, an entire block of addresses is transferred from M2 to M1 including the requested address. By interleaving the memories comprising M2, the block transfer can occur much more rapidly than the access rate of individual memory units. In a similar fashion the cheapest but slowest memory M3 supports M2. Thus if most memory address requests are supported by M1, the memory cache will appear to execute at nearly the speed of M1 but the overall cost of the memory system will be held down.

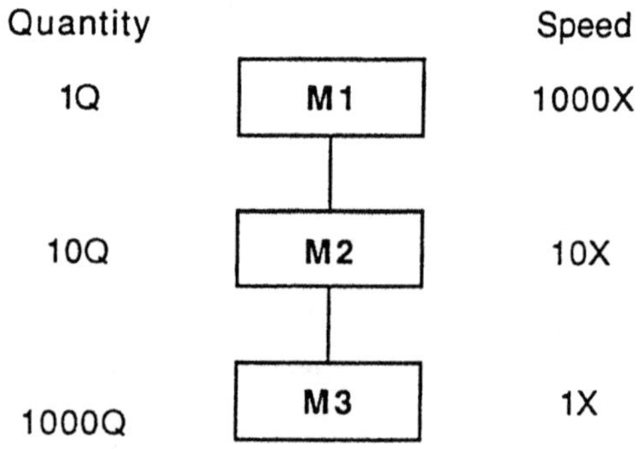

Figure 7. Memory caching system.

If f1 and f2 are the fraction of time M1 and M2, respectively, contain the requested address, then the effective performance speed of the memory cache will appear to be f11000X + f2 10X + (1-f1 - f2) X. Because sequential algorithms tend to access data in a uniform fashion, caching usually results in dramatic performance improvements with little effort, however caching considerations are often important to considering the placement of data in parallel computing algorithms.

3. Designing Parallel Algorithms

In order to achieve a high degree of speedup, data must be correctly positioned so that computations may proceed without waiting for data transfers, i.e. data must be positioned "close" to processing elements and data movement must occur concurrently with computations. Sometimes the design of a parallel algorithm necessarily entails solving a difficult combinatorial optimization problem. To appreciate this aspect of parallel algorithm design, consider how a simple Gauss-Seidel iteration scheme may be parallelized. A dependency graph G=(V,A) is a useful and compact way to think about the communication requirements of an iterative method. Each vertex in V represents an unknown and an arc (i,k) is present in A if and only if function f_k depends on variable x_i. Finding a Gauss-Seidel update ordering that maximizes speedup is equivalent to a graph coloring problem on the dependency graph [3]. This graph coloring problem is NP-complete which means finding an algorithm to optimally solve a particular instance may entail great effort. The implications for parallel computing are that designing good algorithm can necessarily involve significant work.

As an illustration of many of the aspects of parallel algorithm design, consider the two-dimensional heat conduction equation on a rectangular region subject to boundary conditions [3].

$$\frac{\partial^2 T}{\partial x^2}+\frac{\partial^2 T}{\partial y^2}=0$$

such that

$$T(x_0,y)=f_1(y)$$

$$T(x_1,y)=f_2(y)$$

$$T(x,y_0)=f_3(x)$$

$$T(x,y_1)=f_4(x)$$

Suppose the x and y axis are discretized into equally spaced intervals using points $(i\,\Delta x, j\,\Delta y)$, where Δx is the size of the x-interval and Δy is the size of the y-interval and i=0,...,M and j=0,...,N. Furthermore, denote the temperature at space point $(i\,\Delta x, j\,\Delta y)$ as $T_{i,j}$. Finite difference approximation of the derivatives yields:

$$\frac{\partial^2 T}{\partial x^2}=\frac{T_{i-1,j}-2T_{i,j}+T_{i+1,j}}{\Delta x^2}$$

$$\frac{\partial^2 T}{\partial y^2}=\frac{T_{i,j-1}-2T_{i,j}+T_{i,j+1}}{\Delta y^2}$$

which may be combined:

$$\frac{T_{i-1,j}-2T_{i,j}+T_{i+1,j}}{\Delta x^2}+\frac{T_{i,j-1}-2T_{i,j}+T_{i,j+1}}{\Delta y^2}=0$$

Thus $T_{i,j}$ may be computed from the four surrounding space points. The (symmetric) dependency graph for the system of equations implied by this difference equation for M=N=4 is shown in Figure 8. The circles represent the temperature at the space points and the edges represent the symmetric dependency of adjacent space points for updated values in a Gauss-

Seidel solution scheme. The dependency graph may be minimally colored with two colors so that no two adjacent vertices possess the same color. White vertices may be updated simultaneously in a Gauss-Seidel scheme. Likewise for black vertices. In an actual implementation of a Gauss-Seidel algorithm, a processor takes responsibility for updating an equal number of white and black vertices in some localized region of space. Each processor updates white vertices using the information currently stored by the processor for black vertices, exchanges the new white vertex values with neighbors, and then computes new values for black vertices. The process continues until appropriate accuracy is obtained.

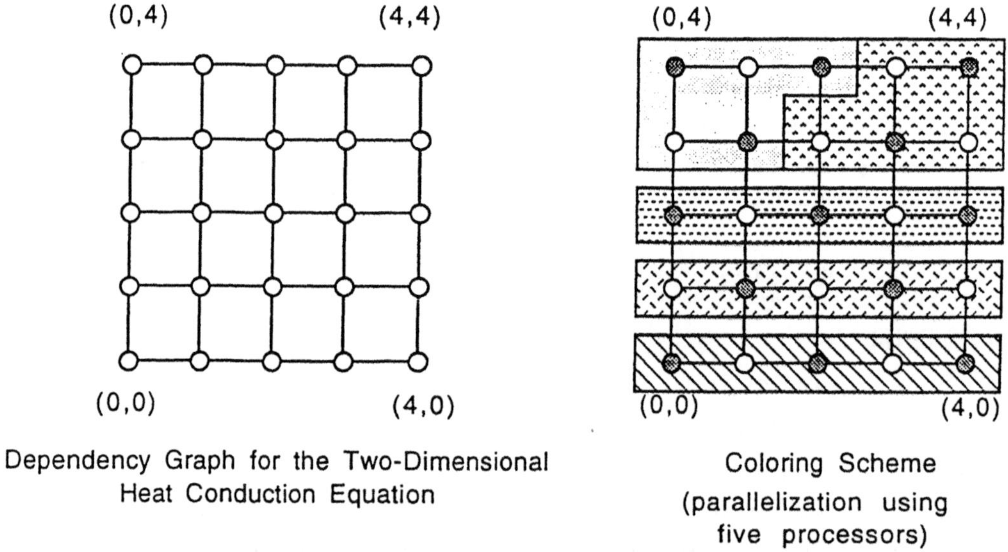

Figure 8. Coloring scheme for finite difference grid.

Because of the localized communication, a number of parallel architectures would be appropriate for the Gauss-Seidel algorithm implied by this dependency graph. A large number of P.D.E.'s are amenable to Gauss-Seidel parallelization via coloring schemes including the Vorticity Transport Equation, Poisson's Equation, Laminar Flow Heat-Exchanger Equation, Telephone Equation, Wave Equation, Biharmonic Equation, Vibrating Beam Equation, and the Ion-Exchange Equation.

As a practical matter, some parallel computer vendors include optimized algorithms for solving linear systems on their hardware. Market pressure should increase this trend. Even so, for large scale problems, applications specialists are still forced to develop their own tailored algorithms for solving linear systems. The need arises because the linear system is very large, ill- conditioned, or the structure of the matrix lends itself to the development of very fast special purpose algorithms. The large community of applied mathematicians and applications specialists in SIAM concern themselves with computational research into quickly solving linear systems. A number of the leading linear algebra packages are being adapted to exploit parallelism [28], e.g. Linear Algebra Package (LAPACK). In addition, a number of special purpose parallel computers already exist for solving linear systems and more are under development [23]. From a process system point of view, the ongoing research into parallel solution of linear systems is important since almost all process system algorithms rely on the solution of linear systems.

3.1. Role of Special Purpose Algorithms

The only purpose of parallel computing is to improve algorithm performance. Tailoring algorithms to particular problems also provides a compatible option for enhancing performance. The form of this tailoring varies considerably, but generally speaking, exploiting structure involves developing theoretical results and data structures specifically tuned to exploiting problem characteristics. For example, very efficient solvers have been developed for sparse systems of linear equations such as tridiagonal systems [13]. As a more detailed measure of the value of special purpose algorithms, consider the assignment problem which is a highly structured linear program.

$$\textit{minimize} \sum_{i=1}^{n}\sum_{j=1}^{n} c_{ij}\, x_{ij}$$

$$\sum_{i=1}^{n} x_{ij} = 1 \quad j=1,\ldots,n$$

$$\sum_{j=1}^{n} x_{ij} = 1 \quad i=1,\ldots,n$$

$$x_{ij} \geq 0 \quad i,j=1,\ldots,n$$

The general purpose simplex method available in GAMS (BDMLP) requires 17.5 seconds of CPU time when applied to an assignment problem of size n=40 (1600 variables) using a Sun 3/280 computer. On the other hand the specialized assignment problem algorithm of [2] requires 0.08 seconds of CPU time to solve the same size problem. In parallel computing terms, the special purpose algorithm achieves a speedup of 218.8 over the general purpose approach. Furthermore, special purpose algorithms often possess easily exploitable parallelism compared to general purpose approaches. Indeed the special purpose assignment problem algorithm of [2] was designed for parallel execution although it is among the best available assignment problem codes when executed in sequential mode. In principle, special purpose algorithms may be developed for every problem. In practice, the expense of developing special purpose algorithms is too great for most applications. This means the development of methods for reducing the costs of special purpose algorithms, such as computer aided software engineering techniques, is as important an area of research as parallel computing. In fact, for enumerative algorithms such as branch and bound in mathematical programming and alpha-beta search in Artificial Intelligence, exploiting parallel computing is clearly of secondary importance when compared to problem structure exploitation since worst case performance of these paradigms results in unreasonably long execution times on any foreseeable computer architecture.

3.2. Hierarchical Parallelism

In engineering computations, opportunities exist for exploiting parallelism at many different levels of granularity. Indeed, the only way to circumvent Amdahl's law for most calculations is to exploit the parallelism available at these different levels. To illustrate the notion of hierarchical parallelism, consider the solution of discrete optimization problems that arise in a number of process systems applications. Exact solutions for the majority of these problems can only be obtained through an enumerative algorithm using some form of branch and bound. Table 1 illustrates the levels of parallelism which can be exploited in a branch and bound algorithm.

Table 1 - Levels of Parallelism for Branch and Bound

Mode of parallelism	**Task granularity**
competing search trees	very high
tree search	high
algorithm components	moderate
function evaluations	low
machine instructions	very low

Parallelization at the three coarsest levels of granularity is the responsibility of the applications expert while the two finest granularity levels are usually the responsibility of computer architects and compiler authors.

In the context of discrete optimization, the branch and bound tree (search tree) represents a means of partitioning the set of feasible solutions. This partitioning is never unique and quite often different partitioning strategies can lead to widely different search times even for identical lower and upper bounding techniques although the best partitioning strategy usually is not known in advance. The availability of multiple partitioning strategies provides an excellent opportunity for exploiting parallelism. In particular, the availability of k partitioning strategies implies that k different search trees can be created. A group of processors can be used to explore each search tree (see below) and all search trees can be exploited whenever one group proves optimality of a solution. As an example of this type of parallelism let $f(x)$ be the probability that a branch and bound algorithm requires time x and $g(x)$ be the probability that k competing trees require time x based on the same lower and upper bounding technique. If the probability of finding a solution in a given time using a particular partitioning strategy is assumed to be independent with respect to other partitioning strategies then $g(x)$ is related to $f(x)$ (a reasonable assumption with a strong problem relaxation).

$$g(x)=k\,[1-F(x)]^{k-1} f(x)$$

where

$$F(x)=\int_0^x f(s)\,ds$$

For example, if

$$f(x)=\frac{x}{12\,e^{\sqrt{x}}}$$

which is a realistic probability distribution function for a branch and bound algorithm with strong bounding techniques, then the expected speedup for searching k branch and bound trees in parallel is given by

$$Speedup,\ S_k=\frac{\int_0^\infty x\,f(x)\,dx}{\int_0^\infty x\,g(x)\,dx}$$

For example, S_2=1.969 and S_4=3.59 so that a parallelization strategy based on competing search trees is quite effective. In practice, the speedups given by this analysis are conservative since competing search trees can share feasible solution information resulting in synergism which can accelerate the search. Additional information on competing search trees may be found in [30].

Each of the competing search trees may also be investigated in parallel. A great deal of research has been done in this area. In particular, the papers by [20, 31] summarize existing strategies for parallelizing the search of a single branch and bound tree. The work reported in [22] outlines conditions under which a parallel search strategy results in superlinear speedup†, near linear speedup, and sublinear speedup. In intuitive terms, the opportunities for parallelism can be related to search tree structure. Superlinear speedup is possible when there is more than one vertex with a relaxation value equal to the optimal solution. Near linear speedup is possible whenever the number of vertices divided by maximum tree depth exceeds the number of processors involved in the search. Sublinear speedup results whenever the number of vertices divided by maximum depth is less than the number of processors.

Finer granularity methods are required to develop parallel algorithms for the components of branch and bound such as the solution of problem relaxations, heuristics, and branching rules. However, parallelization of the branch and bound components complements the concurrency that may be exploited at the competing search tree level and

† Speedup greater than the number of processors.

individual search tree level. The results given in [2, 30, 31] suggest a speedup of at least 70 to be possible for combined parallelization at each of these levels (a product of a factor of 1 to 12 at competing search tree level, factor of 10 to 50 at search tree level, and a factor of 7 to 14 at the component level) for an exact asymmetric traveling salesman problem algorithm based on an assignment problem relaxation. The expected speedup from hierarchical parallelization for this problem is dependent on the ultimate sizes of the search trees with the largest search trees yielding the greatest speedup and virtually no speedup possible for very small search trees.

As another example of a class of process systems algorithms that may exploit hierarchical parallelism, consider decomposition methods for mathematical programming problems, e.g. Bender's Decomposition [11] and the Outer Approximation Method [8]. Both of these decomposition methods iterate through a sequence of master problems and heuristic problems. The master problems invariably rely on enumerative methods, usually branch and bound, so that the hierarchical parallelization described above is applicable in master problem solution. Furthermore, the sequence of iterations provides yet another layer of parallelism to be exploited. Namely, each iteration of the decomposition method produces a set of constraints to further strengthen the bounds produced from master problem solution. Groups of processors could pursue different decomposition paths by, say, using different initial guesses for complicating variables. Each of the decomposition paths would produce different cuts to strengthen the master problem relaxation and the processor groups could share these cuts to create a pool from which they could all draw upon. Additional synergism among processor groups is possible by sharing feasible solutions. Experimental work is necessary to determine the degree to which parallel decomposition could succeed, however experience suggests that the best decomposition algorithms require only a small number of iterations. Thus parallelization of high quality decomposition methods would offer only a small speedup. However, when this small speedup is amplified by parallelization at lower algorithmic levels, the potential overall speedup promises to be quite high.

4. Distributed Computing

The last few years have seen a surge of interest in distributed computing systems. A distributed computing system involves processors a large physical distance apart (say more than 10 meters) and interprocessor communication delays are unpredictable. A set of computers on a corporate or academic site connected using Ethernet technology [39] provides an example of a distributed computing system. Distributed computing systems offer the greatest performance potential and offer the most general view of parallel computing systems.

Engineering computations do not tend to be homogeneous in time in the sense that the workload alternates between disparate types of calculations (e.g. calculation of physical properties, construction of a Hessian matrix, solution of a linear system of equations, etc.). Computational experience over a large number of algorithms shows that different types of calculations yield varying degrees of processor efficiency for any one architecture. Thus no one parallel computing architecture is ideally suited for sophisticated engineering calculations. However, networking technology is evolving to the point where it will soon be possible to decompose a complex calculation into components, distribute the components to the appropriate architecture for calculations, and the reassemble the final result in an appropriate location. Such distributed computing could become a dominant mode of parallelism. This leads to the notion of network based parallelism which is defined as the application of locally distributed and widely distributed computers in the cooperative solution of a given problem over short time scales. Network based parallelism is currently under investigation at a number of locations within the United States, e.g. [19, 41] and preliminary progress suggests that the necessary communication hardware could become more widely available by the middle of this decade with routine application possible by early next century. Of course existing computer networks can support parallelism, however it is severely limited in terms of the amount of information that can be transmitted and the number of algorithms that can simultaneously be supported. As mentioned above, bandwidth and latency are the principle factors in the calculus of network based parallelism. The principle challenge is to construct algorithms that mitigate latency, which cannot be circumvented due to the fundamental speed of light, and exploit bandwidth, which can be made arbitrarily large. The ideal network algorithms communicate infrequently, although when they do, they may exchange enormous quantities of information.

4.1. Controlling Network Communication

The key consideration in the design of network algorithms is how to control communication. The primary choice lies with the degree of control which the algorithm designer wishes to exercise over communication. At the lowest level, algorithms may be implemented using low level paradigms such as TCP/IP socket streams, datagrams, and remote procedure calls [39] which tend to be tedious but offer a high degree of control. At a higher level, packages such as ISIS (Cornell University) and C-Linda (Scientific Computing Associates) are available which hide network communication details through abstract paradigms. The principle drawback of these paradigms is that one must use a given communication model which may only offer awkward support for an algorithm. Existing research promises to offer a number of support mechanisms for network based parallelism. In particular research is proceeding on developing virtual memory across machines in a network whereby a process on one machine could effortlessly reserve and use memory on any of a number of machines. A number of vendors and academic researchers are exploring automatic migration of computational processes through which a process spawned on one machine would quickly locate and use unburdened machines [40]. Provided efficiency issues can be adequately addressed, such paradigms offer great promise to application specialists for implementing network algorithms. Application specific development environments are another area being explored to aid in the implementation of network algorithms. For example, as discussed above, parallel computing offers great promise in reducing the execution times of branch and bound computations [32]. However, the time and effort needed to parallelize algorithms exacerbates the already arduous task of algorithm development. This has prevented the routine use of parallel and distributed computers in solving combinatorial optimization problems. Given that all branch and bound algorithms can utilize the same mode of parallelism, tools can be specifically developed to reduce the burden associated with designing and implementing branch and bound algorithms in a distributed environment [18].

As an example of network based parallelism using existing communication technology, consider the computationally intense task of multiplying two fully dense matrices to produce a third matrix, i.e. C=AB. This calculation can be parallelized in a number of ways but, for simplicity, consider distributing matrix A and B to each of a number of machines on a network. The task of computing C can then be partitioned among the machines so that each machine computes a portion of C according to its relative capability (see Figure 9).

$$C=A*B \quad A,B,C \in R^n \quad n=700$$

Computation Schematic:

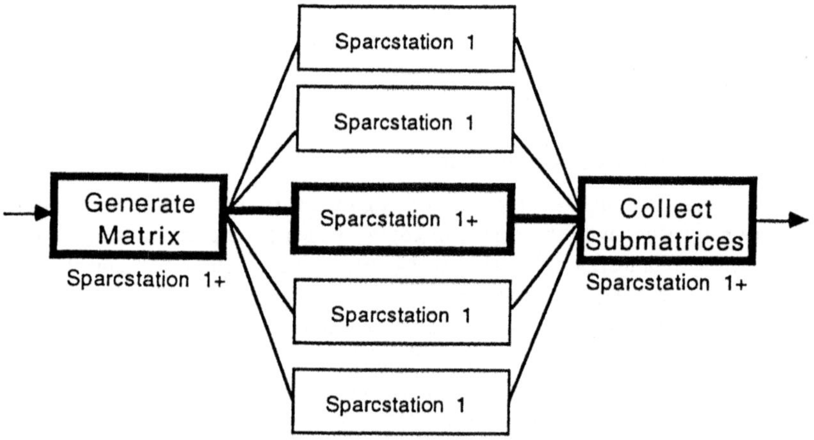

Figure 9. Distributed matrix multiplication scheme.

For square double precision matrices of size 700, the sequential computing time on a Sun Microsystem Sparcstation 1+ is 751.2 seconds. Performance for the network based algorithm depicted in Figure 9 is given in Table 2. In Table 2, the critical path is defined as that string of computations which controlled the wall clock time of the algorithm. In particular, the transmission of the 7.84 megabytes of matrix data (A and B) to each of the four Sparc 1s from the Sparc 1+, the portion of matrix multiplication done on the Sparc 1+, and the collection of the resulting matrix back on the Sparc 1+ controlled the wall clock execution time. The overall speedup for the algorithm computed as the parallel wall clock execution time divided by the Sparcstation 1+ sequential execution time is 3.72 yielding an efficiency of 93% using four processors. This simple example points out a difficulty of using the usual definitions of speedup and efficiency for network calculations.

Table 2 - Distributed Matrix Computation Results	
Operation	time (sec)
Multiplication Time (Sparc 1)	186.7
Multiplication Time (Sparc 1+)	151.0
Latency (along critical computation path)	0.50
Transmission Time (along crit. path)	14.46
Multiplication Time (along crit. path)	186.7
Wall Clock Execution Time	201.7

Namely, the Sparc 1+ is approximately 33% faster than a Sparc 1 while the normal definition of speedup and efficiency assume all processors to be of equal capability. The conservative approach is to perform speedup and efficiency calculations using the fastest processor to collect sequential times.

5. Additional Reading

The last several years have seen a dramatic increase in the number of publications addressing various parallel computing issues [5]. The textbook [3] addresses a number of issues relevant to the design of parallel algorithms. A number of references discuss the trends in computer architecture [14, 29, 38] and implementation technology [6, 24, 37]. Molecular simulation [9] and transport phenomena [16, 35] offer a number of opportunities for applying parallel computing. In addition to the areas discussed above, there has been considerable research into the parallelization of continuous optimization methods [25]. A number of excellent references exist for the parallelization of generic numerical methods [13, 27, 28] and parallel programming languages [1, 4].

References

1. Babb, R. G., *Programming Parallel Processors*, Addision-Wesley, 1988.
2. Balas, E., D. L. Miller, J. F. Pekny, and P. Toth, "A Parallel Shortest Path Algorithm for the Assignment Problem," *Journal of the Association for Computing Machinery*, vol. 38, pp. 985-1004, 1991.
3. Bertsekas, D. P. and J. N. Tsitsiklis, *Parallel and Distributed Computation*, Prentice Hall, Englewood Cliffs, 1989.
4. Brawer, S., *Introduction to Parallel Programming*, Academic Press, 1989.
5. Corcoran, E., "Calculating Reality," *Scientific American*, vol. 264, no. 1, pp. 100-109, 1991.
6. Corcoran, E., "Diminishing Dimensions," *Scientific American*, vol. 263, no. 5, pp. 122-131, 1989.
7. Defense, U. S. Department of, *Critical Technologies Plan (Chapter 3)*, 1990.
8. Duran, M. A. and I. E. Grossmann, "An Outer-Approximation Algorithm for a Class of Mixed-Integer Nonlinear Programs," *Mathematical Programming*, vol. 36, pp. 307-339, 1986.
9. Fincham, D., "Parallel Computers and Molecular Simulation," *Molecular Simulation*, vol. 1, pp. 1-45, 1987.
10. Flynn, M. J., "Very High-Speed Computing Systems," *Proc. IEEE*, vol. 54, pp. 1901-1909, 1966.
11. Geoffrion, A. M., "Generalized Benders Decomposition," *Journal of Optimization Theory and Applications*, vol. 10, no. 4, pp. 237-260, 1972.
12. Ghezzi, C., *Fundamentals of Software Engineering*, Prentice-Hall, Englewood Cliffs, NJ, 1991.

13. Golub, G. H. and C. F. Van Loan, *Matrix Computations (2nd edition)*, John Hopkins, Baltimore, 1989.
14. Hillis, W. D., *The Connection Machine*, MIT Press, 1985.
15. Hwang, K. and F. A. Briggs, *Computer Architecture and Parallel Processing*, McGraw-Hill, New York, 1984.
16. Jespersen, D. C. and C. Levit, "A Computational Fluid Dynamics Algorithm on a Massively Parallel Computer," *International Journal of Supercomputing Applications*, vol. 3, pp. 9-27, 1989.
17. Kim, S. and S. J. Karrila, *Microhydrodynamics: Principles and Selected Applications*, Butterworth-Heinemann, Boston, 1991.
18. Kudva, G. and J.F. Pekny, "DCABB: A Distributed Control Architecture for Branch and Bound Calculations", Computers and Chemical Engineering, vol. 19, pp. 847-865, 1995.
19. Kung, H. T., E. Cooper, and M. Levine, "Gigabit Nectar Testbed," Corporation for National Research Initiatives Grant Proposal (funded), School of Computer Science, Carnegie Mellon University, 1990.
20. Lavallee, I. and C. Roucairol, "Parallel Branch and Bound Algorithms," *MASI Research Report*, EURO VII, Bologna, Italy, 1985.
21. Leebaert, D., *Technology 2001*, MIT Press, 1991.
22. Li, G. and B. W. Wah, "Computational Efficiency of Parallel Approximate Branch-and-Bound Algorithms," *International Conference on Parallel Processing*, pp. 473-480, 1984.
23. McCanny, J. F., J. McWhirter, and E. E. Swartzlander, *Systolic array processors : contributions by speakers at the International Conference on Systolic Arrays, held at Killarney, Co. Kerry, Ireland, 1989*, Prentice Hall, 1989.
24. Meindl, J. D., "Chips for Advanced Computing," *Scientific American*, vol. 257, no. 4, pp. 78-89, 1987.
25. Meyer, R. R. and S. A. Zenios, "Parallel Optimization on Novel Computer Architectures," *Annals of Operations Research*, vol. 14, 1988.
26. Miller, D. L., "Parallel Methods in Combinatorial Optimization," in *Invited Lecture*, Purdue University, West Lafayette, IN, 1991.
27. Modi, J. J., *Parallel Algorithms and Matrix Computation*, Oxford University Press, 1988.
28. Ortega, J. M., *Introduction to Parallel and Vector Solution of Linear Systems*, Plenum Press, 1988.
29. Patterson, D. A., *Computer Architecture: A Quantitative Approach*, Morgan Kaufman Publishers, San Mateo, CA, 1990.
30. Pekny, J. F., "Exact Parallel Algorithms for Some Members of the Traveling Salesman Problem Family," Ph. D. Dissertation, Carnegie Mellon University, Pittsburgh, PA 15213, 1989.
31. Pekny, J. F. and D. L. Miller, "A Parallel Branch and Bound Algorithm For Solving Large Asymmetric Traveling Salesman Problems," *Mathematical Programming*, vol. 55, pp. 17-33, 1992.
32. Pekny, J. F., D. L. Miller, and G. Kudva, "An Exact Algorithm for Resource Constrained Sequencing With Application to Production Scheduling Under an Aggregate Deadline," *Computers and Chemical Engineering*, vol. 17, pp. 671-682, 1993.
33. Pekny, J. F., D. L. Miller, and G. J. McRae, "An Exact Parallel Algorithm for Scheduling When Production Costs Depend on Consecutive System States," *Computers and Chemical Engineering*, vol. 14, pp. 1009-1023, 1990.
34. Pita, J., "Parallel Computing Methods in Chemical Engineering," in *Invited Lecture*, Purdue University, West Lafayette, IN, 1991.
35. Saati, A., S. Biringen, and C. Farhat, "Solving Navier-Stokes Equations on a Massively Parallel Processor: Beyond the One Gigaflop Performance," *International Journal of Supercomputing Applications*, vol. 4, pp. 72-80, 1990.
36. Sierra, H. M., *An Introduction to Direct Access Storage Devices*, Academic Press, 1990.
37. Stix, G., "Second-Generation Silicon," *Scientific American*, vol. 264, no. 1, pp. 110-111, 1991.
38. Stone, H. S., *High-Performance Computer Architecture*, Addison-Wesley, Menlo Park, 1987.
39. Tanenbaum, A. S., *Computer Networks*, Prentice Hall, Englewood Cliffs, 1981.
40. Tazelaar, J. M., "Desktop Supercomputing," *BYTE*, vol. 15, no. 5, pp. 204-258, 1990.
41. Tesler, L. G., "Networked Computing in the 1990s," *Scientific American*, vol. 265, no. 3, pp. 86-93, 1991.
42. Weiser, M., "The Computer for the 21st Century," *Scientific American*, vol. 265, no. 3, pp. 94-104, 1991.

Optimization

Arthur W. Westerberg

Dept. of Chemical Engineering and the Engineering Design Research Center, Carnegie Mellon University, Pittsburgh, PA 15213, USA

Abstract: This paper is a tutorial on optimization theory and methods for continuous variable problems. Its main purpose is to provide geometric insights. We introduce necessary conditions for testing the optimality of proposed solutions for unconstrained, equality constrained and then inequality constrained problems. We draw useful connections between constrained derivatives and Lagrange theory. The geometry of the dual is exploited to explain its properties; we use the ideas to derive the dual for linear programming. We cover pattern search and then show the key ideas behind generalized reduced gradient and sequential quadratic programming methods. Using earlier insights, linear programming becomes a special case of nonlinear programming; we readily explain the Simplex algorithm. The paper ends with presentations on interior point methods and Benders' decomposition. For nonlinear problems, the paper deals only with finding and testing of local solutions.

Keywords: optimization, constrained derivatives, Lagrange multipliers, Kuhn-Tucker multipliers, generalized dual, pattern search, generalized reduced gradient, sequential quadratic programming, linear programming, interior point algorithms, Benders' decomposition.

Introduction

Optimization should be viewed as a tool to aid in decision making. Its purpose is to aid in the selection of the better values for the decisions which can be made by the person in solving a problem. To formulate an optimization problem, one must resolve three issues. First, one must have a representation of the artifact which can be used to determine how the artifact performs in response to the decisions one makes. This representation may be a mathematical model or the artifact itself. Second, one must have a way to evaluate the performance - an objective function - which is used to compare alternative solutions. Third, one must have a method to search for the improvement. In this paper, we shall be concentrating on the third issue, the methods one might use. The first two items are difficult ones, but discussing them at length is outside the scope of this paper.

Example optimization problems are: (1) determine the optimal thickness of pipe insulation; (2) find the best equipment sizes and operating schedules for the design of a new batch process to make a given slate of products; (3) choose the best set of operating conditions for a set of experiments to determine the constants in a kinetic model for a given reaction; (4) find the amounts of a given set of ingredients one should use for making a carbon rod to be used as an electrode in a arc welder.

For problem (1), one will usually write a mathematical model of how insulation of varying thickness restricts the loss of heat from a pipe. Evaluation requires one develop a cost model for the insulation (a capital cost in dollars) and the heat which is lost (an operating cost in dollars/yr). Some method is required to permit these two costs to be compared such as a present worth analysis. Finally, if the model is simple enough, the method one can use is to set the derivative of the evaluation function to zero with respect to wall thickness to find candidate points for the its optimal thickness. For problem (2), selecting a best operating schedule involves discrete decisions which will generally require models that have integer variables. Such problems will be discussed at length in the paper by Grossmann in this ASI.

It may not be possible to develop a mathematical model for problem (4) as we may not know enough to characterize the performance of a rod versus the amounts of the various ingredients used in its manufacture. Here, we may have to manufacture the rods and then judge them by ranking the rods relative to each other, perhaps based partially or totally on opinions. Pattern search methods have been devised to attack problems in this class; we shall consider them briefly later .

For most of this paper, we shall assume a mathematical model is possible for the problem to be solved. The model may be encoded in a subroutine and be known to us only implicitly, or we may know the equations explicitly. A general form for such an optimization problem is:

$$\begin{aligned} &\min F = F(z) \\ &\text{s.t.} \quad h(z) = 0 \\ &\qquad\;\; g(z) \leq 0 \end{aligned}$$

where F represents a specified objective function that is to be minimized. Functions h and g represent equality and inequality constraints which must be satisfied at the final problem solution.

Variables z are used to model such things as flows, mole fractions, physical properties, temperatures and sizes. The objective function F is generally assumed to be a scalar function, one which represents such things as cost, net present value, safety or flexibility. Sometimes several objective functions are specified (e.g. minimize cost while maximizing reliability); these are commonly combined into one function, or else one is selected for the optimization while the others are specified as constraints. Equations $h(z)=0$ are typically algebraic equations, linear or nonlinear, when modeling steady-state processes, or algebraic coupled with ordinary and/or partial differential equations when optimizing time varying processes. Inequalities $g(z) \leq 0$ put limits on the values variables can take such as a minimum and maximum temperature or they restrict one pressure to be greater than another.

One set of issues about practical optimization we shall be unable to address but which is critical if one wishes to solve large problems are numerical analysis issues. We shall not discuss what happens when a matrix is almost singular or how to factor a sparse matrix or how to partition and precedence order a set of equations. Another topic we shall completely exclude is the optimization of distributed systems. This topic is in Biegler et al. [4]. A good text on this topic is by Bryson and Ho [6]. Finally we shall not consider so-called genetic algorithms or those based on simulated annealing. These latter are best suited for solving problems involving a very large number of discrete decisions, the topic of one of the papers by Grossmann.

For further reading on optimization, readers are directed to the following books [17,31].

Packages

There are a number of packages available for optimization. Following is a list of some of them.

(1) Frameworks

- GAMS. This framework is commercially available. It provides a uniform language to access several different optimization packages, many of them listed below. It will convert the model as expressed in "GAMS" into the form needed to run the package chosen.
- AMPL. This framework is by Fourier and co workers [14] at Northwestern University. It is well suited for constructing complex models.

- ASCEND. This framework is our own. Featuring an object-oriented modeling language, it too is well suited for constructing complex models.

(2) Algebraic optimization with equality and inequality constraints

- SQP. A package by Biegler in our Chemical Engineering Department.
- MINOS5.4. A package available from Stanford Research Institute (affiliated with Stanford University). This package is the state of the art for mildly nonlinear programming problems.
- GRG. A package from Lasdon at the U. of Texas, Dept. of Management Science.

(3) Linear programming

- MPSX from IBM
- SCICONIC from the company of that name.
- MINOS5.4
- Cplex. A package by R. Bixby at Rice University and Cplx, Inc.

Most current commercial codes for linear programming extend the Simplex algorithm, and they can typically handle problems with up to 15,000 constraints.

Organization of Paper

Several sections of this paper are based on a chapter on optimization which this author prepared with Biegler and Grossmann and which appears in Ullmann's Encyclopedia of Industrial Chemistry [4]. In that work and here, we partition the presentation on optimization into two parts. There is the theory needed to determine if a candidate point is an optimal one, the theme of the next section of the paper. The second part covers various methods one might use to find candidate points, the theme of the subsequent section. Our goal throughout this paper is to provide physical insight.

In the next section, we state conditions for a point to be a local optimum for an unconstrained problem, then for an equality constrained one, and finally for an inequality constrained one. To obtain the conditions for equality constrained problems, we introduce constrained derivatives as they directly relate the necessary conditions for the constrained problem to those for the unconstrained one. We show a very nice way to compute them. We next introduce Lagrange multipliers and the associated Lagrange function. Lagrange theory and constrained derivatives are elegantly related. This insight aids in explaining the methods we shall describe in the following section to find candidate points.

Contrary to most presentations, we shall present linear programming as a special case of nonlinear programming, making it possible to explain why the Simplex algorithm is as it is. An exciting development in the solving of Linear Programming are the recent interior point algorithms, which we shall discuss.

We end this paper by looking at Benders' decomposition as a method to solve problems having special structure.

Conditions for Optimality

We start by stating both necessary and sufficient conditions for a point to be the minimum for an unconstrained problem.

Local Minimum Point for Unconstrained Problems

Consider the following unconstrained optimization problem

$$\min_u \left\{F(u) \mid u\in \mathbf{R}^r\right\}$$

If F is continuous and has continuous first and second derivatives, it is necessary that F is stationary with respect to all variations in the independent variables u at a point $\hat{u}$ which is proposed as a minimum to F, i.e.,

$$\frac{\partial F}{\partial u_i} = 0\,,\ \ i=1,2,..r \ \text{ or } \ \nabla_u F = 0 \ \text{ at } u = \hat{u} \tag{1}$$

These are only necessary conditions as point $\hat{u}$ may be a minimum, maximum or saddle point.

Sufficient conditions are that any local move away from the optimal point $\hat{u}$ gives rise to an increase in the objective function. We expand F in a Taylor series locally around our candidate point $\hat{u}$ up to second order terms

$$F(u) = F(\hat{u}) + \frac{\partial F}{\partial u_1}(u_1 - \hat{u}_1) + \frac{\partial F}{\partial u_2}(u_2 - \hat{u}_2) + \cdots \frac{\partial F}{\partial u_r}(u_r - \hat{u}_r) +$$

$$\frac{1}{2}\left\langle (u_1 - \hat{u}_1)\frac{\partial^2 F}{\partial u_1 \partial u_1}(u_1 - \hat{u}_1) + (u_1 - \hat{u}_1)\frac{\partial^2 F}{\partial u_1 \partial u_2}(u_2 - \hat{u}_2) + \cdots \ (u_r - \hat{u}_r)\frac{\partial^2 F}{\partial u_r \partial u_r}(u_r - \hat{u}_r)\right\rangle$$

$$+ \cdots$$

$$= F(\hat{u}) + \nabla_u F^T\Big|_{\hat{u}} (u-\hat{u}) + \frac{1}{2}(u-\hat{u})^T \,\nabla^2_{uu} F\Big|_{\hat{u}} (u-\hat{u}) + \ldots$$

If $\hat{u}$ satisfies necessary conditions (1), the second term disappears in this last line. For this case we see that sufficient conditions for the point to be a local minimum are that the matrix of second partial derivatives ∇^2_{uu} is positive definite. This matrix is symmetric so all of its eigenvalues are real; to be positive definite, they must all be greater than zero.

Constrained Derivatives - Equality Constrained Problems

Consider minimizing our objective function F written in terms of n variables z and subject to m equality constraints h(z)=0, i.e.,

$$\underset{z}{\mathrm{Min}} \left\{F(z) \mid h(z) = 0,\ z \in \mathbf{R}^n,\ h{:}\mathbf{R}^n \rightarrow \mathbf{R}^m\right\} \tag{2}$$

We wish to test point $\hat{z}$ to see if it could be a minimum point. It is necessary that F is stationary for all infinitesimal moves for z that satisfy the equality constraints. We discover the appropriate necessary conditions, given this goal, by linearizing the m equality constraints around $\hat{z}$, getting

$$h(\hat{z} + \Delta z) = h(\hat{z}) + \nabla_z h^T \Big|_{\hat{z}} \Delta \tag{3}$$

where $\Delta z = z - \hat{z}$.

We want to characterize all moves Δz such that the linearized equality constraints remain at zero. There are m constraints here so m of the variables are dependent leaving us with r=n-m independent variables. Partition the variables Δz into a set of m dependent variables Δx and r=n-m independent variables Δu. Eqn (3), rearranged and then rewritten in terms of these variables becomes

$$\Delta h = \nabla_x h^T \Big|_{\hat{z}} \Delta x + \nabla_u h^T \Big|_{\hat{z}} \Delta u = 0$$

Solving for dependent variables Δx in terms of the independent variables Δu, we get

$$\Delta x = -\left[\nabla_x h^T \Big|_{\hat{z}}\right]^{-1} \nabla_u h^T \Big|_{\hat{z}} \Delta u \tag{4}$$

Note that, we must choose which variables are the dependent ones to assure that the Jacobian matrix $\nabla_x h$ evaluated at our test point is nonsingular. This partitioning is only possible if the rank of the m by n matrix $\nabla_z h$ is of rank m. Eqn (4) states that the changes in m dependent variables x can be computed once we specify the changes for the r independent variables u.

Linearize the objective function F(z) in terms of the partitioned variables

$$\Delta F = \nabla_x F^T \Big|_{\hat{z}} \Delta x + \nabla_u F^T \Big|_{\hat{z}} \Delta u$$

and substitute out variables Δx using eqn (4).

$$\Delta F = \left\{ \nabla_x F^T - \nabla_u F^T \left[\nabla_x h^T \right]^{-1} \nabla_u h^T \right\}_{\hat{z}} \Delta u$$
$$= \left\{ \frac{dF}{du} \right\}^T_{\Delta h=0} \Delta u = \sum_{i=1}^{r} \left\{ \frac{dF}{du_i} \right\}_{\Delta h=0} \Delta u_i \qquad (5)$$

There is one term for each Δu_i in the row vector which is in the curly braces {}. These terms are called *constrained derivatives.* They tell us how the objective function will change if we change the independent variables u_i while changing the dependent variables x_i to keep the constraints satisfied.

Necessary conditions for optimality are that these constrained derivatives are zero, i.e.,

$$\left\{ \frac{dF}{du_i} \right\}_{\Delta h=0} = 0 \ , \ i=1,2,..r$$

An Easy Way to Compute Constrained Derivatives

Form the Jacobian matrix for the equality constraints h(z) augmented with the objective function with respect to variables z at $\hat{z}$.

$$\nabla_z h^T \Big|_{\hat{z}} \Delta z = 0 \qquad \text{m rows}$$

$$\nabla_z F^T \Big|_{\hat{z}} \Delta z = 0 \qquad \text{1 row}$$

Note that, there are n variables z in these m+1 linearized equations. Perform a forward gauss elimination on these equations including the last row ($\nabla_z F \ \Delta z = 0$) but do not pivot within that row. One will select m pivots. Select them so the m by m pivoted portion of the matrix is nonsingular. The pivoted variables are the dependent variables x for the problem, while the unpivoted are the independent variables u. Fig. 1 shows the structure of the result. The nonzero portion of the last row beneath the variables u contains exactly the numerical evaluation for the *constrained derivatives* given in eqn (5). One can prove this statement by carrying out the elimination symbolically and noting that this part of the matrix is algebraically the constrained derivatives as noted.

Equality Constrained Problems - Lagrange Multipliers

Form a scalar function, which we shall term the Lagrange function, by adding each of the

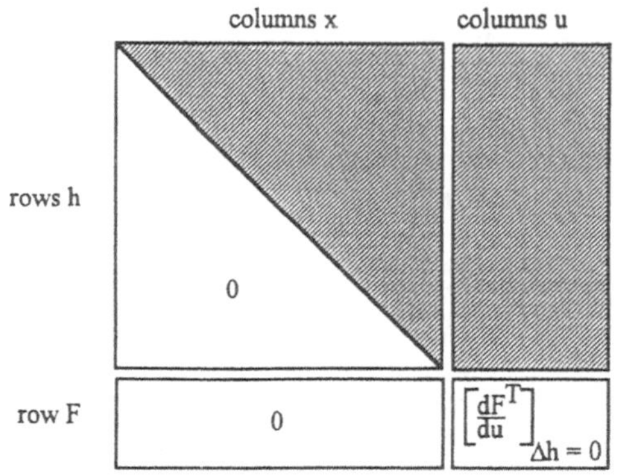

Fig. 1 Partitioning the Variables and Computing Constrained Derivatives in a Single Step Using Gaussian Elimination

equality constraints multiplied by an arbitrary multiplier to the objective function.

$$L(x,u,\lambda) = F(x,u) + \sum_{i=1}^{m} \lambda_i h_i(x,u) = F(x,u) + \lambda^T h(x,u)$$

At any point where the functions h(z) are zero, the Lagrange function equals the objective function.

Next, write the stationarity conditions for L with respect to variables x, u and λ.

$$\nabla_x L^T \Big|_{\hat{z}} = \nabla_x F^T \Big|_{\hat{z}} + \lambda^T \nabla h_x^T \Big|_{\hat{z}} = 0^T \tag{6}$$

$$\nabla_u L^T \Big|_{\hat{z}} = \nabla_u F^T \Big|_{\hat{z}} + \lambda^T \nabla h_u^T \Big|_{\hat{z}} = 0^T \tag{7}$$

$$\nabla_\lambda L^T \Big|_{\hat{z}} = h^T(x,u) = 0^T$$

Solve eqn (6) for the Lagrange multipliers

$$\lambda^T = - \nabla_x F^T \left[\nabla h_x^T\right]^{-1} \tag{8}$$

and then eliminate these multipliers from eqn (7).

$$\nabla_u L^T = \nabla_u F^T - \nabla_x F^T \left[\nabla h_x^T\right] \nabla h_u^T = 0 \tag{9}$$

We see by comparing eqn (9) to eqn (5) that ∇_u are equal to the constrained derivatives for our problem, which, as before, should be zero àt the solution to our problem. Also these stationarity conditions very neatly provide us with the necessary conditions for optimality of an equality constrained problem.

Lagrange multipliers are often referred to as shadow prices, adjoint variables or dual variables, depending on the context. Assume we are at an optimum point for the our problem. Perturb the variables such that only constraint h_i changes. We can write

$$\Delta L = \Delta F + \lambda_i \Delta h_i = 0$$

which is zero because, as just shown, the Lagrange function is at a stationary point at the optimum. Solving for the change in the objective function

$$\Delta F = -\lambda_i \, \Delta h$$

The multiplier tells us how the optimal value of the objective function changes for this small change in the value of a constraint while holding all the other constraints at zero. It is for this reason they are often called shadow prices.

Equality and Inequality Constrained Problems - Kuhn-Tucker Multipliers

In the previous section, we considered only equality constraints. We now add inequality constraints and examine how we might test a point to see if it is an optimum. Our problem is

$$\underset{z}{\text{Min}} \left\{ F(z) \mid h(z) = 0,\ g(z) \le 0,\ z \in \mathbf{R}^n,\ F{:}\mathbf{R}^n \to \mathbf{R}^1,\ h{:}\mathbf{R}^n \to \mathbf{R}^m,\ g{:}\mathbf{R}^n \to \mathbf{R}^p \right\}$$

The Lagrange function here is similar to before.

$$L(z, \lambda, \mu) \equiv F(z) + \lambda^T h(z) + \mu^T g(z)$$

only here, we also add each of the inequality constraints $g_i(z)$ multiplied by what we shall call a Kuhn-Tucker multiplier, μ_i. The necessary conditions for optimality, called the Karush-Kuhn-Tucker conditions for inequality constrained optimization problems, are

$$\nabla_z L\Big|_{\hat{z}} = \nabla_z F\Big|_{\hat{z}} + \nabla_z h\Big|_{\hat{z}} \lambda + \nabla_z g\Big|_{\hat{z}} \mu = 0$$

$$\nabla_\lambda L = h(z) = 0$$

$$g(z) \le 0$$

$$\mu_i g_i(z) = 0\ , \quad i=1,2,\ldots p \qquad (10)$$

$$\mu_i \ge 0\ , \quad i=1,2,\ldots p$$

Conditions (10), called complementary slackness conditions, state that either the constraint $g_i(z)=0$ and/or its corresponding multiplier μ_i is zero. If constraint $g_i(z)$ is zero, it is behaving like an equality constraint, and its multiplier μ_i is exactly the same as a Lagrange multiplier for an equality constraint. If the constraint is away from zero, it is not a part of the problem and should not affect it. Setting its multiplier to zero removes it from the problem.

As our goal is to minimize the objective function, releasing the constraint into the feasible region must not decrease the objective function. Using the shadow price argument above, it is evident that the multiplier must be nonnegative [24].

Constraint Qualifications

The necessary conditions above will not hold if, for example, two nonlinear inequality constraints form a cusp as shown in Fig. 2, and the optimum is exactly at that point. The optimum requires both constraints for its definition, even though they are both collinear at the solution. The equation

$$\nabla F + \mu_1 \nabla g_1 + \mu_2 \nabla g_2 = 0$$

which states that the gradient of the objective function can be written as a linear combination of the gradients of the constraints can be untrue at this point, as Fig. 2 illustrates. One usually states that the constraints have to be independent at the solution when stating the Karush-Kuhn-Tucker conditions, a constraint qualification.

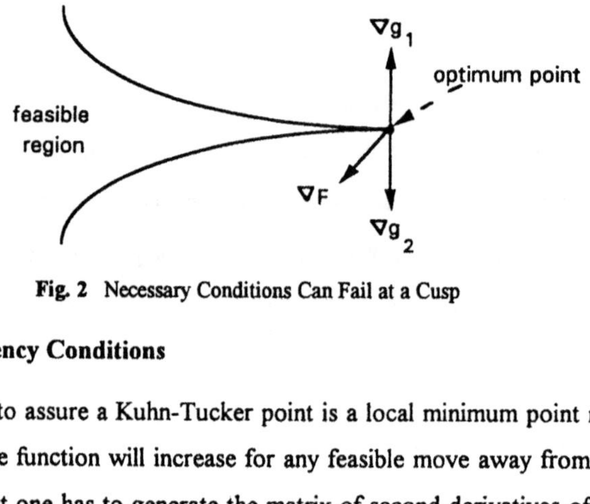

Fig. 2 Necessary Conditions Can Fail at a Cusp

Kuhn-Tucker Sufficiency Conditions

Sufficiency conditions to assure a Kuhn-Tucker point is a local minimum point require one to prove that the objective function will increase for any feasible move away from such a point. To carry out such a test one has to generate the matrix of second derivatives of the Lagrange function with respect to all the variables z evaluated at $\hat{z}$. The test is seldom done as it requires too much work.

The Generalized Dual

Consider the following problem, which we shall call the primal problem.

$$F^* \equiv F(z^*) = \underset{z}{\mathrm{Min}} \left\{ F(z) \mid h(z) = 0, \ z \in S \right\} \tag{11}$$

Let us write the following "restricted" Lagrange function for this problem. We call it restricted as it is defined only for the restricted set of values of $z \in S$.

$$L(z, \lambda) = \{ F(z) + \lambda^T h(z) \mid z \in S \}$$

Pick a point $z \in S$ and plot its corresponding point F(z) versus h(z). Repeat for all $z \in S$, getting the region R shown in Fig. 3. Region R is defined as

$$R = \{ (F(z), h(z) \mid \text{for all } z \in S \}$$

Pass a hyperplane (line) through any point in R with a slope of $-\lambda$, as illustrated. The intercept where h(z) = 0 can be seen to be exactly equal to the Lagrange function for this problem.

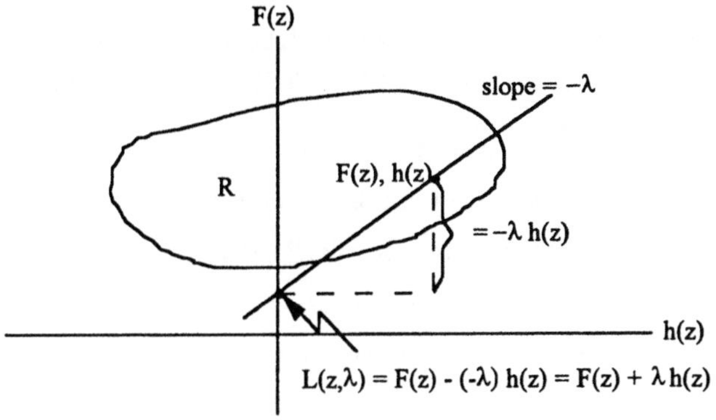

Fig. 3 The Lagrange Function in the Space of F(z) vs. h(z)

If we minimize the Lagrange function over all points in R for a fixed slope $-\lambda$, we obtain the minimum intercept possible for that hyperplane, which is illustrated in Fig. 4. Note that this hyperplane supports (just touches) the region R, with all of R being on only one side of it. This support function

$$D(\lambda) = \underset{z}{\mathrm{Min}} \ \{ L(z, \lambda) \mid z \in S \}$$

is known as the generalized dual function for our original problem. The minimization is carried out over z, and the *global minimum must be found.*

By examining the geometric interpretation for the dual, we immediately see one of its most important properties. It is always below the region R on the intercept where h(z) = 0. As such *it must be a lower bound for the optimal value for our objective function for the primal problem* defined by eqn (11), namely

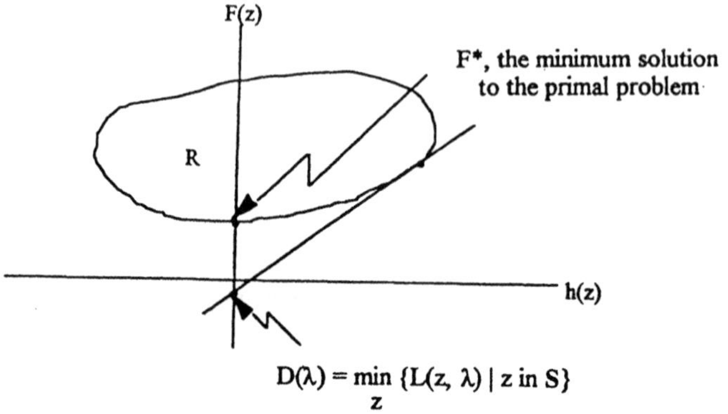

Fig. 4 A Geometrical Interpretation of The Generalized Dual Function, D(λ)

$$D(\lambda) \leq F^*$$

We can now vary the slope λ to find the maximum value that this dual function can attain.

$$D^* \equiv D(\lambda^*) = \underset{\lambda}{\text{Max}} \;\{D(\lambda) \mid \lambda \in R^m\} \leq F^* \equiv F(z^*) = \underset{z}{\text{Min}} \;\{F(z) \mid h(z) = 0,\ z \in S)$$

If a hyperplane can support the region R at the point where h(z) = 0, then we note that this maximum exactly equals the minimum value for our original objective at its optimal solution. It may be that the region R cannot be supported at F*, in which case the optimum of the dual is always less than the optimum for the primal, as illustrated in Fig. 5. Note there are two support points (at least) if this is the case, neither of which satisfies the equality constraints for the primal problem.

Further properties of dual: Examining its geometrical interpretation, we can note some further important properties of the dual.

- First, if the region R fails to cover any part of the axis where h(z) = 0, then the primal problem can have no solution. It is infeasible. For this case, we can find a hyperplane to support the region R that will intersect this vertical axis at any value desired. The maximum value will be positive infinity. *If the original problem is infeasible, the dual is feasible but unbounded.* Conversely, if R covers any part of the axis, the dual cannot be unbounded, so the previous statement is really an if and only if statement.
- Second, consider the case we show in Fig. 6 where the region R is unbounded below but where it does not cover the negative vertical axis. We note there are multipliers(slopes), as the hyperplane labeled 1 illustrates, that will lead from region R to an intersection with vertical axis (h(z)=0) at negative infinity; i.e., for those values the dual function is unbounded

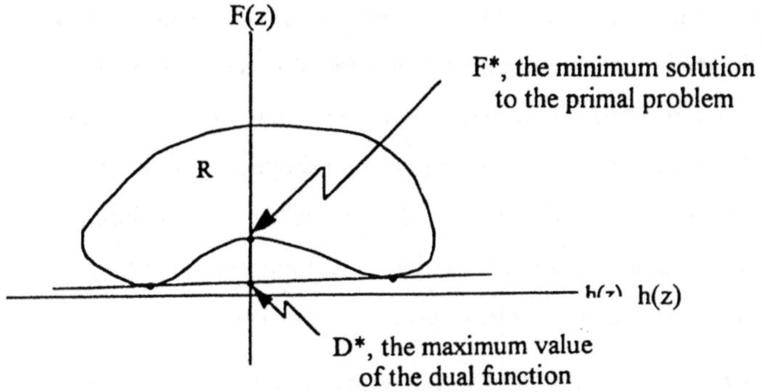

Fig. 5 A Nonconvex Region R with No Support Hyperplane at the Solution to the Primal Problem

below. However, there are also multipliers where the support plane is finite as hyperplane 2 illustrates. Since the dual problem is to search over the hyperplane slopes to find a maximum to the intercept with the vertical axis, we can eliminate looking over multiplier values where the dual function is unbounded below. We shall define the dual function as being infeasible for these values of the multipliers.

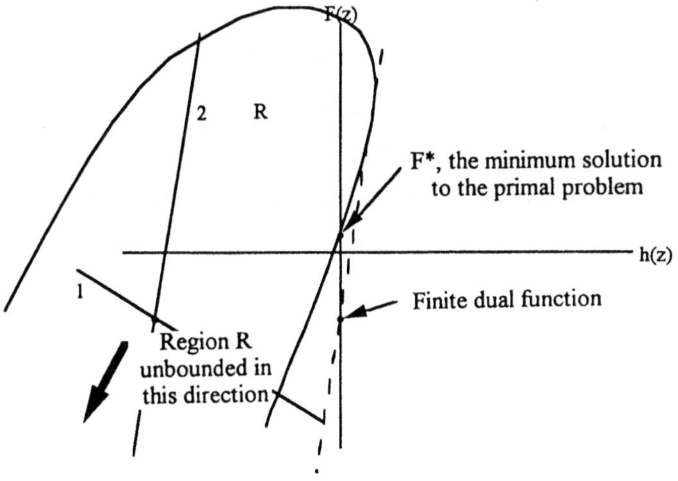

Fig. 6 Case Where Region R Is Unbounded In Negative F(Z) Direction. A Support Hyperplane With Slope Parallel To 1 Yields A Dual Solution Of Negative Infinite While One With Slope Of 2 Yields A Finite Dual Solution.

• Third, if the region R in Fig. 6 covers the entire negative vertical axis where h(z) is zero, then the dual function is negative infinity for all values of the multipliers. Continuing our ideas just above, the dual is infeasible everywhere. Thus we have a symmetry: if the primal is infeasible, the dual is unbounded. If the primal is unbounded, the dual is infeasible.

While it is not immediately obvious, the *dual of the dual* is closely related to the primal problem. It corresponds to an optimization carried out over the convex hull of the region R. It is left as an exercise for the reader to prove this statement.

Example: Let us find the dual for the following problem.

$$\underset{u}{\text{Min}}\ \{ c^T u \mid Au \geq b,\ u \geq 0\}$$

We can introduce slack variables and rewrite this problem as follows

$$\underset{u,s}{\text{Min}}\ \{ c^T u \mid b + s - Au = 0,\ u,\ s \geq 0\}$$

The constrained Lagrange function can then be written as

$$L(u,s,\lambda) = \{c^T u + \lambda^T(b + s - Au) \mid u,s \geq 0\} = \{(c^T - \lambda^T A)\, u + \lambda^T s + \lambda^T b) \mid u,s \geq 0\}$$

from which we derive the dual function

$$D(\lambda) = \underset{u,s}{\text{Min}}\ L(u,s,\lambda)$$

The minimization operation can be carried out term by term, giving

$$\text{Min}\ (c_i - \lambda^T A_i)u_i = \left\{ \begin{array}{ll} 0 & \text{if } c_i - \lambda^T A_i \geq 0 \\ -\infty & \text{if } c_i - \lambda^T A_i < 0 \end{array} \right\} \quad \text{and} \quad \text{Min}\ \lambda_j s_j = \left\{ \begin{array}{ll} 0 & \text{if } \lambda_j \geq 0 \\ -\infty & \text{if } \lambda_j < 0 \end{array} \right\}$$

where A_i is the i-th column of A. For the primal problem to have a bounded solution, we can eliminate looking over multipliers where the dual is unbounded below (see Fig. 6 earlier).

$$A^T\lambda \leq c \quad \text{and} \quad \lambda \geq 0$$

The dual function becomes

$$D(\lambda) = \underset{u,s}{\text{Min}}\ L(u,s,\lambda) =$$

$$= \{0 + 0 + \lambda^T b \mid A^T \lambda \leq c,\ \lambda \geq 0\} = \{b^T\lambda \mid A^T \lambda \leq c,\ \lambda \geq 0\}$$

and the dual optimization problem

$$D^* = \underset{\lambda}{\text{Max}}\ D(\lambda) = \underset{\lambda}{\text{Max}}\ \{b^T\lambda \mid A\lambda \leq c,\ \lambda \geq 0\} \tag{12}$$

We can show that R is convex by noting if the points (u(1),s(1)) and (u(2),s(2)) are both in R then so is their convex combination $\alpha(u(1),s(1)) + (1-\alpha)(u(2),s(2))$ and that it maps precisely into the point

$$\{\alpha c^T u(1) + (1-\alpha)c^T u(2),\ \alpha(b + s(1) - Au(1)) + (1-\alpha)(b + s(2) - Au(2))\}$$

in R. If R is convex, then it has a support hyperplane everywhere. Thus the value of the objective function for both the primal and the dual will be equal at the solution.

Strategies of Optimization

The theory just covered can tell us if a candidate point is, or more precisely, is not the optimum point, but how do we find candidate point? The simplest strategy is to place a grid of points throughout the feasible space, evaluating the objective function at every grid point. If the grid is fine enough, then the point yielding the highest value for the objective function can be selected as the optimum. 20 variables gridded over only 10 points would take place 10^{20} points in our grid, and, at one nanosecond per evaluation, it would take in excess of four thousand years to carry out these evaluations.

Most strategies limit themselves to finding a local minimum point in the vicinity of the starting point for the search. Such a strategy will find the global optimum only if the problem has a single minimum point or a set of "connected" minimum points. A "convex" problem has only a global optimum.

Pattern Search

Suppose the optimization problem is to find the right mix of a given set of ingredients and the proper baking temperature and time to make the best cake possible. A panel of judges can be formed to judge the cakes; assume they are only asked to rank order the cakes and that they can do that task in a consistent manner. Our approach will be to bake several cakes and ask the judges to rank order them. For this type of problem, pattern search methods can be used to find the better conditions for manufacturing the product. We shall only describe the ideas behind this approach. Details on implementing it can be found in Umeda and Ichikawa [35].

The complex method is one such pattern search method, see Fig. 7. First form a "complex" of at least r+1 (r = 2 and we used 4 points in Fig. 7) different points at which to bake the cakes by picking a range of suitable values for the r independent variables for the baking process. Bake the cakes and then ask the judges to identify the worst cake.

For each independent variable, form the average value at which it was run in the complex. Draw a line from the coordinates of the worst cake through the average point - called the centroid - and continue on that line a distance that is twice that between these two points. This point will be the next test point. First decide if it is feasible. If so bake the cake and discover if it leads to a cake that is better than the worst cake from the last set of cakes. If it is not feasible or it is not better, then return half the distance toward the average values from the last test and try again. If it is better, toss out the worst point of the last test and replace it with this new one. Again, ask the judges to find the worst cake. Continue as above until the cakes are all the same quality in the most recent test. It might pay to restart at this point, stopping finally if the restart leads to no improvement. The method takes large steps if the steps are being successful in improving the recipe. It collapses onto a set of point quite close to each other otherwise. It works reasonably well, but it requires one to bake lots of cakes.

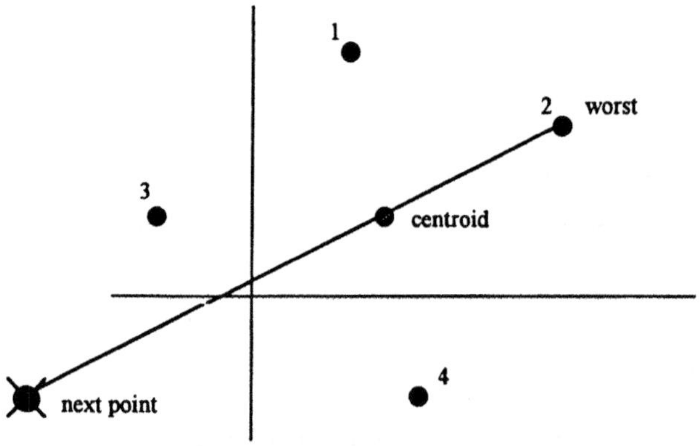

Fig. 7 Complex Method, a Pattern Search Optimization Method

Generalized Reduced Gradient (GRG) Method

We shall develop next a method called the generalized reduced gradient (GRG) method for optimization. We start by developing a numerical approach to optimize an unconstrained problem.

Optimization of Unconstrained Objective: Assume we have an objective function F which is a function of independent variables u_i, $i = 1..r$. Assume we can have a computer program which, when supplied with values for the independent variables, can feed us back both F and its derivatives with respect to each u_i. Assume that F is well approximated as an as yet unknown quadratic function in u.

$$F \approx a + b^T u + \frac{1}{2} u^T Q u$$

where a is a scalar, b a vector and Q an rxr symmetric positive definite matrix. The gradient of our approximate function is

$$\nabla_u F = b + Qu$$

which, when we set it to zero, allows to find an estimate for it minimum

$$u = - Q^{-1} b \tag{13}$$

We do not know Q and b at the start so we can proceed as follows. b contains r unknown coefficients and Q another r(r+1)/2. To estimate b and Q, we can run our computer code repeatedly, getting r equations each time - namely

$$\begin{aligned} (\nabla_u F)(1) &= b + Qu(1) \\ (\nabla_u F)(2) &= b + Qu(2) \\ &\ldots\ldots \\ (\nabla_u F)(t) &= b + Qu(t) \end{aligned} \tag{14}$$

As soon as we have written as many independent equations from these computer runs as there are unknown coefficients, we can solve these *linear* equations for b and Q. A proper choice of the points u(i) will guarantee getting independent equations to solve here.

Given b and Q, eqn (13) provides us with a new estimate for u as a candidate minimum point. We run the subroutine again to obtain the gradient of F at this point. If the gradient is essentially zero, we can stop; we have a point which satisfies the necessary conditions for optimality. If not, we write equations in the form of (14) for this new point, add them to the set while removing the oldest set of equations. We solve these equations for b and Q and continue until we are at a minimum point. If removal of the oldest equations from the set (14) leads to a singular set of equations, then different equations have to be selected for removal. We can keep all the older equations, with the new ones added to the top of the list. Pivoting can be done by proceeding down the list until a nonsingular set of equations is found. We use the older equations only if they have to be. Also, since only one set of equations is being replaced, clever methods are available to find the solution to the equations with much less work than is required to solve the set of equations the first time [10,34].

Quadratic Fit for the Equality Constrained Case: We wish to solve a problem of the form of eqn (2). We proceed as follows. For each iteration k:

1. Enter with values provided for variables u(k).

2. Given values for u(k), solve equations h(x,u) = 0 for x(k). These will be m equations in m unknowns. If the equations are nonlinear, solving can be done using a variant of the Newton-Raphson method.
3. Use eqns (8) to solve for the Lagrange multipliers, λ(k). If we used the Newton-Raphson method (or any or several variants to it) to solve the equations, we will already have generated the Jacobian matrix $\nabla_x^T h|_{z(k)}$ and its L\U factors so solving eqns (8) requires very little effort.
4. Substitute λ(k) into equation (7), which in general will not be zero. The gradient ∇_u (k) computed will be the constrained derivatives of F with respect to the independent variables u(k).
5. Return.

We enter with given values for the independent variables u and exit with the (constrained) derivatives of our objective function with respect to them. We have just described the routine we indicated was needed for the unconstrained problem above where we use a succession of quadratic fits to move toward the optimal point for an unconstrained problem. Apply that method.

This approach is a form of the generalized reduced gradient (GRG) approach to optimizing, one of the better ways to carry out optimization numerically.

Inequality Constrained Problems: To solve inequality constrained problems, we have to develop a strategy that can decide which of the inequality constraints should be treated as equalities. Once we have decided, then a GRG type of approach can be used to solve the resulting equality constrained problem. Solving can be split into two phases: phase 1 where the goal is to find a point that is feasible with respect to the inequality constraints and phase 2 where one seeks the optimum while maintaining feasibility. Phase 1 is often accomplished by ignoring the objective function and using instead

$$F = \sum_{i=1}^{p} \begin{Bmatrix} g_i^2(z) \text{ if } g_i(z) > 0 \\ 0 \quad \text{otherwise} \end{Bmatrix}$$

until all the inequality constraints are satisfied.

Once satisfied, we then proceed as follows. At each point check which of the inequality constraints are active, i.e., exactly equal to zero. These can be placed into the active set and treated as equalities. The remaining can be put aside to be used only for testing. A step can then be proposed using the GRG algorithm. If it does not cause one to violate any of the inactive inequality constraints, the step is taken. Otherwise one can add the closest inactive inequality constraint to the active set. Finding the closet inactive equality will almost certainly require a line search in the direction proposed by the GRG algorithm.

When one comes to a stationary point, one has to test the active inequality constraints at that point to see if they should remain active. This test is done by examining the sign (they should be nonnegative if they are to remain active) of their respective Kuhn-Tucker multipliers. If any should be released, it has to be done carefully as the release of a constraint changes the multipliers for all the constraints. One can find oneself cycling through the testing to decide whether to release the constraints. A correct approach is to add slack variables s to the problem to convert the inequality constraints to equalities and then require the slack variables to remain positive. The multipliers associated with the inequalities $s \geq 0$ all behave independently, and their sign tells one directly to keep or release the constraints. In other words, simultaneously release all the slack variables which have multipliers strictly less than zero. If released, *the slack variables must be treated as a part of the set of independent variables* until one is well away from the associated constraints for this approach to work.

Successive Quadratic Programming (SQP)

The above approach to finding the optimum is called a feasible path method as it attempts at all times to remain feasible with respect to the equality and inequality constraints as it moves to the optimum. A quite different method exists called the Successive Quadratic Programming (SQP) method which only requires one be feasible at the final solution. Tests which compare the GRG and SQP methods generally favor the SQP method so it has the reputation of being one of the best methods known for nonlinear optimization for the type of problems we are considering in this paper.

Assume we can guess which of the inequality constraints will be active at the final solution. The necessary conditions for optimality are

$$\nabla_z L(z,\mu,\lambda) = \nabla F + \nabla g_A \mu + \nabla h \lambda = 0$$

$$g_A(z) = 0$$

$$h(z) = 0$$

Then, one can apply Newton's method to the necessary conditions for optimality, which are a set of simultaneous (non)linear equations. The Newton equations one would write are

$$\begin{bmatrix} \nabla_{zz}L(z(i),u(i),\lambda(i)) & \nabla g_A(z(i)) & \nabla h(z(i)) \\ \nabla g_A(z(i))^T & 0 & 0 \\ \nabla h(z(i))^T & 0 & 0 \end{bmatrix} \begin{bmatrix} \Delta z(i) \\ \Delta\mu(i) \\ \Delta\lambda(i) \end{bmatrix} = - \begin{bmatrix} \nabla_z L(z(i),\mu(i),\lambda(i)) \\ g_A(z(i)) \\ h(z(i)) \end{bmatrix}$$

A sufficient condition for a unique Newton direction is that the matrix of constraint derivatives is of full rank (linear independence of constraints) and the Hessian matrix of the Lagrange function ($\nabla_{zz}L(z,\mu,\lambda)$ *projected into the space of the linearized constraints* is positive definite. The linearized system actually represents the solution of the following quadratic programming problem:

$$\min_{\Delta z} \ \nabla F(z(i))^T \Delta z + \frac{1}{2} \Delta z^T \nabla_{zz} L(z(i),\mu(i),\lambda(i)) \Delta$$

subject to

$$g_A(z(i)) + \nabla g_A(z(i))^T \Delta z = 0$$
$$h(z(i)) + \nabla h(z(i))^T \Delta z = 0$$

Reformulating the necessary conditions as a linear quadratic program has an interesting side effect. We can simply add linearizations of the inactive inequalities to the problem and let the active set be selected by the algorithm used to solve the linear quadratic program.

Problems with calculating second derivatives as well as maintaining positive definiteness of the Hessian matrix can be avoided by approximating this matrix by B(i) using a quasi-Newton formula such as BFGS [5,9,11,12,13,19,34]. One maintains positive definiteness by skipping the update if it causes the matrix to lose this property. Here gradients of the Lagrange function are used to calculate the update formula [22,30]. The resulting quadratic program, which generates the search direction at each iteration i becomes:

$$\min_{\Delta z} \ \nabla F(z(i))^T \Delta z + \frac{1}{2} \Delta z^T B(i) \Delta$$

subject to

$$g(z(i)) + \nabla g(z(i))^T \Delta z \le 0$$
$$h(z(i)) + \nabla h(z(i))^T \Delta z = 0$$

This linear quadratic program will have a unique solution if B(i) is kept positive definite. Efficient solution methods exist for solving it [16,20,25,38].

Finally, to ensure convergence of this algorithm from poor starting points, a step size α is chosen along the search direction so that the point at the next iteration ($z^{i+1}=z^i+\alpha d$) is closer to the solution of the NLP [7,22,32].

These problems get very large as the Lagrange function involves all the variables in the problem. If one has a problem with 5000 variables z and the problem has only 10 degrees of freedom (i.e., the partitioning will select 4990 variables x and only 10 variables u), one is still faced with maintaining a matrix B which is 5000x5000. Berna and Westerberg [3] proposed a method that kept the quasi-Newton updates for B rather than keeping B itself. They were able to reduce the computational and space requirements significantly, exactly reproducing the steps taken by the original algorithm. Later Locke et al [26] proposed a method which permitted B to be approximated only in the space of the degrees of freedom, very significantly reducing the space and computational requirements. More recently a "range and null space" decomposition approach [29,36,37] has been proposed solving the problem. This decomposition loses considerably on sparsity but is numerically more reliable. Lucia and Kumar [27] proposed and tested explicitly creating the second derivative information for the Hessian and then exploited its sparsity. In an attempt to keep the sparsity of the Locke et al approach and improve its numerical reliability, Schmid and Biegler [33] have recently proposed methods to estimate the terms which are missing in the Locke et al algorithm. Finally Schmid and Biegler are developing a linear quadratic programming algorithm based on ideas in Goldfarb and Idnani [20] which is much faster than available library codes.

Linear Programming

If the objective function and all equality and inequality constraints are linear, then a very efficient means is available to solve our optimization problem. Considering only inequalities, we can write

$$\min_{u} \ \{ c^T u \mid Au \geq b, \ u \geq 0\} \qquad \text{(Primal LP)}$$

We label this problem a *primal* linear program as we shall later examine a corresponding dual linear programming when we look at an example problem. Fig. 8 illustrates the appearance of a small linear program.

To solve we change all inequalities into equalities by introducing slack variables

$$\min_{u,s} \ \{ c^T u \mid b + s - Au = 0, \ u, s \geq 0\} \qquad (15)$$

and then observe, as Dantzig [8] did, that our solution can reside at a *corner point* for the feasible region, such as points a, b, c and d in Fig. 8. If the objective exactly parallels one of the boundaries, then the whole boundary - including its corner points - are solutions. If the objective is everywhere equal, then all points are solutions, including again any of the corner points. It is for this reason that we stated the solution *can* always reside at a corner point.

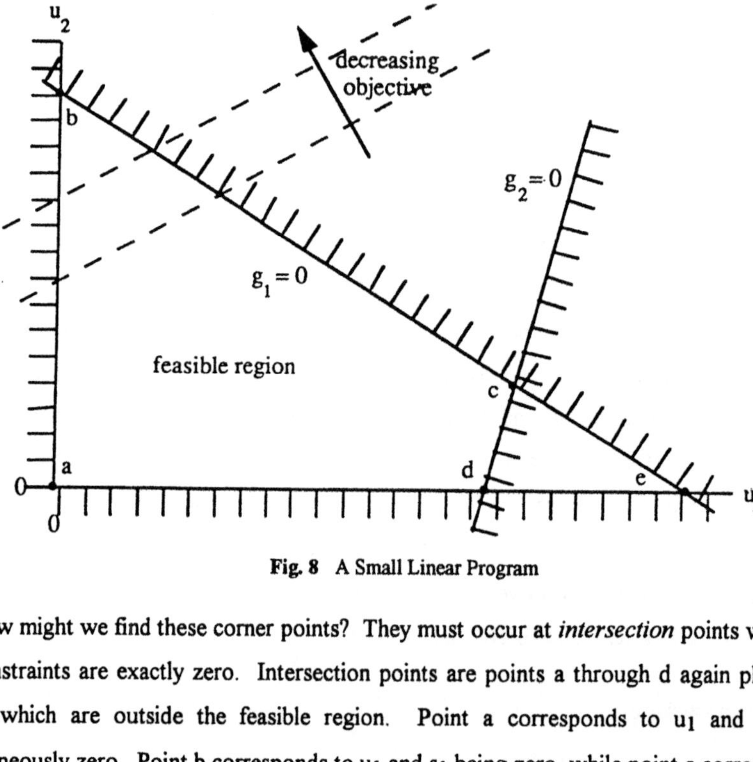

Fig. 8 A Small Linear Program

How might we find these corner points? They must occur at *intersection* points where r of the constraints are exactly zero. Intersection points are points a through d again plus points like e which are outside the feasible region. Point a corresponds to u_1 and u_2 being simultaneously zero. Point b corresponds to u_1 and s_1 being zero, while point c corresponds to s_1 and s_2 being zero. If we examine the degrees of freedom for our problem, we see there are r variables u, p variables s and p equality constraints. Thus, there are r degrees of freedom, the number of variables u that exist in the problem. If we set r variables from the set u and s to zero, solving the equations will provide us with an intersection point. If the remaining variables are nonnegative, then the intersection point is feasible. Otherwise, it is outside the feasible region.

The Simplex Algorithm: Dantzig [8] developed the remarkably effective Simplex algorithm that allows one to move from one feasible intersection point to another, always in a downhill direction. Such a set of moves ultimately leads to the lowest corner point for the feasible region which is then the solution to our problem. Each step in the Simplex algorithm is equivalent to an elimination step in a Gauss elimination for solving linear equations and an examination step of the result to discover which adjacent intersection points are feasible and downhill.

Let us mentally apply the Simplex algorithm to the example in Fig. 8. Suppose we are currently at point d. We examine how the objective function changes if we move along either of the constraints which are active at d. To see what we really are doing, let each of the variables which are zero at the corner point be our independent variables for the problem at this point in time; here we identify variables u_2 and s_2 as our independent variables. Increase one while holding the remaining one(s) at zero. We find we move exactly along an edge whose constraint(s) correspond to the variable(s) being held at zero. In particular release u_2 and move along $g_2 = 0$. Release s_2 and move along $u_2 = 0$.

The constrained derivatives for the degrees of freedom for the problem defined at the current corner point tell us precisely how the objective function will change if we increase one of the independent variables while holding the rest at zero, precisely what we are doing. So we will need constrained derivatives, which we shall see are readily available if we do things correctly.

Next, we need to know how far we can go. As we proceed, we generally encounter other constraints. Suppose we have selected u_2 to increase, moving along constraint g_2. We will encounter the constraint where s_1 becomes zero. In effect, we trade $u_2 = 0$ for $s_1 = 0$ to arrive at the adjacent point. The first variable to become zero as we increase u_2 tells us where to stop.

We are now at point c. We start again with s_1 and s_2 being our independent variables. We must compute the constrained derivatives for them, select one, and move again. This time we move to point b. At point b, the constrained derivatives are all positive indicating there is no downhill direction to move. We have located the optimum point.

Problems occur when the region is unbounded if the objective function decreases in that direction. The examination to discover which constraint is closest in the direction selected finds there is no constraint in the way. The algorithm simply stops, reports an unbounded solution exists and indicates the direction in which it occurs.

Also, there can be *degeneracy* which occurs when more than r variables are zero at a corner point. In Fig. 8, degeneracy would be equivalent to three constraints intersecting at one point. Obviously only two are needed to define the point. This redundancy can cause the Simplex algorithm to cycle if care is not taken. If one encounters degeneracy, the normal solution is to perturb enough of the constraints to have no more than r intersecting at the point. One then solves the perturbed problem to move from the point, if movement is required, removing the perturbations once away from the point.

Example: We shall carry out the solution for the following very small linear program.

$$\text{Min} \quad F = 2u_1 + 3u_2 - u_3$$

subject to

$$g_1: \; u_1 + u_2 \leq 10$$
$$g_2: \; u_3 \leq 4$$
$$h_1: \; 2u_2 - 5u_3 = 6$$
$$u_1, u_2, u_3 \geq 0$$

We shall first put this problem into the form indicated by eqn (15) to identify A, b and c properly. The inequality constraints have to be written in the form $Au = b + s$.

$$-u_1 - u_2 = -10 + s_1$$
$$-u_3 = -4 + s_2$$

We shall write the equality constraint with a special slack variable called an artificial variable, using the same sign convention as for the inequality constraints.

$$-2u_1 + 5u_3 = -6 + a_1$$

The artificial variable a_1 must be zero at the final solution which we can accomplish by giving it a very large cost as follows.

$$F(a) = c^T u + (\text{big number})\, a_1 = [2,3,-1]\begin{bmatrix} u_1 \\ u_2 \\ u_3 \end{bmatrix} + 1000\, a_1$$

The solution to the optimization problem will then make it zero, its least value. If it cannot be removed from the problem in this manner, the original problem is not feasible. The constraints can be put into matrix form, getting

$$Au - s = b \Rightarrow \begin{bmatrix} -1 & -1 & 0 \\ 0 & 0 & -1 \\ 0 & -2 & 5 \end{bmatrix}\begin{bmatrix} u_1 \\ u_2 \\ u_3 \end{bmatrix} - \begin{bmatrix} s_1 \\ s_2 \\ a_1 \end{bmatrix} = \begin{bmatrix} -10 \\ -4 \\ -6 \end{bmatrix}$$

$$u_1, u_2, u_3, s_1, s_2, a_1 \geq 0$$

We choose to transform each of the constraints so each of their RHS terms is positive by multiplying each by -1 to form the following Simplex tableau.

	basic	u_1	u_2	u_3	s_1	s_2	a_1	RHS
g_1:	s_1	1	1		1			10
g_2:	s_2			1		1		4
h_1:	a_1		2	-5			1	6
F(a):		2	3	-1			1000	0
F:		2	3	-1				0

We have included six variables for this problem and three constraints. The first constraint, for example, can be read directly from the tableau; namely, $u_1+u_2+s_1 = 10$. We have set up two objective function rows. The former has a large cost for the artificial variable a_1 while the latter has its cost set to zero. We shall use the row called F(a) as the objective function until all of the artificial variables are removed (i.e., get set to zero) from the problem. Then we shall switch the objective function row to F for the remainder of the problem. Once we have switched, we will no longer allow an artificial variable to be reintroduced back into the problem with nonzero value.

Each variable under the column labeled "basic" is the one which is being solved for using the equation for which it is listed. They are chosen initially to be the slack and artificial variables. All remaining variables are called "nonbasic" variables; they are chosen initially to be all the problem variables (here u_1, u_2 and u_3) and will be treated as the current set of independent variables. As we noted above, we set all independent (nonbasic) variables to zero. The dependent (basic) variables of s_1, s_2 and a_1 have a current value equal to their corresponding RHS value, namely, 10, 6 and 4, respectively. Note that the identity matrix appears under the columns for the basic variables. We put a zero into the RHS position for both of the objective function rows F(a) and F.

If we reduce the row F(a) to all zeros below the dependent (basic) variables, then, as we discussed earlier in our section on constrained derivatives and in Fig. 1, the entries below the independent variables will be the constrained derivatives for the independent variables. To place a zero here requires that we multiply row h_1 by 1000 and subtract it from row F(a), getting the following tableau.

	basic	u_1	u_2	u_3	s_1	s_2	a_1	RHS
g_1:	s_1	1	1		1			10
g_2:	s_2			1		1		4
h_1:	a_1		2	-5			1	6
F(a):		2	-1997	-1			0	-6000
F:		2	3	-1			0	0

The value appearing in the RHS column for F(a) and F are the negative of the respective objective function values for the current solution - namely for $s_1 = 10$, $s_2 = 4$, $a_1 = 6$, and $u_1 = u_2 = u_3 = 0$. The constrained derivatives for u_1, u_2 and u_3 are 2, -1997 and -1 respectively. We see that increasing u_2 by one unit will decrease the objective function by 1997. u_2 has the most negative constrained derivative so we choose to "introduce" it into the "basis" - i.e., to make it a dependent (basic) variable. We now need to select which variable to remove from the basis - i.e., return to zero. We examine each row in turn.

Row g_1: We intend to increase u_2 while making the current basis variable, s_1, go from 10 to 0. u_2 will increase to 10/1 = 10. The "1" used here is the coefficient under the column u_2 in row g_1.

Row g_2: A zero in this row under u_2 tells us that u_2 does not appear in this equation. It is, therefore, impossible to reduce the basis variable for that row to zero and have u_2 increase to compensate.

Row h_1: Here, making the basis variable a_1 go to zero will cause u_2 to increase to 6/2 = 3.

If any of the rows had made the trade by requiring u_2 to take a negative value, we skip this row. It is saying u_2 can be introduced to an infinite positive amount without causing the constraint it represents to be violated.

We can introduce u_2 at most to the value of 3, the lesser of the two numbers 10 and 3. If we go past 3, a_1 will go past zero and become negative in the trade. We introduce u_2 into the basis and remove a_1. To put our tableau into standard form, we want the column under variable u_2 to have a one in row h_1 and zeros in all the other rows. We accomplish this by performing an elimination step corresponding to a Gaussian elimination. We first rescale row h_1 so a 1 appears in it under u_2 by dividing it by 2 throughout. We subtract (1) times this row from row g_1 to put a zero in that row, subtract (-1997) times this row from row F(a) and finally (3) times this row from row F, getting the following tableau.

	basic	u_1	u_2	u_3	s_1	s_2	a_1	RHS
g_1:	s_1	1	0	-2.5	1		-0.5	7
g_2:	s_2			1		1		4
h_1:	u_2		1	-2.5			0.5	3
F(a):		2	0	6.5			998.5	-9
F:		2	0	6.5			-1.5	-9

Note that, we have indicated that u_2 is now the basic variable for row h_1. All the artificial variables are now removed from the problem. We switch our attention from objective function row F(a) to row F from this point on in the algorithm. Constrained derivatives appear under the columns for the independent variables u_1, u_3 and a_1. We ignore the constrained derivative for the artificial variable a_1 as it cannot be reintroduced into the problem. It must have a zero value at the final solution Constrained derivatives for u_1 and u_2 in row F are positive indicating that introducing any of these will increase the objective function. We are at a minimum point. The solution is read straight from the tableau: $s_1 = 7$, $s_2 = 4$, $u_2 = 3$, $u_1 = u_3 = a_1 = 0$. The objective function is the negative of the RHS for row F, i.e., F = 9. Since the artificial variable a_1 is zero, the equality constraint is satisfied, and we are really at the solution.

In the example given to illustrate the generalized dual, we in fact developed the dual to a linear program (as may have been evident to the reader at the time). Eqn. (12) gives the dual formulation for our problem, namely

$$\max_{\lambda} \{b^T\lambda \mid A^T\lambda \leq c, \lambda \geq 0\}$$

which for our example problem is

$$\text{Max} \quad -10\,\lambda_1 - 4\,\lambda_2 - 6\,\lambda_3$$

subject to

$$-\lambda_1 \leq 2$$
$$-\lambda_1 - 2\,\lambda_3 \leq 3$$
$$-\lambda_2 + 5\,\lambda_3 \leq -1$$
$$\lambda_1, \lambda_2 \geq 0$$

The third Lagrange multiplier, λ_3, does not have to be positive as it corresponds to the equality constraint. To convert this problem to a linear program where all variables must be positive, we split λ_3 into two parts, a positive part and a negative part. We can also add in slack variables σ_1 to σ_3 at the same time and write the following equivalent optimization problem.

$$\text{Max} \quad -10\,\lambda_1 - 4\,\lambda_2 + 6\,(\lambda_3^+ - \lambda_3^-)$$

subject to

$$-\lambda_1 + \sigma_1 = 2$$
$$-\lambda_1 + 2\,(\lambda_3^+ - \lambda_3^-) + \sigma_2 = 3$$
$$-\lambda_2 - 5\,(\lambda_3^+ - \lambda_3^-) + \sigma_3 = -1$$
$$\lambda_1, \lambda_2, \lambda_3^+, \lambda_3^-, \sigma_1, \sigma_2, \sigma_3 \geq 0$$

One can show that the values for the constrained derivatives in the final primal tableau provide us with the solution to the corresponding dual problem, namely: $\sigma_1 = 2$, $\sigma_2 = 0$, $\sigma_3 =$

6.5, $\lambda_1 = 0$, $\lambda_2 = 0$ and $\lambda_3 = -1.5$ (i.e., $\lambda_3^+ = 0$, $\lambda_3^- = 1.5$). The numbers are in the objective function row F in the order given here. The reader should verify that this is the solution.

The correspondence is seen as follows. The first column of the original tableau gives us the first equation in the dual; its slack σ_1 corresponds to this column. A similar observation holds for columns 2 and 3. The multipliers λ_1 to λ_3 are the Lagrange multipliers for the equations in the primal problem. Their values are the constrained derivatives for the slack variables for the constraints for primal problem. So, for example, λ_1 appears under the column for s_1.

Interior point algorithms for Linear Programming Problems

There has been considerable excitement in the popular press about so-called interior point algorithms [23] for solving extremely large linear programming problems. Computational demands for these algorithms grow less rapidly than for the Simplex algorithm, with a break-even point being a few thousand constraints. We shall base the presentation in this section on a recent article by MacDonald and Hrymak [28] which itself is based on ideas in references [1,21].

A key idea for an interior method is that one heads across the feasible region to locate the solution rather than around its edges as one does for the Simplex algorithm. This move is found by computing the direction of steepest descent for the objective function with respect to changing the slack variables. Variables u are computed in terms of the slack variables by using the inequality constraints. The direction of steepest descent is a function of the scaling of the variables used for the problem.

A really clever and second key idea for interior point algorithms is the way the problem is scaled. At the start of each iteration one rescales the space one searches so that current point is a unit distance from the constraint boundaries. In general one must rescale all the variables. The algorithm has to guarantee that it will never quite reach the constraint boundaries at the end of an iteration so that one can then rescale the variables at the start of the next to be one unit from the boundary. A zero distance cannot be rescaled to be a unit distance.

One terminates when changes in the unscaled variables become negligible from one iteration to the next. The distances to the constraints which finally define the optimal solution have all been magnified significantly. One will really be close to them.

Consider the following linear program.

$$\underset{u,s}{\text{Min}} \ \{F = c^T u \mid b + s - Au = 0, \ u, \ s \geq 0\}$$

where slack variables have been introduced into the problem to convert inequalities into equalities. Assume matrix A has p rows and r columns and that there are more constraints than there are variables u, i.e., $p \geq r$.

Assume we are at some point in the interior of the feasible region, (u(k), s(k)). Let D_s and D_u be diagonal matrices whose i-th diagonal elements are $s_i(k)$ and $u_i(k)$ respectively. Define rescaled variable vectors as

$$s' = D_s^{-1} s \ \text{ and } u' = D_u^{-1} u$$

and replace the variables in the constraints by these rescaled variables, getting

$$D_s^{-1} b + s' - D_s^{-1} A D_u u' = 0$$

We wish to solve these equations for u' in terms of s'; however, the coefficient matrix $D_s A D_u$ is of dimension mxr. We must first make it square which we do by premultiplying by its transpose (rxp x pxr --> rxr where $p \geq r$ has been assumed above) to get

$$-D_u A^T D_s^{-1}(D_s^{-1} b + s') + (D_u A^T D_s^{-2} A D_u) u' = 0$$

These equations can now be solved, giving

$$u' = (D_u A^T D_s^{-2} A D_u)^{-1} D_u A^T D_s^{-1}(D_s^{-1} b + s')$$

We can now determine the gradient of the objective function, $F = c^T u$, with respect to the rescaled slack variables s', getting

$$\nabla_{s'} F = (\nabla_{s'} u')^T c = D_s^{-1} A D_u (D_u A^T D_s^{-2} A D_u)^{-1} c$$

The direction of steepest descent for changing the rescaled slack variables is the negative of this gradient direction. We let the step in s' be in this direction. The corresponding steps for u' and the unscaled variables s and u follow directly.

$$\Delta s' = -D_s^{-1} A D_u (D_u A^T D_s^{-2} A D_u)^{-1} c$$
$$\Delta u' = -(D_u A^T D_s^{-2} A D_u)^{-1} c$$
$$\Delta s = -A D_u (D_u A^T D_s^{-2} A D_u)^{-1} c$$
$$\Delta u = -D_u (D_u A^T D_s^{-2} A D_u)^{-1} c$$

The above defines the direction to move. We want to move close to but not exactly onto the edge of the feasible region. We encounter the edge when one of the variables in s or u becomes zero while the rest stay positive while taking our step. The variable with the most negative value for $s_i/\Delta s_i$ or $u_i/\Delta u_i$, as appropriate, is the one that will hit the edge first. One typically takes a step that goes more than 99% but not 100% of the way toward the edge.

The last issue to settle is how to get an initial point which is strictly inside the feasible region. By introducing slack variables and artificial variables as we did above for our linear programming example, we can pick a point which is feasible but on the constraint boundary. Pick such a point but then make all the variables which are normally set to zero just slightly positive.

All the work in this algorithm is in factoring the matrix $(D_u A^T D_s^{-2} A D_u)$ which is far less sparse that the original coefficient matrix A. Because the algorithm is useful only when the problems get really large, one must use superb numerical methods to do these computations.

MacDonald and Hrymak show that the Karmarker algorithm (closely related to the above algorithm) is a special case of the Newton Barrier Method [18]. The constraints stating the nonnegativity of the problem variables u and the slack variables s are replaced by adding a term to the objective function that grows to infinity as any one of these variables decreases to zero:

$$F = c^T u - \mu \left(\sum_{i=1}^{r} \ln u_i + \sum_{j=1}^{p} \ln s_j \right)$$

where μ is a positive fixed scalar. One needs a feasible starting point strictly interior to the feasible region.

MacDonald and Hrymak also discuss a method that attacks the problem from both the primal and dual formulations simultaneously. Both forms are written with slack variables where the nonnegativity for all the variables is maintained by erecting barrier functions. The necessary conditions for optimality are stated separately for both the primal and dual with their respective barrier functions in place. The necessary conditions are then combined. Solving them computes a direction in which to move.

Benders' Decomposition

One way to solve the following problem

$$\min_{u,y} \{F(u,y) \mid g(u,y) \le 0,\ h(u,y) = 0\}$$

is to use an approach called projection. Projection breaks the problem into an outer optimization which contains an inner one, such as the following for the above problem

$$\min_{y} \{ \min_{u} \{F(u,y) \mid g(u,y) \le 0,\ h(u,y) = 0\}\}$$

There is no advantage to using projection unless the inner problem possesses a special structure which is very easy to solve. For example the inner problem might be a linear program while the overall problem is a nonlinear problem. Benders [2] presented a method for precisely this case which Geoffrion subsequently generalized [15]. We shall look at the problem Benders presented here, using ideas presented earlier in this paper. Benders' problem was the following

$$\min_{u,y} \{ c^T u + F(y) \mid Au + f(y) \ge b,\ u \ge 0,\ y \in S \}$$

which, when projected, yields the following two-level optimization problem

$$\min_{y} \{ F(y) + \min_{u} \{ c^T u \mid Au \ \ge b - f(y),\ u \ge 0 \} \mid y \in S \}$$

With y fixed the inner problem is a linear program. We can replace the inner problem with its equivalent dual representation

$$\min_{y} \{ F(y) + \max_{\lambda} \{(b - f(y))^T \lambda \mid A^T \lambda \le c,\ \lambda \ge 0 \},\ y \in S \}$$

Solving either form gives us a solution for the original problem; however, there are some advantages to this latter form. The solution to the dual problem occurs at one of the corner points for the feasible region. The feasible region is defined by the constraints $A^T\lambda \le c$, $\lambda \ge 0$, and these constraints do not contain the outer problem variables y so the corner points are *not* a function of y.

If we are given a value for y and if we had a list of all the corner points $\lambda(1), \lambda(2)$ ..., we could solve the inner problem by picking that corner point having the maximum value for its corresponding objective function $(b - f(y))^T \lambda(k)$. If we know only some of the corner points, we can search over just those for the best one, hoping the optimal solution for the given value of y will reside there. Not having all the points, the value found will be less than or at most equal to the maximum; it will provide an lower bound on the maximum. We can subsequently solve the inner problem given y. If we obtain the same corner point, then the inner problem was exactly solved using only the known subset of the corner points. If we do not, then we can add the new corner point to the list of those found so far.

The feasible region may be unbounded for the dual problem which occurs when the primal inner problem is infeasible for the chosen value of y. The directions in which the inner dual problem are unbounded are not a function of y, as we established above. To understand the geometry of an unbounded problem, consider Fig. 9.

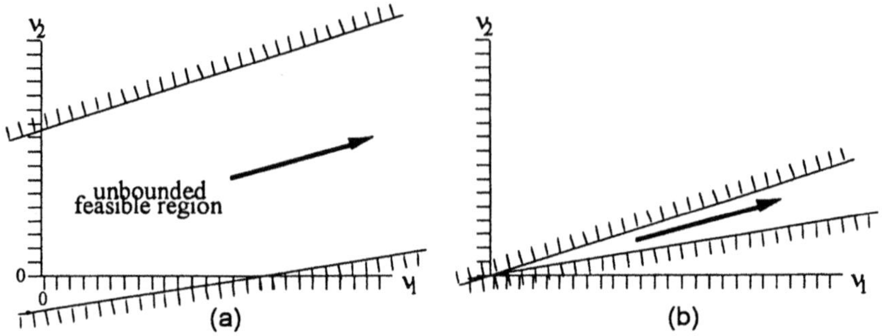

Fig. 9 Unbounded feasible region for the dual. (a) shows the constraints for the original dual problem. (b) shows the constraints shifted to pass through zero.

We shift all the constraints to pass through the origin. Then any convex combination of the constraints bounding the feasible region in (b) represents a direction in which the problem in (a) is unbounded. Any v which satisfies the following set of constraints is an unbounded direction:

$$\left\{A^T v \le 0, v \ge 0, \sum v_i = 1\right\}$$

We can find it by reposing the dual problem as follows: (1) zero the coefficients of the objective function, (2) zero the right-hand-side of the inequality constraints and (3) add in the constraint that the multipliers add to unity.

Such an unbounded direction imposes a constraint on y, and these constraints are called cutting planes. To preclude an infinite solution for the dual - and therefore an infeasible solution for the primal inner problem and thus the problem as a whole, we can state that the inner dual objective function must not increase in this unbounded direction, namely

$$(b - f(y))^T v(k) \le 0$$

We add this constraint to our problem statement. It is a constraint on the value of y, the variables for the outer optimization problem. The inner problem has fed back a constraint to the outer problem on the values of y it should consider when optimizing over them.

The algorithm is as follows.

1. Set the iteration counter k=0. Define empty sets C (for corner points) and U (for unbounded directions).

2. Solve the following optimization problem for y (if set U is empty, set δ equal to zero)

$$\min_{y} \{F(y) + \delta \mid \delta \geq [b-f(y)]^T \lambda(i) \mid \text{for all } \lambda(i) \in C\},\ y \in S,\ [b-f(y)]^T v(j) \leq 0 \text{ for all } v(j) \in U\}$$

This problem tells us to find the value of y that minimizes the sum of F(y) and the maximum of the inner problem objective over all the corner points found so far, subject to y being in the set S and satisfying all the cutting plane constraints found so far. The sets C and U are initially empty so the first time through, one initially finds a value of y that minimizes F(y) subject to y being in the set S. Exit if there is no value of y satisfying all the constraints, indicating that the original problem is infeasible.

3. For this value of y, solve the inner problem

$$\max_{\lambda} \{(b - f(y))^T \lambda \mid A^T \lambda \leq c,\ \lambda \geq 0\}$$

which will give rise to a new corner point $\lambda(k)$ or, if the problem is unbounded, a direction $v(k)$. Place whichever is found in the set C or U as appropriate.

4. If the solution in step 3 is bounded and has the same value for the inner objective as found in step 2, exit with the solution to the original problem. Else increment the iteration counter and return to step 2.

The problem in step 2 grows with each iteration. Thus, these problems can get large.

Geoffrion [15] has generalized Benders' method. The generalized method is used to solve mixed integer programming problems, for example, where the inner problem variables are required to take on values of zero and one only. Such use is indicated in the papers by Grossmann at this Advanced Study Institute.

References

1. Alder, H, N. Karmarker, M. Resende and G. Veigo, "Implementation of Karmarkar's Algorithm for Linear Programming," Mathematical Programming, 44, 297-335 (1989)
2. Benders, J.F., "Partitioning Procedures for Solving Mixed-variables Programming Problems," Numerische Mathematik, 4, 238 (1962)
3. Berna, T.J., Locke, M.H., and A.W. Westerberg, "A New Approach to Optimization of Chemical Processes," AIChE J, 26, 37-43 (1980)
4. Biegler, L.T., I.E. Grossmann, and A.W. Westerberg, "Optimization," in Ullmann's Encyclopedia of Industrial Chemistry, B1, Mathematics in Chemical Engineering, Chapt. 10, 1-106 to 1-128, Weinheim:VCH Verlagsgesellschaft 1990
5. Broyden, C.G., "The Convergence of a Class of Double Rank Minimization Algorithms," J. Inst. Math. Applic., 6, 76 (1970)
6. Bryson, A.E. and Y-C Ho, *Applied Optimal Control*, Washington D. C.:Hemisphere Publishing 1975
7. Chamberlain, R.M., C. Lemarechal, H.C. Pedersen, and M.J.D. Powell, "The Watchdog Technique for Forcing Convergence in Algorithms for Constrained Optimization," Math. Prog. Study 16, Amsterdam:North Holland 1982

8. Dantzig, G., "Linear Programming and Extensions", Princeton:Princeton University Press 1963
9. Davidon, W.C., "Variable Metric Methods for Minimization," AEC R&D Report ANL-5990 (rev.) (1959)
10. Dennis, J.E. and J.J. More,"Quasi-Newton Methods, Motivation and Theory," SIAM Review, 21, 443 (1977)
11. Fletcher, R. and Powell, M.J.D., "A Rapidly Converging Descent Method for Minimization," Computer J., 6, 163 (1963)
12. Fletcher, R., "A New Approach to Variable Metric Algorithms," Computer J., 13, 317 (1970)
13. Fletcher, R., *Practical Methods of Optimization*, New York:Wiley 1987
14 Fourer, R., D.M. Gay and B.W. Kerninghan, "A Modeling Lanugage for Mathematical Programming," Management Science, 36(5), 519-554 (1990).
15. Geoffrion, A.M., "Generalized Benders Decomposition", JOTA, 10, 237 (1972)
16. Gill, P. and W. Murray, "Numerically Stable Methods for Quadratic Programming," Math. Prog., 14, 349 (1978)
17. Gill, P.E., W. Murray and M. Wright, *Practical Optimization*, New York:Academic Press 1981
18. Gill, P.E., W. Murray, M. Saunder, J. Tomlin and M. Wright, "On Projective Newton Barrier Methods for Linear Programming and as Equivalence to Caretakers Projective Method," Math. Prog., 36, 183, (1984)
19. Goldfarb, D., "A Family of Variable Metric Methods Derived by Variational Means," Math. Comp., 24, 23 (1970)
20. Goldfarb, D. and A. Idnani, "A Numerically Stable Dual Method for Solving Strictly Convex Quadratic Programs," Math. Prog., 27, 1 (1983)
21. Goldfarb, D. and M.J. Todd, "Linear Programming", Chapter II in Optimization (eds. G.L. Nemhauser, A.H.G. Rinnoy Kan and M.J. Todd), Amsterdam:North Holland 1989
22. Han, S-P, "A Globally Convergent Method for Nonlinear Programming," J. Opt. Theo. Applics., 22, 297 (1977)
23. Karmarker, N., "A New Polynomial-Time Algorithm for Linear Programming," Combinatorica, 4, 373-395 (1984)
24. Kuhn, H.W., and Tucker, A.W., "Nonlinear Programming," in Neyman, J. (ed), *Proc. Second Berkeley Symp. Mathematical Statistics and Probability*, 402-411, Berkeley, CA:Univ. California Press 1951
25. Lemke, C. E., "A Method of Solution for Quadratic Programs," Management Science, 8, 442 (1962)
26. Locke, M.H., A.W. Westerberg and R.H. Edahl, "An Improved Successive Quadratic Programming Optimization Algorithm for Engineering Design Problems," AIChE J, Vol. 29, pp 871-874 (1983)
27. Lucia, A. and A. Kumar, "Distillation Optimization," Comp. Chem. Engr., 12, 12, 1263 (1988)
28. MacDonald, W.B., A.N. Hrymak and S. Treiber, "Interior Point Algorithms for Refinery Scheduling Problems," in Proc. 4th Annual Symp. Process Systems Engineering, Montibello, Quebec, Canada, pp III.13.1-16, Aug. 5-9, 1991
29. Nocedal, Jorge and Michael L. Overton, "Projected Hessian Updating Algorithms for Nonlinearly Constrained Optimization", *SIAM J. Numer. Anal.*, **22, 5** (1985)
30. Powell, M.J.D., "A Fast Algorithm for Nonlinearly Constrained Optimization Calculations," Lecture Notes in Mathematics 630, Berlin:Springer Verlag 1977
31. Reklaitis, G.V., A. Ravindran, and K.M. Ragsdell, Engineering Optimization Methods and Applications, New York:Wiley 1983
32. Schittkowski, K., "The Nonlinear Programming Algorithm of Wilson, Han and Powell with an Augmented Lagrangian Type Line Search Function," Num. Math., 38, 83 (1982)
33. Schmid, C. and L.T. Biegler, "Acceleration of Reduced Hessian Methods for Large-scale Nonlinear Programming," paper presented at AIChE Annual Meeting, Los Angeles, CA, Nov. 1991
34. Shanno, D.F., "Conditioning of Quasi-Newton methods for function minimization," Math. Comp., Co 24, 647 (1970)
35. Umeda, T. and A. Ichikawa, "I&EC Proc. Design Develop., 10, 229 (1971)
36. Vasantharajan, S. and L.T. Biegler, "Large-Scale Decomposition Strategies for Successive Quadratic Programming," *Computers and Chemical Engineering*, **12, 11**, 1087 (1988)
37. Vasantharajan, S., J. Viswanathan and L.T. Biegler, "Large Scale Development of Reduced Successive Quadratic Programming," presented at CORS/TIMS/ORSA Meeting, Vancouver, BC, May, 1989
38. Wolfe, P., "The Simplex Method for Quadratic Programming," Econometrica, 27, 3, 382 (1959)

Mixed-Integer Optimization Techniques for the Design and Scheduling of Batch Processes

Ignacio E. Grossmann, Ignacio Quesada, Ramesh Raman and Vasilios T. Voudouris

Department of Chemical Engineering and Engineering Design Research Center, Carnegie Mellon University, Pittsburgh, PA 15213, U.S.A.

Abstract: This paper provides a general overview of mixed-integer optimization techniques that are relevant for the design and scheduling of batch processes. A brief review of the recent application of these techniques in batch processing is first presented. The paper then concentrates on general purpose methods for mixed-integer linear (MILP) and mixed-integer nonlinear programming (MINLP) problems. Basic solution methods as well as recent developments are presented. A discussion on modelling and reformulation is also given to highlight the importance of this aspect in mixed-integer programming. Finally, several examples are presented in various areas of application to illustrate the performance of various methods.

Keywords: mathematical programming, mixed-integer linear programming, mixed-integer nonlinear programming, branch and bound, nonconvex optimization, reformulation techniques, batch design and scheduling

Introduction

The design, planning and scheduling of batch processes is a very fertile area for the application of mixed-integer programming techniques. The reason for this is that most of the mathematical optimization models that arise in these problems involve both discrete and continuous variables that must satisfy a set of equality and inequality constraints, and that must be chosen so as to optimize a given objective function. While there has been the recognition that many batch processing problems can be posed as mixed-integer optimization problems, the more extensive application of these techniques has only taken place in the recent past.

It is the purpose of this paper to provide an overview of mixed-integer optimization techniques. We will first present a brief review of the application of these techniques in batch processing. We then provide a brief introduction to mixed-integer programming in order to determine a general classification of major problem types. Next we concentrate in both mixed-integer linear (MILP) and mixed-integer nonlinear programming (MINLP) techniques, introducing first the basic methods and then the recent developments that have taken place. We then present a discussion on modelling and reformulation, and finally, some numerical examples and results in various areas of application.

Review of applications

In this section we will present a brief overview of the application of mixed-integer programming in batch processing. More extensive reviews can be found in [55] and [66,67].

Mixed-integer nonlinear programming techniques have been applied mostly to design problems. Based on the problem considered by Sparrow et al [81], Grossmann and Sargent [28] were the first to formally model the design of multiproduct batch plants with parallel units and with single product campaigns as an MINLP problem. These authors showed that if one relaxes the numbers of parallel units to be continuous, the associated NLP corresponds to a geometric program that has a unique solution. Rather than solving the problem directly as an MINLP, the authors proposed a heuristic rounding scheme for the number of parallel units using nonlinear constraints based on the solution of the relaxed NLP. Since this problem provides a valid lower bound to the cost, optimality was established within the deviation of the rounded solution. This MINLP model was subsequently extended by Knopf et al [36] in order to handle semi-continuous units. A further extension, was the MINLP model for a special type of multipurpose plants by Suhami and Mah [83] in which simultaneous production was only allowed if products did not require the same processing stages. This model was subsequently modified by Vaselenak et al [90] and by Faqir and Karimi [19] to embed the selection of production campaigns. However, all these works did not rigorously solve the MINLP, but they relied on the rounding scheme by Grossmann and Sargent [28] for obtaining an integer number of parallel units.

The first design application in which an MINLP model was rigorously solved was [90] who considered the retrofit design of multiproduct batch plants. These authors applied the outer-approximation method by Duran and Grossmann [18] with a modification to handle separable nonconvex terms in the objective. Recently, [20] removed the assumptions of equal volume for units operating out of phase by Vaselenak et al [90], and formulated a new MINLP model that again was solved by the outer-approximation method. Also, [38] formulated the MINLP model by Grossmann and Sargent [28] in terms of 0-1 variables for the parallel units and solved it rigorously with the outer-approximation method as implemented in DICOPT. Subsequently, [96] applied this computer code to an MINLP model for multiproduct plants under uncertainty with staged expansions.

An important limitation in all the above applications was that convexity of the relaxed MINLP problem was a major requirement. Also, it became apparent that the solution of larger design problems could become expensive. The first difficulty was partially circumvented with the augmented penalty version of the outer-approximation algorithm proposed by Viswanathan and Grossmann [92] and which was implemented in the computer code DICOPT++. This code was applied by Birewar and Grossmann [9] for the simultaneous synthesis, sizing and scheduling of multiproduct batch plants which gives rise to a nonconvex MINLP model.

Papageorgaki and Reklaitis [53,54] developed a comprehensive MINLP model for the design of multipurpose batch plants which involved nonconvex terms. They found that the code DICOPT would get trapped into suboptimal solutions and that the computation time was high. For this reason they proposed a special decomposition method in which the subproblems are NLP's with fixed 0-1 variables and campaign lengths and the master problem corresponds to a simplified MILP. Faqir and Karimi [19] also modelled a special class of multipurpose batch plants with multiple production routes and discrete sizes as an MINLP problem that involves nonconvexities in the form of bilinear constraints. These authors proposed valid underestimators for these constraints and reduced the design problem to a sequence of MILP problems. Recently, [93] have shown that several batch design problems, convex and nonconvex, can in fact be reformulated as MILP problems when they involve discrete sizes. Examples include the design of multiproduct batch plants with single product campaigns and the design of multipurpose batch plants with multiple routes. These authors have also developed a comprehensive MILP synthesis model for multiproduct plants in which the cost of inventories is accounted for [95]. Finally, [65] have reported computational experience in solving a variety of batch design problems as MINLP problems using the computer code DICOPT++, while [82] have applied it in the optimization of flexibility of multiproduct batch plants.

As for scheduling and planning, there have been a large number of MILP models reported in the Operations Research literature. However, in chemical engineering the first major MILP model for batch scheduling was proposed by [68] for the case of multipurpose batch plants in which the products were preassigned to processing units. They used the computer code LINDO [74] to solve this problem, and later extended it to handle the production of a product over several predefined sets of units [69]. Ku and Karimi [42] developed an MILP model for selecting production sequences that minimize the makespan in multiproduct batch plants with one unit per stage. Their model, which can accomodate a variety of storage policies, was also solved with the computer code LINDO.

A very general approach to the scheduling of batch operations was proposed by Kondili et al [40] in which they developed a state-task network representation to model batch operations with complex process network structures. By discretizing the time domain they posed their problem as a multiperiod MILP model that has the flexibility of accomodating variable batch sizes, splitting and mixing of batches, finite, unlimited or no storage, various transfer policies and resource constraints. Furthermore, the model has the flexibility of assigning equipment to different tasks. Recently, [77] have been able to considerably tighten the LP relaxation for this problem and develop a special purpose branch and bound method with which these authors have been able to solve problems with more than one thousand 0-1 variables. These authors have also extended their MILP model to some design and planning problems [76].

For the case of the no-wait flowshop scheduling problem [49] (see also [57]) formulated the problem as an asymmetric traveling salesman problem (see [30]). For this model they developed a parallel .branch and bound method that was coupled with a matching algorithm for detecting Hamiltonian cycles. The specialized implementation of their algorithm has allowed them to solve problems to optimality with more than 10,000 batches, which effectively translates to problems with more than 20,000 constraints and 100,000,000 0-1 variables.

Finally, MINLP models for scheduling of multipurpose batch plants have been formulated by Wellons and Reklaitis [97] to handle flexible allocation of equipment and campaign formations. Due to the large size of these problems, these authors developed a special decomposition strategy for their solution. Sahinidis and Grossmann [70] considered the cyclic scheduling of continuous multiproduct plants with parallel lines and formulated the problem as a large-scale MINLP problem. They developed a solution method based on Generalized Benders decomposition for which they were able to solve problems with up to 800 0-1 variables, 23,000 continuous variables and 3000 constraints.

In summary, what this brief review shows is that both MILP and MINLP techniques are playing an increasingly important role in the modelling and solution of batch processing problems. This review also shows the importance of exploiting the structure of these problems for developing reasonably efficient solution methods. It should also be mentioned that while there might be the temptation to resort to simpler optimization approaches such as simulated annealing, mixed integer programming provides a rigorous and deterministic framework, although it is not always the easiest one to apply. On the other hand, many mixed-integer problems that were regarded as unsolvable 10 years ago are currently being solved to optimality with reasonable computing requirements due to advances in algorithms and increased computer power.

Mixed-integer Programming

In its most general form a mixed-integer program corresponds to the optimization problem,

$$\begin{aligned} \min \quad & Z = f(x,y) && \text{(MIP)} \\ \text{s.t.} \quad & h(x,y) = 0 \\ & g(x,y) \leq 0 \\ & x \in R^n \quad y \in N_+^m \end{aligned}$$

in which x is a vector of continuous variables and y is a vector of integer variables. The above problem (MIP) specializes to the two following cases:

I. *Mixed-integer linear programming (MILP).* The objective function f, and the constraints h and g are linear in x and y in this case. Furthermore, most of the applications of interest are restricted to the case when the integer variables y are binary, i.e. $y \in \{0,1\}^m$. A number of important classes of problems include the pure integer linear programming problem (only integer variables) and a large number of specialized combinatorial optimization problems that include for instance the assignment, knapsack, matching, covering, facility location, networks with fixed charges and traveling salesman problems (see [51]).

II. *Mixed integer nonlinear programming (MINLP).* The objective function and/or constraints are nonlinear in this case. The most common form is linear in the integer variables and nonlinear in the continuous variables [27]. More specialized forms include polynomial 0-1 programs and 0-1 multilinear programs which can be transformed into MILP problems (eg see [6]). The difficulty that arises in the solution of MILP and MINLP problems is that due to the combinatorial nature of these problems, there are no optimality conditions like in the continuous case that can be directly exploited for developing efficient solution methods. In this paper we will concentrate on the modelling and solution of unstructured MILP problems, and MINLP problems that are linear in the 0-1 variables. Both types of problems correspond to the more general type of mixed-integer optimization problems that arise in batch processing. It is very important however, to recognize that if the model has a more specialized structure, general purpose techniques will be inefficient for solving large scale version of these problems, and specialized combinational optimization algorithms should be used in this case.

Mixed-integer Linear Programming (MILP)

We will assume the more common case in which the subset of the integer variables y are restricted to take only 0 or 1 values. This then gives rise to the MILP problem:

$$\begin{aligned} \min Z = \ & c^T x + b^T y \qquad\qquad \text{(MILP)} \\ \text{s.t.} \ & Ax + By \leq d \\ & x \geq 0 \,,\ y \in \{0,1\}^m \end{aligned}$$

In attempting to develop a solution method to solve problem (MILP), the first obvious alternative would be to solve for every combination of 0-1 variables the corresponding LP problem in terms of the variables x, and then pick as the solution the 0-1 combination with lowest objective function. The major drawback with such an approach is that the number of 0-1 combinations is exponential. For example, an MILP problem with 10 0-1 variables would require the solution of $2^{10} = 1024$ LPs, while a problem with 50 0-1 variables would require the solution at $2^{50} = 1.13\text{x}10^{15}$ LPs! Thus, this approach is, in general, computationally infeasible.

A second alternative is to relax the 0-1 requirements and treat the variables y as continuous with bounds, $0 \leq y \leq 1$. The problem with such an approach, however, is that except for few special cases (e.g. assignment problem), there is no guarantee that the variables y will take integer values at the relaxed LP solution. As an example, consider the pure integer programming problem,

$$\begin{aligned} \min Z &= -1.2y_1 - y_2 \\ \text{s.t.} \quad & 1.2y_1 + 0.5y_2 \leq 1 \\ & y_1 + y_2 \leq 1 \\ & y_1, y_2 = 0,1 \end{aligned} \tag{1}$$

By relaxing y_1 and y_2 to be continuous the solution yields the noninteger point y_1=0.715, y_2=0.285, Z= -1.143. Assume we simply round the variables to the nearest integer value, namely $y_1 = 1$, $y_2 = 0$. This, however, is an infeasible solution as it violates the first constraint. In fact, the optimal solution is $y_1 = 0$, $y_2 = 1$, $Z = -1$. Thus, solving the MILP problem by relaxation of the y variables and rounding them to the nearest integer, will in general not lead to the correct solution. Note, however, that the relaxed LP has the property that its optimal objective value provides a lower bound to the integer solution.

In order to obtain a rigorous solution to the problem (MILP) the most common approach is the branch and bound method which originally was proposed by Land and Doig [44] and later formalized by Dakin [16]. In the branch and bound technique the objective is to perform an enumeration without having to examine all the 0-1 combinations. The basic idea is first to represent all the 0-1 combinations through a binary tree such as the example shown in Fig. 1. Here at each node of the tree the solution of the linear program subject to integer constraints for the subset of the y variables that are fixed in previous branches is considered. For example, in node A the root of the tree involves the solution of the relaxed LP, while node B involves the solution of the LP with fixed $y_1 = 0$, $y_2 = 1$ and with $0 \leq y_3 \leq 1$.

In order to avoid the enumeration of all the nodes in the binary tree, we can exploit the following basic properties. Let k denote a descendent node of node l in the tree (e.g. k=B, l = A) and let (P^k) and (P^l) denote the corresponding LP subproblems. Then the following properties can be easily established:

1. If (P^l) is infeasible then (P^k) is also infeasible.

2. If (P^k) is feasible then (P^l) is also feasible, and $(Z^l)^* \leq (Z^k)^*$. That is, the optimal objective of subproblem (P^l) corresponds to a lower bound of the optimal objective at subproblem (P^k).

3. If the optimal solution of subproblem (P^k) is such that y = 0 or 1, then $(Z^k)^* \geq Z^*$. That is, the optimal objective of subproblem (P^k) corresponds to an upper bound of Z^*, the optimal MILP solution.

The above properties can be used to fathom nodes in the tree within an enumeration procedure. The question of how to actually enumerate the tree involves the use of branching rules. Firstly, one does not necessarily have to follow the order of the index of the variables y for branching as might be implied in Fig. 1. A simple alternative is to branch instead on the 0-1 variable that is closest to 0.5. Alternatively, one can specify a priority for the 0-1 variables, or else use a more sophisticated scheme that is based on the use of penalties [17, 86]. Secondly, one has to decide as to what node should be examined next having solved the LP at a given node in the tree. Here the two major alternatives are to use a depth-first (last in-first out) or a breadth-first (best second rule) enumeration. In the former case one of the branches of the most recent node is expanded first; if all of them have been examined we backtrack to another node. In the

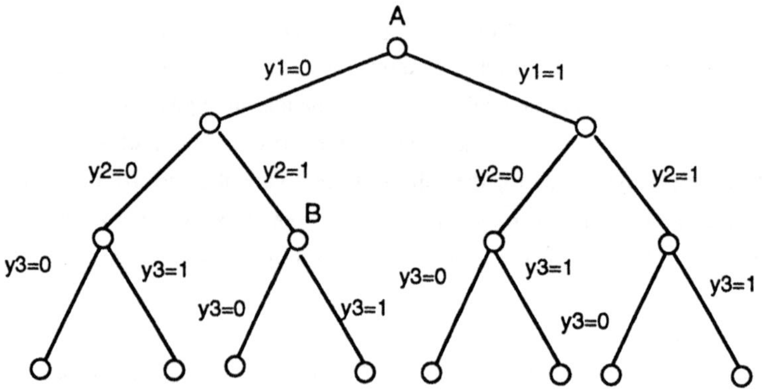

Fig. 1 Binary tree representation for three 0-1 variables

latter case the two branches of the node with the lowest bound are expanded successively; in this case no backtracking is required. While the depth-first enumeration requires less storage, the breadth-first enumeration requires in general an examination of fewer nodes. In practice the most common scheme is to use depth first, but by branching on both the 0 and 1 values of a binary variable at each node.

In summary, the branch and bound method consists in first solving the relaxed LP problem. If y takes integer values we stop. Otherwise we proceed to enumerate the nodes in the tree according to some specified branching rules. At each node the corresponding LP subproblem is solved, typically by updating the dual LP problem of the previous node which requires few pivot operations. By then making use of the properties cited before, we either fathom the node (if infeasible or if lower bound ≥ upper bound) or keep it open for further examination. Clearly the computational efficiency is largely dependent on the quality of the lower bounds of the LP subproblems.

As an example, consider the following MILP problem involving one continuous variable and three 0-1 variables:

$$\begin{aligned} \min Z &= x + y_1 + 3y_2 + 2y_3 \\ \text{s.t.} \quad & -x + 3y_1 + 2y_2 + y_3 \leq 0 \\ & -5y_1 - 8y_2 - 3y_3 \leq -9 \\ & x \geq 0,\ y_1, y_2, y_3 = 0,1 \end{aligned} \tag{2}$$

The branch and bound tree using a breadth-first enumeration is shown in Fig. 2. The number in the circles represents the order in which 9 nodes out of the 15 nodes in the tree are examined to find the optimum. Note that the relaxed solution (node 1) has a lower bound of Z = 5.8, and that the optimum is found in node 9 where Z = 8, y_1=0, y_2=y_3=1, *and x=3.*

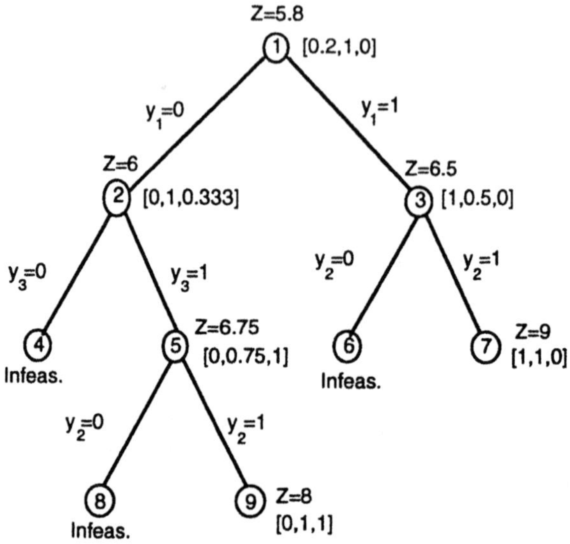

Fig. 2 Branch and bound tree for example problem (2)

The branch and bound method is currently the most common method used for MILP in both academic and commercial computer software (eg. LINDO, ZOOM, SCICONIC, OSL, CPLEX, XA). Some of these codes feature a number of special features that can help to reduce the enumeration in the tree search. Perhaps one of the most noteworthy are the generalized upper bound constraints [7] which are integer constraints of the form,

$$\sum_{i \in I} y_i = 1 \tag{3}$$

In this case instead of performing branching on individual variables, the branching is performed by partitioning the variables into two subsets (commonly of equal size). As a simple example consider the problem:

$$\begin{aligned} \text{min. } & Z = y_1 + 2y_2 + 3y_3 + 4y_4 \\ \text{s.t. } & y_1 + y_2 - y_3 - y_4 \leq 0 \\ & y_1 + y_2 + y_3 + y_4 = 1 \\ & y_i = 0,1 \quad i = 1, 4 \end{aligned} \tag{4}$$

The relaxed LP solution of this problem is $Z = 2$, $y_1 = y_3 = 0.5$, $y_2 = y_4 = 0$. If a standard branch and bound search is performed, 4 nodes are required for the enumeration as shown in Fig. 3a. However, if we instead treat the last constraint as a generalized upper bound constraint, only two nodes are enumerated as shown in Fig. 3b.

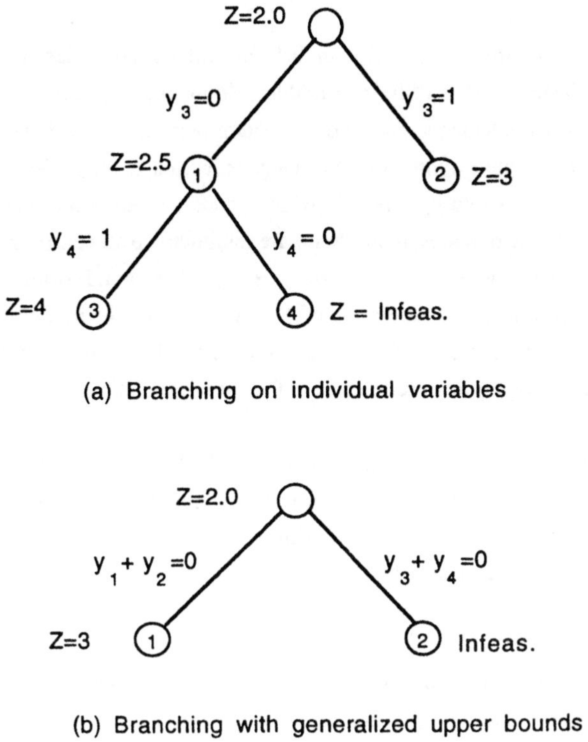

Fig. 3 Standard branching rule and generalized upper bounds

Closely related to the generalized upper bound constraints, are the special ordered sets (see [7, 87]). The most common are the SOS1 constraints that have the form,

$$\sum_{i \in I} y_i = 1, \qquad x = \sum_{i \in I} \alpha_i y_i \tag{5}$$

in which the second constraint is denoted as a reference row where x is a variable and α_i are constants with increasing value. In this case the partitioning of the 0-1 variables at each node is performed according to the placement of the value of the continuous variable x relative to the points α_i. SOS2 constraints are those in which exactly two adjacent 0-1 variables must be nonzero, and they are commonly used to model piecewise linear concave functions. Again, considerable reductions in the enumeration can be achieved with these types of constraints.

Another important capability in branch and bound codes are preprocessing techniques that have the effect of fixing variables, eliminating redundant constraints, adding logical inequalities, tightening variable bounds and/or performing coefficient reduction (see [12, 15, 47]). A simple example of coefficient reduction is for instance converting the inequality $2y_1 + y_2 \geq 1$ into $y_1 + y_2 \geq 1$ which yields a tighter representation in the 0-1 polytope. An example of a logical constraint are minimum cover constraints. For instance, given the constraint $3y_1 + 2y_2 + 4y_3 \leq 6$, $y_1 + y_3 \leq 1$ is a minimum cover since it eliminates the simultaneous selection of $y_1 = y_3 = 1$ which violates this constraint. Preprocessing techniques can often reduce the integrality gap of an MILP but their application is not always guaranteed to reduce the computation time.

Although the LP based branch and bound method is the dominant method for MILP optimization, there are other solution approaches which often complement this method. These can be broadly classified into three major types: cutting plane methods, decomposition methods and logic based methods. Only a very brief overview will be given for these methods.

The basic idea of the original cutting plane methods was to solve a sequence of successively tighter linear programming problems. These are obtained by generating additional inequalities that cut-off the fractional integer solution. Gomory [26] developed a method for generating these cutting planes, but the computational performance tends to be poor due to slow convergence and the large increase in size of the LP subproblems. An alternative approach is to generate strong cutting planes that correspond to facets, or faces of the integer or mixed-integer convex hull. Strong cutting planes are obtained by considering a separation problem to determine the strongest valid inequality that cuts off the fractional solution. This, however, is computationally a difficult problem (it is NP-hard) and for this reason unless one can obtain theoretically these cuts for problems with special structure, only approximate cutting planes are generated. Also, strong cutting planes are generated from the LP relaxation and during the branch and bound search to tighten the LP. Crowder et al [15] developed strong cutting planes for pure integer programming problems by considering each constraint individually and treating each of them as knapsack problems. Van Roy and Wolsey [89] considered special network structures for MILP problems to generate strong cutting planes. In both cases, substantial improvements were obtained in a number of test problems.

A more recent approach for cutting plane methods has been based on the important theoretical result that it is possible to transform an unstructured MILP problem into an equivalent LP problem that corresponds to the convex hull of the MILP. This involves converting the MILP into a nonlinear polynomial mixed integer problem which is subsequently linearized through variable transformations (45, 80]. Unfortunately, the transformation to the LP with the convex hull is exponential. However, these transformations can be used as a basis for generating cutting planes within a "branch and cut" enumeration, and this is for instance an approach that is being explored by Balas et al [5].

As for decomposition methods for MILP, the most common method is Benders decomposition [8]. This method is based on the idea of partitioning the variables into complicating variables (commonly integer variables in the MILP) and noncomplicating variables (continuous variables in MILP). The idea is to solve a sequence of LP subproblems for fixed complicating variables y^k,

$$\begin{aligned} Z^k = \min\ & c^Tx + b^Ty^k \\ \text{s.t.}\ & Ax \le d - By^k \\ & x \ge 0 \end{aligned} \qquad \text{(LPB)}$$

and master problems that correspond to projections in the space of the binary variables and that are based on dual representations of the continuous space. The form of the master problem given K feasible and M infeasible solution points for the subproblems is given by:

$$\begin{aligned} & Z_L^k = \min \alpha \\ & \alpha \ge (\lambda^k)^T(d - By^k) \quad k = 1, .. K \\ & (\mu^m)^T(d - By^k) \le 0 \quad m = 1, .. M \\ & \alpha \in R^1,\ y \in \{0,1\}^m \end{aligned} \qquad \text{(MB)}$$

Since the master problem provides valid lower bounds and the LP subproblems upper bounds, the sequence of problems is solved until equality of the bounds is achieved.

Benders decomposition has been successfully applied in some problems (eg. see [24], but it can also have very slow convergence if the LP relaxation is not tight (see [46]). Nevertheless, this method is in principle attractive in large multiperiod MILP problems. Finally, another type of decomposition techniques are Lagrangean relaxation methods which are applied when complicating constraints destroy the special structure of a problem.

Logic based methods were developed by taking advantage of the analogy between binary and boolean variables. Balas [3] developed Disjunctive Programming as an alternate form of representation of mixed-integer programming problems. MILP problems are formulated as linear programs with disjunctions (sets of constraints of which at least one must be true). Balas [4] characterized the family of all valid cutting planes for a disjunctive program. Using these cuts, disjunctions were re-expressed in terms of binary variables and the resulting mixed-integer problem is solved.

Another class of logic based methods are based on using symbolic inference techniques for the solution of pure integer programming problems. Hooker [32] demonstrated the analogy between unit resolution and first order cutting planes. Jeroslow and Wang [35] solved the satisfiability problem using a numerical branch and bound based scheme but solving the nodal problems using unit resolution. An alternate symbolic based branching rule was also proposed by these authors. Motivated by the above ideas, [62] considered the incorporation of logic in general mixed-integer programming problems in the form of redundant constraints to express with logic propositions the relations among units in superstructures. Here one approach is to convert the logic constraints into inequalities and add them to the MILP. Although this has the effect of reducing the integrality gap, the size of the problem is often greatly increased [63]. Therefore, these authors considered an alternate scheme in which symbolic inference techniques were used on a the set of logical constraints which are expressed in either the disjunctive or conjunctive normal form representations. The idea is to perform symbolic inference at each node during the branch and bound procedure in order to perform branching on the variables so as to fix additional binary variables. Orders of magnitude reductions have been reported by these authors using this approach [64].

Finally, it should be noted that the more recent computer codes for MILP, such as OSL (IBM, 1992) and MINTO [73] have an "open software architecture" that gives the user considerably more flexibility to control the branch and bound search. For instance, these codes allow the addition of cutting planes and modification of branching rules according to procedures supplied by the user.

Mixed-integer Nonlinear Programming (MINLP)

Although the problem (MIP) given earlier in the paper corresponds to an MINLP problem, for most applications the problem is linear in the 0-1 variables and nonlinear in the continuous variables x; that is,

$$\begin{aligned} \min Z = \; & f(x) + c^T y \\ \text{s.t.} \; & h(x) = 0 \\ & g(x) + By \leq 0 \\ & x \in R^n , \; y \in \{0,1\}^m \end{aligned} \qquad \text{(MINLP)}$$

This mixed-integer nonlinear program can in principle also be solved with the branch and bound method presented in the previous section ([31, 50, 11]). The major difference here is that the examination of each node requires the solution of a nonlinear program rather than the solution of an LP. Provided the solution of each NLP subproblem is unique, similar properties as in the case of the MILP would hold with which the rigorous global solution of the MINLP can be guaranteed.

An important drawback of the branch and bound method for MINLP is that the solution of the NLP subproblems can be expensive since they cannot be readily updated as in the case of the MILP. Therefore, in order to reduce the computational expense involved in solving many NLP subproblems, we can resort to two other methods: Generalized Benders decomposition [23] and Outer-Approximation [18]. The basic idea in both methods is to solve an alternating sequence of NLP subproblems and MILP master problems. The NLP subproblems are solved by optimizing the continuous variables x for a given fixed value of y, and their solution yields an upper bound to the optimal solution of (MINLP). The MILP master problems consist of linear approximations that are accumulated as iterations proceed, and they have the objective of predicting new values of the binary variables y as well as a lower bound on the optimal solution. The alternate sequence of NLP subproblems and MILP master problems is continued up to the point where the predicted lower bound of the MILP master is greater or equal than the best upper bound obtained from the NLP subproblems.

The MILP master problem in Generalized Benders decomposition (assuming feasible NLP subproblems) is given at any iteration K by:

$$
\begin{aligned}
Z_{GB}^{K} &= \min \alpha && \text{(MGB)}\\
\text{s.t.}\ \ \alpha &\geq f(x^k) + c^T y + (\mu^k)^T \left[g(x^k) + By\right] \qquad k=1,2...K\\
&\alpha \in R^1 \ , \ y \in \{0,1\}^m
\end{aligned}
$$

where α is the largest Lagrangian approximation obtained from the solution of the K NLP subproblems; x^k and μ^k correspond to the optimal solution and multiplier of the kth NLP subproblem; Z_{GB}^{K} corresponds to the predicted lower bound at iteration K.

In the case of the Outer-Approximation method the MILP master problem is given by:

$$
\begin{aligned}
Z_{OA}^{K} &= \min \alpha && \text{(MOA)}\\
\text{s.t.}\ \ &\left.\begin{aligned}
\alpha &\geq f(x^k) + \nabla f(x^k)^T (x-x^k) + c^T y\\
&T^k \nabla h(x^k)^T (x-x^k) \leq 0\\
&g(x^k) + \nabla g(x^k)^T (x-x^k) + By \leq 0
\end{aligned}\right\} \qquad k=1,2,...K\\
&\alpha \in R^1 \ , x \in R^n \ , y \in \{0,1\}^m
\end{aligned}
$$

where α is the largest linear approximation of the objective subject to linear approximations of the feasible region obtained from the solution of the K NLP subproblems. T^k is a diagonal matrix whose entries $t_{ii}^k = \text{sign}\ (\lambda_i^k)$, where λ_i^k is the Lagrange multiplier of equation h_i at iteration k, and is used to relax the equations in the form of inequalities [37]. This method has been implemented in the computer code DICOPT [38].

Note that in both master problems the predicted lower bounds, Z_{GB}^{K}, and Z_{OA}^{K}, increase monotonically as iterations K proceed since the linear approximations are refined by accumulating the lagrangian (in MGB) or linearizations (in MOA) of previous iterations. It should be noted also that in both cases rigorous lower bounds, and therefore convergence to the global optimum, can only be ensured when certain convexity conditions hold (see [23, 18]).

In comparing the two methods, it should be noted that the lower bounds predicted by the outer approximation method are always greater than or equal to the lower bounds predicted by Generalized Benders decomposition. This follows from the fact that the Lagrangian cut in GBD represents a surrogate constraint from the linearization in the OA algorithm [60]. Hence, the Outer-Approximation method will require the solution of fewer NLP subproblems and MILP master problems (see example 4 later in the paper). On the other hand, the MILP master in Outer-Approximation is more expensive to solve so that Generalized Benders may require less time if the NLP subproblems are inexpensive to solve. As discussed in [72], fast convergence with GBD can only be achieved if the NLP relaxation is tight.

As a simple example of an MINLP consider the problem:

$$\begin{aligned}
\min\ & Z = y_1 + 1.5y_2 + 0.5y_3 + x_1^2 + x_2^2 \\
\text{s.t.}\ & (x_1 - 2)^2 - x_2 \leq 0 \\
& x_1 - 2y_1 \geq 0 \\
& x_1 - x_2 - 4(1-y_2) \leq 0 \\
& x_1 - (1 - y_1) \geq 0 \\
& x_2 - y_2 \geq 0 \\
& x_1 + x_2 \geq 3y_3 \\
& y_1 + y_2 + y_3 \geq 1 \\
& 0 \leq x_1 \leq 4,\ 0 \leq x_2 \leq 4 \\
& y_1, y_2, y_3 = 0, 1
\end{aligned} \tag{6}$$

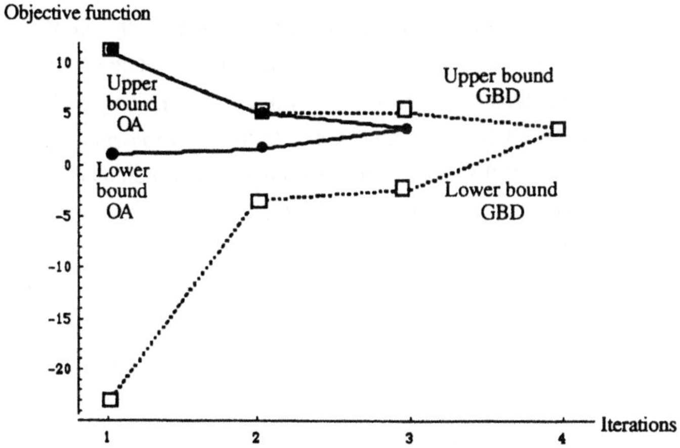

Fig. 4 Progress of iterations of OA and GBD for MINLP in (6)

Note that the nonlinearities involved in problem (6) are convex. Fig. 4 shows the convergence of the OA and the GBD methods to the optimal solution using as a starting point $y_1 = y_2 = y_3 = 1$. The optimal solution is Z =3.5, with $y_1 = 0$, $y_2 = 1$, $y_3 = 0$, $x_1 = 1$, $x_2 = 1$. Note that the OA algorithm requires 3 major iterations, while GBD requires 4, and that the lower bounds of OA are much stronger.

Other related methods for MINLP include the extension of the OA algorithm by Yuan et al [99] who considered nonlinear convex terms for the 0-1 variables, and the feasibility technique by Mawekwang and Murtagh [48] in which a feasible MINLP solution is obtained from the relaxed NLP problem. The latter method has been recently extended by Sugden [84].

In the application of Generalized Benders decomposition and Outer-Approximation, two major difficulties that can arise are the computational expense involved in the master problem if the number of 0-1 variables is large, and non-convergence to the global optimum due to the nonconvexities involved in the nonlinear functions.

To circumvent the first problem, Quesada and Grossmann [59] have proposed for the convex case an LP/NLP based branch and bound method in which the basic idea is to integrate the solution of the MILP master problem and the NLP subproblems which are assumed to be inexpensive to solve. This is accomplished by a tree enumeration in which an NLP is first solved to construct an initial linear approximation to the problem. The LP based branch and bound search is then applied; however when an integer solution is found a new NLP subproblem is solved from which new linear approximations are derived which are then used to update the open nodes in the tree. In this way the cold start for a new branch and bound tree for the MILP master problem is avoided. It should be noted that this computational scheme can be applied to Generalized Benders and Outer-Approximation. As mentioned before the latter will yield stronger lower bounds. However, in this integrated branch and bound method the size of the LP subproblems can potentially become large. To handle this problem Quesada and Grossmann [59] proposed the use of partial surrogates that exploit the linear substructures present in an MINLP problem.

In particular, consider that the MINLP has the following structure,

$$\begin{aligned} Z = &\min\ c^T y + a^T w + r(v) \\ &\text{st}\ \ Cy + Dw + t(v) \le 0 \qquad \text{(MINLP')} \\ &Ey + Fw + Gv \le b \\ &y \in Y,\ w \in W,\ v \in V \end{aligned}$$

in which the equality constraints are relaxed to inequalities according to the matrix T^k and included in the inequality set. Here the continuous variables x have been partitioned in two subsets w and v such that the constraints are divided into linear and nonlinear constraints, and the continuous variables into linear and nonlinear variables. In this representation $f(x)=a^T w + r(v)$, $B^T = [\, C/E]^T$, $g(x)= \begin{bmatrix} Dw + t(v) \\ Fw + Gv \end{bmatrix}$, and X= WxV. By constructing a partial surrogate constraint involving the linearization of the nonlinear terms in the objective and nonlinear constraints, the modified master problem has the form:

$$
\begin{aligned}
& Z_L{}^K = \min \beta && \text{(MMOA)} \\
& \text{st} \quad c^T y + a^T w + b - \alpha = 0 && \\
& \beta \geq r(v^k) + (\lambda^k)^T [\, Cy + Dw + t(v^k)\,] - (\mu^k)^T\, G\,(v - v^k) && k=1,\ldots K \\
& \qquad Ey + Fw + Gv \leq b && \\
& \qquad y \in Y,\ w \in W,\ v \in V,\ \alpha \in R^1,\ \beta \in R^1 &&
\end{aligned}
$$

where λ^k and μ^k are the optimal multipliers of the kth NLP subproblem. It can be seen that as opposed to the Benders cuts, the linearizations are defined in the full space of the variables, requiring only the addition of one new constraint for the nonlinear terms. It can be shown that the lower bound $Z_L{}^K$ predicted by the above master problem is weaker than the one of OA, but stronger than the one by GBD. Computational experience has shown that the predicted lower bounds are in fact not much weaker than the ones by the OA algorithm.

As for the question of nonconvexities, one approach is to modify the definition of the MILP master problem so as to avoid cutting off feasible mixed-integer solutions. Viswanathan and Grossmann [92] proposed an augmented-penalty version of the MILP master problem for outer-approximation, which has the following form:

$$
\begin{aligned}
& Z_{OA}^K = \min \alpha + \sum_{k=1}^{K} (\rho^k)^T(p^k + q^k) && \text{(MOAP)} \\
& \left.\begin{aligned}
& \text{s.t. } \alpha \geq f(x^k) + \nabla f(x^k)^T(x-x^k) + c^T y \\
& \qquad T^k \nabla h(x^k)^T(x-x^k) \leq p^k \\
& g(x^k) + \nabla g(x^k)^T(x-x^k) + By \leq q^k
\end{aligned}\right\} && k=1,2,\ldots K \\
& \alpha \in R^1,\ x \in R^n,\ y \in \{0,1\}^m &&
\end{aligned}
$$

in which the slacks p^k, q^k, have been added to the function linearizations, and in the objective function with weights ρ^k that are sufficiently large but finite. Since in this case one cannot guarantee a rigorous lower bound, the search is terminated when there is no further improvement in the solution of the NLP subproblem. This method has been implemented in the computer code DICOPT++ which has shown to be successful in a number of applications. It should also be noted that if the original MINLP is convex the above master problem reduces to the original OA algorithm since the slacks will take a value of zero.

An important limitation with the above approach is that it does not address the question whether the NLP subproblems may contain multiple local solutions. Recently there has an important effort to address the global optimization of nonconvex nonlinear programming problems. The current methods are either stochastic or deterministic in nature. In the former, generally no assumption about the mathematical structure of the problem is made. Simulated annealing is an example of a method that belongs to this category which in fact has been applied to batch process scheduling[43. 56]. This method however has the disadvantages that no strict guarantee can be given about global optimality and that its computational expense can be high. Deterministic methods require the problem to have some particular mathematical structure that can be exploited to ensure global optimality.

Floudas and Visweswaran [21] developed a global optimization algorithm for the solution of bilinear programming problems. Valid lower and upper bounds on the global optimal solution are obtained through the solution of primal and relaxed dual problems. The primal problem arises by fixing a subset of complicating variables which reduces the bilinear NLP into an LP subproblem. The relaxed dual problems arise from the master problem of GBD but in which the Lagrangian function is linearized and partitioned into subregions to guarantee valid lower bounds. An implicit partition of the feasible space is conducted to reduce the gap between the lower and upper bounds. A potential limitation of this method is that the number of relaxed dual problems to be solved at each iteration can grow exponentially with the number of variables involved in the nonconvex terms.

Another approach for solving nonconvex NLP problems in which the objective function involves bilinear terms is the one presented by Al-Khayyal and Falk [1]. These authors make use of the convex envelopes for the individual bilinear terms to generate a valid lower bound on the global solution. An LP underestimator problem is imbedded in a spatial branch and bound algorithm to find the global optimum. Sherali and Alameddine [78] presented a reformulation-linearization technique which generates tight LP underestimator problems that dominate the ones of Al-Khayyal and Falk. A similar branch and bound search is conducted to find the global solution. Although this method requires the enumeration of few nodes in the branch and bound tree, it has the main disadvantage that the size of the LP underestimator problems grows exponentially with the number of constraints.

Swaney [85] has addressed the problem in which the objective function and constraints are given by bilinear terms and separable concave functions. A comprehensive LP underestimator problem provides valid lower bounds that are used within a branch a bound enumeration scheme in which the partitions do not increase exponentially with the number of variables.

Quesada and Grossmann [60] have considered the global optimization of nonconvex NLP problems in which the feasible region is convex and the objective involves rational and/or bilinear terms in addition to convex functions. The basic idea is based on deriving an NLP underestimator problem that involves both linear and nonlinear estimator functions that provide an exact approximation of the boundary of the feasible region. The linear understimators are similar to the ones by Al-Khayyal and Falk [1], but these are strengthened in the NLP by nonlinear convex underestimators. The NLP underestimator problem, which allows the generation of tight lower bounds of the original problem, is coupled with a spatial branch an bound search procedure for finding the global optimum solution.

Modelling and reformulation

One of the difficulties involved in the application of mixed-integer programming techniques is that problem formulation is not always trivial, and that the way one formulates a problem can have a very large impact in the computational efficiency for the solution. In fact, it is not uncommon that for a given problem one formulation may be essentially unsolvable, while another formulation may make the problem much easier to solve. Thus, model formulation is a crucial step in the application of mixed-integer programming techniques.

While model formulation still remains largely an art, a number of guiding principles are starting to emerge that are based on a better understanding of polyhedral theory in integer programming (see [51]). In this section we will present an overview of modelling techniques and illustrate them with example problems. These techniques can be broadly classified into logic based methods, semi-heuristic guidelines, reformulation techniques and linearization techniques.

It is often the case in mixed-integer programming that it is not obvious how to formulate a constraint in the first place, let alone formulate the best form of that constraint. Here the use of propositional logic and its systematic transformation to inequalities with 0-1 variables can be of great help (e.g. see [98, 14, 62]). In particular, when logic expressions are converted into the conjunctive normal form, each clause has the form,

$$P_1 \vee P_2 \vee \ldots \vee P_m \tag{7}$$

where P_i is a proposition and $\vee$ is the logical operator OR. The above clause can be readily transformed into an inequality by relating a binary variable y_i to the truth value of each proposition P_i (or $1 - y_i$ for its negation). The form of the inequality for the above clause is,

$$y_1 + y_2 + \ldots + y_m \geq 1 \tag{8}$$

As an example consider the logical condition, $P_1 \vee P_2 \Rightarrow P_3$ which when converted into conjuctive normal form yields, $(\neg P_1 \vee P_3) \wedge (\neg P_2 \vee P_3)$. Each of the two clauses can then be translated into the inequalities,

$$\begin{array}{lcl} 1 - y_1 + y_3 \geq 1 & & y_3 \geq y_1 \\ 1 - y_2 + y_3 \geq 1 & \text{or} & y_3 \geq y_2 \end{array} \tag{9}$$

Similar procedures can be applied when deriving mixed-integer constraints.

Once constraints have been formulated for a given problem, the question that arises is whether alternative formulations might be better suited for the computation, and here the first techniques to be considered are semi-heuristic guidelines. These are rules of thumb on how to formulate "good" models. A simple example are variable upper bound constraints for problems with fixed charges,

$$x_i - Uy_i \leq 0 \qquad (10)$$

Here it is well known that although for representation purposes the upper bound U can be large, one should try to select the smallest valid bound in order to avoid a poor LP relaxation. Another well known example is the constraint that often arises in multiperiod MILP problems (selecting unit z implies possible operation y_i in period i, i=1,2,...n),

$$\sum_{i=1}^{n} y_i - n\,z \leq 0 \qquad (11)$$

Here the disaggregation into the constraints,

$$y_i - z \leq 0 \quad i=1,2,..n \qquad (12)$$

will produce a much tighter representation (in fact the convex hull of the 0-1 polytope). These are incidentally the constraints that would be obtained if the logical conditions for this constraint are expressed in conjuctive normal form.

The main problem with the disaggregated form of constraints is the potentially large number of them when compared with the aggregated form. Therefore, one has to balance the trade-offs between model size and tighter relaxations. For example [93] report a model in which the disaggregated variable upper bound constraints were used in the form

$$w_{ijsn} \leq U_{ijsn}\, y_{jsn} \qquad \forall\ i, j, s, n \qquad (13)$$

An equivalent aggregated form of the above constraints is

$$\sum_i w_{ijsn} \leq \sum_i U_{ijsn}\, y_{jsn} \qquad \forall\ j, s, n \qquad (14)$$

When the first set of constraints is used the model involved 708 constraints and required 233 CPUsec using SCICONIC on a VAX 6320. When the second set of constraints is used, the model required only 220 constraints, but the time was increased to 649 sec because of the looser relaxation of (14).

While the above modelling schemes are somewhat obvious, there are some which are not. A very good example arises in the MILP scheduling model of [40]. In this model, the following constaints apply:

(a) At any time t, an idle item of equipment j can only start at most one task $i \in I_j$.

(b) If the item j does start performing a given task $i \in I_j$, then it cannot start any other task until the current one is finished after p_i time units.

Kondili et al [46] formulated the two above constraints as:

$$\sum_{i \in I_j} W_{ijt} \leq 1 \quad \forall j, t$$
$$\sum_{t'=t}^{t+p_i-1} \sum_{i' \in I_j} W_{i'jt'} - 1 \leq M(1 - W_{ijt}) \;\; \forall i, j \in K_i, t \tag{15}$$

where M is a suitably large number. Note that the second constraint has the effect of imposing condition (b) if $W_{ijt} = 1$. As one might expect the second constraint yields a poor relaxation due to the effect of the "big M". Interestingly, [77] found an equivalent representation for the above two constraints, which is not only much tighter, but requires much fewer constraints! These are given by,

$$\sum_{i \in I_j} \sum_{t'=t}^{t-pi+1} W_{ijt'} \leq 1 \quad \forall j, t \tag{16}$$

Thus, this example clearly shows that formulation of a "proper" model is not always trivial or even well understood. Nevertheless, an analysis of the problem based on polyhedral theory can help one understand the reason for the effectiveness of the constraint. A detailed proof for constraint (16) is given in the Appendix.

However, not everything in MILP modelling is an art. A more rational approach that has been emerging is the idea of reformulation techniques that are based on variable disaggregation (e.g. see [61, 33, 34,]), and which have the effect of tightening the LP relaxation. The example *par excellence* is the lot sizing problem that in its "naive" form is given by the MILP (see [51]):

$$\begin{aligned}
&\min \sum_{t=1}^{NT} (p_t x_t + h_t\ s_t + c_t y_t) \\
\text{s.t.}\quad & s_{t-1} + x_t = d_t + s_t \qquad t = 1, NT \\
& x_t \le x^u y_t \qquad t = 1, NT \\
& s_0 = 0 \\
& s_t, x_t \ge 0, \quad y_t \in \{0,1\} \qquad t = 1, NT
\end{aligned} \tag{17}$$

where x_t is the amount to be produced in period t, y_t is the associated 0-1 variable, and s_t is the the inventory for period t; c_t, p_t, h_t, are the set-up, production and storage costs for time period t, t = 1, NT.

As has been shown by Krarup and Bilde [41] the above MILP can be reformulated by disaggregating the production variables x_t into the variables $q_{t\tau}$, to represent the amount produced in period t to satisfy the demand in period $\tau \ge t$; that is,

$$x_t = \sum_{\tau=t}^{NT} q_{t\tau} \tag{18}$$

The MILP is then reformulated as,

$$\begin{aligned}
\min\quad & \sum_{t=1}^{NT} \sum_{\tau=t}^{NT} (p_t + h_t + h_{t+1} + \ldots + h_{\tau-1})\ q_{t\tau} + \sum_{t=1}^{NT} c_t y_t \\
\text{s.t.}\quad & \sum_{\tau=1}^{t} q_{\tau t} = d_t \qquad t = 1, NT \\
& q_{t\tau} \le d_t y_t \qquad t = 1, NT, \quad \tau = t, NT \\
& q_{t\tau} \ge 0,\ y_t = \{0,1\}
\end{aligned} \tag{19}$$

As it turns out this reformulation yields the absolute tightest LP relaxation since it yields 0-1 values for the y variables; thus this problem can be solved as an LP and there is no need to apply a branch and bound search as there is in the original MILP (17). It should be noted that although this example is quite impressive, it is not totally surprising from a theoretical viewpoint. The lot sizing problem is solvable in polynomial time and therefore one would expect that it should be possible to formulate this problem as an LP that is polynomially sized in the number of variables and constraints. It should be noted that the lot sizing problem is often embedded in MILP planning and scheduling problems with which one can reformulate these problems to tighten the LP relaxation as discussed in [71].

Finally, another case that often arises in the modelling of mixed-integer problems are nonlinearities such as bilinear products of 0-1 variables or products of 0-1 with continuous variables. Nonlinearities in binary variables usually involve the transformation of a nonlinear function into a polynomial function of 0-1 variables and then transforming the polynomial function into a linear function of 0-1 variables [79]. For cross products between binary and continuous variables [58] proposed a linearization method which was later extended by Glover [25]. The main idea behind these linearization schemes was the introduction of a new continuous variable to represent the cross product. The equivalence between the bilinear term and the new variable was enforced by introducing a set of equivalence constraints. For the specific case in which the model has a multiple choice structure, an efficient linearization scheme was proposed by Grossmann et al [29]. This scheme compared to the one proposed by Glover gives tighter LP relaxations with fewer number of constraints. The multiple choice structure usually arises in discrete design models, in which the design variables instead of being continuous, take values from a finite set. Batch process design problems often involve discrete sizes and as such the latter linearization scheme is well suited. As an example, consider the the bilinear constraints:

$$\alpha_{ij} \sum_{s=1}^{N(i)} d_{is} y_{is} v_j - \beta_{ij} w_j \leq 0 \qquad j \in J(i)\ ,\ i=1,..n \tag{20}$$

in which y_{is} is a 0-1 variable and v_j is continuous, and where the following constraint holds:

$$\sum_{s=1}^{N(i)} y_{is} = 1 \tag{21}$$

In order to remove the bilinear terms $y_{is}v_j$ in (20), define the continuous variables v_{ijs} such that

$$v_j = \sum_{s=1}^{N(i)} v_{ijs} \qquad j \in J(i)\ ,\ i=1,..n \tag{22}$$

$$v_j^L\, y_{is} \leq v_{ijs} \leq v_j^U\, y_{is} \qquad j \in J(i),\ s=1, N(i),\ i=1..n \tag{23}$$

where v_j^L, v_j^U are valid lower and upper bounds. Using the equations in (21) to (23), the constraints in (20) can be replaced by the linear inequalities

$$\alpha_{ij} \sum_{s=1}^{N(i)} d_{is} v_{ijs} - \beta_{ij} w_j \leq 0 \qquad j \in J(i)\ ,\ i=1,..n \tag{24}$$

The bilinear constraints in (20) can also be linearized by considering in addition to the inequalities in (24), the following constraints proposed by Glover [25]:

$$\begin{aligned} & v_j^L\, y_{is} \le v_{ijs} \le v_j^U\, y_{is} \\ & v_{ijs} \ge v_j - v_j^U(1 - y_{is}) \qquad j \in J(i),\ s=1, N(i),\ i=1..n \\ & v_{ijs} \le v_j - v_j^L(1 - y_{is}) \end{aligned} \tag{25}$$

This linearization, however, requires almost twice as many constraints as (22), (23) and (24) Furthermore, while a point (v_{ijs}, v_j, y_{is}) satisfying (22) and (23) satisfies the inequalities in (25), the converse may not be true. For instance, assume a non-integer point y_{is} such that $v_{ijs} = v_j^U y_{is}$. Using (21) it follows from (25) that

$$v_j^L + (v_j^U - v_j^L)\, y_{is} \le v_j \le v_j^U \tag{26}$$

while (22) yields $v_j = v_j^U$. Thus, the inequalities in (25) may produce a weaker LP relaxation.

For the case when the bilinear constraints in (20) are only inequalities, Torres [88] has shown that it is sufficient to consider the following constraints from (25):

$$\begin{aligned} & v_j^L\, y_{is} \le v_{ijs} \\ & v_{ijs} \ge v_j - v_j^U(1 - y_{is}) \qquad j \in J(i),\ s=1, N(i),\ i=1..n \end{aligned} \tag{27}$$

which requires fewer constraints than the proposed linearization in (22) and (23). However, the above inequalities also can produce a weaker LP relaxation. For instance, setting $v_{ijs} = v_j^L y_{is}$ for a non-integer point y_{is} yields,

$$v_j \le v_j^U - (v_j^U - v_j^L)\, y_{is} \tag{28}$$

while (22) yields $v_j = v_j^L$.

While the modelling techniques described in this section have been mostly aimed at MILP problems, they are of course also applicable to MINLP problems. One aspect however, that is particular to MINLP problems are the modelling of nonlinearities of the continuous variables. In such a case it is important to determine whether the nonlinear constraints are convex or not. If they are not, the first attempt should be to try to convexify the problem. The most common approach is to apply exponential transformations of the form $x = \exp(u)$, where x are the original continuous variables and u the transformed variables; if the original nonlinearities correspond to posynomials these tranformations will lead to convex constraints. A good example is the optimal design of multiproduct plants with single product campaigns [28], with which [39] were able to rigorously solve the MINLP problem to global optimality with the outer-approximation algorithm. When no transformations can be found to convexify a problem, this does not necessarily mean that the relaxed NLP has multiple local optima. However, nonconvexities in the form of bilinear products and rational terms are warning signals that should not be ignored. In this case the application of a global optimization method such as the ones described previously will be the only way to rigorously guarantee the global optimum.

Finally, it should be noted that another aspect in modelling is the computer software that is available for formulating and solving mixed-integer optimization problems. Modelling systems such as GAMS [13] and AMPL [22] have emerged as major tools in which problems can be specified in algebraic form and automatically interfaced with codes for mixed-integer linear and nonlinear optimization (e.g. MINOS, ZOOM, SCICONIC, OSL, CPLEX, DICOPT++). Modelling tools such as these can greatly reduce the time that is required to test and prototype mixed-integer optimization models.

Examples

In this section we will present several examples to illustrate a number of points and the application of techniques for mixed-integer programming in batch processing.

Example 1: The MILP for the State Task Network model for the scheduling of batch operations by Kondili et al. [40] has been used to compare the performance of three MILP codes: ZOOM an academic code, and OSL and SCICONIC which are commercial codes. This example which has 5 tasks, 4 units, 10 time periods and 9 states (see Fig. 5), also demonstrates the effect of modelling schemes on the solution efficiency. The objective is to maximize the production of the two final products. The resulting MILP model, which incorporates the constraints in (16), involves 251 variables (80 binary) and 309 constraints (see [77]). The results of the benchmark comparison between the three codes for this problem are shown in Table 1. The problems were solved to optimality (0% gap) by using GAMS as an interface. As can be seen in Table 1 the performance of the codes is quite different. SCICONIC had the lowest computing requirements: about less than a tenth of the requirements of ZOOM.

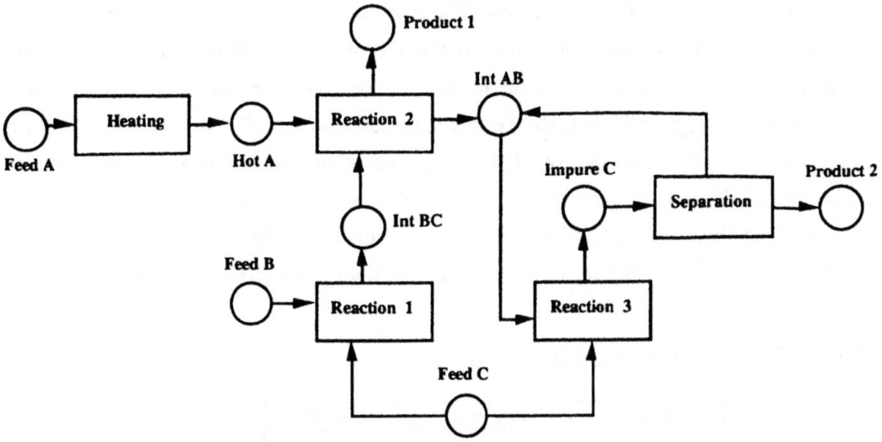

Fig. 5 State-task network for example problem

It should be noted that [77] solved this problem with their own branch and bound method in two forms. In the first the MILP was identical as the one solved in this paper. In this case 1085 nodes and 437 secs on a SUN1- Sparcstation were required to solve the problem within 1% of optimality. In the second form, the authors applied a solution strategy for reducing the size of the relaxed LP and for reducing degeneracies. In this case only 29 nodes and 7 secs were required to solve the problem, which represents a performance comparable to the one by SCICONIC.

To illustrate the effect that alternate formulations may have, two cases were considered and the results are shown in Table 2a. Firstly, realizing that the objective function does not contain any binary variable, the second column involves the addition of a penalty to the objective function in which all the binary variables are multiplied by a very small number so as not to affect the optimum solution. The idea is simply to drive the 0-1 variables to zero to reduce the effect of degeneracies. In the third column, 18 logic cuts in the form of inequalities have been added to the MILP model to reduce the relaxation gap [63]. These logic cuts represent connectivity of units in the state task network. For example, in the problem of Fig. 5, since no storage of impure C is allowed, the separation step has to be immediately performed after reaction 3. As can be seen from Table 2a, both modelling schemes lead to a substantial improvement in the solution efficiency with OSL, while with SCICONIC only the addition of logic cuts improves the solution efficiency. Furthermore, the effect of adding these logic cuts in this problem have been studied for the case of 10, 20, 40 and 50 time periods. The results, shown in Table 2b, demonstrate an increase in the effectiveness of the logic cuts in improving the efficiency of the branch and bound procedure. The reduction in the number of nodes required in the branch and bound search due to the logic cuts increases from a factor of 3 for the 10 period case to a factor of more than 6 for the 40 period case. The 50 time period problem, with 1251 variables (400 binary) and 1509 constraints could not be solved by OSL within 100,000 iterations and 1 hour of CPU time on the IBM POWER 530. With the addition of the logic cuts, the problem is solved in 158.84 sec requiring only 698 nodes and 5017 iterations.

Table 1. Comparison with several MILP codes

	nodes	iterations	CPU time *
ZOOM	410	7866	39.44
OSL	350	918	14.85
SCICONIC	61	318	3.63

* IBM POWER 530

Table 2a. Computational results with modified formulations (OSL / SCICONIC)

	Original Model	Model with altered Objective Function	Model with Logic Cuts
number of nodes	350/61	40/61	108/33
number of iterations	918/318	336/318	620/233
CPU time *	14.85/3.63	2.51/3.65	5.98/2.13
Relaxed Optimum	257.2	257.2	257.2
Integer Optimum	241	241	241

* sec IBM POWER 530

Table 2b. Example 1: Effect of logic cuts for different time periods

	Original Model	Model with Logic Cuts
10 Time Periods 251 variables 80 binary		
Constraints	309	327
Number of nodes	350	108
Number of Iterations	918	620
CPU Time *	14.85	5.98
20 Time Periods 501 variables 160 binary		
Constraints	609	643
Number of nodes	123	67
Number of Iterations	755	658
CPU Time *	10.22	7.81
40 Time Periods 1001 variables 320 binary		
Constraints	1209	1279
Number of nodes	2098	315
Number of Iterations	25964	3423
CPU Time *	424.68	67.17
50 Time Periods 1251 variables 400 binary		
Constraints	1509	1597
Number of nodes	>20,000	698
Number of Iterations	>100,000	5017
CPU Time*	>3,600	158.84

* sec IBM POWER 530

Example 2. In order to illustrate the effect of preprocessing and the use of SOS1 constraints, consider the design of multiproduct batch plants with one unit per stage, operating with single product campaigns, and where the equipment is available in discrete sizes [93]. The MILP model is as follows:

$$\min \sum_j \sum_s c_{js}\, y_{js} \qquad \text{(RP1)}$$

$$\text{s.t.} \quad T_i \geq \sum_s \left(\frac{Q_i S_{ij}}{v_{js}}\right) y_{js}\, T_{Li} \qquad i=1,..,N\ ,\ j=1,..,M$$

$$\sum_i T_i \leq H$$

$$\sum_s y_{js} = 1 \qquad j=1,..,M$$

$$T_i \geq 0 \quad i=1,..,N\ , \quad y_{js} \in \{0,1\} \qquad j=1,..,M\ ,\ s=1,..,ns_j$$

The example considered involves a plant with 6 stages and 5 products. To illustrate the effect that the number of discrete sizes has in the size of model (RP1) as well as in the computational performance, three problems one with 8, one with 15 and another with 29 discrete sizes were considered.The MILP problems were solved using SCICONIC 2.11 (SCICONIC, 1991) through GAMS 2.25 in a Vax-6420.

Table 3. Computational results for example 2

# Dis. Sizes	Constraints	Variables	0-1 Vars's	Cpu-time	Iterations	Nodes
WITHOUT SOS1, DOMAIN REDUCTION AND CUTTOFF						
8	38	54	48	2.93	181	89
15	38	96	90	25.09	985	731
29	38	180	174	44.94	1203	979
WITH SOS1, DOMAIN REDUCTION AND CUTTOFF						
8	38	46	40	1.93	57	53
15	38	82	76	2.85	91	64
29	38	154	148	6.64	182	150
* In VAX- 6420 SECONDS						

As seen from Table 3, the number of discrete sizes has a significant effect in the number of 0-1 variables, and hence in the number of iterations and the CPU time. One can, however, reduce significantly the computational requirements by performing a domain reduction of the 0-1 variables through the use of bounds to fix a subset of them to zero, treating the multiple choice constraints as SOS1 constraints and applying and objective function cutoff as described in [93]. As seen in Table 3, reductions of up to one order of magnitude are achieved.

Example 3. This example will illustrate how strong cutting planes may significantly improve the computational performance of MILP problems with poor continuous relaxations. A good example are jobshop scheduling problems. Consider the case in which one has to schedule a total of 8 batches, 2 for each one of 4 products A, B, C, D so as to minimize the makespan in a plant consisting of 5 stages. The processing times for each product are given in Fig. 6, where it can be seen that not all products require all the stages, and that they all require zero-wait transfer policy.

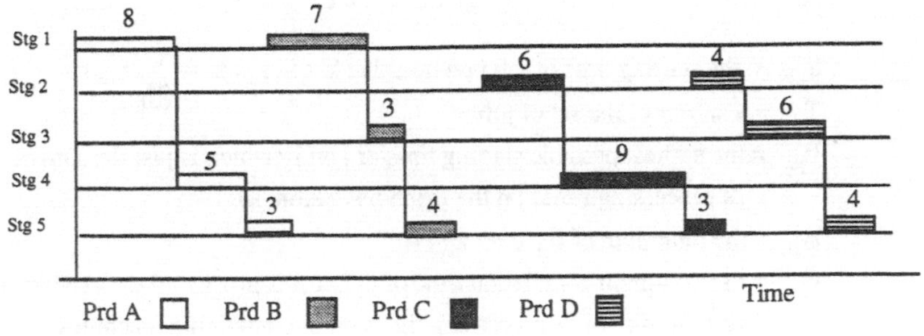

Fig. 6 Processing times of products in various stages

As noted in [94] the makespan minimization problem described above can be formulated as an MILP problem of the form:

$$\min \ Ms \qquad \text{(P1)}$$

$$\text{s.t.} \quad Ms \geq \bar{S}_i + \sum_{k=1}^{M} t_{ik} \qquad \forall i$$

$$\bar{S}_j - \bar{S}_i + W(1-y_{ijk}) \geq \left(\sum_{k=1}^{k'} t_{ik} - \sum_{k=1}^{k'-1} t_{jk}\right) \qquad \forall (i, j, k') \in C$$

$$\bar{S}_i - \bar{S}_j + W y_{ijk} \geq \left(\sum_{k=1}^{k'} t_{jk} - \sum_{k=1}^{k'-1} t_{ik}\right) \qquad \forall (i, j, k') \in C$$

$$y_{ijk} \in \{0,1\} \qquad \forall i, j, k \quad \bar{S}_i \geq 0 \ \ \forall i$$

In the above formulation the potential clashes at every stage are resolved with a pair of disjunctive constraints that involve a large bound W. The difficulty with these constraints is that they are trivially satisfied when the corresponding binary variables are relaxed, which in turn yields a poor LP relaxation. For this example, the LP relaxation had an objective value of 18 compared to the optimal integer solution of 41, which corresponds to a relaxation gap of 56%. The MILP was solved with SCICONIC 2.11 on a Vax-6420 requiring 55 CPUsecs, and the solution is shown in Fig.7. In order to improve the LP relaxation basic-cut inequalities have been recently proposed by Applegate and Cook [2], and they have the form,

$$\sum_{i\in T} t_{ik} S_{ik} \geq E_{Tk} t_{ik} + \sum_{\{i\in T,\, j\in T,\, i<j\}} t_{ik} t_{jk} \qquad \forall k$$

$$\sum_{i\in T} t_{ik} (Ms - S_{ik}) \geq F_{Tk} t_{ik} + \sum_{i\in T} t_{ik}^2 + \sum_{\{i\in T,\, j\in T,\, i<j\}} t_{ik} t_{jk} \qquad \forall k$$

where S_{ik} = the starting time of job i on machine k ($S_{ik} = \overline{S}_i + \sum_{k=1}^{M} t_{ik}$)

T = a subset of the set of jobs

E_{jk} = the earliest possible starting time of j on k (which is just the sum of j's processing times on the machines before k)

E_{Tk} = the minimum of E_{jk} over all $j \in T$

F_{jk} = the minimum completion time of j after it is processed on k (which is just the sum of j's processing times on the remaining machines)

F_{Tk} = the minimum of E_{jk} over all $j \in T$

The impact of these constraints in the MILP was very significant in this example. The LP relaxation increased to 38 which corresponds to a gap of only 7%. In this case the optimal solution was obtained in only 8 CPUsecs.

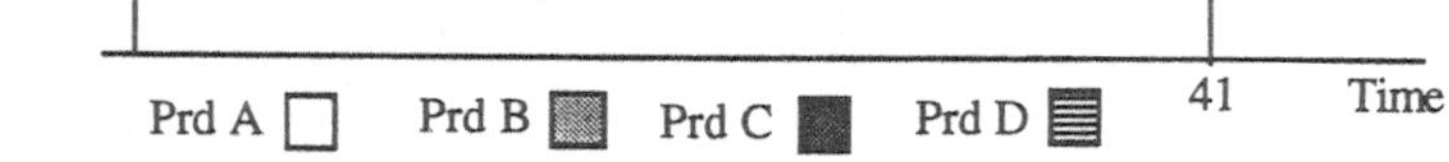

Fig. 7 Optimal schedule for example 3

Example 4. The optimal design of multiproduct batch plants with parallel units operating out of phase (see Fig. 8) will be used to illustrate the computational performance of the different MINLP algorithms.

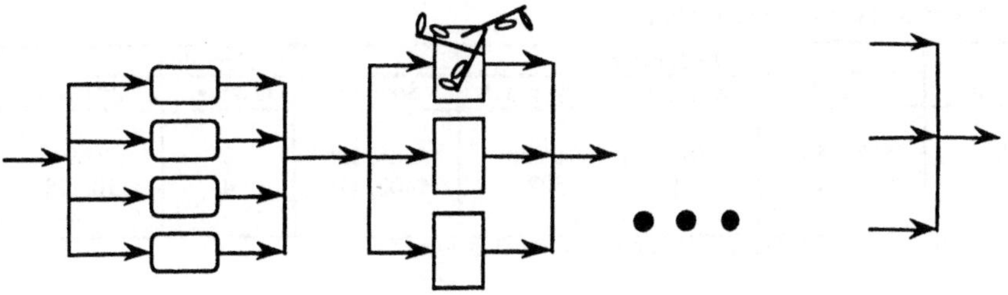

Fig. 8 Multiproduct batch plant with parallel units

Two different cases are considered. One consists of 5 products in a plant with 6 processing stages process with a maximum of 4 parallel units per stage (batch5). The second one has 6 products in a plant with 10 stages and a maximum of 4 parallel units per stage (batch6). The MINLP formulation and the data for these examples are reported in [38] and the size of the problems is given in Table 4. The model is a geometric programming problem that can be transformed into a convex MINLP through exponential transformations.

Table 4. Data for Example 4.

problem	Binary variables	Continuous variables	Constraints
batch5	24	22	73
batch6	40	32	141

The GBD and OA algorithms were used for the solution of both examples and the computational results are given in Table 5. The GBD algorithm was implemented within GAMS, while the version of the OA algorithm used was the one implemented in DICOPT++ with the augmented penalty. MINOS 5.2 was used for the NLP subproblems and SCICONIC 2.11 for the MILP master problems. Note that in both cases the OA algorithm required much fewer iterations than GBD which predicted very weak bounds and a large number of infeasible NLP subproblems during the iterations. For problem batch5 both algorithms found the global optimal solution. For batch6, both algorithms also found the same solution which however is suboptimal since the correct optimum is $398,580. In the case GBD, the algorithm did not converge as it had a large gap between the lower and upper bounds after 66 iterations. In the case of the OA algorithm as implemented in DICOPT ++ the optimal solution was suboptimal due to the termination criterion used in this implementation.

Table 5. Computational results for Example 4.

problem	GBD algorithm Solution	iterations	CPU time*	OA algorithm Solution	iterations	CPU time*
batch5	$285, 506	67	766.88	$285, 506	3	26.94
batch6	$402, 496	66+	2527.2	$402, 496	4	108.58

* sec Vax 6420 + Convergence of bounds was not achieved

In both the above examples the solution of the MILP master problem in the OA algorithm was of the order of 80%. A rigorous implementation of the OA algorithm for the convex case [18] and the LP/NLP based branch and bound algorithm by Quesada and Grossmann [59] were also applied to compare their computational performance with respect to the number of nodes that are required for the MILP master problem. The results are given in Table 6. As can be seen, both algorithms required the solution of 4 and 10 NLP subproblems, respectively, and they both obtained the same optimal solution. However, the LP/NLP based branch and bound required a substantially smaller number of nodes (36% and 16% of the number of nodes required by OA).

Table 6. Results on MILP solution step for problems in Example 4.

problem	optimal solution	Outer Approximation nodes	NLP	LP/NLP branch and bound nodes	NLP
batch 5	$285, 506	90	4	32	4
batch 6	$398, 580	523	10	84	10

Example 5. In order to illustrate the effect of nonconvexities, consider the design and production planning of a multiproduct batch plant with one unit per stage. The objective is to maximize the profit given by the income from the sales of the products minus the investment cost. Lower bounds are specified for the demands of the products and the investment cost is assumed to be given by a linear cost function. Since the sizes of the vessels are assumed to be continuous, this gives rise to the following NLP model:

$$\max \quad P = \sum_i p_i n_i B_i - \sum_j \alpha_j V_j$$

$$\text{s.t.} \quad V_j \geq S_{ij} B_i \quad i=1, N, \; j=1,M \qquad \text{(NLPP)}$$

$$\sum_i n_i T_i \leq H$$

$$\frac{Q_i^L}{n_i} - B_i \leq 0 \quad i=1, N$$

$$V_j, B_i, n_i \geq 0$$

where n_i and B_i is the number and size of the batches for product i, and V_j is the size of the equipment at stage j. The first inequality is the capacity constraint in terms of the size factors S_{ij}, the second is the horizon constraint in terms of the cycle times for each product T_i and the total time H, and the last inequality is the specification on lower bounds for the demands Q_i^L. Note that the objective function is nonconvex as it involves bilinear terms, while the constraints are convex. The data for this example are given in Table 7. A maximum size of 5000 L was specified for the units in each stage.

Table 7. Data for Example 7

Product	T_i (hrs)	p_i (\$/Kg)	Q_i^L (Kg)	S_{ij} (L/kg) 1	2	3
A	16	15	80000	2	3	4
B	12	13	50000	4	6	3
C	13.6	14	50000	3	2	5
D	18.4	17	25000	4	3	4

$\alpha_1 = 50,\ \alpha_2 = 80,\ \alpha_3 = 60$ (\$/L) ; H = 8,000 hrs

When a standard local search algorithm (MINOS 5.2) is used for solving this NLP problem using as a starting point $n_A=n_B=n_C=60$ and $n_D=300$ the predicted optimum profit is \$8,043,800/yr and the corresponding batch sizes and their number are shown in Table 8.

Table 8. Suboptimal solution for Example 8

	A	B	C	D
B	1250	833.33	1000	1250
n	79.15	60	50	289.868

Since the formulation in (NLPP) is nonconvex there is no guarantee that this solution is the global optimum. This problem can be reformulated by replacing the nonconvex terms by underestimator functions to generate a valid NLP underestimator problem as discussed in [60]. The underestimator functions require the solution of LP subproblems to obtain tight bounds on the variables, and yield a convex NLP problem with 8 additional constraints.

The optimal profit predicted by the nonlinear underestimator problem is \$8,128,100/yr with the variables given in Table 9. When the objective function of the original problem (NLPP) is evaluated for this feasible point the same value of the objective function is obtained proving that it corresponds to the global optimal solution. This problem was solved on a IBM/R6000-530 with MINOS 5.2, and 1.6 secs were required to solve the LP bounding problems and 0.26 secs to solve the NLP underestimator problem.

It is interesting to note that both the local and global solutions had the maximum equipment sizes. The only difference was in the number of batches produced for products A and D.

Table 9. Global optimum solution for Example 4.

	A	B	C	D
B	1250	833.33	1000	1250
n	389.5	60	50	20

Concluding Remarks

This paper has given a general overview of mixed-integer optimization techniques for the optimization of batch processing systems. As was shown with the review of previous work, the application of these techniques has increased substantially over the last few years. Also, as was discussed in the review of mixed-integer optimization techniques, a number of new methods are emerging that have the potential of increasing the size and scope of the problems to be solved. While in the case of MILP branch and bound methods continue to play a dominant role, the use of strong cutting planes, reformulation techniques and the integration of symbolic logic hold great promise for reducing the computational expense for solving large scale problems. Also, it will be interesting to see in the future what impact interior point methods will have on MILP optimization (see for instance [10], for preliminary experience). While the solution time of large LP problems can be greatly reduced, the subproblems in branch and bound cannot be readily updated as is the case with simplex based methods. As was also shown with the results, different computer codes for MILP can show very large differences in performance despite the fact that they all rely on similar ideas. This clearly points to the importance of issues such as software and hardware implementation, preprocessing, numerical stability and branching rules. However, care must also be exercised in comparisons because any given method or implementation for MILP may exhibit wide variations in performance by changes in data for the same model.

In the case of MINLP, the application of this type of models is becoming more widespread with the Outer-Approximation and Generalized Benders Decomposition methods. The former has proved to be generally more efficient, although the latter is better suited for exploiting the structure of problems (e.g. see [70]). Aside from the issue of problem size in MINLP optimization, nonconvexities remain a major source of difficulties. However, significant progress is being made in the global optimization of nonconvex NLP problems, and this will surely have a positive effect on MINLP optimization in the future. Finally, as has been emphasized in this paper, problem formulation for MILP and MINLP problems has often a very large impact in the efficiency of the computations, and in many ways still remains an art for the application of these techniques. However, a better understanding of polyhedral theory and establishing firmer links with symbolic logic may have a substantial effect on how to systematically formulate improved models for mixed-integer problems.

Acknowledgment

The authors gratefully acknowledge financial support from the National Science Foundation under Grant CBT-8908735 and CTS-9209671, and from the Engineering Design Research Center at Carnegie Mellon.

Appendix : On the reduced set of inequalities *of [77]*

In order to prove the equivalence of constraint (15) and the one in (16) by Shah *et al* [77], one must first state the following lemma

Lemma : The integer constraint,

$$y_1+y_2+...+y_{K+1} \leq 1 \qquad (1)$$

is equivalent to and sharper than the set of integer constraints

$$y_1+y_2+...+y_K \leq 1 \qquad (A0)$$
$$y_{K+1} + y_1 \leq 1 \qquad (A1)$$
$$y_{K+1} + y_2 \leq 1 \qquad (A2)$$
$$....$$
$$y_{K+1} + y_K \leq 1 \qquad (AK)$$

Proof :

First we note that (1) can easily be seen to be equivalent to the constraints (A0) to (AK) since in these at most one variable y can take an integer value of 1. Multiplying constraint (A0) by (K-1) and adding it with the constraints in (A1)-(AK) yields,

$$K(y_1+y_2+..+y_{K+1}) \leq K\text{-}1 + K$$

$$y_1+y_2+..+y_{K+1} \leq 1 + (K\text{-}1)/K$$

Since $y_i \in \{0,1\}$, the right hand side can be rounded below to obtain the inequality

$$y_1+y_2+..+y_{K+1} \leq 1$$

Proof of constraint :

We know, from (15)

$$\sum_{i \in I_j} W_{i,j,t} \le 1 \qquad \forall\, j, t \qquad\qquad W_{ijt} \in \{0,1\}\ \forall\, i,j,t$$

ie. $\quad W_{i1,j,t}+W_{i2,j,t}+..+W_{inj,j,t} \le 1 \qquad$ for all j,t

If $p_{i1} > 1$, since no unit can process two tasks simultaneously,

$$W_{i1,j,t-1}+W_{i1,j,t} \le 1$$
$$W_{i1,j,t-1}+W_{i2,j,t} \le 1$$
$$..$$
$$W_{i1,j,t-1}+W_{inj,j,t} \le 1$$

From the lemma, we get

$$W_{i1,j,t-1}+W_{i1,j,t}+W_{i2,j,t}+..+W_{inj,j,t} \le 1$$

If $p_{i2} > 1$

$$W_{i2,j,t-1}+W_{i1,j,t} \le 1$$
$$W_{i2,j,t-1}+W_{i2,j,t} \le 1$$
$$..$$
$$W_{i2,j,t-1}+W_{inj,j,t} \le 1$$

Also, $\quad W_{i2,j,t-1}+W_{i1,j,t-1} \le 1$

This leads to $\quad W_{i2,j,t-1}+W_{i1,j,t-1}+W_{i1,j,t}+W_{i2,j,t}+..+W_{inj,j,t} \le 1$

Repeat for all $W_{i',j,t-1}$ where $p_{i'} > 1$ to get

$$W_{i1,j,t-1}+W_{i2,j,t-1}+..+W_{inj,j,t-1}+W_{i1,j,t}+W_{i2,j,t}+..+W_{inj,j,t} \le 1$$

Now, if $p_{i1} > 2$

$$W_{i1,j,t-2}+W_{i1,j,t-1} \le 1$$
$$W_{i1,j,t-2}+W_{i2,j,t-1} \le 1$$
$$...$$
$$W_{i1,j,t-2}+W_{inj,j,t-1} \le 1$$
$$W_{i1,j,t-2}+W_{i1,j,t} \le 1$$
$$W_{i1,j,t-2}+W_{i2,j,t} \le 1$$
$$..$$
$$W_{i1,j,t-2}+W_{inj,j,t} \le 1$$

From the lemma, we get

$$W_{i1,j,t-2}+W_{i1,j,t-1}+W_{i2,j,t-1}+..+W_{inj,j,t-1}+W_{i1,j,t}+W_{i2,j,t}+..+W_{inj,j,t} \le 1$$

Repeat for all $W_{i'j,t-2}$ for $p_{i'} > 2$

Repeat for all $W_{i'j,t-3}$ for $p_{i'} > 3$

...

Repeat for all $W_{i'j,t-pi+1}$ for $p_{i'} > p_i-1$

Finally, we get

$W_{i1,j,t-pi1+1}+.+W_{i1,j,t-1}+W_{i1,j,t}+W_{i2,j,t-pi2+1}+W_{i2,j,t-1}+.+W_{i2,j,t}+W_{inj,j,t-pnj+1}+W_{inj,j,t-1}+.+W_{inj,j,t} \leq 1$

Grouping terms in the above inequality yields

$$\sum_{t'=t-p1+1}^{t} W_{i1,j,t'} + \sum_{t'=t-p2+1}^{t} W_{i2,j,t'} +...+ \sum_{t'=t-pnj+1}^{t} W_{inj,j,t'} \leq 1$$

Further summing over all i, we get the constraint by Shah et al. (1991)

$$\sum_{i\in Ij} \sum_{t'=t-pi+1}^{t} W_{i,j,t'} \leq 1$$

References

1. Al-Khayyal, F.A. and Falk, J.E. (1983) Jointly constrained biconvex programming, *Mathematics of Operations Research* **8**, 273-286.
2. Applegate D. and Cook W. (1991). A Computational Study of the Job-Shop Scheduling Problem, *ORSA Journal on Computing*, **3**, No. 2, pp 149-156.
3. Balas, E (1974). "Disjunctive Programming : Properties of the Convex Hull of Feasible Points." MSRR #348, Carnegie Mellon University.
4. Balas, E. (1975). "Disjunctive Programming : Cutting Planes from Logical Conditions". Nonlinear Programming 2, O. L. Mangasarian et al., eds., Academic Press, 279-312.
5. Balas, E., Ceria, S. and Cornuejols, G. (1993). A Lift-and-Project Cutting Plane Algorithm for Mixed 0-1 Programs. *Mathematical Programming*, **58** (3), 295-324
6. Balas, E., and Mazzola, J.B. (1984). Nonlinear 0-1 Programming: Linearization Techniques. *Mathematical Programming*, **30**, 1-21.
7. Beale, E. M. L, and Tomlin, J. A. (1970). Special Facilities in a Mathematical programming System for Nonconvex problems using Ordered Set of Variables, in Proceedings of the *Fifth International Conference on Operational Research*, J. Lawrence, ed., Tavistock Publications, pp 447-454.
8. Benders, J. F. (1962). Partitioning Procedures for Solving Mixed Integer Variables Programming Problems, *Numerische Mathematik*, **4**, 238-252.
9. Birewar D.B and Grossmann I.E (1990). Simultaneous Synthesis, Sizing and Scheduling of Multiproduct Batch Plants, *Ind. Eng. Chem. Res.*, Vol 29, No11, pp 2242-2251
10. Borchers, B. and Mitchell, J.E. (1991). Using an Interior Point Method in a Branch and Bound Method for Integer Programming, R.P.I. Math. Report No. 195.
11. Borchers, B. and Mitchell, J.E. (1991). An Improved Branch and Bound Algorithm for Mixed-Integer Nonlinear Programs, R.P.I. Math. Report No. 200.
12. Brearly, A.L., Mitra, G. and Williams, H.P. (1975). An Analysis of Mathematical Programming Problems Prior to Applying the Simplex Method, *Mathematical Programming*, **8**, 54-83.
13. Brooke, A., Kendrick, D. and Meeraus, A. (1988). GAMS: A User's Guide. Scientific Press, Palo Alto.
14. Cavalier, T. M. and Soyster, A. L. (1987). Logical Deduction via Linear Programming. IMSE Working Paper 87-147, Dept. of Industrial and Management Systems Engineering, Pennsyvania State University.
15. Crowder, H. P., Johnson, E. L., and Padberg, M. W. (1983). Solving Large-Scale Zero-One Linear Programming Problems, *Operations Research*, **31**, 803-834.
16. Dakin, R. J. (1965). A Tree search Algorithm for Mixed Integer Programming Problems, *Computer Journal*, **8**, 250-255.
17. Driebeek, N., J. (1966). An Algorithm for the solution of Mixed Integer Programming Problems, *Management Science*, **12**, 576-587.
18. Duran, M.A. and Grossmann, I.E. (1986). An Outer-Approximation Algorithm for a Class of Mixed-Integer Nonlinear Programs. *Mathematical Programming* **36**, 307-339.

19. Faqir N.M and Karimi I.A (1990). Design of Multipurpose Batch Plants with Multiple Production Routes, *Proceedings FOCAPD'89*, Snowmass Village CO, pp 451-468
20. Fletcher R, Hall J.A. and Johns W.R. (1991). Flexible Retrofit Design of Multiproduct Batch Plants, *Comp & Chem. Eng.* **15**, 843-852
21. Floudas, C.A. and Visweswaran, V. (1990). A global optimization algorithm (GOP) for certain classes of nonconvex NLPs-I Theory, *Computers chem. Engng.* **14**, 1397-1417
22. Fourer, R., Gay, D.M. and Kernighan, B.W. (1990). A Modeling Language for Mathematical Programming, *Management Science*, **36**, 519-554.
23. Geoffrion, A.M. (1972). Generalized Benders Decomposition. *Journal of Optimization Theory and Applications*, **10**(4), 237-260.
24. Geoffrion,A.M. and Graves, G. (1974). Multicommodity Distribution System Design by Benders Decomposition, *Management Science*, **20**, 822-844.
25. Glover F.(1975). Improved Linear Integer Programming Formulations of Nonlinear Integer Problems, *Management Science*, Vol. 22, No. 4, pp 455-460
26. Gomory, R. E. (1960). An Algorithm for the Mixed Integer Problem, RM-2597, The Rand Corporation..
27. Grossmann, I.E. (1990). Mixed-Integer Nonlinear Programming Techniques for the Synthesis of Engineering Systems, *Research in Eng. Design*, **1**, 205-228.
28. Grossmann I.E and Sargent R.W.H, (1979) Optimum Design of Multipurpose Chemical Plants , *Ind.Eng.Chem.Proc.Des.Dev.* , Vol 18, No. 2, pp 343-348
29. Grossmann I.E, Voudouris V.T., Ghattas O.(1992). Mixed-Integer Linear Programming Reformulation for Some Nonlinear Discrete Design Optimization Problems, Recent Advances in Global Optimization (eds. Floudas, C.A.. and Pardalos, P.M.) , pp.478-512, Princeton University Press
30. Gupta, J.N.D. (1976). Optimal Flowshop Schedules with no Intermediate Storage Space. *Naval Res. Logis. Q.* **23,** 235-243.
31. Gupta, O.K. and Ravindran, V. (1985). Branch and Bound Experiments in Convex Nonlinear Integer Programming. *Management Science*, **31**(12), 1533-1546.
32. Hooker, J. N. (1988). Resolution vs Cutting Plane solution of Inference Problems: some computational experience. *Operations Research Letters*, **7**, 1(1988).
33. Jeroslow, R. G. and Lowe, J. K. (1984). Modelling with Integer Variables. *Mathematical Programming Study*, **22**, 167-184.
34. Jeroslow, R. G. and Lowe, J. K. (1985). Experimental results on the New Techniques for Integer Programming Formulations, *Journal of the Operational Research Society*, **36**(5), 393-403.
35. Jeroslow, R. E. and Wang, J. (1990). Solving propositional satisfiability problems, *Annals of Mathematics and AI*, **1**, 167-187.
36. Knopf F.C, Okos M.R, and Reklaitis G.V. (1982). Optimal Design of Batch/Semicontinuous Processes, *Ind.Eng.Chem.Proc.Des.Dev.* , Vol 21, No. 1, pp 79-86
37. Kocis, G.R. and Grossmann, I.E. (1987). Relaxation Strategy for the Structural Optimization of Process Flowsheets. *Industrial and Engineering Chemistry Research*, **26**(9),1869-1880.
38. Kocis, G.R. and Grossmann, I.E. (1989). Computational Experience with DICOPT Solving MINLP Problems in Process Synthesis Engineering. *Computers and Chem. Eng.* **13**, 307-315.
39. Kocis G.R., Grossmann I.E. (1988) Global Optimization of Nonconvex MINLP Problems in Process Synthesis, *Ind.Engng.Chem.Res.* **27**, 1407-1421.
40. Kondili E, Pantelides C.C and Sargent R.W.H. (1993). A General Algorithm for Short-term Scheduling of Batch Operations. I. MILP Formulation. *Computers and Chem. Eng..*, **17**, 211-228.
41. Krarup, J. and Bilde, O. (1977). Plant Location, Set Covering and Economic Lot Size: An O(mn) Algorithm for Structured Problems in L. Collatz et al. (eds), *Optimierung bei graphentheoretischen und ganzzahligen Problemen*, Int. Series of Numerical Mathematics, 36, 155-180, Birkhauser Verlag, Basel.
42. Ku, H. and Karimi, I. (1988) Scheduling in Serial Multiproduct Batch Processes with Finite Intermediate Storage: A Mixed Integer Linear Program Formulation, *Ind. Eng. Chem. Res.* **27**, 1840-1848.
43. Ku, H. and Karimi, I. (1991) An evaluation of simulated annealing for batch process scheduling, *Ind. Eng. Chem. Res.* **30**, 163-169.
44. Land, A. H., and Doig, A. G.(1960). An Automatic method for solving Discrete Programming Problems, *Econometrica*, **28**, 497-520.
45. Lovacz, L. and Schrijver, A. (1989). Cones of Matrices and Set Functions and 0-1 Optimization, Report BS-R8925, Centrum voor Wiskunde en Informatica.
46. Magnanti, T. L. and Wong, R. T. (1981). Acclerated Benders Decomposition: Algorithm Enhancement and Model Selection Criteria, *Operations Research*, **29**, 464-484.

47. Martin, R.K. and Schrage, L. (1985). Subset Coefficient Reduction Cuts for 0-1 Mixed-Integer Programming, *Operations Research*, 33, 505-526.
48. Mawekwang, H. and Murtagh, B.A. (1986). Solving Nonlinear Integer Programs with Large Scale Optimization Software. *Annals of Operations Research*, 5, 427-437.
49. Miller D.L and Pekny J.F. (1991). Exact solution of large asymmetric traveling salesman problems, *Science*, 251, pp 754-761.
50. Nabar, S.V. and Schrage (1990). Modeling and Solving Nonlinear Integer Programming Problems. Paper No. 22a, Annual AIChE Meeting, Chicago, IL.
51. Nemhauser, G. L and Wolsey, L (1988). Integer and Cominatorial Optimization. Wiley, New York.
52. OSL Release 2 (1991) Guide and Reference, IBM, Kingston, NY.
53. Papageorgaki S. and Reklaitis G.V (1990) Optimal Design of Multipurpose Batch plants-1. Problem Formulation , *Ind.Eng.Chem.Res.*,Vol 29, No. 10, pp 2054-2062
54. Papageorgaki S. and Reklaitis G.V (1990). Optimal Design of Multipurpose Batch plants-2. A Decomposition Solution Strategy , *Ind.Eng.Chem.Res.*,Vol 29, No. 10, pp 2062-2073
55. Papageorgaki S. and Reklaitis G.V. (1990). Mixed Integer Programming Approaches to Batch Chemical Process Design and Scheduling, ORSA/TIMS Meeting, Philadelphia.
56. Patel A.N., Mah R.S.H. and Karimi I.A. (1991). Preliminary design of multiproduct noncontinuous plants using simulted annealing, *Comp & Chem Eng.* 15, 451-470
57. Pekny J.F and Miller D.L. (1991). Exact solution of the No-Wait Flowshop Scheduling Problem with a comparison to heuristic methods, *Comp & Chem. Eng.*, Vol 15, No 11, pp741-748.
58. Petersen C.C.(1991). A Note on Transforming the Product of Variables to Linear Form in Linear Programs, *Working Paper*, Purdue University.
59. Quesada I. and Grossmann I.E. (1992). An LP/NLP based Branch anc Bound Algorithm for Convex MINLP Problems. *Comp. & Chem Eng.*, 16, 937-947.
60. Quesada I. and Grossmann I.E. (1992). Global Optimization Algorithm for Fractional and Bilinear Progams. Submitted for publication.
61. Rardin, R. L. and Choe, U.(1979). Tighter Relaxations of Fixed Charge Network Flow Problems, Georgia Institute of Technology, Industrial and Systems Engineering Report Series, #J-79-18, Atlanta.
62. Raman, R. and Grossmann, I. E. (1991). Relation between MILP modelling and Logical Inference for Process Synthesis, *Computers and Chemical Engineering*, 15(2), 73-84.
63. Raman, R. and Grossmann, I.E. (1992). Integration of Logic and Heuristic Knowledge in MINLP Optimization for Process Synthesis, *Computers and Chemical Engineering*, 16(3), 155-171.
64. Raman, R. and Grossmann, I.E. (1993). Symbolic Integration of Logic in Mixed-Integer Programming Techniques for Process Synthesis, to appear in *Computers and Chemical Engineering*.
65. Ravenmark D. and Rippin D.W.T. (1991). Structure and equipment for Multiproduct Batch Production, Paper No.133a, Presented in AIChE annulal meeting, Los Angeles, CA
66. Reklaitis G.V (1990) Progress and Issues in Computer-Aided Batch Process Design, *FOCAPD Proceedings*, Elsevier, NY, pp 241-275
67. Reklaitis G.V. (1991). "Perspectives on Scheduling and Planning of Process Operations", *Proceedings Fourth Int.Symp. on Proc. Systems Eng.*, Montebello, Quebec, Canada.
68. Rich S.H and Prokopakis G.J. (1986). Scheduling and Sequencing of Batch Operations in a Multipurpose Plant, *Ind.Eng.Chem.Res*, Vol. 25, No. 4, pp 979-988
69. Rich S.H and Prokopakis G.J. (1987). Multiple Routings and Reaction Paths in Project Scheduling, *Ind.Eng.Chem.Res*, Vol. 26, No. 9, pp 1940-1943
70. Sahinidis, N.V. and Grossmann, I.E. (1991). MINLP Model for Cyclic Multiproduct Scheduling on Continuous Parallel Lines, *Computers and Chem. Eng.*, 15, 85-103.
71. Sahinidis, N.V. and Grossmann, I.E. (1991). Reformulation of Multiperiod MILP Models for Planning and Scheduling of Chemical Processes, *Computers and Chem. Eng.*, 15, 255-272.
72. Sahinidis, N.V. and Grossmann, I.E. (1991). Convergence Properties of Generalized Benders Decomposition, *Computers and Chem. Eng.*, 15, 481-491.
73. Savelsbergh, M.W.P., Sigismandi, G.C. and Nemhauser, G.L. (1991) Functional Description of MINTO, a Mixed INTeger Optimizer, Georgia Tech., Atlanta.
74. Schrage, L. (1986). Linear, Integer and Quadratic Programming with LINDO, Scientific Press, Palo Alto.
75. SCICONIC/VM 2.11 (1991). Users Guide", Scicon Ltd, U.K.
76. Shah N. and Pantelides C.C. , (1991). Optimal Long-Term Campaign Planning and Design of Batch Operations, *Ind. Eng. Chem. Res.*, Vol 30, No. 10, pp 2308-2321
77. Shah N., Pantelides C.C. and Sargent, R.W.H. (1993). A General Algorithm for Short-term Scheduling of Batch Operations. II. Computational Issues. *Computers and Chem. Eng.*., 17, 229-244.

78. Sherali, H.D. and Alameddine, A. (1992) A new reformulation-linearization technique for bilinear programming problems, *Journal of Global Optimization*, **2**, 379-410.
79. Sherali H. and Adams W.(1988) A hierarchy of relaxations between the continuous and convex hull representations for zero-one programming problems, Technical Report, Virginia Polytechnic Institute.*t*
80. Sherali, H. D. and Adams, W. P. (1989). Hierarchy of relaxations and convex hull characterizations for mixed integer 0-1 programming problems. Technical Report, Virginia Polytechnic Institute.
81. Sparrow R.E, Forder G.J, Rippin D.W.T (1975) The Choice of Equipment Sizes for Multiproduct Batch Plant. Heuristic vs. Branch and Bound , *Ind.Eng. Chem.Proc.Des.Dev.* , Vol 14, No. 3, pp197-203
82. Straub, D.A. and I.E. Grossmann (1992). Evaluation and Optimization of Stochastic Flexibility in Multiproduct Batch Plants, *Comp.Chem.Eng.*, **16**, 69-87.
83. Suhami I. and Mah R.S.H, (1982) Optimal Design of Multipurpose Batch Plants, *Ind. Eng. Chem. Proc. Des. Dev.*, Vol 21, No. 1, pp 94-100
84. Sugden, S.J. (1992). A Class of Direct Search Methods for Nonlinear Integer Programming. Ph.D. thesis. Bond University, Queensland, Australia.
85. Swaney, R.E. (1990). Global solution of algebraic nonlinear programs. Paper No.22f, AIChE Meeting , Chicago, IL
86. Tomlin, J. A. (1971). An Improved Branch and Bound method for Integer Programming, *Operations Research*, **19**, 1070-1075.
87. Tomlin, J. A. (1988). Special Ordered Sets and an Application to Gas Supply Operations Planning. *Mathematical Programming*, **42**, 69-84.
88. Torres, F. E. (1991). Linearization of Mixed-Integer Products. *Mathematical Programming*, **49**,427-428.
89. Van Roy, T. J., and Wolsey, L. A. (1987). Solving Mixed-Integer Programming Problems Using Automatic Reformulation, *Operations Research*, **35**, pp.45-57.
90. Vaselenak J.A , Grossmann I.E. and Westerberg A.W. (1987). An Embedding Formulation for the Optimal Scheduling and Design of Multipurpose Batch Plants, *Ind.Eng.Chem.Res*,**26**, No1, pp139-148
91. Vaselenak J.A , Grossmann I.E. and Westerberg A.W (1987) Optimal Retrofit Design of multipurpose Batch Plants, *Ind.Eng.Chem.Res*, **26**, No. 4, pp718-726
92. Viswanathan, J. and Grossmann, I.E. (1990). A Combined Penalty Function and Outer-Approximation Method for MINLP Optimization. *Computers and Chem. Eng.* **14**(7),769-782.
93. Voudouris V.T and Grossmann I.E. (1992). Mixed Integer Linear Programming Reformulations for Batch Process Design with Discrete Equipment Sizes, *Ind.Eng.Chem.Res.*, **31**, pp.1314-1326.
94. Voudouris V.T and Grossmann I.E. (1992). MILP Model for the Scheduling and Design of Multipurpose Batch Plants. In preparation.
95. Voudouris, V.T. and Grosmann, I.E. (1993). Optimal Synthesis of Multiproduct Batch Plants with Cyclic Scheduling and Inventory Considerations. To appear in *Ind.Eng.Chem.Res.*
96. Wellons H.S and Reklaitis G.V. (1989). The Design of Multiproduct Batch Plants under Uncertainty with Staged Expansion , *Com . & Chem. Eng.*, **13**, No1/2, pp115-126
97. Wellons M.C and Reklaitis,G.V. (1991). Scheduling of Multipurpose Batch Chemical Plants. 1. Multiple Product Campaign Formation and Production Planning, *Ind.Eng.Chem.Res*, **30**, No. 4, pp688-705
98. Williams, P. (1988). Model Building in Mathematical Programming. Wiley, Chichester.
99. Yuan, X., Piboleau, S., and Domenech, S. (1989). Une Methode d'Optimisation Non Linaire en Variables Mixtes pour La Conception de Procedes. RAIRO Recherche Operationnele

Recent Developments in the Evaluation and Optimization of Flexible Chemical Processes

Ignacio E. Grossmann and David A. Straub

Department of Chemical Engineering, Carnegie Mellon University, Pittsburgh, PA 15213, USA

Abstract: The evaluation and optimization of flexible chemical processes remains one of the most challenging problems in Process Systems Engineering. In this paper an overview of recent methods for quantifying the property of flexibility in chemical plants will be presented. As will be shown, these methods are gradually evolving from deterministic worst-case measures for feasible operation to stochastic measures that account for the distribution functions of the uncertain parameters. Another trend is the simultaneous handling of discrete and continuous uncertainties with the aim of developing measures for flexibility and reliability that can be integrated within a common framework. It will be then shown how some of these measures can be incorporated in the optimization of chemical processes. In particular, the problem of optimization of flexibility for multiproduct batch plants will be discussed.

Keywords: flexibility, design under uncertainty, worst-case analysis, statistical design.

1. Introduction

The problem of accounting for uncertainty at the design stage is clearly a problem of great practical significance due to the variations that are commonly experienced in plant operation (e.g. changes in demands, fluctuations of feed compositions and equipment failure). Furthermore, at the design stage one must rely on values of technical parameters which are unlikely to be realized once a design is actually implemented (e.g. transfer coefficients and efficiencies). Finally, models that are used to predict the performance of a plant at the design stage may not even match the correct behavior trends of the process. In view of all these uncertainties, the common practice is to overdesign processes and/or perform ad-hoc case studies to try to verify the flexibility or robustness of a design. The pitfalls of such approaches, however, are well known and therefore have motivated the study and development of systematic techniques over the last 20 years ([4], [5]).

It is the purpose of this paper to provide an overview of recent techniques that have been developed for evaluating and optimizing flexibility in the face of uncertainties of continuous parameters and discrete states. This paper is in fact an updated version of a recent paper presented by the authors at the COPE Meeting in Barcelona ([7]). In this paper we will emphasize work that has been developed by our group at Carnegie Mellon. This paper will be organized as follows. The problem statements for the evaluation and optimization problems will be given first for deterministic and stochastic approaches. An overview will then be presented for different formulations and solution methods for the evaluation problems, followed by similar items for the optimization design problems. As will be shown, the reason for the recent trend towards the stochastic approach is that it offers a more general framework, especially for integrating continuous and discrete uncertainties which often arise in the design of batch processes. At the same time, however, the stochastic approach also involves a number of major challenges that still need to be overcome, especially for the optimization problems. A specific application to multiproduct batch plants will be presented to illustrate how the problem structure can be exploited in specific instances to simplify the optimization.

2. Problem Statements

It will be assumed that the model of a process is described by equations and inequalities of the form:

$$\begin{aligned} &h(d,z,x,\theta)=0 \\ &g(d,z,x,\theta)\leq 0 \\ &d=Dy \end{aligned} \tag{1}$$

where the variables are defined as follows:

- d - L vector of design variables that defines the structure and equipment sizes of a process
- z - n_z vector of control variables that can be adjusted during plant operation
- x - n_x vector of state variables that describe the behavior of a process
- θ - n_p vector of continuous uncertain parameters
- y - L vector of boolean variables that describes the unavailability (0) or availability (1) of the corresponding design variables d
- D - diagonal matrix whose elements in the diagonal correspond to the design variables d

For convenience in the presentation it will be assumed that the state variables x in (1) are eliminated from the equations $h(d,z,x,\theta)=0$; the model then reduces to

$$\begin{aligned} &f(d,z,\theta)\leq 0 \\ &d=Dy \end{aligned} \tag{2}$$

The evaluation problems that can then be considered for a fixed design D are as follows:

A) **Deterministic Problems**. Let y be fixed, and θ be described by a nominal value θ^N, expected deviations in the positive and negative directions $\Delta\theta^+$, $\Delta\theta^-$, and a set of inequalities $r(\theta) \leq 0$ to represent correlations of the parameters θ:

a) Problem A1: Determine if the design d = Dy, is feasible for every point θ in $T=\{\theta|\theta^N - \Delta\theta^- \leq \theta \leq \theta^N + \Delta\theta^+,\ r(\theta)\leq 0\}$

b) Problem A2: Determine the maximum deviation δ that design d = Dy can tolerate such that every point θ in $T(\delta)=\{\theta|\theta^N - \delta\Delta\theta^- \leq \theta \leq \theta^N + \delta\Delta\theta^+,\ r(\theta)\leq 0\}$ is feasible.

Problem (A1) corresponds to the *feasibility problem* discussed in Halemane and Grossmann [8], while problem (A2) corresponds to the *flexibility index problem* discussed in Swaney and Grossmann [25].

B) **Stochastic Problems** Let θ be described by a joint probability distribution function $j(\theta)$:

a) Problem B1: If y is fixed, determine the probability of feasible operation.

b) Problem B2: If the discrete probability p_ℓ for the availability of each piece of equipment ℓ is given, determine the expected probability of feasible operation.

Problem (B1) corresponds to evaluating the *stochastic flexibility* discussed in Pistikopoulos and Mazzuchi [17], while problem (B2) corresponds to evaluating the *expected stochastic flexibility* discussed in Straub and Grossmann [22].

As for the design optimization problems they will involve the selection of the matrix D so as to minimize cost and either a) satisfy the feasibility test (A1), or b) maximize the flexibility measure as given by (A2), (B1) or (B2), where the latter problem gives rise to a multiobjective optimization problem.

3. Evaluation for Deterministic Case

3.1 Formulations

In order to address problem (A1) for determining the feasibility of a fixed design d, consider first a fixed value of the continuous parameter θ. The feasibility for fixed d at the given θ value is then given by the following optimization problem [8] :

$$\begin{aligned} \psi\left(d,\theta\right) &= \min_{u,z} u \\ \text{s.t.} \quad f_j\left(d,z,\theta\right) &\leq u \quad j\in J \end{aligned} \tag{3}$$

where $\psi \leq 0$ indicates feasibility and $\psi > 0$ infeasibility. Note that the objective in problem (3) is to find a point z^* such that the maximum potential constraint violation is minimized.

In terms of the function $\psi(d,\theta)$, feasibility for every point $\theta \in T$, can be established by the formulation [8]:

$$\chi(d) = \max_{\theta \in T} \psi\left(d, \theta\right) \tag{4}$$

where $\chi(d) \leq 0$ indicates feasibility of the design d for every point in the parameter set T, and $\chi(d) > 0$ indicates infeasibility. Note that the max operator in (4) determines the point θ^* for which the largest potential constraint violation can occur.

As for the flexibility index problem (A2), the formulation is given by (see [25]):

$$\begin{aligned} F &= \max \delta \\ \text{s.t.} \quad & \max_{\theta \in T(\delta)} \psi\left(d, \theta\right) \leq 0 \\ & \delta \geq 0 \end{aligned} \tag{5}$$

where the objective is to inscribe the largest parameter set $T(\delta^*)$ in the feasible region projected in θ-space. An alternative formulation to problem (A2) is,

$$F = \min_{\tilde{\theta} \in \tilde{T}} \delta^*\left(\tilde{\theta}\right) \tag{6}$$

where

$$\begin{aligned} \delta^*\left(\tilde{\theta}\right) &= \max_{\delta, z} \delta \\ \text{s.t.} \quad & f_j\left(d, z, \theta\right) \leq 0 \quad j \in J \\ & \theta = \theta^N + \delta \tilde{\theta} \\ & \delta \geq 0 \end{aligned} \tag{7}$$

and $\tilde{T} = \left\{\tilde{\theta} \mid -\Delta\theta^- \leq \tilde{\theta} \leq \Delta\theta^+\right\}$

The objective in (6) is to find the maximum displacement which is possible along the displacement $\tilde{\theta}$ from the nominal value θ^N. Note that in both (5) and (6) the solution of the critical point θ^* lies at the boundary of the feasible region projected in θ-space.

3.2 Methods

Assume that no constraints $r(\theta) \leq 0$ are present for correlating parameter variations. Then the simplest methods for solving problems (A1) and (A2) are vertex enumeration schemes which rely on the assumption that the critical points θ^* lie at the vertexes of the sets T and $T(\delta^*)$. Such an assumption is only valid provided certain convexity conditions hold (see [25]).

Let $V = \{k\}$ correspond to the set of vertices in $T=\{\theta | \theta^N - \Delta\theta^- \leq \theta \leq \theta^N - \Delta\theta^+\}$. Then, problem (4) can be reformulated as

$$\chi(d) = \max_{k \in V} u^k \tag{8}$$

where

$$\begin{aligned} u^k = &\min_{u,z} u \\ \text{s.t.} \quad & f_j\left(d, z, \theta^k\right) \leq u \qquad j \in J \end{aligned} \tag{9}$$

That is, the problem reduces to solving the 2^{np} optimization problems in (9).

Likewise, problem (6) can be reformulated as

$$F = \min_{k \in V'} \delta^k \tag{10}$$

where

$$\begin{aligned} \delta^k = &\max_{\delta, z} \delta \\ \text{s.t.} \quad & f_j\left(d, z, \theta\right) \leq 0 \quad j \in J \\ & \theta = \theta^N + \Delta\theta^k \\ & \delta \geq 0 \end{aligned} \tag{11}$$

and $\Delta\theta^k$ is the displacement vector to vertex k. This problem again reduces to solving 2^{np} optimization problems in (11).

The problem of avoiding the exhaustive enumeration of all vertices, which increase exponentially with the number of parameters, has been addressed by Swaney and Grossmann [26] and Kabatek and Swaney [10] using implicit enumeration techniques. The latter authors has been able to solve problems with up to 20 parameters with such an approach.

An alternative method that does not rely on the assumption that critical points correspond to vertices, is the active set strategy of Grossmann and Floudas [3]. This method relies on the fact that the feasible region projected into the space of d and θ,

$$R(d,\theta)=\{\theta \mid \psi(d,\theta)\leq 0\} \tag{12}$$

(see Figure 1) can be expressed in terms of active sets of constraints $fj(d,z,\theta)=0$, $j \in J_A^k$ $k=1,...NAS$.

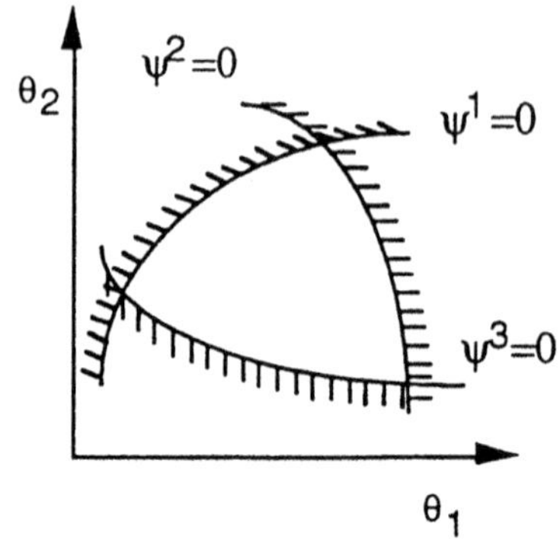

Figure 1 Constraints in the space of d and θ

These active sets are obtained from all subsets of non-zero multipliers that satisfy the Kuhn-Tucker conditions of problem (3)

$$\sum_{j\in J_A^k} \lambda_j^k = 1 \ , \ \sum_{j\in J_A^k} \lambda_j^k \frac{\partial f_j}{\partial z} = 0 \tag{13}$$

Pistikopoulos and Grossmann [15] have proposed a systematic enumeration procedure to identify the NAS active sets of constraints, provided that the corresponding submatrices in (13) are of full rank.

The projected parameter feasible region in (12) can then be expressed as

$$R(d,\theta)=\{\theta \mid \psi^k(d,\theta)\leq 0,\ k=1,\ldots,NAS\} \tag{14}$$

where

$$\begin{aligned} \psi^k(d,\theta) &= \min u \\ \text{s.t.} \quad f_j(d,z,\theta) &= u \quad j\in J_A^k \end{aligned} \tag{15}$$

The above active set strategy by Grossmann and Floudas [3] does not require, however, the *a-priori* identification of constraints ψ^k. This is accomplished by reformulating problem (4) with the Kuhn-Tucker conditions of (3) embedded in it, and expressed in terms of 0-1 variables w_j for modelling the complementarity conditions.

For the case of problem (A1), this leads to the mixed-integer optimization problem

$$
\begin{aligned}
&\chi(d) = \max u \\
\text{s.t.}\quad & s_j + f_j(d, z, \theta) = u \qquad j \in J \\
& \sum_{j \in J} \lambda_j = 1 \\
& \sum_{j \in J} \lambda_j \frac{\partial f_j}{\partial z} = 0 \\
& \left.\begin{aligned} & \lambda_j - w_j \le 1 \\ & s_j - U(1 - w_j) \le 0 \end{aligned}\right\} \quad j \in J \\
& \sum_{j \in J} w_j \le n_z + 1 \\
& \theta^N - \Delta\theta^- \le \theta \le \theta^N + \Delta\theta^+ \\
& r(\theta) \le 0 \\
& w_j = 0,1 \;\; ; \;\; \lambda_j \, ; \, s_j \ge 0 \quad j \in J
\end{aligned} \tag{16}
$$

where U is a valid upper bound for the violation of constraints. For the case of problem (A2), the calculation of the flexibility index can be formulated as the mixed-integer optimization problem (17). In both cases, constraints f_j that are linear in z and θ give rise to MILP problems which can be solved with standard branch and bound methods. For nonlinear constraints models (16) and (17) give rise to MINLP problems which can be solved with Generalized Benders Decomposition [2] or with any of the variants of the outer-approximation method (e.g. [28]). Also, for the case when $n_z + 1$ constraints are assumed to be active and the constraints are monotone in z, Grossmann and Floudas [3] decompose the MINLP into a sequence of NLP optimization problems, each corresponding to an active set which is identified *a-priori* from the stationary conditions of the Lagrangian.

$$
\begin{aligned}
&F = \min \delta \\
\text{s.t.}\quad & s_j + f_j(d, z, \theta) = 0 \qquad j \in J \\
& \sum_{j \in J} \lambda_j = 1 \\
& \sum_{j \in J} \lambda_j \frac{\partial f_j}{\partial z} = 0 \\
& \left.\begin{aligned} & \lambda_j - w_j \le 1 \\ & s_j - U(1 - w_j) \le 0 \end{aligned}\right\} \quad j \in J \\
& \sum_{j \in J} w_j \le n_z + 1 \\
& \theta^N - \delta\Delta\theta^- \le \theta \le \theta^N + \delta\Delta\theta^+ \\
& r(\theta) \le 0 \\
& \delta \ge 0 \;\; ; \;\; w_j = 0,1 \;\; , \;\; \lambda_j \, , \, s_j \ge 0 \quad j \in J
\end{aligned} \tag{17}
$$

4. Evaluation for Stochastic Case

4.1 Formulations

In order to formulate problem (B1), the probability of feasible operation given a joint distribution for θ, $j(\theta)$, this involves the evaluation of the multiple integral

$$SF(d) = \int_{\theta:\psi(d,\theta)\leq 0} j(\theta)\, d\theta \tag{18}$$

where SF(d) is the stochastic flexibility for a given design (see [17], [22]). Note that this integral must be evaluated over the feasible region projected in θ space (see eqn. (12) and Figure 2). In Figure 2 the circles represent the contours of the joint distribution function j.

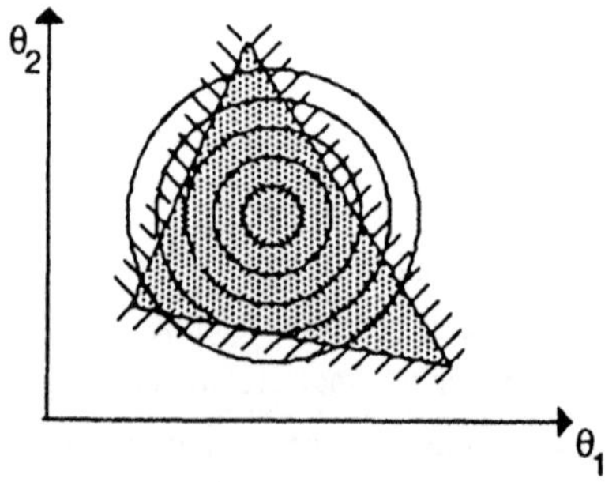

Figure 2.SF is evaluated by integration over the shaded area.

For the case when uncertainties are also involved in the equipment, discrete states result from all the combinations of the vector y. It is convenient to define for each state s the index sets

$$Y_1^s = \{\ell \mid y_\ell^s = 1\}\ ,\ Y_0^s = \{\ell \mid y_\ell^s = 0\} \tag{19}$$

to denote the identity of available and unavailable equipment. Note that state s is defined by a particular choice of y^s which in turn determines the design variables for that state, $d^s = Dy^s$. Also, denoting by p_ℓ the probability that equipment ℓ be available, the probability of each state P(s) is given by:

$$P(s) = \prod_{\ell \in Y_1^s} p_\ell \prod_{\ell \in Y_0^s} (1 - p_\ell) \quad s=1,...2^L \tag{20}$$

In this way the probability of feasible operation over both the discrete and continuous uncertainties (i.e. problem (B2)) is given by

$$E(SF) = \sum_{s=1}^{2^L} SF(s)\, P(s) \tag{21}$$

where E(SF) is the expected stochastic flexibility as proposed by Straub and Grossmann [22].

4.2 Methods

The solution of problems (18) and (21) poses great computational challenges. Firstly, because (18) involves a multiple integral over an implicitly defined domain. Secondly, (19) involves the evaluation of these integrals for 2^L states. For this reason solution methods for these problems have been only reported for the case of linear constraints:

$$f_j\left(d, z, \theta\right) = a_j^T d + b_j^T z + c_j^T \theta + c_j^O \tag{22}$$

Pistikopoulos and Mazzuchi [17] have proposed the computation of bounds for the stochastic flexibility, SF(d) by assuming that j(θ) is a normal distribution. Firstly, expressing the feasibility function $\psi^k(d,\theta)$ as given in (15) through the Lagrangian, this yields for (22) the linear equation

$$\psi^k\left(d, \theta\right) = \sum_{j \in J_A^k} \lambda_j^k \left[c_j^T \theta + c_j'\right] \tag{23}$$

where $c_j' = c_j^O + a_j^T d$. Since (23) is linear in θ and these are normally distributed $N(\mu_\theta, \Sigma_\theta)$, then the distribution function $\phi(\psi^k)$ is also normally distributed with mean and variance,

$$\mu_{\psi^k} = \sum_{j \in J_A^k} \lambda_j^k \left[c_j^T \mu_\theta + c_j'\right] \tag{24}$$

$$\sigma_{\psi^k}^2 = \left[\sum_{j \in J_A^k} \lambda_j^k c_j^T\right] \Sigma_\theta \left[\sum_{j \in J_A^k} \lambda_j^k c_j^T\right]^T \tag{25}$$

where Σ_θ is the variance-covariance matrix for the parameters θ.

The probability of feasible operation for the above set k is then given by the one-dimensional integral

$$SF^k = \int_{-\infty}^{0} \phi\left(\psi^k\right) d\psi^k \tag{26}$$

which can be readily evaluated. For multiple sets the probability is defined as follows (shown for 2 sets),

$$SF^{k\text{-}\ell} = \int_{-\infty}^{0} \int_{-\infty}^{0} \phi_{MVN}\left(\psi^k \psi^\ell\right) d\psi^\ell d\psi^k \tag{26b}$$

where ϕ_{MVN} is the multivariate normal probability distribution function.

Lower and upper bounds of the stochastic flexibility SF(d) are then given by

$$SF^{L}(d) = \sum_{k=1}^{NAS} SF^{k} - \sum_{k<l} SF^{kl} + \sum_{k<l<m} SF^{k+m} \dots \tag{27}$$

$$SF^{U}(d) = \min_{q=1,Q} \left\{ \prod_{k \in JA(q)} SF^{k} \right\} \tag{28}$$

where JA(q) $\subseteq$ JA, q=1,...Q, are all possible subsets of the inequalities $\psi^{k}(d,\theta) \leq 0$, k=1,...NAS. It should be noted that the bounds in (27) and (28) are often quite tight providing good estimates for the stochastic flexibility.

Straub and Grossmann [22] have proposed a numerical approximation scheme for arbitrary distribution functions using Gaussian quadrature within the feasible region of the projected region R(d,θ) (see Fig. 3).

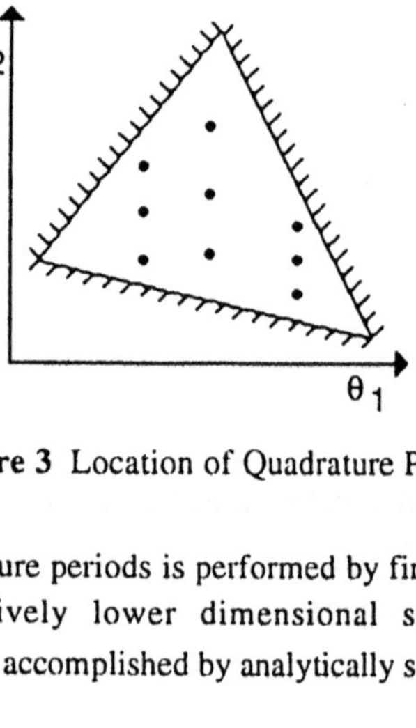

Figure 3 Location of Quadrature Points.

The location of the quadrature periods is performed by first projecting the functions $\psi^{k}(d,\theta)$, k=1...NAS into successively lower dimensional spaces in θ ; i.e. :$[\theta_1, \theta_2, \dots \theta_M] \rightarrow [\theta_1, \theta_2, \dots \theta_{M-1}] \dots \rightarrow [\theta_1]$ This is accomplished by analytically solving the problems r=1,2,...,M-1:

$$\begin{aligned} \psi^{r+1,k}\left(d, \theta_1, \theta_2, \dots \theta_{M-r}\right) &= \min u \\ \text{s.t.} \quad \psi^{r,k}\left(d, \theta_1, \theta_2, \dots\right) &\leq u \quad k=1 \dots NAS(r) \end{aligned} \tag{29}$$

where $\psi^{1,k} = \psi^{k}(d,\theta) = \sum_{j \in J_A^k} f_j(d,z,\theta)$, and NAS(r) is the number of active sets at the rth state of the projection.

In the next step, lower and upper bounds are generated together with the quadrature points for each θ_i component in the order $\theta_1 \rightarrow \theta_2 \ldots \rightarrow \theta_M$. This is accomplished by using the analytical expressions $\psi^{r,k}(d,\theta_1,\theta_2,\ldots,\theta_{M+r-1})$ in the order r=M, M-1,..., to determine the bounds. For instance, the bounds θ_1^L and θ_1^U are determined from the linear inequalities $\psi^{M,k}(d,\theta_1) \leq 0$, k=1,...NAS(M). The quadrature points θ_1^{q1} then are given by:

$$\theta_1^{q_1} = \frac{v_{q1}\left(\theta_1^U - \theta_1^L\right) + \theta_1^U + \theta_1^L}{2} \qquad q_1=1,\ldots QP_1 \tag{30}$$

where v_{q1}, q1=1,..., QP_1 represent the location of QP_1 quadrature points in [-1,1]. In the next step, bounds for θ_2 are computed for each θ_1^{q1} from $\psi^{M-1,k}(d,\theta_1,\theta_2) \leq 0$, k=1,...NAS(M-1). These bounds are denoted as $\theta_2^L(\theta_1^{q1})$ since they depend on the value of θ_1^{q1}. Quadrature points are then computed as in (30) and the procedure continues until the the bounds $\theta_M^L\left(\theta_1^{q_1},\theta_2^{q_1q_2}..\theta_{M-1}^{q_1q_2q_{M-1}}\right)$, $\theta_M^U\left(\theta_1^{q_1},\theta_2^{q_1q_2}\ldots\theta_{M-1}^{q_1q_2q_{M-1}}\right)$ and quadrature points $\theta_M^{q_1q_2\ldots q_M}$ are determined.

The numerical approximation to (18) is then given by

$$SF(d) = \frac{\theta_1^U - \theta_1^L}{2} \sum_{q=1}^{QP1} w_{q2} \frac{\left[\theta_2^U(\theta_1^{q_1} - \theta_2^L(\theta_1^{q_1})\right]}{2} \sum_{q2=1}^{QP2} w_{q2} \frac{\left[\theta_3^U(\theta_1^{q_1},\theta_2^{q_1q_2}) - \theta_3^L(\theta_1^{q_1},\theta_2^{q_1q_2})\right]}{2} \cdots \frac{\theta_M^U(\theta_1^{q_1},\ldots\theta_{M-1}^{q_1q_2\ldots q_{M-1}}) - \theta_M^L(\theta_1^{q_1},\ldots\theta_{M-1}^{q_1q_2\ldots q_{M-1}})}{2}$$
$$\times \sum_{q_M=1}^{QPM} w_{qM}\, j\left(\theta_1^{q_1},\theta_2^{q_1q_2},\ldots\theta_M^{q_1q_2\ldots q_M}\right) \tag{31}$$

where w_{q_i} are the weights corresponding to each quadrature point.

It should be noted that equation (31) becomes computationally more expensive as the number of parameters θ increases which is not the case with the bounds in (27) and (28). However, as pointed out before, (31) can be applied to any distribution function (e.g. normal, beta, log) while the bounds can only be applied to normal distribution. Also, both methods require the identification of active sets which may become large if the number of constraints is large.

As for the solution of equation (21) for the expected flexibility, Straub and Grossmann [22] have developed a bounding scheme that requires the examination of relatively few states despite the fact that these can become quite large in number. They represent the states through a network as shown in Fig. 4.

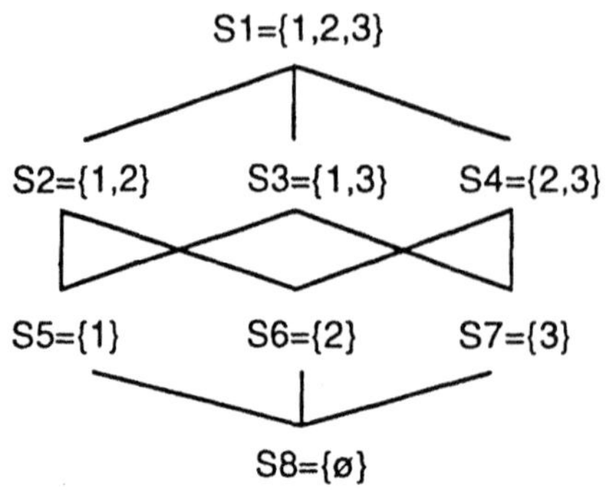

Figure 4 State Network showing different possible sets of active units.

Here the top state has all the units active (i.e. $y_\ell = 1$), while the bottom state has all units inactive. Since the states with active units will usually have the higher probability, the evaluation starts with the top state.

At any point of the search the following index sets are defined:

$$E=\{s|SF(s) \text{ is evaluated}\}$$
$$U=\{s|SF(s) \text{ is not evaluated}\} \tag{32}$$

The lower and upper bounds are then given as follows:

$$E(SF)^L = \sum_{s\in E} SF(s)\ P(s)$$
$$E(SF)^U = \sum_{s\in E} SF(s)\ P(s) + \sum_{s\in U} BSF(s)\ P(s) \tag{33}$$

where BSF(s) are valid upper bounds that are propagated through the subnetwork from higher states that have been evaluated. Convergence with this scheme for a small tolerance is normally achieved within 5 to 6 state evaluations (see Figure 5) provided the discrete probabilities $p_\ell > 0.5$. The significance of this method is that it allows the evaluation of flexibility and reliability within a single measure accounting for the interactions of the two.

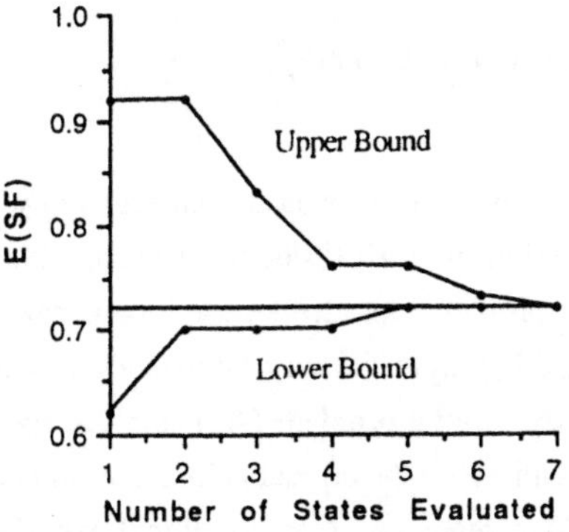

Figure 5 Example of the progression of the bounds.

5. Design Optimization

Most of the previous work ([9], [11]) has only considered the effect of the continuous uncertain parameters θ for the design optimization, and for which the minimization of the expected value of the cost function has been considered using a two-stage strategy:

$$\min_{d} E_{\theta}\left[\min_{z} C\left(d, z, \theta\right) \mid f\left(d, z, \theta\right) \leq 0\right] \tag{34}$$

In order to handle infeasibilities in the inner minimization, one approach is to assign penalties for the violation of constraints (e.g. $C(d,z,\theta)=\overline{C}$ if $f(d,z,\theta) > 0$. This however can lead to discontinuities. The other approach is to enforce feasibility for a specified flexibility index F (e.g. [8]) through the parameter set $T(F)=\{\theta \mid \theta^L - F\Delta\theta^- \leq \theta \leq \theta^U + F\Delta\theta^+,\ r(\theta) \leq 0\}$. In this case (34) is formulated as

$$\begin{aligned} &\min_{d} \ E_{\theta \in T(F)} \left[\min_{z} C\left(d, z, \theta\right) \Big| f\left(d, z, \theta\right) \leq 0\right] \\ &\quad \text{s.t.} \quad \max_{\theta \in T(F)} \psi\left(d, \theta\right) \leq 0 \end{aligned} \tag{35}$$

A particular case of (35) is when only a discrete set of points θ^k, k=1..K are specified which then gives rise to the problem

$$\begin{aligned} &\min_{d, z1, .. zk} \sum_{k=1}^{K} w_k C\left(d, z^k, \theta^k\right) \\ &\text{s.t.} \quad f\left(d, z^k, \theta^k\right) \leq 0 \qquad k=1..K \end{aligned} \tag{36}$$

where w_k are weights that are assigned to each point θ^k, and $\sum_{k=1}^{K} w_k = 1$.

Problem (36) can be interpreted as a multiperiod design problem which is an important problem in its own right for the design of flexible chemical plants. However, as shown by Halemane and Grossmann [8] this problem can also be used to approximate the solution of (35). This is accomplished by selecting an initial set of points θ^k, solving problem (36) and verifying its feasibility over T(F) by solving problem (A1) as given by (4). If the design is feasible the procedure terminates. Otherwise the critical point from (4) is included to the set of K θ points and the solution of (36) is repeated. Computational experience has shown that commonly one or two major iterations must be performed to achieve feasibility with this method (e.g. see [3]).

While the above procedure can be applied to general linear and nonlinear problems, one can exploit the structure for specialized cases. For instance, consider the case of constraints that are linear in d, z, and θ, and where the objective function only involves the design variables d. This case commonly arises in retrofit design problems.

As shown by Pistikopoulos and Grossmann [13], equation (23) holds for linear constraints. Therefore, the constraint in (35) can be simplified into NAS inequalities as shown in the following model:

$$\begin{aligned} &\min_{d} \; C(d) \\ \text{s.t.} \quad &\sum_{j \in J_A^k} \lambda_j^k \left[c_j^T \theta^{ck} + c_j^o + a_j^T d \right] \le 0 \qquad k=1..NAS \\ &d^L \le d \le d^u \end{aligned} \tag{37}$$

where

$$\theta_i^{ck} = \begin{cases} \theta^N + F\Delta\theta^+ & \text{if } \dfrac{\partial \psi^k}{\partial \theta_i} > 0 \\[2ex] \theta^N - F\Delta\theta^- & \text{if } \dfrac{\partial \psi^k}{\partial \theta_i} < 0 \end{cases}$$

The significance of problem (37) is that the optimal design can be obtained through one single optimization which however requires prior identification of the NAS active sets.

Pistikopoulos and Grossmann [13] have presented an alternative formulation to (37) in which one can easily derive the trade-off curve of cost versus the flexibility index. The formulation is given by

$$\begin{aligned} &\min_{\Delta d} \; C\left(d^E + \Delta d\right) \\ &\text{s.t.} \;\; \delta^k \geq F \\ &\delta^k = \delta_E^k + \sum_{l=1}^{L} \sigma_l^k \, \Delta d_l \end{aligned} \Bigg\} \; k=1,..NAS \qquad (38)$$

$$\Delta d^u \leq \Delta d \leq \Delta d^U \; , \; \delta^k \geq 0$$

where δ_E^k is the flexibility index for active set k at the base design d_E and $\sigma_l^k = \frac{\partial \delta^k}{\partial \psi^k} \frac{\partial \psi^k}{\partial d_l}$ are sensitivity coefficients that can be determined explicitly; Δd are design changes with respect to the existing design d_E.

Also, these authors extended the formulation in (37) to the case of nonlinear constraints. Here, the inequalities in (37) are augmented within an iterative procedure similar to the scheme based on the use of the multiperiod design problem, except that problem (15) is solved for each active set to determine the critical points and multipliers.

Finally, the determination of the optimal degree of flexibility can be formulated for the case of linear constraints as

$$\max Z = \mathop{E}_{\theta \in T(F)} \left\{ \max_z p(z, \theta) \,|\, f(d, z, \theta) \leq 0 \right\} - C(\Delta d)$$

$$\begin{aligned} &\text{st.} \;\; \delta^k \geq F \\ &\delta^k = \delta_E^k + \sum_{l=1}^{L} \sigma_l^k \Delta d_l \end{aligned} \Bigg\} \; k=1,..NAS \qquad (39)$$

$$d = d_E + \Delta d \; , \; \delta^k \geq 0$$

where $p(z,\theta)$ is a profit function.

Pistikopoulos and Grossmann [14] simplified this problem as maximizing the revenue subject to minimizing the investment cost; that is (see Fig. 6):

$$
\begin{aligned}
&\max_{F} \quad Z = R(F) - C(F) \\
&\text{s.t.} \quad C(F) = \min C\,(\Delta d) \\
&\qquad\qquad \delta^{k} \geq F \\
&\qquad \delta^{k} = \delta_{E}^{k} + \sum_{\ell=1}^{L} \sigma_{\ell}^{k} \Delta d_{\ell}
\end{aligned}
\tag{40}
$$

and where

$$
\begin{aligned}
&R(F) = \underset{\theta \in T}{E} \left\{ \max_{z} p\,(z, \theta) \,|\, f\,(d, z, \theta) \leq 0 \right\} \\
&\qquad\qquad \Delta d = \arg\left[C(F)\right]
\end{aligned}
\tag{41}
$$

which is solved by a modified Cartesian integration method. Since problem (40) is expressed in terms of only the flexibility index F, its optimal value is found by a direct search method.

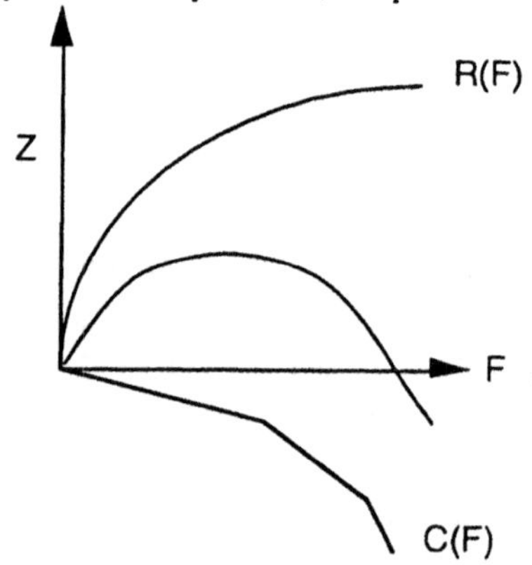

Figure 6 Curves for Determination of Optimal Flexibility

6. Application to Multiproduct Batch Design

The methods presented in the previous section have been only applied to continuous processes. On the other hand batch processes offer also an interesting application since these plants are built because of their flexibility for manufacturing several products. Reinhardt and Rippin [19], [20] have reported a design method when demand uncertainties are described by distribution functions. Wellons and Reklaitis [29] have developed a design method for staged expansions for the same type of uncertainties. In this section we will summarize the recent work by Straub and Grossmann [23] which accounts for uncertainties in the demands (continuous parameter) and equipment failure (states). This will serve to illustrate some of the concepts of Section 4 and show how the structure of the problem can be exploited to simplify the calculations; particularly the optimization of the stochastic flexibility. Consider the model for the design of multiproduct batch plants with single product campaigns (see [6]):

$$
\begin{aligned}
&\min_{NV} \sum_{j=1}^{M} \alpha_j N_j V_j^{\beta_j} \\
&\text{s.t.} \quad \left.\begin{aligned} V_j &\geq S_{ij} B_i \\ T_{Li} &\geq t_{ij}/N_j \end{aligned}\right\} \quad i=1,NP,\ j=1,M \\
&\sum_{i=1}^{N} \frac{Q_i}{B_i} T_{Li} \leq H \\
&V_j^L \leq V_j \leq V_j^U \ , \quad N_j=1,2.. \quad j=1..M \\
&B_i \ , \ T_{Li} \geq 0 \qquad i=1,NP
\end{aligned} \tag{42}
$$

Although problem (42) is nonlinear, for fixed design variables V_j (sizes), N_j (number of parallel units), the feasible region can be described by the linear inequality

$$\sum_{i=1}^{NP} Q_i \gamma_i \leq H \tag{43}$$

where $\quad \gamma_i = \max_j \{t_{ij}/N_j\}/\min_j \{V_j/S_{ij}\}$.

If we define

$$H_A = \sum_{i=1}^{NP} Q_i \gamma_i \tag{44}$$

then the problem of calculating the probability of feasible operation for uncertain demands Q_i, i=1,N, can be expressed through the one-dimensional integral

$$SF = \int_0^H \phi(H_A)\, dH_A \tag{45}$$

which avoids the direct solution of the multiple integral in (18). Furthermore, the distribution $\phi(H_A)$ can be easily determined if normal distributions are assumed for the product demands with mean μ_{Q_i} and variance $\sigma^2_{Q_i}$. Then proceeding in a similar way as in (24) and (25) the mean and the variance of $\phi(H_A)$ are given by

$$
\begin{aligned}
\mu_{H_A} &= \sum_{i=1}^{NP} \gamma_i \mu_{Q_i} \\
\sigma^2_{H_A} &= \sum_{i=1}^{NP} \gamma_i^2 \sigma^2_{Q_i}
\end{aligned} \tag{46}
$$

with which the integral in (45) can be readily evaluated for the stochastic flexibility.

As for the expected stochastic flexibility, let p_j be the probability that a unit in stage j is available. Also let n_j^s, j=1,M be the number of units that are available for any given state s. Then it can be shown that the number of feasible states where at least some production can be obtained is given by $TFS = \prod_{j=1}^{M} (N_j)$, and that the probability of each state is given by

$$P(s) = \prod_{j=1}^{M} \frac{N_j!}{(n_j^s)!\,(N_j-n_j^s)!}\, p_j^{n_j^s} (1-p_j)^{(N_j-n_j^s)} \tag{47}$$

In this way the expected stochastic flexibility can be expressed as

$$E(SF) = \sum_{s=1}^{TFS} SF(s)\; P(s) \tag{48}$$

where SF(s) and P(s) are given by (45) and (47), respectively. The value of E(SF) can then be obtained by applying the bounding procedure at the end of Section 4 (eqtns. (32) and (33)).

In order to determine the sizes V_j and number of parallel units N_j that maximize the stochastic flexibility (i.e. only uncertainties in the demands) given a limit for the capital investment, $\overline{C}$, one would have to optimize in principle the integral in (45) over the constraint set in (42). However, this can be avoided in view of the fact that maximizing the normal deviate $z=(H-\mu_{HA})/\sigma_{HA}$, is equivalent to maximizing the integral. Thus by applying appropriate exponential transformations to (42) to convexify the problem, the optimal design that maximizes the stochastic flexibility for a limit in the investment cost, can be formulated as the MINLP (see [23]):

$$\max\ z=(H-\mu_{HA})/\sigma_{HA}$$

$$\text{s.t.}\quad \left.\begin{aligned} & b_i \le v_j - \log(S_{ij}) \\ & t_{Li} \ge \log(t_{ij}) - \eta_j \end{aligned}\right\} \begin{aligned} & i=1,\dots,NP \\ & j=1,\dots,M \end{aligned}$$

$$\left.\begin{aligned} & \eta_j = \sum_r w_{jr} \log(r) \\ & \sum_r w_{jr} = 1 \end{aligned}\right\}\ j=1,\dots,M$$

$$\sum_j \alpha_j \exp(\eta_j + \beta_j v_j) \le C$$

$$\xi_i = t_{Li} - b_i \quad i=1,\dots,NP$$

$$\mu_{HA} = \sum_i \exp(\xi_i)\, \mu_{Q_i} \tag{49}$$

$$\sigma^2_{HA} = \sum_i \exp(2\xi_i)\, \sigma^2_{Q_i}$$

$$\ln (B_i^L) \le b_i \le \ln (B_i^U) \quad i=1,\ldots,NP$$
$$\ln (V_j^L) \le v_j \le \ln (V_j^U) \quad j=1,\ldots,M$$
$$\ln (T_{Li}^L) \le tl_i \le \ln (T_{Li}^U)$$
$$-\infty \le \xi_i \le \infty \qquad i=1,\ldots,NP$$
$$H, \mu_{H_A}, \sigma_{H_A}^2 \ge 0$$
$$\ln (V_j^L) \le v_j \le \ln (V_j^U) \quad j=1,\ldots,M$$
$$w_{jr}=0,1 \quad j=1,\ldots,M \; r=1,\ldots, N_j^U$$

By solving this MINLP for different values of $\overline{C}$ one can then determine trade-off curves of the expected stochastic flexibility versus cost (see Figure 7). Also, note that if the number of parallel units is fixed (49) reduces to an NLP problem.

As for the optimization of expected flexibility, problem (49) can be extended as a multiperiod design problem if N_j is fixed, where each period corresponds to a given state s. There is no need to solve, however, for all the states since a preanalysis can easily establish the relative magnitudes of (47) and valid upper bounds for SF(s). The optimization of N_j and V_j is in principle considerably more complicated. However, here Straub and Grossmann [22] have developed an enumeration procedure that relies on the state network representation and which minimizes the number of multiperiod optimization problems that need to be examined. The details of this method can be found in their paper.

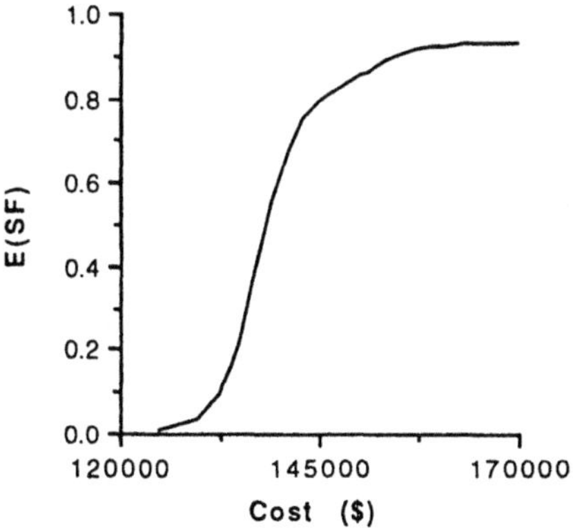

Figure 7 Trade-off Curve

7. General Nonlinear Models

The ideas developed in Straub and Grossmann [22] to evaluate the SF with linear models have recently been extended to nonlinear models, see Straub and Grossmann [24] and Straub

[21]. In addition the new concepts allow a straightforward extension to the design optimization problem.

Observe that eqn (18) can be written as follows for a specified design d,

$$SF=\int_{\theta_1^L}^{\theta_1^U}\int_{\theta_2^L(\theta_1)}^{\theta_2^U(\theta_1)}\cdots\int_{\theta_M^L(\theta_1,\theta_2,\ldots,\theta_{M-1})}^{\theta_M^U(\theta_1,\theta_2,\ldots,\theta_{M-1})} j(\theta)\; d\theta_M \ldots d\theta_2\, d\theta_1 \tag{50}$$

With nonlinear models in is difficult to analytically determine the constraints ψ that lead to the bounds on θ in the linear case. Thus the bounds on each θ need to be determined by a nonlinear programming problem: max $\theta_1{}^U$ s.t. h=0, g≤0. The difficulty here is that this may potentially involve a large number of optimization problems. However, it has been shown that a single NLP can be used to determine the bounds in place of all of the individual NLP's. The benefit of the single NLP is that to extend the SF evaluation problem to that of the design optimization problem one need only consider the design variables as decision variables instead of parameters. The NLP for design optimization is shown below,

$$\max SF=\frac{\theta_1^U-\theta_1^L}{2}\sum_{q1=1}^{Q1} w_1^{q1}\frac{(\theta_2^{U\,q1}-\theta_2^{L\,q1})}{2}\sum_{q2=1}^{Q2} w_2^{q2}\, j(\theta_1^{q1},\theta_2^{q1,q2})$$

$$\begin{aligned}
\text{s.t.}\quad & g_1^L(d,z^{(\cdot)},x^{(\cdot)},\theta_1^L,\theta_2^{(\cdot)})\le 0\\
& g_1^U(d,z^{(\cdot)},x^{(\cdot)},\theta_1^U,\theta_2^{(\cdot)})\le 0\\
& \theta_1^{q1}=0.5\,[\theta_1^U(1+v_1^{q1})+\theta_1^L(1-v_1^{q1})] \qquad q_1=1,\ldots Q_1\\
& g_2^{L\,q1}(d,z^{(\cdot)},x^{(\cdot)},\theta_1^{q1},\theta_2^{L\,q1})\le 0 \quad q_1=1,\ldots Q_1\\
& g_2^{U\,q1}(d,z^{(\cdot)},x^{(\cdot)},\theta_1^{q1},\theta_2^{U\,q1})\le 0 \quad q_1=1,\ldots Q_1\\
& \theta_2^{q1\,q2}=0.5\,[\theta_2^{U\,q1}(1+v_2^{q2})+\theta_2^{L\,q1}(1-v_2^{q2})] \qquad q_1=1,\ldots Q_1\;\; q_2=1,\ldots Q_2\\
& \text{cost}(d)\le\alpha\\
& \theta_1^{MIN}\le\theta_1^L\le\theta_1^U\le\theta_1^{MAX}\\
& \theta_2^{MIN}\le\theta_2^{L\,q1}\le\theta_2^{U\,q1}\le\theta_2^{MAX} \quad q_1=1,\ldots Q_1\\
& \theta_2^{MIN}\le\theta_2^{(\cdot)}\le\theta_2^{MAX}\\
& d\in D
\end{aligned} \tag{51}$$

Note that only the inequality constraints are shown for simplicity in presentation. Also note that the quadrature approximation to the SF , eqn. (31), is used as the objective function and the constraints defining the location of the quadrature points are also included. Finally, this NLP model incorporates explicitly the design variables d with which one can optimize the stochastic flexibility subject to a cost constraint and in that way generate trade-off curves as the one in Fig. 7.

The authors in [24] also address the computational difficulties of the program above. They propose a solution method based on Generalized Benders Decomposition [2]. The solution technique involves decomposing the large NLP into a series of smaller NLP's that are easier to solve. The techniques have effectively been applied to a reactor design problem and to a small flowsheet problem to determine the optimal size of the various pieces of equipment.

8. Conclusions

This paper has given an overview of methods and formulations for evaluating and optimizing flexibility in chemical processes. As has been shown, deterministic methods have reached a stage of maturity whereby their wider application, like in process simulators, should be technically feasible, although not necessarily trivial (e.g. computation of flexibility index and multiperiod design problems). Stochastic methods on the other hand are in principle computationally more expensive, except for few specific cases (e.g. bounds for linear models, optimization of multiproduct batch plants). However, the major advantage with the stochastic approach is that it offers the possibility of integrating flexibility and reliability under a common measure as has been discussed in this paper. It is clear that major advances are required to make computationally feasible the optimization with the stochastic approach.

Acknowledgments

The authors would like to acknowledge financial support from the Department of Energy under Grant DOE: DE-FG-02-85ER13396

References

1. Floudas, C.A. and I.E. Grossmann: Synthesis of Flexible Heat Exchanger Networks with Uncertain Flowrates and Temperatures. Comp. Chem. Eng. 11, 319 (1987).

2. Geoffrion, A.M.: Generalized Benders Decomposition. JOTA 10, 237-260 (1972).

3. Grossmann, I.E. and C.A. Floudas: Active Constraint Strategy for Flexibility Analysis in Chemical Processes. Comp. Chem. Eng. 11, 675 (1987).

4. Grossmann, I.E., K.P. Halemane and R.E. Swaney: Optimization Strategies for Flexible Chemical Processes. Comp. Chem. Eng. 7, 439 (1983).

5. Grossmann, I.E. and M. Morari: Operability, Resiliency and Flexibility-Process Design Objectives for a Changing World. Proc. 2nd Int. Conf. Foundations Computer Aided Process Design (Westerberg and Chien Eds.). CACHE, 937 (1984).

6. Grossmann, I.E. and R.W.H. Sargent: Optimum Design of Multipurpose Chemical Plants. Ind.Eng.Chem. Process Design Dev. 18, 343-348 (1978).

7. Grossmann, I.E. and D.A. Straub: Recent Developments in the Evaluation and Optimization of Flexible Chemical Processes. Proceedings of COPE-91 (eds. Puigjaner, L. and A. Espuna), Barcelona, Spain,49-59 (1991)

8. Halemane, K.P. and Grossmann, I.E.: Optimal Process Design under Uncertainty. AIChE J. 29, 425, (1983).

9. Johns, W.R., G. Marketos and D.W.T. Rippin: The Optimal Design of Chemical Plant to Meet Time-varying Demands in the Presence of Technical and Commercial Uncertainty. Design Congress 76, F1, (1976).

10. Kabatek, U. and R.E. Swaney: Worst-Case Identification in Structured Process Systems. Comp. Chem. Eng. 16, 1063-1072 (1992).

11. Malik, R.K. and R.R. Hughes: Optimal Design of Flexible Chemical Processes. Comp. Chem. Eng. 3, 473 (1979).

12. Marketos, G.: The Optimal Design of Chemical Plant Considering Uncertainty and Changing Circumstances. Doctoral Dissertation No 5607 ETH Zurich, (1975).

13. Pistikopoulos, E.N. and Grossmann, I.E.: Optimal Retrofit Design for Improving Process Flexibility in Linear Systems. Comp. Chem. Engng. 12, 719 (1988).

14. Pistikopoulos, E.N. and Grossmann, I.E.: Stochastic Optimization of Flexibility in Retrofit Design of Linear Systems. Comp. Chem. Engng. 12, 1215 (1988).

15. Pistikopoulos, E.N. and I.E.Grossmann: Optimal Retrofit Design for Improving Process Flexibility in Nonlinear Systems- I Fixed Degree of Flexibility. Comp. Chem. Engng. 13, 1003 (1989).

16. Pistikopoulos, E.N. and Grossmann, I.E.: Optimal Retrofit Design for Improving Process Flexibility in Nonlinear Systems- II Optimal Level of Flexibility. Comp. Chem. Eng. 13, 1087 (1989).

17. Pistikopoulos, E.N. and T.A. Mazzuchi, "A Novel Flexibility Analysis Approach for Processes with Stochastic Parameters," Comp. Chem. Eng., Vol. 14, No. 9, pp.991-1010 (1990).

18. Pistikopoulos, E.N., T.A. Mazzuchi and C.F.H. Van Rijn: Flexibility, Reliability, and Availability Analysis of Manufacturing Processes: A Unified Approach. Comp. App. in Chem. Eng. (eds. Bussemaker, H. Th. and Iedema, P.D.) Elsevier, Amsterdam (1990).

19. Reinhart, H.J. and D.W.T. Rippin: The Design of Flexible Batch Chemical Plants. Annual AIChE Meeting, Paper 50e, New Orleans (1986).

20. Reinhart, H.J. and D.W.T. Rippin: Design of Flexible Multi-Product Plants A New Procedure for Optimal Equipment Sizing Under Uncertainty. Annual AIChE Meeting, Paper 92f, New York (1987).

21. Straub, D.A.: Evaluation and Optimization of Process Systems with Discrete and Continuous Uncertainties. PhD Thesis, Carnegie-Mellon University (1992).

22. Straub, D.A. and Grossmann, I.E.: Integrated Statistical Metric of Flexibility for Systems with Discrete State and Continuous Parameter Uncertainties. Comp. Chem. Eng.14, 967, (1990).

23. Straub, D.A. and Grossmann, I.E.: Evaluation and Optimization of Flexibility in Multiproduct Batch Plants. Comp. Chem. Eng. 16, 69-87 (1992).

24. Straub, D.A. and Grossmann, I.E.: Design Optimization of Stochastic Flexibility. Comp. Chem. Eng., 17, 339-354 (1993).

25. Swaney, R.E. and Grossmann, I.E.: An Index for Operational Flexibility in Chemical Process Design. Part 1 Formulation and Theory. AIChE J. 31, 621 (1985).

26. Swaney, R.E. and Grossmann, I.E.: An Index for Operational Flexibility in Chemical Process Design. Part 2 Computational Algorithms. AIChE J. 31, 631 (1985).

27. Van Rijn, C.F.H.: A Systems Engineering Approach to Reliability, Availability and Mantainance. Presented at the Conference on Foundations of Computer Aided Operations, FOCAPO, Salt Lake City (1987).

28. Viswanathan, J. and Grossmann, I.E.: Combined Penalty Function and Outer-Approximation Method for MINLP Optimization. Comp. Chem. Eng. 14, 769-782 (1990).

29. Wellons, H.S. and G.V. Reklaitis: The Design of Multiproduct Batch Plants under Uncertainty with Staged Expansion. Comp. Chem. Eng.13, 115, (1989).

Artificial Intelligence Techniques in Batch Process Systems Engineering

Jack W. Ponton

Department of Chemical Engineering, University of Edinburgh King's Buildings, Edinburgh EH 9 3JL, Scotland

Abstract: The discipline of Process Systems Engineering has evolved largely in the area of continuous chemical processing. However, systematic developments in the batch area have taken place in a number of centres, and a brief overview is provided of the present state of batch PSE

An introduction to some relevant techniques from the field of artificial intelligence focusses on their use in process systems. Most applications, once again, have been in the continuous, but extrapolation to batch processing is discussed.

Finally, some speculative applications of ideas from other areas of manufacturing are outlined.

Keywords: Process systems engineering, batch processing, artificial intelligence

1 Introduction and Overview

The author is uneasy about writing this paper. Firstly, he has little or no experience of batch processing. Secondly, despite having coauthored a number of papers on the theme of 'artificial intelligence in process engineering', he has never really used any of the techniques of AI himself! The work for which he has received vicarious credit has been performed by his colleagues. What follows is thus necessarily highly derivative, and the reader who wishes to follow it up must refer to the bibliography.

In the course of writing the author has in fact managed to clarify some of his own ideas about a discipline of batch process systems engineering. A key difference between the systems engineering of continuous and of batch processes is that the former focusses on *design*, and the latter on *operation*. The serious reader would do well to read an excellent and wide ranging paper by George Stephanopoulos on AI in plant operations, [30]. A review of AI in process engineering, which the reader should be warned was not written for an audience of processes engineers, but which contains a useful overview and many references, particularly in the important area of safety, may be found in [11].

The outline of the paper is as follows:

- An overview of the process systems engineering of batch processes.
- Relevant artificial intelligence techniques and their application:
  - Rule based systems
  - Qualitative modelling
  - Object oriented techniques
- Design environments and databases
- A bit of blue sky

2 The Process Systems Engineeering of Batch Processes

Most work in process systems engineering PSE has concentrated, explicitly or implicitly, on techniques for continuous processes. Batch processing has had a small, but enthusiastic following in the areas particularly of scheduling and in the application of techniques from general manufacturing, see e.g. [29, 26, 13]. There have been occasional industrial papers, [27], but it is only recently that a wider range of ideas from the 'mainstream of PSE have been specifically directed to the batch processing field, e.g. [9, 34, 16, 28].

In this section, I shall look at some characteristics of batch processes and look for needs and opportunities in the application of PSE techniques.

2.1 Batch *versus* Continuous

In discussing the choice between batch and continuos processes, Douglas, [8] identifies the following characteristics for candidate batch processes:

1. Modest production rates of less than 1000 to 10,000 te/year.
2. Seasonal or limited demand for products.
3. Multiproduct operation.
4. 'Difficult' operations, e.g.
   - solids or slurries,
   - long reaction times,
   - fouling.

Characteristic 3. can probably be regarded as both an opportunity and a consequence arising from 1. and 2., the plant not being fully occupied at all times.

The items under 4. are indicative of the inability to design certain types of operation, arising largely from a lack of fundamental understanding of these operations. To them we could add. particularly in the context of biological processes:

- reactions liable to stop or deviate after extended operation, and
- processes sensitive to contamination.

Some properties of batch processes suggested by the above and other observations are as follows.

Batch plants are in general 'less designed' than continuous plants. They tend to be assembled from standard items of equipment, whose size is often not seen to be particularly critical. This noncriticality is partly because capital equipment charges are typically a smaller proportion of production cost than for continuous plants due to higher cost feedstocks and higher labour charges, see below. It is also the case that flexibility may be introduced by adjusting batch times rather than by changing equipment sizes.

There is thus a general shift from concentration on the *design* of the process to its *operation.*

Flexibility in operation is often obtained at the expense of increased labour, resulting from the choice of equipment. For example, batch filters are more flexible than continuous, but involve higher labour costs. There is however an overall increase in the rôle of the operator as compared with continuous processes. As well as the desire, or need, to maintain flexibility, a poorer understanding of basic process principles, e.g. in the handling of wet, sticky solids, or of complex biological reactions, regularly calls for operator intervention.

2.2 PSE Tools and Benefits in Continuous Processes

For continuous process the following PSE tools and techniques have yielded benefits.

1. Steady state mass and energy balancing programs: better sensitivity analysis and easier evaluation of alternative designs and operating policies.
2. Large Scale Optimisation: reduced design margins, better and cheaper plants and/or increased throughput.
3. Thermodynamic analysis: more energy efficient plants.
4. Detailed modelling and dynamic simulation: better understanding and hence more effective use of key operations, e.g. distillation and reactors.
5. Advanced control: better regulation of certain difficult operations and processes.
6. Methodology of process synthesis and design: better ways of carrying out design and developing alternative and new processes.

7. Design and management data systems and databases: faster designs and faster response to operational needs.

8. Analysis of safety critical systems: the 'Hazop' method and safer plants.

Most of these techniques have brought benefits to both design and operation. However, the emphasis of most work has been towards design

2.3 Application to Batch Processing

With the exception of the steady state mass and energy balancing, or 'flowsheeting' program, all these techniques have already been used effectively in batch processing, although not necessarily in quite the same way, nor with the same benefits.

1. The steady state 'flowsheeting' program, as a flexible and general purpose design aid, has no real counterpart in the unsteady state world of batch processing. [1] Discrete event simulators, e.g. [26], developed for use in a variety of other fields, have never quite filled the same rôle. This suggests an opportunity, because the flowsheeting program has arguably the most practically effective tool to have come from PSE. It is possible that some development of the ideas behind the state-task network, [16], might lead to an appropriate tool.

2. Optimisation is a powerful tool in batch processing. As noted earlier, operation, particularly scheduling, tends to feature rather than design, although retrofit design, [34, 9], is now assuming significant importance.

3. Although some work has been done on thermodynamic analysis of batch plants, the incentives are less than for continuous processing, energy being usually a less significant part of production cost. There are also fewer opportunities since there is no convenient way of moving energy between the time phases of batch processing.

4. Our ability to model, for example solids and biological operations is less advanced than for fluid and chemical systems. This is less because of mathematical intractability than because the basic processes themselves are less well understood. We thus see empirical or 'black box' models, such as neural nets, used where a fundamental or mechanistic model is really required.

   The importance of the *operator* in batch processing suggests that we need operator as well as process models.

5. The control requirements of batch processes are rather different from continuous processes. The emphasis shifts from regulation to servos, and most particularly to sequencing. The global control problems of continuous processing are primarily to do with *structure*, while those of batch plant are concerned with *sequence*. This is not to say that the structure of control systems is unimportant, especially in

[1] Flowsheeting programs can be, and indeed are, used to establish time averaged material balances in batch processing. However, in the continuous domain they are nowadays used as more general design tools. It is this aspect of their use which has no clear batch counterpart.

multiproduct plants which may require major reconfiguration of the control system for different products. However the structure of a batch processing operation tends to be mapped into the time domain as well as, and to a large extent, instead of, the spatial domain.

6. There is no methodology of batch process design which corresponds to any of the hierarchical schemes, e.g. [8], for continuous processes. In particular there is no corresponding hierarchy for *operation* which the different emphasis of batch processing suggests may be needed.

7. The use of advanced database techniques in the process industries is still at an early stage of its development. Some applications to both design and to operation have been reported. In the latter area there seem to be many opportunities for batch processing.

8. The Hazop methodology and techniques of quantitative hazard assessment are applicable to both batch and continuous processes, and indeed outside the chemical industries. They are already extensively applied to batch processes for both design and operation. However, Hazop, in particular, is a very demanding and time consuming tool. Any way of making it easier or more effective would be of great value.

2.4 Summary

In an attempt to draw some conclusions from these comments and comparisons, I suggest the following as ideas on which to focus in examining PSE techniques, including those of AI, for application to batch processes.

- Operation rather than Design
- Sequence instead of Structure
- Ill-defined rather than well understood
- Operators as well as Operations

Some specific areas of likely payoff seem to be:

- A batch process engineering 'flowsheeting' tool
- Techniques for operator modelling
- A hierarchical approach to design of batch plants and operating strategies
- Hazop improvements
- Advanced data systems

These are not exactly neglected areas, and some, e.g. the last two, are equally applicable to continuous processes. As it happens, however, they are areas where ideas from the AI community may be helpful.

3 Some Relevant AI Techniques

AI has provided us with two things: programming languages and tools, and concepts. These often appear to be closely linked, in that particular concepts are associated with tools, but they are in fact independent. AI concepts may be implemented in any language, even FORTRAN, and AI languages may be used to write more general kinds of program.

3.1 Rule Based Programming

Rule based programming is conceptually little more than the FORTRAN `if` construction:

`if` (condition) action

This forms the basis of the most familiar AI product, the so-called 'expert' system. AI languages include special rule based programming tools called expert system 'shells'. A complete system may be written in such a shell, and consists of a set of rules and an interpreter which applies some search strategy to see which rules match a given problem. The final rule which 'fires' is the solution to the problem. Thus a system to analyse some aspect of process safety might include rules such as the following:

```
1:if P<P_design then No_pressure_hazard
2:if (C<LEL and C>UEL) then No_flam-hazard
3:if (No_flam_hazard and No_pressure_hazard) then Safe else Unsafe
```

The system can also 'explain' its conclusions by reporting which rules fired. Such explanations frequently leave much to be desired. The above set of rules might produce the explanation:

```
Unsafe by Rule 3 because not No_flam_hazard by Rule 2 not C<LEL
```

...which is logical but scarcely intelligent!

Although apparently rather successful in some areas of engineering, rule based systems as such have had limited success in most areas of process engineering, where few concepts are sufficiently straightforward to be described adequately by reasonably sized sets of rules.

One area of success has been in the modelling of operator actions to replace or augment the plant operator. It would appear that operators work with a set of rules and apply them, informally and often inconsistently. By formalising the actions of the best process operators, rule based systems have been able to improve plant operation in some cases significantly. The same effect could have been obtained, without the use of a computer, simply by writing down the rules as operating procedures and requiring that all operators follow them. Automating the rules ensures that they *are* followed, and is of course much faster.

An obvious problem with a book of rules, however implemented, is that it may not cover all eventualities. There is no 'understanding' of the process and there is thus no means of extrapolating to new situations. In an attempt to overcome this limitation, rules have been combined with various sorts of model.

See, for example [24, 17, 18]

It would be contentious to claim that rules form the basis of most other AI techniques, but to a large extent this is so. There are other ways of writing rules which make them more tractable for use in different circumstances. Their properties, however, including their limitations, remain the same.

3.1.1 Blackboard Systems

A blackboard system may be thought of as a communicating collection of separate rule based systems, each working on a different aspect of the same problem. Such systems have been succesful, [3, 4] in handling problems beyond the scope of any single rule based system. Difficulties do arise in the coordination of these individual 'experts', none of which understands the basis on which the others are working. Some means of global management is required. Unfortunately, there is no obvious way of doing this without restoring the difficulty, a very large and therefore unmanageable rule base, which caused the problem to be decomposed in the first place.

3.1.2 Planning

A possible candidate for the blackboard manager is often seen to be the AI planner. Extensive work has been carried out in this area by the AI community, and some has been applied in the process systems area.

Particularly relevant to our interest is work on planning operating procedures, [21, 22]. At the end of the day, however, the basis of these planners is once again the use of rules.

The way in which these rules operate may vary. In *linear* planning the rule interpreter attempts to progress towards a given goal, backtracking when it fails and trying again by a different route. *Nonlinear* planners construct partial solutions and when these fail, return to a previous partial solution and attempt to modify it to prevent the observed failure. The nonlinear planner thus works both backwards and forwards. The difference lies in how efficiently the solution is found, and should not affect what the solution is. A useful illustration of the operation of these strategies is given in [22].

The area of operating procedure synthesis, with particular reference to startup, and thus highly relevant to batch operation, has been investigated with these and related techniques, see also [10]. The extent to which procedures developed in this way have actually been used is unclear.

3.2 Qualitative Simulation

Within the AI community there has been considerable interest in the ideas of qualitative simulation, see e.g [14]. The reason for this is, presumably, scientific curiosity, since no particular claims of utility appear to have been advanced.

Qualitative simulation models the behaviour ofa system using a range of non numerical measures: HIGH, LOW, RISING, FALLING, etc. Corresponding models can either be steady state or dynamic, the latter being qualitative differential equations or *confluence equations.*

Within the process engineering community there has been some interest in these techniques, largely in the areas of safety, e.g. [32], and fault diagnosis, e.g. [19]. The justification for this interest is twofold. Firstly, it may be possible to infer 'universal truths' which do not depend on specific numerical values. In this context, it is interesting to note that qualitative simulation has been long established in the process industries in the line-by-line hazard and operability study, [15], which uses the concepts HIGH, LOW, NONE, DIFFERENT, etc., applied to process parameters FLOW, PRESSURE, TEMPERATURE, etc. in what is essentially a simulation procedure carried out by a group of people.

The technique thus clearly has possibilities. Another expectation, relevant particularly to real time fault diagnosis, is that qualitative models may be easier to construct and to solve than full quantitative models. As far as construction is concerned, this is probably true. It is probably of comparable difficulty to write down equations for both types of model, but the qualitative model will not require numerical values for all its parameters, a major saving. However, experience suggests that qualitative models are often more timeconsuming, sometimes by orders of magnitude, than conventional models!

There appear to have been no applications of qualitative modelling in the batch processing field. However, this may represent a missed opportunity, since discrete event simulation, frequently used to model scheduling, is more 'qualitative' than continuous dynamic modelling. Adding the concepts BEFORE, DURING and AFTER would provide an interesting framework for batch process qualitative simulation.

It is likely that qualitative models will be most useful when combined in some way with conventional quantitative equation models. The two approaches are in a sense complementary. Qualiative models embody the ideas of 'cause and effect', while equations represent only relationships. It is also possible to refine ideas of qualitative representation so that they provide almost as much information as numbers, by including qualitative measures of first and second derivatives [6].

3.3 Object Orientated Programming Techniques and Concepts

One of the most powerful ideas to emanate from the AI and computer science community is that of *Object Oriented* programming and the concepts which underly it. Conventional programming languages, such as FORTRAN, provided representational facilities for a limited number of ideas, initially only numbers, either integral or general, and simple groupings of these into vectors and arrays. Later languages, such as Pascal and Algol'68, added a few further concepts, character strings and logical variables, as well as more general groupings of these entities into *structures.*

Object oriented programs allow the user to define *anything* as an 'object' and, within the other constraints of the language, to create and manipulate these objects. Objects may contain other objects, and by a controllable *inheritance mechanism*, contained objects may automatically possess properties of the object which contains them. This leads naturally to a hierarchical representation which is highly appropriate to physical objects, such as processes which are objects containing other objects, namely streams and units, the latter possessing parts such as vessels and flanges.

More important, the representation is also appropriate to process engineering procedures, such as the design of a plant in the hierarchical manner, [8]. Unlike the structures of conventional languages, objects are not just passive. They may contain active computer code which carries out some activity when the object is referenced. For example, a process unit may be redesigned whenever its specification is changed.

These ideas have been used in process engineering both for physical items and for conceptual activities. For an extensive example of the first, and an excellent overview of objected oriented programming, see [31].

The representation of the design process itself, which in the context of batch processing would certainly include the design of the operating procedure, is discussed below.

4 Design Environments and 'Intelligent' Databases

There are many similarities between the object oriented paradigm and developments in database technology. Indeed, the coming generation of data bases will be object oriented databases, [5]. The ability to represent complex objects is clearly crucial to any computer based system for carrying out process design, whether of continuous or of batch processes. More subtle is the need to represent the *process* of design: the designers intentions, the history of activities which embody the explanation of why design decisions were taken, the assumptions used by the designer and all the design tools which he used, and so forth. Many of these are subtle and complex, and techniques for their representation, let alone manipulation, remain to be developed.

The complexity of the artifacts to be designed, the process of design and the tools used by the designer has pointed to the need for a unified *design environment* within which all activities may be conducted. The environment must provide a means of representing all the entities described above within what must be, at the very least, an extremely flexible system.

An example of the use of AI object oriented techniques as a database to keep track of designers' decisions and to check these for violation of either imposed constraints, e.g. for safety, or for violation of the designer's original intentions, may be found in [33]. This particular application is interesting to the batch process systems engineer in that it has been applied to the design of a batch process, although it was conceived initially in the context of continuous processing, [20].

Ideas of modelling the process of design are described in [2]. From that particular work has grown a prototype *design environment*, intended for use by a group of designers working in a cooperative manner on different aspects of the same design. An illustration of the environment in use is shown in figure 1.

The activities in the numbered windows are as follows.

1. This is the top, system manager, window from which the design was started. It identifies the Project, the designer and the process, `block4`.

2. This is the design history and hierarchy. The main process is decomposed into a reaction section `reac-sec1` and a recycle separation section `rsep-sec1`. For the

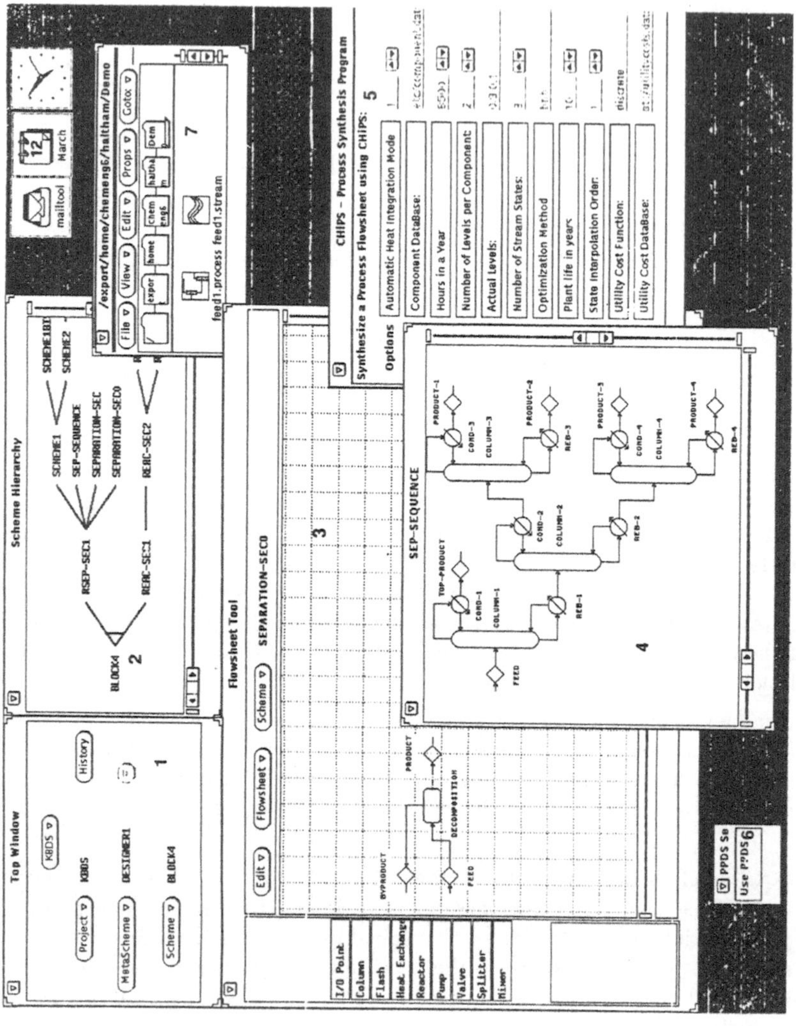

Figure 1. Process Design Environment

latter, a range of alternative designs are being examined and are at different stages of detail.

3. The Flowsheet Tool contains one of these alternatives which is at the BFD level of detail. A particular separation task, for example from within this BFD, has been dropped into CHiPS ...

4. ... which has synthesised a flowsheet and designed the columns, displaying them in this flowsheet window. The units are 'mouse sensitive' and can be selected and examined or changed. If the user were not satisfied with the automatically generated process it could be dropped into the Flowsheet Tool for manual editing.

5. This menu window was popped up by CHiPS in the course of its operation. The options are 'greyed out' as they are not active at this point.

6. The PPDS server is available both to programs, for example, it would have been called in for the column design, or for direct access by dropping into it a stream from a flowsheet, or elsewhere.

7. Finally, this is a *épée* manager window. It shows a process and a stream object which can be pulled into compatible tools or windows.

In the course of developing the environment, the need for a high level representation of process engineering concepts was reinforced. Seeing this as the 'top level' of a database system not as yet available, an object oriented *messaging system* was developed. This is intended to serve as a means of communication within an environment such as that shown above, but using an object orientated database as the actual 'firmware' communication mechanism [1].

5 Blue Sky: Robots and Railroads

Finally, here are two ideas quite different from anything above, but perhaps of importance to batch process engineering.

It is easy to move fluids around between processing steps in either batch or continuous processes. Solids when dry can be air conveyed, and when slurried are relatively easily handled. There is a difficult gap between suspended solids, which is how they are usually formed, and dry solids, as they are generally sold, in which the materials are wet and sticky. The plate-and-frame batch filter remains the most common method of primary solid-liquid separation. Unlike most processing operations it is highly labour intensive. Labour costs are thus significant, but the need to have an operator present at all is the major drawback, for example in toxic or sterile conditions.

We have recently succeeded, [12] in developing a laser visioned robot operated filter press. The vision system in particular turns out to be very useful in identifying lumps of material where they are not wanted, for example in vessel emptying. Techniques from conventional, non-chemical manufacturing have limited applicability, since these assume the artifacts being handled have a fixed shape and size. A new methodology of robotics seems to be required for dealing with amorphous products.

An interesting example of lateral thinking is involved in a batch process, [25], which instead of moving chemicals between processing units, moves the processing units, on rail tracks! Initial reaction to this proposal, when photographs of a plant in use in Japan were shown, was concern about safety. However, the idea is intriguing, and raises the question as to whether there are any other ways of radically rethinking a the philosophy of chemical production. While not wishing to dampen enthusiasm for such radical developments, it is worth noting that, for fluids, pipes do have many fundamental advantages over tank cars. However, in the area of wet solids processing referred to above, the issue may well not be so clear cut.

References

1. G Ballinger et al, Epee: A Process Engineering Environment, Proceedings of ESCAPE 3, Graz, Austria, 1993
2. R Banares-Alcantara, Representing the Engineering Design Process: Two Hypotheses, prize winning paper at AI in Design'91, proceedings, ed J Gero, published by Butterworth Heinemann, 1991. Also in Computer-Aided Design, 595-603, 23, 9 1991
3. R Banares-Alcantara, EI Ko, AW Westerberg and MD Rychener, DECADE: a Hybrid Expert System for Catalyst Selection: Part I, Expert System Considerations, Comp and Chem Eng 11, 267-277, 1987
4 ... Part II- Final Architecture and Results, ibid, 12, 923-938, 1988
5. MR Blaha, J Mehta and RL Motard, Structure and Methodology in Engineering Information Management, AIChE Fall Meeting, 137e, Los Angeles, 1991
6. JT-Y Cheung and G Stephanopoulos, Representation of Process Trends: Parts I and II, Comp and Chem Eng 14, 495-510 and 511-540, 1990
7 BJ Cott and S Machietto, An Integrated Approach to Computer Aided Operation of Batch Chemical Plants, Comp and Chem. Eng. 413, 11/12, 1263-1273, 1989
8. JM Douglas, The Conceptual Design of Chemical Processes, McGraw-Hill, 1988
9. R Fletcher, JAJ Hall and WR Johns, Flexible Retrofit Design of Multiproduct Batch Plants. Comp and Chem., Eng 15, 12, 843, 852, 1991
10. RH Fusillo and GJ Powers, Operating Procedure Synthesis using Local Models and Distributed Goals, ibid, 12, 1023-1034, 1988
11. D Hutton, JW Ponton and A Waters, AI Applications in the Process Industries, The Knowledge Engineering Review, 5, 2 69-95, 1990
12. A Jaffrey, N Macleod and JW Ponton, The development of a laser guided robotic discharge system for filter presses, Proc IChemE Research Event, 143, 564-566, 1992
13. LS Karimi and GV Reklaitis, Intermediate Storage in Noncontinuous Processing, in 'Foundations of Computer Aided Process Design', eds. Westerberg and Chien, CACHE publications, 1984
14. J de Kleer and JS Brown, A Qualitative Physics Based on Confluences, Artificial Intelligence, 24, 7-83, 1984
15. T Kletz, Bazop and Hazan, IChemE, London, 1986
16. E Kondili, CC Pantelides and RWH Sargent, A General Algorithm for Scheduling Batch Operations, Proc PSE'88, 62-75, Sydney, 1988
17. MA Kramer, Malfunction Diagnosis Using Quantitative models and non-Boolean Reasoning in Expert Systems, AICHE J 33, 130, 1987
18. MA Kramer, Expert Systems for Fault Diagnosis: a General Framework;, in Foundations of Computer Aided Process Operations', eds Reklaitis and Spriggs, CACHE, Austin, Texas, 1987
19. MA Kramer and BL Palowitch, A Rule based Approach to Fault Diagnosis Using the Signed Directed Graph, AICHE J, 33, 1067-1078, 1987
20. EM Lakie, Evaluation of a Tool for the Representation and Checking of Design Constraints by its Continuous Use in a Student Process Design Project, Project Report, Department of Chemical Engineering, University of Edinburgh, June 1989
21. A Lakshmanan and G Stephanopoulos, Synthesis of Operating Procedures for Complete Chemical Plants. I: Hierarchical Structured Modeling for Nonlinear Planning, Comp and Chem. Eng. 12, 985-1002, 1988
22 ... II: a Nonlinear Planning Methodology, ibid 1003-1021

23. J Love and B Shiel and N Drakos, A Declarative Approach to Applications Software for Batch Process Control, proceedings of IChemE Research Event 680-682, IChemE Rugby, 1992
24. DR Myers, JF Davis and DJ Herman, A Task Oriented Approach to Knowledge Based Systems for Process Engineering Design, Comp and Chem. Eng. 12, 959-972, 1988
25. T Niwa, Transferable Vessel Type Multipurpose Batch Plant, Proc PSE'91, IV, 2.1-2.15, 1991
26. BW Overturf, G. V. Reklaitis, JM Woods, GASP-IV and the Simulation of Batch and Semicontinuous Operations: Single Train Processes, IEC Proc Des Dev, 17, 161, 1978
27. ML Preston and DH Cherry, Batch Process Scheduling - ICI's BatchMASTER, in "Process Systems Engineering: PSE'85", Institution of Chemical Engineers, Rugby, 1985
28. DWT Rippin, Simulation of Single and A Multiproduct Batch Chemical Plants for Optimal Design and Operation, Comp and Chem. Eng. 7, 3, 137-156, 1983
29. RE Sparrow, GJ Forder and DWT Rippin, The Choice of Equipment Sizes for Multiproduct Batch Plants: Heuristics vs Branch and Bound, IEC Proc Des Dev 14, 3, 197, 1975
30. G Stephanopoulos, The Scope of Artificial Intelligence in Plant- Wide Operations, in 'Foundations of Computer Aided Process Operations', 505-555, eds Reklaitis and Spriggs, CACHE, Austin, Texas, 1987
31. G Stephanopoulos et al, Design Kit: an Object Orientated Environment for Process Engineering, Comp and Chem,. Eng. ό311, 6, 655-674, 1987
32. A Waters and JW Ponton, Qualitative Simulation and Fault Propagation in Process Plants, ChERD 67, 407-422, 1989
33. A Waters and JW Ponton, Managing Constraints in Design: Using an AI Toolkit as a DBMS, Computers and Chemical Engineering, 16, 10/11, 987-1006, 1992
34. JA Vaselnak, IE, Grossman and AW Westerberg, Optimal Retrofit Design in Multiproduct Batch Plants, IEC Res 26, 718-726, 1987

Elements of Knowledge Based Systems
Representation and Inference

Kristian M. Lien

Department of Chemical Engineering, The University of Trondheim, Norway

Abstract: Over the last decade, the use of Knowledge Based techniques have become popular in Process Systems Engineering. This paper will give an overview of some of the main issues involved in the application of such techniques. It does not emphasize batch systems in particular, but does rather present a collection of useful general concepts in search, representation, inference and systems architectures, applicable as well in batch process systems as in e.g. synthesis of process flowsheet. After all, there is quite a number of similarities between batch process systems engineering and process synthesis, since both may be viewed as configuration problems: In batch processing, a major task is to configure a processing plan configuration of a set of actions in time. Search, representation formalisms, inference mechanisms and problem solving systems architectures are the main ingredients of this paper, where the latter of these three is particularly focused towards blackboard systems. The last part of the paper describes the development of a particular blackboard system - AKORN D, its successor AKORN DT and an application system for synthesis of separation systems built on top of AKORN DT.

Keywords: Knowledge Based Systems, Search, Representation, Inference

Search

Search is a fundamental concept in computer problem solving. From an initial situation - *the initial state*- it is desired to get to another situation- *the goal state*. A set of operators can be applied to states. States are transformed into new states when operators are applied to them; operators facilitate state transitions. In order to determine whether the search should continue or be stopped when a new state is reached, a *termination criterion* is applied to every new state generated. If the criterion is satisfied- for example if the new state is a goal state - the search terminates. Otherwise, the search continues.

```
State ← initial state;
Until State satisfies Termination-Criterion do
    begin
        Select an operator Op from the set of operators applicable to State;
        State ← result of applying Op to State;
    end
```

A sequence of applied operators resulting in a goal state constitute a solution to the search problem - *a solution path*. The set of possible states that can be reached is termed the problem's *search space*.

There are numerous ways the search among the possible states can be performed. The two major kinds of search - *search regimes* - that are distinguished are termed *irrevocable* and *tentative* search regimes. In an *irrevocable search regime*, operators are selected and applied *without provision for reconsideration* later. There may be alternative operators applicable to a particular state, but one is chosen, and the state it is applied to may never be returned to later. The pseudo algorithm above is of this type. Since states are never reconsidered in irrevocable search regimes, these do not need to save descriptions of former states. In *tentative search regimes*, previous *states may be returned to* so that alternative sequences of operators may be applied. This implies that states must be saved so that they may be returned to at some point later in the search process.

When may an irrevocable search regime be used, and when is a tentative regime to be preferred ? That depends on the *nature of the search space*: If all sequences of operators lead to goal states - no partial sequence ends in a blind alley - and any sequence is as good as any other, then irrevocable regimes may be good enough. Application of *hill climbing techniques*, e.g. 'steepest ascent' methods, to determine local maxima of mathematical functions is an example in case. However, if the global maximum is desired and the function of concern has more than one local maximum, the hill climbing procedure may easily get stuck on one of these, failing to find the global maximum solution.

Tentative search regimes may be classified as *backtracking* or *graph search* regimes. *Backtracking search regimes* insist on exploring every possible path from the most recently generated state before tracking back to previously generated states to explore alternative paths from there. Backtracking regimes need therefore only to store the states on the current path. Graph search regimes do not put such an emphasis on finding extensions to the current path. Several partial paths may coexist, and one may switch among these.

Evaluation of Solutions and Paths

There must of course be a rationale behind switching from one partial path to another; a belief that one partial path is *better* than another in some sense. In order to determine the *goodness* of a path, some *goodness criterion* must be present. A path from the start to a goal state may often be assigned a value or cost, but for the goodness criterion to be useful in guiding the search process- deciding which of the existing partial paths to extend - it must be applicable to *partial* solutions as well as to completed ones.

The value or cost of a partial path from start to an already generated state S is usually the easiest one to calculate - either, if the cost or value is additive, by summation of the cost of each operator application on the path- or, if the cost is not additive, by an evaluation of the cost or value associated with the state S. The problem lies in the estimation of what the cost or value of extending the path from S to a goal state is going to be, since that part of the path is yet unknown; it remains to be generated.

To illustrate the problem, suppose that the problem is to find a path to travel from New York City (NYC) to Los Angeles (LA). Being in New York City there is the choice between going to Boston and going to Chicago. ("Go to Boston" and "Go to Chicago" are the operators applicable to the state "Being in NYC") If it is not known where Boston or Chicago is relative to LA, the goal state, one may as well go to Boston first, because that is the closer of the two, less expensive than going to Chicago. Being in Boston, there is the choice between going to Montreal and going to Cleveland. Not knowing where Cleveland or Montreal is relative to LA, one may as well go to Montreal first, because that is closer than Cleveland. All the way, the least expensive next move is selected, minimizing local cost, but the overall result may be getting nowhere near the goal state (LA) until all other possibilities are explored. If it is possible to estimate remaining distances from the next possible location to the goal, efficient routes may be found with much less effort: Go to Chicago first, because the distance

$$\text{NYC} \rightarrow \text{Chicago} + \text{Chicago} \rightarrow \text{LA}$$

is less than

$$\text{NYC} \rightarrow \text{Boston} + \text{Boston} \rightarrow \text{LA}.$$

It may be shown that if the following three conditions are satisfied:

a) The goodness criterion is additive:

Cost of going from A to B through S

= Cost of going from A to S

\+ Cost of going from S to B

b) The cost of the remaining part of any partial path is never overestimated, that is, the real cost will never turn out to be lower than the estimated cost.

c) The path extensions are always selected so that the path which has the lowest combined cost (Cost of already generated part of the path + Estimated cost of remaining distance to the goal) is expanded next.

then the first complete path to the goal state will comprise an *optimal* path: *No other path leading to the goal state generated subsequently can have a lower cost than the path already found.*

The A^* search algorithm is an example of an optimal search algorithm . A description of A^* and an informal proof of its optimality may be found in Nillson[48]. If it cannot be guaranteed that the estimated cost is always an underestimate, it cannot, in general, be guaranteed that an optimal path has been found until all alternatives have been explored.

Suppose that there are two alternative estimation functions for remaining paths, *F1* and *F2*. Neither overestimate remaining cost, but *F1* consistently estimates higher costs than *F2*. Then, it may be shown that a larger number of states will be generated with search procedures using *F2* than with search procedures using *F1*. It may also be shown that any state generated using *F1* will also be generated using *F2*. This does not necessarily mean, however, that a search procedure using *F1* will be more efficient than one using *F2*, because efficiency also will have to include how long it takes to evaluate the estimation function on a state or partial path. But, the search tree generated using *F1* will be less "bushy" than the search tree generated using *F2*.

The best conceivable estimation function is one that estimates remaining costs exactly. If such a function can be devised, every state generated in the search will be on the optimal path.

The above statements about optimality and *heuristic power* (heuristic power is a term reflecting the "bushiness" of the search tree generated: the less bushy the tree, the more heuristically powerful the underlying search procedure.) hold only if the underlying assumptions hold; that the goodness function is additive, that estimated costs never exceed real costs, and that the path with lowest estimated *total* cost is always extended next.

The above description assumes that there is one goal state or that any goal state is as good as any other. But often the problem is not to find an optimal path to a goal but to find a path to an optimal goal. (Referring to the traveling example, the problem is not to find the optimal path to LA but rather to find the city on the west coast with the best opportunities for getting into the movie business. At first it seems unimportant how long it takes to get there, because being there is what counts. In reality that is not entirely true. You may spend so much time looking for the place that when you finally get there you are too old to become an actor.) Optimal goal problems can often be addressed by the same means as optimal path problems. This is particularly true for *configuration* or *assembly* problems: The goal state is described in terms of functionality, the function that the configuration must be able to perform. The search space is described in terms of parts that may or may not be part of a solution configuration. The problem is to select a subset of those parts and combine them so that the combination of parts performs the specified function better than any other combination of parts present in the search space with respect to some goodness criterion.

It is of course implicit in this formulation that *optimal solutions* are optimal in a context: the present search space and the present goodness function. Another goodness criterion or search space might have termed a totally different solution optimal.

An underlying assumption here is that it is possible to decompose the specified goal functionality into subfunctionalities that can be implemented by devices present in the search space that can be combined to a device that performs the goal function. The naive approach to solving the problem would be to try every possible configuration of the search space elements, evaluate the goodness of each and pick the best. This would be naive because only very small search spaces could be handled that way. To illustrate, assume that the search space contains M parts. The optimal solution to the problem could contain one of those parts, two of them, three . . or at most all M. Assuming that parts can only be configured serially, one part after the other, a solution comprising N parts can be configured in

$$\binom{M}{N} * N! = \frac{M!}{(M-N)!}$$

ways from the M parts in the search space: There are $\binom{M}{N}$ ways to pick N different parts from a set of M, and each of these can be configured in N ! ways if the configurations are restricted to linear strings of parts. The total number of string-like configurations to consider (of which only a small fraction will constitute valid solutions, most likely, but that cannot be known before they are analyzed to check validity) will thus be

$$\sum_{i=1}^{M} \frac{M!}{(M-i)!} = M! * \sum_{i=1}^{M} \frac{1}{i!} \cong 2 * M!$$

For $M = 5$, this amounts to approximately 250 possible combinations, which is not an unreasonably large number, but already when $M = 10$ the number of possibilities approaches one million ! And, even this is a gross underestimation, since only string -like configurations are allowed.

This does of course not mean that all these alternatives are going to be explored, the example is included only to demonstrate how hopelessly large search problems become in the absence of more specific knowledge about the problem. Problem domain specific *decompositions, classifications and constraint logics* enable the configurer to avoid taking the naive approach: The goal functionality can be decomposed into subfunctionalities. A few of the search space elements can be classified as being able to perform a given subfunction (that is why they were included in the search space in the first place; to be alternatives for performing a subfunction that likely would have to be performed given the goal functionality) so that the search space can be viewed as partitioned into functionality groups. And within and between the functionality groups choices are often constrained by domain considerations: Inclusion of one element into a partial solution makes it unnecessary to consider certain other elements or groups. For example, if a particular pump has been selected to increase pressure and the pump has sufficient capacity to increase the pressure enough, it would not make any sense to select another pump to increase the pressure since that functionality already has been implemented. Nor would it make any sense to include a pressure decreasing device - say, a valve or expander - into the partial solution immediately before or after the selected pump.

An optimal search approach may be taken towards the configuration problem if it is possible to estimate the goodness of a partial solution by calculation of the cost of parts already included in the solution plus an estimated cost of remaining parts to include that is not an overestimate.

Unfortunately, it is often hard to find good underestimator. However, it may still be possible to use a modified approach: Partition the cost of the parts into two classes, a class for which the optimal search assumption holds and a "residue". Then, use optimal search on the part of the costs for which the assumption holds.

Example

The dominant costs associated with a chemical plant are investment costs and energy costs. The investment costs can be considered to be additive, but energy cost is a function of how well energy integrated the plant is. If the energy consumption of each piece of equipment is summed, one may get an overestimate of the energy consumption of the plant. Then, if the investment cost is the dominant cost factor, ignore the energy cost and use optimal search on the investment part. Alternatively, if it is possible to associate an energy cost with a piece of equipment that does not reflect the entire energy consumption of the unit, but is close to what its final share in the total energy cost is going to be, include that part of the cost into the modified optimal search procedure.

The first modification has the consequence that the search procedure loses some of its heuristic power as a result of the systematic underestimation of real costs. This will result in a more 'bushy' search tree, but the procedure will finally terminate with a solution that is optimal with respect to investment since no overestimation takes place. If investment costs are much larger than energy costs, then this solution will also be close to the optimum for combined investment and energy costs.

The second modification may lead to overestimation of costs. Harris [24] has shown that if the maximum possible overestimation k of an estimation function can be determined, it can be guaranteed that no solution will be generated costing more than k more than the optimal solution before the optimal solution is generated. It may thus be tempting to try to use an overestimating function, keep track of all estimated costs, compare the estimated costs with the real cost when the first solution is found and claim that the solution found is no more than a more expensive than the optimal solution, where a is the maximum overestimation of all the cost estimates on the way to the found solution. However, this does not work unless it can be demonstrated that the function overestimates maximally on the way to that particular solution, that is, $a = k$ for that particular solution.

The basic problem in finding the optimal solution in general is that the cost of the optimal solution is unknown initially. So in principle one may risk an exhaustive search before the optimal solution can be recognized. In practice, this is often avoidable: Sometimes one may predict an attainable lower bound on the cost of an optimal solution, sometimes one may find predictive cost functions whose maximum possible overestimation can be determined, and sometimes both these sources of information may be exploited:

a) If a lower bound L on the optimal solution can be predicted, then no partial solutions with (partial) cost above L need to be explored.

b) If a predictive cost function has k as its maximum over-estimation, only the partial solutions with the lowest estimated total cost and those partial solution with an estimated total cost within k above the lowest one need to be explored.

c) If an attainable lower bound L on the cost of an optimal solution can be predicted and a predictive cost function exists with a maximum overestimate k, no partial solutions need to be explored that has an estimated total cost above $L + k$.

Case c) stated above allows the expansion of partial solutions up to the point where the cost of already generated parts plus estimated remaining cost exceed $L + k$. Case a) allows the expansion of partial solutions up to the point where the cost of the already generated part of the solutions exceed L. When estimated remaining costs exceed k, unfruitful partial solutions will thus be kept longer with a) than with c); c) is a stronger exclusion criterion than a), even though it first may seem to be the other way around.

Optimal solutions may be hard to find. Often a major search effort is required to find the one best solution, particularly if many solutions cost little more than the optimal one. However, when cost differences are minute, finding the best solution may be more of an academic exercise than an issue of real importance. Often solutions 'reasonably close' to optimum will be accepted. In most cases, the search is performed within a search space that only approximates the set of alternatives available in reality. Furthermore, the criteria used to evaluate solutions are frequently simplified versions of real world considerations. Issues like human factors, politics, company policies, etc., are rarely easily quantifiable.

The context in which the search is performed is important. Even though optimal search procedures may be devised within a given search space, it is hardly a too strong claim that the goodness of sets of alternatives to explore and evaluation criteria to use strongly depend on familiarity with characteristics of the problem. Unless search space and rating function are carefully selected, the optimal solution may be no more than an excellent answer to the wrong question.

Representation

In the previous section of this paper, problem solving was viewed as search within a space of possibilities. Each possible situation on the path from initial problem to solution was termed a state. Operators could change one situation or state into another. This chapter addresses representational issues; how to describe states and how to describe operators so that computers can transform initial problem state descriptions into solution state descriptions.

In computer problem solving, states may be viewed as data and operators as procedures. The operator descriptions can be given procedural interpretations, the interpretation of a state description relies entirely on the correspondence between the symbols comprising it and the external world the symbols represent.

A state may be described in terms of the objects present in the state, their properties and the relations existing among the present objects. A collection of symbols can be said to be a valid representation of a state if every distinguishable object, property and relation of concern is represented by a unique symbol in the state representation. A collection of symbols can be said to be a valid representation of an operator or state transition between two states S1 and S2 *iff* the collection of symbols has a procedural interpretation generating behavior capable of symbol manipulation characterized by

a) leaving every symbol corresponding to objects, properties and relations existing both in S1 and in S2 unchanged,

b) destroying every symbol corresponding to objects, properties and relations present in S1 and absent in S2, and

c) creating new symbols uniquely corresponding to any object, property and relation of concern absent in S1 and present in S2.

Representations of valid solutions may be mechanically produced by application of operators to state representations when such symbol structures representing states and state transitions can be devised. However, note that *the meaning is in the transformations*: The interpretation of such a collection of symbols relies entirely on the procedures manipulating it.

There is more than one approach to representing application of a set of operators to a set of states: One might explicitly code the operators in some procedural computer language as a set of functions that take a state description as input, where every such function explicitly is called to check whether it is applicable to the current state description, for example representation as a global block of data. If it is, execute the function and let its result be the new current state. Iterate explicitly until no function is applicable, or an explicit exit from the loop is encountered. Alternatively, one could imagine the operator application cycle partitioned into

a) a background processing mechanism that 'invisibly' selects 'operators proper' for execution and handles the details of state description management and update, and

b) a higher level description of the 'operators proper' that is interpreted and executed by the background mechanism.

Rule Based Programming

Rule-based programming systems incorporate such a partition. Frame-based or object oriented systems are also examples of similar partitions of representation and processing into layers, as are many process simulators, equation solvers, network communication protocols and robot programming languages, to name a few. A rule-based program consists of three parts:

1) A global collection of symbols serving as state representation. This collection of data is also frequently termed the system's 'dynamic database' or 'working memory'.

2) A collection of conditional statements or rules serving as operators representation. A rule is on the form:

    If <antecedent statements>
    Then <consequent statements>

3) A background process or interpreter. The interpreter, which is invisible in the explicit code comprising the rule based program, iteratively

    a) finds the rules that can be applied to the current state

b) selects a rule to apply among these

c) applies the rule.

Application of a rule may change the current state description, enabling some rules not applicable in a given iteration cycle n to be applied in iteration n + 1.

Rule-based systems are often classified as being either *forward chaining* or *backward chaining*. A forward chaining system compares the antecedent statements with the current state description, and if antecedent statements and state description statements are compatible, given some compatibility criterion, the rule is applicable and may be selected. If it gets selected, the consequent statements are processed. The processing of the consequent statements change the state description. In other words, the forward chaining system produces symbol structures that should be understood as consequences of applying a sequence of rules to an initial state description.

A backward chaining system focuses on a particular statement and tries to establish a path backwards through rule applications to statements present in the state description. It tries to prove that this particular statement will be a consequence of application of a sequence of rules from the current state. The statement of concern, frequently termed the goal statement, is compared to consequent statements of rules. All rules that have consequent statements compatible with the goal statement are considered to be on the path of rule applications from the initial state to states containing the goal statement. One of these rules is selected and now the antecedent statements of this rule become new statements of concern, new goal statements - subgoals. This is illustrated below:

```
Database ← initial state description;
Goal ← statement of concern;
Rules ← the set of rules;
Result ← Backchain (Goal, Database, Rules).
Function Backchain (Goal, Database, Rules);
    Result ← Success;
        If Goal is not matched by any statement in Database, then
            If a rule R exists in Rules with a consequent statement matching Goal, then
                For every antecedent statement A in R OR until Result = Fail do
                    Result ← Backchain (A, Database, Rules);
```

else Result ← Fail;

end if;

Backchain ← Result;

The function Backchain as described above employs an irrevocable search regime for identification of a rule sequence that will generate the initial goal statement as a consequence of the statements in the initial state description. This is done for the sake of simplicity in the presentation; a tentative search regime may of course be used, so that when a rule R is selected from Rules that is a blind alley, alternative rules may be chosen until one succeeds or all alternatives are explored.

It should be emphasized at this point that rules can serve two distinct purposes as representational elements. Rules as representations of state transitions has already been described. However, rules may also serve as a representational mechanism for explication of implicit statements of state descriptions. Consider the statements:

C1 is a Big Compressor.

C1 is Expensive Equipment.

C2 is a Big Compressor.

C2 is Expensive Equipment.

Using rules a more compact description is possible:

C1 is a Big Compressor.

C2 is a Big Compressor.

If X is a Big Compressor, then X is Expensive Equipment.

Clearly, this rule does not represent any state transition. The world the statements describe does not change depending on whether the rule is applied or not. The only thing that changes is *the amount of explicit information* about that same world. Making decision (state transitions) and deducing logical implication are totally different activities and should be treated accordingly in any rule-based system:

Logic programming = logical implications.

Production systems programming = state transitions.

Logic programming is a rule-based approach to programming principally concerned with deducing logical implications of state descriptions. Production systems programming is another rule-based approach to programming oriented more towards decision making. These two approaches will now be presented and exemplified by a logic programming language; PROLOG, and a production system language; OPS.

Logic Programming and Mechanical Theorem Proving

In the early days of Artificial Intelligence, there was a dominant view focusing on formal logic as the foundation for computerization of human intelligence. Mechanical derivation of theorems and proofs from a set of initial axioms by means of one or a few formal rules of inference has therefore had significant attention in this research area. The foundation of much of this work has been the Predicate Calculus, a formal representation of classical logic.

To represent a natural language (here English) sentence in the predicate calculus, focus on the relations and entities the sentence describes. There are many possible ways to represent an English sentence in terms of predicate calculus formulae:

"The topstream output from column 1 is the feed to column 2"

may, for example, be represented as:

Same-stream (topstream (Col1), feedstream (Col2))

or as:

Same-stream (stream (Col1, Top), stream (Col2, Feed))

or as two connected formulae:

Top-stream (Streaml, Col1) $\cap$ Feed-stream (Streaml, Col2)

etc.

Atomic formulae are the building blocks of the predicate calculus . *Connectives*, such as '$\cap$' (logical 'and'), '$\cup$' (logical 'or') and '$\rightarrow$' (logical implication), are used to combine these into more complex formulae. '$\neg$' (logical negation) is also often termed a connective, even though it is really not used to combine formulae, but to negate them. Formulae connected by '$\cap$' (and) are termed *conjunctions*, formulae connected by '$\cup$' (or) are termed *disjunctions*, and formulae prefixed by '$\neg$' (not) are termed negations. In addition, we do also have *implications*, e.g.:

"IF column1 is a distillation column

and it doesn't have a partial condenser

Then it has a total condenser"

which may be represented as:

Distillation-column (Col1)

∩¬Condensertype (Col1, Partial)

→Condensertype (Col1, Total)

The part of the last formula preceding the '→' is termed its *antecedent*. It is composed of *antecedent statements*. The part of the formula succeeding the '→' is termed its *consequent*. It is composed of *consequent statements*. It should be noted that implications may be expressed in another equivalent form. "A implies B" is e.g. logically equivalent to "Not A or B".

When variables occur in a predicate calculus formula, there are alternative ways to interpret the formula: It may be true for whatever object one may imagine in the domain of discourse (universal quantification), or it may be true for at least one such object. (existential quantification). The symbol '∀' denotes universal quantification, and the symbol '∃' denotes existential quantification. The meaning of

∀ (x) [Distillation-column (x)

→ Condenser-type x, Partial)]

("All distillation columns have partial condensers")

is obviously different from the meaning of

∃ (x) [Distillation-column (x)

→ Condenser-type (x, Partial)]

("Some distillation columns have partial condensers").

Resolution-A Widely Used Role of Inference

In the predicate calculus, *rules of inference* may be applied to formulae *to produce new formulae mechanically*. Resolution is a widely used rule of inference on predicate calculus formulae for mechanical reasoning purposes [51]. Resolution may only be applied to a certain class of formulae called *clauses*, defined as formulae consisting of disjunctions of atomic (or negated atomic) formulae only. But this is no real restriction, since any formula in the predicate calculus may be converted to a set of clauses. To explain how is a fairly lengthy process, so for brevity the reader is referred to Nilsson [48].

Resolution may be described as the combination of a pair of *parent clauses* into a new clause termed the *resolvent* of the parent clauses. The resolvent is computed by taking the disjunction of the two parent clauses, eliminating any complementary pair. (A complementary pair is the disjunction of a formula and the same formula negated, for example $P \cup \neg P$)

Example Parent clauses:

$$P$$

$$\neg P \cup Q$$

Resolvent:

$$P \cup [\neg P \cup Q]$$

$$\equiv [P \cup \neg P] \cup Q$$

$$\equiv Q$$

Since $P \cup \neg P$ is a complementary pair (with truth value always true, independent of P's truth value), eliminate the pair. It should be noted that the expression $\neg P \cup Q$ is logically equivalent to the expression $P \rightarrow Q$.

Atomic formulae can make up complementary pairs even when they contain variables. For this to be true, the variables in the two formulae must match, their bindings must be made to be the same.

Example

(A is a constant symbol denoting some named object, x, y and z are variable symbols.)

Parent clauses:

$$P(x, A, y)$$

$$\neg P(A, x, w) \cup Q(x, w)$$

Resolvent:

$$Q(A, z)$$

where $x = A$, $w = z$, $y = z$

The process of finding variable substitutions to make expressions identical is termed *unification*. Nilsson gives a description of unification [48]. Resolution refutation systems are a particularly widespread type of resolution system. The basic difference between a resolution system and a resolution refutation system is that instead of inferring the formulae inferable from the initial set of formulae (the axiom formulae), one attempts to demonstrate that from the *negation* of a particular formula - the *hypothesis formula* - combined with the set of axiom formulae, logical contradiction will be inferred by means of resolution, thereby demonstrating that the negated hypothesis is inconsistent with the set of axiom formulae.

The Programming Language PROLOG

PROLOG is a logic programming language of increasing popularity [9]. The language is a resolution refutation system restricted to a particular class of clauses - Horn clauses. While general clauses are arbitrary disjunctions of atomic predicate calculus formulae, Horn clauses are restricted to having at most one unnegated atomic formula per clause. This does not restrict the expressive power of Horn clauses, however, since any unrestricted clause can be equivalently represented by a set of Horn clauses. An example of a Prolog statement follows:

```
Simple-column (col):-
    Nr-of-feeds(col, 1)
    Nr-of-topstreams (col, 1 ),
    Nr-of-bottomstreams (col, 1 ) .
```

This corresponds to the Horn clause:

Simple-column (col) $\cup \neg$ Nr-of-feeds(col, 1)

$\cup \neg$ Nr-of-topstreams (col, 1) $\cup \neg$ Nr-of-bottomstreams(col, 1)

which may be phrased in English in several ways:

1) Either col is a simple column
   or it does not have one feed
   or it does not have one topstream
   or it does not have one bottomstream.
2) If col has one feed, one topstream and one bottomstream
   then it is a simple column.
3) In order to prove that col is a simple column,
   It must be demonstrated that:
   col has one feed,
   col has one topstream, and that
   col has one bottomstream.

The last paraphrase is the one closest resembling the way Prolog works.

Resolution refutation systems may in principle consider every possible pair of clauses in the role as parent clauses in each resolution step. If there are m initial clauses, and n resolution steps are performed, there will be m + n clauses at the end of the process, and the total number of pairs of clauses considered as parent clauses during the entire resolution process will be:

$$\sum_{i=0}^{n} [(m+i)*(m+i-1)] = m^2n + n^2m + (n^3 - n)/3$$

since each of the m + i clauses present after i resolution steps at least in principle can be paired with any of the other m + i -1 clauses present.

In order to reduce the number of pairs of clauses to consider, Prolog uses a variant of a search strategy called *linear input resolution*. Using a linear input resolution strategy, the hypothesis to be proved is initially resolved with one of the initial axioms resulting in a new clause, the resolvent. In the next resolution step, this particular resolvent clause is resolved with another present clause, and so on. This implies that in every resolution step one of the parent clauses is already selected, only the other parent clause needs to be determined. The number of possible pairs to be considered is consequently reduced to

$$\sum_{i=0}^{n} (m+i-1) = mn + n(n-1)/2$$

The use of such a strategy imposes a strictly hypothesis-oriented, goal-driven nature on the behavior of the search process: There are no means of inferring consequences of the axioms beside those explicitly asked for. The variant of linear input resolution Prolog uses is further restricted in two ways.

1) The search proceeds in a *depth first* manner: If there are several clauses {*S*} in a given resolution step i that can be resolved with the resolvent from the previous step one of these will be attempted to produce a new resolvent. Only after every possible search path from the new resolvent has failed to demonstrate logical contradiction, another clause from {S} will be attempted. This is handled automatically by the chronological backtracking search mechanism built into the Prolog interpreter.
2) Alternative clauses to be resolved with the resolvent from the previous resolution step will be attempted in the order they occur textually in the Prolog program. The ordering

of clauses in a Prolog program may therefore have a substantial effect on the time it takes to find a proof.

A Prolog program consists of a set of conditional or unconditional axioms and an hypothesis to be shown to be a logical consequence of the axioms. The schematic example below is without variables for brevity. Prolog's termination criterion is the production of the empty clause- a clause with no formulae in it.

	Prolog form	Horn Clause form
1)	A.	A
2)	B.	B
3)	C:–A, B.	$C \cup \neg[A \cap B] \equiv C \cup \neg A \cup \neg B$
4)	D:–C.	$D \cup \neg C$

1) and 2) are unconditional axioms stating that A and B are true. 3) and 4) are conditional axioms. 3) states that C is true if A is true *and* B is true. 4) states that D is true if C is true.

Suppose now that we want to demonstrate the truth of D; the logical consequence of the statements 1) through 4): Add the negation of D to the set of clauses and attempt to demonstrate that this is inconsistent with the axioms present:

5) :–D. $\neg D$

Once this is done, the resolution refutation process starts searching for the empty clause, denoted ':-' in Prolog form, 'NIL' in Horn clause form. (The 'or' denotes the logical disjunctive combination of two Prolog form Horn clauses, and the '=>' is followed by the resulting resolvent, the new clause with complementary pairs removed): First, Combine 5) and 4):

:–D. or D:–C. $\neg D \cup [D \cup \neg C]$

=> 6) :–C. $\neg C$

Then, combine 6) and 3):

:–C. or C:–A, B. $\neg[C \cup [C \cup \neg[A \cap B]] \equiv [\neg C \cup C] \cup \neg[A \cap B]$

=> 7) :–A, B. $\neg[A \cap B]$

Thereafter, combine 7) and 1):

:–A, B. or A. $\neg[A \cap B] \cup A \equiv [\neg A \cup A] \cup \neg B$

=> 8) :–B. $\neg B$

Finally, combine 8) and 2):

:–B. or B. $\neg B \cup B$

=> 9) :– NIL

The empty clause is produced after four resolution steps: ¬D is shown to be inconsistent with the database 1) through 4) of clauses. D may thus be said to be a consequence of the initial statements present in the program.

Production Systems

Production system programming is another widespread approach to rule based programming. This approach differs substantially from logic programming approaches:

The information processing is focused towards responding to the data present in the current database rather than towards proving that a current hypothesis may be deduced from these data.

There is no formal logic underlying the manipulation of data representations; the style of problem solving is more associative than deductive.

While a logic program typically involves one rule 'calling' another in order to prove sub-hypotheses of the current hypothesis, there is no explicit rule interaction in production systems; rules do not 'call' one another, but should be viewed as independent, interacting through the current data in the database only.

Allen Newell [39] has explained the concept that later turned into the realization of production systems as follows:

"*Metaphorically we can think of a set of workers, all looking at the same blackboard: Each is able to read everything that is on it, and to judge when he has something worthwhile to add to it. This concept is .... a set of demons, each independently looking at the total situation and shrieking in proportions to what they see fit their natures [39].*"

Newell was concerned with the structure of the existing problem solving programs when he wrote this. The programs existing at that time were mostly organized along a generate-and-test search model, and the primary difficulties with this organization were inflexible control of the search and restricted data accessibility. The blackboard solution cited above that Newell proposed was later termed production system. In a production system, each 'worker' or 'demon' is represented as a condition-action rule, and data are globally accessible in a working memory.

One of the many 'shrieking demons' (those which have their conditions satisfied) is selected through a conflict resolution process emulating the selection of the loudest shrieking demon.

As an illustration of the production system concept, consider the following informal simple **example**.

Initial working memory

1) current temperature is 300 K
2) boiling point of n-C6 is 342 K
3) boiling point of n-C7 is 371 K

Rules

a) If <x> is liquid, and <y> is liquid,
   Then consider distillation to separate <x> from <y>.
b) If <x> is vapor, and <y> is vapor,
   Then consider adsorption to separate <x> from <y>.
c) If <x> has its boiling point above the current temperature,
   Then <x> is liquid.
d) If <x> has its boiling point below the current temperature,
   Then <x> is vapor.

Given the above working memory and set of rules the following steps may take place:

Step 1: Rule (c) matches working memory elements (1) and (2).
Insert '**4) n-C6 is liquid**' into working memory.

Step 2: Rule (c) matches working memory elements (1) and (3).
Insert '**5) n-C7 is liquid**' into working memory.

Step 3: Rule (a) matches working memory elements (4) and (5).
Insert '**6) consider distillation to separate n-C6 and n-C7**' into working memory.

After step 3, working memory includes statements (1) through (6).

The execution of a production system is an iterative process consisting of three steps: Match, Select and Execute. In the Match step, every rule is inspected (its condition part is) to see whether the rule is applicable. The set of applicable rules from this step is termed 'the conflict set'. The Select step decides which of the rules in the conflict set to apply next. Obviously this choice makes a difference in the behavior of the system, and a wide variety of mechanisms have been provided in different systems. Such rule selection mechanisms are termed 'conflict resolution strategies'. In the Execute step, the selected rule is executed - or more precisely - the statements comprising its action part are. Rule action statements typically specify patterns of change to take place in the working memory, changes that will enable rules not present in the current conflict set to be present in the conflict set in the next cycle. This way a forward chain of rule applications take place.

In the following section, the main features of OPS5, the currently most widely used production system programming language, will be described.

OPS5, A Production System Language

OPS5 programs consist of working memory elements (data) and productions (rules). A working memory element is, to some extent, similar to a Record in Pascal. Working memory elements (wmes) may be of different types (belong to different wme classes), where each type has a number of predefined attributes (record fields) where attribute values may be stored. For example the sentence "Col1 is a distillation column with partial condenser" may be represented in OPS5 as:

(distillation-column ↑name col1 ↑condenser-type partial)

Wmes cannot (as Pascal Records can) be explicitly linked with pointers to form complex data structures. In this respect, wmes are more like entries in a relational database table: Links are present only implicitly, through shared attribute values.

Example

"Col1 is a distillation column with condenser cond1, whose surface area is A." may be represented as:

(distillation-column ↑name col1 ↑condenser cond1)

(condenser ↑name cond1 ↑surface-area A)

The link between the two wmes is implicit in that the former and latter have the same symbol-cond1-referred to as value of one of their attributes. The symbol NIL is the default value for all attributes in OPS5 working memory elements.

Each wme has associated with it a *time-tag*. This is a unique identifier, any wme may be identified by its timetag. Whenever a wme is modified, it gets a new time-tag. OPS5 productions operate on the database of wmes. It should be noted that the entire set of wmes and productions is memory-resident. A production's If-part - termed its left hand side (lhs) - may be viewed as a query, using database terminology, into the database of wmes. This query is in the form of a pattern, and it is matched against the contents of the wme database. The Then-part - termed the right hand side (rhs) of the production - suggests patterns of change to be performed on subsets of wmes that match the left hand side query.

Example

"If the distillation column *col1* has a condenser *cond1* and *cond1* is not explicitly associated with any column then associate *cond1* with *col1*"

may be represented as the production:

(p create-inverse-link-from-cond1-to-col1
distillation-column ↑name col1 ↑condenser cond1)
condenser ↑name cond1 ↑column NIL)
→
(modify 2 ↑column col1))

The above production (where 2 is a cursor to the 2nd wme referenced in the production's lhs) is not particularly useful, it contains no variables in its lhs, so the pattern matching involved here is a simple equality test. A more useful and general production is possible if variables are substituted for constant symbols:

(p create-inverse-link-from-a-condenser-to-the-column-it-belongs-to
(distillation-column ↑name <col> ↑condenser <cond>)
(condenser ↑name <cond> ↑column NIL)
→
(modify 2 ↑column <col>))

OPS5 productions may also match against missing data - verify the absence of a piece of data in a particular data pattern. It should be noted that this is not the same as the logical negation of the data pattern.

Example

"If there is a distillation column C associated with no particular condenser, and there is no condenser which is associated with this particular column,

Then create a new unique symbol X to denote the name of the condenser, associate X with C and C with X. "

```
(p      make-new-condenser-wme-and-associate-bidirectionally-with-its-column
        (distillation-column ↑name <c> ↑condenser NIL)
    -   (condenser ↑column <c>)
        →
        (bind <x>)
        (make condenser ↑name <x> ↑column <c>)
        (modify 1 ↑condenser <x>))
```

As any production system, OPS5 works in cycles, each involving the three steps Match, Select and Execute. Step one, Match, identifies all *production instantiations* matching the current wme database: Multiple sets of data may match a production consistently (a variable must be consistently matched throughout the entire production; all its occurrences must denote the same constant symbol throughout the production), and these sets must be distinguished among. The pair of such a set of wmes and the identity of the production (its name) comprise a production instantiation. The set of production instantiations comprises the current cycle's *conflict set*.

In *conflict resolution* - the Select step following the conflict set identification step- one production instantiation is selected as being *dominating* according to OPS5's conflict resolution strategy. The main principles in OPS's strategy are *Refraction, Recency* and *Specificity. Refraction* implies that if a production instantiation has been *selected before*, it *cannot be selected again.* This is to avoid infinite iterative execution of the same production instantiation. OPS5 identifies instantiations by means of the wmes' time-tags, so if a wme gets a new time-tag, it and all production instantiations it is part of are considered to be new, and may thus be allowed to execute again. Recency specifies that if a particular wme referred to by a production instantiation *Z* is more recently created or modified than any other wme referred to by any other production instantiation in the conflict set, then *Z* will dominate in conflict resolution. If several instantiations have the same recency rating- they refer to the same set of wmes - then the one associated with the production with the largest number of condition elements will dominate. This is termed the Specificity principle in OPS5.

The rationale underlying the Recency principle is an attempt to extend the 'current line of processing' - a focusing of attention: Give preference to the processes operating on the most recently created or modified data. The rationale underlying the Specificity principle is: The more conditions a production lhs contains, the more specific is the situation the production is intended to address likely to be, and specific prescriptions are given priority over more general ones.

When the dominating production instantiation is identified, the rhs of the production it refers to is executed with variable bindings identical to the bindings from its lhs. (These were stored as part of the production instantiation.)

OPS5 is an entire programming language, this brief description of its basic characteristics is only an overview. More thorough descriptions may be found in the OPS5 manual [19] and in the textbook by Brownston et. al. [6]. Two points should, however, be emphasized with respect to production systems programming:

1) The textual occurrence of a rule within a production system program is insignificant.
2) There is no explicit sign of a rule's interaction with other rules present in the system, no direct references to any rule from any other.

These features have been marketed as a blessing, since they make it possible to introduce new pieces of program code into an existing program without having to insert it in a particular place in the program code. No references to the new piece of code is necessary, since productions do not 'call one another' as in procedural languages or as in Prolog, so in principle the existing program does not have to be revised. However, in practice this may be a mixed blessing: The new rule may execute in situations where it was not intended to, or it may fail to execute in places where it was supposed to. This is due to the fact that it is not sufficient for a rule to execute that it is present in the current conflict set - that its conditions for execution are satisfied. It must also be the winner of the conflict resolution step - achieve the highest rating in the conflict set as calculated by the conflict resolution strategy.

In principle, it should be possible to formulate a rule so that it specifies what should be done in a given situation without concern for when this situation might occur in the execution course of the program. However, in reality the situation description triggering this rule, call it rule A, may coexist with a situation description triggering another rule, rule B, whose action part disables the execution of rule A. (Formally, rule B may negate at least one of A's condition statements).

If the production system programmer does not foresee this situation, and rule B has a higher rating than A according to the conflict resolution mechanism in charge, then rule A will not get a chance to execute. However, in order to foresee such a situation the programmer must have the knowledge that there is a rule B that may prevent rule A from executing: He must be concerned with what other rules might execute in the situation where rule A is intended to, and he may also have to change some of the existing rules when new ones are added to prevent the old ones from executing in situations that were unforeseen before the new rule was added to the program.

A rule may be explicitly prevented from executing by adding statements to its condition part that limit the range of situations where it is satisfied. (Rule specialization) It is important that this is done in such a way that the added statements can reasonably be related to the domain of discourse. Otherwise, the rule may easily become a mixture of a rule representing domain knowledge and an execution control element, its interpretation becomes unclear, and the modularity of domain knowledge representation that a production system offers may become very difficult to maintain as the system grows. Flexibility turns into chaos.

Structured Objects Representations; Frames

In 1975 Marvin Minsky wrote one of the more influential papers on knowledge representation. The paper introduced the notion of 'frames', one of the most popular representation styles in the years to follow [37]. Minsky's basic idea was that:

> *"The ingredients of most theories both in Artificial Intelligence and in Psychology have been on the whole too minute, local, and unstructured to account - either practically or - for the effectiveness of commonsense thought. The 'chunks' of reasoning, language, memory, and perception ought to be larger and more structured; their factual and procedural contents must be more intimately connected in order to explain the apparent power and speed of mental activities [38]."*

Minsky pointed to other recent works too, as

> *"moving away from the traditional attempts . . . in trying to represent knowledge as collections if separate, simple fragments."* (an implicit reference to expressions in predicate calculus)

Brachman and Levesque later pointed to the core of Minsky's argumentation:

"That the pursuit of formal logic in AI has been very misleading, and that some of the demands of logic, such as consistency and completeness, are probably not even desirable in a representation language [4]."

The extraordinary influence of frame representations led Patrick Hayes in 1979 to take a closer look at what this movement had produced since Minsky's original work in 1975:

"A frame is a data structure . . . intended to represent a 'stereotypical situation'. It contains named 'slots', which can be filled with other expressions - 'fillers', which may themselves be frames, or presumably simple names or identifiers (which may themselves be somehow associated with other frames,) .... For example, we might have a frame representing a typical house, with slots called kitchen, bathroom, bedrooms . . . A particular house is then to be represented by an instance of this house frame, obtained by filling in the slots with specifications of the corresponding parts of the particular house, so that, for example, the kitchen slot may be filled with an instance of the frame 'contemporary-kitchen' which has slots 'cooker', 'floorcovering', 'sink', 'cleanliness', etc. . . Not all slots in an instance need be filled, so that we can express doubt. . . *Looked at this way, frames are essentially bundles of properties*.... Thus far, then, working only at a very intuitive level, it seems that frames are simply an alternative syntax for expressing relationships between individuals, i.e. for predicate calculus. But we should be careful, since although the meanings may appear to be the same, *the inferences sanctioned by frames may differ in some crucial way from those sanctioned by logic. In order to get more insight into what frames are supposed to mean we should examine the ways in which it is suggested that they be used [25].*"

Minsky argues that

"traditional formal logic is a technical tool for discussing either everything that can be deduced from some data or whether a certain consequence can be deduced; it cannot discuss at all what *ought to* be deduced under ordinary circumstances [38]".

This emphasis on *control of the inference process in frame systems* points to what Hayes terms 'the heuristic interpretation':

"that frames are a computational device for organizing stored representations in computer memory, and perhaps also for organizing the processes of retrieval and inference which manipulate these stored representations [25]."

Frame systems have been said to involve two fundamental issues that distinguish them from other representational frameworks like for example the predicate calculus. The first is the use *of procedural attachment* in the retrieval and update processes, enabling very specialized inference procedures to be used efficiently, because they are explicitly connected to the sets of data they are meant to be applied to. The second is the ability to represent and manipulate descriptions of and inferences on the representation itself: frames, slots and values may all have attached meta-information, frames describing the frames/slots/values. Hayes has emphasized this latter point, pointing out that frames may be used as a medium for *reflexive reasoning: reasoning involving description of the self.* In fact, Hayes claims that the emphasis on the analysis of processes of reflexive reasoning is the only thing that is *conceptually* new. Apart from this he claims that the frames idea's real force is not on the representational level at all, but on the implementational level:

"How to organize large memories; that assertions should be stored in namable 'bundles' which can be retrieved via some kind of indexing mechanism on their names. In fact, that assertions should be stored in non-clausal form [25]".

The following section describes some central features of a frame-based representation system called CRL - Carnegie Representation Language [7].

CRL -A Frame-Based Representation Language

CRL is a frame representation language developed at Carnegie Mellon University / Carnegie Group Inc. It is, like most AI knowledge representation languages, based on Lisp. (Common Lisp) Frames are termed schemata in CRL, and the schema is a particular type of Lisp object. Schemata have slots (attributes), and slots have values. Slot values may be arbitrary Lisp objects unless otherwise explicitly specified.

Schemata will here be displayed in the form:

{{schema-name

slot-1: value (or values; multiple values are allowed by default)

slot-2: value or values

....

slot-n: value}}

Example

"Solution X1 consists of the distillation columns *C1* and *C2*, and it has a high rating." may be represented in CRL as:

```
{{solution}}
{{X1
  instance: solution
  solution-elements: C1 C2
  rating: high}}
{{distillation-column}}
{{C1
  instance: distillation-column}}
{{C2
  instance: distillation-column}}
```

A set of schemata comprises a schema database, and this database may be queried or manipulated using access-functions, such as:

(get-value *schema slot*) , which returns the value of *slot* in *schema*

(new-value *schema slot value*) , which sets the *value* of *slot* in *schema*

(add-value *schema slot value*), which adds the *value* to already existing ones

(delete-value *schema slot value*) , which removes *value* from *slot* in *schema*

etc.

Values may be *inherited*. Suppose that the database is:

```
{{separator
  purpose: separate}}
{{distillation-column
  is-a: separator}}
{{C1
  instance: distillation-column
  number-of-feeds: 1}}
```

Then, the following query results may be obtained:

```
(get-value 'C1 'number-of-feeds) => 1
```

(get-value 'C1 'purpose) => separate

The value 1 is the *resident value* of slot *number-of-feeds* in schema *C1*, but the value separate is inherited from schema *separator*. Inheritance is possible because the special slots is-a and instance belong to a set of special slots in CRL termed *relations*.

The behavior of any slot in a CRL schema database may be specified using a *slot control schema* for the slot. This is a schema which has the same name as the slot, where additional information about it is stored. It is not necessary for a slot to have a slot control schema, but if it has one, it typically contains the following information:

{{<slot-name>

domain: some value or expression

range: some value or expression>

cardinality: two integers

demon: a sequence of names of demon schemata

...... }}

The *domain* slot of the slot control schema makes it possible to restrict the use of the slot to a set of schemata by use of a system-defined restriction grammar. The *range* slot makes it accordingly possible to define the set of legal values a slot may have. *Cardinality* specifies an upper and lower bound on the number of values the slot is allowed to have. If the *demon* slot is used, it specifies pre-and/or post-processing procedures that will be executed automatically, without being explicitly called (thus, the name) when the slot is accessed in a particular way. The values of the demon slot must be schemata that describe the nature of the demons: There must be one schema for every demon, and such a schema must contain the information:

{{<demon's schema-name>

instance: demon

access: the name of the access function(s) triggering this demon

when: before or after the access-function executes?

action: name of the Lisp function that constitutes the 'demon proper'

. . . . . . }}

If the same kind of pre - or post-processing is needed every time a particular slot is accessed using a particular access-function - for example, delete-value - then the use of demons is a convenient way of hiding behavior specifications from main program code.

Example

{{vital-information
  demon: you-blew-it}}
{{you-blew-it
  instance: demon
  access: delete-value
  when: after action: (lambda (schema slot value)
    (tell-user schema slot value "is gone. Too bad!"))}}

This is a sketch of a demon executed every time a value in slot *vital-information* is deleted by means of access-function *delete-value*. The demon executes *after* the value is deleted, and its purpose is to tell the user that the value is gone.

If a slot is to be used as a relation, its slot control schema must be is-a or instance linked to the system-defined schema *relation*, where the necessary 'relation-handling machinery' is stored and may be inherited from.

Relations do frequently have additional slots in their slot control schemata. These - termed *inheritance specifications* - allow specification of how to inherit information over the relation link: *Inclusion-specs* specify information that is allowed to be inherited. *Exclusion-specs* specify information that is not allowed inherited. *Introduction-specs* may introduce new information into a schema when the relation slot is created in that schema. *Map-specs* specify information transformations during inheritance (English-metric unit conversion, for example).

The slot control schema is a *global* definition of slot behavior, the definition applies to *all occurrences* of the slot in the entire schema database. But local specialization of slot behavior is allowed using *meta-slots*: Any entity in a CRL database may have *meta-information* attached; information describing information. A schema may have a *meta-schema* attached, a particular slot occurrence in a schema may have a *meta-slot* attached (this is NOT the same as the slot control schema, which has a global scope), and any slot value may have a *meta-value* attached. Both meta-schemata, meta-slots and meta-values are represented as schemata, they may have an arbitrary number of slots and values, and these may again have meta-information attached; arbitrary levels of meta-information is representable.

Meta-slots may have the same slots as slot control schemata - domain, range, demon, etc., and if they have, these function as local overrides of the slot control schema. A meta-slot may be entirely local to the slot occurrence it is a meta-slot for, or it may be inherited down through a hierarchy of schemata, thus making is possible to define localized slot behavior for arbitrary classes of slot occurrences.

The value(s) of a slot in a CRL schema database may be arbitrary Lisp objects (unless otherwise specified in the slot's slot control schema or meta-slot). Functions are valid Lisp objects, so functions or function names are also valid slot values: Pieces of Lisp code may be stored in CRL schemata, and results of CRL database retrieval may therefore be executed as programs:

```
{{any-schema
    slot-with-function-name-as-value: plus}}
(funcall
    (get-value 'any-schema 'slot-with-function-name-as-value)
    2
    2)
=> 4
```

A CRL schema database may be partitioned in a tree structure using *contexts*. This is an advanced feature of CRL that may be used to implement hypothetical reasoning (multiple worlds) in a convenient way: Suppose that a decision point is reached in the problem solving where several alternative decisions should be considered. Then a new context may be created for each of these, and changes in the CRL representation of these decisions and their consequences are stored in their respective contexts separated from each other. Only changes are stored, incrementally: Unchanged information is inherited in any context *C* from the set of contexts comprising a path from *C* up to the root-context, so that any such path comprises a *consistent world view*.

Object Oriented Programming

Object-oriented programming is a programming concept closely related to frame-based representation systems. (Actually, it is the other way around; historically the concept of object oriented languages was there almost a decade before Minsky's influential paper on frames: Simula [10], the first object oriented language, was designed and implemented as early as in the mid-sixties by the Norwegian researchers O.J. Dahl and K. Nygaard.) Today, C++ and Smalltalk are probably the most widespread object oriented languages.

An object oriented program consists of a collection of *objects* and *methods*. Objects are similar to frames; they have attributes with values, they are organized in hierarchies, and attribute values may be inherited down through the hierarchy as described for frame systems. Methods are procedures attached to particular objects' attributes. They may be equated with procedural attachment to slots in frame systems. Methods are inheritable like attribute values.

Objects communicate by *message passing*. A message is a data structure consisting of three parts; the *verb*, the *receiver object* and *optional arguments*. When the message is received by the receiver object, the method associated with the verb is retrieved and executed. (Either there is a procedure attached to the verb attribute in the receiver object, or the receiver object inherits such a procedure from higher up in the object hierarchy it is part of) Sending messages to other objects is typically part of any method's execution.

The philosophy underlying object oriented programming is that as much as possible of the inner workings of an object should be hidden to the outside world. The outside world must know that the object exists, that certain operations (verbs) can be performed on it, but the details of how that is done is of no concern. It is the object's own responsibility to manage itself. This approach to programming makes it convenient to write very compact and high-level code once the building blocks (the collection of objects and methods) are established. Object oriented programming makes procedural abstraction very convenient and natural: Several distinct procedures may be said to do conceptually 'the same thing', but to different types of objects and consequently in different ways, without having to specify such details in code calling on the execution of these procedures. As an example, assume that the chemical components A and B may be separated using either distillation, extraction or absorption. The goal is to simulate the separation, and the simulation procedures 'dist-sep-simulation', 'extr-sep-simulation' and 'abs-sep-simulation' exist. Traditionally, one would then say:

call dist-sep-simulation (A, B, --specifications data), or

call extr-sep-simulation (A, B, --specifications data), or

call abs-sep-simulation (A, B, --specifications data).

Adopting an object oriented approach instead, one would create objects representing prototypical distillation-columns, etc., and attach the procedures to these objects as methods, for example with method name 'Simulate separation'. Then, one would create particular instances of these objects, with attribute values reflecting e.g. what solvent is used in the extraction unit to be simulated, its size, operating pressure and temperature, etc. Finally, to trigger the simulation, one would send the message:

Simulate-separation (<instance>, A, B),

where <instance> would be the name of an instance of a distillation column object, an extraction unit object or any other object that has a procedure associated with the verb (method) 'Simulate separation'.

The primary distinction between object-oriented programming systems and frame-based representation systems seems to lie more in how the systems are used than in how they differ in their inner workings: While people who have used structured representation techniques as an augmentation to procedural programming have termed their systems 'object-oriented', people with a rooting in representation and inference have tended to use the phrase 'frame-based'.

The current tendency is integration of different styles of programming; logic-based, production-based, procedural and structured-representation programming all present in the same system. In such a context, it seems like objects and frames are evolving in the same direction.

Problem Solving Systems Architectures

This section addresses organizational frameworks for 'Expert-level' problem solving computer systems - different logical architectures.

Hayes-Roth et al. provide a categorization of Expert Systems applications [28]. It is proposed that each of their categories have particular problem class characteristics, and that these characteristics should be reflected in the organization of problem solving systems attacking these classes of problems. Their classification is shown below:

Expert System Application Types

Interpretation	Inferring situations from descriptions.
Prediction	Inferring likely consequences of given situations.
Diagnosis	Inferring system malfunctions from observables.
Design	Configuring objects under constraints.
Planning	Designing actions.

Monitoring	Comparing observations to plan vulnerabilities.
Debugging	Prescribing remedies for malfunctions.
Repair	Executing a plan to administer a prescribed remedy.
Instruction	Diagnosis, debugging and repair of student behavior.
Control	Interpreting, predicting, repairing and monitoring system behaviors.

However, systems architectures developed within one of these application areas may very well be suitable for application in other areas. The blackboard architecture is an example in case. This architecture was originally designed for speech interpretation [17], but over the last years variants of the blackboard architecture have been used also in other domains; crystal structure determination [57], planning [27], design .

While the early problem solving systems tended to be built from scratch, a number of system building tools exist today, making it possible to start system design and implementation on a higher lever than before. Ranging from simple PC-based rule-interpreters to powerful hybrid representation and reasoning frameworks, it may often be hard to determine which system to use for the problem at hand. Particularly, because flexible hybrid systems often will be harder to learn and use than simpler, but more restricted ones. To guide system designers in their selection, a number of 'Guides to Expert Systems' have been published [22,23,31,32,60].

The architecture of a system should reflect the characteristics of the problem it addresses. While simple, uniform architectures may suffice for relatively limited problems, more complex and heterogeneous architectures may be required when the problem becomes less structured and less understood. (Apart from this, it should be stated, present system architectures naturally also reflect the way the system designer believes the problem *should* be solved; subjective preferences for particular problem solving strategies and formalisms.)

Stefik et al. (in [28] Ch. 4) give an overview of Expert Systems architectures. They classify problems along three dimensions-*problem size, reliability of data, and constancy of data*- and they suggest systems architecture guidelines for each of eleven classes of problems.

Problem Size; Decomposition, Precedence Ordering And Constraints:

Large problems must be decomposed to be solved. If one does not decompose such a problem, then simple statistical arguments [55] demonstrate that they will be almost impossible to solve.

There are many ways to decompose a problem. *Functional decompositions* decompose the problem into subproblems that address different functional aspects of the problem. For example, the problem 'design a chemical plant' may be decomposed into 'design the reactor system', 'design the separation system', 'design the heat exchanger network', etc.

Decomposition into *different levels of abstraction* is another frequently used decomposition strategy. For example, the problem 'design a chemical plant' may be decomposed into 'design the Process Flow Diagram', 'design the Piping & Instrumentation Diagram {P&ID)', 'perform the mechanical and thermal design', etc.

A third frequently used decomposition is *decomposition into views*, for example, 'a distillation column as viewed by mechanical engineers', 'a distillation column as viewed by chemical engineers', 'a reactor viewed as a heat source', ' a reactor viewed as a chemical components converter', etc. Real world decompositions often involve these three types, and probably others as well, although this may not be explicitly recognized, because some of the decompositions are implicit in the educational and professional training of human problem solvers. As an illustration, when a chemical engineer is given a problem of the type 'design a process for production of methanol from natural gas', he will typically come up with a Process Flow Diagram (PFD) and consider this to be a solution to the problem stated. However, if you give this solution to an instrumentation engineer and he will view it as part of a problem specification. ("One's roof is another's floor", as the idiom goes.) It is rare that any given decomposition decomposes the problem into completely independent subproblems; subproblems usually interact, weakly or strongly. When subproblems interact, there are constraints on either the solution of the problem, the solution process itself, or both, depending on the nature of the interactions: A P&ID requires that a PFD exists - there is a strict precedence ordering constraint on the design of PFDs and P&IDs, and the results of the design on the PFD level forms part of the problem specification for the design on the P&ID level. In other cases, there is perhaps no strict precedence ordering, but the order in which the subproblems are addressed has effect on the final solutions: If a car is designed by designing the engine first and then the body, it is very likely that the results is going to be different from what it would have been if the body were designed first.

The initial problem has an initial number of degrees of freedom. When subproblems are solved they consume some of these, leaving fewer degrees of freedom to the remaining yet unsolved subproblems. The consequence is that the remaining subproblems have to be solved within a limited space of alternatives. If the precedence ordering is selected poorly, there may not even be a single valid alternative left for the last subproblems, or there are only poor solutions left for them. (If the validity of this claim is doubted, one should try to put on one's tightest pair of jeans after having put on one's biggest pair of boots.)

The intuitive heuristic on precedence orderings coming out of this discussion is that if the problem can be decomposed into subproblems of varying importance to solution of the overall problem, then the most important ones should be addressed first, before the space of alternatives gets too constrained by already made decisions.

Uncertainty

The reliability of data and inferences are important parameters in the design of system architectures. In classical logic, statements may only be either true or false, and consequently inferences on such statements are assumed to be either true or false, too. When premises are said to be more or less probably true, and inferences more or less likely follow from the premises, classical logic is no longer sufficient.

Mycin, an early medical diagnosis Expert System [53] and Prospector, an early geological data interpretation Expert System [14], are examples of systems designed to work with uncertain data and probabilistic inferences.

Formal methods exist for the solution of such problems - Bayes statistics- but Expert Systems tend to use other less formal certainty factors and certainty propagation procedures. The rationale for this has been that the experts providing the information both factual and judgmental - present in these systems "do not use the information comparable to implemented standard statistical methods. However, the concept of certainty factors did appear to fit . . . their judgment of how they weighted factors, strong of weak, in decision making" [3]. Such subjective probabilistic approaches to problem solving have become popular particularly in diagnosis and interpretation applications. The main characteristic of such problems is that:

1) There exists a finite number of predetermined hypotheses, a set of more or less certain observed data, and a set of probabilistic conditional statements (rules).
2) The task is to identify how probable the hypotheses are, given the data and their certainty This is done by quasi-statistical ('Bayes-like') inference combination of data and rules with an update of probability- or rather degree of subjective belief - similar to the

statistical calculation of posterior probabilities from observed events (data) and a priori probabilities on conditional statements (rules) involving these observed data.

Besides giving an estimate of how strongly an hypothesis is believed be true, certainty factors are also used as control parameters in systems taking this approach. During the inference process, the various inferences produced by the system get their certainty factors updated. These are propagated through the system so that the hypotheses' degree of belief also will change. There are typically many hypotheses to be considered, and a heuristic frequently used is to focus on the one with the highest current degree of belief. (Rule applications may increase on decrease this degree of belief). Shortcliffe [53], Davis [11] and Duda [14] describe such certainty propagation procedures. Fuzzy logic is one of several other approaches to reasoning with uncertainty [63].

Assumptions and Expectations

Logical deduction is a necessary but not sufficient condition for problem solving. 'Educated guessing' often plays a major role - making the appropriate assumptions and solving the problem assuming these hold. Assumptions may of course turn out to be wrong, and in that case experienced human problem solvers will be able to steps back, review and modify assumptions, retract the parts of the solution relying on invalidated premises and start over again, but not from scratch; even the solutions based on faulty assumptions have given some insight into the problem.

Problem solving is performed within a framework of assumptions and beliefs, and the sooner the appropriate framework can be established the more efficient will the problem solving be; assumption making is indeed another form of problem decomposition.

Assumptions may be verified up front by analysis or experiments; 'assume that no azeotrope can be formed from a mixture of components A and B', they may be unverifiable in principle, which is the case e.g. with predictions and forecasts; 'assume that the product market price over the next five years will be X', or they may be verifiable only after decisions are made and solutions are analyzed; 'assume that all the distillation columns may be run using low pressure steam'.

If an assumption is verifiable, that does not necessarily mean that it should be verified immediately. That depends on how strongly the assumption is believed to hold, and it depends on the effect an assumption violation would have on the solution. The important thing is that they should be visible as assumptions as long as they are unverified - explicitly represented so that they may be inspected later if possible and desirable. Dependencies should also be explicitly represented when reasoning is performed on assumed premises: A derived results that depends on an assumption may loose its foundation when the assumption is retracted - the result itself might have to be retracted. In order to retract it when necessary, it must be explicitly represented how it was derived. For a discussion of dependencies see [8].

So-called *hypothetical reasoning* - 'What if ...'- reasoning, also termed 'Multiple worlds'-reasoning, is an approach to reasoning with uncertainty, based on assumptions and expectations instead of certainty factors and statements that are true to a certain degree. Systems incorporating hypothetical reasoning allow inconsistent statements to coexist in *different worlds:* In one world, a statement may be considered to be true, and inferences depending on the truth of this statement are derived in this world. In another coexisting world the same statement may be considered to be false. This world will therefore produce other inferences than the former.

It is the present author's opinion that decision making systems (as opposed to systems where the truth of an hypothesis is to be determined) lend themselves more naturally to the hypothetical reasoning approach than to the certainty factor approach. The justification for this position is that *decision systems are primarily concerned with consequence and utility - not truth*: A decision is not primarily true or false - it may be more or less appropriate, better or worse. A medical doctor may be 20% or 70% certain that Mr. Smith has a particular disease, and this degree of belief will hopefully affect the treatment prescribed for him. However, when it comes to *acting on the belief - deciding* what kind of treatment to give, there are only two alternatives - either assume that he has the disease, or assume that he dose not. And, a chemical engineer may be 30% or 60% certain that a particular separation task in a petrochemical process may be performed using a simple flash drum instead of a more expensive distillation column, but in order to verify or falsify the belief, an analysis must be carried out where *either* the drum *or* the column is present.

Meta-Level Problem Solving

In the section addressing search, it was emphasized that several competing operators are typically applicable to a particular state. Selection among these could be guided by the evaluation of a rating function. The rating function was therefore said to control the search. However, in complex decision making problem domains, typically involving uncertainty and assumption making, where analysis and decision making both are part of the problem solving process, it is frequently hard to identify a single simple rating function capable of rating properly all the actions that could conceivably be carried out as part of the problem solving process: Alternative decision making operators must compete among themselves for priority, but they must also share resources with analysis operators - operators capable of identification, verification or falsification of critical assumptions. The control problem - determination of what to do next in a complex problem solving system - may therefore itself become a complex task.

Operators may be divided into distinct *base-level* and *meta-level* categories [20], where base-level operators are those which when strung together solve the domain problem the system is intended to solve, and meta-level operators are those involved in deciding which base-level operators to apply next. Genesereth [20] emphasizes the importance of keeping base-level and meta-level components as 'separate agents'; not letting an operator have both these roles. Inclusion of a meta-operator layer makes it possible to bridge the gap between a general purpose interpreter and domain specific base-operators without confusing control and domain knowledge in the base-operators.

Why Is It Important Not To Confuse Control And Domain Knowledge? The main issues are modularity and flexibility. Complex systems tend to grow incrementally, operators are added one at a time. If a particular set of operators contain information assuming the presence or absence of other operators, addition of new operators or removal of existing ones may have wide-ranging effects on those left unchanged.

Hayes-Roth et. al. argue that "An intelligent system reasons about its actions", and that "In order to perform effectively, an intelligent system must have knowledge about its actions" [30]. Such reasoning on knowledge about actions implies that the system must have -or be able to generate - information describing the purpose and consequences of operator applications. The consequences of applying an operator should of course not have to be known in full detail in advance; if that were the case it would not be necessary to apply the operator, it is superfluous. However, basic characteristics of it could be described so that a meta-level layer of operators would be able to forecast an approximate potential utility of applying it, like for example the typical time it needs to execute, the expected quality, accuracy and level of detail of the results it produces, etc.

Goal- and State- Driven Integration of Independent Operators

The collection of operators comprising a system reflects an implicit decomposition of the problem the system is intended to solve- an operator addresses a task that is part of the problem. The purpose of the operator is to perform the task. How the different tasks in a problem are related can be represented either explicitly or implicitly and with or without *subgoaling*.

Subgoaling is a mechanism where operators respond to goal statements present in the current state description, where a goal corresponds to a request for a task to be performed, and the operator responding to the goal statement either performs the task to completion or posts subgoals, thereby letting operators capable of performing subtasks of its task execute to their completion before completing its own task. Every operator must have encoded in it what kind of tasks it may solve.

With such a subgoaling scheme, there is no need to represent the purpose of an operator explicitly. Its purpose will be to respond to a goal statement, and if goal statements are labeled with the identity of the goal statement activating the operator posting these goals, there will be a link upwards through all currently suspended operators from the currently active one all the way up to the top level goal.

It is important to note the interaction between goal statement, task and operator. The operator does not refer to any particular sub-operator. It posts a goal, and once the goal is posted the sub-operator 'sees' that the goal refers to a task that it can respond to. It is therefore in principle not necessary that anything but goals are communicated among operators in this type of system.

There may of course be many operators capable of solving the same type of task. In that case, one of these will have to be selected among those applicable. This is typically the case in problems where there are alternative solution methods to solve a problem, possibly with varying accuracy, detail and computational resource requirements. In such cases, the goal statement may be used to specify restrictions on the current range these parameters are allowed to take. The operator performing the goal-task matching these restrictions better than any of the others performing the goal-task may than be selected for execution and executed. Meta-level operators may be used to determine what constitutes a 'better match'.

The above assumes that there is only one active goal at any time-it is a depth-first goal-oriented architecture. Real world decision problems are frequently more complex: Many different tasks may be performed as the next one, leading to a situation where many goals should be considered as the one to pursue next. Now, the *relevance* of achieving a particular goal to the overall solution of the problem will have to be considered, too: Which of the currently active goals will promote a good solution and an efficient solution process the most?

Two approaches may be taken: try to first establish which of the uninstantiated goals to pursue, and after that decide which operator is best suited to achieve the selected goal. Alternatively collapse these two decision phases into one: Decide directly which applicable operator should be selected without any *separate* concern for the goal it achieves.

It may be argued against the latter that this looks a lot like adding apples and oranges. To some extent that is a valid objection, but in the alternative approach- select goals first and then ways to achieve them - information available on the expected characteristics of potentially applicable operators is not considered; information that could have an impact on the selection of the next goal to pursue is not used.

Architectures without subgoaling are also possible: Problems can be decomposed so that an operator that requires subgoaling is decomposed into

a) the part of the operator that is executed before subgoaling would occur.

b) the part of the operator that is executed after.

It is not even necessary to represent the subgoaling explicitly: The state description at the instance where subgoaling would occur may be used instead of the goal statement as the statement telling the sub-operator that it should signal its applicability. So instead of looking out for a goal statement to appear, the operators look out for specified data patterns in the state description. This is termed data-driven processing: The state description determines the set of applicable operators.

The major difference between goal- and data-driven architectures may be phrased as follows: In a goal-driven, scheme operators performing particular tasks are requested. In a data-driven scheme, there are no requests, operators 'volunteer' their bids for execution.

The terms goal- and data-driven are frequently confused with the terms backward- and forward-chaining, but these two concept pairs are orthogonal: Backward-and forward-chaining are concepts describing how a system implementation is operated, while goal-and data-driven are concepts describing how the problem is decomposed into operators and whether operators are invoked voluntarily or on request. The stereotypical view is that data-driven systems chain forward while goal-driven systems chain backwards. However, goal-driven systems may also chain forward, as in:

"If Goal X is currently active

then do something procedural...

and post subgoal Y

and do something more procedural..."

In both cases - data-driven or goal-driven - the applicable operators should submit a signal, globally visible to the system, that they are applicable. They should not be allowed to execute before the system realizes system wide that they are applicable. If they were, there would be no way eventual meta-operators could submit their 'opinion' on whether an operator should be allowed to execute or not. Such a globally visible signal - a bid for execution - should contain as much information about the operator submitting it as possible, because the more the system gets to know about the operators the better is the basis for deciding which to choose.

Goal-driven system architectures, where several goals are allowed to be active at any instance, may be viewed as a special case of a data-driven system architecture. The rationale for this claim is that the relevance of any active goal needs to be computed in order to determine which of the alternative goals to pursue as the next. The goal is there, but this does not any longer have an imperative interpretation in the sense that one in the set of operators able to address this particular goal is guaranteed to execute next. The set of operators have to compete with other sets of operators trying to achieve other active goals.

Hearsay II - An Early Example of Integration of Goal-and Data Driven Processing

The Hearsay II speech understanding system [17] was an early system addressing this kind of problem. Hearsay II's domain was 'close-to-real-time' interpretation of connected speech natural language sentences in a restricted vocabulary containing approximately 1000 words. The system consisted of three parts; a 'Blackboard' -the global data structure holding the representation of the current state, a set of 'Knowledge Sources' (KSes) - independent pieces of code performing the different domain tasks the system could perform, and a scheduler-a mechanism for selection of Knowledge Sources to execute next.

The Hearsay II blackboard was partitioned into a 2-dimensional grid, one dimension comprising the different levels of abstraction that the speech understanding system had to work on, the other dimension representing the time interval during which the sentence was uttered. Most of the Hearsay II KSes worked on the border between two of the Blackboard levels, forming hypotheses on the upper level from the data on the level below or generating data on the lower level under the assumption that hypotheses on the upper level were true, extending the set of hypotheses along the time dimension and comparing and rating hypotheses originating from different KSes. In other words, both the KSes and the Blackboard data structure reflected a particular decomposition of problem and solution methods.

Hearsay II can neither be classified as goal-driven nor as data-driven. It can best be described as applying an *opportunistic reasoning* model - applying pieces of knowledge in the direction towards either goals or data in an opportune manner. The input data to Hearsay II- parameterized voice signals- is the basis of hypotheses formed on the syllable level. Signal patterns may indicate that a particular syllable has been uttered in a given time interval in the sentence. Combinations of recognized syllables may further indicate that a particular word has been uttered, etc. This kind of processing is data-driven; data patterns suggest hypotheses to verify. When a number of words have been hypothesized as possible explanations of segments of the voice signal, further processing may be done on alternative interpretations of such a segment, establishing a relative likelihood that a particular hypothesized word is the correct one by comparison of an expected signal pattern (such expected patterns are stored within the system for each word it is able to recognize) with the observed pattern. This is goal-driven processing; attempting to identify which in a set of alternative hypotheses to believe the most. Alternatively, combinations of words may be combined into word phrases, for example a verb phase if the hypothesized word is a verb, word phrases may be combined into full sentences, and from the assumption that every utterance conforms to a specified grammar, a grammatical analysis may then be applied to determine whether an interpretation of a segment of the utterance constitutes a legal statement in the language described by this grammar. Formation of hypothesized word phases and sentences are data-driven activities, analysis and verification of these are goal-driven activities. Hearsay II switches opportunistically between these two modes of reasoning, all the time trying to:

1) move up the levels from signal description (bottom level) to semantic content (top level) of the sentence,
2) expand the time interval of the currently most strongly supported hypothesized sentences,

3) discriminate among alternative hypothesized interpretations.

The meta-level or control reasoning in Hearsay II- deciding which KS to execute next- is all contained in the systems' scheduler, a single, complex fixed piece of procedural code whose task it is to predict where on the blackboard grid it is most profitable to focus, what kind of processing is needed and which KS is the best one suited to do this.

The scheduling mechanism is described in [26], and Hearsay II as a structure for problem solving is described in [15]. A description of the implementation may be found in [17].

Hearsay II originated a new class of problem solving systems - *blackboard systems*. Nii [46,47] gives an overview of "The blackboard Model of problem solving and the Evolution of Blackboard Architectures." She states that "Subsequently, many application programs have been implemented whose solutions were formulated using the blackboard model. Because of the different characteristics of the application problems and because the interpretation of the blackboard model varied, the design of these programs differed considerably." The HASP/SIAP system (military system for identification of various types of ocean vessels from digitized hydrophone data; [45]), the Chrysalis system (structure interpretation of X-ray diffraction data for protein crystals; [57] and OPM (simulation of human errand running protocols; [25]) are examples in case.

The essence of the control (meta-level) organization in OPM has been extended and modified into the domain independent blackboard system BB1 [29]. The core idea underlying BB1 is that domain and control problem solving are intertwined, and therefore the domain and the control problem should be solved simultaneously. It is suggested that the control problem should be approached as a real-time planning problem.

The basic control scheme in BB1 is very simple; three small and simple basic control KSes are activated iteratively. These three basic KSes update the agenda of KSes that may be executed, select one for execution, and when selected, interpret it. The reason that the scheduling mechanism may be kept so simple is that additional control knowledge may be encoded in more or less domain specific control KSes that, when selected, alter the criteria for KS selection in a way reflecting general or domain specific problem solving strategies. The control problem and the domain problem are solved simultaneously, but BB1 specifies two separate blackboards - a control blackboard and a domain blackboard. The structure, contents and vocabulary of the domain blackboard will vary from one application to another, but the structure, contents and vocabulary of the control blackboard are determined by BB1. A thorough (but unfortunately not too readily comprehensible) description of the control blackboard may be found in [29]- it is too lengthy and complex to be accounted for in any detail here. The key point is that the control blackboard specifies *a control plan* for the solution of the domain problem *on several levels of abstraction and detail:*

Desirable actions or classes of actions according to some explicitly represented (named) strategy.

The current focus of problem solving.

Past and future foci.

Policy decisions determining weighs in a multiobjective rating function.

The rating function itself

The set of actions that may be immediately performed and their variable bindings.

The KS currently being interpreted.

All this information is represented declaratively on the control blackboard, and any piece of it may be modified by control KSes. This makes the BB1 architecture extremely flexible and potentially applicable to a wide range of problem domains.

The generality of BB1 makes it a complex tool to use, and instead of building application systems directly on top of BB1, work has later been focused towards providing higher level system building tools on a level between BB1 and specific applications.

ACCORD is one such problem solving framework for *arrangement assembly tasks* built on top of BB1 [30]. It is hypothesized that assembly problems constitute a distinguishable class of problems that may be approached specifically with a set of control structures and 'semi-domain-independent' concepts tailored to problems of this class. The concept of *planning islands*, a problem solving strategy particularly emphasized in Stefik's system *MOLGEN* [56] is underlying the design of the ACCORD framework. ACCORD also provides generic verbs to use in the description of problems modeled as arrangement assembly tasks, and the system 'knows' how to interpret these, thereby providing a higher level description language than the underlying BB1 representation that ACCORD statements map into.

The major benefit of having such a layered structure is that control and representation paradigms proven to be useful in a problem are made available to similar problems without having to redesign and reimplement it all down to the lowest level of the system building tool.

AKORN D -A Blackboard Model For Process Synthesis.

A problem solving system may be viewed from at least two points of view: The structure and representation of the domain specific information the system is supposed to manipulate, and the structure and representation of the manipulating mechanism itself- the system architecture. It is naturally desired that the problem solving architecture to a large extent should reflect the basic characteristics of the domain problem to be solved. One of the basic characteristics of process design and synthesis problems is their open-endedness, there is hardly ever one correct solution, but a lot of better and worse ones. Another basic characteristic is that such problems are usually configurative; the solutions of the problem do not pre-exist in form of a fixed set of hypotheses to be verified or falsified, it is a major part of the problem solving process to generate them. A third characteristic is that the overall problem solving strategy seems sensitive to special characteristics of the domain problem, for example the identification of bottlenecks in the design - dominant features of the problem pointing to essential decompositions and precedence orderings once the features are recognized. A fourth characteristic is the mixed use of analysis and assumption making used to limit the search space. A fifth characteristic is the use of multiple approaches to the problem- more or less accurate analysis methods, heuristics based on one point of view or another, and several levels of abstraction on which the problem may be described.

These characteristics are typical for problems that have been approached using blackboard models. It was therefore a natural choice to attempt to use a blackboard model approach here, too. The proposed architecture, which was given the name AKORN D for obscure reasons, attempts to be general and flexible but at the same time simple. Concepts were borrowed from BB1, but AKORN D distinguishes itself from BB1 in that it is much less complex: In BB1, a substantial effort is put into representing declaratively extensive control plans on multiple levels of abstraction independent of application problem domain. In the design of AKORN D, it was felt that such complexity may, when it comes to representing domain control knowledge, be counterproductive - it requires substantial insights in the domain problem to be able to formulate and represent extended control plans, and it may not even be fruitful to any worthwhile extent to do so in an early development phase. So the opposite approach was taken - it was assumed to be sufficient to be able to decide what to do next in a qualified manner. Crucial prerequisites to be able to do this are:

The system has a system-wide awareness of what can be done next.

It is able to apply control knowledge to select among the options available.

As in BB1, it was decided to parameterize the basic KS selection function and to keep the parameters represented on a separate control blackboard distinguishable from a domain solution blackboard, where emerging domain solutions are represented. It was also decided to retain the BB1 concept of separate domain and control KSes.

The KSes may be viewed as (simulated) independent processes, each constantly monitoring a common data area looking for data patterns indicating their relevance and responding to these. When a KS finds a pattern it can respond to, it posts a data element called a bid on the control blackboard. The bid is a signal to the rest of the system that a particular KS considers itself to be useful. It contains information about the characteristics of the KS posting it and about this KS's own opinion of the importance of being executed in the current situation.

Knowledge Sources: In AKORN D, KS consists of three parts:

1) A *monitor* looking for characteristic patterns: The monitor submits a bid when it detects 'its' pattern. This bid is put onto the control blackboard.
2) A *declarative representation of the task* the KS performs and characteristics of the way the KS achieves it.
3) An *action part* that when executed performs the actual task and typically manipulates one of the blackboards.

Domain KSes monitor and manipulate the domain blackboard only. They comprise the available domain actions- the set of actions that may be performed next to promote or extend an emerging solution represented on the domain blackboard. *Control* KSes essentially monitor and manipulate the control blackboard. They do also have monitoring access to the domain blackboard, but they are not allowed to manipulate it.

Scheduler: The *scheduler* is a low-priority independent process whose purpose it is to select a bid from the collection of bids present on the control blackboard and hand over execution control to the action part of the KS that posted this winning bid. The winning bid is selected by means of a selection function and the parameters currently present on the control blackboard. Figure 1 illustrates this information flow.

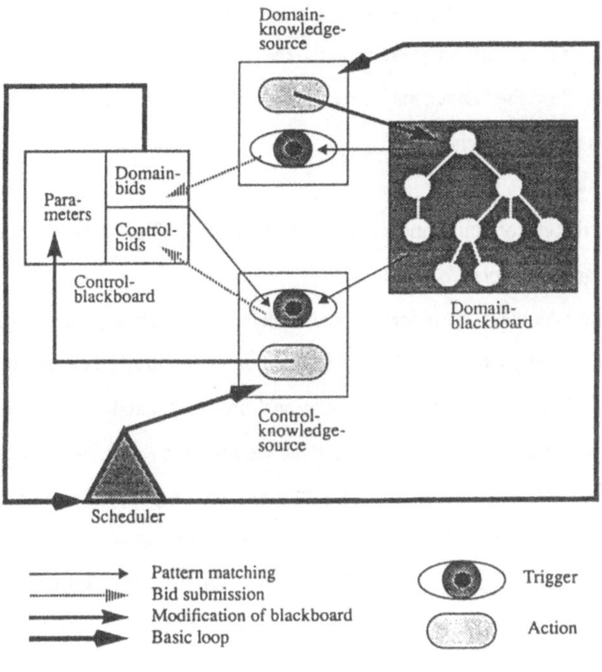

Figure 1. AKORN D information flow

AKORN D operates in cycles. In one cycle,

a) All KSes that detect their potential for application submit a bid which is placed on the control blackboard.

b) When the bid submission halts the low priority scheduling process takes control and selects one of the bids on the blackboard.

c) Execution control is handed over to the action, part of the KS that submitted the selected bid.

When the selected action is completed a new cycle is started. In the meantime, a blackboard has typically been modified so that there are new data patterns to respond to for the set of KSes.

Execution may be terminated either explicitly by a KS executing a halt command or by the scheduling process if no bids are present on the blackboard.

AKORN D as described above may be termed an extended production system:

It is based on pattern matching

It operates in cycles

There is no direct communication among rules (KSes). They communicate solely through working memory (the blackboards)

In every cycle a 'conflict set' is formed and 'conflict resolution' is performed on it in the sense that bids are submitted and KS selection is based on information contained in these bids.

Yet, AKORN D is different from a conventional production system: *The conflict set (the set of bids) is an integral part of the problem state because it is explicitly represented in working memory (on the control blackboard.* This is an important feature in that it *allows the system to respond to signals about what it considers to do before actually doing it,* and does so independently of proposing it. For example, a control KS containing knowledge about situation where one task is preferred over another may trigger when these two task have bids signaling their potential application on the blackboard. This is readily done without having the domain KSes contain references to the conflict situations or to each other. Domain knowledge (which options are available) and control knowledge (which actions are preferred in the current situation) is clearly separated.

Using previously defined terms, domain KSes may be viewed as base operators, while control KSes are meta-level operators (or, if a control KS triggers on the presence of a control bid-meta-level, etc.). An interesting point is that all levels manifest themselves in the same basic cycle.

Conflict Resolution in AKORN D: Nothing has yet been said about the conflict resolution mechanism in AKORN D- the basic mechanism for selection of a bid and thereby of the next action to perform. This basic mechanism is a parameterized selection function, where the parameters are explicitly represented on the control blackboard and thus modifiable by control KS actions. The selection function parameters describe the current problem solving focus of attention in a domain independent way. The following are the key points justifying their form and existence: The design of AKORN D attempts to allow several partial solution to the problem at the same time on the blackboard, so the system has to decide which of the present solutions to work on. The KSes thus have to be explicit about which solution they propose to extend. Every solution must be provided with a rating describing how promising it is currently believed to be. The system may also decide to focus on a particular set of task. Since it is desirable to be able to add new task without having to modify existing code, there is no explicit reference to task names. It is instead assumed that the problem task can be said to exist on a *level*, so that a reference to the level is an implicit reference to a set of task. The KSes have to know their own level. The bids submitted by a KS on a particular level will thus represent potential actions on that level. Within a level there may exist alternative actions. A domain KS recognizing its applicability will also to some extent be able to predict the goodness of the action it proposes.

Some problems have more strongly predefined precedence ordering than others. The amount of search needed to find a good solution depends very much on the problem characteristics, so it should be possible to *modify the width of the search*, both in terms of moving among alternative solutions and in terms of moving around on different levels in any particular solution. AKORN D distinguishes among three different classes of bids: ***Decision*** bids are bids from KSes that make decision: A partial solution gets closer to completion by application of decision KSes. ***Analysis*** bids are bids from KSes that perform analyses (generate information) without making any decision. ***Control*** bids are bids from control KSes. These propose to modify selection function parameters. Control KSes do not exist on any level.

The selection function parameters present on the control blackboard are:

Current-ps: The partial solution currently in focus. A partial solution name.

Current-level: The level currently in focus. A level name (a positive integer).

Current-rigor: (Applicable to analysis bids only)
The current importance of accurate analyses.
(a number in the range from 0 (unimportant) to 1 (important))

Current-speed: (Applicable to analysis bids only)
The current importance of fast analysis.
(a number in the range from 0 (unimportant) to 1)

Ps-dop: Degree of focusing on the current partial solution.
(number ranging from 0 (no focusing) to 1 (strong focusing))

Level-dop: Degree of focusing on the current level.
(number ranging from 0 (no focusing) to 1 (strong focusing))

Four additional parameters are used by the scheduler to compute bid priorities:

Ps-support: A binary switch with value 1 if the bid addresses the current partial solution, -1 otherwise

Level-support: A binary switch with value 1 if the bid addresses the current level, -1 otherwise

Opgood: The 'subjective opinion' of the KS submitting the bid on the potential value of its application.
(number ranging from 0 (not very promising) to 1 (very promising))

Psgood: The 'subjective opinion' of a partial solution on its relative value.
(number ranging from 0 (not very promising) to 1 (very promising))

Opgood and Psgood will clearly be strongly domain dependent. AKORN D does therefore not specify *how* to compute these numbers, but a mechanism is specified for their determination: It is assumed that every partial solution and KS is represented as an object with a method attached or inheritable which computes these numbers. These methods are allowed to have read-access to the control blackboard.

Domain bids (decision or analysis bids) are rated according to the following selection function:

DBP =[2 + Ps-dop * Ps-support]
* -2 + Level-dop * min {Level-support, Ps-support * Level-support}]
* 1 + Opgood]
* 1 + Psgood]

DBP is a number in the range from 1 to 36. It should be pointed out that level here becomes subordinate to partial solution: It rarely makes any sense to promote actions on solutions out of focus even if they are on the current level. Level focusing is thus viewed as a refinement of solution focusing. Control bids are rated according to:

$$CBP = 1 + 6^{(1 + Opgood)}$$

CBP is a number in the range from 7 to 37.

The rationale for these rating functions is an attempt to represent desired cross-effects of the parameters involved: Control KSes represent control heuristics which may address more or less specific situations. Although any KS signals its application through an active assessment of the situation - it does not suggest itself unless it identifies the current situation as interesting, there may be situations where both general knowledge and more specialized knowledge compete for application. The general guideline is that the more specific set of situations a KS is intended to handle, the higher should the Opgood values of the bids it submits be. If high rated domain bids are present-the actions they represent seem locally very promising, they address a promising solution, this solution is currently in focus, the level is the current one and the focusing is very strong-then only very specialized control knowledge is allowed to interfere. (for example to discriminate between two particular (named) very promising actions, if such control knowledge is present in the system.)

When the rating of domain bids drop, a larger set of control KSes become competitive-that is, if the situation is interesting to them; they have submitted bids. But still the more specialized of them will dominate. There are several reasons for domain bids to get low rating. Often these reasons may themselves indicate the proper control actions to get back to high ratings:

There may not be any more good bids on the current level in the current solution.

If good bids are present on another level in the same solution and the solution is believed to be good then this suggests to focus on this other level.

If good bids are present in another solution on a particular level and this other solution is believed to be comparable with the current one, then this suggests to switch focus to the solution and level where the good bid is.

If it is unclear what level and solution to focus on, this may indicate that search is needed to identify where to focus: Reduce the emphasis on focus-lower the values for Ps-dop and/or Level-dop and work where it locally seems to be good until a more specific control pattern is recognized and acted upon.

Bids representing defocusing control actions should in general probably have lower Opgood values than bids representing focusing ones. This because focusing aims to reduce search effort, while defocusing increases it.

AKORN D attempts to provide an additional degree of freedom in representing domain knowledge as compared with a traditional production system: A KS may have its conditions for execution partitioned among

1) the KS trigger, which determines whether the current situation is interesting at all to this particular KS,
2) the selection function and its current parameters, which decide whether this KS in the current situation is interesting *enough* to allow its execution,

3) (an arbitrary number of) control KSes which are capable of changing the current situation (the focusing on a particular solution, level, and also the degree of such focusing) so that a locally interesting action may become interesting enough to be selected.

The original AKORN D architecture was is implemented using the development system Knowledge Craft as a set of CRL schemata, a set of Lisp functions and an OPS5 rule driving the basic loop. An in-depth description of the implementation, as well as an example of how the architecture was used to generate heat integrated distillation sequences using mainly qualitative information, may be found in [35].

S6 - Automatic Synthesis of Separation Sequences with a Blackboard Based System combined with a Mathematical Programming Model

The S6 program combines Knowledge Based and Operations Research techniques in the generation of solutions to sharp split heat integrated distillation sequence problems. *The KBS parts of the system* ***construct superstructures automatically. These superstructures are subsequently pruned down to sets of cost optimal solutions by combinatorial optimization***. KBS techniques are furthermore used to represent complete and emerging solutions, as well as to allow ***some extent of user interaction*** in the problem solving process.

The overall structure of S6 is illustrated in Figure 2. The 'knowledge based' parts of the program are implemented on a Symbolics Lisp machine running ***Knowledge Craft,*** while other parts of the program (e.g. parts of GAMS input files and the GAMS optimization system itself) reside on a SUN workstation connected to the Symbolics via Ethernet. The knowledge based parts of the program are implemented using the ***AKORN DT*** blackboard architecture, an extension of AKORN D. The parts of the program residing on the SUN workstation are invoked on demand from the Symbolics computer. Communication between the two machines is controlled by a separate communication module. Some user interaction is allowed via a graphical interface.

AKORN DT is an extension and reimplementation of the AKORN D blackboard architecture. The 'T' indicates that tentative search has been included in the new version of the architecture. Both AKORN DT and its predecessor are opportunistic control architectures, meaning that there is no explicit representation of a hierarchy of goals to be fulfilled. Any KS may in principle be activated at any time, providing a 'broad focus of attention'. The bids submitted onto the control blackboard serve as a globally visible and explicitly represented set of alternative next moves. This set resembles the conflict set in production systems, but its explicit representation as part of the problem solving state provides improved opportunities for the use of domain specific information in the conflict resolution.

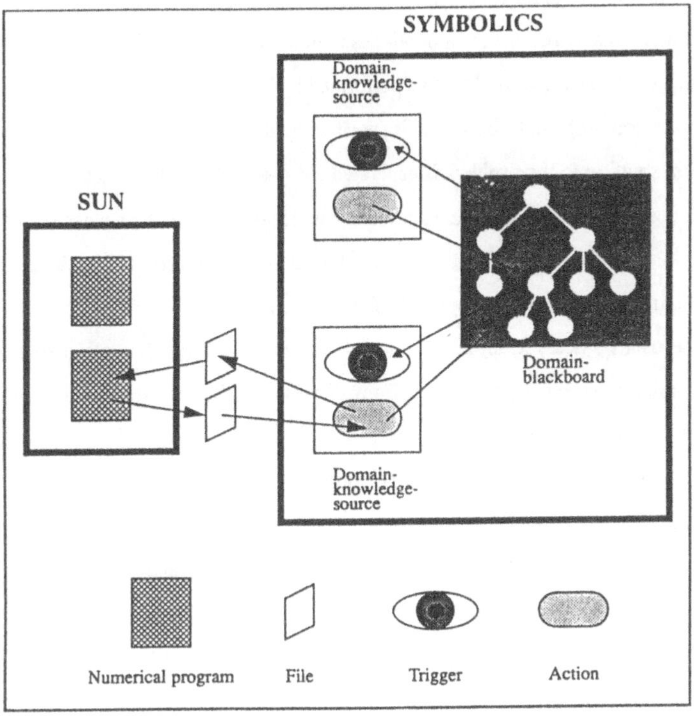

Figure 2. S6 overall structure

Some basic modifications had to be made to AKORN D in order to facilitate tentative search: Incremental representation of partial solutions, combined with explicit representation of backtracking points and alternative next moves. In AKORN DT, incremental representation of partial solutions has been achieved by use of the Context Mechanism provided in Knowledge Craft: When the action part of a KS is activated, the results generated by this KS are stored in a new context, which is the 'child' context of the context where the KS triggered. Only new or modified information is stored in the new context, unaltered information is made available through inheritance from ancestor contexts. This is a representation form which makes it easy to preserve consistency during tentative search, and the set of possible backtracking points may be defined in a very general way: Every context is a potential backtracking point. Furthermore, if previously unused bids from domain KSes are stored with an association to the context where they were submitted, then the possible next moves from any context may also be defined in a very general way: Every unused bid associated with a context represents a possible next move from that context. Wahl presents an in-depth description of AKORN DT [58].

On a high level of abstraction, the tasks performed by S6 may be described as follows:

1) A separation problem to be solved is specified by the user on the Symbolics machine.
2) A first solution to the problem is generated automatically by the program and displayed.
3) The user may decide that the solution is the desired one, or that parts of the proposed solution should be modified. (The modifications are limited to modifications of the superstructure from which optimal solutions are extracted)
4) If the user requests it, an alternative solution may be generated automatically, and the new solution is displayed. Repeat from step 3) until the user accepts the solution or no more solutions can be generated.

Automatic generation of the first solution involves the following steps:

a) Generate the set of possible separation tasks needed to separate the mixture described in the separation problem specification. Store this in a semantic network (e.g. a frame structure) on the domain blackboard.
b) Simulate, size and cost each of the separation tasks. Augment the semantic network with this information.
c) Generate a logical superstructure consisting of separation columns on different levels. Store this superstructure in the semantic network.
d) Construct an input file to the combinatorial optimizer GAMS from the information contained in the semantic network on the domain blackboard. Perform the optimization, extract a description of the optimal solution from the GAMS output file and store this information in the semantic network on the domain blackboard.

As currently implemented, S6 may use two different mathematical programming formulations of the problem - the Andrecowich Transshipment formulation or the corresponding Transportation formulation.

All information about the emerging design solutions is stored in a semantic network (e.g. a frame structure) on the domain blackboard. The network consists of schemata (the Knowledge Craft notion for 'frames' or 'objects') describing entities such as columns, streams, and hot and cold utilities, or representing decisions / alternatives and relations among these. A simplified graphical presentation of the network is given in Figure 3.

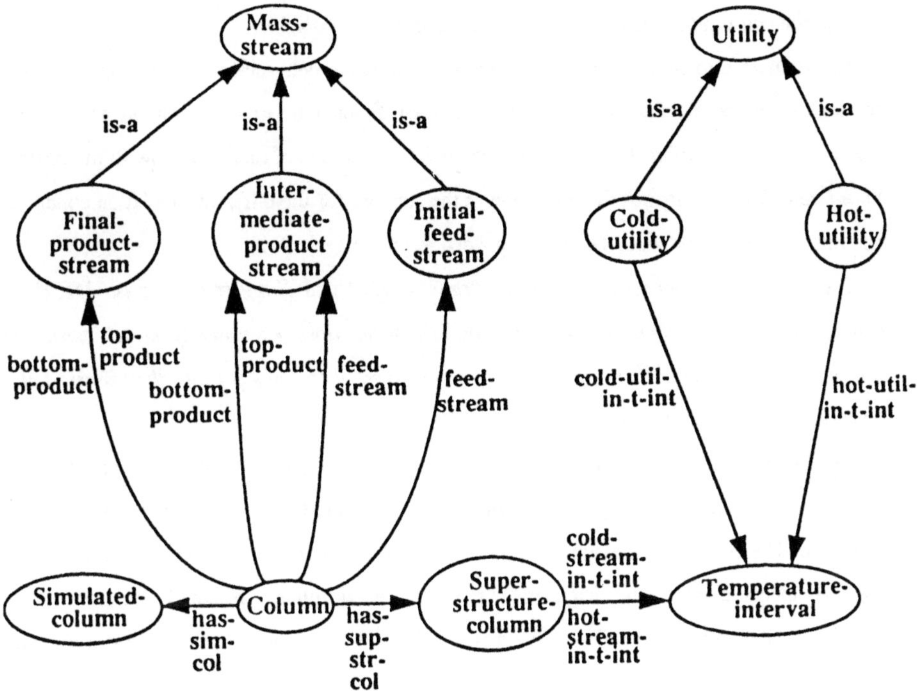

Figure 3. Part of the semantic network in AKORN DT

The knowledge sources currently implemented in S6 are basic knowledge sources which are general and rather lean on heuristic information. Other knowledge sources implementing specific heuristics can be added to compete with or supplement the basic ones.

The domain knowledge source ***Generate-Columns*** generates a network of streams and conceptual columns. This knowledge source does not include any heuristic knowledge, it generates all possibilities. Later, other knowledge sources may be added that either create another network or modify this network. This reflects a basic principle adapted by us, believed to be rather important when the purpose of the knowledge based parts of a system is to provide a superstructure that is subsequently going to be pruned down to one or a small set of solutions by OR methods: *Heuristic elimination of a subset of the set of logically possible alternatives will usually be based on a weaker foundation than a cost based elimination in the subsequent optimization of the superstructure. Therefore, within the limits imposed by computational tractability (e.g. the size of the superstructure), the superstructure should to the extent possible contain the entire set of logically possible solutions to the problem.*

The domain knowledge source ***Simulation-Conditions*** determines the pressure and flow conditions for the simulation of the columns. Four simulations of each column are required -simulations at two different pressures for two different amounts of total feed flow to the column. ***Simulate-columns*** performs the simulation of the columns for the different simulation conditions determined from the previous described knowledge source.

The knowledge sources ***Suggest-HI- Transh-MILP*** and ***Suggest-HI-Transp-MILP*** are conceptually equal: "*If there is a separation problem which does not have any models associated with it and there exist columns that have been simulated, then suggest to we the transportation (transshipment) model for the problem.*"

Use-Char-Temp-Alg uses the characteristic temperature algorithm to create a superstructure of columns [1]. The knowledge source generates an input file to the characteristic temperature algorithm. The contents of the input file is created on the Symbolics machine and transferred to the SUN machine, where the characteristic temperature algorithm is executed. The result file is transferred back to the Symbolics, where it is transformed to another format, to be stored in the semantic network.

The two domain knowledge sources ***Use-Gams-Hi-Transh-Milp*** and ***Use-Gams-Hi-Transp-Milp*** use the optimizer GAMS to extract the optimal solution from the ones embedded in the superstructure. They are almost equal, so they can be explained together ."*If there is a separation problem with the transportation/transshipment model and there exists a set of superstructure columns and no solution exists, then suggest to we GAMS to solve the problem.*" First, this knowledge source creates the 'dynamic' parts of the GAMS input file, including e.g. sizes of sets, values of parameters and so on. 'Static' parts of the input, e.g. indexed model equations, are stored in other files. A series of Unix commands for the SUN machine are then invoked from the Symbolics. These commands merge the different input files, start GAMS, and extract the relevant information from the GAMS output file. The extraction of information from the output file is done using AWK, an efficient UNIX pattern matching utility program. These results are retrieved to the semantic network and the solution is displayed.

It is up to the user to decide what to do with the displayed solution - four alternatives are available through a menu in the user interface. Each of the four choices are carried out by activation of a control knowledge source: ***User-Response-Halt*** stops further execution of the system. It is used when the user indicates satisfaction with the solution. ***User-Response-Ok*** "*If there is a user response with the value OK and there exists a unused bid (other alternatives are possible), then suggest to explore this alternative bid.*" The knowledge source finds the context where this bid was submitted and makes it the default context (i.e. moves to this context) and lets the bid execute. ***User-Response-Next*** adds an integer cut - the current solution is not allowed in the next attempt. This causes the second best solution to be generated. ***User-Response-Edit*** signals that the user has chosen to edit the superstructure.

The domain knowledge source ***Edit-Formulation*** is triggered when the user has signaled that he wants to edit the current superstructure. It pops up a graphical superstructure editor, where the user may move the temperature / pressure level a superstructure column is defined on, delete an existing column, or add a new column.

The Graphical User Interface: The main purpose of the graphical interface is to display the solutions and the superstructures in a convenient format when the user has the possibilities to interact with the system. The graphical interface has been implemented using the Knowledge Craft Window/Graphics System. The screen is divided into different areas (viewports and windows). In the upper left corner is a list of icons, in the upper right corner is the name, at the bottom there is a command window. The rest of the screen consists of a viewport for display of costs for the selected solution, one for display of the superstructure, one for display of the labels used in the superstructure, one for display of the heat integration pattern, and one for display of the mass flow pattern. This is illustrated in Figure 4 and described in some detail below:

The ***Command Window*** allows the user to type in Lisp commands instead of using icons and menus. The ***Icon Window*** consists of five icons: Quit, Problem Specification, Run, Display, and Edit. All of theses are activated by clicking the mouse bringing up a menu. *Quit* makes it possible to pause (temporarily leave and return) or to exit (leave with no return). *Problem Specification* brings up a slot filling editor where the specifications for the separation problem can be filled in. Alternatively, one has the option to read the problem specification from a file. *Run* starts or restarts the system (loading of files + a few commands). Display gives a menu for the display of the viewports Cost, *Superstructure, Heat Integration and Mass Flow*. Edit is used for modifying the superstructure. The possibilities are: *Delete Column*, which removes a superstructure column, *Edit Column*, which allows the user to modify the temperatures for the condenser and reboiler by the use of a slot filling editor, and *Create Column*, which allows the user to generate another superstructure column by use of the same slot filling editor.

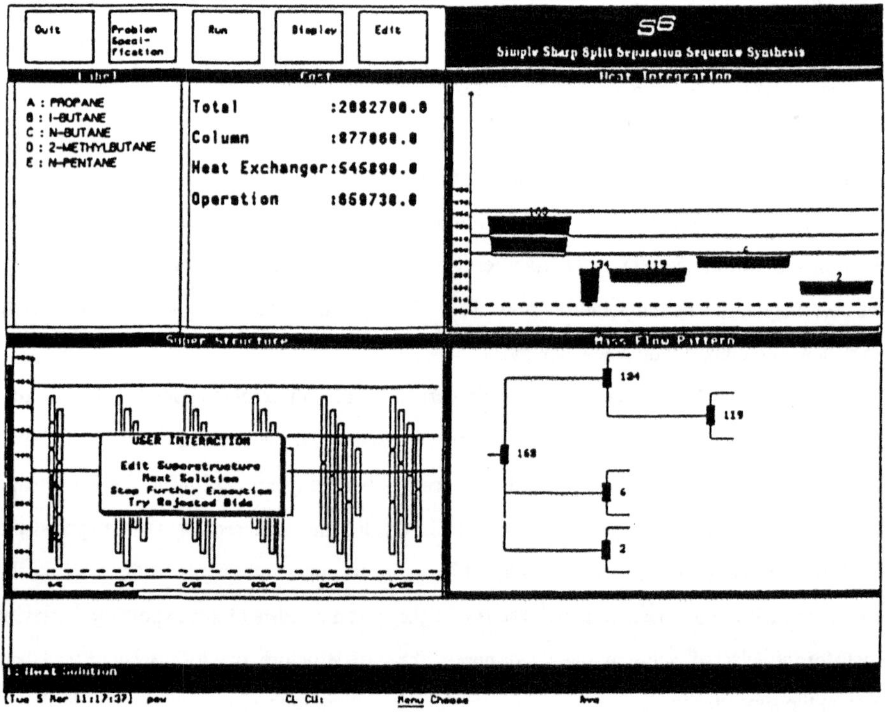

Figure 4. S6 user interface

The ***Cost Viewport*** displays some cost variables returned from GAMS annualized cost, which is the objective function of the math. programming model, investment cost for the columns and the heat exchangers, and operating costs cold and hot utilities). The purpose of this window is to give the user information about the distribution of the cost and thus hopefully make it easier to evaluate the returned solution or decide whether to modify the superstructure or not. The ***Superstructure Viewport*** displays the superstructure generated by the characteristic temperature algorithm. The superstructure columns are represented as white boxes. When a solution has been generated, the selected superstructure columns are displayed as black boxes with an index. Utilities are shown as horizontal lines -hot utilities are solid and the cold utilities are dashed. The splits are labeled by letters for compactness in the graphical representation, and displayed labels are described in an own window above the superstructure window.

In the ***Mass Flow Viewport***, each column is represented as a black box with an index (which also occurs in the superstructure viewport) and the name of the light and heavy key. The ***Heat Flow Viewport*** describes the heat integration of the selected solution. Columns are here represented as trapezoids in a temperature-energy coordinate system, resembling the Andrecowich box representation. (Trapezoids are used instead of rectangles here, because a column's reboiler and condenser duties are not necessarily equal.) When the transportation formulation is used, one trapezoid may be displayed above another, to represent graphically the heat integration of the corresponding columns. When the transshipment formulation is used, on the other hand, the optimization results do not explicitly tell which columns to integrate. The graphical representation will in these cases not display any trapezoid on top of another.

There are also two other windows, which will normally be buried by the others: One for the OPS process (OPS is used for generation of bids and for triggering of the AKORN DT scheduler) and other messages from the Symbolics system, and one terminal window for processes running on the remote Unix machine. These are shown in Figure 5. The former is useful for debugging purposes. The latter may be useful e.g. for the monitoring of progress in time-consuming optimization runs.

Chemical components are organized in a hierarchical network structure with multiple inheritance. Components and component classes are implemented as schemata (objects) where each slot can have values or methods attached. This gives rise to possibilities for object oriented programming. Physical property parameters for each compound are stored in each schema. Mainly the slots of the schemata are filled with quantitative data taken from [50]. Physical properties, e.g. vapor pressures, may be computed by alternative equations, where a particular

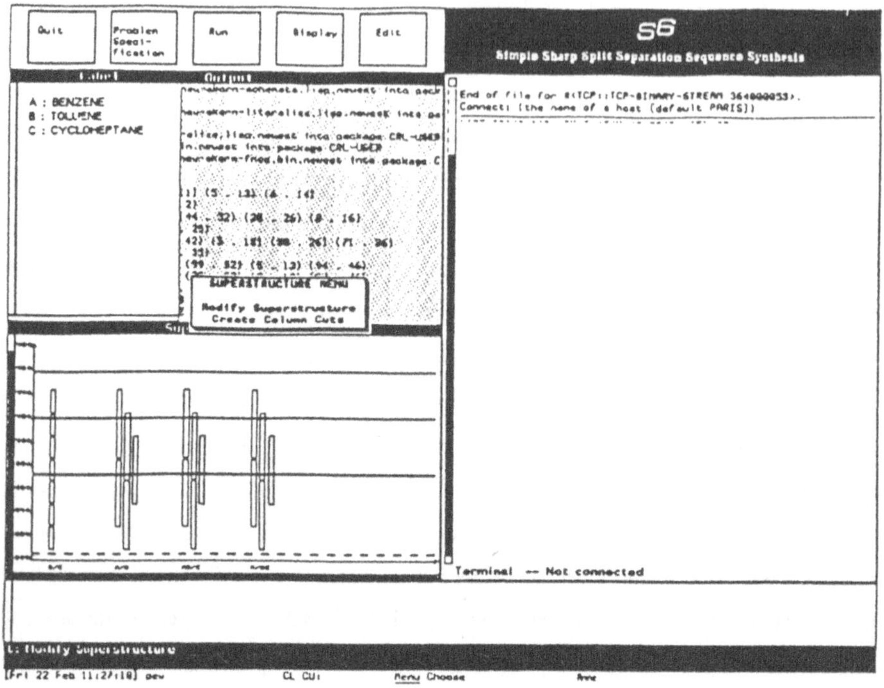

Figure 5. Hidden trace and debugging/monitoring windows

equation is better suited for some components or components classes than others. In S6, each of the components (or component classes) have stored references to which of these methods to apply, so that the calculation method needs not be known 'outside' the component. This gives the calculations a flavor of object oriented programming: When the program needs e.g. the vapor pressure of a component, a message is sent to the component, saying that the vapor pressure at a given temperature is needed. The choice of method is then determined by the component 'itself'.

The functions used in S6 for the simulation and cost estimation of the columns and heat exchangers are to a large extent the shortcut methods reported in [13]. Interfacing of S6 to a commercial simulator has been considered, but since the main purpose of the program is to demonstrate the integration of AI and OR techniques, we were not willing to spend the required effort. The required minimum set of shortcut methods was instead implemented in Lisp in the version of the system reported here. It should however be mentioned that in an even more recent reduced implementation of S6 in C++ [59]. The program has been interfaced to the commercial simulator ASPEN+ for generation of simulation, sizing and costing data.

Final Remarks

This paper has presented an overview of some important elements of knowledge based systems. The overview has by no means been intended to be complete. Instead, we have tried to focus on what the current author believes to be the central issues. It is expected that knowledge based systems techniques will continue to play an important role in the research and development of useful computer aids for process engineers.

Two key issues described here -pattern recognition and search, are expected to have a profound influence on the development of such systems, but it is difficult to predict what form this will take. Our - and others' -experience with pattern directed knowledge based methods in the form of rule based systems indicate that there is a need for more powerful pattern recognition methods than those offered by symbol oriented rule based languages, e.g. pattern detection in multivariable continuous variable spaces. Currently, such isolated pattern matching features are offered both by e.g. Neural Networks and by several more established statistics based pattern recognition algorithms. It may therefore be expected that the future may bring hybrid systems where symbolic problem solving and search is integrated with this kind of advanced pattern matchers. This would in turn have a profound effect on representation techniques: The various representation techniques described in this paper are essentially symbol oriented, and this is most like going to be insufficient if more advanced pattern matching techniques are going to be used. What is needed under these circumstances is a *subsymbolic* representation - a representation where explicit symbols are not necessarily explicit in the representation but may possibly be generated as needed as a kind of abstraction.

In search, it may be expected that knowledge based search techniques may be combined with the more powerful analysis methods of mathematical programming, because mathematical programming allows a wide range of powerful tools from mathematics to be applied. Object oriented representations will naturally also play a role in this development process, but in the design of object oriented representations one should be aware of the fact that the information hiding characteristic of object oriented programming may in fact in many cases make advanced pattern matching more difficult rather than easier, simply because the data to match are hidden and not globally visible to the pattern matcher.

Acknowledgments

The work reported here has been funded in parts by the Norwegian Science Foundation (NFR), Norsk Hydro, Statoil and the Nordic Petroleum Research Program. Thanks to Prof. A. W. Westerberg, Carnegie Mellon Univ., for many inspiring discussions during the development of the AKORN blackboard architectures, and to Dr. P.E. Wahl who developed the AKORN DT architecture and S6 as parts of his Ph.D.

References

1. Andrecowich, M.: Synthesis of Heat Integrated Distillation Sequences, Ph.D. Dissertation Thesis, Chem. Eng. Dept., Carnegie-Mellon Univ., PA, (1983).
2. Andrecowich, M. and Westerberg, A. W.: A Simple Synthesis Method based on Utility Bounding for Heat-Integrated Distillation Sequences, AIChE J. 31, p.363, (1985).
3. Barr, F. and Feigenbaum, E.: The Handbook of Artificial Intelligence, W. Kaufmann, Los Altos, CA, (1982).
4. Brachman, R. and Levesque, EI.: Readings in Knowledge Representation, Morgan Kaufmann, Los Altos, CA (1985).
5. Brooke, A., D. Kendrick and A. Meeraus: "GAMS - A User's Guide", The Scientific Press, Redwood City, CA (1988).
6. Brownston, L. et. al.: Programming Expert Systems in OPS5, Addison-Wesley, Reading, MA, (1985).
7. Carnegie Group: Knowledge Craft 3.1 Reference Manual, Carnegie Group Inc., Pittsburgh, PA, (1986).
8. Charniak, E., Riesbeck, C. K. and McDermott, D. W.: Artificial Intelligence Programming, Lawrence Erlbaum Assoc., Hillsdale, NJ, (1980).
9. Clocksin, W. F. and Mellish, C. S.: Programming in Prolog, Springer-Verlag, Berlin, (1981).
10. Dahl, O. J. and Nygaard, K.: Simul-an Algol-Based Simulation Language, ACM Comm. 9, p.671, (1966).
11. Davis, R. and Lenat, D.: Knowledge-Based Systems in Artificial Intelligence, McGraw-Hill, NY, (1982).
12. Douglas, J.: A Hierarchical Decomposition Procedure for Process Synthesis AIChE J. 31, p.353, (1981).
13. Douglas, J. M.: "Conceptual Design of Chemical Processes", McGraw-Hill, NY (1988).
14. Duda, R. O. et. al.: Final Report, SRI Projects 5821 and 6415, SRI International Inc., Menlo Park, CA, (1978).
15. Erman, L. D. and Lesser, V. R.: A Multi-Level Organization for Problem Solving using many, diverse, cooperating Sources of Knowledge, Proc. 4th IJCAI, Tsibili, USSR (1975).
16. Erman, L. D. and Lesser, V. R.: System Engineering Techniques for Artificial Intelligence Systems, in Hanson, A. and Riseman, E.: Computer Vision Systems, Academic Press, NY, (1977).
17. Erman, L. D. and Lesser, V. R.: The Hearsay-II System: A Tutorial, in Lea, W. A. (ed): Trends in Speech Recognition, Prentice-Hall, Englewood Cliffs, NJ, (1978).
18. Forgy, C. L. and McDermott, J.-: OPS, a domain independent production system language, Proc. 5th IJCAI, MIT AI Lab., Cambridge, MA (1977).
19. Forgy, C. L.: OPS5 User's Manual, Tech. Rep., Dept. of Comp. Sci., Carnegie Mellon Univ., PA, (1981).
20. Genesereth, M.: An overviews of meta-level architecture, Proc. 3rd Ann. Conf. on AI, Los Altos, CA (1983).
21. Goldberg, A. and Robson, D.: Smallta1k-80: The Language and its Implementation, Addison-Wesley, Reading, MA, (1980).

22. Goodall, A.: The Guide to Expert Systems, Learned Information Ltd., Oxford, England, (1987).
23. Harmon, P. and King, D.: Artificial Intelligence in Business. Expert Systems., John Wiley, NY, (1985).
24. Harris, L. R.: The heuristic search under conditions of error, Artificial Intelligence, 5, No. 3, p.217, (1974).
25. Hayes, P. J.: The Logic of Frames, in Metzing, D. (ed): Frame Conceptions and Text Understanding, Walter de Gruyter & Co., Berlin (1979).
26. Hayes-Roth, F. and Lesser, V. R.: Focus of Attention in a Distributed-Logic Speech Understanding System, Tech. Rep., Comp. Sci. Dept., Carnegie-Mellon Univ., PA (1977).
27. Hayes-Roth, B.: Modeling Planning as an Incremental, Opportunistic Process, Proc. 6th IJCAI, Los Altos CA, (1979).
28. Hayes-Roth, F. (ed): Building Expert Systems, Addison-Wesley, Reading, MA, (1983).
29. Hayes-Roth, B.: A Blackboard Architecture for Control, Artificial Intelligence, 26, p.251, (1985).
30. Hayes-Roth, B. el. al.: A Modular and Layered Environment for Reasoning about Action, Tech. Rep. KSL-86-38, Stanford Univ., CA (1987).
31. Hewett, J. and Sasson, R.: Expert Systems 1986, Vol 1: the USA and Canada, Ovum, (1986).
32. Hewett, J. and Sasson, R.: Commercial Expert Systems in Europe, Ovum, (1986).
33. Kraft, A.: XCON: An Expert Configuration System at Digital Equipment Corporation, in Winston, P. and Prendergast, K. A. (Eds): The AI Bwiness: The Commercial Uses of Artificial Intelligence, MIT Press, Cambridge, MA (1984).
34. Lien, K. Suzuki, G. and Westerberg, A. W.: The Role of Expert Systems Technology in Design, Chem. Eng. Sci., 42, No. 5, p.1049, (1987).
35. Lien, K. M.: "Expert Systems Technology in Synthesis of Distillation Sequences", Dr. Ing. Thesis, The Norwegian Institute of Technology, Trondheim, Norway (1988).
36. Lien, K. M.: "A Framework for Opportunistic Problem Solving, Computers and Chemical Engineering, vol. 13, no. 4/5 (1989).
37. Minsky, M.: A Framework for Representing Knowledge, in Winston, P. (ed): The Psychology of Computer Vision, McGraw-Hill, NY, (1975).
38. Minsky, M., in Haugeland, J. (ed): Mind Design, The MIT Press, Cambridge, MA, (1981).
39. Newell, A.: Some Problems of Basic Organization of Problem Solving Programs, in Yovits, M. C. (ed): Proc. Conf. on Self-Organizing Systems, Wash. D.C., (1962).
40. Newell, A., Shaw, J. C. and Simon, H.: Empirical Explorations of the Logic Theory Machine: A case study in heuristics, in Feigenbaum, E. and Feldman, J. (Eds): Computers and Thought, McGraw-Hill, NY (1963).
41. Newell, A., et. al.: Information Processing Language V Manual, Prentice-Hall, NY, (1964).
42. Newell, A, and Simon, H.: Human Problem Solving, Prentice-Hall, NY, (1972).
43. Newell, A.: Production Systems: Models of Control Structures, in Chase, W. C. (Ed): Visual Information Processing, Academic Press, NY, (1973).
44. Newell, A.: The Knowledge Level, Artificial Intelligence, 18 No. 1, p.87 (1982).
45. Nii, H. P.: HASP / SIAP Case Study, The AI Mag., Spring, p.23 (1982).
46. Nii, H. P.: Blackboard Systems: The Blackboard Model of Problem Solving and the Evolution of Blackboard Architectures, The AI Mag., Summer, p.38, (1986).
47. Nii, E. P.: Blackboard Application Systems, Blackboard Systems from a Knowledge Engineering Perspective, The AI Mag., August p.82, (1986).
48. Nilsson, N.: Principles of Artificial Intelligence, Tioga Publ. Co., Palo Alto, CA, (1980).
49. Pearl, J.: Heuristics, Addison-Wesley, Reading, MA, (1984).
50. Reid, R. C., J. M. Prausnitz and B. E. Poling: "The Properties of Gases and Liquids", McGraw-Hill, NY (1987).
51. Robinson, J. A.: A Machine Oriented Logic Based on the Resolution Principle, JACM, 12, No. 1, p.23, (1965).
52. SanGiovanni, J. P. and Romans, H. C.: Expert Systems in Industry: A Survey, Chem. Eng. Progr., Sept.-87, p.52 (1987).
53. Shortcliffe, E. H.: Computer Based Medical Consultations, Elsevier, NY, (1976).
54. Simulation Sciences Inc.: PROCESS Reference Manual, Simulation Sciences Inc., Fullerton, CA, (1984).
55. Simon, E.: The Sciences of the Artificial, MIT Press, MA, (1968).
56. Stefik, M.: Planning with Constraints, Ph.D. dissertation Thesis, Comp. Sci. Dept., Stanford Univ., CA, (1980).
57. Terry, A.: The Chrysalis Project-Hierarchical Control of Production Systems, Tech. Rep. HPP-83-19, Stanford Univ., CA, (1983).

58. Wahl, P. E. "Synthesis of Heat Integrated Distillation Systems - Approaches Combining Artificial Intelligence and Operations Research Methods", Dr. Ing. Thesis, The Norwegian Institute of Technology, The University of Trondheim, Trondheim, Norway (1991).
59. Wahl P. E. and K. M. Lien: "S6 - A Computer Program for Automated Synthesis of Heat Integrated Distillation Systems", Sintef Tech. Report, The Sintef Group, Trondheim, Norway (1992).
60. Waterman, D. A.: A Guide to Expert Systems, Addison-Wesley, Reading, MA, (1986).
61. Wehe, R, Lien, K. and Westerberg, A. W.: Control Architecture Considerations for a Separations Systems Design Expert, Proc. NFS-AAAI Workshop on Artificial Intelligence in Process Engineering, Columbia Univ., NY, March-87 (1987).
62. Winston, P. and Eorn, B. K. P.: LISP, Addison-Wesley, Reading, MA, (1984).
63. Winston, P.H.: "Artificial Intelligence", Addison-Wesley, Reading, MA (1984).
64. Zadeh, L.: A Theory of Approximate Reasoning, in Michie, D. (Ed.): Machine Intelligence Vol. 9, Ellis Horwood, Chichester, (1979).

Selected Topics in Artificial Intelligence for Planning and Scheduling Problems, Knowledge Acquisition, and Machine Learning

Aydin K. Sunol, Muzaffer Kapanoglu, Praveen K. Mogili

College of Engineering, University of South Florida Tampa, FL 33620, USA

Abstract: Application of artificial intelligence techniques to planning and scheduling problems are briefly reviewed, issues involved in knowledge acquisition are discussed, and methods for machine learning are introduced.

More in depth treatise of selected approaches, with processing systems engineering examples, are used to illustrate some of the issues involved. The examples utilize genetic programming for batch scheduling and design problems, and a hybrid of symbolic and connectionist approaches to automated knowledge acquisition (machine learning).

Concepts and issues involved in Genetic Programming are discussed within the context of batch processing systems examples. A scheduling example is used to illustrate discrete variable based decision models and associated terminology while a design problem with continuous decision variables and constraints is also optimized. Possibility of using genetic programming as an integrating tool for computer integrated manufacturing is also discussed.

A novel instance based learning algorithm that allows symbolic information to be encoded into a connectionist representation is introduced. Selection of complex distillation column sequencing designs is used as the example. However, the approach is ideally suited for fault diagnosis and structured selection problems. The knowledge/rules extracted from the learning algorithm are very similar to the design heuristics proposed in literature. The performance of the Symbolic Connectionist network (SC-net) is studied based on its knowledge extraction capabilities and the classification accuracy in the test case.

Keywords: Machine learning, planning, genetic programming, artificial intelligence, batch process scheduling, optimal design of batch processes

Planning And Scheduling

Introduction

Planning is a difficult task for both natural and artificial intelligence. A non-exhaustive list of its various dimensions, with some overlap of the items, include: achieving goals, satisfying constraints, minimizing cost, maximizing profit, resolving conflicting goals, handling undesirable side-effects, scheduling, learning from failure, generalizing and reusing old plans, and planning in a dynamic environment. These dimensions bring forth the complex and heterogenous nature of planning which is a NP-hard problem where the components are usually tackled hierarchically, in part through a portfolio of techniques.

Batch production or manufacturing problem is often tackled in two phases, process planning and scheduling. In the planning phase, a partial ordering of batch production operations or activities is generated to achieve one or more goals. The scheduling problem is that of assigning resources and times for each activity in a plan such that the ordering relations imposed by the plan and the capacity limitations of shared resources are met.

The planning and scheduling problem could be tackled through ***optimization based***, ***heuristic based***, and ***knowledge based*** approaches. Each class of methods has inherent advantages and shortcomings over the other two. Optimization-based schedulers guarantee optimality in a static way. Combinatorial optimization techniques and graph-theoretic approaches are effective for small-size schedules. However, as the search space increases, heuristic methods become necessary for reduction of the search space and result in a tractable problem. Unfortunately, the heuristic methods do not guarantee the best solution, or indeed any solution. For complex problems, testing heuristic methods is difficult and only comparative tests are possible. In the relevant literature, including the articles in this book, many recent surveys of the optimization and heuristic methods are available for the manufacturing scheduling problem. Knowledge based schedulers use expertise in dealing with uncertainty and complexity. The approach has comparative advantages in capturing human expertise and intermeshing efficiently both the heuristic and the algorithmic methods. Knowledge acquisition step appears to be a serious bottleneck in AI-based approaches. Recent surveys of the knowledge based approaches to scheduling include a review by Noronha and Sarma [25]. The current thinking on use of artificial intelligence techniques in manufacturing may be best found in a book [8] which is an outgrowth of several workshops held by the American Association of Artificial Intelligence (AAAI) special interest group on manufacturing. For researchers in Artificial Intelligence in Planning, a recent monograph that contains a collection of articles [22] may also be of interest.

Knowledge Based Approaches to Planning And Scheduling

A comparison of discrete manufacturing systems is shown in Figure 1. At one extreme, we have mass production which is characterized by high production rate, high production volume, reduced cost, special tooling, minimal labor skill, product flow type, fixed automation, and very limited product variety. At the other extreme, we have job-shop manufacturing with low production rate, low production volume, higher cost, general tooling, high-level labor skill, process-based plant layout, programmable automation, and high product variety. The batch production may be considered an intermediate within these extremes. The trade-offs between the approaches at the extremes may be addressed in part through Flexible Manufacturing Systems (FMS). Modular Manufacturing Systems (MMS) have been conceptualized to further expand the flexibility [15]. The motivation here is to be able address the scheduling and planning problems of MMS, which in turn would hopefully facilitate effective solution of the batch scheduling and design problems.

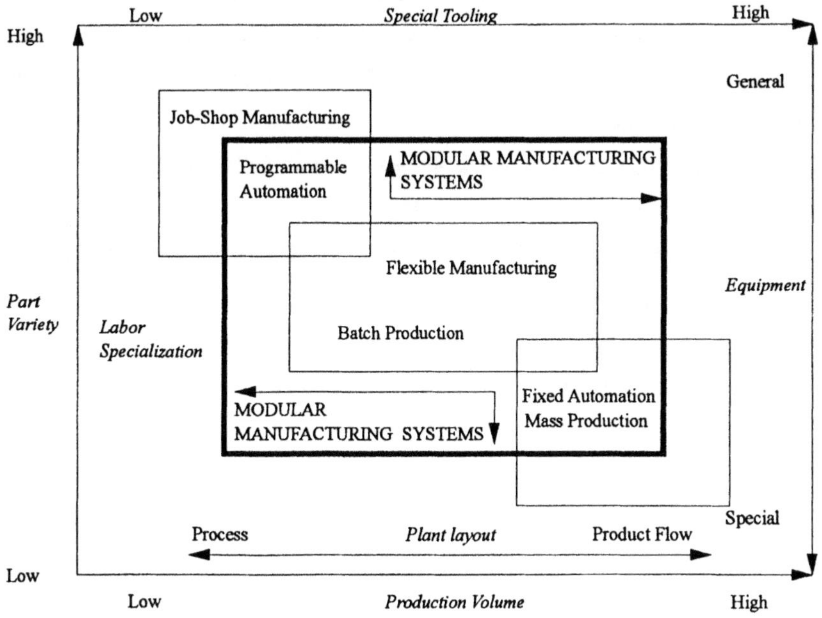

Figure 1. Characteristics of Discrete Manufacturing Systems

General Approaches and Considerations

There are many issues to consider in the development of an AI-based scheduler. One dimension of scheduler classification [16] include ***Expert Systems*** [17,33], ***Deeper Method-Based systems,* and *Interactive*** schedulers [4,33]. Batch-Kit [18] presented in this NATO ASI falls between the categories of Deeper Method-Based and Interactive.

Another dimension is whether the scheduler is ***predictive (deliberative)*** or ***reactive***. Most optimization methods and simulation provide static (predictive) schedules that may quickly become obsolete over time. A completely different approach to the aforementioned ***deliberative*** planning techniques is ***the reactive systems*** [3]. The idea here is to avoid planning altogether, and instead use the observable situation as a clue to which one must react. The system must have access to a knowledge base in order to know what actions to take under which circumstances. The reactive approach is instrumental in interleaving planning and plan execution. Deliberative planners that learn to be reactive over time seems to be the direction in which practical planning systems may emerge [23].

Equipment break-down, operator availability, raw material delivery, continual arrival of new orders, and product quality assurance are typical elements that contribute to the dynamic environment of a batch plant. As the dynamism of the environment increases, it becomes more difficult to distinguish the difference between the predictive and reactive behavior, so the need for ***real-time systems*** arise [35].

Another consideration involves, ***distributed*** versus ***centralized*** batch scheduling systems. The advantages of the former to the latter include [8]:

a) The distributed systems address the problem of coordinating a number of local schedulers which may each be responsible for scheduling a part of the work of a larger project better.
b) The underlying scheduling model's organization, if represented as a network of scheduling agents, can easily be changed dynamically to adapt to changes on the plant (e.g. addition or removal or rearrangement of the equipment).

The solution to planning problems often involve decomposition. ***Hierarchical planning*** involves first solving the problem completely, considering only preconditions whose criticality value (a measure of expected difficulty of satisfying preconditions) and augmenting the plan with operators that satisfy the preconditions. It could be viewed as a length-first search. As the computational power keeps increasing, popularity of ***concurrency*** and ***aggregated*** approaches will grow.

One shortcoming of rule based systems is the context dependency of the production rules and definitions, limiting the effectiveness of the approach to narrow domain applications. ***Case Based Reasoning*** aim to remedy the aforementioned shortcomings. The application of the idea in scheduling problems would be ***Case-Based Planning*** [12] where the old plans are re-used to make new ones.

In the planning stage, the design and production information is translated into instructions for production of either intermediates (parts) or end-products through a combination of human planners and software packages. Since these plans are not always executed as planned, the system must therefore monitor the execution of the plans, learn from the successful as well as unsuccessful plans, and replan the process. These issues, ***knowledge acquisition and machine learning*** are discussed in more depth in the second part of this paper.

Search Methods

In order to solve most non-trivial problems, it is necessary to combine some of the basic problem solving strategies (search methods) with one or more of the knowledge representation mechanisms. Although search and knowledge representation issues are discussed as another paper in this book, applicable search related practice will be briefly discussed.

Once the problem is defined as a state space search problem [27], search algorithms such as A* could be used to find a solution. A* provides a way of conducting best first search through a graph representation of the problem. Each node that is examined by the algorithm represents a description of a complete problem state, and each operator describes a way of changing the total state description. For more complicated problems, decomposition into smaller subproblems and ways of avoiding recomputing entire problem state, as it changes, is necessary. AO* algorithm [24] provides a way to decompose the problem when it is completely separable.

One of the earliest planning methods is ***goal-stack*** planning where individual goals are solved one at a time in order. This approach can not handle cases where goals interact. ***Constraint Posting*** is an alternative for such ***nonlinear planning*** problems. The basic idea here is to build up a plan by incrementally hypothesizing operators, partial orderings between the operators, and bindings of variables within the operators. The differences between state space search and constraint posting could be found in most text books on AI techniques [31].

In order to solve more complicated problems, it is important to be able to eliminate some of the detail of the problem until a solution that addresses the main issues is found. Subsequently, the appropriate details are filled in. These ***hierarchical planning*** approaches differ in the way the search problem is decomposed and the operators used.

Although many knowledge-based approaches and mathematical programming techniques have been suggested to solve scheduling problems, a single universally acceptable technique that could solve the problem is yet to emerge. In the recent years, Genetic Algorithms attracted considerable interest [9,10,11,14,19]. It seems to be a quite robust search technique and an effective tool for Machine Learning in Knowledge Based Systems [13]. The literature in GA is quite scarce, particularly in Chemical Engineering literature. The recent interest to the field, its promise for scheduling problems, the robustness of the approach as a search technique, and utility in machine learning are responsible in the extended coverage of the subject.

Genetic Programming

Genetic Programming (GP) is composed of genetic modelling (GM) and genetic algorithms (GA). Genetic Modeling involves description of a system using genetic structures, such as genes, chromosomes, alleles and so on. For instance, any number between 0 and 7 can be modeled as a chromosome with three genes by using a binary alphabet (0 and 1). If the number is 6, it could then be genetically modeled by using the following chromosome:

0 1 1 0

The differences between genetic algorithms and conventional search methods could be summarized as follows [6]:

a) GAs work with a coding of the parameter set, not the parameters themselves.
b) GAs search from a population of points not a single point.
c) GAs use payoff (objective function) information, not gradients or other auxiliary knowledge.
d) GAs use probabilistic transition rules, not deterministic rules.

Genetic Algorithms manipulate the genetic models to obtain optimal solutions to the problem represented by this model. This manipulation utilizes genetic operators such as crossover, mutation, mating, and reproduction.

Since GA was invented to mimic some of the processes observed in natural evaluation, its description is inevitably based on the biological vocabulary. Table 1 provides the equivalence of traditional and GA search terminology. The flow-chart of a simple genetic algorithm is presented as Figure 2.

Table 1. Comparison of the Terminologies of Genetic Programming and Operations Research

GENETIC PROGRAMMING	MATHEMATICAL PROGRAMMING (Decision Theory)
Genetic Model (Genotype, Structure)	Mathematical Model-Decision Model
Genetic Algorithm (Reproduction Technique)	Mathematical Algorithm- Solution Technique, Decision Making
String (Chromosome)	A Mathematical Statement- A Constraint, an Objective Function, etc.
Gene (Feature, Character, Detector)	Mathematical Variable- Decision Variable
Allele (Feature Value)	Value of a Mathematical Variable- Value of a Decision Variable
Locus (String Position)	-
A Solution of Genetic Model (Phenotype, a Decoded Structure)	A Feasible Solution of the Model- A Decision
Generation	Iteration- Next (Improved) Decision
Epistasis	Nonlinearity, Multi-modality- Complex Decision
Alphabet	Variable Types- Decision Types
Population	A Set of Feasible Solutions

The number of genetic operators that participate in reproducing an organism determines the complexity of that organism. Hence, more sophisticated problems require more advanced genetic operators. This leads to finer models and, thus, better solutions (Dominance, diploidy, haploidy). In this study, simple genetic programming, hence, basic genetic operators are introduced and

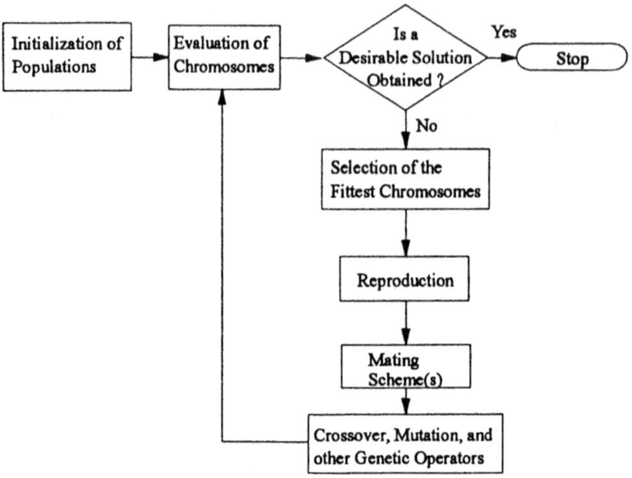

Figure 2. Flowchart of a Simple Genetic Algorithm

discussed in the context of batch process systems engineering.

Selection: Selection process is performed based on fitness values (a low fitness value means poor adaptability to the environment or large deviation from the goal) of chromosomes (structures, alternative policies,...etc.).

Reproduction: This operation determines the number of replicates of each selected chromosome used in the new generation (iteration). A higher chance to contribute to reproduction is given to those chromosomes with higher fitness values (lower cost, higher profit, briefly, better adaptability to the environment). Usually, reproduction and selection are implemented at the same computational stage.

Example: It is very common to use roulette-wheel to allot off-spring strings.

Parent #	Fitness	Expected #	Actual Count (next generation)
1	6	1.8	2
2	1	0.30	0
3	3	0.90	1
Total	10	3	3
Average		3.33	

Crossover: The crossover operator provides the discovery mechanism for GAs. Therefore, when modeling a problem, the type of crossover operator to be applied should be considered carefully. A comprehensive coverage of the possible crossover types will not be attempted, a rather simple one-point crossover will be given for illustration.

Parent #1 BBBBBB Parent #2 AAAAAA

Initially, a crossover point is selected randomly, all chromosomes are then cut and spliced. If the crossover points are in the middle, then following siblings are obtained:

Child #1 AAABBB Child #2 BBBAAA

This operator provides information exchange between strings. Crossover enables combination of ***building blocks*** of better solutions from parent chromosomes. Since the new population is generated from superior parents (or bit strings with higher fitness value), crossover allows information exchange among the elite. It is expected that some superior parents may have many features in common. These common characteristics (similarities) of some parents are reflected on similarity templates. Each similarity template is called a ***schema***. Holland's Schema Theorem asserts that genetic algorithms manipulate ***schemata*** during its execution [14]. Since GAs allow only better parents to reproduce, one can predict the relative increase or decrease of a certain schema in the next generation. The expected number of occurrences of schema H in the next generation of the population is n*r/a minus disruptions caused by mutation and crossover [2]. Here, r is the average fitness of all chromosomes in the population containing the schema H; n is the number of chromosomes; and "a" is the average fitness of all chromosomes.

The reproduction ends with the growth and decay of the several schemata contained in a population. How the growth and decay of some schemata may lead to an improvement is illustrated next. The bit string [0 1 1 1 1 1 0 0 0 1 0 0 1 1 0 0 0 1 0 1] which can symbolically be written as $A = [a_1 \; a_2 \ldots a_{20}]$ consists of genes (a_i, i = 1, 2,... 20), each of which represent a single binary feature, that may take a value of either 1 or 0. A_j's (j = 1, 2,... n) represent individual strings and A(t) symbolizes the population of A_j's at time t.

For the description of a schema, one may use a wild card "*" which may take a value of either 1 or 0 at a given particular position. If one considers the first ten digits of the bit string given above, the schema H = [0 1 1 * * * 0 0 * 1] represents the modified bit string A' = [0 1 1 1 1 1 0 0 0 1]. It is apparent that there are more schemata than bit strings. More specifically, there are 3^L schemata defined over a binary string of length L.

There are two important concepts yet to be defined, schema order and its defining length. The order of a schema is the number of fixed positions in the template. For example, the order of the schema H given above, is o (H) = 6. The defining length of a schema is the distance between the first and last specific position (i.e. the defining length of the schema H, d (H) = 10 - 1 = 9). Schemata and their properties provide the means for analyzing the net effect of reproduction and genetic operators on building blocks of the population [6]. A particular schema grows as the ratio of the average fitness of the schema to the average fitness of the population increase. Of course, one can not generate any new (better) schema by copying old strings without any change. Therefore, at this stage, crossover may come into picture. Crossover is nothing but information exchange between the newly reproduced strings. Since the number of undesirable strings decrease during the reproduction process, one may expect some improvement with this exchange. There is no guarantee that crossover will improve the fitness of the pool of bit strings. Therefore, since each string is representative of some schemata, the effect of crossover on the schemata has to be analyzed. For the bit string and schemata given below:

A = [0 1 0 1 1 1 0 1 0 1 0 1 0 1 0 0 1 0 0 0]
H_1 = [0 1 0 1 1 * * * * * * * * * * * 1** 0]
H_2 = [* * * * * * * 0 1 0 1 0 1 * * * 1 * * 0]

String A can be represented by either schemata H_1 or H_2. However, in order to mate and crossover A with another string, H_1 and H_2 has to be disrupted in a different fashion. For a uniformly distributed random number between 1 and 19, the probability of destruction for schema H_1 is

$$P_d = d(H_1) / L\text{-}1 = 19/19 = 1.0$$

whereas, same probability for the schema H_2 is

$$P_d = d(H_2) / L\text{-}1 = 12/19 = 0.6315$$

This example can be generalized as, "the schema with a shorter defining length has a higher probability to survive". Then, for any schema H, the probability of survival in the next generation is given as

$$P_s = 1 - P_d$$

and the expected number of schema in the next generation becomes

$$n(t+1) = (n(t)*r/a)*P_s.$$

So far, the crossover rate (P_c) did not come into the picture. When P_c=1, all strings are mated and the crossover operation is performed. The crossover rate controls the frequency with which the crossover operator is applied. For each new population, crossover is applied on P_c*N strings.

As the crossover rate increase, the rate at which new individuals are introduced into the population increases. Crossover rate determines the exploration rate. While higher crossover rates may disrupt the fittest chromosomes, lower P_c may reduce the improvement rate. Usually, a crossover rate of 0.6 is considered appropriate for many applications.

Mutation: Mutation is a secondary operator. After selection, each bit position of each string in the new population is applied a random change with a probability equal to the mutation rate P_m. Consequently, approximately P_m*N*L mutations occur at each generation, where N is the number of strings and L is the number of bits at each string. Mutation operator helps to recover lost strings that had high-performance. However, high mutation rate causes an increased level of randomness. For P_m=.5, GA becomes a random search technique. With A = [0 1 **0** 1 1 1 0 1 0 1 0 1 0 1 0 0 1 0 0 0], if mutation is performed on the third gene of the chromosome, then one gets A=[0 1 **1** 1 1 1 0 1 0 1 0 1 0 1 0 0 1 0 0 0].

Performance Measures: In an effort to quantify the effectiveness of GAs, DeJong devised two measures, one to monitor convergence characteristics and the other for evaluation of ongoing performance [7]. These measures are called off-line (convergence) and on-line (ongoing) performance, respectively. The off-line performance is a running average of the best performance values over a particular time period and is given as

$$X_e^*(s) = \frac{1}{T}\sum_1^T f_e^*(t),$$

The on-line performance is an average of all function evaluations up to the current trial and is given as

$$X_e(S) = \frac{1}{T}\sum_1^T f_e(t).$$

Genetic Programming for Batch Process Systems Engineering

The application of Genetic Algorithms to Batch Process Systems Engineering are fairly recent and the authors are aware of only a set of internal reports [29,30]. Simulated annealing, which is often considered an alternative to GA, received more attention for both design [26] and scheduling [5] of Batch Processes. The following example illustrates the use of GA to solve a scheduling problem is included primarily for its pedagogical value.

Application 1. A Scheduling Problem [20]:

This is a 4-Product 3-Stage Flowshop problem which will be used to illustrate the genetic algorithm for a case that involves discrete variables. The processing times for the problem are given in the following table and is followed by its genetic model:

Stage	Product A	Product B	Product C	Product D
1	2	4	5	6
2	4	4	2	4
3	6	4	5	2

Chromosome Syntax: Each product is represented with a gene where alleles are the product types (A,B,C,D).

Chromosome Interpretation: The locus of each gene corresponds to the order of that specific product. The part corresponding to the most left gene is processed first.

Chromosome Evaluation: Population contains four chromosomes. Makespan time is used to evaluate the fitness of each chromosome. Selections are made through roulette-wheel.

Operators: Order-based crossover is used on all chromosomes (P_c=1.0). Order-based mutation probability (P_m) is 0.03. Chromosomes are initialized randomly (Generation # 0).

Population #	Population	Fitness (makespan)	Actual Count
1	A B D C	24	2
2	D A B C	28 (die off)	0
3	C B A D	25	1
4	B A C D	25	1
On-Line (total)		102	4
Off-line (best)		24	

According to actual count values, mating and crossover operators are used on the following four parent strings:

P1 = [A B^{*} D^{*} C], P2 = [A^{*} B^{*} D C], P3 = [C B A D], P4 = [B A C D]

Initially, P1-P4 and P2-P3 are mated. While applying order-based crossover, a set of locations is randomly picked and this sequence is imposed on the other parent. Mutation is not used at this iteration.

P2-P3 has the following two children C#1 = [A C D B] and C# 2 = [C A B D]

P1-P4 has the following two children C#3 = [A B D C] and C# 4 = [B A C D]

Population #	Population	Fitness	Actual Count
1	A C D B	26 (die off)	0
2	C A B D	24	2
3	A B D C	24	1
4	B A C D	25	1
Total		99	4
On-line fitness = (102+99)/2 =101.5		Total	99
		Improvement	3
Off-line Performance	3	Improvement	0

After reproduction, mating and crossover operators applied between P1&P3 and P2&P4

P#1 = [C^{*} A B D^{*}] P#2 = [C A^{*} B^{*} D] P#3 = [A B D C] P#4 = [B A C D]

the following siblings are obtained

C#1 = [A B C D] C#3 = [A C B D] C#2 = [A C B D] C#4 = [A B C D]

Suppose that mutation is applied to the first child of the generation, to explore some overlooked or lost sequences and to avoid premature convergence. If the location is selected randomly, then C#1 = [A B C^{*} D^{*}] becomes C#1 = [A B D C] and the following new generation (#2) is obtained:

Population #	Population	Fitness
C # 1	A B D C	24
C #2	A C B D	23
C #3	A C B D	23
C #4	A B C D	23

Thus, two alternative solutions, [A C D B] and [A B C D], are obtained. The optimum makespan time is 23.

Criteria	Performance	Improvement
Total fitness	93	6
On-line	97.5	4
Off-line	23.6	0.6
Best	23	1

An exhaustive search would have required evaluation of 4!=24 sequences to obtain the optimal, instead of 8 different sequences were evaluated with GA.

Application 2. A Batch Plant Design Problem [1]

The following problem has been widely used in the batch plant design literature for testing the efficiency of various algorithms. The problem is formulated as follows while the design that is modeled below is depicted in Figure 3 and the parameter values used are given in Table 2. :

0. The Objective Function

$$\text{Minimize} \quad \sum_{j\in B} m_j a_j (V_j)^{\alpha_j} + \sum_{j\in S} \sum_{k\in K(j)} b_k (R_k)^{\beta_k}$$

1. The batch equipment capacity constraint

$$V_j \geq S_{ij} B_i, \qquad j\in B$$

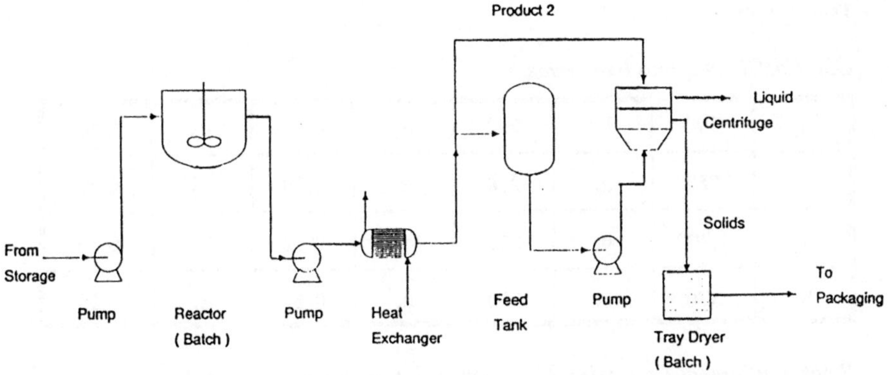

Figure 3. A Batch Plant

2. The continuous equipment capacity constraint

$$R_k \geq \frac{D_{ik} B_i}{t_{ij}}, \qquad k \in K(j), \quad j \in S$$

3. The cycle time constraints

$$\sum_{i=1}^{n} \frac{Q_i T_i}{B_i} \leq H$$

4. The production time constraints

$$T_i \geq t_{ij}, \qquad j \in S$$

$$T_i \geq \frac{(t_{i,j-1} + P_{i,j} + t_{i,j+1})}{m_j}, \qquad j \in B,$$

5. The equipment capacity bounds

$$V_j^{max} \geq V_j \geq V_j^{min}$$

$$R_k^{max} \geq R_k \geq R_k^{min}$$

Table 2. Data for Application 2

Cost Coefficients and Exponents

a_1	592	b_2	250	α_1	0.65	β_2	0.40
a_2	582	b_3	210	α_2	0.39	β_3	0.62
a_3	1,200	b_4	250	α_3	0.52	β_4	0.40
b_1	370	b_5	200	β_1	0.22	β_5	0.85

Batch Equipment Processing Times for each product, P^0_{ij} (hours)

Products	P^0_{ij} *(hours)*		
1	3	1	4
2	6	-	8
3	2	2	4

Production Requirements (Q_i) and Size Factors (S_{ij})

Products	**Q_i (lbs/year)**	**Size Factors (S_{ij})**		
1	400,000	1.2	1.4	1.0
2	300,000	1.5	-	1.0
3	100,000	1.1	1.2	1.0

Duty Factors (D_{ik})

Product	**Duty Factors**				
1	1.0	1.2	1.2	1.4	1.4
2	1.0	1.5	1.5	-	1.5
3	1.0	1.1	1.1	1.2	1.2

After relaxation of some of the constraints, the revised model reduces to 23 decision variables and 30 constraints.

The literature on genetic algorithms is composed of either applications in various domains or analysis of its performance for optimization of simple unconstrained functions with continuous variables. For these relatively simple test cases, GAs are relatively well understood and accepted as a search technique. However, GM still needs to be handcrafted for successful deployment of GPs in real world problems. One practical problem is how to incorporate constraints into the model.

There are two approaches for incorporation of constraints in the GM. One is to apply a penalty for the constraints that are violated. In this case, the chromosome corresponding to the objective function doesn't change; however, a penalty is charged to the fitness value of this structure. This may, and frequently does, cause the algorithm to slowdown due to resulting ***infeasible solutions***. If this method is used, initial population should be setup properly. Several ***feasible solutions*** has to be inserted into the initial population to reduce the execution time. This method is insensitive to how close an infeasible solution is to the feasible region. One advantage of this approach is that it keeps the structure shorter. The shorter the defining length, the more efficient is the information exchange. However, the population size would be larger than usual in order to explore legal solutions. In this approach, the time to obtain a legal solution may be excessive due to the number of constraints, the way of setting the initial population, the size of the search space, and the penalty parameters.

The other approach that can be adapted is to form a penalty function and include all the constraints in the structure (chromosome). The basic difference between the two approaches lies in the fact that the latter tracks the deviation from the feasible region [6]. In this case, as the length of a chromosome increases, the efficiency of the crossover operator decreases due to the reduction in quality of the information exchanged. Therefore, one may need to apply multi-point crossover in lieu of one-point crossover. The parameters that are recommended for each of the approaches are discussed next.

The generic parameter setting is given as $P_c = 0.6$, $P_m = 0.001$ and $N = 50$. The current literature on the effect of parameters on the search is limited to small size and/or benign problems. For the problem given above, the following experiments were run:

N	P_c	P_m	Policy	# of trials
100	0.4/0.6/0.8	0.001/0.03/0.1	Elitist	5000

The following table shows the effect of mutation and crossover rate for Application 2. The values listed are for the penalty function with hundred units of penalty for each unit deviation from any constraint.

The values are negative due to the effect of penalty parameters. However, this shouldn't hide the convergence properties of each parameter setting. It can be seen that highest mutation rate produces the slowest convergence. The same behavior could be also seen when the average characteristics are inspected. This result is consistent with the expected effect of the mutation

Parameters	$P_c = 0.4$	$P_c = 0.6$	$P_c = 0.8$	Average
$P_m = 0.001$	-5.11	-5.14	-5.15	-5.13
$P_m = 0.03$	-5.13	-5.12	-5.11	-5.12
$P_m = 0.01$	-5.11	-5.0	-4.90	-5.00
Average	-5.11	-5.08	-5.05	

operator. Principally, it is the crossover operator that makes the GAs different from the random search techniques. As can be seen from the column labeled average, lower crossover rates result in faster convergence. This behavior is contrary to the general wisdom. A formal question that should be answered is why lower rate of information exchange is better? At the extreme, this can be construed as random search being superior to the GA, for the problem tackled here. The relation between crossover rate and convergence can be better attributed to the simple GA form used. This GA was deficient in producing legal (feasible) solutions (unless some legal solutions are introduced to the initial population). One possibility is to update the crossover operator so that crossover produces only legal structures-feasible solutions (including initial population). For constraint models with continuous decision variables, it is easy to apply a crossover method that generates feasible-legal structures.

For the problem given above, a fitness function that adapts some sort of lagrangian relaxation was used. This fitness function is given below:

$$\phi = f(x) + \mu(p(b-g(x)))^2$$

where p is a penalty parameter that balances the effect of the square power on numbers between -1 and 1. "μ", unlike lagrange multiplier, is a binary parameter that takes the value of 1 when the constraint is violated and 0 otherwise. Thus, a penalty proportional to the deviation from the feasible region is applied to the fitness value of the corresponding chromosome. This approach appears to be superior to the two simple approaches mentioned before. Using this approach, the effect of control parameters on the efficiency of genetic algorithms for multi-constraint models were analyzed. More widely-accepted control parameter settings for GAs were tested with this design problem. Representative results are shown in Figure 4a, b. As can be seen in Figure 4a., the optimal population size is 50, although the sensitivity with respect to this parameter is not that significant. On the other hand, as can be seen in Figure 4b. the higher crossover rates (e.g. 0.7) beneficial for this problem. This conclusion conflicts with the earlier observations discussed during the analysis of interrelation between the mutation operator and crossover rate. The cross interaction of GM and control parameters is a fruitful area for future research.

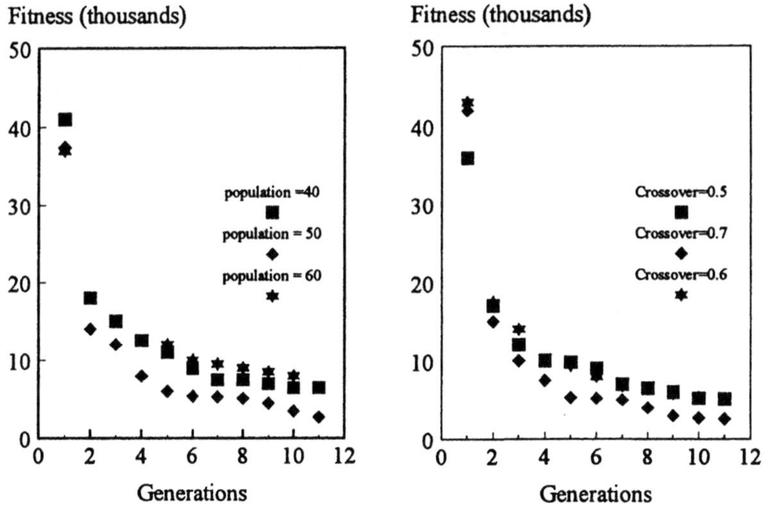

Figure 4 a, b. Convergence Characteristics of Genetic Algorithms

Knowledge Acquisition and Machine Learning

Background

The development of knowledge based systems for planning and scheduling has been limited by the knowledge acquisition bottleneck. Formalizing knowledge and implementing within knowledge bases are major tasks in the construction of large hybrid systems that efficiently integrate algorithmic and symbolic methods. The hundreds of rules and thousands of facts have to be obtained through various means shown in Figure 5. Some of the modes of knowledge acquisition involve human, i.e. "Domain Expert and Knowledge Engineer" while some of the knowledge acquisition modes utilize machines. The interactive and incremental development of knowledge based systems allow experimentation and validation of knowledge, structure, and methods at various levels. The stages and type of information involved in acquisition of knowledge are summarized in Figure 6.

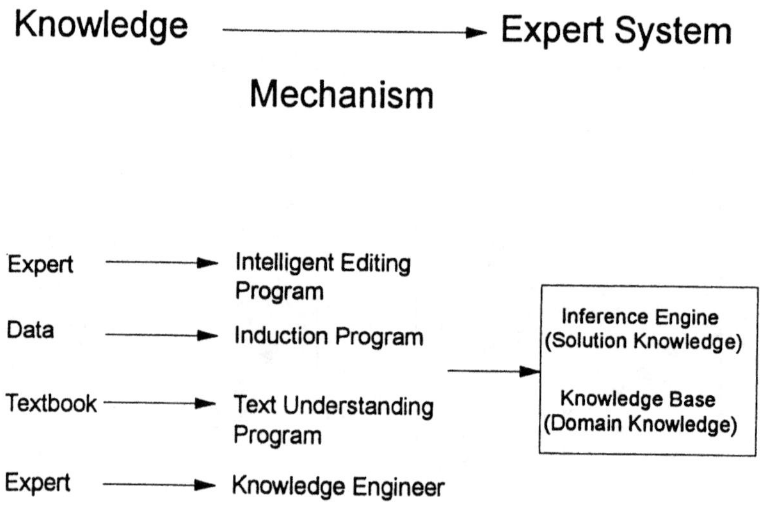

Figure 5. Mechanisms of Knowledge Acquisition

Advances for automation of the knowledge acquisition process would constitute a major

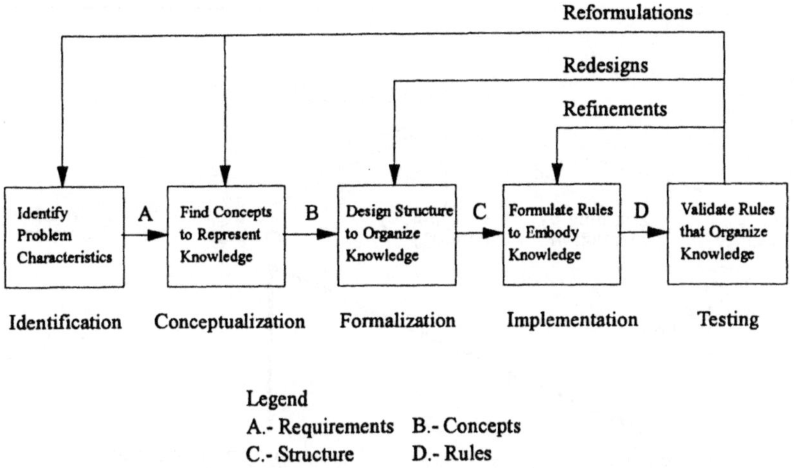

Figure 6. Stages of Knowledge Acquisition

contribution to AI technology and such techniques are within the domain of machine learning. There is no generally accepted categorical definition of learning and maybe it is best to give a prototype instead. A learning system can be modified by itself to improve its performance over an extended period of time. A machine learning program will allow more efficient execution of computer software, in comparison to ones with no machine learning component.

The literature on machine learning is vast and rapidly expanding. However, these techniques are conveniently summarized within the two volume handbook on machine learning [21]. A taxonomy of various symbolic machine learning topics is shown in Figure 7, where the paths of the topics discussed here are highlighted. One of the earliest approaches to learning is called ***Rote Learning***. In rote learning, computed solutions are stored along with the corresponding attributes that contribute. It is quite effective whenever computation is more expensive than recall. Learning element in computer programs that played games such as chess utilized rote learning. Today, ***Explanation Based Learning*** systems seem to capture the spotlight. They attempts to learn from a single example by explaining why it is an example of the target concept. ***Genetic Algorithms*** discussed earlier would be useful in learning through observation and discovery.

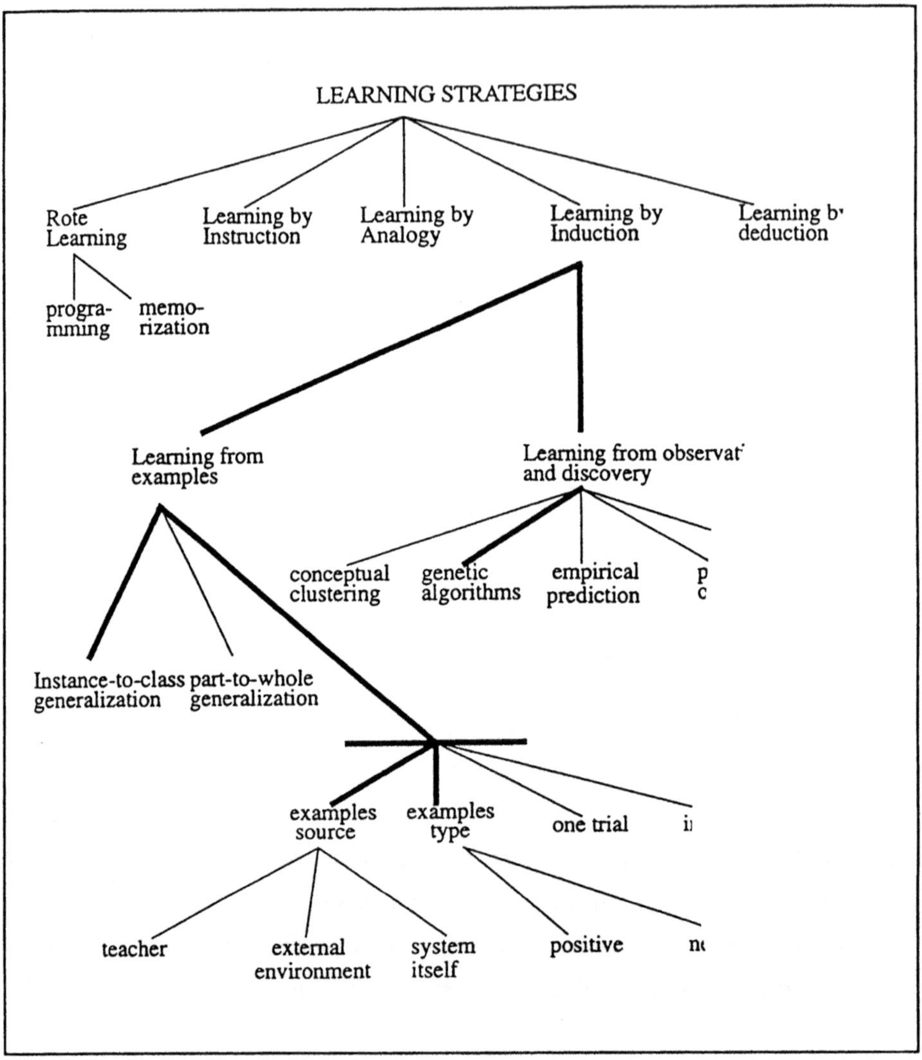

Figure 7. Classification of Learning Mechanisms

Symbolic approaches to learning, connectionist learning approaches, and genetic programming all offer significantly different points of view, each with strong and weak points. Genetic Algorithms were discussed earlier, connectionist approaches and their comparison with statistical methods are readily available in literature including the papers presented at this NATO ASI [28,

36]. Our focus here will be to give a hybrid approach that combines the symbolic and connectionist approaches to machine learning.

A Hybrid Approach to Machine Learning

The knowledge extraction and background knowledge encoding problems associated with connectionist learning algorithms and inability of the symbolic learning algorithms to handle with continuous variables effectively, and allow parallel knowledge representation have prompted the researchers in AI to seek a hybrid approach. One development in this direction is Symbolic Connectionist network (SC-net) [32]. The SC-net, designed specifically for constructing Expert Systems, is based on a hybrid of Symbolic and Connectionist architectures. The system allows extraction of knowledge in the form of rules and can handle both scalar and fuzzy variables. **The following features of SC-net make it a connectionist method**

1) A highly parallel and uniform representation of knowledge
2) Fault tolerance and noise resistance.
3) A built-in ability to deal with non-crisp inputs and outputs.

and the following features make it a symbolic method

1) The ability to encode rules to support knowledge refinement.
2) Allow for rule extraction as a direct means to elicit learned knowledge and support the implementation of Expert System standards such as consultation and explanation facilities.
3) Provides means to represent symbolic constructs such as variables, comparators and quantifiers. This leads to a more powerful language for describtion of knowledge.

The network topology in SC-net is based on the training examples unlike the user specified ones in connectionist approaches, promising an optimal representation. Fuzzy logic is used to deal with uncertainty. The next section describes the different learning stages involved in SC-net with enough emphasis on each learning component.

Learning Mechanism

Different learning steps involved in SC-net are shown in Figure 8. Each time SC-net is given a training set that consists of input and corresponding output information, it invokes a call to the Recruitment of Cells Algorithm (RCA). The main function of RCA is to map the training set into

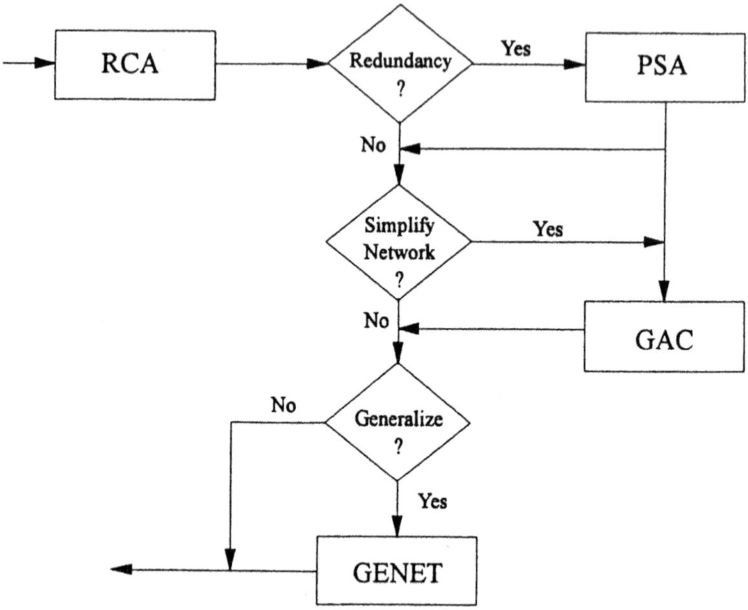

Figure 8. Learning with SC-net

a network representation for further processing through the other algorithms.

The network generated consists of Input Cells (IC), Information Collector Cells (ICC), Negative Collector cells (NC), Positive Collector cells (PC), UnKnown (UK) cells, and Output Cells (OC). The input cells represent the attributes of the training set, information collector cells transform the input information into an intermediate form, and the NC and PC cells collect negative and positive evidence present in the training set for a conclusion. The NC and PC cells are connected to all output cells and IC cells. The UK cells act as threshold for the NC and PC cells, enabling them to propagate. The activation is based on the type of evidence (positive, negative, unknown). The output cells represent the output classes.

In the SC-net, the network structure consists of cells modelling the min and max operators of fuzzy logic. Every cell contains a bias value CB_i, which indicates the type of fuzzy function the cell models and its value lies between -1 and +1. The min cells have a bias value of -1 and the max cells have 1. The absolute value of CB_i represents an upper threshold on the cell activation, CA_i. These cells are connected by links which carry weights. The link weights represent the degree of contribution the lower cell makes for the upper cell's concept. If cell C_i (with CA_i) and cell C_j (with CA_j) are connected then the weight of the connecting link (CW_{ij}) is finite, otherwise (i.e., not connected) the corresponding weight is zero. The new cell activation CA'_i is computed according to the formula given below.

$$CA_i' = \begin{cases} CB_i * \min\{ CA_j * CW_{ij}\} & \text{if } C_i \text{ is a min cell.} \\ CB_i * \max\{ CA_j * CW_{ij}\} & \text{if } C_i \text{ is a max cell.} \\ CA_{i\ positive} + CA_{i\ negative} - \tfrac{1}{2} & \text{if } C_i \text{ is a final output cell.} \\ 1 - (CA_j * CW_{ij}) & \text{if } C_i \text{ is a negate cell.} \end{cases}$$

Here, a cell is a min cell if there is negative evidence, a max cell if there is a positive evidence, and negate cell if there is no evidence.

The final output of cell C_i is always bounded by [0,1] range using the following relationship,

$O_i = \min\{ 1, \max\{ 0, CA_i'\} \}$

Every training instance is presented to the network as a single feed forward pass. After the pass has been completed, the actual and the expected activation for each output are compared. Three possible conditions may result from this comparison:

1) If the example was correctly identified (error is below some epsilon, $\epsilon < 0.02$), then no modifications are made to the network.
2) If the example was similar to at least one previously seen and stored instance (error within 5ϵ), the bias of these cells are adjusted to incorporate the new instance.
3) If the example was not identified by the network, then this results in the recruitment of a new cell or information collector cell (ICC). Appropriate connections from the network to the ICC are created. The ICC cell itself is connected to either positive collector (PC) or negative collector (NC) cell. The PC is used to collect positive evidence, whereas the NC accumulates the negative evidence. The unknown cell always propagates an activation of 0.5, guaranteeing the PC cell only attains activation greater than or equal to 0.5, whereas the NC cell propagates activations of less than or equal to 0.5.

The network generated by the aforementioned procedure is next processed through the Pre-Selection of Attributes (PSA) algorithm that introduces redundancy into the network. Since the RCA provides a minimized representation of the input domain, certain input features may be incorrectly identified as important. This is especially true when the number of training cases is insufficient. PSA algorithm addresses this problem by selecting all the input attributes that represents a concept and was found to be quite effective in pattern-matching.

The network topology of SC-net is set based on the training examples, unlike the user specified ones in other connectionist approaches. The cells are recruited at a single level and the cell growth proceeds horizontally, suitable for paralelization. One of the major problems with the RCA learning component is the linear growth of the network, under worst case conditions, each example may result in recruitment of one to three new cells. This problem may be rectified through injection of background knowledge (i.e. if then rules) in the training set. In doing so, the learning system makes use of fuzzy logic language hedges (low, medium, high), quantifiers (never, sometimes, always), and operators (max, min). As a consequence, one ends up with a natural language like interface. The network growth is further controlled, independently, through the global attribute covering (GAC) algorithm.

Given the RCA (with/without PSA) generated network as its input, GAC generates (except for contradictions and inconsistencies) an equivalent network, minimized both in the number of cells and links. As a side effect, extraction of highly general and simple rules are possible for explanation purposes. At the first step of GAC, all links to information collector (IC) cells are disconnected, $CW_{ij} = 0$. This forces IC cells to propagate an activation of 1 (firing state), regardless of the inputs presented to the network, during the feed forward pass. Since all IC cells fire, they will also activate the corresponding output cells they are connected to. All IC cells that are incorrectly activated are identified and form an conflict list. Furthermore, all links entering the IC cell as inputs will be considered as potential inhibitors for network minimization. By reconnecting specific links, IC cells can be prevented from firing for the same inputs. An "undesirability" measure associated with each link is used to prune the network.

The GAC generated network can be further generalized through use of the GEneralization of NETwork (GENET) learning component. In symbolic terms, generalization is accomplished through replacing single conditions by a disjunction of similar conditions, allowing extraction of rules that cover a larger domain. This can be hard coded into the network, through use of domain specific knowledge.

The process of generalization involves grouping of the set of links entering an IC cell that have a common parent cell. Each of these possible groups (containing more than one link) are first represented in the form of a max-cell, and the links within the group form the input connections to this cell. Here, there is a possibility of over generalization i.e., it is possible for output cells to fire even when they should not have fired. Once again, this is controlled by the "undesirability" measure which is applied to the conflict vector. After each pass through this conflict vector, the link with the highest undesirability index is selected and removed from the max-cell. This process is continued until the conflict vector set is empty and the resulting network is more general. The generalized network may be able to handle an input pattern that does not exactly match the one presented for training. The SC-net system can handle crisp variables as well as fuzzy quantifiers. Figure 9 shows a sample program that includes fuzzy quantifiers and operators. This representation helps in dealing with uncertain/fuzzy domains and for generating rules which are similar to that of a natural language.

Range	0.00- 0.05	# completely_false #	
	0.05- 0.20	# kind_of_false #	
	0.20- 0.40	# sometimes_false #	
	0.40- 0.60	# unknown #	
	0.60- 0.80	# kind_of_true #	
	0.80- 0.95	# mostly_true #	
	0.95- 1.00	# completely_true #	
Quantifiers	almostall:	0.90-1.0	
	always:	1.00-1.0	/* same as always true */
	sometimes	0.70-0.9	
	never	0.00-0.0	/* same as not */
Operators	almost	:= **they** almostall are completely_true	
	incorrect	:= **they** always are completely_false	
	hardly_false	:= **they** are never completely_false	
Rules	**if** almost (u1,u2,u3,u4,u5) **then** u7		(1)
	if incorrect (u1,u2,u3,u4,u5) **then** u8		(1)
	if never (greater (u1,0.8)) **then** u9		(0)
	if hardly_false (u1,u2,u3,u4,u5) **then** u10		(1)
	if and (sometimes (u1), never (u2), always (u3)) **then** u7		(1)

Figure 9. Sample Program Containing Quantifiers and Quantifying Operators

Another powerful feature of SC-net is its ability to incorporate background knowledge. Background knowledge could be provided before, during and after the learning process. This background knowledge could be domain specific and/or meta-knowledge. This feature is particularly helpful in cases where there are infrequent examples that may never be covered by the learning algorithms.

Before the learning component of SC-net is invoked, it is possible to encode a set of rules into the network. This can affect both the network growth, the training time, and the number of rules generated. The background knowledge base can be constructed either through constructing a prototype or via the Life-Cycle approach. In the first method, a set of rules may be used as background knowledge along with training examples. In the second approach, the rules generated from examples are modified by the domain expert(s), whenever needed and maintained in the learning system The latter approach is illustrated in Figure 10.

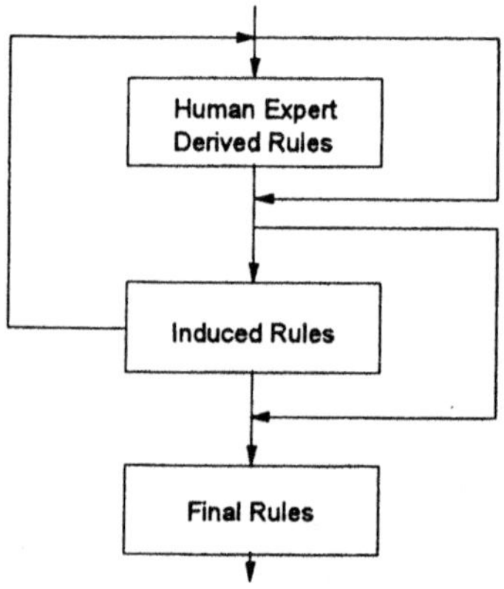

Figure 10. Life Cycle Approach to Knowledge Refinement

During the learning phase, the background knowledge can be utilized through GAC algorithm to select the best groups of links. This background knowledge can enter SC-net in three ways:

1) Unknown value heuristic: This allows the specification of attribute values, which are considered to represent concepts such as doesn't apply, unknown, etc. This feature is helpful when a domain exhibits a strong degree of ignorance i.e., where many of the features are deemed not important or redundant.
2) Complementary value heuristic: In cases where the attributes contain complementary values such as (high,low), (good,bad), (strong,weak) etc., this information can be used to direct the GAC algorithm by temporarily discarding groups of links in that emanate from cells representing complementary value heuristic.
3) Important attribute heuristic: Here, the background knowledge is entered in the form of

description. They specify, for each output concept (class), only the attribute of importance. This helps to reduce the search space learning algorithm operates on.

The GENET algorithm allow encoding background knowledge after learning as well. This process can also be done through injecting the meta-knowledge after each learning cycle, resulting in informed learning.

In the next section, an example that illustrates SC-net's learning ability is presented.

Example: Design of Complex Column Sequences

The design of complex distillation column sequences has been of interest to chemical engineers. We used the method proposed by Tedder & Rudd [34] to test the capability of SC-net. The optimal/best column sequencing is based on separation and cost efficiency as given by Tedder and Rudd. The sequences used are shown in Figure 11 and are abbreviated as

O_1 - Direct sequence O_2 - Indirect sequence O_3 - Side stream rectifier

O_4 - Side stream stripper O_5 - Complex design O_6 - Side stream columns

O_7 - Side stream columns

For training purposes, the SC-net is presented with a set examples covering as many different permutations of the selected attributes as possible. The training set used for this example is given in Table 3. Only 15 training cases were used for learning. The input variables in the training vector are very general and represented using fuzzy numbers. This fuzzy representation helps us to represent a region (pi-shaped) as opposed to a single point. The following notation is used for input variables:

ESI - Ease of Separation Index

MP - Middle Product

OH - Over Heads

BP - Bottom Product

OHBP - |OH-BP| indicates the how close overheads and bottoms are

Design - The type of Column Sequence.

These fuzzy variables are defined in the following range

lt_n - less than n, any number in $[o,n-\epsilon]$

ge_n - greater than or equal to n, any number in [n,100]

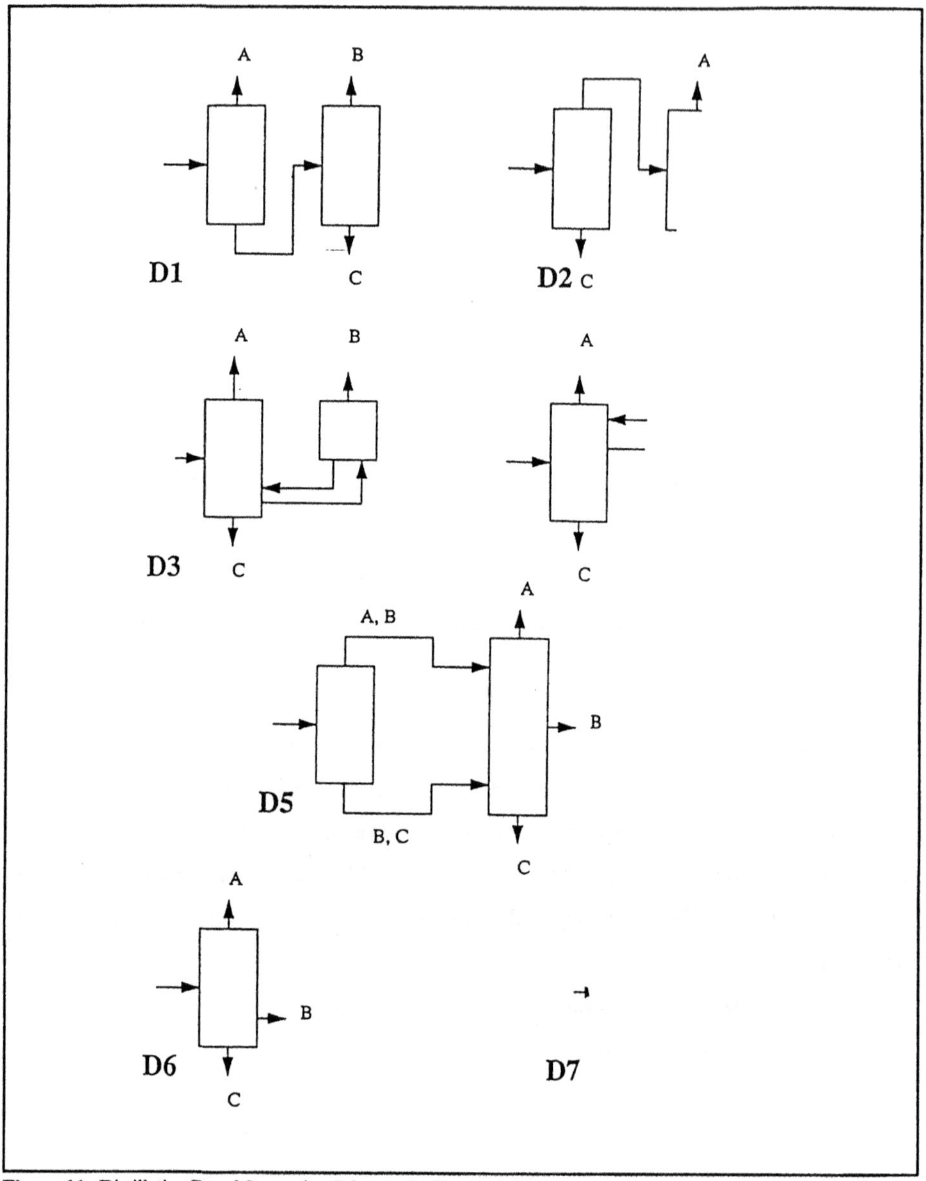

Figure 11. Distillation Based Separation Schemes for Ternary Mixtures

b_n1_n2 - between n1 and n2, any number in [n1,n2]

any - any number [0,100].

"n", "n1", and "n2" may have any value in the region R = [0,100] and ϵ is a small real number O(0.01). The output class has values D1, D2, D3, D5, D6, D7 which correspond to the type of

Table 3. Training Set for the Complex Column Sequencing Example

ESI	MP	OH	BP	OHBP	Class
lt_1.6	b_40_90	any	any	lt_10	D5
ge_1.6	ge_50	any	b_5_20	any	D5
lt_1.6	ge_50	any	lt_5	any	D6
lt_1.6	lt_50	any	lt_5	any	D1
ge_1.6	ge_50	any	lt_5	any	D6
ge_1.6	lt_50	any	lt_5	any	D3
lt_1.6	ge_50	lt_5	any	any	D7
lt_1.6	lt_50	lt_5	any	any	D2
ge_1.6	ge_50	lt_5	any	any	D7
ge_1.6	lt_50	lt_5	any	any	D2
lt_1.6	lt_50	lt_50	ge_50	any	D2
ge_1.6	lt_50	lt_50	ge_50	any	D2
lt_1.6	lt_50	ge_50	lt_50	any	D1
lt_1.6	lt_50	any	any	lt_10	D3
ge_1.6	any	ge_50	lt_50	any	D3

column sequencing. Figure 12 shows the network generated by SC-net for this example, the input attributes ESI, MP, OH, BP and OHBP are in the input cell layer and the interconnection network consists of the network created for defining each of the attributes values. The output cells from the interconnection network will be the input attributes with each of its values presented in the example case, only few of them are shown in the figure. These cells are linked to the IC cells depending on the example cases presented and the IC cells connect to the PC/NC cells depending on the type of evidence available (Positive/Negative) in examples and the outer layer has all the outputs presented in the examples.

The knowledge extracted from the SC-net is shown in Table 4. In terms of simplicity and accuracy, these rules are very similar to the ones proposed by Tedder and Rudd[34] .

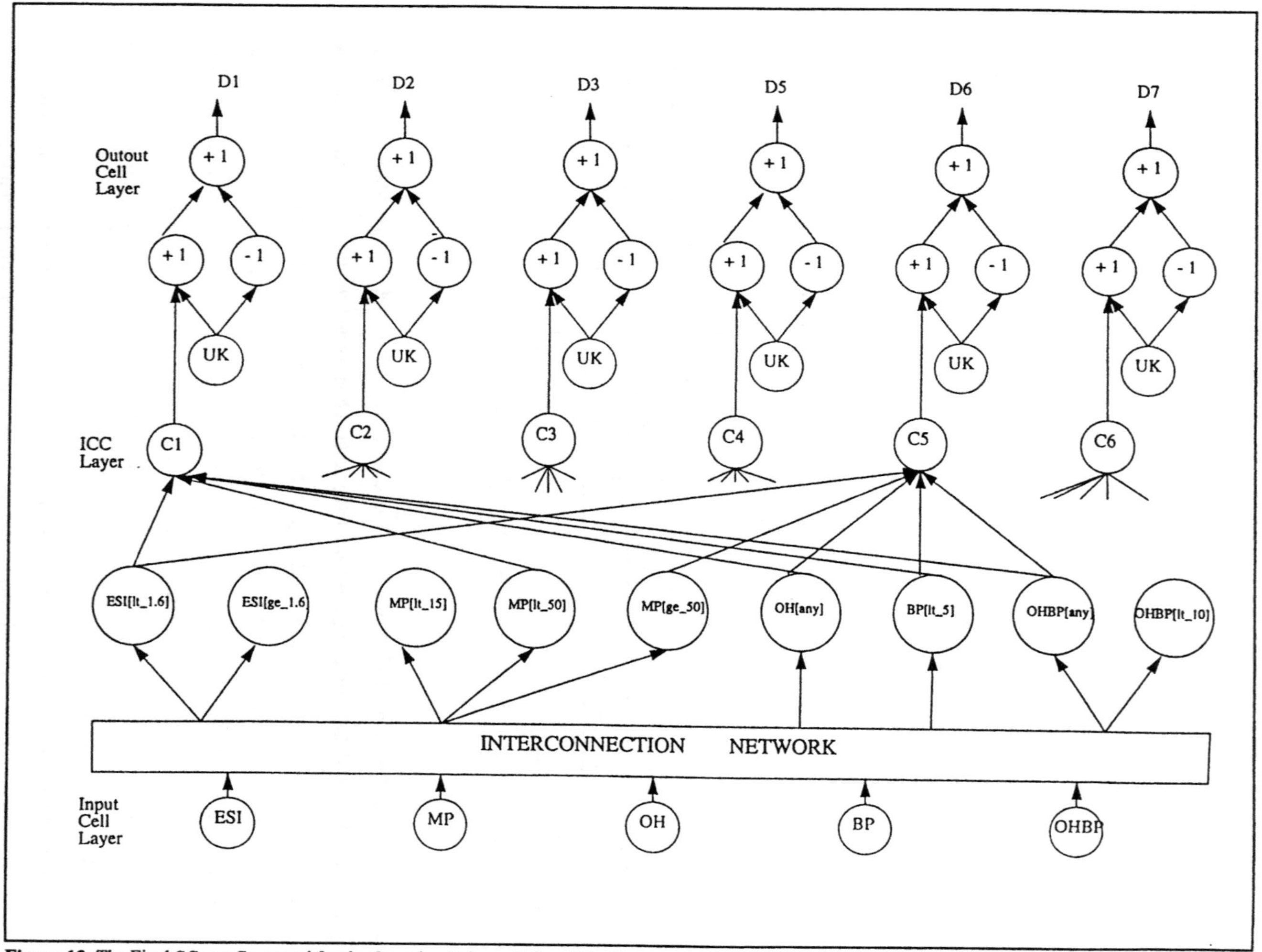

Figure 12. The Final SC-net Generated for the Complex Column System Selection

Table 4. Rules Generated by SC-net

If and (ESI [lt_1.6], MP[lt_50], BP[lt_5]) then D1 (1.00)
If and (ESI [lt_1.6], MP[lt_50], OH [ge_50], BP[lt_50]) then D1 (1.00)
If and (MP[lt_50], OH [lt_5]) then D2 (1.00)
If and (MP[lt_50], OH [lt_50], BP [ge_50]) then D2 (1.00)
If and (ESI [ge_1.6], MP[lt_50], OH [ge_50], BP[lt_5]) then D3 (1.00)
If and (ESI [lt_1.6], MP[lt_50], OH [ge_50]) then D3 (1.00)
If and (ESI [ge_1.6], OH [ge_50]) then D3 (1.00)
If and (ESI [lt_1.6], MP[b_40_90], OHBP [lt_10]) then D5 (1.00)
If and (ESI [ge_1.6], MP[ge_50], BP [b_5_20]) then D5 (1.00)
If and (MP[ge_50], BP [lt_5]) then D6 (1.00)
If and (MP[ge_50], OH [lt_5]) then D7 (1.00)

After the training phase was over, the SC-net was given some new cases. The new cases used are shown in Table 5. The classification accuracy is very encouraging as it is able to identify all the classes correctly. Due to the fuzzy nature of the input attributes, in some cases, more than one design was identified as optimal. This feature is particularly desirable in uncertain domains where there are large number of attributes with interdependency and the attributes are best represented in a qualitative fashion.

Conclusion

The objective of this paper was to provide an overview of various approaches to planning/scheduling problems and knowledge acquisition/machine learning. The evaluation of these techniques within the context of batch processing systems engineering will naturally take several years. However, in each topical area, a more promising and new approach is discussed in more depth with motivating examples. Thus, the conclusions of this chapter will be limited to these cases.

Table 5. Comparison of Sc-net Performance with Simulation

ESI	MP	OH	BP	OHBP	Expected	SC-net
1.3	65	17	18	1	**D5**	***D5***
1.8	70	15	15	0	**D5**	***D5***
1.3	80	18	2	16	**D6**	***D6***
1.3	20	78	2	76	**D1**	***D1,D6***
1.8	65	31	4	27	**D6**	***D6***
1.8	20	78	2	76	**D3**	***D3,D6***
1.3	65	4	31	27	**D7**	***D7***
1.3	20	2	78	76	**D2**	***D2,D7***
1.8	65	4	31	27	**D7**	***D7***
1.8	20	2	78	76	**D2**	***D2,D7***
1.3	25	20	55	35	**D2**	***D2***
1.8	25	20	55	35	**D2**	***D2***
1.3	25	55	20	35	**D1**	***D1***
1.3	10	45	45	0	**D3**	***D3***
1.8	15	60	25	35	**D3**	***D3***

Genetic programming is a robust search technique that has been applied effectively in several different solution spaces. This robustness of Genetic Programming could be exploited in interfacing distinct decision modules of a batch production decision environment. The promising performance of GP is expected to increase its application in NP-hard combinatorial optimization problems such as scheduling. The examples shown illustrate this potential as well as significance of the modelling element. One of the strong elements of Genetic Programming is its robustness in allowing the user great flexibility to bring insight both in modeling and search.

A hybrid learning mechanism which has the features of both symbolic and connectionist approaches has been analyzed and its applicability for a chemical engineering design problem is illustrated. The ability of learning system to extract knowledge in the form of expert/natural language type rules makes it a very good choice for building expert systems. Its performance accuracy, despite generalization of the network during the learning phase, makes it a valid choice for overcoming the knowledge acquisition problems. Due to the background knowledge encoding feature, it can be used to generate rules in domains where there is a lack of expert knowledge and systems which require some a priori knowledge.

References

1. Biegler,L.T., I.E. Grossmann and G.V. Reklaitis: Application of Operations Research Techniques in Chemical Engineering. In: Engineering Design, Better Results through Operations Research Methods (R.Levary, ed.), Elsevier Science Publishing Co., Inc. 1988
2. Booker,L.: Improving Search in Genetic Algorithms. In: Genetic Algorithms and Simulated Annealing (L. Davis, ed.) Morgan Kaufman Publishers, Inc., Los Altos, CA, 1987
3. Brooks, R.A.: A Robust Layered Control System for a Mobile Robot. IEEE Journal of Robotics and Automation RA-2 1 (1986)
4. Cukierman, D., R. Owans, and S. Sloseris: Interactivity Activity Scheduling with Object Oriented Constraint Logic Programming. In: Application of Artificial Intelligence in Engineering VIII. Vol:II Computational Mechanics Publications/Elsevier Applied Science. (G. Rzevski et al. ed.), 1993
5. Das, H., P.T. Cummings, and M.D. LeVan: Scheduling of Serial Multiproduct Batch Processes via Simulated Annealing, Computers in Chemical Engineering, 14 (12), 1351-1362, (1990)
6. Davis, L: Handbook of Genetic Algorithms. NY: Van Nostrand Reinhold, 1991
7. DeJong,K.A.: An Analysis of the Behavior of a Class of Genetic Adaptive Systems. Doctoral dissertation, University of Michigan, 1975
8. Famili, A., D. Nau, and S. Kim: Artificial Intelligence Applications in Manufacturing, AAAI Press/MIT Press, 1992
9. Fuernsinn, M.,and G. Meyer : The Configuration of Automobile Manufacturing Plants Using FAKIR. In: Application of Artificial Intelligence in Engineering VIII. Vol:1 Computational Mechanics Publications/Elsevier Applied Science. (G. Rzevski et al. ed.), 1993
10. Goldberg, D.E.: Genetic Algorithms in Search, Optimization, and Machine Learning. Reading, MA: Addison-Wesley 1989
11. Grefenstette, J.J., L.Davis, and D. Cerys: GENESIS & OOGA - Two Genetic Algorithm Systems, Melrose MA: TSP 1991
12. Hammond, K., Chef: A Model of Case-Based Planning. In: Proceedings of AAAI-86 (1986)
13. Hillard, M.R., G.E. Liepens, M. Palmer, M. Morrow, and J. Richardson: A Classifier Based System for Discovering Scheduling Heuristics In: Genetic Algorithms and Their Applications. Proceedings of the Second International Conference on Gnetic Algorithms, 231-235 (1987)
14. Holland,J.H., :Adaptation in Natural and Artificial Systems, The University of Michigan:Ann Arbor 1975
15. Kapanoglu, M., Genetic Intelligence in Scheduling of Modular Manufacturing Systems, PhD dissertation (in preparation), U.of South Florida, 1995
16. Kempf, K. C. LePape, S. Smith, and B. Fox: Issues in The Design of AI-Based Schedulers: A Workshop Report. AI Magazine 11 (5) (1991)
17. Lambrou, S.K., and A.J. Dentsoras: A Machine Based System for Valuation and Consultation on Machine Assemblies during Design for Configuration. In: Application of Artificial Intelligence in Engineering VIII. Vol:1 Computational Mechanics Publications/Elsevier Applied Science. (G. Rzevski et al. ed.) 1993
18. Laszlo, H, M, Hoffmeister, D.W.T. Rippin: Gantt: an Interactive tool, This volume, p. 706
19. Liepins, G.E., M.R. Hilliard, J. Richardson, and M.Palmer: Genetic Algorithms Applications to Set Covering and Traveling Salesman Problems. In: Operations Research and Artificial Intelligence: The Integration of Problem Solving Strategies (1990)
20. Mah,R.S.: Chemical Process Structures and Information Flows, Boston: Butterworths 1990
21. Michalski, J., G. Carbonell, and T.M. Mitchell.: Machine Learning, Volumes I and II, Los Altos: Morgan Kaufman 1986
22. Minton, S. Machine Learning Methods for Planning, Morgan Kaufmann, 1993
23. Mitchell, T.M.: Becoming Increasingly Reactive. In: Proceedings AAAI-90
24. Nilsson, N.J.: Principles of Artificial Intelligence Palo Alto: Morgan Kaufmann 1980
25. Noronha, S.J., and V.V.S. Sarma: Knowledge Based Approaches for Scheduling Problems: A Survey, IEEE Transactions on Knowledge and Data Engineering, 3, 160-171 (1991)
26. Patel, A.N., R.S.H. Mah, and I.A. Karimi. Preliminary Design of Multi-product Noncontinuous Plants Using

Simulated Annealing, Computers and Chemical Engineering, 15, 451-469 (1991)

27. Pearl, Judea: Heuristics. Addison-Wesley, 1985
28. Raich, A., X. Wu, and A. Cinar: Comparison of Neural Networks and Nonlinear Time Series Techniques for Dynamic Modeling of Chemical Processes. This volume, p. 309
29. Realff, M,: Learning Localized Heuristics in Batch Scheduling. Internal report LISPE 88-053, MIT, 1989
30. Realff, M,: An Analysis of the Chemical Batch Production Problem and a Detailed Methodology for Scheduling. Internal report LISPE 88-054, MIT, 1989
31. Rich, E and K. Knight: Artificial Intelligence, McGraw-Hill, 1991
32. Romaniuk, S.G. Extracting Knowledge from a hybrid Symbolic, Connectionist network. Ph.D thesis, Univ. of South Florida 1991
33. Smith, P., E. Fletcher, F. Gronskov, L. Malmgren-Hansen, K.Ho, B. Gaze: CIM.REFLEX- Using Expert System Technology to Achieve Real Time Shop Floor Scheduling. In: Application of Artificial Intelligence in Engineering VIII. Vol:II Computational Mechanics Publications/Elsevier Applied Science. (G. Rzevski et al. ed.) 1993
34. Tedder, D.W., Rudd, D.F. : Parametric Studies in Industrial Distillation. AICHE J. 24, No. 2, (1978)
35. Tsukiyama, M., K. Mori, and T. Fukuda: Strategic Level Interactive Scheduling and Operational Level Real-Time Scheduling for Flexible Manufacturing Systems. In: Application of Artificial Intelligence in Engineering VIII. Vol:II Computational Mechanics Publications/Elsevier Applied Science. (G. Rzevski et al. ed.) 1993
36. Venkatasubramanian, V.: Fault Diagnosis Through Neural Networks. This volume, p. 631

Integrating Unsupervised and Supervised Learning in Neural Networks for Fault Diagnosis

Venkat Venkatasubramanian*, Surya N. Kavuri

Laboratory for Intelligent Process Systems, School of Chemical Engineering, Purdue University, West Lafayette, IN 47907, USA

Abstract: Recently, there has been considerable interest in the use of neural networks for fault diagnosis applications. To overcome the main limitations of the neural networks approach, improvements are sought mainly in two respects: (a) a better understanding of the nature of decision boundaries (b) determining the network structure without the usual arbitrary trial and error schemes. In this perspective, we have compared different neural network paradigms and developed an appropriate integrated approach. A feedforward network with ellipsoidal units has been shown to be superior to other architectures. Two different types of learning strategies are compared for training neural networks : unsupervised and supervised learning. Their relative merits and demerits are discussed and a combination has been proposed to develop a network that meets our diagnosis requirements. Unsupervised learning component serves to identify the features and establish the network structure. Supervised learning serves to fine-tune the resulting network. We present results from a reactor-distillation column case study to demonstrate the structure of the measurement pattern distribution and the suitability of ellipsoidal units approach. By considering the transient behavior in the diagnosis framework, we point out that the problem of fault diagnosis can be treated on the same footing for both batch and continuous processes.

Keywords: Fault Diagnosis, Neural Networks, Artificial Intelligence, Learning

1. Introduction

Managing process operations requires that the process situation be properly identified. This requires the ability to answer [14] : What is the current state of the plant ?

---

* Author to whom all correspondence should be addressed.

Before we proceed further, it is useful to illustrate the problem of fault diagnosis in a pictorial manner that brings out the essential characteristics of this problem. Figure 1 shows the schematic of a typical distribution of the fault classes in the space of sensor measurements.

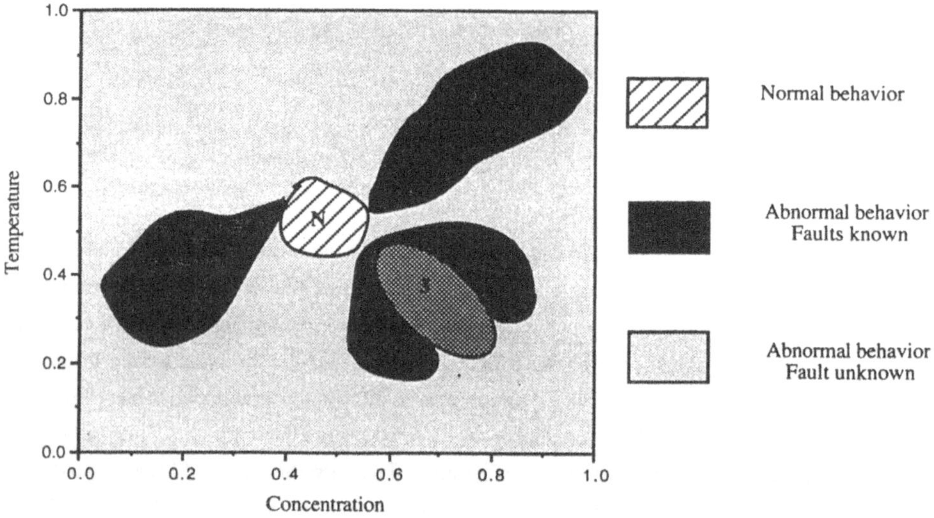

Figure 1. Distribution of fault classes in the measurement space

Examples of such distributions in the space of two significant sensors may be found in [20, 18]. From the day-to-day operations of a process plant, sufficient patterns can be acquired for the normal behavior which is therefore well defined. In Figure 1, the central region represents the region where the measurements corresponding to the normal process operation fall. When a pattern falls in the normal region, the fault classifier should be able to identify that the process behavior is *normal.* Otherwise, the process behavior is *abnormal.* For the abnormal behavior, only a few patterns may be available either from simulations or process history, covering only portions of the abnormal region. These regions define the known faults (causes of abnormality) as shown in Figure 1. In the case of abnormal behavior, the classifier should identify whether the behavior falls under one of the *known* fault classes and if so identify the fault class. If the behavior has a novel symptom pattern, the classifier should accordingly identify the behavior as abnormal while indicating that its cause is *unknown.* Fault classifier should be able to do this using only the examples of known classes.

This question is not easy to answer because of the following reasons:

- the measurements may be insufficient
- the measurements may be unreliable - sensor biases or failures
- process unit malfunctions - unnoticed leakages or blockages
- process unit degradation - catalyst deactivation, heat exchanger fouling
- large unknown external disturbances - Increase in the coolant temperature

Data reconciliation techniques traditionally serve to rectify data or measurement inconsistencies by using the redundancy provided by the process model. If the process model is in itself of doubt for its validity, this framework needs to be extended. In this perspective, fault diagnosis is a complementary subdiscipline of data reconciliation where the goal is to determine what is wrong without assuming either the process structure (model) or the parameters are correctly known. In the general context of fault diagnosis, the definition of a fault includes :

- gross parameter changes in a model (e.g.change in the feed concentration)
- structural changes (change in the model itself, e.g. failure of a controller, valve stuck, pipe broken etc.)
- measurements are grossly corrupted (e.g. sensor biases, failure, saturation etc.)

Feedforward neural networks have been an attractive alternative for the design of classifiers for fault diagnosis. The interest in them stems mainly from the fact that they provide a means to parameterize the classifier without a specification of the functional form of the discriminant function. This has an advantage over other schemes such as linear and quadratic classifiers [5], used commonly in statistical pattern recognition, where a commitment is made *a priori* to a particular functional form. Neural networks have been successfully demonstrated in their ability to approximate arbitrary functions [3]. This generality also comes with a lack of control and sometimes undesirable effects such as extrapolation problems. Application of neural networks to fault diagnosis has been a very active area of research in recent years [7, 19, 17, 15, 12] . Attempts were made to characterize their classification and generalization performance on one hand, and the speed of their training on the other. Use of networks using ellipsoidal units [9, 10] have been proposed as a viable alternative to the standard backpropagation network on two accounts. First, the standard backpropagation network uses linear activation function that classifies regions by generating intersecting hyperplanes. These hyperplanes of infinite extent have been noticed to cause (a) erroneous extrapolation (b) place decision boundaries close to the class boundaries (c) sometimes generate unintuitive classification of some regions of the input space. Second, the training time of backpropagation networks can be very high for large networks. This requires the use of decomposition techniques that help speed up the training. Networks with ellipsoidal units permit the development of decomposition techniques which result in the training of small networks with few training patterns.

2. Fault Diagnosis of Continuous and Batch Processes

Traditionally, fault diagnosis of continuous processes has been concerned with steady state models and the deviation of the process from the nominal steady state. The situation with batch processes is different in that the process is in unsteady state. Detection and diagnosis of faults in batch processes has to be done using dynamic models instead of steady state models and dynamic trends instead of steady state values. These distinctions would, however, blur when one considers the transient dynamics resulting from a fault in a continuous process as well. In such a general context, the problem of diagnosis becomes one of detecting and isolating a fault from the time trajectories of measurements for the different faults. The problem of diagnosis is to partition the state space into various fault classes and the normal region. Partitioning may be performed by the generation of envelopes that enclose possible trajectories for each of the fault classes. Envelopes so generated serve to isolate each trajectory to a possible fault class. Figure 2 shows the state-space trajectories of a process transient for two different faults.

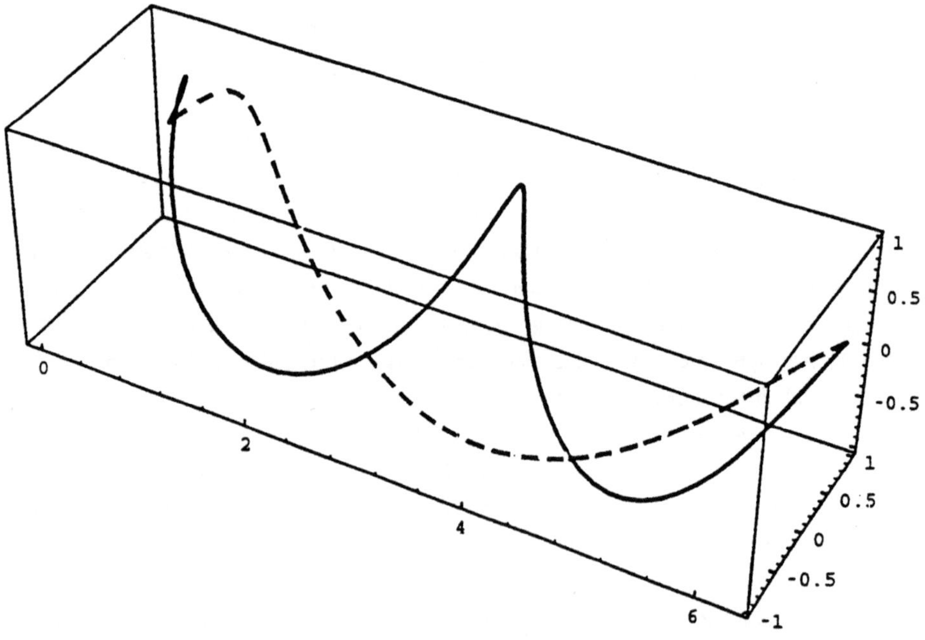

Figure 2. Transient state-space trajectories of two different faults

The classifier approximating the different fault distributions in the state-space should develop envelopes around the region where the sample trajectories for that fault exist. When trying to classify a trajectory to a particular fault class, two checks have to be made: (a) nearness of the trajectory to envelopes; (b) similarity in the dominant directions in which the trajectory and the envelopes spread in measurement space. Classifiers which use Euclidean distances can identify whether the trajectory is close to a given envelope. However, modeling the direction in which the envelope spreads is not possible with the use of radially symmetrical measure of distance such as the Euclidean distance. Use of an asymmetric distance metric is warranted for such situations. The required asymmetry is brought into the network by the ellipsoidal activation as will be discussed next.

3. Linear Versus Ellipsoidal Activation

The standard backpropagation network uses a linear activation function, a sigmoid squashing function and RMS error minimization. The ellipsoidal activation function [9] $E(\mathbf{x})$ is the following:

$$E(\mathbf{x}) = -(\mathbf{x}-\mathbf{m})^T(\mathbf{D}^T\mathbf{D})^{-1}(\mathbf{x}-\mathbf{m}) + 1 \qquad (1)$$

$E(\mathbf{x}) = 0$ defines an n-dimensional ellipsoid oriented along the reference axes, where $\mathbf{m}$ is the center of the ellipsoid and $\mathbf{D}$ is the diagonal matrix of the half-lengths of the principal axes. Activation $E(\mathbf{x})$ is incorporated in the squashing function to determine the bounded network value. Since the squashing function, a sigmoid, is a monotonic function, the node value would also have ellipsoidal contours of constant value. The node value for a sigmoid with the range (0 1) is given by:

$$\frac{1}{1+\exp(-E(\mathbf{x}))} \qquad (2)$$

- When $\mathbf{x}$ is on the ellipsoid, $E(\mathbf{x}) = 0$, and the node value is 0.5.
- When $\mathbf{x}$ is inside the ellipsoid, $E(\mathbf{x}) > 0$ and the node value is in the range (0.5 1). As $\mathbf{x}$ moves away from the ellipsoid towards the center, the node value tends closer to 1.
- When $\mathbf{x}$ is outside the ellipsoid, $E(\mathbf{x}) < 0$ and the node value is in the range (0 0.5). As $\mathbf{x}$ moves away from the ellipsoid, the node value tends closer to zero.

Figure 3, schematically shows the difference between linear and ellipsoidal activation networks. In networks using linear activation functions, decision regions are formed by the intersection of hyperplanes generated at the hidden nodes. Decision regions so carved out may include regions of input space which are not represented by training patterns. In the case of linear activation, because of the infinite extension of these hyperplanes, the network classifies the entire input space into the various classes. This causes these networks to display unsatisfactory generalization characteristics, and make them not suitable for fault diagnosis.

Ellipsoidal activation functions, on the other hand, generate closed decision regions. This avoids arbitrary inclusion of input regions not represented by training patterns.

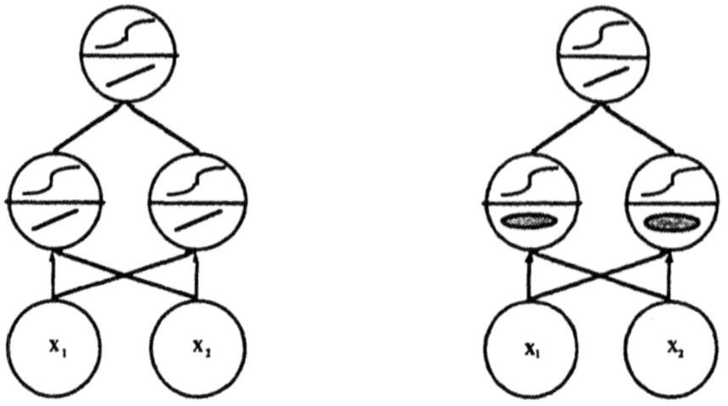

Figure 3. Schematic representation of activation functions in the network

Figure 4a shows a schematic of the classification by linear activation networks for the fault diagnosis problem shown in Figure 1. Figure 4b shows the classification using ellipsoidal activation networks. The following discussion compares these networks and explains why ellipsoidal activation is more desirable for diagnostic applications.

Linear Activation Networks

- *Decision regions are formed by hyperplanes.*
  Hyperplanes have an infinite extent, resulting in arbitrary classification of regions of input space not covered by training patterns.
- *There is no region of indecision.*
  The network would classify all measurement patterns to one of the known classes. A pattern that is far from the training patterns may belong to a known fault class or to some unknown fault class. In Figure 4a, we see that all of the measurement space has been apportioned to the known classes. This does not allow any scope for indicating that the cause for an abnormal measurement pattern cannot be ascertained.
- *Decision regions for fault classes can be unbounded.*
  In Figure 4a, all the fault classes have unbounded decision regions though the fault classes themselves are bounded.
- *Hyperplanes are not unique.*
  Hyperplanes formed by the network can be spatially translated to some extent without affecting the classification of the training patterns. However, this translation changes the decision regions formed. While such a translation of hyperplanes may not affect the correct classification of the training patterns, it would change the class membership

of some of the regions in the measurement space. As a result, generalization results of the network are arbitrary and are sensitive to the network training. Different initial weights or learning coefficients can result in different generalization results. This arbitrariness results in a non-robust classification.

- *No notion of distance is used by the classifier.*
  All the patterns that fall on the positive-side of the hyperplane get high activation, close to one. This activation provides no information on the distance of these patterns from the training patterns. If a hyperplane is adjacent to the class boundary, patterns immediately on the negative-side of the hyperplane would have low activation even though they are very close to the training patterns.

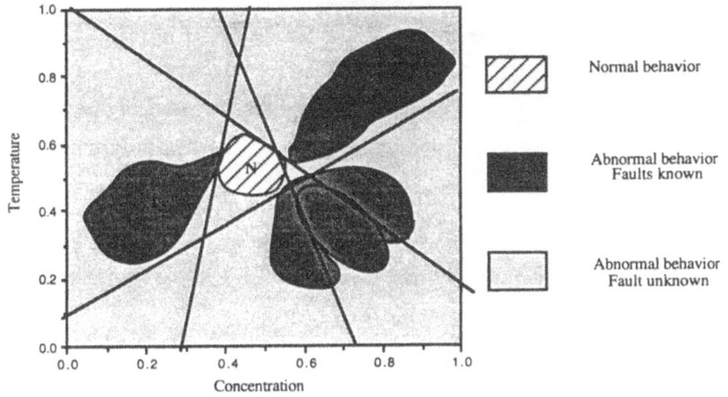

Figure 4a. Fault space classification of linear activation networks

Ellipsoidal Activation Networks

- *Decision regions are formed by ellipsoids.*
  Ellipsoids are bounded regions. Their size can be controlled so that the decision regions do not include regions of input space far from the training patterns.
- *Identifies regions of indecision.*
  Ellipsoidal units have a decreasing output value as one moves away from their centers radially. Since these units are placed on the training patterns, they have lower values when they are far from these patterns. As a result, patterns which are far from all the training patterns are clearly identified as not belonging to any of the known classes including the normal region. They are interpreted as an indication of abnormal behavior of the process whose cause cannot be determined from the available information.

- *Decision regions for fault classes are always bounded.*
  This is because the ellipsoids are bounded regions. The intention is to generate decision boundaries close to class boundaries, allowing scope for reasonable generalization but preventing erroneous extrapolations.
- *Ellipsoids use a notion of distance.*
  Generalization results are thus based on this notion and hence robust. A pattern that is closer is more likely to belong to a class than a pattern that is farther.

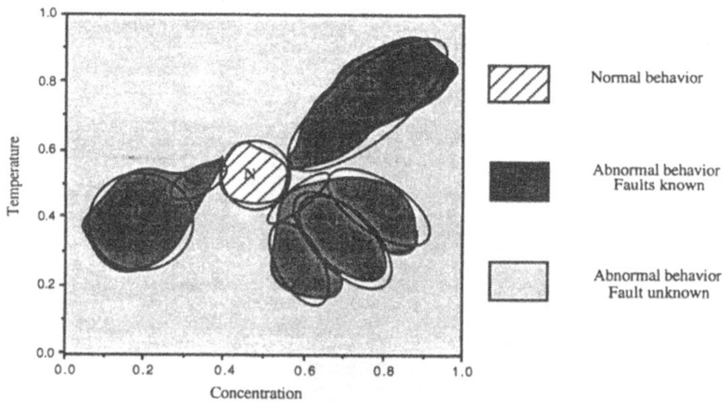

Figure 4b. Using ellipsoidal activation functions for fault diagnosis problem

From Figure 4b, we see that a single ellipsoid is sufficient to represent the normal region and the fault classes 2 and 3. However, to represent fault classes 1 and 4, we need more than one ellipsoid. From Figure 4b, it is clear that the network training algorithm has to address three issues carefully in order to make such a representation possible: (a) it has to determine the number of ellipsoids required to approximate each fault class. This is crucial since too few ellipsoids would result in a poor representation of the fault class and too many would increase the complexity of network training; (b) It has to locate these ellipsoids correctly and determine a shape based on the local distribution of training set; and (c) it has to ensure that the size of the ellipsoids generated is contained.

In the following sections, we address these issues related to network architecture and training. We characterize the problems faced in the training of these networks and propose solutions. The main issues of focus are (a) robust generalization characteristics and (b) compact representation so that a class is represented by as few ellipsoids as possible.

4. Unsupervised and Supervised Learning Strategies

A satisfactory solution to these problems require the consideration of two different approaches to learning, namely, unsupervised and supervised learning. We compare these two approaches in this section and then go on to propose a strategy that integrates both.

4.1. Unsupervised Learning

- Unsupervised learning is used when the class information is not available but the patterns are to be grouped based on similarity.
- Methods used for this purpose are familiarly known as self-organized learning or competitive learning algorithms. The idea is to partition the training patterns into disjoint groups so that each group has patterns similar in some respect.
- The features extracted by these methods are such that they identify which patterns are similar. The class information of these patterns is of no relevance and is not used in the construction of these features.
- Objective of these methods is to give credit for similar patterns to group together. No penalty is given for grouping patterns, belonging to different classes, as being similar.
- Since the objective uses a notion of distance (similarity), patterns far away from a group can be accordingly judged in terms of the distance metric. Thus, in Figure 1, patterns far from fault classes 1 and 2 would be considered as belonging to the region of indecision or to other classes.
- In Figure 1, fault classes 3 and 4 are very close and appear similar. A simple measure of distance has no reason to view them as distinct. Without the class information, they are treated as belonging to the same or similar group.

Competitive learning (or self-organized learning) is used as a means to group the training data by similarity. The grouping procedures give credit for patterns which are similar but do not pay any attention to their membership to different classes. In Figure 1 for example, all patterns of classes 1 and 2 are properly categorized just by grouping by similarity. However, a simple measure of similarity will not solve the classification problem for classes 3 and 4. Even a representation by many clusters may not solve the problem of misclassification.

4.2. Supervised Learning

- Supervised learning is used for classification problems where the class information is available.

- Typically, these methods use an error criteria such as RMS error minimization. Objective of these methods is to penalize patterns which are wrongly classified. No credit is, however, given for detecting similarity between a group of patterns.
- The features extracted by these methods are such that they can be useful for discriminating between the patterns of different classes. Notion of similarity plays no direct role.
- The objective of these methods does not penalize for including regions which are not similar to the patterns in a class. This is one of the causes for the extrapolation problems which arise in classification. Since the objective does not consider the effect on patterns which are not a part of the training set, there is no built-in mechanism for avoiding extrapolation completely. Thus, in Figure 1, there is no reason why the decision regions for fault classes 1 and 2 should not include the regions of indecision. This results in poor generalization.
- Performs credit assignment to various features so that a finer classification can be brought about even when two classes are close and sharing nonlinear decision boundaries.

4.3. Integrating Unsupervised and Supervised Learning Strategies

Supervised learning tries to determine features which serve to result in a maximum separation between the classes. Features determined solely on this criteria do not consider the notion of similarity. As a result classes may be arbitrarily assigned regions of measurement space where no information (training patterns) is available. Unsupervised learning tries to determine features by grouping patterns based on similarity. However, no information of class is considered during the feature determination and the resulting classification can be poor. To avoid the arbitrariness and get good classification, what is necessary is a method that uses features which use a notion of similarity and then perform a credit assignment to these features to achieve good classification. Unsupervised learning provides a way to group patterns by similarity thus identifying required features. Supervised learning would assign relative credit to these features to achieve finer classification. This integrated learning strategy brings together the good characteristics of both methods while overcoming their undesirable characteristics. In Figure 1, clearly neither unsupervised or supervised learning will independently do the proper isolation of classes 3 and 4. However, their combination will solve the problem. Unsupervised learning can be used to identify the necessary features for each of these classes separately, without the information of other classes. These features can be used to determine the structure of a network [e.g. each feature forms a hidden node] and thus avoid the problem of arbitrarily determining the number of hidden nodes as practiced in the standard approach. Then, supervised learning can be used to perform a credit assignment to these features, this time with the use of class information, so that a proper classification is achieved.

Use of this integrated learning strategy also offers solutions to other problems [9, 10] :

- solve the hidden node problem avoiding the unsatisfying and expensive trail-and-error methods
- overcome local minima traps which are severe when supervised learning alone is used
- allow the development of decomposition strategies which make the fault diagnosis of large-scale industrial processes tractable (discussed later in section 7)
- avoid extrapolation problems arising through the introduction of bounding schemes

5. Clustering as Unsupervised Learning

Unsupervised learning is also known as associative learning, competitive learning or simply clustering. The underlying purpose of these approaches is as follows: In the input space, there may be regions where data samples are heavily concentrated. These regions are called clusters. The centers of the clusters are the representative points of the training patterns and are much fewer in number than the training patterns. The idea pursued here is that a cluster center along with a local measure of distance should define the local density of the data distribution sufficiently well. Clustering saves storage space since only a few cluster centers have to be stored instead of a large number of training patterns. It saves computational time since the computations (such as distance measures) have to be done only with cluster centers and not all the training patterns. It gives good generalization because it captures the general characteristics of the underlying distribution.

5.1. Why Mixture Density Representation?

If a density function has local regions where the densities are high, then it is important to know the covariance of the density in each local region. The local covariance is used in the local region for the purpose of measuring distance with a non-Euclidean distance measure. When the density function of the class concerned has multiple modes, it should be represented by a mixture density. The basic motivation behind this approach is that the modes of a continuous density function are important in describing and analyzing the underlying data structure for recognition purposes. The problem of isolating the modes is called *feature extraction* or *clustering*. A mixture of density functions from a particular family, such as the Gaussian, can be used to approximate the density function of the underlying distribution. Upon detecting the modes, one could approximate the local region by a Gaussian and the overall density by a mixture of Gaussians.

5.2. Scope of Clustering

Clustering makes sense only in terms of *a priori* knowledge used. For example, the results of clustering may not make sense when interpreted by a person who is using a field of knowledge not incorporated in the clustering procedure. Clustering methods use some notion of a metric and this can be validated only by assumptions made on the distribution of the input patterns. Clusters could be very complex in the measurement space. For example, clusters for two different classes could be interwoven as hyper-coil springs in a multi-dimensional space. Without the knowledge that the respective classes have complex loci, the data points for different classes will appear[1] close together and it will not be possible to separate points from different classes. There may be a need to define locii such as hyper-spirals, hyper-oscillations etc. in order to determine the appropriate distance metric a priori. Choice of a distance metric used for clustering would thus presume something about the class distributions. Empirical studies have shown that decision regions of many classification problems are ellipsoidal and many class distributions are approximately Gaussian [1]. Most often, a multivariate normal distribution (or a mixture of them) is assumed for each class, and although this assumption is not often met, the resulting features are still considerably better than those that are randomly derived.

A clustering procedure is practicable only when it forms a minimal number of cluster centers. Otherwise, a tight upper bound should be available for the number of clusters a priori, but this is generally not possible. What is desirable is a clustering technique which, once given a particular choice of distance metric, determines a conservative number of clusters. Even if the preset upper limit on the number of clusters is high, the algorithm should utilize only the required number of clusters.

5.3 K-means Clustering

K-means clustering [4] is a simple and powerful clustering algorithm. K-means clustering presupposes the number of clusters needed and would cluster the data accordingly. It utilizes all the cluster centers so that each of the clusters is guaranteed at least one pattern. We present

[1] The look will be with a Euclidean distance measure.

the k-means clustering algorithm in an adaptive form as a network [13]. This form makes it convenient to compare with other methods to be discussed later. The network has a layer of input nodes connected to a layer of output nodes. Each of the output nodes is a cluster. The weights connecting the input nodes to an output node j give the cluster center coordinates.

Notation

c = number of cluster centers chosen

n = total number of inputs

N_p = total number of patterns

W_{ji} : weight connecting input node i to output node j

$[W_{j1}, W_{j2}, ...W_{jn}]$ = cluster center j

d_{kj} = Euclidean distance of jth cluster center from the kth pattern

K-means Clustering Algorithm

1. Choose *a priori* the number of clusters c. Randomly choose c training patterns and assign these as the initial cluster centers $\mathbf{W}_j$.
2. For each training pattern k, find the distance d_{kj} from all the cluster centers j. Let j^* be the cluster center that is closest to the pattern.

$$\Delta W_{j^*i} = (X_{ki} - W_{j^*i})$$

$$W_{j^*i} = W_{j^*i} + Lrate*\Delta W_{j^*i}$$

3. Go to 2 if there are changes in the cluster centers $\mathbf{W}_j$.

Note that in step 2, a pattern is allowed to belong to *only one cluster* - the cluster that is closest to the pattern. Each pattern is grouped to a cluster center that is closest. The cluster centers are then moved to the mean of the corresponding groups. This process is repeated until the cluster centers stabilize. The main drawback of this method is that utilizes all the cluster centers in the formation of clusters. Since the exact number of clusters in a distribution is in general not known, one will have to overestimate the number of clusters in order to properly use this method.

Figure 5 shows K-means clustering in the form of a network. For each training pattern, distances are computed from each of the cluster centers and the pattern is included only in the closest cluster. This is called a winner-take-all strategy where a cluster which is closest gets all the credit for a pattern. This is a form of competitive learning which introduces a lateral inhibition between the output nodes (or clusters). This is also sometimes referred to as hard competitive learning, since it lets only one winner considering the rest to be losers.

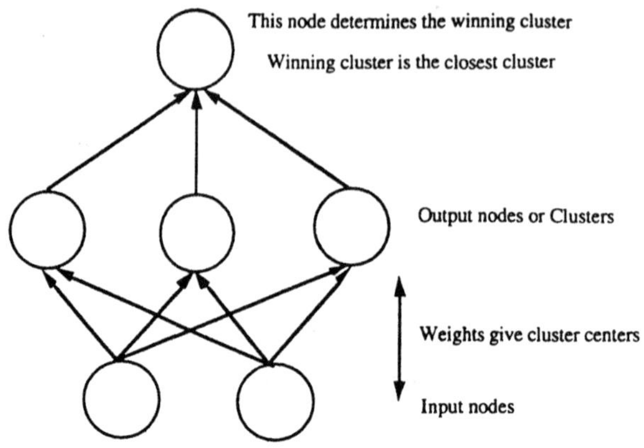

Winner-take-all Competitive Learning

Figure 5. Network structure of K-means clustering

A generalization to this, called soft competitive learning, can come in two forms:

- There are k winning clusters and the rest loose.
- All clusters have a fuzzy membership which specifies to what degree they win or loose a pattern, instead of simply specifying whether they win or loose a pattern.

5.4. Kohonen's Self-organizing Feature Maps

Kohonen's self organizing maps is related in principle to k-means clustering. Kohonen's model also consists of two layers. The first layer contains the input nodes and the second layer contains the output nodes. Adjustable weights completely connect all the input nodes to the output nodes. Though the original Kohonen's algorithm [11] used output nodes in a 2-dimensional array, others have used it in a linear array [8]. The neighborhood number is set initially to $\frac{c}{2}$ and gradually reduced to 0. Topological neighbors are considered on either side of the linear array of output nodes.

Kohonen's Algorithm

1. Randomly initialize weights W_{ji}. Set the neighborhood size N_c to $\frac{c}{2}$.

2. For each training pattern k, select the output node j^* such that d_{kj^*} is minimum.

Update W_{ji} using the following rule:

$$\Delta W_{ji} = (x_{ki} - W_{ji})$$

$$W_{ji} = W_{ji} + Lrate * \Delta W_{ji}$$

where j includes j^* and each of its N_c neighboring nodes on either side.

3. Go to step II if there are significant changes in weights.
4. If $(N_c = 0)$ stop

   else $N_c = N_c - 1$ and Go to 2.

Notice that the above algorithm uses the notion of topological neighborhood. The topology is in reference to the ordering of the output nodes (cluster centers) in the network. The algorithm identifies the cluster center that is closest to a training pattern, and updates this cluster center and all its topological neighbors. Kohonen's algorithm (known familiarly as the self-organizing map) is based on the idea that the neighboring cells in a neural network compete in their activities by means of mutual lateral interactions, and develop adaptively into specific detectors of patterns. The idea is that the locations of the responses tend to become ordered. Kohonen's method differs from the k-means clustering in that there is no single winner anymore. Instead, all the clusters in the neighborhood are winners. K-means clustering can be shown to be a special case of Kohonen's clustering algorithm. If we set the neighborhood size $N_c = 0$ in step I of the Kohonen's algorithm, we get the k-means algorithm.

After a neighborhood number is decided, the algorithm makes all the clusters in the neighborhood to be the winners of the pattern. In fact, it gives the same weightage to all the clusters, in the topological neighborhood, in updating their centers for a pattern k. However, it is better to have the weightage for updates dependent on the relative distances of their centers from pattern k. This is essential because topological neighbors may not be neighbors by distance and it makes sense to give a higher weightage to a closer cluster. A fuzzy-membership scheme may be used such that a cluster center which is relatively far off from a pattern has a very little movement towards the pattern. There are other advantages to the use of fuzzy membership such as reducing the effect of gravity from dense regions which will be discussed later.

6. Notion of a Fuzzy Membership

When a cluster belongs to the neighborhood, Kohonen's method updates the center, otherwise it does not. This is a case of hard partitioning where a pattern has either a membership value of 0 or 1 to a cluster. A fuzzy partitioning, on the other hand, allows us to give a soft partitioning by allowing any value in the closed interval between 0 and 1. We use below a membership function due to Bezdek[2] in his fuzzy-c-means algorithm. Huntsberger and Ajjimarangsee [8], used this membership function in a self-organizing model for unsupervised pattern recognition based on Kohonen's self-organizing maps.

u_{kj} : membership value of kth pattern in the jth cluster

$$u_{kj} = 1 \text{ if } d_{kj} = 0$$

$$= 0 \text{ if } d_{kl} = 0,\ (l \neq j,\ 0 \leq l, j \leq c\text{-}1)$$

$$= \sum_{l=0}^{c-1} \left(\frac{d_{kj}}{d_{kl}}\right)^{-1} \text{ otherwise}$$

This fuzzy membership value is in the range [0 1]. Membership value is inversely proportional to the distance of kth pattern from j^{th} cluster, d_{kj}. Membership values are normalized. Membership of any pattern in all the clusters add up to 1. This means that membership considers only the relative distances of the pattern from the cluster centers.

6.1. Modified Kohonen's Algorithm with Fuzzy Membership

1. Randomly initialize weights $\mathbf{W}_j$. Set the neighborhood size N_c to $\frac{c}{2}$. Set Lrate[2] value.
2. For each training pattern k, select the output node j^* such that d_{kj^*} is minimum.

   Update W_{ji} using the following rule:

   $$\Delta W_{ji} = u_{kj} * (x_{ki} - W_{ji})$$

   $$W_{ji} = W_{ji} + \text{Lrate} * \Delta W_{ji}$$

   where j includes j^* and each of its N_c neighbors on either side.
3. Go to step II if there are significant changes in weights.
4. If $(N_c = 0)$ stop

   else $N_c = N_c - 1$ and Go to 2.

Figure 6 shows the clustering procedure in the form of a network. From a comparison of Figures 5 & 6, it is clear that there is no single winning cluster for any pattern when membership function is used. Instead, the membership specifies which clusters are close and which are far. In the case of Kohonen's method, one could consider that there is a hard membership function which gives a membership of 1 for the closest cluster and zero for the rest of the clusters.

Use of fuzzy membership serves in four respects:

- The fuzzy membership moves a cluster center towards a pattern by an increment inversely proportional to the distance of the cluster center from that pattern. As a result, a cluster center which is relatively far off from a pattern gets a smaller move towards the pattern than a cluster center that is closer.

---

[2]In all the case studies considered, Lrate was set at 0.001. It has been recommended in literature that the value of Lrate be decreased with iterations. Larger Lrate means that cluster centers get larger movements.

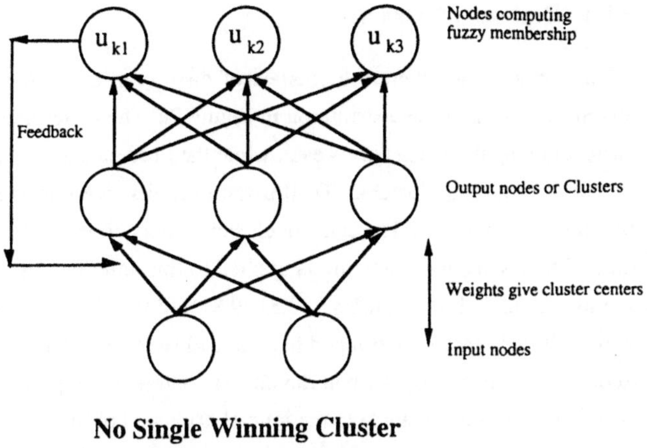

Figure 6. Network structure for Kohonen clustering with fuzzy membership

- Cluster centers may tend to move close when there are dense regions of patterns. There is no mechanism in the clustering algorithm to prevent cluster centers from moving towards each other until they coincide. Fuzzy membership provides such a facility indirectly. Direct repulsion by penalty could be an alternative to avoid cluster centers from moving closer to each other. However repulsion may inhibit the movements of cluster centers thus resulting in local minima problems. When a pattern is relatively very close to a particular cluster center, the fuzzy membership is close to one for that cluster and is near zero for all other clusters. This particular feature discourages cluster centers from moving close to each other. It prevents the clusters from getting attracted to regions of high gravity.
- Fuzzy-membership also helps avoid excessive usage of cluster centers - i.e., cluster centers are sparingly used. Redundant cluster centers do not get utilized because their fuzzy-membership values drop to zero for all training patterns. This conservative approach to the use of cluster centers results in a minimal set of features (cluster centers) needed for the description of the distribution of the patterns. This is essential to classification problems because the number of features determines the complexity of the supervised learning to be performed for classification.
- Fuzzy membership provides a mechanism to allow *overlapping* clusters. When the fuzzy membership of two clusters is high for a pattern, the pattern is a part of both the clusters. The final clusters developed are thus allowed to overlap. The Kohonen's basic algorithm like other clustering methods does not allow clusters to overlap.

6.2. Avoiding Regions of Gravity

Gravity problem arises when there are clusters of disproportionate sizes. To illustrate the gravity problem, we consider the distribution in Figure 7a. There are three clusters in the data with two clusters having nine data points each while the third cluster has only two. Figure 7a shows three nonoverlapping clusters. To illustrate gravity, we consider the initial cluster centers within the large clusters. The starting cluster centers chosen are : (0.8, 0.9) (0.9, 0.9) and (0.0, 1.0). When K-means clustering is applied to this situation, the algorithm identified the following as the final cluster centers - (0.81, 0.81), (0.91, 0.91) and (0.77, 0.11). Notice that the first two cluster centers are trapped in a large cluster while the smaller cluster was left unrepresented. This problem arises when the cluster centers get trapped in a region of higher "gravity". This problem is attributable to the mutual exclusion of the clusters considered in the K-means clustering.

Figure 7b shows the trajectory of the cluster centers for Kohonen's algorithm without the fuzzy membership. Because of the notion of neighborhood, clusters are allowed to share patterns and the cluster center is able to move out of the local cluster but is not able to steer away from the gravity due to the larger clusters to move towards the smaller cluster. It performs better than K-means clustering because the clusters are not localized with mutual exclusion. Instead, the cluster sizes are started with large neighborhoods which is then gradually reduced. However, the method still suffers from the influence of high gravity - though there are cluster centers present in the gravity areas, other cluster centers are still attracted to these denser regions.

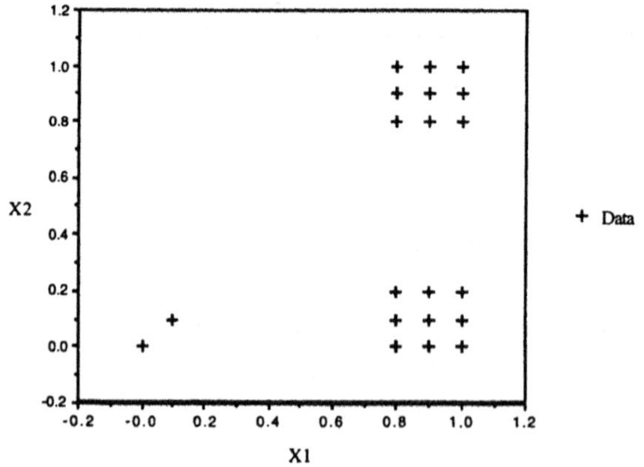

Figure 7a. Clusters with disproportionate number of points

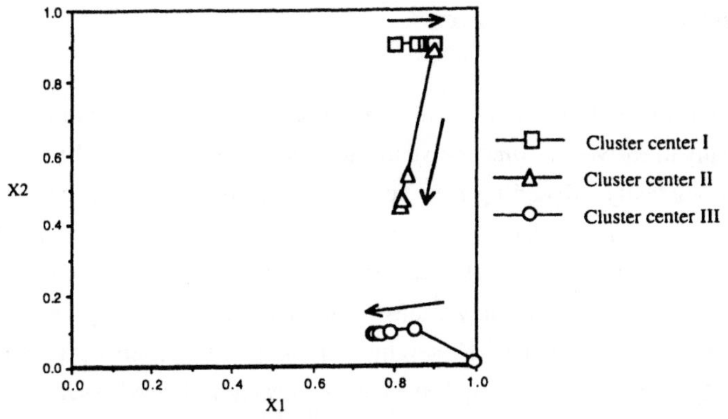

Figure 7b. Trajectory of clusters centers - Standard algorithm

One could circumvent this problem if one has a scheme to give less importance to regions (dense or otherwise) which are already represented by cluster centers. Figure 7c shows the trajectory of the cluster centers when fuzzy membership is used. Since one cluster center provides for the representation at each of the denser regions, the membership value for the third cluster center is low at these regions of higher gravity. The third cluster center is thus able to steer its way to the third smaller cluster. By using fuzzy membership, provision was made for repulsion between the cluster centers. This repulsion helps the cluster center to steer against the gravity of the larger centers towards the smaller cluster.

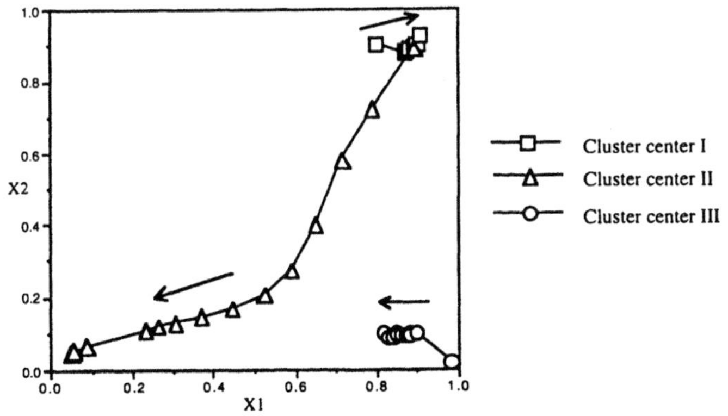

Figure 7c. Trajectory of clusters centers - Algorithm with fuzzy membership

7. Decomposition Strategies

For the diagnosis of large-scale industrial processes, network training task becomes computationally intensive. It is imperative that decomposition techniques be used for achieving practicable complexity. Speed up may be achieved by the following three decomposition strategies which result in smaller networks with few weights and few training patterns :

(a) ***Network decomposition:*** The network is broken into smaller networks so that each of the subnetworks can be independently trained. Since the network training complexity is at least $O(N^3)$, where N is the number of weights, training the subnetworks independently is computationally more efficient. If N_k is the number of weights in the kth subnetwork, then :

$$\sum_{k=1}^{m} N_k = N$$

and $$\sum_{k=1}^{m} (N_k)^3 \ll N^3$$

(b) ***Training set decomposition:*** When the training set is very large, it can dominate the computational complexity as well. Training set decomposition exploits the fact that only a part of the training set is relevant to the training of each of the subnetworks. Training patterns which have negligible influence on the training of a subnetwork are identified and eliminated. This results in small training set sizes.

(c) ***Input space decomposition:*** In a typical industrial process, the number of sensors is quite large. In such a case, the input space dimension or the number of input nodes becomes a dominating factor. Dimensionality reduction can be achieved by exploiting such factors as sensor redundancy. Restricting to the use of local information (with respect to process structure) is another possibility where only the close by relevant sensors are considered for each fault class.

7.1. Network Decomposition

In networks with ellipsoidal units, hidden nodes are not shared between the output nodes, i.e. each of the hidden nodes is connected to only one output node corresponding to a particular class. This is because the ellipsoids formed by the hidden nodes are dedicated to a particular class. Not sharing the hidden nodes amounts to reducing the number of connections substantially thus decreasing the dimensionality (number of weights) of the training problem. The network architecture is inherently decomposable since the hidden nodes are not shared between the output classes in the network. Each of the subnetworks has one output node and its own dedicated hidden nodes. Such a decomposition implies that there is no communication required between the subnetworks and hence their training can be done asynchronously on different processors. Since the complexity of network training is at least $O(N^3)$, a linear decomposition of the network would result in a faster network training i.e., the sum of the training times for the subnetworks would be much smaller than that of the completely connected network. Further, the decomposition allows for an asynchronous computation on the subnetworks, resulting in a parallelism of 100% efficiency.

7.2. Training Set Decomposition

The training of feedforward neural networks exploits the parallelization that comes from localized computation. The fact that computations can be localized at the processors allows for a fast computation. When localized representation is used, as in the case of Gaussian units or ellipsoidal activation functions, the network still retains this property of localized computation. In addition, the network training can be modified so that it exploits the localized representation of units in the input space. The value of a node with ellipsoidal activation function drops radially to zero as one moves away from its center. For patterns far from its center, its value is near zero. When the standard backpropagation is used in network training, the error backpropagated to a hidden node by a pattern far from it, is near zero. As a result, the contribution of this pattern is negligible to the modification of the weights connecting the input nodes to the hidden nodes. If this near zero error is not backpropagated to the hidden node, its weight updates are negligibly affected. This idea is being exploited in partitioning the training patterns in the training of the network. All the patterns which give near zero value to a hidden node are removed from the training set of that node. This considerably reduces the training task.

7.3. Input Space Decomposition

So far, we had considered feedforward neural networks with ellipsoidal activation functions and speedup of their training by network decomposition as well as training set decomposition. However, when the number of input nodes is very high, the complexity of network training will be dominated by the training of input-hidden nodes weights. One can reduce the dimensionality by considering only the structural close sensor measurements. There is an alternative method that utilizes only the distribution information of the training patterns. Recall that each of the hidden nodes describes an ellipsoid in the input space and approximates a part of the distribution of a particular class. Ideally, each of the principal axes of the ellipsoid should show how this data is distributed locally along those directions. If the distribution is oriented to the axes of reference, approximating this with a set of unoriented ellipsoids is not efficient. To use the power of representation that ellipsoids provide, we need to allow for the orientation of the ellipsoids generated. For representing an oriented ellipsoid in an n-dimensional space, the number of parameters N (center coordinates + axes lengths + axes orientation vectors) grows as $O(n^2)$. Since the network training is at least $O(N^3)$, it is imperative that we limit the growth of N with problem size. For large n, the problem of network training (determining the parameters of oriented ellipsoids) would be very time-consuming. To avoid this complexity, the number of parameters is traditionally kept at O(n) by fixing the orientation of the ellipsoids to be along the reference axes [6, 9]. Fixing the orientation of ellipsoids a priori reduces the advantage of using ellipsoids instead of spherical units. For example, in approximating an oriented Gaussian distribution, we would need many fixed orientation (along reference axes) ellipsoids. We introduce a method of computing the orientation, without increasing the complexity, by virtue of the fact that the network training is performed in two phases: namely (1) clustering phase (2) ellipsoidal unit tuning. Clustering phase determines the clusters present in a given distribution. Each of these clusters is then approximated by an ellipsoid and tuned. If the cluster is oriented, a fixed orientation ellipsoid is a poor approximation. If we can determine the cluster orientation, this orientation can be used as the fixed orientation of the corresponding ellipsoidal unit. This is a more sensible choice than the arbitrary choice of reference axes.

Let $x_1, x_2, ..., x_m$ be the original variables defining the cluster patterns. Clusters would be oriented, with respect to the reference axes in the **x**-space, if these variables are correlated in the patterns of the cluster. Orientation of the cluster can be determined through *principal component analysis* (PCA). It is essentially a mathematical technique involving transformation of the original variables to a new set of uncorrelated variables called principal components. Principal components are linear combinations of the original variables and can sometimes be identified as being meaningful in their own right. Let $y_1, y_2, ..., y_m$ be the principal components. Then:

$$y_j = a_{j1}x_1 + a_{j2}x_2 + ... + a_{jm}x_m \quad (j = 1,2, ..., m) \quad \text{or } \mathbf{y} = \mathbf{A}\mathbf{x}$$

with the condition that A is an orthonormal matrix i.e., $A^TA = I$.

Normalization of A ensures that distances are preserved between the patterns in the cluster so that transformation by A only results in orientation of the cluster. Further, orthogonality ensures that the new variables **y** are uncorrelated in the patterns describing the cluster. By performing this transformation from **x** to **y**, we have determined a new frame of reference where the variables are uncorrelated ie., the cluster which is oriented to the reference axes in the **x**-space is now oriented along the reference axes in **y**-space. For example, an oriented Gaussian distribution in **x** would be unoriented in **y**. Therefore, we can approximate the distribution in **y**-space by an ellipsoid oriented along the reference axes. Characterization of the transformation can be done the following way: Let Σ_x be the covariance of the cluster in **x**-space. Then:

$$\Sigma_x \Phi = \Phi \Lambda$$

where:

Φ = matrix of the eigenvectors of Σ_x

Λ = diagonal matrix of the eigenvalues of Σ_x

Now, consider the transformation: $\mathbf{y} = \Phi \mathbf{x}$.

$\mathbf{m}_x$ = mean of $\mathbf{x}$

$\mathbf{m}_y$ = mean of $\mathbf{y}$

$\Sigma_y = \Lambda$

Since Σ_y is diagonal, the y variables are uncorrelated. Thus, if we choose $A = \Phi$, we achieve the required transformation. This transformation to **y**-space chooses the directions of eigenvectors of Σ_x as the new reference axes.

There is another advantage to this transformation besides determining the orientation of the cluster. Because the measurements in a process are often redundant, ie. they are correlated, the number of "independent pieces" of information is much fewer in number than the number of sensors. Among these "independent pieces", only a few carry the significant information. Each of the eigenvectors is an "independent piece" of information. Magnitude of the corresponding eigenvalues gives the quantity of information (variance) associated with it. These remarks can be quantified as follows: Let us assume, without a loss of generality, that the eigenvectors are ordered in decreasing order of eigenvalues. Variance of the distribution along the jth principal component y_j is given by the corresponding eigenvalue : $\lambda_j = \mathrm{var}(y_j)$.

$$\text{Total variance in the data} = \sum_{j=1}^{m} \mathrm{var}(y_j) = \sum_{j=1}^{m} \mathrm{var}(x_j)$$

The proportion of variance accounted for by the first p principal components is :

$$\frac{\sum_{j=1}^{p} \mathrm{var}(y_j)}{\sum_{j=1}^{m} \mathrm{var}(y_j)}$$

If this proportion is reasonably large (say 95%), then the first p principal components provide an adequate description of the data. The rest of the components can be conjectured to consist mostly of noise. Not using the components with small variance serves to eliminate noise. Further, if the original variables are redundant, Σ_x is singular, some of the y_j would have zero variance ie., have no information along these directions. Considering only significant components considerably reduces the dimensionality of the problem (from m to p, where p is much smaller than m).

8. Fault Diagnosis of a Reactor-Distillation System

We will consider the fault diagnosis of a reactor-distillation process system shown in Figure 8, for illustrating the application of the techniques discussed so far. The CSTR processes reactant A to result in a mixture of A and product B. Two P-controllers are used to control the reactor temperature and holdup by manipulating the coolant flowrate and the product stream flowrate respectively. The reactor product stream containing the binary mixture of A and B is fed into a distillation column where it is separated in to a distillate stream with 98% B and a bottoms stream containing 2% B. Two PI-controllers are used to control the overhead and bottom product compositions by manipulating the reflux rate and the vapor-boilup rate, respectively. A total of ten sensor measurements are available. Details of the process model and the listing of faults considered may be found in Vaidyanathan [16]. The purpose of this case study is twofold. One, to perform fault diagnosis on a system that is of interesting complexity. Second, to examine the underlying structure of the fault space spanned by process measurements and to bring out the asymmetric nature of fault classes that is often found in large scale problems. We show how the redundancy in the data can be used to reduce the dimensionality of the data and why ellipsoids are well suited for representing the distribution.

A total of eight faults were considered for training the network. Training patterns were chosen from the simulation 40 minutes after the inception of the fault so that the fault signature is significant in the measurements. Training data was next analyzed for its correlation structure. For illustration purposes, we considered fault F_4 (increase in the temperature of the coolant feed) for the analysis. Data is analyzed for each fault by finding the principal components. Principal components are the eigenvectors of the covariance matrix of the distribution. To explain geometrically, they correspond to the direction of orientation of the distribution. For example, if the distribution is an oriented Gaussian distribution, principal components correspond to the directions of its principal axes. Corresponding eigenvalues give the variance along each of these directions. To allow for oriented ellipsoids, the number of parameters grow as $O(n^2)$, where 'n' is the number of measurements. If the orientation is fixed, the number of parameters grows only as $O(n)$. The first advantage of principal components is that they identify the data orientation so that by fixing the orientation of the ellipsoids along them, we not only use the orientation information but also limit the number of parameters to $O(n)$. The second advantage of using principal components is that they are uncorrelated unlike the measurements. This means that one can see how the data spreads along each of these directions independently. The spread on the direction of a principal component is given by the variance (corresponding eigenvalue of the covariance matrix). If the sensor measurements are correlated, and they usually are because of their redundancy, then the variance of the distribution is significant only along some of these directions. Along the rest, the variance is negligible or zero. This means that one can approximate the distribution very well by only considering the few significant principal components. By mapping the distribution from measurement space to the principal components, we can not only identify the orientation of the distribution but also reduce the dimensionality by ignoring the insignificant components.

Principal components were identified for the distribution of fault F_4. These components were ordered in the data of decreasing variance along those directions. The first three components are shown in Table 1. These three components account for 92% of the total variance in the data, which can be interpreted as 92% of the information in the distribution of the data. This means that the data in the distribution predominantly exists in the subspace spanned by these three dimensions. The remaining space (7-dimensional) has nearly no information on the distribution of the data. Distribution of the training patterns for fault class F_4 are shown in the space of these three significant components in Figure 9a.

In using a network with ellipsoidal units, a single ellipsoidal unit is used to approximate the distribution for each fault class, in the space of components. That is, the principal axes of the ellipsoid are now oriented along the principal components and not the axes of the measurement space. Because the ellipsoidal unit can have different lengths along its different principal axes, it is possible to limit the unit to the subspace spanned by the significant components with near zero spread along the remaining. If one were to use hyper-spheres instead of hyper-ellipsoids, we would be stuck with a large ten-dimensional sphere in this example as the hyper-sphere, due to its spherical symmetry, would span the measurement space with equal variance (i.e. equal radius) in all the directions. This leads to serious generalization problems. Figure 9b shows the distribution in Figure 9a being approximated with a single ellipsoid. In this case, clearly a single ellipsoidal unit approximated the class distribution. It is to be emphasized that the ellipsoid is actually in the 10 dimensions spanned by the components with near zero or zero variance along 7 of the dimensions. The largest axis of the ellipsoid is along the principal component with the largest variance, the second largest axis along the second component with the second largest variance and so on.

Table 1. Significant directions in data distribution for Fault 4

Significant Principal Components	Variance
$[0.091\ \ -0.075\ \ 0.206\ \ 0.203\ \ -0.127\ \ 0.932\ \ 0.134\ \ 0.019\ \ 0.013]^T$	0.981
$[0.414\ \ 0.352\ \ -0.508\ \ 0.046\ \ 0.154\ \ 0.041\ \ 0.5\ \ -0.196\ \ 0.354\ \ -0.092]^T$	0.125
$[-0.213\ \ 0.422\ \ -0.112\ \ -0.3\ \ -0.097\ \ 0.073\ \ 0.346\ \ 0.36\ \ -0.42\ \ 0.483]^T$	0.084

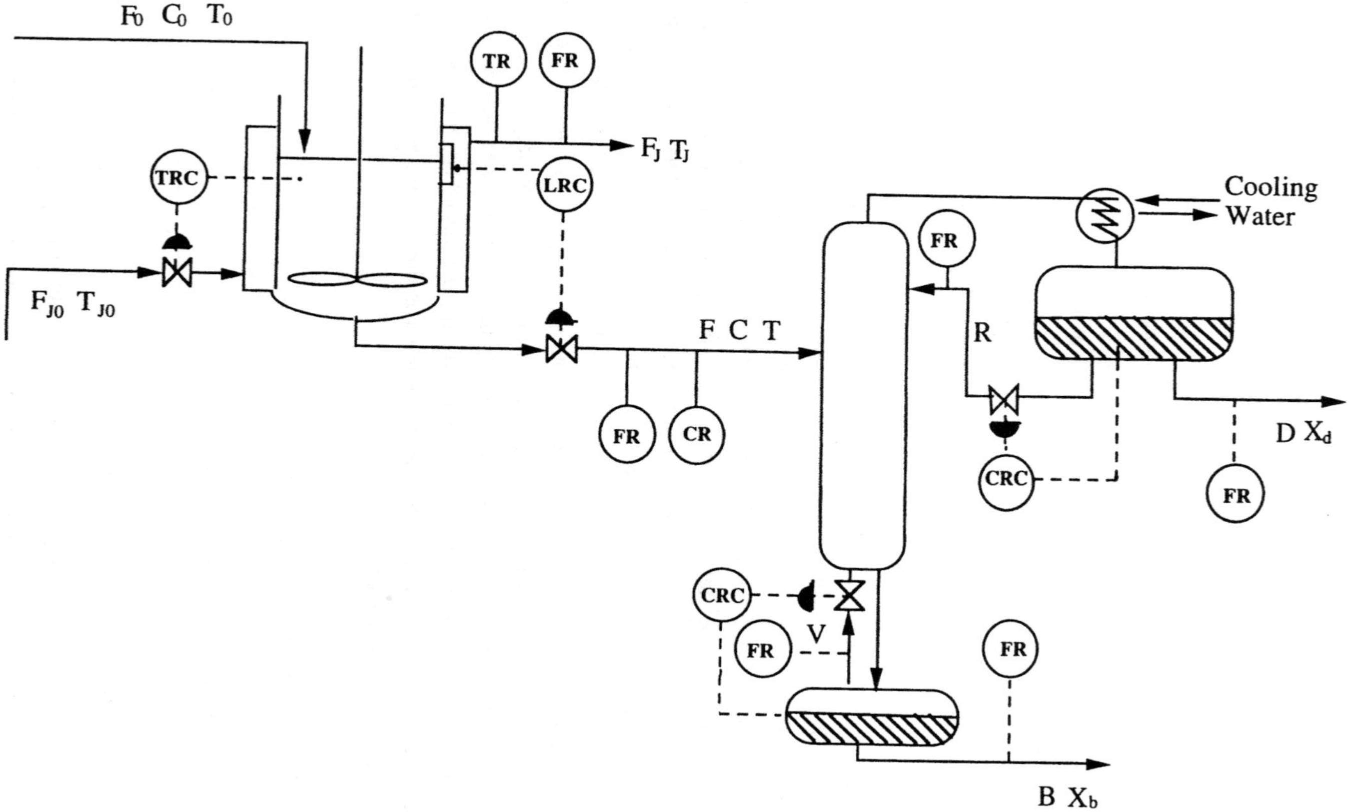

Figure 8. Reactor-distillation column case study

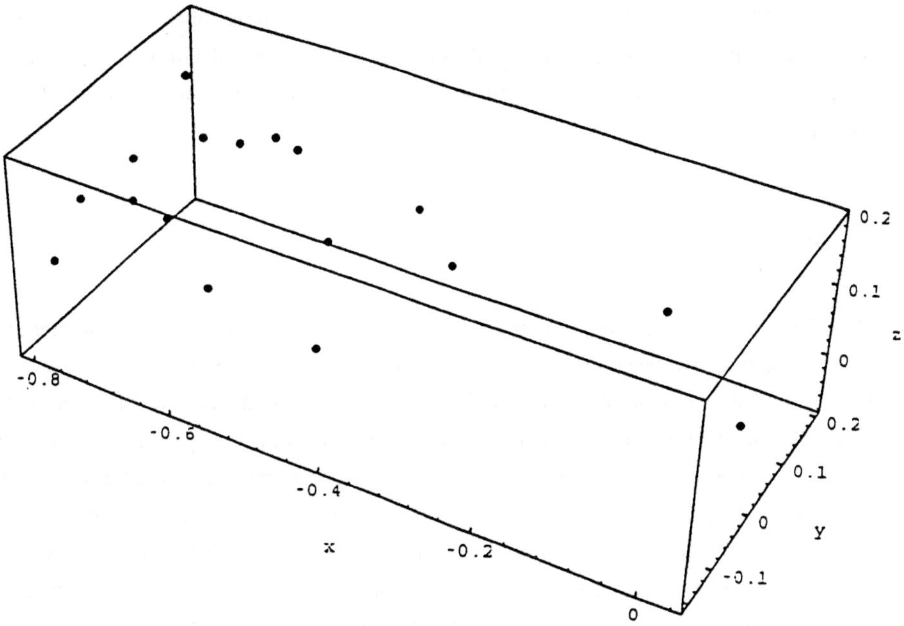

Figure 9a. Distribution of training patterns in the space of principal components for fault F4

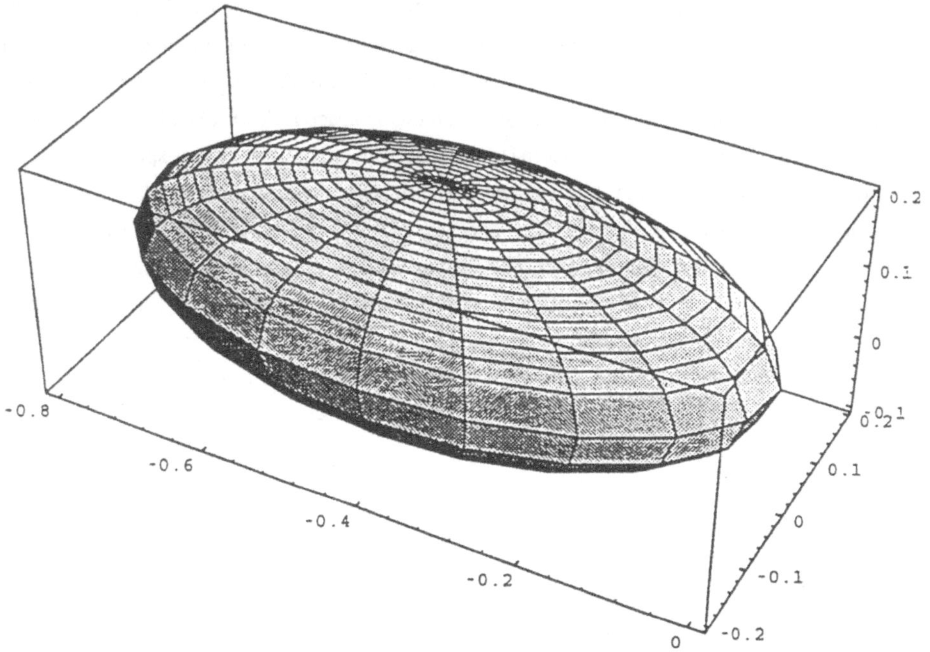

Figure 9b. Ellipsoidal approximation of the training patterns for fault F4

The same procedure was performed for all the other fault classes and a single ellipsoid was found sufficient to approximate each of the classes. Only 100 epochs of tuning were used for 100% correct classification. This demonstrates the efficiency of the representation of data by ellipsoids.

9. Summary

Classifiers have to meet the minimum requirements of robust classification and reasonable generalization for successful fault diagnosis applications. In this paper, we have shown that linear activation neural networks have serious problems in meeting these requirements and the use of ellipsoidal activation is more appropriate. After comparing unsupervised and supervised learning strategies, we propose an integrated learning scheme which has the advantages of both the approaches and avoids the problems of each. In this integrated approach, a fuzzy unsupervised learning is used to determine the features and the structure of the network. Then, supervised learning is used to perform credit assignment to various features by fine-tuning the network parameters. Decomposition techniques have been introduced to result in smaller networks with fewer training patterns. This makes the network suitable for large-scale process plants. Since the hidden nodes are not shared between classes, the network decomposes into smaller networks one per class. Since the ellipsoidal units are localized in the input space, only the training patterns in its proximity need be considered for tuning its parameters. This results in training set decomposition. Finally, the measurement space is decorrelated through principal component analysis. This helps decrease the dimensionality of the input space. A reactor-distillation column process system has been used to show the structure of the distributions in the measurement space for various faults and the suitability of ellipsoidal units was addressed in this context. We also point out that fault diagnosis of continuous and batch processes can be treated uniformly if the transient behavior is considered, ie. using trends to perform diagnosis instead of steady state values. Ellipsoidal units provide a means to develop envelopes which enclose the trends for individual fault classes. Network output values give a measure of how close or far we are from the envelopes of different fault classes.

Acknowledgments

The authors gratefully acknowledge the National Science Foundation (Grant ECS-9013349) and the National Institute of Occupational Safety and Health (Grant OH02740-A1) for their support of this work.

References

1. Batchelor, B., *Practical approaches to pattern recognition*, Plenum Press, New York, 1974.
2. Bezdek, J. C., *Pattern recognition with fuzzy objective function algorithms*, Plenum Press, New York, 1981.
3. Cybenko, B., "Approximation by superposition of a sigmoidal function, *Math. Control. Signals Systems*, **2**, 303-314, 1989.
4. Duda, R. O. and Hart, P. E., *Pattern classification and scene analysis*, Wiley, New York, 1973.
5. Fukunaga, K., *Introduction to pattern recognition*, Academic Press, New York, 1972.
6. Holcomb, T. and Morari, M. "Local training for radial basis function networks: Towards solving the hidden unit problem", 2331-2336, *American Control Conference*, 1991.
7. Hoskins, J. C. and Himmelblau, D. M., "Artificial neural network models of knowledge representation in chemical engineering", *Comput. Chem. Engng.*, **12**, 881-890, 1988.
8. Huntsberger, T. L. and Ajjimarangsee, P. "Parallel self-organizing feature maps for unsupervised pattern recognition", *Int. J. General Systems*, **16**, 357-372, 1990.
9. Kavuri, S. N. and Venkatasubramanian, V., "Using Fuzzy Clustering with Ellipsoidal Units in Neural Networks for Robust Fault Classification", *Comput. & Chem. Engng*, **17**, 8, pp. 765-784, 1993.
10. Kavuri , S. N. and Venkatasubramanian, V.,"Solving the hidden node problem in networks with ellipsoidal units and related issues", *International Joint Conference on Neural Networks*, Baltimore; June 1992.
11. Kohonen, T., *Self-organization and associative memory*, Springer-Verlag, Berlin, 1984.
12. Leonard, J. A. and Kramer, M. A. "Limitations of the backpropagation approach to fault diagnosis and improvement with radial basis functions", *presented at the AICHE Annual Meeting*, Chicago; November, 1990.
13. Moody, T. J. and Darken, C. J., "Fast learning in networks of locally tuned processing units", *Neural Computation*, **1**, 281-294, 1989.
14. Rippin, D. W., "The future of process operations", presented at the workshop Computer Aided Process Engineering, *World Congress of Chemical Engineering*, Karlsruhe, Germany, June 1991.
15. Ungar, L. H., Powell, B. A. and Kamens, S. N., "Adaptive networks for fault diagnosis and process control, *Comput. Chem. Engng.*, **14**, 561-573, 1990.
16. Vaidyanathan R, *Process fault detection and diagnosis using neural networks*, Doctoral Thesis, Department of Chemical Engineering, Purdue University, December 1991.
17. Vaidyanathan, R., and Venkatasubramanian, V., "Representing and Diagnosing Dynamic Process Data Using Neural Networks", *Engineering Applications of Artificial Intelligence Journal*, **5**, 1, pp. 11-21, 1992a.
18. Vaidyanathan, R., and Venkatasubramanian, V., "On the Nature of Fault Space Classification Structure Developed by Neural Networks",*Engineering Applications of Artificial Intelligence Journal*, **5**, 4, pp. 289-297, 1992b.
19. Venkatasubramanian, V., Vaidyanathan, R. and Yamamoto, Y., "Process fault detection and diagnosis using neural networks: I. Steady state processes", *Comput. Chem. Engng.*, **14**, 699-712, 1990.
20. Venkatasubramanian, V., "Recall and Generalization Performances of Neural Networks for Process Fault Diagnosis ", in *Proceedings of the Fourth International Conference on Chemical Process Control*, South Padre Island, Texas, Feb 17-22, 1991.

Overview of Scheduling and Planning of Batch Process Operations[1]

G.V. Reklaitis

School of Chemical Engineering, Purdue University, West Lafayette, IN 47907, USA

Abstract: Scheduling of batch operations is an important area of batch process systems engineering which has been receiving increasing attention in the last decade, especially in its role within computer integrated process operations. In this paper, we review the basic issues which scheduling methodology seeks to address and outline some of the reasons for the growth of interest in this field. The components of the scheduling problem are described and the main threads of the available recent solution methodology are reviewed.

Keywords: Batch size, campaign, cycle time, equipment, multiproduct plant, multipurpose plant, network flowshop, operating strategy, planning, reactive scheduling, recipe, resource constraints, storage, time discretization.

Introduction

The period of the last fifteen years has seen an increasing level of interest, research, and publication in production planning and scheduling methodology for the chemical processing industry. In this paper, we will examine this growth first by describing the basic issues which this methodology seeks to address and then outlining some of the reasons for this growth of interest. The components of the batch process scheduling problem will be described in detail and the main threads of the relevant research directed at its solution will be summarized.

---

[1]This paper draws extensively on a plenary lecture entitled "Perspectives on Scheduling and Planning of Process Operations", presented by the author at the Fourth International Symposium on Process Systems Engineering, Montebello, Quebec, Canada, August 5-9, 1991.

Distinction between Planning & Scheduling

In the chemical processing context, production planning and scheduling collectively refer to the procedures and processes of allocating resources and equipment over time to execute the chemical and physical processing tasks required to manufacture chemical products. Usually, the production planning component is directed at goal setting and aggregate allocation decisions over longer time scales measured in months, quarters or a year, while scheduling focuses on the shorter time scale allocation, timing, and sequencing decisions required to execute the plan on the plant floor. The division between these two decision problems is to a large degree a matter of tradition, reflecting the conventional hierarchical distinctions between corporate level planning and plant or production line level decisions to execute those plans. The difficulty with that division has been the absence of effective methods for aggregating/disaggregating information and decisions originating at one level and exploiting them at the other level. The conventional solution to this problem of coordination of the planning and scheduling levels in the presence of uncertain information has been to employ the rolling horizon strategy under which planning is performed for several time periods in the future, scheduling is performed for only the first period of the plan, reconciliation of demands and output are made when the period is completed, and the process is repeated.

While it is often convenient to make reference to various decision levels in planning/scheduling, the trend in computer integrated manufacturing is to fully integrate all levels of decision making, thus focusing on the interactions rather than the distinctions between levels. Therefore, for purposes of the present discussion, we will not differentiate between planning and scheduling, referring to the whole as the scheduling problem. We simply recognize that scheduling can be performed over different times scales from planning a campaign of runs, to developing a master schedule for a given campaign, to modifying the master schedule to respond to changes and unexpected perturbations which inevitably arise as the schedule is executed.

Scheduling is required whenever there is competition among activities for limited resources which are available over a finite time period. It involves three key elements: ***assignment*** of resources, ***sequencing*** of activities, and determination of the ***timing*** of the utilization of resources by those activities. The assignment component involves the selection of the appropriate set of resources for a given activity. The sequencing component concerns the ordering of the execution of activities assigned to resources, while the timing component involves the determination of specific start and stop times for each of the activities undergoing scheduling. Consider a plant with two reactors and six different products each requiring a batch to be produced. The assignment step may involve selecting the first reactor for the first three batches and the second reactor for the other three. The sequencing step would involves electing the order in which the batches are to be produced on each line and the timing component would, of course, involve the determination of the exact start and stop times for the production of each batch. These three components of scheduling lend themselves to mathematical description and computer based solution approaches. Indeed, scheduling techniques have been extensively investigated as a core area of the operations research domain since the 1950's, as documented in the classical books by Baker [2], Coffman et al [11], and the text by French [19]. Because the need for scheduling arises in so may different commercial and manufacturing contexts, various forms of the methodology have been widely investigated in many fields of engineering, computer science, management science, and business. The literature is thus quite large, diverse, and diffuse. In this paper, we will primarily focus on that portion of the literature most relevant to batch chemical processing applications.

Stimulus for Scheduling in the CPI

In contrast to the long term concern with scheduling in the discrete parts manufacturing industries, the attention devoted to computer aided scheduling methodology in the chemical processing industry and the chemical engineering profession is much more recent, beginning in the middle 1970's (see reviews [56] and [35]). The principal exceptions are refinery scheduling and distribution logistics LP applications which were already being investigated in the 195O's. The accelerating interest in scheduling methodology in the chemical processing context, particularly in the last decade, has been stimulated in part by major changes in the business environment faced by the industry as well as changes in the information and computer technologies which can be deployed to solve these highly complex decision problems. We briefly review these factors since they provide the environment within which advances in the field of chemical process scheduling must be pursued.

First, within the last decade, the CPI of the industrialized nations has changed dramatically and, perhaps irreversibly, as enterprises have recognized that they must operate on a world-wide basis to be competitive. Manufacturing operations must be globally coordinated, production must be highly cost-efficient with tighter inventory control and high capital facilities utilization. The quality movement in the industry has focused on responsiveness to customer needs, predictable product delivery, and consistency in meeting product specifications. Furthermore, the strategic plans of the leaders of the industry have identified more sophisticated chemical and biochemical products, with high value-added and short product life-cycles as the future of the CPI. The increased tailoring of products to specific customer needs has led to more product types and grades and lower production requirements for individual products. This has led to increased sharing and coordination of production resources to meet the needs of multiple products and less reliance on dedicated facilities. As a result, multi product batch and semicontinuous operations have become, if not more prevalent, (they have been historically widely used throughout the industry) more central to corporate long term business plans. Shared production facilities are no longer a temporary expedient to be replaced by a continuous dedicated plant when the market expands, instead, they have become a permanent reality.

Secondly, rapid changes in information technology have made essential enterprise information quickly and widely available. Automated, on-line plant information systems, no longer mere vendor novelties, are available to provide instantaneous equipment status reports. Computerized data base systems for handling customer orders, inventory status, and work in progress have made timely and informed order processing decisions possible. The CIM movement in the discrete parts manufacturing industries has drummed home the message that the electronically available enterprise information can and should be used to do something more than generate management reports and archive hourly, daily, or monthly production averages. Since unit level control systems have already been widely implemented in the chemical industry, it has become abundantly clear that the next levels of application lie in supervisory control and scheduling.

The advances in information technology have, if anything, been outstripped by those in computer systems. Continuously improving, low cost computing power and memory are beginning to make established but previously too costly solution approaches for scheduling problems such as minced integer linear programming or branch and bound methods, feasible for selected applications. Developments in graphical user interfaces offer promise in making comprehensible the extensive information required to capture the details of scheduling problems and make software easier to use for non-programmers. The advent of affordable high performance desk top computing make it possible to deliver effective interactive scheduling tools to plant personnel even in smaller production facilities. Finally, the realities and prospects of advanced architecture machines make feasible the rigorous solution of schedule optimization problems heretofore deemed intractable.

Clearly, the stage has been set for significant and rapid developments in the exciting field of process scheduling in the next five years. In the following sections, we will seek to define more precisely the directions of the most promising avenues. We begin with a more detailed examination of the scheduling problem and its various components and forms.

The Scheduling Problem

In its most general form, the scheduling problem consists of an operating strategy, a set of plant equipment, a set of resources such as operators, utilities, and materials, a set of product recipes and product precedence relations, specifications of resource availabilities and product requirements, and one or more criteria which must be optimized. In the following, we characterize each of these problem components:

Operating Strategy

A key decision which underlies any scheduling application is the strategy which has been selected for organizing the manufacturing process. As shown in Figure 1, [55], the choices of operating strategies for batch processing can be represented in terms of a two dimensional space which is characterized by degree of product recipe similarity and by the relative length of time of individual production runs or campaigns. Plants which must deal with products which have low recipe similarity and short runs (one or two batches) require different modes of organization than those which have high recipe similarity and long runs or campaigns (dozens of batches). It is reasonable to assume that the issue of selection of operating strategy will have been resolved at plant design or at the time of the most recent retrofit. Thus, for scheduling purposes, it will be assumed that the strategy has already been determined and therefore is a problem input. Additional more detailed aspects of operating policy which are implied by specific choices of operating parameters such as inventory targets and tardiness penalties also have to be treated as input parameters.

Equipment

The equipment items of the plant consist of processing units, storage tanks, transfer units, and connecting networks.

1) Processing equipment: These are the individual units available to the plant, often grouped into types which have equivalent processing characteristics. Equipment items are characterized by a nominal processing capacity and, possibly, a range of capacities over which operation is feasible. Batch equipment is characterized by volume and semicontinuous equipment by rates.

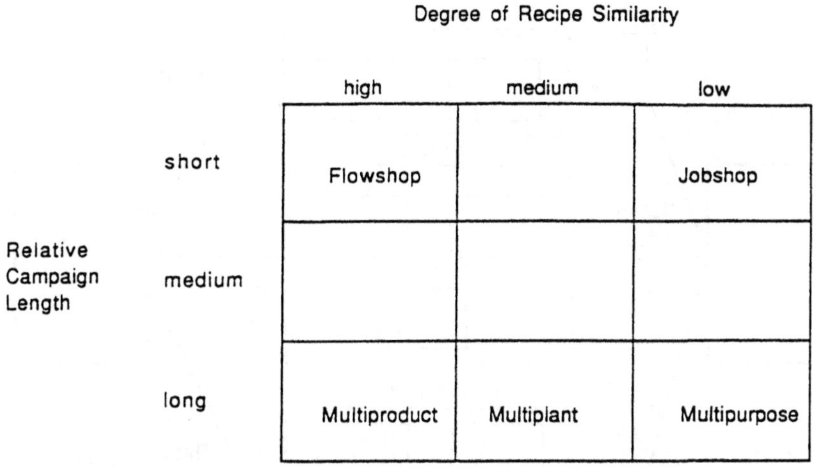

Figure 1. Operating strategy space

2) Storage: Intermediate, raw materials, and product storage tanks may be of different type and are characterized by volume.

3) Connections: the individual plant equipment items are linked through a network of lines which may be fixed or flexible. The network represents all feasible connections between equipment. Equipment is often organized into groups consisting of units used simultaneously (in-phase). Groups performing the same function are organized into stages, where groups in a stage are used sequentially (out-of-phase) as shown in Figure 2.

4) Transfer units: Transfer lines are devices for moving materials between process equipment and are characterized by a rate or possibly by an allowable range of rates. They may include conveyors, blowers, manned transfer vehicles or even autonomous vehicles. Transfer lines are associated with a specific set of connections, may be used on a shared basis, and thus may be entities subject to scheduling.

All of the equipment items may have release times and lists of block-out times associated with them. The release time for an item sets the time at which that item first becomes available for scheduling. Block out times are time periods during which an item is unavailable for processing because of maintenance or other requirements.

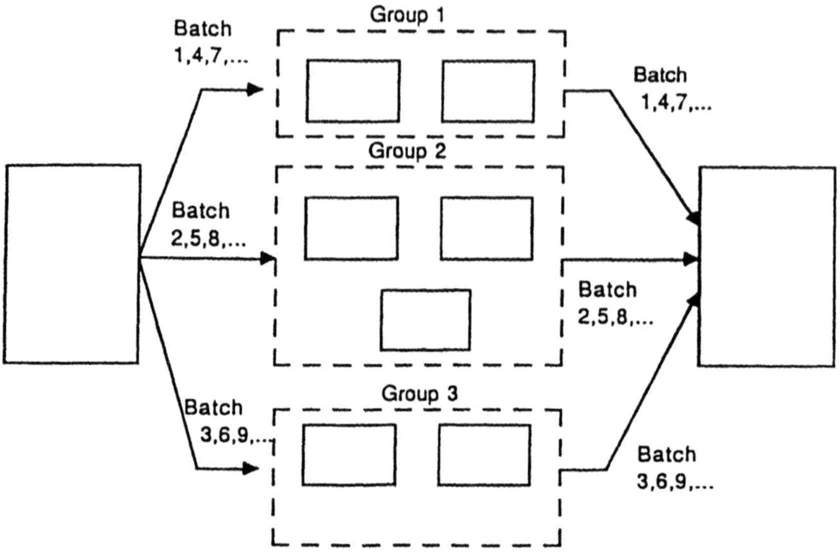

Figure 2. Generalized structure of a stage

Resources

Secondary resources are globally shared production inputs which may be either renewable or nonrenewable. Renewable resources are those inputs whose availability levels are restored to original levels immediately after usage. For example, upon completion of a task, an operator become available for a subsequent task. The most common renewable resources are manpower, electricity, heating and cooling utilities. Nonrenewable resources are those that are consumed and depleted in the process and thus must be replenished after usage. For instance, when a raw material charge is made to a unit, the inventory of that material is reduced and can only be restored through specific action, e.g., delivery of new raw material.

Recipe

A product recipe can be viewed as a directed network of processing tasks which must be performed to manufacture a given product. The nodes are the tasks while the directed arcs denote the precedence order among the tasks. While usually the network structure is fixed and independent of the system state, the possibility exists of conditional tasks, that is, tasks whose execution depends upon a state dependent decision. For instance, a reprocessing task may be initiated if a composition is outside of desired ranges or a devolatilization task may be executed if the need arises for temporary storage of the output of a given task.

Task

A task is a collection of elementary chemical and physical processing operations which must be completed in a serial fashion within a certain type or set of process equipment. Each task has associated with it a required processing time, a set of material balance or size factors, a final state, a stability condition, a storage condition, change over times, resource requirements, and resource substitution relations. Each these components may depend on the equipment and resources which are selected for that task.

1) Processing time: The task processing time (for batch operations) or rate (for semicontinuous operations) in general may be a function of the specific type of equipment employed, the amount of material processed, the level of assigned resources, as well as of selected state variables. The overall processing time of a batch task may contain transfer time components to reflect the time associated with material transfer steps which may be batch size dependent.

2) Feasible equipment list: a set of equipment types which can be used to execute the given task. In general, the list may be ordered by priority.

3) Size factor: the mass or volume which must be processed in a given task in order to produce a unit amount of final product. In general, there may be a size factor associated with each member of the feasible equipment list for a task.

4) State: The final state of a task is characterized by composition, phase, temperature, and other appropriate system variable values.

5) Stability condition: The stability condition indicates whether the material is unstable and thus must be immediately processed (Zero Wait, ZW), can be held for a specified period of time (Finite Wait, FW), can be held indefinitely (Unlimited Wait, UW), or can be made UW upon the execution of a further processing task (Conditionally Stable, CS).

6) Storage condition: The storage condition reflects the material holding options available upon task completion: storage only within the processing unit (No Intermediate Storage, NIS), storage only within an external storage vessel or specified set of vessels (Finite Intermediate Storage, FIS), both of the above options (Conditional Finite Intermediate Storage, CFIS), or

unlimited external storage (UIS).

7) Change-over times: At the completion of a task there is a clean-out time which is incurred before the equipment assigned to that task can be employed to process another task. In general, for a given task, the change-over time will depend on the successor task and the specific equipment item employed.

8) Resource requirements: The resource requirements are the amounts or rates at which the various resources which are necessary for the execution of the task are consumed. Although often assumed to be constant, these requirements may vary over the duration of the task in a specified fashion.

9) Resource substitution relations: The requirements for specific resources can sometimes be met through several alternate means according to specified priorities. For instance, a task may require one of a certain class of operators but, if none is available, a more highly skilled, swing operator may be employed. Substitution relations can be expressed in the form of a set of ordered lists where the first member of the list is the normal choice of resource type while the subsequent members are the allowed ordered substitutes.

In general, each task will itself be composed of a series of individual chemical and physical steps, such as filling, mixing, heating, addition of a reagent, reaction. cooling, decanting, addition of a solvent, and emptying. Each of these individual steps may be defined in terms of a set of specific characteristic values of the above eight components. Thus, the recipe description can extend beyond the network of tasks to a detailed description of each task with multiple input and output streams associated with each task.

Product Precedence Relations

Product precedence relations are directed graphs consisting of nodes representing the products and directed arcs denoting the required (partial) order in which products need to be produced. For instance, a given product may be an intermediate which is required in the production of one or more other products and thus may need to be produced prior to those products.

Resource Specifications

The availability level of each resource type is a function of time, which may be constant over the entire time period under consideration or could be a periodically repeating or a general piecewise nonlinear function.

Demand Specifications

Product requirements are given as a set of orders, where each order has associated with it the identity of the product, the required amount, an earliest start time for the production of the order, a due date, and a priority in case of tardiness.

Performance Measures

The most general measure of the effectiveness of a schedule is an economic one which includes components such as sales revenue, task specific operating costs based on the consumption of resources, labor, and materials, inventory carrying charges, change-over cost, and order tardiness penalties. Often, direct plant performance measures such as time to complete all orders or the sum of the tardiness of all orders are used for simplicity.

Scheduling Problem Solution

The solution to the batch/semicontinuous plant scheduling problem will in general consist of three components:

1) A production plan which will indicate the sequence in which the orders will be processed and the time frames within which orders or campaigns will be completed.

2) The specific assignment of equipment items and secondary resource levels to individual production tasks.

3) The detailed schedule of operations expressed as starting and completion times of each task, the predicted distribution of inventory levels over time, and the detailed profile of resource utilization levels over the scheduling period.

The specific details of these three components will differ considerably depending upon the particular strategy under which the plant is operated.

Special Problem Forms

To facilitate the subsequent review of methodology, it is expedient at this time to outline some of the more important special cases and forms of the resource constrained scheduling problem (RCSP). The classification of these special cases can be made according to the presence or absence of active constraints on resources other than equipment and the mechanism employed for dealing with uncertainty in the scheduling data. Within each of these subdivisions, the cases can be further divided on the basis of the operating strategy which is employed. Finally, one can also consider an extension of the general RCSP, the multiplant problem, in which scheduling encompasses the coordination of a number of geographically distributed and interacting production facilities.

Equipment vs Resource Dominance: While the resource constrained problem is clearly the more general, there does exist a universe of applications in which the principal limiting resource is the process equipment. This so called equipment dominant case has received the bulk of the attention in the scheduling literature, including that devoted to chemical processing applications. The next level of complexity is the case involving shared intermediate storage. The presence of shared storage, for instance, a set of storage vessels which can be employed for material from several different products and tasks, introduces the need to consider the impact of the availability of sufficient storage resource on the start and stop times of tasks which will require storage. Since the time over which storage is required is not fixed as a property of that resource or of the product recipe but is driven by the needs of processing units, shared storage introduces a coupling between processing units and tasks that is not present in the equipment dominant case. Conceptually, this coupling requires that the usage of storage be checked at each point in time at which there is a change in the assignment of an equipment resource. The coupling induced by multiple shared resources makes such checking even more burdensome and contributes significantly to the complexity of the general resource constrained problem.

Rolling Horizon vs Reactive: The classical scheduling problem form is that in which all of the production requirements, equipment status, and resource availability profiles are assumed known at the time the schedule is generated. In many manufacturing environments, however, the equipment and resource availabilities change frequently and unpredictably, task processing times are subject to variations, order information is amended and new orders accepted continuously. Thus, while a master schedule generated based on a priori information is an important planning tool, the schedule must be almost continuously updated and revised. In principle, these situations can be addressed using two limiting approaches: a moving horizon rescheduling strategy and a reactive scheduling strategy.

In the first approach, an initial schedule is generated and then a form of moving horizon approach is used in which the schedule is periodically regenerated whenever a sufficiently large number of new events or inputs are encountered. Rescheduling can be done using an a priori scheduling procedure which only considers tasks which have not yet been completed. Each new schedule generated in this fashion will generally result in some resequencing and reassignment of equipment and resources and, thus, will lead to some degree of schedule "chatter" or "nervousness" as orders with intermediate due dates are repeatedly juggled in successive schedules. Such repeated schedule changes can be undesirable because, rather than mitigating the effects of process upsets and order changes, they, in fact, induce further changes. For instance, materials requirements plans, work assignments, shipping schedules, and other associated supporting functions, which have lead times of their own, may have been established based on an earlier schedule and could be significantly impacted upon rescheduling. In many chemical industry applications, it is highly desirable or even essential to minimize undue changes in the plans for preparatory tasks and supporting activities to avoid the associated costs and delays.

The second approach seeks to reduce system nervousness by using a master schedule as the basis for planning followed by a reactive scheduling strategy which responds to each processing variation, equipment or batch failure, or order change by appropriately readjusting the master schedule in a least cost or least change way. A range of adjustment modes can be employed: readjustment of only task start and stop times, reassignment of intermediate storage capacity, reassignment of equipment items or resource types, and resequencing of tasks or orders. In specific instances, one mode will be preferred to another. Thus, the selection of modes might be made so as to minimize the overall cost of the change or to minimize the difference in the completion times of orders between the master schedule and the revised schedule .

Operating Strategy Alternatives: The three limiting cases of operating strategies which have received the most attention are the multiproduct plant, the network flowshop, and the multipurpose plant.

As shown in Figure 3, the multiproduct strategy is employed for products with highly similar recipes and long product campaigns. Since the equipment network structure is fixed and all products essentially visit the same equipment in the same order, the scheduling solution consists of only the first and third components. The production plan consists of the determination of the order in which campaigns of individual products are made and their length (lot size), allowing multiple campaigns of the same product in order to reduce inventory charges. At the lowest level,

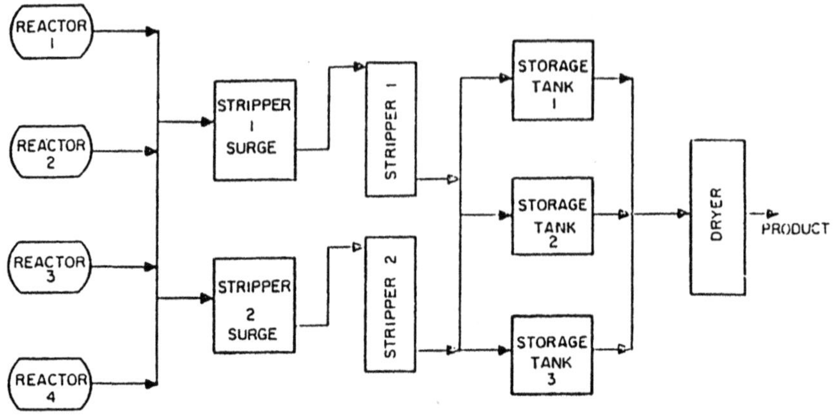

Figure 3. Multiproduct plant

the detailed schedule simply reduces to a set of single product line schedules. Each line schedule consists of the start and stop times for a sequence of batches of the same product sweeping through the fixed equipment network.

The network flowshop is appropriate for the case with similar product recipes and short campaigns. The schedule for this case is similar to the multiproduct strategy, except that there is no campaign structure, rather each product batch is treated as a separately scheduled entity. Thus, the schedule consists of the order in which the individual batches are started in the plant together with start and stop times for each batch on each of the units that it visits.

The multipurpose plant strategy is used for products with dissimilar recipes and longer campaign lengths. Schedules under this strategy will involve all three solution components: the production plan, the production line configurations used in each campaign, and the detailed production line schedules. The production plan identifies the campaign structure, that is, the sets of products which are to be produced at the same time, the campaign sequence, and the campaign durations. The production line configuration or campaign formation portion involves the assignment of specific equipment items to specific product tasks. These assignments serve to define the individual production lines which will be used in executing the campaigns. The line schedules themselves again reduce to single product schedules. This organization is illustrated in Figure 4. As shown, the first campaign involves products A and B, while the second processes A and C. In each campaign the plant equipment is reorganized into two different production lines.

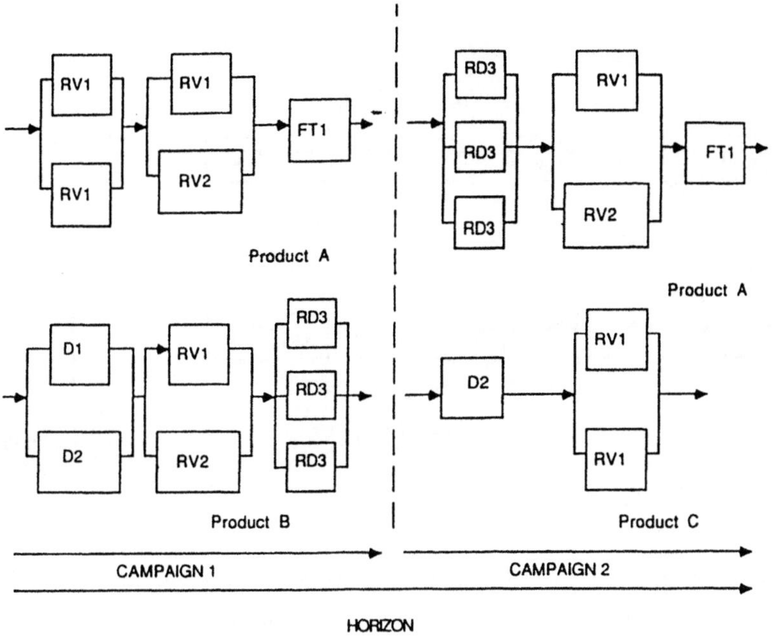

Figure 4. Multipurpose plant

Note that in both the network flowshop and the multiproduct cases it is not essential that all products employ all equipment. Specific paths may be assigned to specific products or product families. For instance in the batch resin process studies by Ishikawa et al [25] and shown in Figure 5, products may follow five different paths which use units 1 through 7; 8 and 2 through 7; 9 through 16; 8, 10 through 16; and 8, 10, 11, 12, 17, and 16.

In practice, it is not uncommon to find mixed strategies in which the above limiting strategies are combined. For instance, a portion of the plant may operate using dedicated lines, another portion may be operated in the multipurpose mode, while the last portion may operate like a multiproduct plant. Alternatively, the particular strategy used at one point in time may change as demands and product mix change.

The Multiplant Extension: In many instances, products and major intermediates are made using multiple plant sites which may be geographically widely distributed and may exchange intermediates. However, each plant is operated under its own strategy and schedule. In this case, the scheduling problem must include an extra decision level which makes the assignment of the

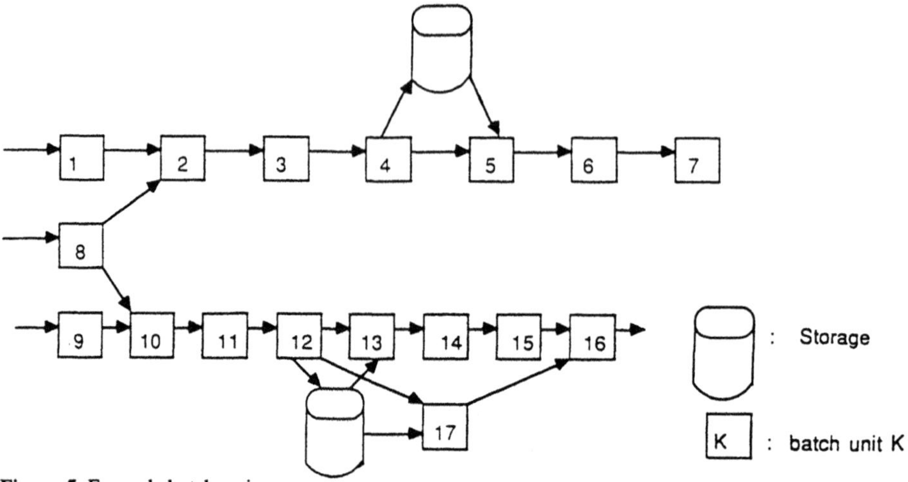

Figure 5. Example batch resin process

different products and intermediates to different plants, tracks inventories, and accounts for the time losses due to shipping of materials between the plants. However, this extra decision level must be closely coupled to the specific plant and production line schedules, since the production rates achieved in the individual lines will affect the overall assignment of products to specific lines in specific plants and vice versa.

Before we begin with a review of the approaches to solving these various forms of the general problem, we briefly summarize some of the difficulties attendant to the scheduling problem and its solution from both a computational and a practical point of view.

Impediments to Scheduling Applications

Scheduling applications have very large information demands and lead to optimization problems of considerable computational complexity. Engineers attempting scheduling studies are often stymied because of the great diversity of reported problem forms, specialized solution approaches, and nonstandardized terminology, generally are handicapped by the lack of convenient and reliable off-the-shelf solution tools, and can be hamstrung by organizational impediments.

Information Load: Large amounts of information are required for the specification of the problem: commercial data, equipment data, recipe data, and plant status information. This information load directly contributes to the complexity of the problem and has materially impeded the adoption of scheduling methodology in practice. Unquestionably, the key first step towards routine computer assisted scheduling is having the appropriate database of current and reliable plant and order status information.

High complexity and dimensionality: The resource constrained scheduling problem (RCSP) is inherently a high dimensionality, mixed discrete/continuous domain optimization problem (MINLP) which, in its general and even in most of its simpler forms, remains among the most severe challenges for optimization methodology. Mathematical programming formulations for resource constrained scheduling problems have been under study since the middle sixties but solution approaches to problems of practical scope have in the past been stymied because of the formidable combinatorial complexity of the problem.

Many forms and special cases: In practice, the RCSP exists in many different forms, with very particular structural constraints and site specific operating restrictions and rules. As noted above, the research community in turn has proposed for investigation a number of idealized problem forms as a means of identifying and exploring elements of solution procedures for more realistic forms of the RCSP. Unfortunately, these two sets of problem forms have not always enjoyed significant commonality. In addition, a very diverse, nonstandardized terminology has arisen for describing problem features. These factors have led to undue confusion and entropy in the field and have served to discourage practitioners from exploring and exploiting the substantial knowledge base that does exist.

Paucity of quality software: Although in recent years there has been much research on solution approaches, the development of high quality process scheduling software has lagged considerably. Available commercial quality scheduling software typically only offer LP, at best MILP, or heuristic dispatching/assignment rules as solution vehicles. MILP and even heuristic methods are quite restricted in the size of applications which can be successfully attacked. Several good graphical display tools to support scheduling have been announced recently but these either leave the main scheduling decisions to the user or offer only very simple heuristic schedule construction mechanisms.

Organizational barriers: A team undertaking a scheduling study often encounters considerable impediments arising from the need to cross various traditional jurisdictional boundaries between the corporate planning and logistics functions, the sales and marketing arm, plant management, and direct plant supervision. Management support can be difficult to garner because the benefits from better scheduling are difficult to predict a priori. Operating supervisors are often too busy fighting fires to invest the time on another plant wide application and may be reluctant to allow that the operation, in fact, requires improvement. Operating staff are suspicious of "help" to improve their productivity, viewing such efforts as interference in their well-established routines.

Although much remains to be done, these difficulties can all be mitigated sufficiently so that at the present time, using existing methodology, practical scheduling applications can be expeditiously implemented and made operationally successful. However, the implementation of such applications does require strong management support and a fair level of expertise in both scheduling problem formulation and its solution.

Status of Methodology

In this section, the main approaches proposed for the solution of various forms of the general scheduling problem will be reviewed. No claims are made of providing an exhaustive catalogue of all references on the subject. Instead the focus is on contributions which are representative of the available approaches to solving an interesting class of problems and are relevant to chemical processing applications. We first consider deterministic scheduling, beginning with the equipment dominant case and continuing with the resource constrained case. This is followed by a discussion of reactive scheduling approaches.

Equipment Dominant Scheduling

The literature on deterministic scheduling of equipment dominated plants can be separated into approaches dealing with the network flowshop, the multiproduct plant, and the multipurpose plant. By far the greatest portion of the literature has focussed on vanous forms of the flowshop. The three limiting forms of this problem, the single unit problem, the simple serial flowshop and the singlc stage parallel system have been studied quite extensively as descnbed in the survey papers by Gupta and Kypansis 124], Graham et al. [21], [Graves, [231, and Lawler, et al. [40]. Results from complexity theory [20] indicate that most instances of these limiting problems belong to the class of NP complete optimization problems. Consequently, it is conventionally argued that effective solutions to these problems and the more general network flowshop can only be expected via heuristic methods. This generally has been the case, although recent work by Miller and Pekny [44] with the classical traveling salesman problem offers clear evidence that rigorous solution of NP complete problems is possible for problem instances of practical size by careful exploitation of problem structure, effective bounding procedures, and the use of parallelization. As shown by Pekny et al. [52], the single machine shop and the serial flowshop under zero wait restrictions, both with sequence dependent change-over costs, transform to asymmetric traveling salesman problems and, thus, large instances of both problems (100 plus batches/jobs) could be routinely solved. The ATSP solution approaches developed by Pekny and Miller are now routinely used by DuPont to sequence processing lines to considerable economic advantage.

Network Flowshop: The general network flowshop problem has to date received relatively little attention, however, various special cases have been investigated, especially in recent years. Recall that under this operating policy it is assumed that each product is only produced in a few batches and, thus, scheduling involves consideration of each batch as a separate entity. Multiple batches of the same product are thus effectively treated no differently than those of other products. Furthermore, since under this strategy all products are assumed to follow the same recipe structure, then under the assumption of no reuse of any equipment item by different tasks of the same product, tasks of a recipe and the stages of the network become equivalent structures. Consequently, the task stability and storage conditions are merged and described as properties of the stages of the equipment network Thus, for example, the UIS, NIS, ZW and FIS conditions are all treated as stage and/or network properties.

The available literature on network flowshop scheduling can be divided into efforts focussed on networks operating under UIS or ZW conditions and those operating with finite storage.

UIS/ZW Networks: Most of the work in this area has been confined to networks in which the parallel units in a stage are identical and are employed out-of-phase. The first effort in this category is that due to Salvador [62], whose investigation appears to have been motivated by a nylon manufacturing application. In this work, an exact algorithm for minimizing makespan for ZW networks with identical processors within each stage was proposed. A branch and bound algorithm was used to determine the optimal sequence, while a dynamic programming subproblem was solved at each node to determine the completion time. No computational experience was reported, but clearly the computation time can be expected to be severe for problems of practical significance.

Kuriyan and Reklaitis [37,38] have investigated approximate algorithms for the network flowshop operating under UIS or ZW conditions, with the objective of minimizing makespan. A two level approach was proposed which decomposes the problem into a simplified sequencing subproblem followed by a sequence evaluation subproblem. The batch processing sequence is determined using heuristic sequencing algorithms drawn from those available for simple serial flowshop and parallel network problems, namely, dispatching rules, bottleneck sequencing, local search sequencing, and best fit sequencing. The sequence evaluation subproblem is solved using two different strategies: one product at a time (PAT) and one stage at a time (SAT). In the former case, a product (batch) is assigned to a unit at every processing stage before another product (batch) is considered. In the SAT scheme, every product is assigned a unit at a stage before assignments for subsequent stages are determined. Computational tests with a broad series of numerical test problems in which 20 products were scheduled on 2, 3, and 4 stage networks, "hourglass" networks, and parallel networks with randomly selected processing times, support the following conclusions. First, the best fit and local search procedures are the most effective sequencing procedures. Second, the quality of the initial sequence can significantly affect the performance. Since the difference between the PAT and SAT methods are not significant, the former is recommended because it can be used for both UIS and ZW networks.

Citing the results of a survey of 20 companies, Musier and Evans [45] proposed the investigation of multistage network flowshops with intermediate storage between each stage of parallel units, operating under the objective of minimizing total tardiness of orders. As first step towards the solution of this form of the network flowshop problem, an approximate procedure, called the Heuristic Improvement Method, was developed for dealing with the single stage parallel system. The procedure considered all single and pairwise interchanges of products (batches) to improve an initial sequence. Results were reported for problems with up to 100 batches and 12 parallel units with a maximum deviation of 1396 from optimality. The performance of the method declined as the number of batches and parallel units increased, presumably because of the significant growth of local optima (with respect to pairwise exchanges).

The extension of the method to multistage processes with UIS between stages and the total tardiness criterion has been reported by Musier and Evans [46]. The extension involves the repeated application of the single stage heuristics in a reverse flow direction for a specified number passes through the network. As is true with virtually all heuristic approaches to parallel and network flowshops, the method can readily accommodate nonidentical parallel units and restriction of units to specific products. Computational results with randomized problems involving four stages with up to seven units per stage, up to 280 batches, and integer problem data indicated that good solutions can be obtained, although clearly optimality can not be claimed. Overall the quality of the solutions is difficult to assess since no solution lower bounds are reported.

Flowshops with Finite Intermediate Storage: Most batch chemical plants operate under finite storage (FIS) capacity constraints and may require different storage conditions at each stage of the process. Storage represents the most direct example of a shared resource and thus investigations of even simplified forms of scheduling problems with storage restrictions are important for insights that they offer to the general RCSP. To date most of the finite storage work has focussed on the simple serial case.

Early work on FIS scheduling addresses the two unit serial system, where storage capacity is measured in number of batches. Dutta and Cunningham [15] proposed an exact algorithm using dynamic programming. However, because of heavy computational demands, two approximate methods were also developed. Papadimitriou and Kanelakis [51] showed that the two unit FIS problem is NP-complete and proposed a two step solution strategy in which the product sequence is determined by solving the ZW problem and the FIS schedule is determined using an unspecified completion time algorithm. Wiede et al [77], developed an optimal polynomial time algorithm for the completion time calculation of the two-stage FIS case. When combined with approximate methods for determining product sequencing using the related UIS or ZW problem, a two step solution approach resulted which yielded makespans on average within 2.4% of the optimum makespan.

Wiede and Reklaitis [75] further investigated the multi-unit FIS problem where the available storage capacity is shared by all stages of the production line. An approximate solution method which finds the storage assignment resulting in the earliest completion time for each product was proposed for the makespan minimization problem. Products were scheduled one at a time but in the worst case the storage feasibility check required multiple passes thus raising the computational complexity of the completion time procedure to $O(M^2N^2)$ where M denotes the number of stages and N the number of products. Wiede and Reklaitis [76] extended the basic logic of this approach to accommodate the Mixed Intermediate Storage (MIS) case in which different stages operate under different ZW, NIS, UIS, or FIS strategies. Ku and Karimi [32] reported a completion time algorithm which reduces the computational complexity to O(MN) by only making feasible storage assignments and thus eliminating the storage feasibility check. This approach was extended to accommodate the mixed intermediate storage case. Ku and Karimi [34] also examined the less general FIS problem in which finite but dedicated storage units occur between process stages as well as the CFIS case. An exact solution approach using recurrence relations for the completion time calculation and an MILP for sequence determination were proposed. Because of excessive computational demands, an approximate approach was also developed using local search procedures. These ideas have been extended to the MIS case with transfer and set-up times by Rajagopalan and Karimi [54].

The only study to date to deal with the scheduling of multi-unit flowshops under a tardiness based performance function is that of Ku and Karimi [33]. In this work, a branch and bound solution method is proposed for the UIS serial problem with the goal of minimizing weighted tardiness. Computational costs for this approach were found to increase exponentially so that solution of problems with more than 15 products was found to be impractical.

Finally, a study of a specific network flowshop consisting of two stages of parallel processors with nonidentical units and finite intermediate storage was reported by Kuriyan et al [39]. In this work, the storage contents was updated in terms of actual quantities rather than as integer number of batches. The problem was treated by using a simplified model for sequence determination and a discrete/continuous plant simulation to predict the completion time of the schedule. It was shown that the best fit heuristic using as initial list the products ranked by decreasing order of differences between first and second stage processing times was most effective among a suite of sequencing heuristics. The study demonstrated that a simulation can be effectively used as a practical completion time calculation mechanism which can readily accommodate plant specific constraints and monitoring of in-process inventory levels.

Multiproduct Plant: The key distinction of the multiproduct plant from the network flowshop is that many batches of each product are assumed to be required, thus operation is organized into longer campaigns. The principal advantage of longer campaign operation is that it can be assumed that, within a given campaign, production will occur in regularly repeating patterns. Such repeating patterns obviate the need for scheduling of individual batches. Instead, the structure and parameters of one instance of the repeating pattern can be used to characterize the production rate for the entire campaign. Thus, the scheduling problem is reduced to determining the characteristic parameters of the repeating batch processing pattern for each campaign and the duration of the individual campaigns. The available literature can be divided into those approaches in which the repeating pattern involves batches of the same product (single product campaign) and those in which the repeating pattern involves batches of more than one product (mixed campaigns). Within each category, the work can be differentiated based on whether or not the determination of campaign length is explicitly considered as part of the scheduling procedure.

Single Product Campaigns: The classical form of the multiproduct plant involves operation in long single product campaigns where within each campaign batches are processed in a periodic fashion described by a characteristic cycle time and batch size. The repeating pattern is thus one batch of the given product. The ratio of batch size to cycle time for each product constitutes the production rate for that product and therefore campaign lengths can be determined from product demands by simple calculation. If sequence dependent change-over losses must be considered, then the order of the campaigns can be determined by using single facility sequencing methods [24] such as solving a mixed integer linear programming formulation. These models can also readily accommodate inventory considerations and, thus, lot sizing decisions, that is, breaking of long single product campaigns into several shorter campaigns so as to optimally balance inventory holding costs against change-over costs.

The key parameters of batch size and cycle time for each product are readily computed for the simple case of constant task/stage processing times, no reuse of equipment items by multiple tasks of the same product, and identical parallel groups for each stage. However, in the presence of batch size dependent processing times and transfer rates and unequal groups, as first noted by Mauderli [43], the batch size and cycle time of a batch can become dependent upon the particular set of groups that are assigned to that batch. If the groups in each stage are used in a repeated cycle, with each group used once in each cycle in fixed order, then batches will be produced sequentially along NP paths of the production line, where NP is the least common multiple of the number of groups in each stage. The repeating pattern in this case becomes the sequence of NP paths called a simple path sequence. In general, depending upon the specific order in which groups on the different stages are assigned to batches, different simple path sequences can be generated, each with a different overall average processing rate. For instance, in the four stage example shown in Figure 6, the second and third stages each contain two groups, the B and C units constituting equipment used in-phase. Since any simple path sequence will contain two paths,

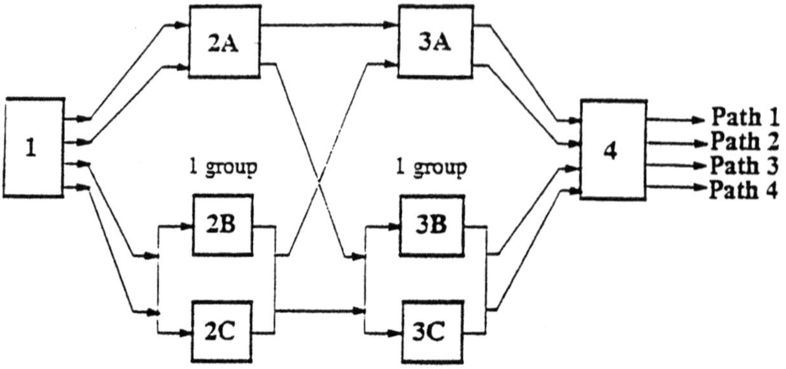

Figure 6. Path sequencing

paths 1 and 4 and paths 2 and 3 constitute two simple path sequences.

Wellons and Reklaitis [69] developed a mixed integer nonlinear programming formulation for the problem of determining the simple path sequence for the general single product production line which maximizes the average production rate of the line. Since combinatorial analysis showed that the formulation is subject to a large number of degenerate solutions because of rotational and operational equivalences, (Wellons and Reklaitis [70]), an explicit enumeration procedure was developed for identifying only unique simple path sequences. Furthermore, a reformulation of the MINLP was reported which restricts the feasible region of the model to distinct path sequences and is much more efficient for larger applications, (Wellons and Reklaitis [73]). Interesting general observations derived from this study are first that the production rate of lines with unequal groups can be significantly increased through the use of simple path sequences as the basic repeating unit for the campaign. Second, in the presence of variable processing times, the optimum batch size for a line, whether using a single batch size or path dependent batch sizes, can in general be less than the maximum batch size allowed by the capacity of the units in the line. Thus, significant production rate enhancements can be obtained by proper choice of repeating batch pattern and optimized batch sizes.

Mixed Product Campaigns: As noted by Wittrock [78] in the case of flexible manufacturing systems, the productivity of a flowshop can be increased by replacing a series of long single product campaigns with combinations of batches of several products which are repeated in a periodic fashion. Specifically, consider three products, A, B, and C, which are to be produced in a three stage network in campaigns of three batches each (Birewar and Grossmann [4]). If the time-limiting stage is different for each product and if there are no time losses due to changeovers, then as, shown in Figure 7, production in a cycle of one batch each of products C, B, and A, results in a reduction in the overall makespan to 42 hours in the case of ZW operation and to 38 hours in the case of UIS operation. Of course, if time losses due to change-overs/clean-ups are large enough, then the benefits of this so-called mixed product campaign (MPC) mode can be lost.

The implications of this strategy on the design and operation of multiproduct plants were investigated by Birewar and Grossmann [4], who considered the plant with a single unit per stage and Cerda, et al [8], who report heuristic procedures to guide the introduction of parallel units. In the former work, it is shown that for sufficiently large numbers of batches, makespan minimization can be replaced by overall cycle time minimization. MILP formulations suitable for determining the optimal cycle which included one batch of each product were reported. This work

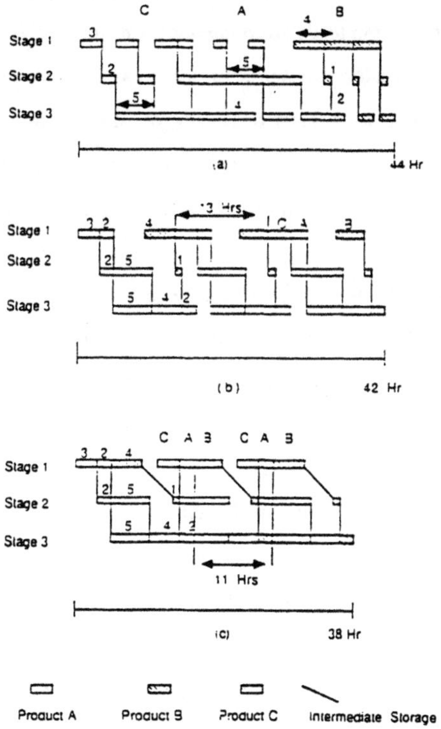

Figure 7. Scheduling with alternate cycle structures

was extended in [3] to allow consideration of more general repeating patterns involving multiple instances of batches of the same product by using a representation in terms of pairs of consecutive batches of products. This led to a compact LP formulation for determining the number of pairs of each type present in the optimal cycle and to graph constructions for devising specific cyclic schedules. While the work reported to date is effectively confined to single unit process trains and cases where the number of products is small compared to the total number of batches to be produced, the MPC concept merits further investigation as it may well be an effective production strategy for new batch plant concepts such as plants using autonomous transfer vehicles for which change-over costs are less significant.

Scheduling with Lot Sizing: In the classical multiproduct case, campaign lengths are governed by aggregate demands and, thus, only the order of campaigns is significant. The combined slack times and clean-out times between each potential pair of successive campaigns can readily be precomputed given the cycle schedules for each campaign. Thus, the optimal campaign sequence can be obtained as the solution of a modestly sized integer program. In the more general form of the multiproduct plant problem, inventory costs are considered and processing occurs in response to orders which have associated due dates. Thus, consideration of multiple campaigns of the same product and choices of campaign length (often referred to as lot sizing) must be added to choices of campaign sequences. Only a limited amount of work has been reported to date on this more general form of the problem.

Specifically, Birewar and Grossmann [5] considered the case of one or more nonidentical parallel production lines, consisting of a single unit per stage and processing multiple products. A multi-time period mixed integer programming formulation was presented in which the integer variables are the number of batches of a given product which are to be produced on a given line in a given time period. Average inventory levels over a time period were considered but change-over and set-up costs and times were neglected. The formulation accommodated both single product campaigns and mixed product campaigns by embedding the MPC formulation of [3]. If the integrality condition is relaxed, then the problem reduces to an LP. A two phase solution strategy is, thus, proposed in which the relaxed problem is solved first, the variables violating integrality conditions are rounded down to the nearest integer and the LP is resolved with the integer, variables maintained at fixed values.

Musier and Evans [47], investigated a single stage process consisting of nonidentical parallel units taking into account order quantities and due dates, inventories of finished products, and sequence dependent clean-out requirements between products. The scheduling goal was to determine the assignment of orders to units and the sequencing of orders on units so as to minimize the sum of the times during which the net inventory of a product is negative. The solution approach involved several heuristic components. First, an estimate was made of the maximum number of batches of each product which need to be produced. Next, an initial feasible schedule was determined using a composite best fit heuristic. Finally, the resulting initial schedule was subjected to a local improvement procedure, consisting of single batch repositioning and pairwise interchanges of batch positions. Each heuristic improvement candidate was tested for inventory feasibility and the process was terminated when a local optimum is reached. Multiple randomly selected starting points were used to improve the likelihood of finding the global optimum to the problem. Computational results showed that the number of local optima increased with the number of batches being processed and thus the likelihood of attaining global optima decreased dramatically with problem size. The main limitations of the approach are the restrictive network structure and the total stockout time criterion which does not differentiate between products/orders or the amount of the stockout.

Recently, Sahinidis and Grossmann [60] addressed the closely related problem of cyclic multiproduct scheduling of semi-continuous parallel lines for constant product demands, It is assumed that production occurs on nonidentical parallel lines, each operated using a characteristic fixed cyclic schedule which involves a subset of the products. A cost based objective function is used which accounts for production, sequence dependent transition, and inventory holding costs as well as a credit for freeing-up capacity. The result is an MINLP which, after reformulation to eliminate some inconvenient nonlinearities, is solved rigorously using a Benders Decomposition approach. The key assumptions of the approach are that the production lines share no resources, hence operate independently, that demand rates are constant over time, and that inventory costs can be computed independently for each production line. The key assumption which avoids the need to discretize time is that of cyclic operation. In that sense, the representation of time is analogous to that used in the mixed product campaign case.

Finally, Kudva et al. [36] consider a more general form of the network flowshop, shown in Figure 8, with multiple nonidentical units in each stage in which not all products can be processed on all units. The formulation accommodates intermediate product draw-offs, raw materials feeds to any stage, blending of intermediates to form specific final products, finite intermediate storage,

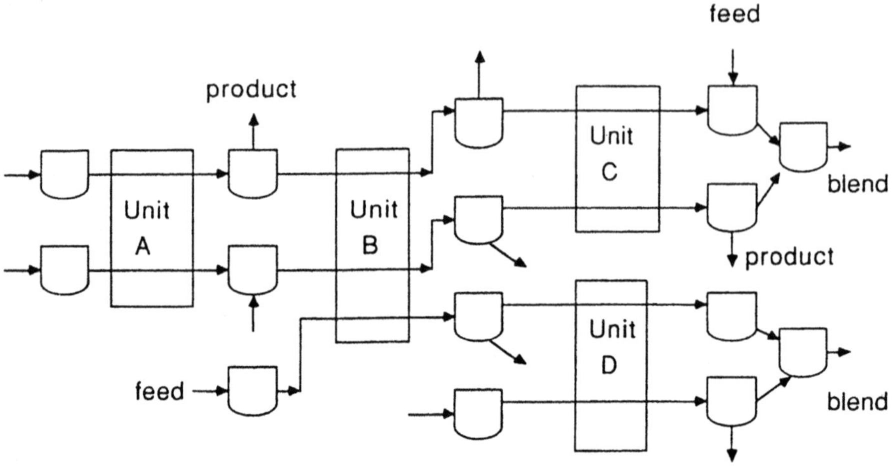

Figure 8. General network flowshop

and order deadlines. The solution strategy uses a fixed discretization of the scheduling horizon and a two step approach to schedule generation. In the first step, orders are scheduled sequentially according to priorities as close to due dates as possible while ensuring that inventories are maintained above safety levels. In the second step, schedule heuristics are employed to improve the schedule by aggregating orders and coalescing runs of the same task. Products are deblended via a backwards component explosion and the required intermediates are scheduled as separate orders, if not available from inventories. Different schedules are generated by changing order priorities and evaluated using a cost based objective function which includes inventory costs, changeover costs and deadline violations. The algorithm has been successfully tested on historical data from an existing multiproduct plant, was found to give significantly better schedules than those manually generated by plant staff, and is now implemented on the plant site for routine use.

Multipurpose Plant: The multipurpose plant problem is the most general form of the scheduling problem. Production occurs in campaigns where each campaign involves one or more parallel production lines and each production line itself will contain out-of-phase groups in a stage, multiple in-phase units in a group, and, possibly, intermediate storage. Thus, planning of campaigns, formation of campaigns, and scheduling of production lines must all be considered. We will divide the review of the literature into those efforts which address only the production planning aspects of the problem and those which deal with all three levels of decision making. The latter category is especially interesting because it represents an area of applications where integration of all three levels is essential because of the strong interaction among the decisions made at each level.

Production Planning: In contrast to the other operating strategies, in the multipurpose plant case the production planning problem involves not only the selection of campaign sequence and length so as to balance set-up and inventory charges but also involves the decision of selecting among alternative campaigns so as to meet production needs in the most efficient way. The selection of campaigns in effect is equivalent to assigning the production rates of the various products, where these assignment are mutually constrained by the availability of a limited set of equipment. In principle, to fully reflect the inherent flexibility of this operating mode, any production planning approach must, thus, address all of these decision aspects. Unfortunately, many of the reported formulations do not.

The production planning models proposed by Suhami and Mah, [65] and Rich and Prokopakis [58] assume that a fixed set of equipment is used for each product and that the product batch size and processing times are fixed. Suhami and Mah use a preprocessing procedure to assign due dates and lot sizes for final and intermediate products given the due dates for the final products. Based on the results of the preprocessing procedure, alternative production sequences are generated and the tardiness of each schedule is determined by solving a linear program. The schedule with least tardiness is retained.

Rich and Prokopakis [58] use an MILP formulation to sequence and schedule production runs. A variety of scheduling objectives, such as minimize tardiness and minimize makespan are investigated. A key assumption of their formulation is that every order due date must have associated with it a corresponding production run, so that the campaign lengths are implicitly fixed by the due dates. In effect, the due dates become the means of fixing the discretization of the time domain. The main result of the solution of the MILP model is thus the sequencing of the orders on the available equipment in the plant. Rich and Prokopakis [57] extend this approach to allow production of a product over several different predefined sets of processors.

Both of these approaches concentrate on the precedence relationships among the products to ensure the timely production of intermediates. This is indeed very important because in many multipurpose plant applications, product recipes are long and thus it is expedient to break long recipes into several subrecipes associated with intermediates so as to allow the use of the same equipment for several of the intermediates. This is, in effect, a means of reuse of the equipment for several tasks of the overall product recipe. However, a major deficiency of both approaches is that only one production line, and, thus, one production rate, is considered for the production of each product While this does simplify the planning problem, it does neglect the inherent flexibility of a multipurpose plant to accommodate a variety of production rates through a large number of possible production line configurations.

Combined Planning and Scheduling: Only Mauderli and Rippin [42], Lazaro and Puigjaner [41], Wellons and Reklaitis [71,72] have addressed all three components of the multipurpose plant scheduling problem, especially the generation and evaluation of alternative campaigns from an existing set of equipment items. In this section, we outline the essential elements of the three approaches and note their limitations.

Mauderli and Rippin [41] employed evolutionary enumerative techniques to generate alternative single product production lines. They first enumerated all possible equipment arrangements with only one equipment group on each stage. After eliminating inefficient candidates, these set-ups were then combined into all possible single product lines. Heuristic rules were used to control the potentially explosive number of alternatives that were generated. The effective production rates of these lines were determined assuming arbitrary sequencing of the groups on each stage, ZW transfer between stages, and all batch sizes set at maximum path capacity. The single product lines were then combined in an enumerative fashion to form a combinatorially large number of alternative single and multiple product campaigns. An LP screening procedure was then used to extract a much smaller subset of dominant campaigns. The LP procedure obtains these dominant campaigns by identifying the campaigns forming the convex hull of the set of feasible campaigns. A campaign is dominant if its vector of rates can not be expressed as a linear combination of the rates of other campaigns. This concept is illustrated in Figure 9 for two products A and B, whose production rates are shown on the axes. Campaign 2 is a dominant campaign since its production rates can not be obtained as any linear combination of the rates of two other campaigns. Campaign 3 on the other hand is inferior as it could be exceeded by campaign 2, for example. Given the set of dominant campaigns, a multi-time period MILP production planning model is used to allocate the available production time in each period to specific dominant campaigns. Constant change over times are assumed between campaigns and

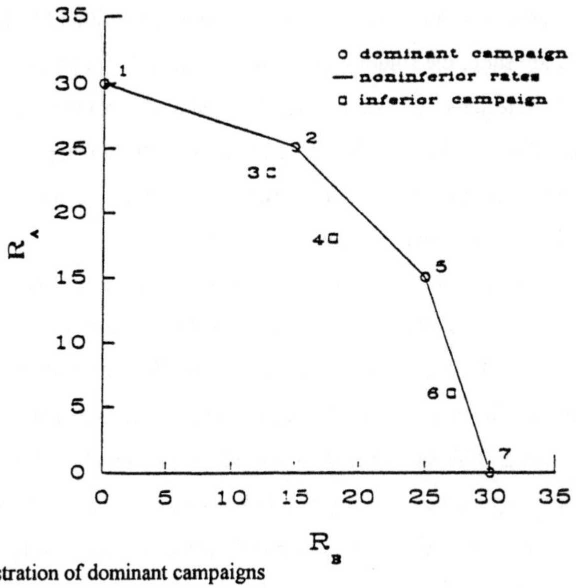

Figure 9. Illustration of dominant campaigns

the time losses due to start-up of the production lines are neglected.

Although this is the pioneering work on the subject, the work does have some notable limitations. Specifically, since the campaign generation procedure is a heuristically controlled partial enumeration, there are no assurances that the dominant campaigns are indeed truly dominant over the set of all possible campaigns. Furthermore, the production rates determined for the individual candidate lines are not optimized and, hence, may lead to poor campaign selection decisions. Finally, the campaign screening procedure simplifies lost production time due to changeover and start-up times by using average values, thus potentially leading to infeasible production plans because the effective production rates may be either under or overestimated. Nonetheless, the work is noteworthy for its recognition of the importance of all three decision levels on the efficient operation of a multipurpose plant.

Lazaro and Puigjaner [41] used an exhaustive enumeration procedure to generate alternative single-product production lines, allowing only in-phase operation of parallel equipment items on a stage. A set of good production lines was chosen based on a heuristic selection index which incorporates the production rate, equipment utilization fraction, and production cost per unit of final product. Production planning was performed in a hierarchical fashion. A single product planning problem was solved for each product to determine the number of batches to be produced in each dominant production line in each planning period. The task sequencing and scheduling for each period were accomplished with a heuristic job shop scheduling algorithm [2]. Since the plant is treated as a classical job shop, the solution procedure assumed UIS operation for all stable intermediates and relaxes the requirement that products be produced in campaigns. However, the job shop scheduling algorithm is able to take into account constraints on manpower and utilities. Since the production plan for each product is determined independently, the planning problem is applied recursively until a feasible production plan is achieved.

The main limitation of this work is that the scheduling procedure treats the plant like a job shop and assumes unlimited intermediate storage. The resulting schedules can be inefficient because out-of-phase operation of units is not considered or difficult to implement because unfinished batches of several different products effectively have to be moved in and out of storage as the plant produces several different products at the same time on shared equipment. The campaign structure of multipurpose plant operation is thus obscured if not lost. Furthermore, the production lines for each product are chosen on a greedy principle based on the value of the selection index. Consequently, the lines selected for different products may require many of the same units, making simultaneous production of these products difficult or inefficient.

Wellons and Reklaitis [71,72] revisited the multipurpose plant scheduling problem, developing rigorous MINLP formulations for all three decisions levels and solving them using mathematical programing techniques. The solution to the problem is approached via three components: the single product campaign formation problem, the multiple product campaign formation problem, and the overall production planning problem. The selection of the equipment items to form single or multiple production lines and the determination of the most efficient schedule for each line are tightly coupled decisions which do not lend themselves to simple decomposition. Thus, the MINLP formulations for these problems include not only the 0-1 variables, which assign specific items to specific equipment groups in the stages of production lines, but also incorporate the single product scheduling formulation of Wellons and Reklaitis [69,70,73]. The production planning problem on the other hand, relies only on the gross parameters of the dominant campaigns and thus can be solved decoupled from the campaign formation and line scheduling problems. We briefly summarize the salient features of each of these three component problems.

The single product campaign formation problem incorporates the following assumptions and conditions:

1) The production stages may contain nonidentical groups
2) Each equipment item is used only once in a production line

3) Batch aggregation is not allowed and batch splitting is only allowed within an equipment group

4) ZW and NIS operating policies are considered but not intermediate storage

5) Transfer times from/to batch units to/from semicontinuous units are accommodated

6) Processing and transfer times may be batch size dependent.

The problem is posed as an MINLP in which the equipment and path assignment decisions are represented by three key 0-1 variables and the objective is to maximize the production rate. The first variable type assigns an equipment of a given type to a specific position in a specific equipment group on a specific stage. The second determines the number of equipment groups on each stage and the third denotes the existence of specific paths of a simple path sequence. The resulting problem is nonconvex and potentially has large dimensionality. A Benders Decomposition approach is developed in which the complicating variables are the number of groups in each stage. The two resulting subproblems are both MINLPs but one is significantly smaller and the other has substantially fewer degeneracies. Comparisons with campaigns generated in Mauderli's work shows that the proposed formulation yields better or equivalent results in all cases tested, with improvements of up to 12.5% in the production rate.

The multiple product campaign formation problem involves the creation of multiple parallel lines which have the highest production rate for a given product ratio. It is thus a multicriterion problem in which one seeks campaigns such that the production rate of each line is as large as possible. The desired set of solutions to this problem will be the dominant campaigns in the sense of [41]. However, rather than generating a large number of possible campaigns and then extracting from these those that are dominant, Wellons introduced the use of the Noninferior Set Estimation method [12]. The NISE method allows the set of dominant campaigns to be generated in a sequential fashion starting with the set of optimized single product campaigns by repeated solution of a suitably modified form of the single product campaign formation problem. As in that case, the solution of the resulting MINLP is best obtained using a Benders Decomposition strategy analogous to that developed for the single product case. Computational results using the Mauderli and Rippin [41] problems show that the proposed approach yielded campaigns that dominate Mauderli's results in 22 of 38 cases, yielding as much as 20% improvement in the production rates, and are equivalent to Mauderli's campaigns in the remaining cases.

The multi-time period production planning problem formulated by Wellons is an MILP which selects from among the dominant campaigns those which maximize net profit subject to upper and lower bounds on sales. The formulation takes into account both the changeover time between campaigns and the start-up time for each campaign. As shown in Figure 10, the change over time is denoted by COT while the start-up time is denoted by SU.

Clearly if the actual production time T of a campaign in a given period is relatively short, then change over and start-up times can be quite significant. The key 0-1 variable in the formulation selects a given campaign for a given period. The number of these variables clearly grows with the number of campaigns and, especially, with the number of multiple product campaigns.

There are three significant limitations of all three of these comprehensive approaches to the multipurpose plant problem. First, the formulations do not explicitly include the decision of how to subdivide the production of a product into several separately scheduled intermediates. Instead, that decision must be made manually. Second, the structure of the individual lines does not permit the inclusion of intermediate storage vessels which would allow changes in batch size in the train. This can be done in principle using the approach of Yeh and Reklaitis [79] at the cost of some additional complexity. The third and key limitation, however, lies in the need to construct the set of dominant campaigns. In essence, the campaign formation component seeks to identify the most efficient campaigns for all of the different combinations of products and for all possible different production scenarios. Given P products to be produced and allowing for the possibility of campaigns involving from 1 up to P products, the number of distinct product combinations is 2P.

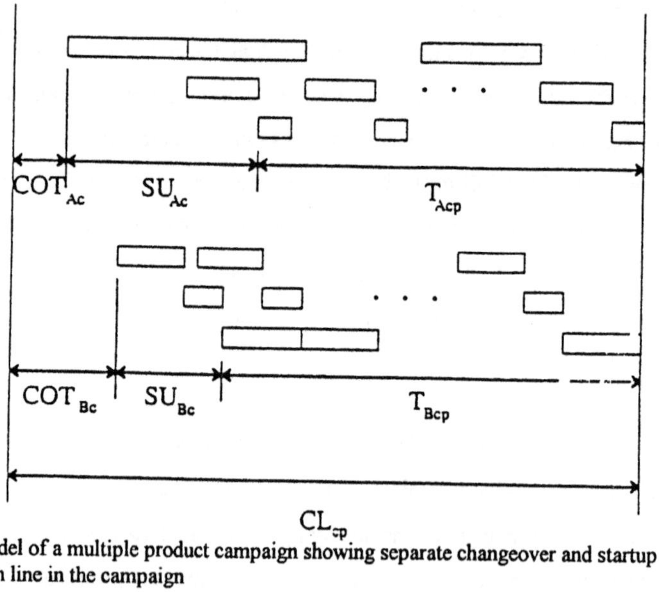

Figure 10. Model of a multiple product campaign showing separate changeover and startup times for each production line in the campaign

In principle, a separate multiple product campaign formation problem would have to be solved to identify the dominant campaigns from among these combinations. If the problem formulation is extended to incorporate consideration of resource constraints, then the structure of the dominant campaigns will also depend on the levels of resource availability. Thus, the campaign formation subproblems would in principle have to be repeatedly solved for different resource level scenarios. It is, thus, apparent that, in order to accommodate the consideration of limited resources, an approach is needed which only forms campaigns as and when they are required for specific resource and production needs.

Resource Constrained Scheduling

The subject of resource constrained scheduling has seen limited, although increasing, attention by workers in the chemical engineering research community but has been more extensively investigated in the operations research and artificial intelligence communities. As noted earlier, the resource constrained problem is inherently more difficult than the equipment dominant case because in contrast to process equipment, resources such as materials and utilities are divisible. Thus, in addition to sequencing of tasks in a temporal dimension, it is also necessary to consider the feasible grouping of simultaneously executed tasks so as to jointly utilize resources in a feasible fashion.

Chemical Process Oriented Contributions

Since the work of Lazaro and Puigjaner [41] who reported on the adaptation of a job-shop algorithm for scheduling multipurpose plants with resource constraints, a growing series of papers has emerged to address various forms of the RCSP. The work has largely resulted from the efforts of research groups at ETH, Imperial College, and Purdue University.

Egli and Rippin [16] considered the problem of determining a short term schedule for a multiproduct batch plant subject to resource constraints which meets a specified demand pattern while minimizing costs associated with storage, changeovers, and utility consumption. The scheduling procedure accounts for constraints on working patterns, limited availability of shared resources, inventory requirements for intermediate products, and raw material deliveries. The procedure is enumerative in nature: all possible production sequences are generated and then progressively eliminated by imposing the problem constraints so that only the favorable sequences are fully evaluated. Feasible schedules are generated by shifting each batch schedule forward or backward in time until all resource constraints are satisfied. Results are reported for a relatively small example with four products and eleven equipment items. The work is notable for its compilation of all relevant problem elements but offers a solution methodology of very limited scope.

Tsirukis and Reklaitis [67,68] considered the resource constrained form of the multipurpose plant scheduling problem and proposed a strategy which uses global approximation methods to reduce the solution domain sufficiently so that math programming approaches can be more effective. In this solution approach, the problem is decomposed into two main decision levels. The first decision level is concerned with the efficient sequencing of actions that may not overlap in time and, therefore, occupy disjoint intervals on the time axis. The second decision level is involved with actions that must be executed within the same time interval. The first level corresponds to the assignment of orders to campaigns and the allocation of a time frame for each campaign. This decision problem is represented as a relaxed MINLP formulation which is solved as a generalized Hopfield network. The second level involves the allocation of equipment and production resources to the specific orders within a given campaign. These allocation decisions are highly interdependent and constrained and are approached using a novel feature extraction search scheme which serves to narrow the search domain, followed by detailed optimization over the reduced decision domain. Figure 11 shows the hierarchical organization of the MBP scheduling system. The MBP system has been used to successfully solve scheduling problems with nine products, forty equipment items, and four resource types whose direct formulation would involve up to 1120 binary variables [68].

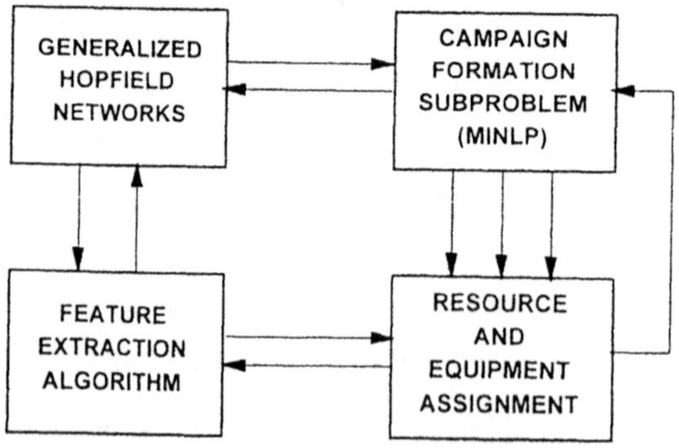

Figure 11. MBP scheduling system

Each task is allowed to have associated with it a constant processing time, a constant resource requirement, and a set of feasible equipment. To model the use of limited plant resources over time, the scheduling horizon is discretized into a number of uniform time intervals. The basic time quantum for the discretization is selected to be sufficiently small so that the processing and other characteristic times can be sufficiently accurately represented. Under this representation, the key set of decision variables are those which control the allocation of a unit to a task in a specific time period, that is, binary variable W_{ijt} takes on the value 1 if unit j is assigned to processing task i in time quantum t and zero otherwise. By assuming constant processing times and resource utilization rates during the occurrence of a task, the material balance, resource utilization, capacity bounds, and unit allocation constraints can be expressed in linear form. The result is a mixed integer-linear programming problem whose solution can be sought using general purpose MILP solvers or specialized branch and bound strategies which exploit the particular problem structure. The authors have reported variations of the formulation to handle sequence dependent set-up/clean-out constraints, several types of storage restrictions and both short term and campaign type operating modes. Continuous processing step are treated by discretizing runs into discrete increments.

The time discretization approach employed in this work is the classical means of dealing with resource allocation problems in the presence of globally shared resources whose utilization is distributed over time. The well-known difficulty with uniform time discretization is that the number of integer variables required to represent practical problems with realistic processing times, numbers of tasks, resources, and equipment can be quite large. The classical dilemma, which then arises in practice, is the following. If the problem data is rounded so that the time quantum can be increased (and thus the number of decision variables decreased), then the solution which will be obtained will be approximate and either too conservative or infeasible. If the problem data are accurately treated, then the MILP can become too large for routine solution. In addition, if sequence dependent factors are taken into account, the number of constraints grows quite large, making the LP relaxations which must be solved as the branch and bound solution progresses large and often resulting in large integrality gaps. On the positive side, the MILP framework is quite flexible and offers considerable expressive power in accommodating application specific features, such as equipment connectivity restrictions, finite shared storage, and conditional post processing steps. To date investigations of various forms of single product, multiproduct, and multipurpose

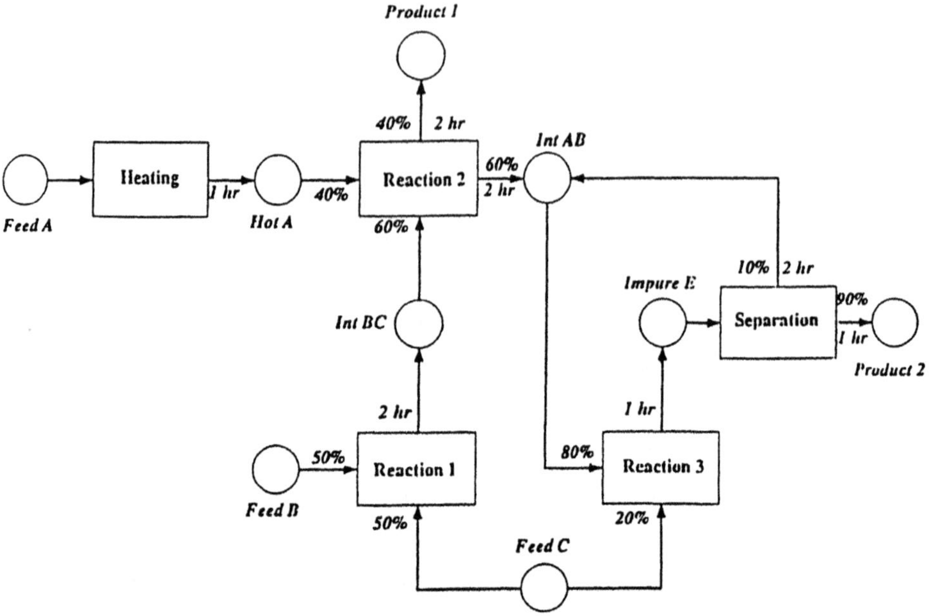

Figure 12. State-task network example (after Kondili et al)

operations have been reported with successful solution of applications which are of modest size in terms of the operational problem but large in terms of variable dimensionality (up to 4000 variables).

From a conceptual point of view, the large number of uniformly spaced time discretization points is a modeling artifact which is introduced to simplify the verification that resource constraints remain satisfied throughout the scheduling horizon as task to equipment assignments are made. Yet, assuming constant processing times and constant resource utilization rates during task execution, the actual times at which checking of resource utilization is required is limited to the points in time at which the resource availability changes and the start and stop times of the individual tasks. These times correspond to the original input data of the scheduling problem and, thus, ideally should be the only event times which ought to be explicitly considered. As shown in Figure 13, the number of the data-related resource utilization event times and the model-related event times can differ significantly. Effectively, the data-driven event representation leads to a sparser, nonuniform discretization which can potentially be a much more efficient approach to the RCSP.

An initial attempt to exploit this concept was given in [80,81] where the development of an

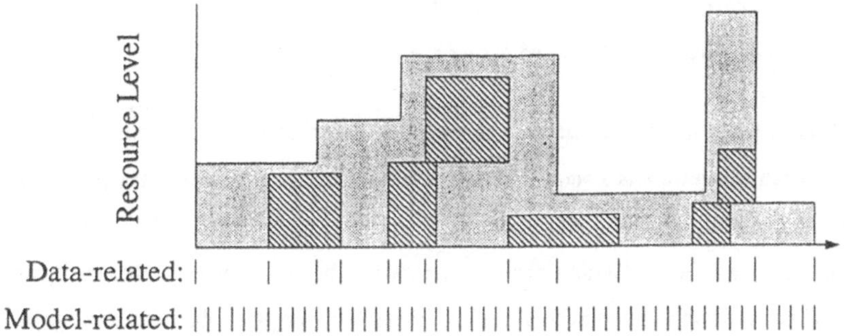

Figure 13. Data and model related resource time events

enumerative search scheme for handling a general form of the short term scheduling problem is reported. The key component of that approach is an interval processing framework which employs tree structures to generate a nonuniform set of time intervals, one for each recipe task during which the task may float freely and not violate resource constraints. Given this set of intervals and the appropriate set of precedence constraints, the actual assignment of exact start times for tasks can be accomplished through a linear program. The approach allows a very flexible treatment of various forms of material stability conditions, intermediate storage, renewable and nonrenewable resources, as well as detailed step descriptions for each task. However, the implementation using the production system language OPS83 proved too inefficient for large scale application. In [82], an MILP formulation of the interval processing framework is reported which overcomes most of the limitations of the direct enumerative search approach. A preliminary comparison of the nonuniform formulation with the conventional uniform discretization representation suggests that for a significant range of problems substantial reductions in the number of 0-1 decision variables can be achieved. Considerable work remains to be undertaken to develop efficient solution algorithms for both formulations since with MILP/MINLP problems it is only through rigorous computational experiments that questions concerning relative solution effectiveness can be answered.

Operations Research Contributions: The operations research literature on the subject of resource constrained scheduling is quite extensive. A comprehensive summary of the main avenues of attack can be found in [7] and an updated review in [6]. As in the equipment dominated case, most of the approaches reported address only severely simplified resource constrained problems and employ either very large scale multi-time period MILP formulations or, in a few specific instances, polynomial time algorithms. The opportunities for direct transfer of solution approaches applicable to the problem structures discussed earlier are, thus, limited.

Artificial Intelligence Contributions: Alternatively, the schedule for a resource constrained batch process can be viewed as a sequence of actions in time that transform the plant from an initial state to a final state in some optimal fashion. Under this perspective, the RCSP can be considered as an instance of the planning problem which has been extensively studied by the artificial intelligence research community. Planning research investigates the automatic generation of a course of action that transform the world from an initial state to a desired goal state. The research is focussed on two major problems: (i) the action representation in which a convenient medium is sought for describing the interactions among problem variables and (ii) the action sequencing problem where efficient solution methods are sought that can sequence the actions and produce the desired result.

The most important action representation vehicle has employed first order logic. The world is described by a set of logical variable whose interactions are represented by appropriate sets of propositions. The actions are represented by operators that transfer the state of the world by manipulating the truth value of the logical variables and the propositions. This action representation was introduced by Fikes and Nilsson [17] and has been used with variations in most subsequent efforts. The solution methods were initially based on intuitive rules that minimized the differences of the actual world state from the goal state. However, the inability of these approaches to guarantee feasibility let to more sophistical heuristic methodologies, such as hierarchical decompositions, least commitment strategies [59] and temporal reasoning models [1].

More recently the planning problem was shown to belong to the class of NP complete problems [9] and, thus, research has reoriented from seeking general systems for solving planning problems to approaches tailored to more narrowly defined domains, such as discrete manufacturing scheduling systems [18] and [50]. These efforts effectively are compilations of problem dependent heuristic rules and intuitive decomposition strategies which are not transported readily to other domains. Overall, the representational ability of planning methodology, linear or nonlinear, is inadequate for representing chemical process scheduling problems which are described by continuously varying quantities, discrete variables, and nonlinear relations among variables. In addition, the RCSP requires solutions that are not only feasible but also, if not optimal, then at least very good as measured by a performance function. It thus appears that the most effective approach to RCSP is the mathematical programming approach supplemented by heuristics, approximation methods, and search strategies, as the practicalities of obtaining working solutions dictate.

Reactive Scheduling

Although the need for frequent schedule readjustment has been actively discussed in the process systems engineering literature [48], [35], and [74], concrete developments have been sparse. However, the issue of rescheduling in a dynamic environment has been an active area of research in the discrete parts manufacturing arena for some years and, thus, an extensive literature does exist in that domain. We first consider contributions in the chemical processing domain.

Chemical Process Oriented Contributions: The first contribution to focus on the issue of reactive scheduling of batch chemical plants is that of Cott and Macchietto [13 & 14]. This work reports on the development of a system for real-time rescheduling of a batch plant subject to processing time variations. The underlying methodology is a completion time calculation algorithm which considers storage, processing units, and secondary resource limitations, and responds to perturbations by readjusting the start and stop times of operations so that feasibility of the schedule is restored. The start and stop time redetermination amounts to an earliest start heuristic for the completion time calculation for a flowshop under the restrictions that no task to resource reassignments are allowed. Tongo and Reklaitis [66] presents a heuristic completion time calculation approach for the resource constrained multiproduct plant. The formulation accommodates splitting and merging of batches, multiple storage options, including CFIS, product-unit combination dependent processing, transfer and changeover times, as well as order release and due dates. The approach selects earliest completion of all orders while satisfying resource limitations. Effectively, both developments can be viewed as simplified forms of discrete event simulation. Discrete/continuous simulators, such as BATCHES [10], which are designed to monitor resource requirements and proceed in the execution of tasks based on dispatching rules such as earliest start, could in principle accomplish the same task, although with some additional overhead. Unfortunately, any reassignment decisions which can be carried out within a simulation are inherently myopic in nature and thus also suffer from the limitations of the earliest start strategy.

In addition to this fundamental limitation, these rescheduling strategies also do not consider the full range of schedule modification alternatives, which are available in practice, from simple time shifts to unit reassignments, to resource substitutions, and eventual complete rescheduling. By way of example, Figure 14a shows two parallel two stage production lines, each processing two product batches. Suppose that, as shown in Figure 14b, a processing time deviation takes place on unit ul which leads to a time conflict in the schedule for unit u2. The simple time shift solution to this conflict would be to delay the start of batch b2 on ul, thus disrupting preparations for that batch which may have already taken place and propagating further delays to any subsequent processing or product shipping steps. An alternate resolution of the conflict, which causes no such disruption, is to assign batch bl to unit u3 which is idle during that time interval. Of course, this may require an extra clean-out of unit u3, with its attendant costs.

A framework for balancing time shifts against unit reassignments which uses a beam search

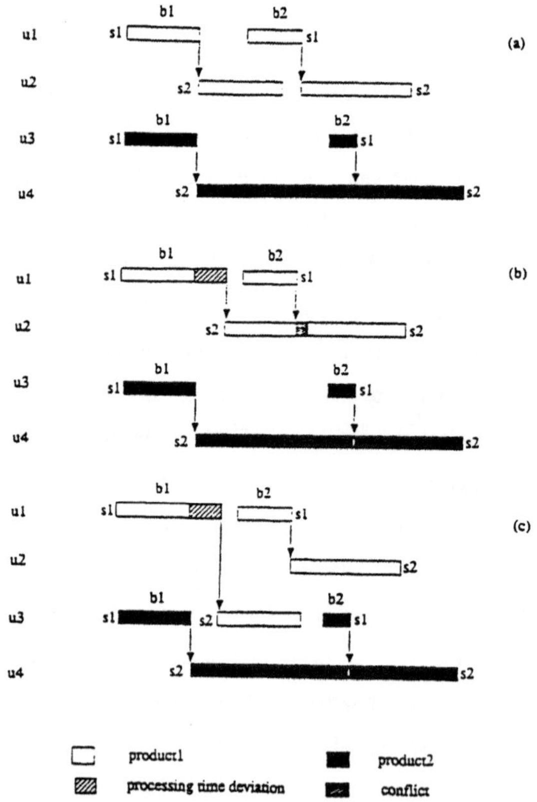

Figure 14. Conflict resolution (a) Original schedule (b) Simple shifting (c) Replacement

with heuristic pruning was reported by in [30]. The approach employs the conflict detection/conflict resolution paradigm under which a two level decision tree is constructed by tracking each batch and tracing each task for each batch undergoing conflict until all conflicts are resolved. The tree is pruned using heuristic pruning criteria as well as Zentner's multiple time interval propagation technique for eliminating infeasible choices. The decision tree is subjected to a beam search which seeks to minimize the weighted deviation from the master schedule. The choice between time shifting or unit replacement is made based on the earliest completion of the task which is subject to conflicting demands. Results reported using a three product, 51 batch, 26 equipment item test case show significant improvement over the use of a myopic criterion for eliminating conflicts. The approach, nonetheless, suffers from the limitations of rapid growth in computation time with increases in the beam search parameter and the suboptimal solutions which can be introduced by virtue of the sequential nature of the batch conflict resolution mechanism. Clearly, additional work is required to develop a reactive scheduling strategy which accommodates the full range of rescheduling decision alternatives.

Discrete Parts Oriented Contributions : Nof et al [49] provide a survey of literature on dynamic, real time scheduling and rescheduling for job shops and flowshops. Most of the systems developed so far are directed at issues connected with integration of real time monitoring, scheduling, and schedule modification in an interactive scheduling framework.

Ow et al [50] developed a system, OPIS, that performs reactive schedule revisions by focussing on a short "conflict horizon". Based on the nature of the scheduling conflict, the conflict is classified into a conflict class. Using a heuristic rule base, the system chooses from a set of predefined remedial actions and applies one such action to the schedule in the "conflict horizon". These alternative remedial actions are investigated using a beam search. No attempt is made to predict or account for the impact of a given decision on the remainder of the schedule. Rather, any additional conflicts are merely imposed on the rest of the schedule. Kanai et al [29] developed a system which does reactive modification of a job shop schedule. The schedule is represented as a constrained network of individual steps and a simple constraint propagation scheme is used for propagating the effects of specific deviations through the rest of the schedule. Rerouting of jobs in the face of machine breakdowns is not considered: all corrective actions are confined to time delays.

Prosser [53] describes a scheduling system which has a combination of predictive and reactive components. The rescheduling problem is treated hierarchically in three layers. At the lowest level, conflicts are resolved by time-shifting. If the system finds a conflict that cannot be resolved by time shifting, a higher level tactical agent is engaged to attempt load balancing. If this fails to resolve the conflict, control is passed to the highest level which employs job resequencing. A key limitation of this system is that it chooses only one solution among the possible alternatives at a given level and backtracks to the next level if that alternative fails. There are no provisions made for interlevel backtracking. Grant and Nof [22] developed an automatic adaptive scheduling system for a jobshop which similarly involves the simple components of time shifting and resequeneing unscheduled orders. As deviations occur in the schedule over time, these are accommodated through time shifting. As these shifts occur, the difference between the expected and the actual completion times of tasks are monitored. When the difference exceeds some tolerance limit, resequencing is invoked. With the exception of OPIS, none of these developments perform a search over the decision space to find the best decision, rather, they proceed with the first feasible alterative that is identified.

The concept of reactive plan modification has also been investigated by researchers in the AI planning domain. For instance, [26,27,28] presents a formal theory of reactive plan modification within the framework of hierarchical nonlinear planning. During the rescheduling process, an attempt is made to minimize the disturbance caused to the applicable portions of the original plan. At each node of the task network, the number of plausible choices is controlled using a heuristic strategy which orders the choices with the aim of minimizing interactions. As in the basic deterministic scheduling case, the main limitation of planning theory approaches derives from the fact that the focus is entirely on obtaining feasible schedules rather than conducting a search for the best (or at least very good) solution. Furthermore, an underlying premise is that enough expert knowledge is available about the domain of interest for the development of a comprehensive body of rules. In the case of highly combinatorial scheduling problems, the effectiveness of expert knowledge can be severely limited.

Conclusions

In this paper, we have sought to define the essential elements of scheduling problems characteristic of chemical manufacturing. The available methodology, especially the developments of the last five years were reviewed. Much has truly been accomplished and a number of research groups in the academic chemical engineering world now are active in the field. However, further work is clearly required. Specifically, further investigation is needed of alternative formulations and solution methods for resource constrained scheduling problems, systematic treatment of reactive scheduling within an overall framework for handling uncertainty in scheduling, and effective approaches to the coordinated scheduling of multiple plant sites.

References

1. Allen, J.F.: Maintaining Knowledge about Temporal Intervals. Communications of the ACM, 26, 11, (1983)
2. Baker, K. R.: Introduction to Sequencing and Scheduling. Wiley, New York, (1974)
3. Birewar, D. B., and I. E. Grossmann: Efficient Optimization Algorithms for Zero Wait Scheduling of Multiproduct Batch Plants. Ind. Eng. Chem. Res., 28, 1333-1345 (1989)
4. Birewar, D. B., and I. E. Grossmann: Incorporating Scheduling in the Optimal Design of Multiproduct Batch Plants. Comput, & Chem. Engng., 13 (1/2), 141-161 (1989)
5. Birewar, D. B., and I. E. Grossmann: Simultaneous Production Planning and Scheduling in Multiproduct Batch Plants. -Ind. Eng. Chem. Res., 29, 570-580 (1990)
6. Blazewicz, J., G. Finke, R. Haupt, and G. Schmidt: New Trends in Machine Scheduling. Europ. J. of Operat. Res., 37, 303-317 (1988)
7. Blazewicz, J., W. Cellary, R. Slowinski, and J. Weglarz: Scheduling under Resource Constraints - Determinate Models. Baltzer, Basel, (1987)
8. Cerda, J., M. Vincente, J. Gutierrez, S. Esplugas, and J. Mata: Optimal Production Strategy and Design of Multiproduct Batch Plants. Indust. & Eng. Chem., 29 (4), 590- 600 (1990)
9. Chapman, D.: Planning for Conjunctive Goals. Artificial Intelligence, 32, 333-377 (1987)
10. Clark, S. M., and Kuriyan, K.: BATCHES-Simulation Software for Managing Semicontinuous and Batch Process Paper 32f, AICHE National Mtg, Houston, TX (April, 1989).
11. Coffmann, E. G., Jr., (ed): Computer and Job/Shop Scheduling. Wiley, New York, (1976)
12. Cohon, J. L., R. L. Church, D. P. Sheer: Generating Multiobjective Trade -Offs: An Algorithm for Bicriterion Problems. Water Resour. Res., 15, 1001-1010, (1979)
13. Cott, B. J., and S. Macchietto: A General Completion-Time Determination Algorithm for Batch Processes. Presented at Annual AIChE Meeting, San Francisco, (Nov. 1989)
14. Cott, B. J., and S. Macchietto: Minimizing the Effects of Batch Process Variability Using On-Line Schedule Modification. Comput. Chem. Engng., 13 (1/2), 105-113 (1989)
15. Dutta S. K.,and A. Cunningham: Sequencing Two Machine Flowshops with Finite Intermediate Storage. Manag. Sci., 21, 989-996 (1975)
16. Egli, U.M., and D.W. T. Rippin: Short-term Scheduling for Multiproduct Batch Chemical Plants. Comput. Chem. Engng., 10(4), 303-325 (1986)
17. Fikes, R. E. and N. J. Nillson: STRIPS: A New Approach to the Application of Theorem Proving to Problem Solving. Artificial Intelligence, 2 (3/4), 189 208 (1971)
18. Fox, M. S., and Smith, S. F.: ISIS - a Knowledge-Based System for Factory Scheduling. Expert Systems Journal 1, (1), 25-49, (1984)
19. French, S.: Sequencing and Scheduling: an Introduction to the Mathematics of the Job-Shop. Horwood, Chichester, (1982)
20. Garey, M. R. and D.S. Johnson: Computers and Intractability: A Guide to the Theory of NP-Completeness. Freeman and Co. (1979).
21. Graham, R.L., E L. Lawler, J. K. Lenstra and A. H. G. Rinnooy Kan: Optimization and Approximation in Deterministic Sequencing and Scheduling: A Survey. Ann. Discrete Math., 5, 287-326 (1979)
22. Grant, F. H., and S.Y. Nof: Automatic Adaptive Scheduling of Multiprocessor Cells. Presented at ICPR, Nottingham, U.K, (August 1989)
23. Graves, S.C.: A Review of Production Scheduling. Operations Res., 29(4), 646 675 (1981)
24. Gupta, S. K., and J. Kyparisis: Single Machine Scheduling Research. Omega Int. J. Of Mgmt. Sci., 15 (3), 207-227 (1987)
25. Ishikawa A., H. Shinji and I. Hashimoto: Module-Based Scheduling Algorithm for a Batch Resin Process. Proceedings of the ISA/90 International Conference on Advances in Instrumentation and Control New Orleans, LA, (Oct. 1990)

26. Kambhampati, S. and J.A. Hendler: Control of Refitting During Plan Reuse. Proceedings of International Joint Conference on AI, 943-948 (1989)
27. Kambhatnpati, S.: A Theory of Plan Modification. Proceedings of AAAI Conference, 176-182 (1990)
28. Kambhampati, S.: Mapping and Retrieval During Plan Reuse: A Validation Structure Based Approach. Proceedings of AAAI Conference, 170-175 (1990)
29. Kanai, N., S. Yakai, K Fukunaga, and Y. Tozawa An Expert System to Assist Production Planning. Proceedings of International Workshop on AI for Industrial Applications, 219-224 (1988)
30. Kanakamedala, K. B., V. Venkatasubramanian, and G.V. Reklaitis: Reactive Schedule Modification in Batch Chemical Plants, Ind. Eng. Chem. Res. 33 (1), 77-80 (1994)
31. Kondili, E., Pantelides, C. C., and Sargent, R.W.H.: A General Algorithm for Scheduling Batch Operations. Proceedings for the Third International Symposium an Process Systems Engineering, Sydney, Australia, 62-75, (1988). See also, Comput. & Chem. Engng., 17 (2), 211-229 (1993)
32. Ku, H.-M., and I. Karimi. Completion Time Algorithms for Serial Multiproduct Batch Processes with Shared Storage. Comput. & Chem. Engng., 14 (1), 49-69 (1990)
33. Ku, H.-M. and I. Karimi: Scheduling in Serial Multiproduct Batch Processes with Due-Date Penalties. Ind. Eng. Chem. Res., 29, 580-590 (1990)
34. Ku, H.-M. and I. Karimi: Scheduling in Serial Multiproduct Batch Processes with Finite Interstage Storage: A Mixed Integer Linear Program Formulation. Ind. Eng. Chem. Res., 27, 1840-1848 (1988)
35. Ku, H.-M., D. Rajagopalan, and I. Karimi: Scheduling in Batch Processes. Chem. Engng. Prog., 83 (8), 35-34 (1987).
36. Kudva, G., A. Elkamel, J. F. Pekny, and G. V. Reklaitis: A Heuristic Algorithm for Scheduling Multi-Product Plants with Production Deadlines, Intermediate Storage Limitations, and Equipment Changeover Costs. Fourth International Conference on Process Systems Engineering, Montebello, Canada, (Aug. 1991). See also, Comput. & Chem. Engng. 18 (9), 859-876 (1994)
37. Kuriyan, K., and G. V. Reklaitis: Approximate Scheduling Algorithms for Network Flowshops. PSE '85: The Use of Computers in Chemical Engineering, IChE Symposium Series 92, 79-90, Pergamon, Oxford, U. K, (1985)
38. Kuriyan K, and G.V. Reklaitis: Scheduling Network Flowshops so as to Minimize Makespan. Comput. Chem. Engng., 13, 187-200 (1989)
39. Kuriyan, K, G. Joglekar, and G.V. Reklaitis: Multiproduct Plant Scheduling Studies using BOSS Ind. Eng. Chem. Res., 26, 1551-1558 (1987)
40. Lawler, E. L.., Lenstra. J. K., and Rinnooy Kan, A. H. G.: Recent Developments in Deterministic Sequencing and Scheduling: a Survey. In M A H Demster, J. K. Lenstral and A H. G. Rinnooy Kan (eds.) Deterministic and Stochastic Scheduling, Reidel Dordrecht, 35-73 (1982)
41. Lazaro, M., and L. Puigjaner Simulation and Optimization of Multi-Product Plants for Batch and Semi-Batch Operations. I. Chem. Symp. Series, 92, 209-222 (1985)
42. Mauderli, A.M., and D.W.T. Rippin Production Planning and Scheduling for Multi-Purpose Batch Chemical Plants Comput. Chem. Engng., 3, 199-206 (1979)
43. Mauderli, A.M. : Computer-Aided Process Scheduling and Production Planning for Multi-Purpose Batch Chemical Plants. Ph.D. thesis; E T H Zurich Nr. 6451 (1979)
44. Miller, D L., and J F. Pekny: Exact Solutions of Large Asymmetric Traveling Salesman Problems. Science, Vol 251, pp. 754-761, (1991)
45. Musier, R F. R, and L B. Evans: An Approximate Method for the Production Scheduling of Industrial Batch Processes with Parallel Units. Comput. & Chem. Engng., 13 (½), 229-238 (1989)
46. Musier, R F. H., and L.B. Evans: Schedule Optimization for Multi-Stage Chemical Processes with Multiple Units at Each Stage. AIChE Annual Meeting, Chicago, (November, 1990)
47. Musier, R F.., and L. B. Evans Schedule Optimization with Simultaneous Lot-Sizing in Chemical Process Plants. AIChE J., 37, pp. 886-896 (1991)
48. Musier, R.F.H., and L.B. Evans: Batch Process Management. Chem. Eng. Prog., 86 (6), 66-77 (1990b)
49. Nof, S. Y., V. N. Rajan, and S.W. Frederick: Knowledge-based, Dynamic, Real-time Scheduling and Rescheduling: A Review and Some Annotated References. Research Memorandum No. 89-16, School of Indus. Engineering, Purdue University, West Lafayette, N (1990)
50. Ow, P.S., S F. Smith and A. Thiriez: Reactive Plan Revision. Proceedings of AAAI Conference, 77-82 (1988)
51. Papadimitriou, C. H., and P.C. Kanellakis: Flowshop Scheduling with Limited Temporary Storage. J. Assoc. Comput. Mach., 27, 533-549 (1980)
52. Pekny, J.F., D.L. Miller and G. J. McRae: An Exact Parallel Algorithm for Scheduling When Production Costs Depend on Consecutive System States. Comput. & Chem. Engng., 14 (9), 1009-1023 (1990)
53. Prosser, P: A Reactive Scheduling Agent. Proceedings of AAAI Conference, 1004-1009 (1988)
54. Rajagopalan, D. and I.A. Karimi: Completion Times in Serial Mixed-Storage Multiproduct Processing with

Transfer and Set-up Times. Comput. & Chem. Engng., 13 (½), 175-186 (1989)
55. Reklaitis, G.V.: Progress and Issues in Computer Aided Batch Process Design. Proceedings of the Third Int. Conference on Foundations of Computer-Aided Process Design, CACHE-Elsevier, New York:, pp. 241-276 (1990)
56. Reklaitis, G.V.: Review of Scheduling of Process Operations. AIChE Symposium Series, Vol. 78, No. 214, 119-133 (1982)
57. Rich, S.H., and G. J. Prokapakis: Multiple Routings and Reaction Paths in Project Scheduling. Ind. Eng. Chem. Res., 26(9), 1940-1943 (1987)
58. Rich, S.H., and G. J. Prokapakis: Scheduling and Sequencing of Batch Operations in a Multipurpose Plant. Ind. Eng. Chem. Process Des. Dev., 25(4), 979-988 (1986)
59. Sacerdoti, E.D.: The Nonlinear Nature of Plans. Advance Papers of the Fourth Jnternational Joint Conference on Artificial Intelligence, Morgan Kaufmann, Los Altos, CA, 206-214 (1975)
60. Sahinidis, N.V., and I.E. Grossmann: MINLP Model for Cyclic Multiproduct Scheduling on Continuous Parallel Lines. Comput. & Chem. Engng, 5 (2), 85-103 (1991)
61. Sahinidis, N.V., and I. E. Grossmann: Reformulation of Multiperiod MILP Models for Planning and Scheduling of Chemical Processes. Comput. & Chem. Engng., 15 (4), 255-272 (1991)
62. Salvador, M.S., Ph.D. Thesis: Case Western Reserve University, Cleveland, OH (1978)
63. Shah, N., C.C. Pantelides and R.W.H. Sargent: A General Algorithm for Short-term Scheduling of Batch Operations - II. Computational Issues. Comput. & Chem. Engng., 17 (2), 224-244 (1993)
64. Rapacoulias, C., N. Shah, and C. C. Pantelides: Optimal Scheduling of Order-driven Batch Chemical Plants. In L. Puigjaner and A. Espuna (eds), Computer Oriented Process Eng., Elsevier, Amsterdam, pp. 145-160 (1991)
65. Suhami, I., and R.H.S. Mah: Scheduling of Multipurpose Batch Plants with Product Precedence Constraints. Proc. of the Second FOCAPD Conference, A.W. Westerberg and H H. Chien, Eds; Amer Institute of Chemical Engineers: New York, (1984)
66. Tongo, G.O. and G.V. Reklaitis: Completion Time Calculation of a General Multipurpose Batch Plant with Resource Constraints. Paper 102a, AIChE National Meeting, Orlando, (March, 1990)
67. Tsirukis, T., and G.V. Reklaitis: A Comprehensive Framework for the Scheduling of Resource Constrained Multipurpose Batch Plants. Proc. of the Fourth International Symposium of Process Systems Engineering, Montebello, Canada, (August, 1991)
68. Tsikuris, A., and G.V. Reklaitis: Feature Extraction Algorithms for Constrained and Global Optimization - I Mathematical Foundations and II. Batch Process Scheduling Applications, Ann. Opns. Res.42, 275-312 (1993)
69. Wellons, M. C., and G.V. Reklaitis: Optimal Schedule Generation for a Single-Product Production Line - I. Problem Formulation. Comput. & Chem. Engng. 13 (1/3), 201-212 (1989)
70. Wellons, M.C., and G.V. Reklaitis . Optimal Schedule Generation for a Single-Product Production Line - II. Identification of Dominant Unique Path Sequences. Comput. & Chem. Engng., 13 (½), 213-227 (1989)
71. Wellons, M. C., and G.V. Reklaitis: Scheduling of Multipurpose Batch Chemical Plants, 1. Formation of Single -Product Campaigns. Ind. Eng. Chem. Res. 30, 671-688 (1991)
72. Wellons, M C., and G.V. Reklaitis: Scheduling of Multipurpose Batch Chemical Plants. 2. Multiple-Product Campaign Formation and Production Planning. Ind. Eng. Chem. Res. 30, 688-705 (1991)
73. Wellons, M.C.: Scheduling of Multipurpose Batch Chemical Plants, PhD thesis, Purdue University, West Lafayette, IN, December, 1989.
74. White, C.H.: Productivity Analysis of a Large Multiproduct Batch Processing Facility. Comput. Chem. Engng, 13 (1/2), 239-245 (1989)
75. Wiede, W., and G.V. Reklaitis: Determination of Completion Times for Serial Multi-product Processes - 2. A Multiunit Finite Intermediate Storage System. Comput. Chem. Engng., 11(4), 345-356 (1987)
76. Wiede, W., and G.V. Reklaitis: Determination of Completion times for Serial Multiproduct Processes - 3. Mixed Intermediate Storage Systems. Comput. Chem. Engng., 11(4), 357-368 (1987)
77. Wiede, W., K. Kuriyan and G.V. Reklaitis: Determination of Completion Times for Serial Multiproduct Processes - 1. A Two Unit Finite Intermediate Storage System. Comput. Chem. Engng., 11(4), 337-344 (1987)
78. Wittrock, RJ.: Scheduling Algorithms for Flexible Flowlines. IBM J. Res. Develop., 29, 401-412 (1985)
79. Yeh, N.C., and G.V. Reklaitis: Synthesis and Sizing of Batch Semicontinuous Process: Single Product Plants. Comput. Chem. Engng., 11, 639-654 (1987)
80. Zentner, M., and G.V. Reklaitis: An Interval Based Approach for Resource Constrained Batch Processing Scheduling, Part I: Interval Processing Framework:. COPE-91, Barcelona, Spain, (Oct. 1991)
81. Zentner, M., and G.V. Reklaitis: An Interval Based Approach for Resource Constrained Batch Process Scheduling, Part II: Assignment and Adaptive Storage Retrofitting. Paper 140d, AIChE Annual Mt., Los Angeles, (Nov. 1991)
82. Zentner, M., and G.V. Reklaitis: An Interval-based Mathematical Model for the Scheduling of Resource Constrained Batch Chemical Processes. This volume p. 779

GanttKit - An Interactive Scheduling Tool

L. Halasz, M. Hofmeister, D.W.T. Rippin

TCL, Eidgenossische Technische Hochschule, CH-8092 Zurich, Switzerland

Abstract: A prototype interactive scheduling capability has been developed as part of the BatchKit knowledge based system for planning and operation of batch chemical processes.

The main goal of the work was the verification of the potential of knowledge-integration techniques in developing a flexible toolkit for a wide variety of tasks:

- Entity-relationship schema for the representation of pertinent complex objects and their relationship
- Active graphics support for man-machine cooperation
- Version management techniques to support case-based reasoning and learning from problem solving experience
- Generalization of branch-and-bound search algorithm

For each product one or more types of batches are defined, each of which represents the allocation of the necessary processing tasks for that product to appropriate equipment items. The schedule is constructed to satisfy the demand for a number of products for which due dates within the planning period and corresponding demands are specified. Various constraints can be imposed between successive tasks and successive batches of the same or different products.

A simple branch-and-bound procedure is available for rapidly generating plausible schedules and the merits of a schedule can be assessed by a performance criterion based on how closely the cumulative demand curves for each of the products can be matched by the scheduled production. Gantt charts of the schedules are displayed on the screen where they can be interactively manipulated.

Keywords: Batch processing, scheduling, production planning

1. Introduction

Many problems in the design and operation of batch processes have been solved in recent years. Increasingly sophisticated methods of mathematical programming are being applied to the solution of larger and more general problems.

However, although special-purpose programs may be available to solve different problems of batch processing such as design, medium term capacity planning or short term scheduling considerable effort is still needed to transfer data from one to the other. Furthermore, however general the formulation, practical use of the programs almost always calls for some additional features to be incorporated, which even if feasible can commonly only be realised with substantial programming effort.

In addition there are still wide differences of practice in batch processing. Because of different historical experience or the special circumstances of different batch environments companies have different criteria, different practices and different calculation methods.

This situation favours the provision of a general batch processing environment in which a variety of batch processing calculations can be carried out under the control of the user who is kept informed and is free to intervene or interact with the programs.

Rather than a set of programs, we prefer to consider that the user has at his disposal a *toolkit* which he can learn to use effectively and creatively. The BatchKit project has been launched to investigate the potential of new technologies such as knowledge engineering and artificial intelligence to provide the type of flexible environment or toolkit needed to aid batch processing.

A related objective of the BatchKit project has been to gain experience in integrating a relatively large set of problem formulations and solutions as opposed to dealing with completely new problems. Thus some of the first capabilities to be realised in BatchKit are similar to those previously available in earlier special purpose programs ([22], [16], [14], [6]) but in simplified form for demonstration purposes and with interactive facilities. The individual modules (of which GanttKit is one example) will share conceptual knowledge of the BatchKit system, as designed in the BatchKB knowledge base).

Research experience has shown that the main problem of such an integration, independently of the domain, is the design of a common representation based on a unified data model.

For data modelling in BatchKit, first a relational data modelling system (AXI) has been developed on top of IntelliCorp's KEE[10]. AXI has proved to be useful for the formalisation of the general structure which currently consists of two subsystems. The first subsystem has been built around the SimKit discrete event simulation package in KEE. It made use of the extensive graphical capabilities of this package and makes provision for the solution of some problems of equipment allocation, batch sizing and capacity planning as described in[9] .

The second subsystem, GanttKit, a short term interactive scheduling system is described here. The complete integration of the two systems is possible, however, they are separated so that workstations which have KEE but not SimKit can still make use of GanttKit. The GanttKit system was demonstrated during the NATO ASI workshop, June 1992, in Antalya, Turkey.

1.1 Related Work

Current work in the field of production planning can benefit from the foundations laid down in other fields :

- Determination of the optimal sequence of operations (jobs) has been of interest in *operations research*. Recent advances in the development of optimization algorithms allow e.g. the formulation of some types of planning problems as MINLPs — see e.g. [18] for review.

- Systems for production scheduling have been created for medium-term *industrial* needs (to our knowledge, e.g. Sandoz, Ciba-Geigy, Rhone-Poulenc and ICI companies and certainly many more are using such systems, see [17]), mostly based on simplified formulations adapted to particular needs, many of them from the machine or aerospace industries.
- Planning (i.e. finding an optimal sequence of operations (such as robot movements) to reach a given goal) has been a research topic in *artificial intelligence* research. AI research also established methods of *temporal reasoning* and the formulation and methods for solution of the *Constraint Satisfaction Problem* [24].
- In Computer Science, the establishment of constraint programming as a generalisation of logic programming has opened up a new perspective in knowledge integration in engineering domains[11].

An example of a system for the study of batch processes is the BATCHES system [4] - mainly used for simulation. A planning system ISIS based on constraints formulation has been designed at Carnegie-Mellon University [8] and later extended to OPIS (Opportunistic Intelligent Scheduler) [21].Other systems have been reported in [25] where e.g. the concept of "plan critics" and local plan repair has been introduced to arrive at a final feasible and acceptable, though suboptimal, plan.

An overview of the work in the area of scheduling and planning of batch process operations can be found in [18].

1.2 Structure of the Paper

The general characteristics of a toolkit are described in Section 2, with some reference to the BatchKit environment for the treatment of general problems in batch processing.

Attention is then directed to the GanttKit aid to scheduling which operates within the BatchKit environment.

In Section 3 distinctive features of the problem are identified and an objective function representing deviations from an ideal schedule is defined.

In Section 4 a partly heuristic solution to the scheduling problem is presented. This generates favourable sequences of production which are then checked for feasibility.

Section 5 describes the structure of the knowledge base and the graphical representation and in Sections 6 and 7 various operations on the graphical representations of the schedules are presented.

Section 8 is about the DATAIMAGE system which is a powerful tool for interactive data editing.

Section 9 provides an illustrative example showing the most important features of GanttKit such as optimization, reconciliation of inconsistencies, manual modifications, sensitivity analysis and version management.

Some conclusions are drawn in Section 10.

2 Tool Kit

A tool kit is a collection of programs which operate in a common environment while maintaining consistency throughout the operations. In contrast to a single program which solves a well defined problem, a tool kit contains a set of problem solvers of varying capabilities which may provide partial or complete solutions to the original problem formulation or enhance the information available to the user in other ways.

GanttKit consists of the following main components:

1. KEE system implemented in the LISP language system and the graphical environment of the X-Windows system.
2. AXI system — the relational tool.
3. BatchKB — The conceptual model of common concepts of batch processing in terms of object classes and relations.
4. The DimageKB — The knowledge base of the general data editing subsystem.
5. GanttKB — The knowledge base providing the GanttKit-specific objects, relations and functions.
6. The individual problem knowledge base storing the problem-specific data such as plant structure, recipes, orders, products etc.
7. The problem solving algorithms :
    a. General-purpose branch-and-bound and its parametrisation for the short-cut method described below.
    b. The schedule reconciliation function implementing the shifting algorithm described below.

The most important issues of a toolkit design will be treated in the following sections.

2.1 Hierarchy of Problem Solving Levels

Typical complex problems such as production planning can be formulated and solved (both by man and by the computer) at various levels of abstraction - ranging from detailed calculations of single production steps to complete production plan optimization.

Maintaining problem state consistency while supporting the possibility of operating at different levels of detail poses a difficult problem for which no general solution currently exists. E.g., the low-level manipulation of the production plan - such as the shifting of individual production steps - may violate assumptions made about the problem state by higher-level problem solvers or by the user.

2.2 Problem Solution by Man-Machine Cooperation

The necessity of cooperation between man and the computer seems to be characteristic of the very complex problems arising e.g. in planning. The *first* reason is the impossibility of completely capturing in models the reality of a dynamically changing environment. The *second* reason is the difficulty arising in solving problems formulated using more complete, complex models. As a result, the modelling detail must be traded against the cost or time needed to formulate the problem and to find the "best" solution under constraints on available resources (including time). Therefore, the domain expert who may be in possession of information not represented in the model (or models) should be included in the solution process. His knowledge may allow important decisions to be made when partial solutions have been acquired, or complete solutions are to be compared.

2.3 The Role of Representation

The cooperative solution of a problem by combination of user and program operations at varying levels of abstraction necessitates the maintenance of equivalence of representations of the current problem state in all cooperating problem solving components (including the human expert or agent).

This equivalence is most easily maintained if there is one common representation based on a sufficiently general and powerful data model. In BatchKit, this task is fulfilled by the AXI system - a relational extension of the KEE object-based representation which supports the development of an Entity-Relationship schema [23]. The conversational language and the retrieval system provided with AXI allows flexible manipulation of the problem state representation using addition, deletion and modification operations. The overhead which the KEE-AXI representation carries with it may call for transformation into a more efficient form when the problem formulation is ready for solution.

2.4 Advanced User Interface

The cooperative problem solution requires intensive communication between man and machine. The user interface design must provide representations which facilitate this communication within the limitations of human perception and comprehension. The information should be presented in a form which is easy to interpret for a human expert. Graphical representation offers a suitable and powerful tool for manipulation by a human expert because human comprehension is in many cases better with a graphical than with a textual representation (a textual representation of a medium size graph is difficult to understand and to manipulate when compared to its graphical equivalent).

The KEE software development environment uses a graphical interface for user interaction. It also provides a sophisticated set of tools for the development of user interfaces.

E.g., the active graphics possibility provided by the KEE system allows to immediately propagate changes to internal data towards the graphical interface and vice versa. This facility can be used to maintain the equivalence between the graphical and the internal problem representation by bidirectional propagation of changes also in the course of problem solution.

2.5 Version Maintenance

If the toolkit efficiently supports the maintenance of multiple problem versions, the user can learn from problem solving experience and postpone the selection of final solution.

The KEE-AXI system uses three different representation techniques to maintain multiple problem versions:

Knowledge bases — suitable when the different problem versions are intended to be independent of each other.

Object class instances allow simultaneous work with several versions and easy transition between them.

KEE worlds: The reachability relations between versions (i.e., the possibility to reach a version by incremental changes to another version) are expressed explicitly in the form of the KEE world graph which can also be represented graphically using the KEE worlds browser.

3 Problem Analysis

As in other problem domains, the development of methods and algorithms for scheduling of chemical batch plants pursues two conflicting objectives :

- Generalisation - Successively more general problem formulations are being proposed, resulting in increased efficiency problems.
- Efficiency - The efficiency problems are countered by proposals of new, more efficient search methods.

The efficiency problems are addressed by the following methods :

- General efficient methods - which meet the original problem requirements but are more efficient.
- Relaxation - The requirements on the solution are relaxed in some sense (constraints or optimality) and *heuristic* methods are proposed.
- Specialisation - Within the general formulation, knowledge of special problem properties (implicit in the class of formulations) is utilised to select or to construct a specialised problem solver.

Problems of real life scheduling in both the chemical and machine industries are particularly difficult to represent completely and solve rigorously. A recent meeting of experts (AAAI Special Interest Group on Manufacturing (SIGMAN) [13]) made among others the following statements :

- Optimality is hard, if not impossible, to define in a realistic scheduling environment.
- The computation of optimal schedules is a futile enterprise, given the unpredictability of factory operations.
- It is preferable to produce and maintain satisfactory schedules.

In view of these difficulties the scheduling feature of BatchKit does not attempt to provide a rigorous optimization but explores a different approach. A partly heuristic optimization procedure rapidly provides feasible schedules with some favourable characteristics. If desired, this optimization procedure can be applied repeatedly to partially completed schedules, in arbitrary alternation with manual adjustments to the Gantt chart displayed on the screen, where an indication of the deviation from a hypothetical ideal schedule by cumulative product balance curves is also always available.

The scheduling problems of multipurpose batch chemical plants have distinctive features which call for a generalisation of earlier formalisations of the scheduling problem.

An early formalisation - a flow-shop problem for 2 machines - is due to Johnson [12]. Gradually, as more powerful computers and algorithms became available, the problem was extended and generalised. [20] present a job-shop scheduling problem and propose the representation in a directed disjunctive graph.

A definition of the job-shop problem according to Bellman [2], p48 is: When *n* jobs or *n* items are processed by *m* machines, under a given ordering that is, given the machine order to process each job, and given processing times of each job on each machine, the problem is to determine the sequence of *n* jobs on each machine in order to minimize a given objective function (performance measure).

In the present paper, this formulation is further generalised in that *processes* are not characterised by a fixed, linear sequence of operations but by a set of *precedence constraints* which may be represented by an acyclic digraph.

Further, the *plant* is assumed to be a set of equipment items with possibly restricted interconnections. Therefore, the *assignment problem* also has to be solved before scheduling can be performed. In GanttKit, the process-to-plant assignment is solved separately and a set of selected solutions is given, represented by batch patterns.

3.1 Characterization of Batch Production

In this section, some characteristic differences between the chemical and machine industry relevant to the scheduling problem are pointed out.

The assumption of treating the material as individual items can be made more frequently in the machine industry than in the chemical industry where the materials treated are a continuum and batches can be split and merged to any degree and thus lose their identity.

The nature of the physical and chemical processes taking place in chemical batch plants allows some simplifications which are not generally applicable in the machine industry :

- Chemical processes normally cannot be interrupted — no preemption and no inserted idle time can be stipulated.
- Each batch processing equipment item can process only one batch at a time —no overlap is allowed.

A batch chemical product is produced in a series of process steps. The allocation of each step to an equipment item (or machine) for a specified time defines a **batch production pattern**. Different batch patterns may be defined by the assignment of the process steps to different equipment items.

The definition of a batch pattern also includes the **transfer times** between equipment items. These may significantly influence the timing of the schedule, particularly if large volumes of material have to be transferred or there are long connections between successive equipment items.

Constraints on the permitted delays between operations may be imposed when the materials at the end of a step are *unstable*. The extreme case where no delays are permitted between operations is designated as zero wait (ZW). The opposite extreme in which a batch on which an operation has been completed may be allowed to remain in the equipment item indefinitely until the succeeding item becomes available is designated as no intermediate storage (NIS). For intermediate cases minimum and maximum permitted delays between operations may be specified.

The availability and role of **intermediate storage** will also influence the batch pattern and may permit sections of a pattern to be decoupled and treated relatively or completely independently of one another. The treatment of finite intermediate storage (FIS) is not considered in the current prototype GanttKit. However unlimited intermediate storage (UIS) can be specified.

Batches can be merged and split if some equipment items are used in parallel. Different raw materials might be required and also different products of the same batch might be produced at different points in time.

Product demand may be specified in different ways [19]. It is subsequently assumed that orders are identified as triplets of product, quantity and due-date. A common feature of multipurpose batch production is that the feeds for some products may themselves be intermediate products produced in the same plant leading to a product precedence structure.

The **planning horizon** is a time span given by start and end times. The planning horizon can be closed, or open at one end or at both ends. Some intervals in the planning horizon can be disabled to represent holidays or special working conditions.

Constraints on the **availability of resources**, such as equipment (machines), utilities, labour and raw materials are common. These are not currently modelled in GanttKit.

The **changeover** time needed between products on the same equipment item is likely to be sequence dependent, i.e., determined by both the preceding and the succeeding products (a vessel which has just produced black dyestuff needs a thorough cleaning before it can be used to produce white dyestuff, if the production of the two dyestuffs is carried out in reverse order, the cleaning time required may be much shorter). Changeover times but not their associated costs are considered in GanttKit.

The list of constraints could be continued further which suggests that the constraints need a general treatment which will be formalized in the following sections.

3.2 The Objective Function

Many alternative objective functions have been defined to measure the performance of a schedule e.g. completion time, lateness, tardiness and machine idle time ([2], p64;[3], p8).

For the prototype scheduling system in GanttKit an objective function was devised which measures how well different schedules are able to match a given set of product demands.

It is assumed that with an ideal schedule the correct amount of each product will be produced at exactly the right time to satisfy each order of the demand. The total profit earned by this ideal schedule is the sum over all products of the amount of each product supplied, multiplied by its profit per unit. In practice the production pattern will not exactly match the demand pattern. In the objective function, penalties are imposed for deviations from the schedule. When products are produced too early they have to be kept in stock until required by the order, thus incurring a storage cost. When products are produced too late the delivery of the order has to be delayed, thus incurring a lateness charge. Both of these penalties are charged per unit amount and per unit time. The charges are not generally equal, the lateness charge is likely to be higher.

A convenient way of recording deviations from the ideal schedule is by the cumulative difference between production and demand over the planning period, also used in FMS/FAS systems [1] and in other reported work on batch chemical production [26].

The production functions for all the products can be obtained from the set of currently scheduled batches. It is assumed that all feeds are available before the start and all products produced appear after the completion of the batch (see section 5.1). Thus if a terminal (initial) step of a batch ending (starting) at time t produces (consumes) x mass units of product (feed) i, a positive (negative) product increment of magnitude x becomes a set member of the definition of $p_i(t)$, the production function of product (feed) i at time t. (Increments which occur simultaneously are of course added.)

The production function for product i is defined over the set of time points T_p^i as $p_i(t_1^i)$, $p_i(t_2^i), \ldots, p_i(t_{n_p}^i)$, with $t_{j+1}^i > t_j^i$ and $t_0 \leq t_j^i \leq t_e$ $\forall j$ where t_0 is the start time, t_e is the end time of the planning horizon. $p_i(t_j^i)$ is the amount of product i which becomes available at time t_j^i and is of positive sign, any amount of the same product consumed as feed for any batch appears in the production function with a negative sign. A surplus or shortage may also be recorded at the starting time t_0.

The demand function of product i is similarly defined as amount increments over the set of time points T_q^i as $q_i(t_1^i)$, $q_i(t_2^i), \ldots, q_i(t_{n_q}^i)$, where $t_{j+1}^i > t_j^i$ and $t_0 \leq t_j^i \leq t_e$ $\forall j$.

The cumulative balance of product i :

$$R_i(t) = \sum_{t_j^i \leq t} p_i(t_j^i) - \sum_{t_j^i \leq t} q_i\left(t_j^k\right)$$

The cumulative production and demand can be analogously defined at all points including the end point of the planning horizon.

$$P_i(t) = \sum_{t_j^i \leq t} p_i(t_j^i)$$

$$Q_i(t) = \sum_{t_j^i \leq t} q_i(t_j^i)$$

Differentiated penalties c_i^+, c_i^- are imposed for positive or negative deviations from the ideal production schedule reflected in positive or negative values of the cumulative balance. The differential penalty at time t is a function of the cumulative balance at that time, $C_i(R_i(t))$ where the (asymmetric) function $C_i(\tau)$ of real value τ (quantity of product missing or in excess of demand in our interpretation) may be defined e.g. as :

$$C_i(\tau) = \begin{cases} c_i^+ \cdot \tau & : \tau > 0 \\ -c_i^- \cdot \tau & : \tau \leq 0 \end{cases}$$

The cumulative penalty until time t is then

$$L_i(t) = \int_0^t C_i(R_i(\vartheta)) \cdot d\vartheta$$

The symbols c_i^+, c_i^- are the penalties per unit amount of product i per unit time for surplus and deficiency respectively.

The maximum attainable profit for product i is $\Pi_i^{max} = f_i \times Q_i(t_e)$, where f_i is the profit per unit mass of product i supplied. Since it may not be possible to produce the whole of the demand, the actual gross profit is

$$\Pi_i^* = f_i \times (\min(Q_i(t_e), P_i(t_e)))$$

The net profit for product i is

$$\Pi_i = \Pi_i^* - L_i(t_e)$$

and the total net profit is

$$\Pi = \sum_i \Pi_i$$

4. The Scheduling Problem

In this section, the formulation of the scheduling problem and the principles of its solution are presented in general terms in the introductory subsection. The principles of the solution as they are implemented in the GanttKit system are presented in the following subsections.

4.1 Problem Formulation

For the scheduling problem the following are given:

1. Partial plan comprising batches already allocated to equipment items, their timing and other resource utilization.
2. Batch production patterns given by precedence graphs including constraints on timing of successive steps.
3. Demands given by product - quantity - due-date triplets.
4. Change-over constraints between different batches.
5. Holiday constraints.
6. Equipment item break-down constraints.

The problem is to complete the given partial plan so as to fulfil all unfulfilled orders and maximize the total net profit.

4.2 Principles of Solution

In the first approximation, the solution of the scheduling optimization problem can be seen as a two stage, generate-and-test procedure: the first stage generates feasible solutions and the second stage tests them for improvement with respect to the current best solution. In logic programming notation (which is very expressive for the formulation of chronological backtracking, see [15]) :

```
(best-plan ?x)←(possible-plan ?x) ;; Generate
               (improved-plan ?x) ;; Test
               (asserta (best-so-far ?x)) ;; Store
               (fail). ;; Repeat loop
(best-plan ?x)←(best-so-far ?x) ;;Collect result
(improved-plan ?x)←(not (best-so-far ?y)). ;; The first
(improved-plan ?x)←(best-so-far ?y)
                   (better-plan ?x ?y)
                   (retract (best-so-far ?y).
```

In the above program, the predicate possible-plan generates and binds to the variable ?x the possible solutions, i.e., feasible plans. (An attentive reader will have noticed the simplification made above: The set of feasible plans will be infinite e.g. in presence of continuous variables e.g. when the time is modelled by a continuous variable. In such a case, the assumption has to be made that ?x is bound to a suitably selected representation of subsets of feasible plans — e.g., all solutions with a given sequencing of steps but without the actual absolute position on the time axis.)

The identification of the optimal solution by the above program necessitates that the union of the plans generated by the predicate `possible-plan` and bound to the variable `?x` covers the complete set of solutions, i.e., all feasible plans. Due to the infeasibility of the full enumeration of all feasible schedules, incompleteness is introduced in two ways into the above procedure :

1. Only a subset of all possible feasible schedules is generated
2. An allowance is made in the optimality test in favour of efficiency.

This approach would work well if the set of solutions were representative, i.e., if we were able to produce a sufficient number of essentially different solutions (not just in the vicinity of a single local optimum).

The above approach is rather inefficient for scheduling problems of realistic size for two reasons: firstly, the solution space is very large, and secondly strong bounding functions are difficult to design.

4.3 Principles of Solution in GanttKit

The problem formulation in GanttKit is restricted in the following ways :

- Batch splitting and merging can be introduced by explicitly defining batch patterns (recipes) for the same product but with the allocation of more than one equipment item to a process task. The relative timing of the steps of the batch pattern which process the parts of a split batch must reflect the nature of production (e.g. in-phase or out-of-phase parallel production at one stage).

- General resource constraints are not considered.

In GanttKit, the solution of the short-term scheduling problem has been decomposed into two subproblems:

1. Optimization of a relaxed problem (short-cut optimization): Determine the number and types of the batches required to satisfy the given orders and the order in which the individual batches will be allocated to the plan.
2. Allocate sequentially the batches generated by the short-cut optimization to the current state of the plan. This problem is solved using the batch allocation procedure which is capable of allocating a single batch to a partial plan.

An advantage of this decomposition is that the batch allocation procedure can be used also for other purposes. E.g. the user can add a batch "manually" to an existing partial plan using the batch allocation procedure which determines a feasible timing for the batch; or an infeasible plan can easily be reconciled, allowing fast generation of feasible plans under different constraints; a development of this idea is the sensitivity calculation module which is also based on the batch allocation procedure, where the change of the objective function value is recorded under changing constraints.

In the"short-cut" optimization procedure, the maximum total load on an equipment item is used as the objective function, which can be strongly bounded and is independent of most of the constraints of the original problem.

The suggested procedure can be given in logic programming notation as follows:

```
(good-plan ?x)←(best-plan ?y)(feasible-plan ?y ?x).
```

The first subgoal of the procedure generates an optimal solution ?y of the relaxed problem.

The second subgoal generates a feasible solution for the original problem ?x from ?y. It is assumed that an optimal solution of the reduced problem results in a "good" solution for the original problem.

On the other hand, the solution to the relaxed problem may result in a batch sequencing for which no feasible schedule can be found by the second stage, although a feasible solution exists. This may be the case in strongly constrained problems where batch sequencing becomes critical.

4.4 "Short-cut" Optimization

The "short-cut" optimization is a branch and bound procedure which determines a series of batches (with their number, type and desired latest starting time) to be added to the current schedule which are required to fulfil the remaining outstanding orders.

It uses two heuristics :

1. Batches producing the most urgently required products are added first.
2. It seems to be appropriate to assume that only those schedule candidates can be easily timed which evenly distribute the load between different equipment items, i.e., if the maximal equipment item load is minimized.

4.4.1 Product urgency The cumulative difference curves for each products yield the information needed to determine which product is needed most urgently. By considering the missing quantities of products and the corresponding times, the latest time at which the production must start to deliver the required quantity can be calculated for each type of batch which produces the product. The product for which this time is the earliest is considered to be the most urgent (or, the corresponding orders are the hardest to fulfill).

4.4.2 Equipment occupancy The maximum of the workload over the equipment items is independent of the timing of batches. It can be calculated from the specification of the batch patterns to be implemented. Thus a lower bound can be calculated for each partial solution by estimating the smallest total equipment item load (for each equipment item) resulting from fulfilling the remaining orders.

For each equipment item additional loads for unfulfilled orders are added to loads for batches already assigned to give a lower bound for the workload on that item. The overall bound is the largest of these workloads over all equipment items.

It should be noted that in estimating the additional workload needed to satisfy the outstanding orders, additional loads are placed only on equipment items whose use is unavoidable for at least one product. If a product can be manufactured by several batch patterns the workload will be increased only for equipment items required in all the batch patterns. Nothing is added to the load of any other item.

For minimization, the lower bound for a partial solution is an estimate of the objective function value which is lower than the objective function value of any complete solution derived from this partial solution. If the lower bound is greater than the objective function value of an already established complete solution then the corresponding partial solution (and thus all complete solutions which could be derived from it) will be eliminated. This may lead to a substantial reduction of the search space.

The procedure for calculating the lower bound for any partial solution is as follows.

Let $E = \{e_1, e_2, ..., e_{n_e}\}$ be the set of equipment items and let λ_i be the current load of equipment item e_i. Let $P = \{p_1, p_2, ..., p_{n_c}\}$ be the set of products, let $D = \{d_1, d_2, ..., d_{n_c}\}$ be the set of total unfulfilled demands, so that d_i is the total unfulfilled demand of product p_i. Let $B = \{b_1, b_2, ..., b_{n_b}\}$ be the set of batch patterns.

Let $L = L(e_i, d_j, b_k)$ be a function defining the load on equipment item $e_i \epsilon E$ if the demanded amount $d_j \epsilon D$ of product $p_j \epsilon P$ is fulfilled with batch pattern $b_k \epsilon B$.

If some b_k does not produce product p_j then $L(e_i, d_j, b_k)$ is taken as infinity. If b_k produces product p_j then $L(e_i, d_j, b_k)$ is taken to be zero for all equipment items which are not used by b_k otherwise it can be calculated from the mass balance and the equipment item occupancy of the batch pattern.

The minimum load on e_i to satisfy the unfulfilled demand d_j of product p_j can be given by

$$l_i(d_j) = \min_B L(e_i, d_j, b_k); i = 1, 2, ..., n_e; j = 1, 2, ...n_c$$

The minimum load on e_i to satisfy the unfulfilled demand of all products can be given by

$$l_i^{min} = \sum_D l_i(d_j); i = 1, 2, ..., n_e$$

The lower bound value is

$$\max_E(\lambda_i + l_i^{min})$$

4.4.3 The Algorithm The short-cut optimizer works as follows:

1) For all products cumulative difference curves are calculated resulting from the present existing (consistent) schedule and the partial solution, i.e., the set of batches already added.

2) The remaining orders for which the quantity in stock plus the production predicted for the current partial solution is insufficient to satisfy all orders are ranked by their urgency value described in the preceding subsection. The most urgent product is selected.

3) All batch patterns for the most urgent product are considered for adding to the production plan to produce a new short-cut partial solution.

4) Partial solutions are evaluated in terms of a bounding function estimating the minimal necessary total load on equipment (described in the preceding subsection). Partial solutions for which the bound on the load is higher than the value for a complete solution already identified are pruned, i.e., eliminated from further consideration. The remaining partial solutions are ranked according to their bound value and the batch of the partial solution with lowest bound is added to the partial schedule. The partial schedule is not checked for consistency. The choice between partial solutions with the same bounding value is arbitrary.

5) The first four steps are repeated until all orders are fulfilled, i.e., the cumulative difference curves for all products are non-negative at the end of the planning horizon.

At this point a solution is obtained for the sequence of batches to be manufactured to meet the total product demand. This solution is then compared with the stored best solution.

Notes :

1. The partial solution contains only the latest required start time for each required batch without establishing its feasibility.
2. During the subsequent adding of batches to the schedule the actual start time of each batch may be changed by the shifting due to the constraints.
3. The short-cut algorithm therefore uses an approximation or prediction of the cumulative curves.

The logic programming formulation is given by defining the predicate `possible-plan` in the preceding section "Principles of Solution" as folows :

```
(possible-plan
 (?sched
  )
 (or (and
      ;; Solve by finding completion of an empty partial schedule
       (schedule-completion NIL ?sched)
       )

       )
  )
```

```
(schedule-completion
 ;; Find the completion of partial schedule
 (?partial-schedule
  ?complete-schedule
  )
 (or
  (and (cumulative-curve ?partial-schedule
                          ?cumulative-curves)
       (find-hardest-order ?cumulative-curves
                           ?hardest-order)
       (batch-alternatives ?hardest-order
                           ?alternative-batches)
      ;; Select the batch alternative with the
         ;; lowest max. equipment occupancy
       (select-batch-alternative ?partial-schedule
                                 ?alternative-batches
                                 ?batch)
       (better-than-bound ?partial-sched ?batch)
       ;; Schedule is better than current bound
       (add-batch-to-schedule ?partial-schedule
                              ?batch
                              ?new-schedule)
       (extend-if-incomplete ?new-schedule
                             ?complete-schedule)
       )
  )
 )

(extend-if-incomplete
 (?new-schedule
  ?complete-schedule
  )
 (or (and (complete-schedule ?new-schedule)
   ;; No further production required
   (is ?complete-schedule ?new-schedule)
   )
    ;; Unsatisfied orders remain
     (and
      (schedule-completion ?new-schedule
                           ?complete-schedule)
      )
     )
 )
```

Notes :

1. The enumeration of alternative hardest orders is by the predicate `find-hardest-order` which successively selects the alternative orders in the order of increasing latest possible start time. This predicate currently does NOT backtrack, i.e., the sequence of products is given by the recursively applied product urgency criterion.
2. The enumeration of batch alternatives is by the predicate `select-batch-alternative` which successively selects the alternative batches in the order of increasing maximum equipment occupancy.
3. The predicate `schedule-completion` is recursively called in the second clause of the predicate `extend-if-incomplete`.

Optimality in terms of equipment load is guaranteed if the complete search is performed but the program allows the computational effort to be reduced if necessary by, for example, reducing the number of alternatives considered at each level or generating a solution which, while not necessarily optimal, is guaranteed to be within a specified interval from the optimum.

The "short-cut" optimizer returns a list of order - batch pattern pairs, so that the orders are sorted according to their urgency, from left to right, e.g.,

```
((ORDER-3 A-BATCH) (ORDER-4 B-BATCH)
 (ORDER-3 A-BATCH) (ORDER-2 B-BATCH))
```

In the partial schedule for which the optimization has been carried out, `ORDER-2`, `ORDER-3` and `ORDER-4` were unfulfilled. Note that `ORDER-3` appears twice because one batch of `A-BATCH` could not fulfil it completely and with its remaining unfulfilled quantity it became the 3rd most urgent demand.

4.5 Batch Allocation

The batch allocation procedure processes the list of order/batch pattern pairs returned by the short-cut optimizer and progressively extends the partial plan by adding each batch to the list. After adding each batch, a new, feasible plan results.

The timing of the batch is subject to two types of constraints :

- Plan constraints (global and inter-batch).
- Precedence constraints (intra-batch).

The plan constraints are those by which the current state of the plan constrains the allocation of steps of a batch. They include the partial occupancy of available resources (equipment, labour, utilities etc.) by already allocated batches and general constraints such as no overlapping, no preemption, availability of resources, holidays, limits on utilities, labour etc.

Precedence constraints are imposed upon the sequence of processing steps by which each batch is processed and are internal to the batch. They may reflect e.g. transfer time and material stability between successive steps. The precedence relation can be represented as a connected precedence graph from which the predecessors and successors of each step can be determined. Steps without predecessors are called start steps, steps without successors are called terminal steps.

A precedence constraint determines the earliest and latest starting times of a step relative to the end of a predecessor step. If a step has more than one predecessor then its earliest and latest relative start times are constrained by the conjunction of all its precedence constraints. If the interval between the earliest and latest start time for a step is empty then the batch cannot be allocated.

4.5.1 Feasible Allocation The batch to be allocated is defined as a graph $G(S, P)$ where $S = \{s_1, s_2, ..., s_n\}$ is the set of steps to be allocated, P is a subset of $S \times S$ defining precedences for some pairs of S. If $(s_i, s_j)\epsilon P$ then s_j is a successor step of s_i or conversely s_i is predecessor of s_j. Note : $G(S, P)$ can be represented by a directed acyclic graph (precedence graph).

Let $T = \{t_1, t_2, ..., t_n\}$ so that t_i is the start time of s_i, $U = \{u_1, u_2, ..., u_n\}$ so that u_i is the duration of s_i.

A precedence constraint $\mathcal{D}_{ij}^{\pm}$ is a pair $(\mathcal{D}_{ij}^{-}, \mathcal{D}_{ij}^{+})$. $\mathcal{D}_{ij}^{-}$ and $\mathcal{D}_{ij}^{+}$ constrain the earliest and the latest starting time of s_j, respectively, in terms of its predecessor s_i, as follows:

1. $\mathcal{D}_{ij}^{-} : t_j \geq t_i + u_i + d_{ij}^{-}$
2. $\mathcal{D}_{ij}^{+} : t_j \leq t_i + u_i + d_{ij}^{+}$

where d_{ij}^{-} and d_{ij}^{+} are predefined values, $d_{ij}^{-} \leq d_{ij}^{+}$.

The relative timing of s_i and s_j is feasible if the constraint pair $\mathcal{D}_{ij}^{\pm}$ is fulfilled. A predicate

$$F^G(\mathcal{D}_{ij}^{\pm})$$

can be given which is true if and only if the allocation of s_j is feasible in terms of one of its predecessors s_i. A step s_j is feasible in terms of all its precedence constraints iff

$$\forall i(s_i, s_j)\epsilon P \longrightarrow F^G(\mathcal{D}_{ij}^{\pm})$$

4.5.2 The Partial Plan The partial plan is defined as a set of constraints $\mathcal{C}^P$. To determine how the partial plan constrains the allocation of s_j, a predicate function F^P can be given

$$F^P(s_j, \mathcal{C}^P)$$

which is true if and only if the allocation of s_j (i.e., including the starting time) is feasible in terms of $\mathcal{C}^P$.

It is assumed that the $\mathcal{C}^P$ is constant, i.e., during the allocation of a batch it does not change. This assumption allows the separation of the treatment of precedence and plan constraints.

4.5.3 The Problem The problem is to find a feasible set of timings $T' = \{t_1', t_2', ..., t_n'\}$ so that

$$\forall j((\forall i(s_i, s_j) \epsilon P \longrightarrow F^G(\mathcal{D}_{ij}^{\pm}) \wedge F^P(s_j, \mathcal{C}^P))$$

4.5.4 The Solution The algorithm used to allocate a batch in an existing schedule in a feasible manner is similar to those presented e.g. in [6] or [5]. The algorithm searches for the feasible timing of all steps of a batch to give a completion time which is latest before or earliest after the given starting position which was determined e.g. by the short-cut optimization. The feasible timing is obtained by repeated shifting of steps - always in the same direction - by the minimal amount to obtain a new starting time which makes a step feasible with respect to all plan constraints and one of its precedence constraints. Because this new location might violate other precedence constraints of the step considered, backtracking is made until no constraint is violated. It is, however, possible that no feasible timing exists between the start and end time of the planning horizon.

If a production unit has to fulfil a given order by some due date then an initial timing for all of its constituent batches can be given which satisfies the order just-in-time. In the case that this just-in-time allocation is not feasible, a backward shifting is tried (it is better to produce too early rather than too late). If the start time of the planning horizon prevents finding a feasible solution using backward shifting then forward shifting can be tried starting from the just-in-time allocation.

4.5.5 Implementation of plan constraints Let s_i and s_j be two steps already allocated to the same equipment item so that s_i precedes s_j, i.e., $t_j \geq t_i$ and there is no step between s_i and s_j. The constraint $\mathcal{O}_{ij} \epsilon \mathcal{O}$ can be given as follows:

$$t_j \geq t_i + u_i + o_{ij}$$

where o_{ij} is the changeover time required after s_i when it is followed by s_j. Normally $o_{ij} \neq o_{ji}$.

The problem is to allocate an additional step s_k between two already allocated consecutive steps — s_i followed by s_j on the same equipment item - so that the changeover constraints will be satisfied. The feasible location, t_k for s_k can be calculated as follows:

$$t_k \geq t_i + u_i + o_{ik}$$

$$t_j \geq t_k + u_k + o_{kj}$$

The above formulation implicitly contains two other assumptions:

- Each step of a batch can be allocated to only one equipment item.
- After finishing a step the equipment item becomes immediately free for further use subject to the corresponding changeover time.

The work pattern or holiday constraints $\mathcal{H} \subset \mathcal{C}^P$ restrict the use of equipment in certain periods of time. A holiday constraint $\mathcal{H}_i \epsilon \mathcal{H}$ can be expressed as an inequality inhibiting the overlapping of all steps with a predefined time interval.

The planning horizon represents a hard constraint. If, during shifting, any step of the batch to be allocated has reached the end of the planning horizon, i.e., if $t_i + u_i > t_\epsilon$ then the corresponding allocation is infeasible.

5. Representation of Scheduling

The main objective was to find a representation which enables a realistic and flexible description of short—term scheduling problems and which allows flexible manipulation of schedules both by programs and by the user through the (graphical) interface.

To fulfil this loosely formulated main objective the system has to be capable of

- Representing batch production patterns as building blocks of schedules.
- Maintaining any number of (partial) solutions of a scheduling problem.
- Allowing both graphical and programmed manipulation of the schedule, while maintaining the equivalence between the two forms of representation. The graphical and programmed manipulations include creation, modification and deletion of objects.
- Representing the constraints as global objects which are to be applied to all scheduling variants.

With these objectives an entity-relationship schema has been developed (Figure 1). Note that the actual model in GanttKit contains many attributes and functions and also some object classes which are not denoted here, however the structure is essentially the same.

The meaning of symbols in Fig.1 is as follows :

- Rectangles represent object classes.
- Rounded rectangles represent relations (among members of object classes).
- Ovals represent attributes of the members of object classes.

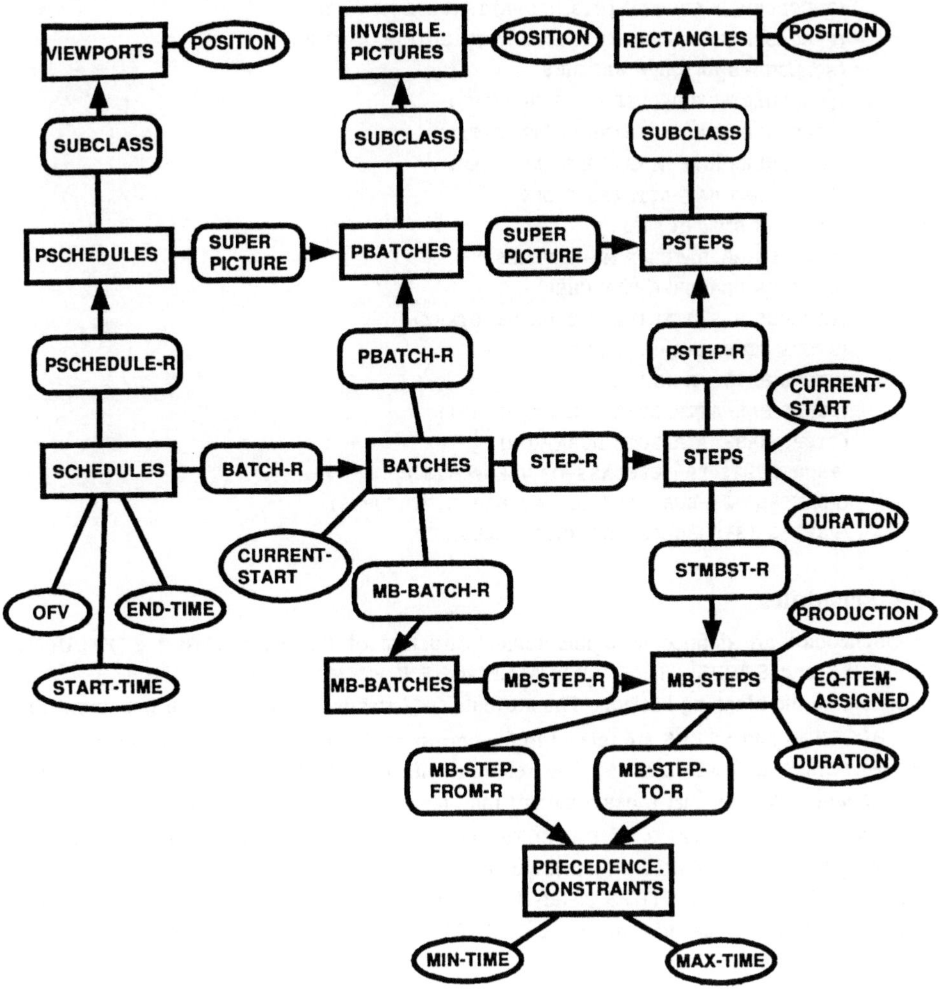

Figure 1: ER-Graph

The E-R diagram of Figure 1 can be transformed into a set of AXI assertions as follows:

```
(MB-BATCHES MB-STEP-R MB-STEPS)
(MB-STEPS MB-STEP-FROM-R PRECEDENCE-CONSTRAINTS)
(MB-STEPS MB-STEP-TO-R PRECEDENCE-CONSTRAINTS)
(MB-STEPS HAS-ATTRIBUTE EQ-ITEM-ASSIGNED)
(MB-STEPS HAS-ATTRIBUTE DURATION)
(MB-STEPS HAS-ATTRIBUTE PRODUCTION)
```

```
(PRECEDENCE-CONSTRAINTS HAS-ATTRIBUTE MIN-TIME)
(PRECEDENCE-CONSTRAINTS HAS-ATTRIBUTE MAX-TIME)
(SCHEDULES BATCH-R BATCHES)
(SCHEDULES PSCHEDULE-R PSCHEDULES)
(SCHEDULES HAS-ATTRIBUTE START-TIME)
(SCHEDULES HAS-ATTRIBUTE END-TIME)
(SCHEDULES HAS-ATTRIBUTE OFV)
(BATCHES STEP-R STEPS)
(BATCHES MB-BATCH-R MB-BATCHES)
(BATCHES PBATCH-R PBATCHES)
(BATCHES HAS-ATTRIBUTE CURRENT-START)
(STEPS STMBST-R MB-STEPS)
(STEPS PSTEP-R PSTEPS)
(STEPS HAS-ATTRIBUTE CURRENT-START)
(STEPS HAS-ATTRIBUTE DURATION)
(PSCHEDULES IS-SUBCLASS-OF VIEWPORTS )
(PBATCHES IS-SUBCLASS-OF INVISIBLE.PICTURES)
(PSTEPS IS-SUBCLASS-OF RECTANGLES)
```

5.1 Schedules

Schedules are defined as a hierarchical structure of the object classes SCHEDULES, BATCHES and STEPS. The START-TIME and END-TIME attributes of the SCHEDULES class define the planning horizon. The attribute OFV stands for the objective function value.

BATCHES and STEPS are related to the corresponding production pattern object classes: MB-BATCHES and MB-STEPS. The relations MB-BATCH-R and STMBST-R have N:1 functionality showing that a batch pattern may have any number of instances.

The CURRENT-START of BATCHES means absolute time in the planning horizon, whereas the CURRENT-START attribute of STEPS is relative to that of BATCHES. The DURATION attribute of STEPS stands for the occupancy of the corresponding equipment item. Its value is equal to that of the DURATION of the related MB-STEPS object when the step is created.

Both the user and the scheduling algorithm can independently and simultaneously manipulate any number of SCHEDULES - versions of a scheduling problem. This allows backtracking to earlier solutions when the version currently pursued proves to be infeasible.

5.2 Batch Production Patterns

The notion of **batch pattern** plays a central role in the GanttKit system — a schedule is a set of allocated batches, each of which is derived from one of the batch patterns by assigning a starting time to each step.

Precedence graphs have been found to be adequate for the representation of **batch patterns**. As an example consider Figure 2. The nodes correspond to the operations carried out in order to transform the batches. Each node has a unique identifier (Step0 ... Step8), an equipment item (e.g. Vessel1) and (except the terminal nodes) a duration associated with it. Terminal nodes describe the sources of raw materials (e.g., Step0) and final destination of products (e.g., Step7 and Step8). The equipment item associated with a terminal node can be final or intermediate storage. The components required and produced by a batch are also described by terminal nodes. The batch pattern in Figure 2. e.g. requires 500 mass units of A and 500 mass units of B at Step0 and produces 100 mass units of B and 900 mass units of C at Step7 and Step8, respectively.

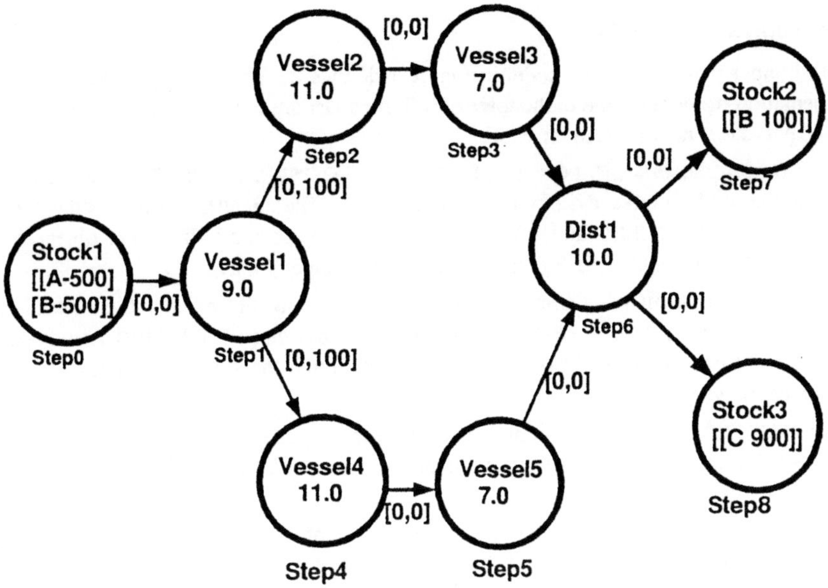

Figure 2: Precedence Graph

The arcs of the batch pattern precedence graph are labeled by the earliest and latest start times of the step at the end of the arc relative to the end time of the step at the beginning of the arc. E.g. the pair $[0,0]$ expresses a zero wait policy between two successive operations.

The precedence graph $G(S,P)$ is represented by the object classes `MB-BATCHES`, `MB-STEPS` and `PRECEDENCE-CONSTRAINTS`. The relation `MB-STEP-R` with functionality 1:N between `MB-BATCHES` and `MB-STEPS`, defines that a batch consists of any number of steps. A step is always carried out in one equipment item, described by the `EQ-ITEM-ASSIGNED` attribute. The `DURATION` attribute defines the time required to carry out the step on the equipment item. The PRODUCTION attribute which is used only for terminal steps describes the required or produced materials in the form of a list of pairs, e.g.

```
( (A-PROD -500.0) (B-PROD -500.0) )
```

The first element of each pair is the name of the material, the second one is the quantity which is negative for feeds and positive for products.

The precedence constraints between batch steps are described by the PRECEDENCE-CONSTRAINTS object class. MIN-TIME constrains the minimal, MAX-TIME the maximal time lag between the finish time of a step and the start time of one of its successors.

5.3 Graphical Objects

The upper part of the E-R model in Figure 1 defines the objects supporting the graphical representation of the schedule in the form of a Gantt chart implemented using the KEEPictures graphical system incorporated in KEE.

The object classes VIEWPORTS, INVISIBLE.PICTURES and RECTANGLES are defined in the KEEPictures system knowledge base. The GanttKit picture object classes PSCHEDULES, PBATCHES and PSTEPS are subclasses of the KEEPicture object classes; they inherit the attributes and functionalities of their superclasses.

Only two levels in the picture hierarchy are really visible, namely the objects of the PSCHEDULES and PSTEPS object classes, whereas the class PBATCHES is defined as a subclass of INVISIBLE.PICTURES; its role is to define the structure of graphical objects without being visible.

The SUPERPICTURE relation and its inverse the SUBPICTURES relation are defined by the KEEPictures system. The structure of the picture object classes corresponds to that of the schedule.

In accordance with the equivalence between the graphical and internal representation, the relations between the internal and graphical objects have a 1:1 functionality, i.e., exactly one graphical object corresponds to each schedule object. The internal representation objects can be used independently of the graphical ones : a schedule may exist without the graphical representation and the user may decide when s/he wants to display it graphically.

An important attribute of the grapical objects is POSITION which defines the horizontal and vertical location of the object picture relative to its superpicture in terms of pixels.

6. Interactive Operations

The GanttKit Panel is the pictorial representation of a schedule in the form of a Gantt chart. A panel is implemented as a KEEPictures viewport, or more precisely as an object of the PSCHEDULES object class which is defined as follows:

```
(PSCHEDULES    IS-SUBCLASS-OF    VIEWPORTS)
```

Similarly to any other viewport, the viewed picture of a GanttKit panel is the topmost pictorial object of all pictures inside the panel.

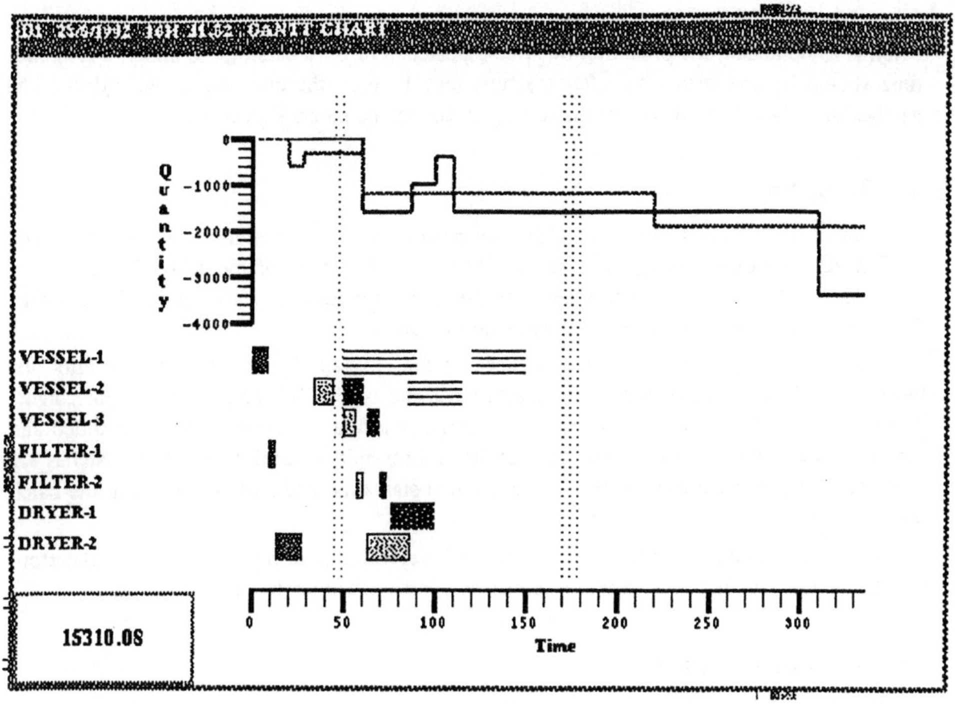

Figure 3: GanttKit panel with three manually allocated batches

A GanttKit schedule panel (such as the one in Fig.3) consists of the following components:

- The panel itself with a frame.
- The **time axis** (x-axis) representing the planning horizon.
- **Equipment item axis** (y-axis) labeled by the equipment items.
- **Objective function value field** appears in the left bottom corner of the panel. In constrast to other picture objects of the GanttKit system, it is not a KEEPicture object but an active image.
- **Cumulative curve plots** are displayed in the upper part of the panels. Depending on the value of the APLOT-TYPE attribute of a schedule which is either demand, or production, or difference, the cumulative demand, or production, or difference of one or more products are shown. The cumulative difference curve is the default.
- **Schedule** appears as a set of boxes in GanttKit panels. Section 7 contains a detailed discussion of schedules.

6.2 Add batches

The **Add batches** operation is the implementation of the batch allocation algorithm defined earlier. It allows one or more batches to be added to an existing schedule. Each batch added is an instance of one of the predefined batch patterns of the current set-up. If more than one batch is to be added, the operation must be repeated. For each batch to be added a reference time should be specified (the reference time may be e.g. the due date of an order). The batches already allocated will not be changed during the batch allocation.

6.3 Reconcile

Manual operations or changing of the constraints may make existing schedules infeasible. In GanttKit, the exact optimum is not sought but the feasiblity of the schedules is a basic requirement. Therefore, a mechanism is needed which makes infeasible schedules feasible. This mechanism in Ganttkit is the Reconcile operation.

There are many ways to reconcile a schedule, e.g., trivially by deleting all of its allocated batches. GanttKit's Reconcile leaves the number and type of allocated batches unchanged. Their timing will be modified as follows: first the schedule to be reconciled is emptied and the batches are temporarily stored (this results in a feasible schedule). Then the batches are taken one by one, according to increasing original start time and reallocated using the batch timing algorithm described earlier.

The batch timing algorithm uses the original step timings as initial start times, therefore the Reconcile is efficient. A feasible schedule is left unchanged by reconciliation.

6.4 Cumulative Curve

The **Cumulative Curve** menu item allows the cumulative difference curve, which was defined earlier, to be displayed.

6.5 Fulfil Rest

The **Fulfil Rest** is the principal operation of the GanttKit tool. It fulfils the currently unsatisfied product demand using the B&B scheduling algorithm described earlier.

6.6 Panel

The **Panel** menu item has only one subitem: "Delete". It allows deletion of the panel (the pictorial representation of the schedule) without deleting the schedule itself.

Any schedule currently not displayed can be displayed again any time later using the **Schedule - Display** menu item of the global GanttKit menu bar.

6.7 Schedule

The **Schedule** menu item has two subitems : **Copy** and **Delete**.

Copy makes a duplicate of a schedule under a new name. **Delete** can delete an empty schedule, i.e., one which does not contain any batches. A non-empty schedule should first be emptied (see next section) before it can be deleted.

6.8 Unfulfilled Orders

The cumulative difference curve makes it possible to identify the unfulfilled orders which can be displayed using this menu item.

7. Operations on Schedules

A schedule appearing in a GanttKit panel is a complex structure consisting of a hierarchy of SCHEDULES, BATCHES and STEPS objects and all related objects, including those implementing the graphical representation.

This structure enables us to **navigate** among objects. The navigation allows related objects to be reached starting from a given object along paths composed of relations and inverse relations. On the user interface only the PSTEPS objects, i.e., the graphical representations of the STEPS are directly accessible with the mouse.

Because the operations on many different objects are very similar to each other the following subsections are presented in terms of operations, always noting how the particular operation applies to different objects.

7.1 Highlight

Highlighting is the way to emphasize selected parts of the structure of a schedule by graphical means, e.g. by changing the fill pattern of a picture. Any object at each level of the object hierarchy can be highlighted to reveal the object structure of the schedule.

7.2 Move

The **Move** operation allows parts of the schedule (STEPS, BATCHES) to be shifted along the time axis. When moving an aggregated object, all its subobjects are moved simultaneously.

Inherently, many constraints apply to moves along the time axis. The immediate propagation of their effects is not used for several reasons. Firstly, it may be sometimes necessary to allow inconsistent states of the schedule which can be reconciled later. Secondly, the propagation of too many time constraints would slow down the move operation. Thirdly, many of these constraint violations may be obvious to the user.

A move operation includes the following steps:

1. Identifying the object to move.
2. Collecting all component objects to be moved together.
3. Moving the selected objects with the mouse.
4. Calculating the new start time for each moved object.
5. Recalculating the objective function and updating the product plots.
6. Shading the panel which marks potential inconsistency.

7.3 Move+

The **Move+** operation allows any user defined set of steps (of the same or different batches) to be moved simultaneously. To accomplish this, the steps to be moved together should first be designated with a **Mark** operation which is available for batches and steps.

Steps or sets of steps which have been marked by mistake can be "unmarked" with the **Demark** operation. The colour of demarked steps will be set back to the colour before the marking.

7.4 Delete

A complete schedule or any batch of a schedule can be deleted. Because batches are the smallest independent parts of a schedule, objects lower in the hierarchy, such as steps, cannot be deleted. Deleting a batch means the physical deletion of the object, whereas deleting a schedule deletes only its lower level objects and leaves an empty schedule.

After a deletion the objective function value is recalculated and the contributions of the products are displayed in the KEE Typescript window.

7.5 Verify

The **Verify** operation, which is available for steps and schedules, carries out the following tests:

1. Equipment items can carry out only one step at a time (no overlapping).
2. For each step of the schedule :
    a. Changeover constraints.
    b. Transfer time constraints.
    c. Material stability constraints.
    d. Holiday constraints.
    e. Equipment break down constraints.

8. Interactive Problem Editing

Scheduling problems normally require a large amount of data which may be prepared in an input file or, in the best case, are gathered from a company-wide database. A third possibility, implemented in the GanttKit system, is to use an interactive data editor. For this purpose, the interactive data editing DATAIMAGE tool was developed which can be attached to any KEE knowledge base.

A DATAIMAGE control unit which can be attached to any KEE object class allows the user to define which member units of the class and which of their attributes should be displayed and how they should be treated. More than one DATAIMAGE control unit can be attached to any object class which allows the same object class to be visualised in different ways.

Figure 13 shows a DATAIMAGE panel which is attached to the **ORDERS** object class. Here, all member units are displayed, sorted by product name and due-date.

A DATAIMAGE panel can be used to add or to delete member units, or to change attribute values. The possibility of changing a unit's attribute is indicated by a heavily lined box. In Figure 13 e.g. columns 1, 3, and 4 are mouse sensitive and can be changed but 2 and 5 are not. The first column contains the object (unit) names.

8.1 Sensitivity

It is often of interest to know how the optimal (or suboptimal) solution of a system changes with variations of the parameters in the formulation. Even if the global optimum is not found it is meaningful to ask how a suboptimal solution changes if a variable is changed and a new suboptimal solution is found by the given problem solver, which in our case is the **Reconcile** operation.

The DATAIMAGE system allows sensitivity analysis to be carried out using as the independent variable any numerical data appearing in a mouse sensitive DATAIMAGE box. A bound value is to be defined for the independent variable and GanttKit reoptimizes the problem for 10 equidistant values of the independent variable between the current value and the bound value. A sensitivity plot is made which shows the independent variable on the x-axis and the change of the objective function value relative to the current value on the y-axis.

The use of the **Reconcile** operation for sensitivity analysis has important advantages but also certain limitations. It is advantageous in that the results of the sensitivity analysis correspond to those obtained by the use of the **Reconcile** operation. Moreover, the solution technique is efficient. However, because the **Reconcile** operation modifies only infeasible schedules, the problem formulation at the bound value of the examined variable should be more constrained than at the current value. Otherwise the indicated change of the objective function will be zero. Because both the GanttKit optimizer and the sensitivity analysis compute suboptimal solutions, the objective function value obtained may be greater for a more constrained problem than for a less constrained one. Further details about the sensitivity analysis are given in the example.

8.2 Versions

GanttKit allows many problem versions to be maintained simultaneously. This feature allows the "what - if" type of questions which are often of interest in ill-defined and dynamic problems such as scheduling.

The problem versions are represented by KEE's **KEEworlds** system which allows different versions of a single problem to be built by incremental changes starting from a root variant (background) and their interdependencies to be subsequently maintained as nodes of an acyclic, directed graph.

9. Illustrative Example

This section illustrates some of the important features of the GanttKit system but is not intended to be a comprehensive numerical example.

A short time scheduling problem should be solved with a planning horizon of two weeks, i.e., 336 hours, starting on Monday at 0 o'clock and finishing next week on Sunday at 24 o'clock.

The plant represented in a SimKit design window (Figure 4) consists of 3 vessels, 2 filters and 2 dryers and facilities to supply raw material and store final products. The plant works in three shifts, round the clock. Two plant shut-downs are defined, the first one from 45 to 50 hours and the second one from 170 to 180 hours. It is assumed that VESSEL-1 is not available between 50-90 and 120-150 and VESSEL-2 between 85-115.

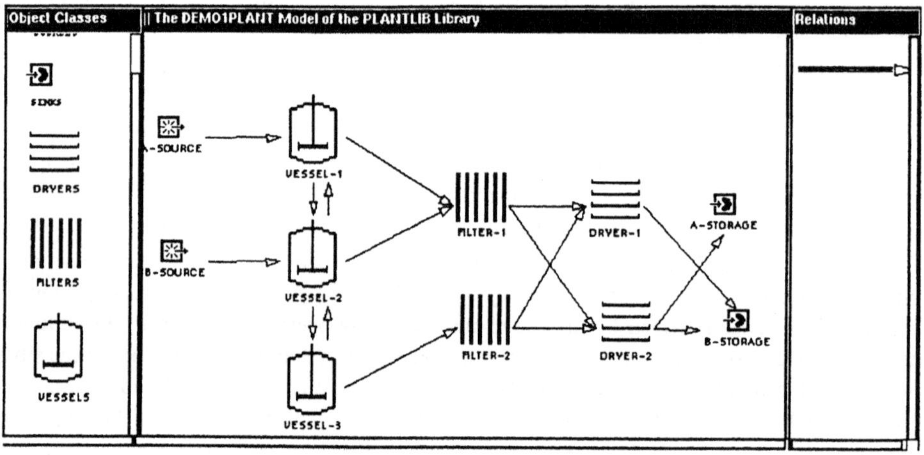

Figure 4: Plant in SimKit design window

Two products — RED-PROD and BLUE-PROD — can be produced according to Table 1, i.e., there is one production variant (batch pattern) for RED-PROD and two for BLUE-PROD. Table 2 together with Table 3 and 4 define three batch patterns RED-BATCH, BLUE-BATCH-A and BLUE-BATCH-B. For the sake of simplicity the terminal nodes are omitted and it is assumed that the products appear always at the end of the last batch step.

Table 1: Batch patterns and production quantities

Product	**Batch Pattern**	**Batch size [kg]**
RED-PROD	RED-BATCH	300.0
BLUE-PROD	BLUE-BATCH-A	600.0
	BLUE-BATCH-B	600.0

Table 2: Batch patterns and step durations

Batch Pattern	Step	Equipment	Dur	Successors
RED-BATCH	RED-STEP-1	VESSEL-1	9.0	RED-STEP-2
	RED-STEP-2	FILTER-1	4.0	RED-STEP-3
	RED-STEP-3	DRYER-2	15.0	
BLUE-BATCH-A	BLUE-STEP-A1	VESSEL-2	11.0	BLUE-STEP-A2
	BLUE-STEP-A2	VESSEL-3	7.0	BLUE-STEP-A3
	BLUE-STEP-A3	FILTER-2	4.0	BLUE-STEP-A4
	BLUE-STEP-A4	DRYER-1	24.0	
BLUE-BATCH-B	BLUE-STEP-B1	VESSEL-2	11.0	BLUE-STEP-B2
	BLUE-STEP-B2	VESSEL-3	7.0	BLUE-STEP-B3
	BLUE-STEP-B3	FILTER-2	4.0	BLUE-STEP-B4
	BLUE-STEP-B4	DRYER-2	24.0	

Table 3: Precedence constraints

Batch Pattern	FROM	TO	MIN	MAX
RED-BATCH	RED-STEP-1	RED-STEP-2	2.0	7.0
	RED-STEP-2	RED-STEP-3	1.0	3.0
BLUE-BATCH-A	BLUE-STEP-A1	BLUE-STEP-A2	2.0	5.0
	BLUE-STEP-A2	BLUE-STEP-A3	0.0	2.0
	BLUE-STEP-A3	BLUE-STEP-A4	2.0	3.0
BLUE-BATCH-B	BLUE-STEP-B1	BLUE-STEP-B2	2.0	5.0
	BLUE-STEP-B2	BLUE-STEP-B3	0.0	2.0
	BLUE-STEP-B3	BLUE-STEP-B4	2.0	3.0

Changeover times are defined in Table 4 for those batch steps which are assigned to the same equipment item.

Table 4: Changeover constaints

Equipment	From step	To step	Time
VESSEL-1	RED-STEP-1	RED-STEP-1	4.0
VESSEL-2	BLUE-STEP-A1	BLUE-STEP-A1	1.0
	BLUE-STEP-B1	BLUE-STEP-B1	4.0
	BLUE-STEP-A1	BLUE-STEP-B1	4.0
	BLUE-STEP-B1	BLUE-STEP-A1	4.0
VESSEL-3	BLUE-STEP-A2	BLUE-STEP-A2	4.0
	BLUE-STEP-B2	BLUE-STEP-B2	4.0
	BLUE-STEP-A2	BLUE-STEP-B2	4.0
	BLUE-STEP-B2	BLUE-STEP-A2	4.0
FILTER-1	RED-STEP-2	RED-STEP-2	4.0
FILTER-2	BLUE-STEP-A3	BLUE-STEP-A3	4.0
	BLUE-STEP-B3	BLUE-STEP-B3	4.0
	BLUE-STEP-A3	BLUE-STEP-B3	4.0
	BLUE-STEP-B3	BLUE-STEP-A3	4.0
DRYER-1	BLUE-STEP-A4	BLUE-STEP-A4	5.0
DRYER-2	RED-STEP-3	RED-STEP-3	2.0
	BLUE-STEP-B4	BLUE-STEP-B4	5.0
	RED-STEP-3	BLUE-STEP-B4	3.0
	BLUE-STEP-B4	RED-STEP-3	3.0

Table 5: Profits and penalties

Product	Profit [SFr/kg]	Penalty+ [SFr/100kg*h]	Penalty- [SFr/100kg*h]
RED-PROD	25.0	0.285	2.85
BLUE-PROD	30.0	0.342	3.42

The profit and penalty for over- and underproduction are defined in Table 5, the initial stock for each product is assumed to be 0.0. The penalty for underproduction is 10 times higher than that for overproduction.

The orders for products `RED-PROD`, `BLUE-PROD` with due dates and quantities are given in Table 6.

Table 6: Demand in Kg.

Hour	RED-PROD	BLUE-PROD
20	600.0	
60	900.0	1600.0
110		1200.0
220	700.0	
310		1800.0

To solve the problem, first a new, empty panel is opened (Figure 5), consisting of a Gantt chart in the lower part and a curve plot in the upper part.

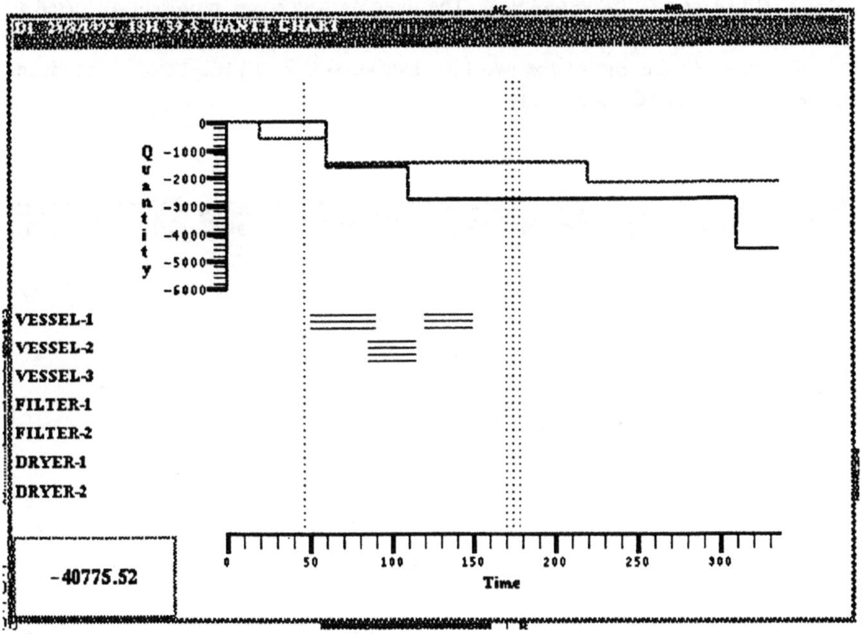

Figure 5: Empty GanttKit panel

The time axis of the Gantt chart represents the planning horizon of 0 to 336 hours, the Y-axis the equipment items. The two vertical shaded bars represent the two plant shut-down constraints, the horizontal bars represent the three equipment item break-downs. The field in the bottom left corner of the panel shows the objective function value recalculated whenever the schedule is modified. The contribution of each demanded product to the objective function value is typed out in the typeout window.

When an empty panel is created as in Figure 5, e.g. the following message appears:

```
Current and maximal profit for schedule D1
 BLUE-PRO -25978.32 138000.00
 RED-PROD -14797.20 55000.00
-------------------------------------------
                 -40775.52. 193000.00
```

The maximal profit for BLUE-PROD of SFr 138000.00 results from the total demand of BLUE-PROD of 4600 kg and the profit per unit amount of BLUE-PROD of 30.00 SFr/kg. The 2200 kg total demand of RED-PROD with 25.00 SFr/kg unit profit yields SFr 55000.00, i.e., the total maximal profit is SFr 193000.00. The initial profit for both products is negative due to the penalty because no orders can as yet be fulfilled.

The upper part of the panel shows the cumulative difference curve for both products. Because no batches have as yet been allocated, the curve for both products is equal to the cumulative demand given in Table 6.

Figure 3 is an example of the batch allocation procedure and its implementation by the **Add batches** menu item. One batch was allocated for each batch pattern RED-BATCH, BLUE-BATCH-A and BLUE-BATCH-B. The three dashed boxes represent allocated RED-BATCH steps, the light grey boxes are BLUE-BATCH-A steps, the dark boxes are BLUE-BATCH-B steps. At the end of the two blue batches 600 kg BLUE-PROD is produced, so the CD curve is updated accordingly.

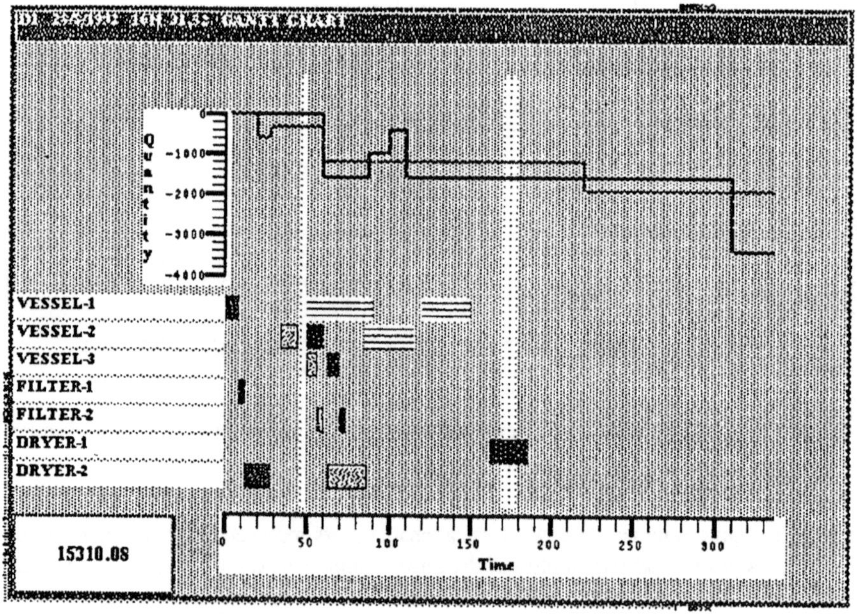

Figure 6: GanttKit panel after manual operation

Figure 6 is an illustration of a manual operation which has been carried out on the panel in Figure 3. The last step of one of the blue batches has been shifted forward to a time where it violates some of the constraints, because it overlaps with a holiday and also violates the precedence constraints. The panel is shaded indicating possible inconsistencies.

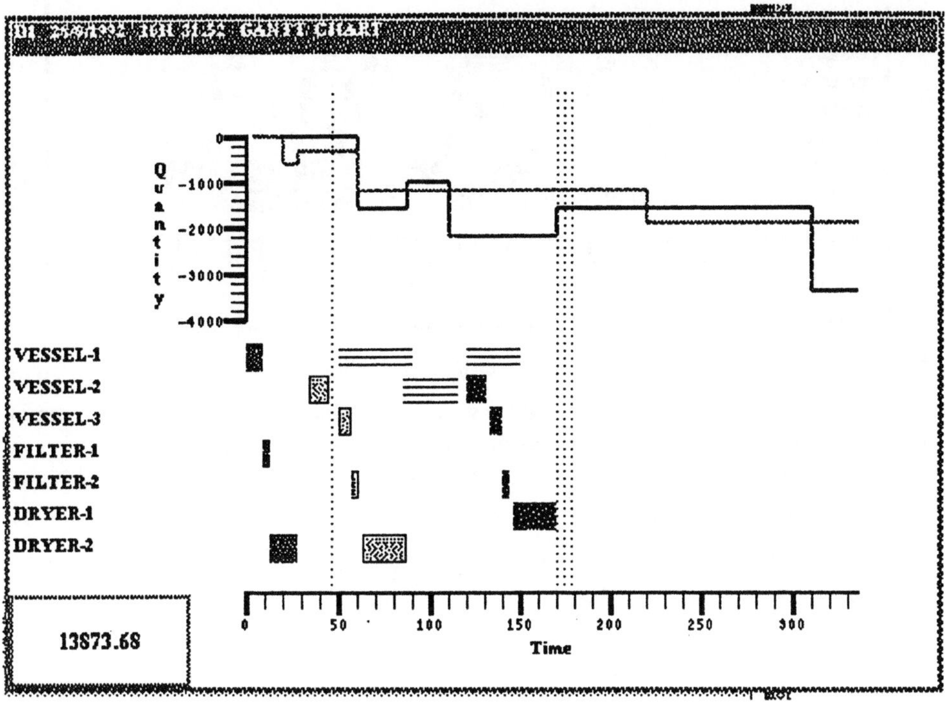

Figure 7: GanttKit panel after restoration of consistency using Reconcile

The **Reconcile** operation can be used to remove the inconsistencies. Figure 7 shows the new state after reconciling the schedule of Figure 6. The step which was previously manually moved has been shifted backward to the latest feasible time and the steps belonging to the same batch have also been adjusted in order to satisfy the precedence constraints. The resulting panel is feasible, the cumulative difference curve has been updated, the objective function recalculated and the background set to white.

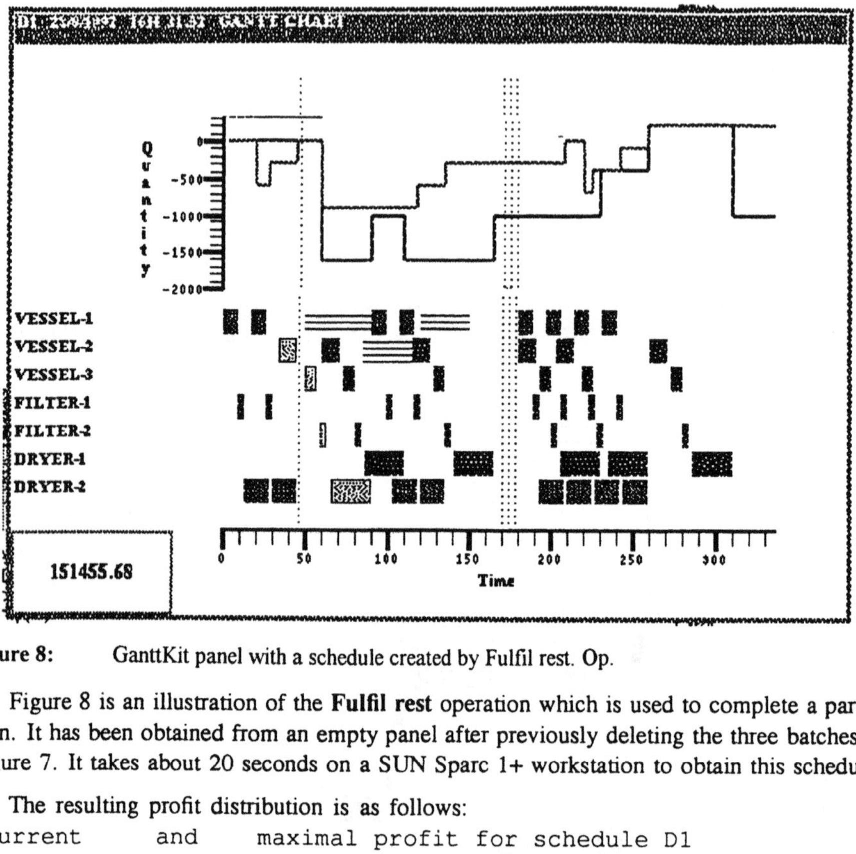

Figure 8: GanttKit panel with a schedule created by Fulfil rest. Op.

Figure 8 is an illustration of the **Fulfil rest** operation which is used to complete a partial plan. It has been obtained from an empty panel after previously deleting the three batches of Figure 7. It takes about 20 seconds on a SUN Sparc 1+ workstation to obtain this schedule.

The resulting profit distribution is as follows:

Current	and maximal	profit for schedule D1
BLUE-PROD	129739.33	138000.00
RED-PROD	50392.80	55000.00
	180132.13	193000.00

The 8 batches of BLUE-PROD each of size 600 kg produce 200 kg more than the demanded 4600 kg, the 8 RED-PROD batches each of size 300 kg yield 200 kg of product in excess of the 2200 kg total demand. The BLUE-PROD CD curve is in the negative region between about 60 - 260 hours which means that orders with due dates falling in this interval cannot be fulfilled on time. Also, the last order of BLUE-PROD cannot be completed within the planning period which is indicated by the negative CD curve between hours 300 to 310, showing a deficit of 1000.0 at the end of the planning period.

HOLIDAYS

UNIT-NAME	FROM	TO
HOLIDAY-1	45	50
HOLIDAY-2	170	180

Equipment Break-downs

UNIT-NAME	EQITEM	FROM	TO
EQBREAK-1	VESSEL-1	50	90
EQBREAK-2	VESSEL-1	120	150
EQBREAK-3	VESSEL-2	85	115

Figure 9: Holiday and equipment breakdown constraints

The equipment item break-down DATAIMAGE panel is shown on Figure 9. All fields enclosed in thick lines can be accessed using the mouse and the corresponding values can be changed. Also existing units (which is always a row in a DATAIMAGE panel) can be deleted or new units can be added. All these changes are propagated to the schedules.

A new equipment item break-down is added to `FILTER-1` from 50 to 100 hours, the modified DATAIMAGE panel is shown on Figure 10.

Equipment Break-downs

UNIT-NAME	EQITEM	FROM	TO
EQBREAK-267	FILTER-1	50	100
EQBREAK-1	VESSEL-1	50	90
EQBREAK-2	VESSEL-1	120	150
EQBREAK-3	VESSEL-2	85	115

Figure 10: A DATAIMAGE panel after addition of a fourth equipment breakdown constraint

This change is also propagated to the schedule panel D1, however the newly added break-down constraint does not violate the feasibility of the existing schedule, so it is left unchanged (Figure 11.).

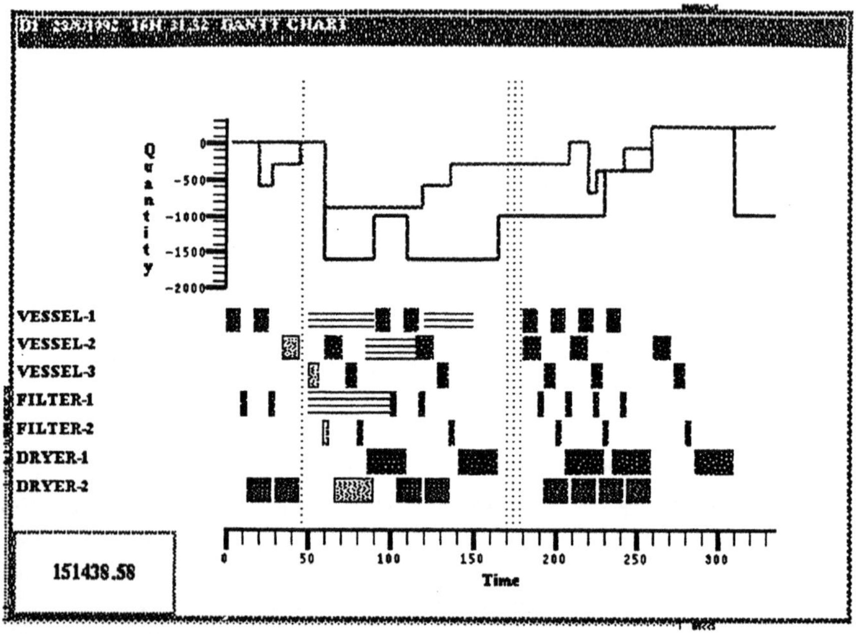

Figure 11: New schedule after addition of a fourth equipment breakdown

Sensitivity analysis can be carried out on any numerical value in the active fields of DATAIMAGE panels. Figure 12 shows the change of the objective function value when the upper bound of the last added equipment item break-down is varied between 100 and 125 hours with a step size of 2.5 hours. This change, as can be seen in Figure 11, affects only the second step of the third RED-BATCH directly but indirectly the entire third RED-BATCH is affected. Because of other constraints the batch cannot be moved backward, therefore it is shifted forward. A small drop of the objective function value can be observed from 100 to 102.5 hours because small forward shifts of the third RED-BATCH can balance the increasing length of the equipment break-down. However, at 102.5 hours these small shifts cannot help anymore and the batch must be shifted to the end of the planning horizon which results in a sharp drop of the objective function value.

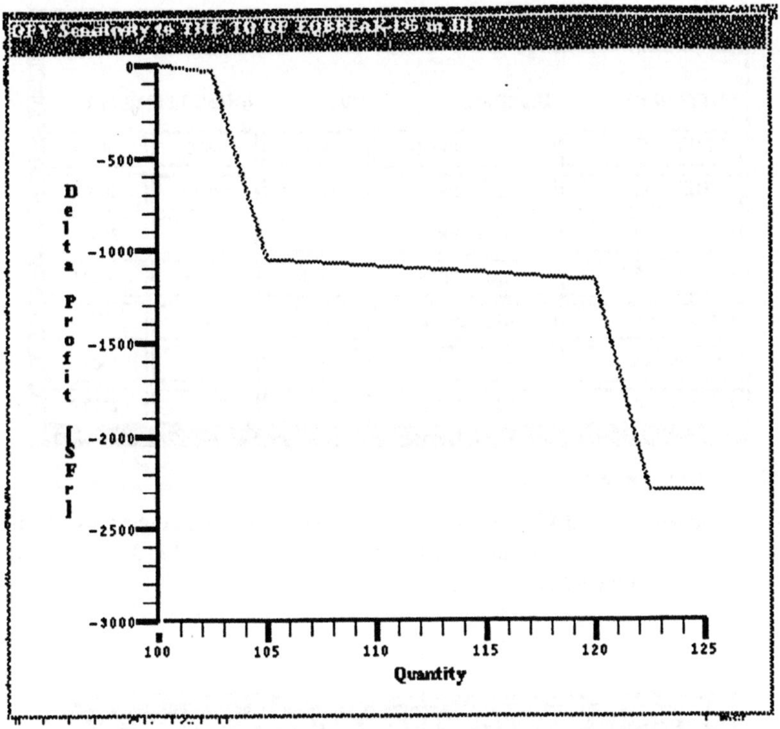

Figure 12: Sensitivity curve

Version management makes it possible to try out different variants of the same basic problem. In our example we have created a world W1 as a descendant of the background which means that the problem definition in W1 is the same as the one defined so far. In W1, a new problem formulation is developed with changes in the orders for BLUE-PROD. First the demand of BLUE-PROD is changed as shown in Figure 13, i.e., it is assumed that 1000 kg out of ORDER-B1 is required no earlier than 320 hours.

ORDERS in world #[World: W1]

UNIT-NAME	PRODUCT	DUE	QUANTITY	CUMUL
ORDER-B1	BLUE-PROD	60.0	600.0	600.0
ORDER-B2	BLUE-PROD	110.0	1200.0	1800.0
ORDER-B3	BLUE-PROD	310.0	1800.0	3600.0
ORDER-363	BLUE-PROD	320.0	1000.0	4600.0
ORDER-A1	RED-PROD	20.0	600.0	600.0
ORDER-A2	RED-PROD	60.0	900.0	1500.0
ORDER-A3	RED-PROD	220.0	700.0	2200.0

Figure 13: Orders in W1

The facts added to world W1 after this modification are shown in Figure 14. Additionally, the first holiday constraint is deleted and the first equipment item break down constraint is shortened to the interval 80-90.

||| (Output) Facts in W1 World

Primitive Facts:
(A WORLD-UNITS OF ORDERS IS ORDER-363)
(A QUANTITY OF ORDER-B1 IS 600.0)
(A CUMULATIVE-QUANTITY OF ORDER-B1 IS 600.0)
(A CUMULATIVE-QUANTITY OF ORDER-B2 IS 1800.0)
(A CUMULATIVE-QUANTITY OF ORDER-B3 IS 3600.0)
(A CUMULATIVE-QUANTITY OF ORDER-363 IS 4600.0)

Deduced Facts:
None

Figure 14: Additional facts in World W1

After these modifications an empty panel is opened in world W1 which reflects this new problem formulation (Figure 15).

The objective function distribution of this panel is

```
 Current      and     maximal profit for schedule D1
BLUE-PROD       -17086.32          138000.00
RED-PROD        -14797.20           55000.00
------------------------------------------------
                -31883.52          193000.00
```

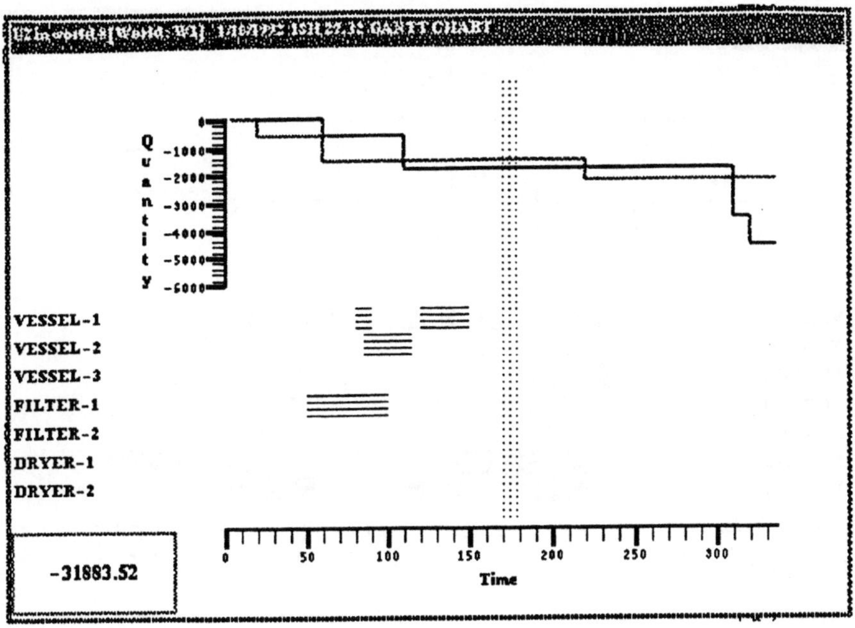

Figure 15: Empty panel in W1

Figure 16 shows the result of the **Fulfil rest** operation carried out on the empty panel. The CD curve of BLUE-PROD has become positive during most of the planning period, i.e., the first three BLUE-PROD orders can be fulfilled on time (note that a surplus is built up towards the end of the planning period to do this). However, the last order for the BLUE-PROD (of 1000.0 at 320 hours) still cannot be satisfied before the planning horizon and a deficit of 1000.0 remains at the end of planning period. The RED-PROD orders cannot be fulfilled on time but it is easy to see that the schedule cannot be improved for RED-PROD under the given constraints. For the two first RED-PROD orders whose early due dates (20 and 60 hours) are too restrictive the production starts as early as possible and for their 1500 kg total amount exactly 5 RED-BATCH batches are allocated each with 300 kg batch size. For the third RED-PROD order with due date 220 hours and 700 kg amount, 3 RED-BATCH batches are allocated, i.e., there will be 200 kg overproduction at the end of the planning period. These 3 RED-BATCH batches are allocated from the due date 220 shifting them backward in time.

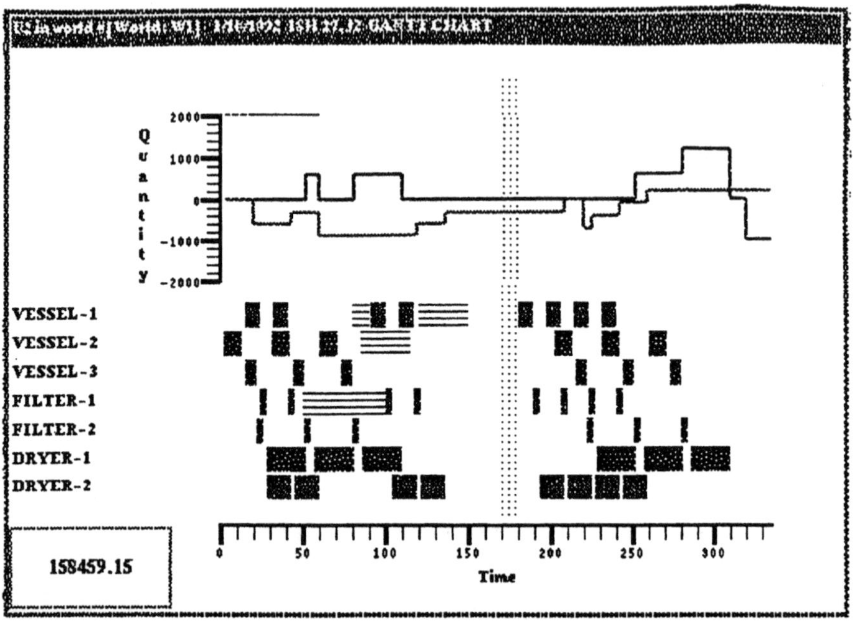

Figure 16: Schedule in W1

The profit distribution is as follows:

Current	and	maximal profit for schedule D1
BLUE-PROD	137658.68	138000.00
RED-PROD	53461.29	55000.00
	191119.97	193000.00

10. Conclusions

As mentioned in the introduction, GanttKit was developed as a pilot project within a longer-term project — BatchKit. The completion of this work allows the following conclusions to be drawn :

- Batch processing : Only a limited number of problem formulations were solved. This may be explained by the fact that the work was dominated by the overall project's goals — more attention has been paid to general aspects and experimentation with new methodology and less effort was devoted to comprehensive coverage of even a limited subfield of batch processing (such as scheduling).
- General problem solving techniques :

  a. The relational representation has confirmed its value as a basic technique.

b. Generalised branch-and-bound algorithm has been investigated but an algorithm of only limited generality was finally used, mainly for efficiency reasons, however the principle used allows an easy adaptation to other problem formulations.
c. Version management is currently based on a mixture of techniques. A synthesis and a more uniform approach is desirable. A necessary pre-requisite of such a generalisation would be a sufficiently general problem representation framework, so that the versions can be defined in terms of sufficiently homogeneous increments.
d. The graphical interactive interface of the scheduling system is its most impressive component, especially considering the small proportion of the total effort actually spent in its development. This has confirmed the value of KEE as a good development toolkit.
e. Generalisation of problem solution for short-term scheduling poses a difficult problem — the program specialisations and transformations which are almost always needed even in simple cases to achieve efficiency in problem solving are as yet by no means sufficiently understood.

- Knowledge integration : The GanttKit as a pilot project did not attain generality to a degree sufficient for an integration of a larger body of knowledge from even the limited field of short-term scheduling, although it offers a meaningful selection of problem-solving techniques. A more expressive representation is needed to describe further aspects such as :

  a. Quantified constraints in terms of complex nested object structures
  b. Enumeration strategies in the sequential explication of the search space and in the application of alternative problem solving methods to subproblems.

  Only significant extensions to the representation can help to overcome the limitations of the current techniques.

The above results appear to be modest considering the ambitious goals of the BatchKit project [HHR89]. The modest effort invested so far did not lead to achieving the "competence level" threshold postulated by Feigenbaum [7]— confirming the general recognition that a large body of application domain knowledge is needed to treat even apparently trivial problems. The effort which could lead to such a system will probably be larger by orders of magnitude.

On the other hand, the development has confirmed the importance of knowledge integration as a goal and also the benefits gained by using the latest software development technology. The benefits can however only be achieved if the requirements of know-how and interdisciplinary education (e.g. in operations research, computer science, chemical engineering) imposed by these techniques are met.

Further Work The current basic relational representation technique is being extended to a fully deductive system. This work has been completed since the presentation of this paper. The new representation is based on the integration of object and logic programming in the LinX system which replaces the AXI system described here.

Introduction of plant and process concept (also already completed) will allow other types of problems to be addressed, such as e.g. the automatic equipment allocation and batch pattern generation, design, mass balance calculations or multiplant design.

The introduction of constraints will allow the inclusion of further standard constraints (e.g., more general types of resource constraints).

Further generalisation of branch-and-bound together with the introduction of constraints will allow easier transition between MINLP formulations and the default branch and bound problem solving, while retaining the advantage of user-friendly problem formulation.

This effort will also be supported by the extension of the relational system towards a fully deductive knowledge base system integrating object-oriented and logic programming.

Concluding remarks The GanttKit system is the first fully interactive system with a graphical interface for the short-term scheduling of batch chemical plants based on the presented type of technology. The technology utilised holds a promise that the ease of operation experienced with GanttKit in short-term scheduling can be extended to other areas.

Due to the variability of problem formulations and to the combinatorial complexity of the resulting optimization problems, the methodology needed to develop useful systems for the treatment of problems in the batch processing area in general will continue to present new challenges.

For this research to become practically relevant for industry and for the benefits of rationalisation to be achieved it is necessary for the information and production automation systems in industry to reach a critical lowest integration level.

From the point of view of an academic research group, the challenge can probably be met at the problem solution level by a combination of strategies and at the toolkit development level by continuously adjusting the knowledge acquired so far to new emerging technologies.

References

1. Akella, IL *Operations Research Models in Flexible Manufacturing Systems (Archietti F., Ed.).* ch. Real Time Part Dispatch in Flexible Assembly, Test and Manufacturing Systems, Spring, 1989.
2. Bellman, R., Esogbue, A., and Nabeshima, I. *Mathematical Aspects of Scheduling and* Applications. Pergamon Press, 1982.
3. Brucker, P. *Scheduling.* Akad. Verlagsgem. Wiesbaden, 1981.
4. Clark, S., and Kuriyan, K. Batches - simulation software for managing semicontinuous and batch processes. In *Proc. AIChE Nat. Mg.*, Houston (1989).
5. Cott, B., and Macchietto, S. Minimizing the effects of batch process variability using on-line schedule modification. *Computers and Chem. Eng. 13,* 1, 105-113, (1989).
6. Egli, M., and Rippin, D. Short-term scheduling for multiproduct batch chemical plants. *Comp. and Chem. Eng. 10,* 4, 303-325, (1986).
7. Feigenbaum, E. A., and Lenat, A. On the thresholds of knowledge. In *IJCAI'87*, vol. 10, (1987).
8. Fox, M., and Smith, S.F. ISIS - A knowledge-based system for factory scheduling. *Expert Systems,* 25, 49, (1984).
9. Halasz, L., Hofmeister, M., and Rippin, D. A flexible knowledge based toolkit in batch processing. In *AIChE Annual Meeting* (1989).
10. IntelliCorp, X *KEE User's Guide.* IntelliCorp Inc., 1988.
11. Jaffar, J., and Lassez, J.L. Constraint logic programming. In *Proc. Conf. on Principles of Programming Languages*, 1987.
12. Johnson, S. Optimal two- and three stage production with setup times included. Naval Res.Logistic Quarterly 1, 61-68, (1954).
13. Kempf, K, LePape, C., Smith, S., and Fox, B. Issues in the design of ai-based schedulers:A workshop report. *AI Magazine 11*, Jan (1991).
14. Klossner, J. *Computer-gestuetzter Entwurf von absatzweise arbeitenden chemischen Mehrproduktanlagen.* PhD thesis, ETH Zuerich, 1985.
15. Kowalski, IL *Logic for Problem Solving*, North-Holland, 1979.
16. Mauderli, A., and Rippin, D. Production planning and scheduling for multi-purpose batch chemical plants. *Compt. and Chem. Eng. 3*, 199-206, (1979).
17. Musier, R., and Evans, L. Batch process management. *Chem. Eng. Progress*, 66-72 , June (1990).
18. Reklaitis, G.V. Overview of scheduling and planning of batch process operations. This volume, p. 660
19. Rippin, D.W. Design and operation of multiproduct and multipurpose batch chemical plants - an analysis of problem structure. *Computers and Chem. Eng. 7*, 463-481, (1983).
20. Roy, B., and Sussmann, B. Les problemes avec constraintes disjonctifs. Tech. Rep. 9, SEMA Montrouge, 1964.
21. Smith, S.F., Ow, P.S., Potvin, J.Y, and Nicola, M. An integrated framework for generating and revising factory schedules. *J. Operations Res. Soc. 41,* 6, 539-552, (1990).
22. Sparrow, R. Multibatch: A computer package for the design of multi-product batch plants. *The Chemical Engineer 289*, 520-525, Sept (1974).
23. Ullman, J.D. *Database and Knowledge-Base Systems.* Computer Science Press, 1988.
24. Van Hentenryck, P. *Constraint Satisfaction in Logic Programming.* MIT Press, 1989.
25. Wilkins, D.E. *Practical Planning: Extending the Classical Al Planning Paradigm.* Morgan Kauhmann Publ., San Mateo CA, 1988.
26. Yamasaki, Y., Morikawa, H., and Nishitani, H. Production scheduling system for a batch process and its application to retrofit design. In *Proc. PSE88, Sydney* (1988).

An Integrated System for Batch Processing

S. Macchietto, C.A. Crooks and K. Kuriyan

Centre for Process Systems Engineering
Imperial College, London SW7 2BY, UK

Abstract: The engineering activities accompanying a batch plant development project are reviewed and the requirements for an integrated batch system environment are detailed. The role of some of the currently available supporting tools is discussed. Finally, a design is presented of a system aimed at integrating batch plant and sequence control system design, off-line operations scheduling and on-line operations management. Preliminary results are presented with particular reference to a multipurpose batch pilot plant example.

Keywords: Batch plant, design, scheduling, production management, sequence control, modeling, integration.

1. Introduction

The need for better integration of process engineering functions in the batch processing field has been discussed several times recently [32,39,43,45,47,48]. Over the last year, within the Batch Processing Project in the Centre for Process Systems Engineering we carried out a collaborative study with several industrial partners aimed at identifying; in some detail, the major technological and research opportunities, needs and obstacles associated with an integrated system for batch processing. Three main industrial partners participated in the study, representing a wide range of industries, backgrounds and present and future demands. APV Baker is a leading supplier of equipment and processes to the food industry. This industry is characterized by rapid evolution in market conditions, high product innovation, fierce competition and often tight production margins. It is not unusual for some food plants to produce 100-200 product varieties with exceedingly short demand and delivery horizons. The closeness to the end-consumer market makes food processing representative of the highly dynamic and responsive production requirements expected in other industries. John Brown is a engineering company with strong presence in the more mainstream chemicals, pharmaceuticals and biotechnology industries and with specific experience in batch plants. Both these companies are contractors, rather than operating companies. As such, they contributed the distilled experience and requirements of many projects and end-users. BNFL builds and operates plants in the nuclear sector, where many important operations (fuel preparation, decommissioning, etc.) are discontinuous. Thus, they reflect both the plant designer and plant operator viewpoints. For obvious reasons, this industry has particular concerns regarding lifecycle engineering, safety, formal development procedures and validation. These are already very important in other segments of batch processing (e.g. pharmaceuticals) and are expected to become ever more central to all batch processing projects, as regulatory and environmental demands increase and the responsibility is imposed on plant suppliers and operators to demonstrate compliance. In addition to these major partners, discussions were held with a large number of other industrial collaborators, in particular fine chemicals producers.

As part of the project, we conducted a detailed analysis of the engineering activities involved in typical batch processing projects. The goal was to identify the main requirements for a set of tools and an integration environment to support all such activities. The analysis involved in-depth discussions with engineering, design and automation teams and included retracing the evolution of several representative projects from initial preliminary design to commissioning and operation. To gain first hand experience, we also built our own fully automated, multipurpose batch pilot plant facility. The design was commissioned to APV Baker and Centre's personnel was seconded for several months to them to work on the project as integral part of the development team.

In the following, first we will briefly summarize some conclusions of the industrial survey, and then detail some of the background integration "infrastructure" issues, in particular with respect to approaches to integration, special modeling requirements posed by batch processing, data base management systems and modeling environments. The second part of the paper deals more specifically with some models, techniques and software for preliminary batch design/scheduling and batch management/control. Some developments are then presented in the area of automatic synthesis of batch management and control procedures, which have the effect of bringing together the design and operation aspects. Finally, preliminary results are presented, with particular reference to our pilot plant.

2. Batch Process Engineering Activities

In spite of the large differences in type of processes and industries covered by the various industrial partners, the approaches to design and operate batch plants were remarkably similar.

The main engineering activities, schematically shown in Figure 1, consist of:

Preliminary Process/Plant Design the goal of which is to identify the major equipment, simple flowsheets and layouts, descriptions of all processes and major costs. For batch plants, the demand for equipment is clearly tied to its utilization, hence operational decisions (e.g. operation in campaign mode, use of distinct lines for different products, operating schedules, etc.) must also be taken at this stage. Design and operational decision which may have large effect on the achievable production rates and costs will have to be considered. This includes type, number and size of process equipment required, their connection network, utility systems and manpower levels. Broad operations specifications are drawn up and the main control system requirements are established at this stage. Contracts are typically won and budgets fixed on the basis of this work.

Detailed Design. Detailed line diagrams (including control hardware) are produced starting from the sketches generated at the preliminary design stage, together with utility line diagrams, a detailed plant layout and a more detailed specification of the operating procedures. Mechanical design (vessels, pipes etc.) is typically done using CAD tools.

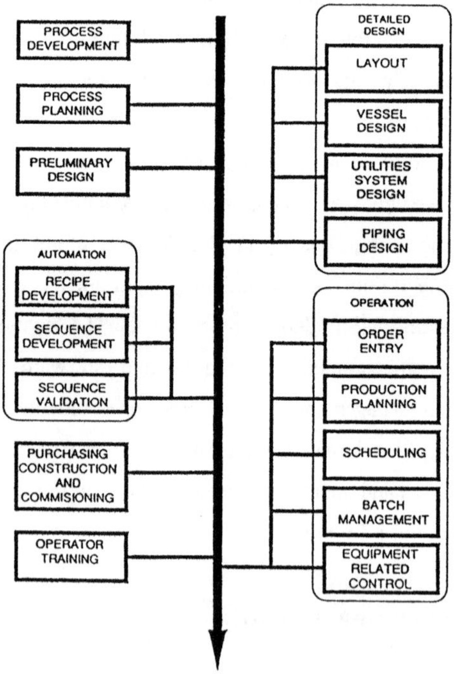

Figure 1. Lifecycles Activities for a Multipurpose Batch Plant

Automation. Starting from the functional specification and broad operating procedures, the control software is further specified through various stages of detail, down to individual sequence control code, and implemented from the bottom up.

Operation. Once built, the usual range of operational activities for a batch plant include production planning and scheduling, production management, batch execution, etc.

The activities described cover a very wide range of expertise and technologies, typically involving several organisations over a plant lifecycle. Time horizons may range from two weeks for preparation of a preliminary design bid to several years between initial design, commissioning and a major plant retrofit [16,17]. Increasingly, decommissioning aspects will also have to be considered as part of a plant lifecycle.

Some of the major conclusions of the study are summarized here:

I) The preliminary design decisions by and large determine the eventual plant throughput range and performance. Techniques which permit exploring in some detail the interaction between design and operation decisions are required.

ii) Detailed design activities are essentially the same as for continuous plants and are reasonably well covered by existing CAD systems.

iii) There is considerable overlap between activities of different departments (process, automation, drafting, etc.) making decisions which are highly interactive. There is much interest in having a consistent set of information easily accessible to all functions at all stages of development. This should include the description of "operations" (functional and procedural aspects) as well as those of a plant flowsheet, from the early stages of a design. These descriptions should evolve in parallel with equipment design.

iv) The development of batch control and automation systems is an engineering intensive activity. A major problem area lies in the interface between process engineering and control/automation groups. As the process and plant are refined many iterations are often required to clarify an operating sequence or implement it correctly. Apparently, minor changes in equipment or process detail may have major implications on the automation systems and vice versa, with costly reconciliation iterations. There is considerable interest in techniques for improving automation development productivity, reducing errors and validating control sequences. Interfaces to general purpose simulations are desirable for "what if" studies and operator training.

v) It is becoming more common for contractors to deliver detailed 3D "design models" to plant operating companies, but the latter typically have to develop their own operations models for the supervisory control functions (scheduling and batch management). There is a large interest in formal but generic "production models" which can be used at the design stage to prove the overall plant performance and then transferred directly to operations.

vi) The need to frequently revise product formulations (recipes and control sequences), supervisory production control strategies and plant configurations requires that all techniques used for grassroots plant and operation design should be model driven and also applicable to the retrofit case. Particular safety requirements arise in the maintenance and retrofit of control code.

vii) The "models" of the process and plant used in the process engineering, automation and operations scheduling are often informally defined, incomplete, ambiguous, and incompatible. A major challenge consists in obtaining consistency between models used for various purposes.

viii) There is a tremendous scope for computer based integration of process design and process operations activities.

ix) The major economic benefits for the design activity are to be found in shorter design-to-operation cycles, lower capital and operating costs made possible by increased confidence in plant operability, lower redesign and commissioning costs associated with increased quality of the control system development process and lifetime maintenance of control code. On the operations side, major benefits are associated with tighter integration between scheduling, batch management and batch control functions. Depending on specific circumstances, benefits are to be gained from increased plant and raw materials utilization, reduction of stocks and wastes, more competitive marketing positions achievable through sharing of more accurate and timely information between order acquisition, planning/scheduling and production execution functions, shorter order-to-delivery cycles and overall more responsive plant operation.

3. Batch Integration Issues

Forms of Integration

The term integration has several different connotations. For example, Pekny et al. [39] identify two different forms of integration: problem integration, which refers to the integration of software tools used to solve different problems, and tool integration, which refers to the integration of different software tools for solving the same problem. Perkins [40] differentiates between software integration and the integration of life cycles aspects during process design. An example of the latter form of integration is the simultaneous design of a continuous process and its control system. Here, the following classification will be used:

I) *Integration by Communication.* Computer programs may be linked to each other by messages transmitted over a computer network. These messages may be requests for services, or responses to such requests from service providers. This form of integration is appropriate when there is a clear division of functions between the software tools being integrated. Typically, the volume of information to be shared is small or of a transient nature. The closer coordination of the modules being integrated should reduce delays and improve the responsiveness of the system.

ii) *Integration through Shared Databases*. Integration by communication over computer networks is inappropriate when a permanent, consistent record of shared data is required. This is the case in process design where, after one group of engineers has finished a design task, they must pass on the design to another set of engineers using a different set of software tools. In process operations, sales/marketing, scheduling and batch management functions need to share data on the current and expected levels of stocks, etc. Use of a common database eliminates manual transfers and the resulting inconsistencies.

iii) *Integration of Analytical Methods*. This refers to the integration of methods which permit qualitatively different types of analysis. Some examples are the combined use of mathematical programming and logical inference in process synthesis [44] and the use of pattern recognition techniques in process control [54].

iv) *Task integration*. This refers to the merging of multiple design or operation tasks, e.g. simultaneous production planning and detailed scheduling, simultaneous batch plant design and scheduling [6] or simultaneous plant layout and equipment network synthesis. Task integration can lead to improved designs and schedules by taking advantage of the interactions between plant and operation aspects which would otherwise be considered separately.

For computer integration of batch processing operations, an architecture has been proposed (ISA, 1992, discussed in more detail in a later section) which is due to become an international standard by the end of 1993 (Figures 2a, 2b). This architecture could in principle implemented in a variety of ways. We can envisage a shared database to provide common recipe information to planning, scheduling and batch management functions. Communication links are required to coordinate production scheduling and process control in real-time. Since most optimal scheduling algorithms require a significant amount of computation time, a fast heuristic search method may be used for on-line scheduling [13], while a more rigorous algorithm may be used for major off-line schedule decisions. A batch management system may also be required to co-ordinate dynamic scheduling with process monitoring tasks using, for example, statistical methods or neural networks [55] for fault detection and classification. Thus, an integrated system for batch processing may employ all four forms of integration described above.

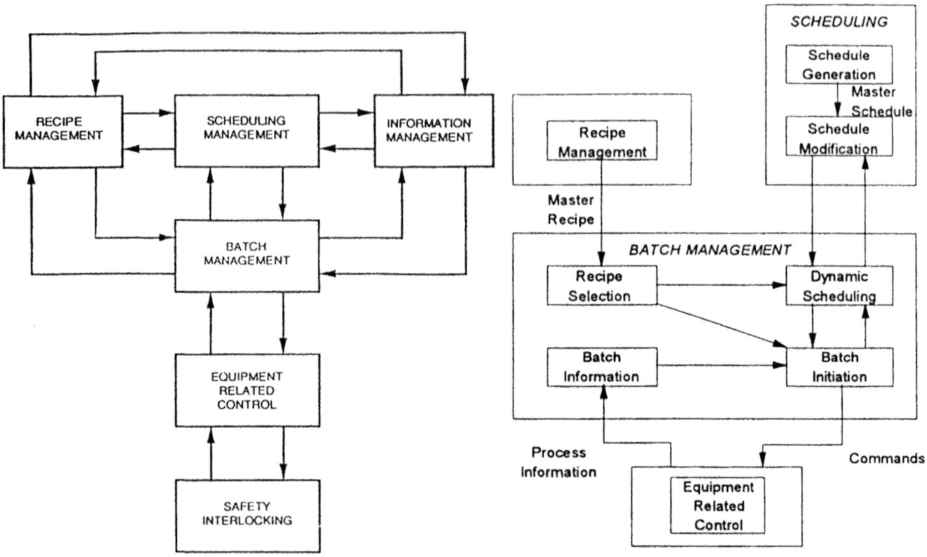

Figure 2a. SP88 Architecture for Integrated Operation of Batch Processes **2b.** SP88 Architecture for Batch Management

Modeling requirements

This section focuses on the requirements for representing batch processes on a shared database. Existing process engineering databases [1,14,42] employ a Process Flow Diagram as the focal point for integration. This, however, does not provide an adequate description of operations and, therefore, cannot be used to support most batch engineering applications. The adequate representation of operating procedures is crucial to the design of a batch process engineering database. Several aspects should be taken into account

i) The flexibility of batch operations results in a variety of plants ranging from dedicated single product facilities to flexible multipurpose plants and pipeless plants. The use of multipurpose equipment means that there is not a one-to-one mapping between unit operations and physical equipment items. Hence, product recipes and operations must be represented separately. A key issue is the representation of operations networks in a way which reflects the flexibility of the batch mode of operation. Recent developments in incorporating layout considerations in preliminary design [27] make it desirable to extend the simple equipment network models that are commonly used for preliminary design to include information on the spatial location of equipment.

ii) A process may be successively viewed at different levels of resolution. The determination of campaign lengths in multipurpose plants is typically carried out by aggregating groups of batches into cycles with nominal production rates. Papageorgiu and Pantelides [38] show that the performance of individual cycles may then be improved by allowing the reuse of equipment and by exploiting intermediate storage. This step requires detailed scheduling within each cycle of batches. The retiming of cycles can then trigger a recalculation of campaign lengths. Similarly, Hasebe and Hashimoto [24] describe a scheduling procedure where groups of batches are successively disaggregated and desegregated to create an overall production schedule.

iii) A further complication is that individual unit operations in a process may themselves be viewed at many different levels of resolution. A unit operation is typically represented as a simple elapsed time model for production schedule optimization, as a mixed discrete-event/differential-algebraic model for dynamic simulation and optimal control studies [37], or as a finite state machine for sequence control purposes.

iv) In batch processing, it is often not possible to completely separate the process representation from the algorithms used for simulation, scheduling or optimization, since solution methods often apply to narrow domains. For example, models may have to be simplified (e.g. by ignoring transfer times, transfer lines, etc.) to reduce computation times for schedule optimization. This imposes restrictions on product recipes and equipment networks and results in the need to modify the model when transferring from the domain of one solver to that of another.

v) There are significant differences in the number and nature of database objects between preliminary and detailed design stages [14]. During preliminary design, a number of alternative plant and operation networks may have to be considered, with several alternatives maintained for evaluation. The number of objects is small but the relationship between equipment topology, recipes and operations networks and models of the unit operations is very complex, in particular, if we wish to explicitly represent such models and the assumptions on which they are based [52]. Such models are represented by complex semantic networks and cannot easily be stored in relational databases. During detailed design, on the other hand, the number of objects increases rapidly as individual piping, instrumentation, electrical wiring, control devices, etc. are added in. Many software tools exist, such as 3D drawing packages, typically with specialized data storage formats to speed-up search and retrieval of objects. Again, a database of operations and detailed control strategies, sequences, control instructions, will have to be defined and tied to the equipment and control data. While databases for operations/control have been developed for specialized in-house use (in particular by automation suppliers), there are at present no generic integrated tools available to support the whole spectrum of batch engineering functions.

Data Base Management Systems

Here, we consider the issue of database management systems suitable for supporting integrated batch process engineering. Existing process engineering databases use a variety of data models [5]. PRODABAS [1] uses an extended relational model, while Huang and Fan [25] advocate the use of a hybrid object-relational model. Motard [36] reports on the relative merits of relational and object-oriented database management systems. Marquardt [34] proposes the use of an object-oriented database management system to support a process modeling environment. A disadvantage of relational database management systems (RDBMS) is the difficulty in accessing and manipulating complex hierarchies of objects, particularly if each object can have multiple versions. The transaction model that is used to regulate the updating of records in a multi-user setting is also inappropriate for design environments where transactions can take place over days or weeks. Object oriented database management systems (ODBMS) have been developed explicitly to support "non standard" application areas such as engineering design. In addition to supporting object-oriented programming concepts such as object classes, inheritance and encapsulation, these database management systems often provide storage for multiple versions of objects and a locking mechanism appropriate for use in a design environment [29]. They are usually designed to run on a distributed computing network of multiple file servers and client workstations. Many ODBMS now provide interfaces to relational store managers, thus allowing the user to employ object oriented data models while still being able to access data in a RDBMS. Another option is to structure information as a knowledge base using a frame based system such as KEE, which was used to implement MODEL.LA [52,53]. Frame based systems provide excellent facilities for modeling complex relationships between data items through frames, slots, daemons, etc. and are well suited for preliminary design. However, they are not well suited to storing the large volumes of information that must be managed during the later phases of design.

Finally, special requirements arise during operations to share large amounts of data between functions in real time. Now what is "real time" is clearly subjective and application dependent. A practical definition is "so that results are produced just faster than they are needed". At the lowest control level, this means milliseconds, at the unit and plant wide scheduling level, seconds/minutes, at the planning/scheduling from minutes upwards. These requirements imply that specialized architecture/software/hardware are typically needed for the more detailed control levels so as to ensure such "real time" performance over ever wider domains and transaction volumes. From a practical point of view, this demands that on-line applications must be adapted (sometimes extensively) to these architecture. On the other hand, bridges between a real time environment and general purpose database management systems (e.g. a relational data base) are often available. It is thus necessary to make a-priory decisions regarding which data and applications will reside in the real time environment and which will not. These decisions clearly fundamentally affect portability etc. across systems. There is much work underway in the computing science and software communities on the development of ODBMS with real time performance and cross systems portability. Although not much is presently available commercially, it can be expected that significant progress will be made in this direction over a five years horizon.

Existing Modeling Environments

Most existing modeling environments for process engineering are oriented towards continuous processes. Much effort has recently gone into developing frameworks for creating and maintaining complex process models. Some examples are MODEL.LA [52,53], ASCEND [41] and VEDA [34]. MODEL.LA is notable for supporting multifaceted modeling [58] and in particular I) the representation of processing systems at several levels of abstraction ii) the automatic generation of basic relationship (equilibrium equations, balances, etc.) iii) the representation of qualitative or semiqualitative relationships and iv) the documentation of underlying assumptions. This language addresses one of the issues raised in the previous section, namely the development of process models at varying levels of resolution. However, for use in batch process modeling further extensions are needed: I) to allow separate description of recipe and equipment networks, and the permitted mappings between the two ii) additional semantic relationships to represent the temporal sequencing of operations iii) representation of production requirements and schedules and iv) the representation of combined discrete event/continuous systems.

While modeling environments for continuous processes support both simulation and optimization, environments for batch process modeling are specialized into two categories. Packages such as UNIBATCH [21], BATCHES [8] and gPROMS [4] allow the process engineer to build combined discrete-event/differential algebraic models for simulation studies. On the other hand, a number of packages provide simplified models without process dynamics for use in production scheduling and preliminary design. These include GANTT_KIT [22], BatchMaster [9], SUPERBATCH [12] and gBSS [51]. The first three systems employ constraint satisfaction strategies to generate a feasible schedule and perform some local optimization. They are especially useful for performing schedule updates in a dynamic environment [11]. The gBSS system, on the other hand, creates optimal schedules through the use of mixed integer linear programming techniques. No system is presently available that can address both design and scheduling with plant wide batch production models and simulation or optimization with detailed dynamic models.

4. Models and Software for Batch Processing

In this section, attention is focused on specific models for integrated batch processing, preliminary design and scheduling, and batch management/control. No attempt is made to present a comprehensive review of all the possible approaches and solution techniques. Instead, some ideas and specific tools are described which have been developed over the last few years at Imperial College and which will be used as "building blocks" for our integration project, described in the last section.

Reference models for integrated batch processing

A number of organizations and individuals have analyzed the above problems or at least some of its component parts and produced "reference models" for an integration architecture. For more general considerations on integrated processing, the book of Rijnsdorp [46] and the work of Williams [56] are useful. Specifically, for batch processing operation, a very useful analysis and set of recommendations were proposed by the NAMUR committee. This organization, with membership by a large set of European companies (mainly users), is aimed at the definition of standard industrial practices in process measurements and control. A subcommittee formed in 1979 produced a set of recommendation specifically for batch control [49], presently being revised. Much of their results were taken on board by the Instrument Society of America (ISA) SP88 Committee (mainly suppliers), which is developing a standard terminology and architecture for batch control operations. This is due to produce a final report by the end of 1993 with the view of establishing an international ISO standard. Detailed interim documents have been produced (ISA, 1992).

The major driving force for both activities is the need to agree standards so as to permit the development of batch control and application software which is modular, highly flexible, can be tested, maintained and easily evolve with changes in processes, plant and products. The major NAMUR recommendations can be briefly summarized in the use of I) a hierarchical structure for batch management and control systems (definitions of functions and data flows are given) ii) standard terminology iii) the definition of "operations" independently from equipment, with a hierarchical organization defined for operations and recipes iv) all application development by configuration rather than programming. The ISA SP88 committee extends the scope somewhat at the supervisory control level which include scheduling (Figures 2a, 2b). The analysis produced by both bodies is much more detailed for the lower level control functions, while the interactions with off-line planning and scheduling remains a bit fuzzy. Nonetheless, the models produced have had significant impact with users and vendors, with many commercial batch control systems already offering compliance and user companies demanding it.

All the above "reference models" concentrate mainly on integrated operations or design of control systems for a given batch plant with hardly any attention to the design of the plant itself.

Modeling for scheduling and preliminary design

State Task Networks. It is common to employ simple "elapsed time" models of unit operations for preliminary design and scheduling of batch processes. Typically, the process is represented as a directed graph where nodes represent unit operations and arcs represent precedence relationships between unit operations. The State Task Network representation developed by Kondili et al. [30] uses a directed graph with two kind of nodes: "state" nodes, representing unit operations and "task" nodes representing material states. The explicit inclusion of state nodes in the network allows the description of several characteristic features of batch chemical processes which are not manifested in discrete parts manufacturing: alternative pathways for producing the same material; accumulation of material from successive batches; batch splitting, recycles, etc. State Task Networks have been used as the representation framework for defining a variety of optimization problems with formulations based on uniform time discretization. These result in mixed integer linear programming problems. Problems addressed include short term and cyclic scheduling [51], the determination of optimal campaign cycles and schedules for each cycle [38] and the preliminary design of multipurpose plants [51].

Maximal State Task Networks. In its original form, the STN models have a number of ambiguities, resulting from a coarse view of the scheduling problem (assignment of main process vessels to tasks). For example, limitation due to potential connections between main processing vessels are not considered and "states" are ambiguously identified with storage. An extension of the concept which resolve these ambiguities was developed by Crooks [15] who proposed a procedure for generating a maximal State Task Network (mSTN). This includes information about connectivity and the representation of storage options: storage of input states in a processing vessel; storage of output states after processing in a vessel; dedicated storage for input and/or output states; explicit "storage" tasks which can be allocated to processing vessels. The first step in the construction of a mSTN is to represent any legal movement of material from one processing vessel to another by a sequence of an output state associated to the upstream vessel, a transfer task and an input state associated to the receiving vessel. The generation of the mSTN is completed by defining one instance of each task, along with the corresponding input and output states, for every piece of equipment in the plant on which the task may be performed (legal assignment). Thus, the mSTN takes into account both the processing network and the plant network.

Unit State Task Networks. A description of batch processes must also include information regarding the state of processing equipment and the effect of carrying out operations on the equipment. For example, Batch A may leave Unit U1 in a dirty state, necessitating a cleaning operation before Batch B may be processed in the same vessel. Such sequence constraints may be important for detailed scheduling and are essential for control purposes. Interactions between batches and processing units could be represented by Petri Nets [23,57] and various finite state automata. Crooks [15] further extended the above mixed integer linear programming formulation by introducing the explicit definition the legal states in which a processing unit may be (uStates) and of the transitions between these uStates caused by processing and ancillary (e.g. cleaning) tasks. This gives for each vessel a finite state automaton (a unit State Task Network, uSTN) which may be used to represent efficiently a great variety of state and task precedence constraints, still formulated as extended MILP problems.

The above scheduling models (described in detail in Crooks et al. [18,19,20]) may be extended to design by introducing additional discrete variables to represent the existence or otherwise of individual processing vessels [51], and connectivity networks [2], with continuous variables representing size (of vessels or connections). With the addition of suitable equipment and operating cost models (and modifications of the constraints), we can then address problems related to minimum design (or retrofit) cost with preliminary or more detailed models. All of the above models based on the STN representation, utilize a discretization of time into intervals of fixed duration, with any events happening only at interval boundaries. While this may be useful for preliminary design and scheduling, clearly discretization approximations are introduced, especially when operations involving widely different time scales have to be considered. Thus, short operations are normally neglected.

gBSS Overview

A software program (gBSS, for general Batch Scheduling System) which performs both preliminary design and scheduling based on the STN approach described above was developed and is described by [51]. It uses a number of techniques to improve the efficiency of a branch and bound method for the solution of the resulting MILP programs which enables the solution of relatively large problems.

Modeling for batch operation and control

The type of models required for batch execution ranges for approximately the same level of detail as used for plant wide scheduling (discussed in the previous section) to much more detailed models for control of individual sequences and steps. A key aspect here is that while for design and scheduling a single recipe, procedure, etc. define the processing of all batches of the same type, execution of two instances of exactly the same recipe on exactly the same equipment will give two "batches" with distinct identities, since they may be made up by distinct supplies of the same ingredients, will typically have slightly different processing history, quality, etc. In addition, basic recipes may have to be adjusted for individual batches, for example, to compensate for known variations in raw material quality, reduced equipment performance, etc. Batch control and management systems must then be able to execute, track and monitor individual instances of each "batch", for all batches of all products. The ISA recommendations include the definition of a batch recipe model with generic master recipes, parameterised according to equipment type and batch size, and for which the operating procedures are defined in terms of generic control sequences. Control sequences are similarly defined in a generic, equipment independent form and instantiated for specific equipment. Thus, a transfer from an upstream tank to one of 4 similar downstream tanks may be described as a generic transfer phase (with associated parameters such as rate, amount transferred, etc. and a list of upstream and downstream equipment types) of which there are four specific instances. A master recipe is product and process dependent but equipment independent. For individual batches, a master recipe is then instantiated into a *control recipe*, which includes allocation of specific equipment, batch size, and with the generic control sequences instantiated into specific sequences for the target equipment (control *phases*). The NAMUR and ISA models represent many years of effort by the batch control vendors and operators towards achieving control code which is structured, modular, easily reconfigured to reflect recipe, equipment and production plans changes, portable across platforms and maintainable in the long run. It is worthwhile noting that this effort generated a model based on concepts essentially equivalent to classes and hierarchical inheritance which make it a natural candidate for object oriented implementation. It may also be observed that if we defined the parameters for master recipes and generic control phases in a suitable way, in term of required resources, batch size, control parameters and if we additionally assigned a time estimate for the completion of each generic control phase, this would allow their use for scheduling purposes as well as for control. However, models for execution and control purposes must be more comprehensive and include all processing steps, including short transfers of material, etc. Additionally, control models must be based on a continuous time representation. Thus, a number of differences exist between the models required for preliminary design/scheduling and execution/control.

SUPERBATCH Overview

A system, SUPERBATCH, has been developed at Imperial College over the last few years which implements the supervisory batch management and scheduling functions of the ISA model. Based on the ideas described in Cott and Macchietto [12], the system is broadly structured as in Figure 3. It is assumed that major economic decisions are made by some higher level planning/scheduling function, which establishes the (possibly optimal) assignment of batches to processing units, batch sizes and the sequence of batches. If these decisions have been made by

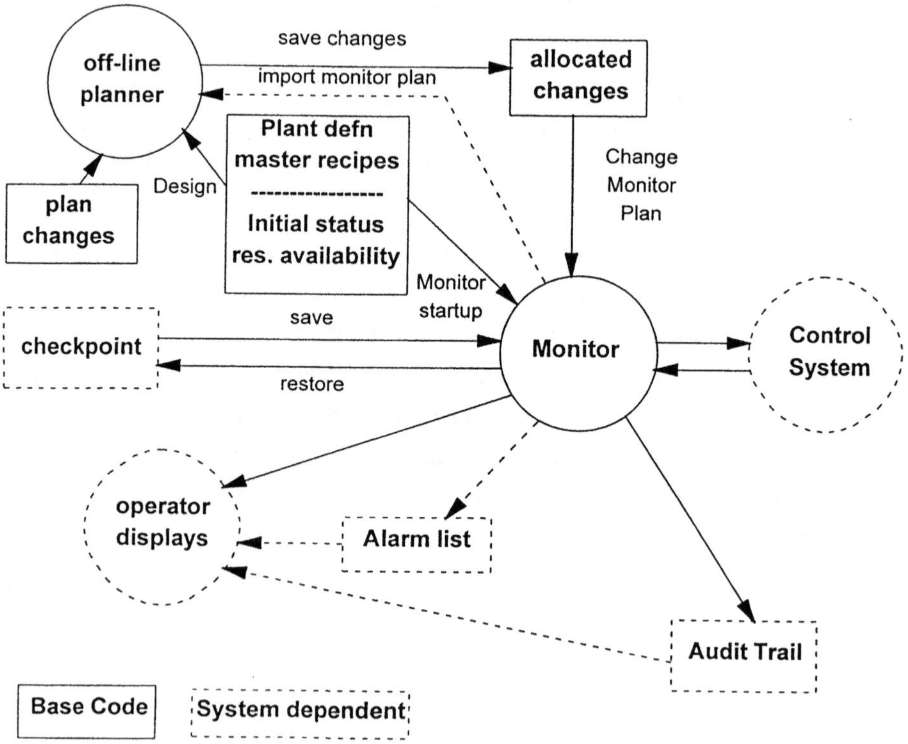

Figure 3. SUPERBATCH - system overview

some external scheduler (either manually or algorithmically), they are imported to the system and verified for feasibility, not just at the initial time but for the entire production horizon, utilizing the current plant, procedure and constraint models within SUPERBATCH and information on the state of the current production in the plant, expected availability of all resources, etc. Any missing decisions or conflicts may also be resolved or changed interactively in an off-line scheduling mode (using the "off-line planner" in Figure 3), by viewing the resulting schedules, utilization statistics, etc. until a satisfactory "production plan" is obtained. This can then be passed on to the on-line portion of the system (the "monitor" in Figure 3) which implements the batch management functions of the ISA model. The monitor runs in a real-time environment and processes any new requests as changes to the current schedule. It sets the parameters for and initiates control phases as and when required by sending commands to (possibly distributed) control system(s). It then monitors their external execution and revises the schedule for all remaining production in view of actual as opposed to planned duration, amounts transferred, etc. The schedule revisions are done at a frequency of, typically, one minute or less utilizing a completion time algorithm based on minimal algebra [13]. In order to insure fast solutions (essential for real-time execution) the algorithm deals with only a subset of all possible scheduling decisions. In particular, it retains (with minor exceptions) the batch sequence order, sizes and allocations last specified, and revises start and finish times to meet a set of detailed operational constraints. Thus, it can ensure detailed operational feasibility but not economic optimality. At any one time, however, the complete production schedule can be made available to both the off-line planner and external scheduler for deciding on major scheduling decision changes. The same algorithms are used in the on-line monitor and off-line planner.

The system is completely driven by a hierarchical model which first defines the plant, recipes and control phases in terms of generic resources (for both equipment and operations). Each resource is then parameterised and is instantiated as individual equipment units, control phases, etc. (some of the terminology used was based on earlier versions of the ISA model, thus ISA's master and control recipes are called master procedures and batch procedures, respectively, in SUPERBATCH, etc. - the concepts are however respected). The above design assumes that phases are delegated for execution to external controllers and this defines the level of supervisory vs. distributed control and the "granularity' of the supervisory management models (Figure 4). The interface between supervisory system and external controllers (e.g. the mapping between phase parameters and specific control tags, devices, etc.) is also part of the SUPERBATCH model definition and therefore external controllers may be easily changed.

An implementation of this design (into an IBM product called ASSYST, for Adaptive Scheduling System) within a fully integrated plant wide industrial control environment (IBM's RTPMS) has been produced which permits access to all portions of the models and production data at any time from a proprietary shared real time database. Thus, multiple users may simultaneously access the current actual production status (in various forms, including Gantt charts) for all executing and scheduled batches, individually examine alternative scenarios in an off-line mode (changes may range form simple additions and deletions of batches, to changes in the recipes and plant configurations) and, subject to authorization, promote changes to the on-line system for execution. It is possible to use all the standard industrial system functions to produce, for example, user interfaces, detailed history, alarms, etc., to communicate with external relational databases for corporate accounting or other functions, and to communicate with external "user programs" running synchronously for heavier numerical work. The external communication mechanism may be used for interfacing more sophisticated off-line schedule optimisers (to carry out optimal scheduling functions) and dynamic simulators (so as to enable running a simulation instead of the actual plant), as described in Macchietto et al. [33]. Some applications of this type were developed by Cott [10] and Kassianides [28], who demonstrated the execution of supervisory batch control function in conjunction with both Speedup simulations and actual equipment (an older pilot plant in the Department of Chemical Engineering).

The basic ideas and algorithms were also incorporated into a software product (APV Baker's

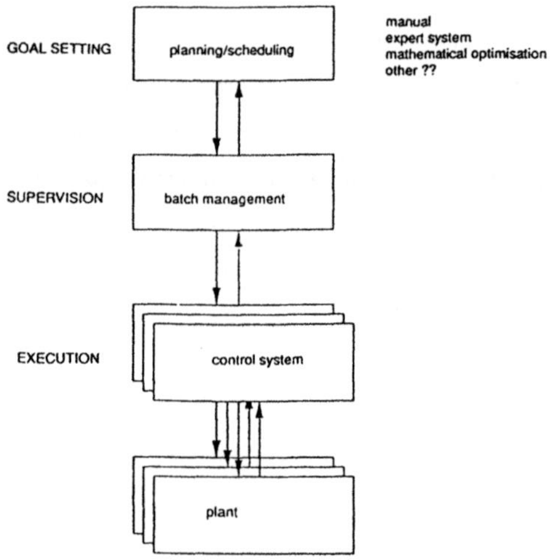

Figure 4. Control/Management Environment

BatchManager), which provides graphic interfaces for interactive use in a UNIX environment, with all models definitions held in a relational database and with interfaces to order acceptance systems. At present, Batch Manager only works in an off-line mode, but work is under way to integrate it with the ACCOS plant wide automation and control systems which APV supplies mainly to the food industry.

5. Integration Project

In this section, we describe initial work towards the development of a system which addresses in particular three of the major problems identified in the industrial survey study: integration of preliminary plant/process design with operations models, integration of automation and control system development with plant/process design and integration of batch design tools with batch management/execution tools. The focus will be on the general aspects and on an example of recent results. Details on individual aspects will be presented elsewhere.

The approach has been as follows:

I) For preliminary design, the approach based on the STN representation provides a good basis for simultaneous consideration of plant, processes and market interactions, for a sufficiently wide class of problems. In the very early stages, simple models of plant and operations may be considered (major processing vessels). These can be evolved into rather more detailed models to include the actual connectivity of the plant, much more detailed operational constraints, etc. using the extensions of Crooks [15] and Barbosa and Macchietto [2], outlined in a previous section (mSTN, uSTN, etc.).

ii) For batch operations management/control, the NAMUR and ISA SP88 models define a design basis which is both appropriate and which cannot be ignored, since it has been and will be adopted by most industrial batch control systems. Most of the interesting work lies at the supervisory batch management level, in the dynamic scheduling and in the coordination with off-line planning/scheduling. For these activities, the approach of Cott and Macchietto [12] is adopted.

iii) Techniques for the development of automation and control systems should utilize initially the same plant/process/constraint models used for process/plant design (supplemented later by details of piping, valves, instrumentation, etc. arrangements and more detailed control objectives, strategies, constraints) and produce eventually models for batch control and supervision in a form compliant with NAMUR/SP88. This should guarantee that the control structures and sequences generated can be passed on to and executed by standard industrial batch operation systems. The required machinery was developed by Crooks [15] and will be briefly described in the next section [16,17,18,19,20]. A prototype software as developed (CAPS) which implements those ideas.

iv) Different representation are clearly best for addressing different problems. The conversations of mode presentations between different domains and data aggregation/desegregation were given particular attention so as to achieve consistency.

CAPS overview

A functional overview of the CAPS (Computer Aided Procedure Synthesis) system of Crooks [15] is given in Figure 5. The problem is posed as follows. Given: a description of the plant and processes involved, operating goals and constraints identify I) the control sequences (phases), ii) the master procedure networks, made up by those phases and iii) detailed sequence control instructions for each phase so as to achieve one or more desired operating goal(s), subject to all restrictions posed by equipment and operations configurations and other constraint specifications. Where parallel operations are possible, control phases and master procedures should be produced both in a generic, parameterised form and also in instantiated form for each of the possible equipment allocations. Since there is often more than one way of achieving a goal, a choice should be made according to some objective function. Typical goals may be to transfer a quantity of material from one end of the plant to another while leaving any equipment used in a clean state,

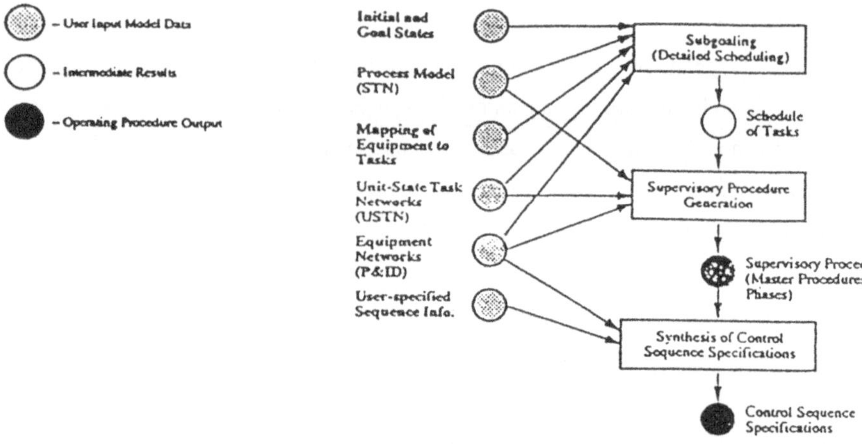

Figure 5. CAPS - system overview

to produce an intermediate product according to a specified recipe, etc. The solution approach followed consists in a sequence of three main steps. First, the problem is formulated as a detailed task scheduling problem, using the STN representations and modeling extensions previously discussed, and solved as a MILP problem. Plant items which are functionally equivalent are represented at this stage as a single generic unit type (a resource). This produces a generalization of all possible schedules of the tasks required to achieve the goals. We may note that this can be done even before the exact number of equipment units in a resource is fixed, i.e. very early in the design stage. This information is then processed in a second step where generic control phases and master procedures are identified, according to the SP88 guidelines and parameterised according to equipment resources, control parameters, etc. The boundaries between phases are identified so as to enable distributed control and distinct master procedures are chosen so as to enable monitoring of batch identity (usually destroyed by batch merging and splitting operations). Tasks may be aggregated into phases in various ways as required for control purposes. For example, all the tasks involved in a recirculation operation, or in the simultaneous feed of several ingredients must be carried out together and will constitute a single control phase. Any additional control phases which were not present in the original gBSS model (e.g. small transfers) are identified and added. More detailed flowsheet information, when it becomes available from design, enables identifying multiple instances of the generic control phases. These first two steps broadly accomplish what has been called "subgoaling" in the procedure synthesis literature.

Other researchers have approached this problem for continuous plants using a variety of methods for "planning" the required operation sequences. The main features of the CAPS approach are the use of an operations model which is sufficiently rich so as to reflect all important interactions, the use of a numerical technique which is very good at considering complex interactions and interacting goals, the choice of balance between use of information about desired operations which are easily supplied by the user and information which are more easily generated by the computer, and the use of industrial guidelines to help in decomposing the problem into sensible components.

Finally, the third step of CAPS utilizes much more detailed information about the plant network (a process and instrumentation diagram, P&ID) to develop rather detailed sequence control specifications for each phase in terms of typical discrete and continuous control primitives (e.g. open/close valve, check item, put control loop on automatic at setpoint x, monitor logical condition y, etc.). Because individual control phases are now largely independent (their interactions having already been identified), very detailed searches and extensive practical control knowledge and constraints may be applied to this step, which is accomplished using a rule-based method. The final output from this step is a specification for each control phase. For execution, individual specifications will have to be translated into actual sequence control code for a target control system. In principle, this step could also be automated, although we have not yet done it (all control systems are slightly different, requiring one translator each). The specifications, however, are sufficiently low level that automatic translation should be rather easy. In practice, we have manually implemented some of the control phases on two industrial control systems to which we have access (RTPMS, by IBM and ACCOS, by APV Baker).

The output of CAPS can be produced in principle in a variety of ways. One implementation which we have developed produces all the definitions needed to define a SUPERBATCH model of the plant, master procedures and control phases, in the required formats. These can be inputted directly to the supervisory system. It is then possible for an operator to introduce a production plan (however generated), involving any number of batches for which the corresponding recipes and control phases have been synthesized, and ask SUPERBATCH to schedule it. These production plans may of course include different and much more complex operation scenarios than were initially used in the optimal plant design. If individual control phases have been defined (by CAPS), but these have not yet been implemented in a control system, then only off-line scheduling exercises can be carried out. This will enable assessing the actual schedules, detailed plant operations, performance measures, etc. obtained in a continuous time representation. As already noted, this can be done very early on in the design stage. When the control phases are implemented in a control system, SUPERBATCH can then initiate their execution and carry out all its other on-line supervisory control and dynamic scheduling functions. The controllers can manipulate actual control devices and receive actual signals from the plants, but, initially, it is also possible to drive just a simulation (either a simple or a more detailed dynamic simulation, possibly with random variations also modeled, as described by Cott and Macchietto [18]. The performance of the actual batch and supervisory control systems can then be assessed early on with a simulated plant. Later on, when the plant is built, actual end control devices and signals may be connected to each control phase and exactly the same supervisory and batch control systems will drive the actual plant. This switch is made very simple since all control variables and parameters used for each control phase are not hardwired but explicitly declared as part of a SUPERBATCH model.

Overall, the approach described permits going almost automatically all the way from preliminary design of a batch plant taking into account operations models, to detailed design of batch operation sequences and control procedures at various levels of abstraction, to execution of the same procedures in an on-line system which features supervisory batch management and dynamic scheduling.

If we now assume that a plant configuration is fixed and turn the attention to the usual operations problem of generating and executing an optimal production schedule, we may in principle use again the gBSS plant and operations models already developed at the design stage, specify a new production objective, new cost parameters, etc. and calculate an optimal schedule for the new operation problem. If we go again through step 2 of CAPS, we will have one of two situations: either the new production can be carried out using master procedures and control phases which had been already identified, in which case CAPS will just determine the new parameters (batch sizes, transfer rates, etc.) to be used with the existing procedures, or some novel processing arrangement is to be done which was never done before (this may be the case if a product recipe variation is introduced, or an alternative route is proposed as optimal). In this case, new control phases may have to be developed, and possibly entire new master procedures. CAPS was designed so that these could be incrementally added to its procedure data base. A similar situation will arise if the equipment is modified (say, a new reactor is added in parallel to an existing one). In some cases, it is only sufficient to generate new instances of existing generic control phases. In others, it is possible to generate new master procedures as a different network of existing phases. Finally, for more radical changes, it will be necessary to generate some completely new control phases and master procedures. It should be noted that even the simple addition of a new procedure may require changing several other ones, for example to include interlocks.

6. Application Example - Batch Pilot Plant

In this section, an example of the ideas outlined above is given, which is used to illustrate the current progress. The example refers to our own batch pilot plant (described in Macchietto [31]), a schematic diagram of which is given in Figure 6. This plant is simple in terms of its basic processing units, but a great deal of complexity is added by the presence of cleaning operations (partially sharing the same pipework used by the process), operations involving external recirculation loops, two way flow in some pipes, transfer paths which can be configured through flexible valve manifolds (so called Flow Separation Plates), as is common in the food industry. In the initial design stages, a simplified view of the plant may be considered with only the major equipment items, and representative processes used for preliminary sizing, etc. This is well within the capabilities of the optimal gBSS software. As the requirements are refined, details may be

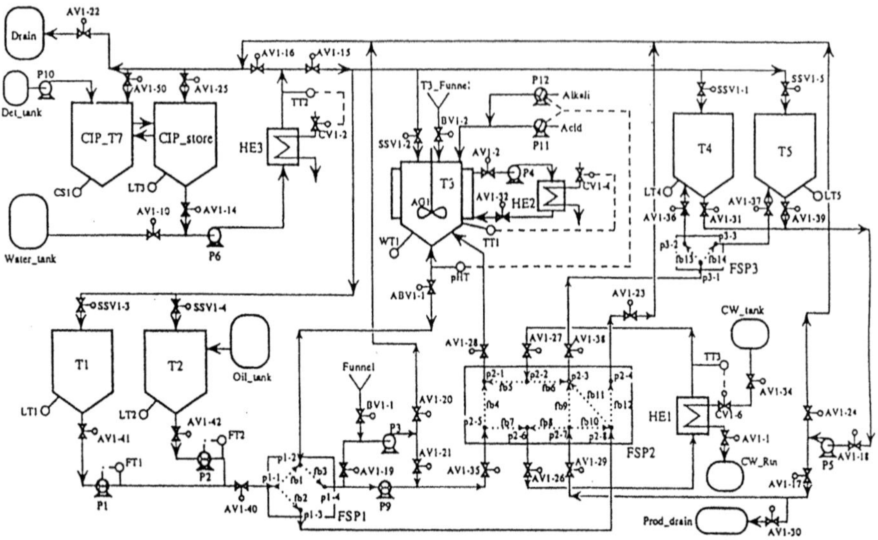

Figure 6. Batch Pilot Plant Process and Instrumentation Diagram

added to the model to include connections and more detailed operational constraints. For scheduling purposes, such a model was developed by Crooks [15] which could be used to optimally schedule a small number of batches in greater detail. This model was used as input to CAPS to generate the control procedures for batch management and detailed sequence specification. One of the basic processes carried out in the plant is an enzymatic starch conversion to sugars. With a goal specified of producing one batch of glucose, a single master procedure was identified by CAPS, with 14 control phases (with associated resource requirements, precedence network between phases, phase parameters, etc.) which corresponded very well with the design of the actual control system (one of this is an external recirculation of the starch from reactor T3 through heat exchanger H1 for temperature control purposes). This exercise was then carried out for the rather complex cleaning operations, defined by an STN model. Here, the synthesis goal specified was simply to change the state of a target processing vessel from "dirty" to 'clean". The initial goal of cleaning T3, for example, resulted in 4 control procedures (preparation of hot cleaning-in-place (CIP) solution, water rinse T3, detergent wash T3 and final check of cleaning solution strength). To carry out a T3 clean operation alone, these sequences are scheduled sequentially as CIP preparation, water rinse T3, detergent wash T3, water rinse T3, final check. The water rinse sequence is only generated once but used twice with different parameters, for pre and post detergent wash, as required by the cleaning STN definition. If we continue the exercise

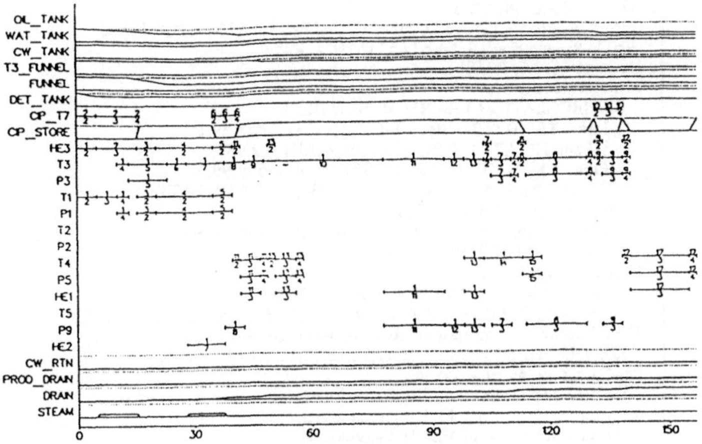

Figure 7. SUPERBATCH Detailed Schedule for Glucose Production and Cleaning-in-Place (CIP) Operations (on horizonal axis, time in minutes for start)

and specify, as new synthesis goal, the cleaning of T1-T2 (modeled as a single resource type), CAPS generates two additional procedures only, water rinse T1-2 and detergent wash T1-2. The CIP preparation and final check sequences are also identified as necessary, but these are recognized as being already available in the CAPS procedure database. To carry out a T1-T2 clean alone the sequences are scheduled in the sequence CIP preparation, water rinse T1-2, detergent rinse T1-2, water rinse T1-2, final check. With a flowsheet showing two vessels for the resource T1-2, CAPS will produce two instances for each of the generic water and detergent rinse phases. We can now prepare production plans consisting of batches of glucose and cleaning cycles, as required, and produce a corresponding operations schedule with SUPERBATCH. For example, a detailed schedule (of the flexible connecting paths as well as of the main and service vessels) for a single batch of glucose with final cleaning of the main reactor T3 is shown in Figure 7. The complex resource contentions during all processing and transfer steps are well handled.

In the next step, the detailed P&ID diagram of the plant is utilized by CAPS to generate the detailed control specifications required to carry out each of the control phases so far identified. A typical output of CAPS is shown in Figure 8 for one of the glucose production phases (the addition of starch through a funnel with external recirculation of the reactor contents). Again, the specification produced compared well with those for the corresponding sequence in the actual plant. This example is discussed in detail in a forthcoming publication (Crooks and Macchietto, 1993b). To actually execute the sequence, we would need to translate this specification into

```
Sequence SEQ_U_PHASE5
  Corresponds to Unit Phase U_PHASE5 and Resource Phase R_PHASE5
  Weight Parameter = 9.9, based on FUNNEL loss
  Duration = [10.0, TIME]
  Material Transfers:
    RECIRCULATION TRANSFER BLOCK:
      From: T3   To: P3   State: MESS2_1   Amount: 89.1
        Path: [T3,ABV1_1,P1_2,FB3,P1_4,AV1_19,P3]
      From: P3   To: T3   State: MESS1_1   Amount: 99.0
        Path: [P3,AV1_21,AV1_35,P2_5,FB4,P2_1,AV1_28,T3]
    LINEAR TRANSFER BLOCK:
      From: Funnel   To: P3   State: STARCH   Amount: 9.9
        Path: [FUNNEL,BV1_1,P3]
  STN Tasks performed:
    Task: RECIRC_ST   Unit: T3   Amount: 99.0
    Task: ADD_STARCH   Unit: P3   Constant Rate from subgoaler: 9.9
  The following sequences are inhibited by this sequence:
    [SEQ_U_PHASE34,23,22,21,17,16,15,12,11,10,9,8,7,6,4,1]
  Pre-Checks for the sequence are:
    Ensure flowplate bend: FB7 is disconnected.
    Ensure flowplate bend: FB5 is disconnected.
    Ensure flowplate bend: FB1 is disconnected.
    Ensure flowplate bend: FB3 is in place.
    Ensure flowplate bend: FB4 is in place.
    Unit Funnel minimum level: 9.9
    Unit T3 pre-UState #1: CLEAN
  Close and Inhibit the following valves:
    [AV1_20,AV1_2,AV1_32,BV1_2,SSV1_2]
  Inhibit the following flowplate bends which must not be used during the sequence:
    [FB1,FB5,FB7]
  Inhibit the following items for use in the sequence:
    [AV1_21,AV1_35,P2_5,FB4,P2_1,AV1_28,T3,ABV1_1,P1_2,FB3,P1_4,AV1_19,
    FUNNEL,BV1_1,P3]
  Open valves: [AV1_21,AV1_35,AV1_28,ABV1_1,AV1_19,BV1_1]
  Start the following devices:
    P3 with a nominal flowrate of: 20
    Agitator AG1
  Termination Condition(s) for the sequence:
    1.  After nominal time: 10.0
  Termination Instructions for the sequence:
    Stop agitator AG1
    Stop flow device: P3
    Close valves: [AV1_21,AV1_35,AV1_28,ABV1_1,AV1_19,BV1_1]
  Uninhibit the following plant items:
    [AV1_21,AV1_35,P2_5,FB4,P2_1,AV1_28,T3,ABV1_1,P1_2,FB3,P1_4,AV1_19,
    FUNNEL,BV1_1,P3,FB1,FB5,FB7,AV1_20,AV1_2,AV1_32,BV1_2,SSV1_2]
  Post-conditions are:
    Unit T3 post-UState #1: DIRTY
```

Figure 8. Control Sequence Specification produced by CAPS for recirculation Addition of Starch

control code. The pilot plant is controlled by an ACCOS system, by APV, with its own sequence control language (PARACODE). Assuming this done for all the required sequences, we could then run the plant from the same SUPERBATCH model. As already mentioned, we are presently working on an ACCOS/SUPERBATCH interface, and plan to complete this demonstration shortly. Two other application demonstrations have been developed, where the plants are simulated by a dynamic model (running under the control of the RTPMS system). Here, we were able to show the final execution and supervisory management of a complete production plan.

These applications, involving an industrial dairy plant and a multigrade PVC process, are presented by Crooks and Macchietto [19] and Bretelle, Chua and Macchietto [7]. Some details of the dairy plant application are also given in Crooks et al. [15,16,17].

Finally, the retrofit design of our pilot plant is discussed by Barbosa and Macchietto [2]. The MILP model of Crooks is utilized again, with additional design variables, to explore the minimum cost modification of the plant so as to permit the inclusion of a novel process. The optimal solution, involving the installation of double seat valves instead of the flow separation plates and some rerouting of pipes, is presently being implemented in the plant.

7. Discussion and Conclusions

Clearly, proper integration of all batch processing activities is quite a formidable undertaking and achieving it is still quite a way away. We still need far better and more powerful tools to deal with each of the individual problems involved. Optimal design and scheduling techniques are still haunted by combinatorial complexity, although much progress has been achieved recently. On-line dynamic scheduling could do with more powerful algorithms so as to deal with more (all?) scheduling and control decisions in real time. Techniques for dealing with the automatic development of operating and control procedures from basic design information, in spite of my personal excitement at the results discussed above, are still rudimentary if compared to the level of sophistication required, say, to pass a strict safety audit. Proper and formal validation techniques will no doubt have to be used, and some work in this direction is indeed underway [35]. Finally, the tools required to support a flexible and powerful integration infrastructure (object oriented data base management systems, modeling environments, etc.) still need further development before they can be routinely used for the job at hand. More comprehensive models and representations are still needed at pretty much all levels to cover in particular the boundaries between classic, well defined but isolated problems.

The collaborative study with our industrial friends confirmed that the separation on-line management/control and off-line design/scheduling is a major source of problems (or, conversely, a very rich ground for potential advances). The recent work [15] on procedure synthesis, in particular, has forced us to take a very hard look at these two aspects of batch processing, at the underlying assumptions and at development the consistent models. Integration techniques utilized ranged from the use of shared databases, task integration and integration of analytical techniques. The results obtained indicate that, perhaps for a limited set of problems and activities, some interesting steps can already be achieved towards an integrated system for batch processing.

Acknowledgment

This work was supported by SERC/AFRC and a Commonwealth Institute scholarship (CAC). Industrial partners of the Batch Processing Project, in particular APV Baker, BNFL Ltd. IBM UK and John Brown Ltd. are warmly thanked.

References

1. Angus, C. J. and P. Winter (1985). An Engineering Database for Process Design. IChemE Symp. Series No. 92, Pergamon Press, Oxford.
2. Barbosa-Povoa, A. and S. Macchietto (1992). Design of Multipurpose Batch Plants - I. Problem Formulation, Computers Chem. Engng., 17S, 33-38.
3. Barbosa-Povoa, A. and S. Macchietto (1993) Redesign of a Multipurpose Batch Pilot Plant with Cleaning in Place (CIP) Integration. Proc. ESCAPE 3, Graz (Austria), July 5-7.
4. Barton, P. and C. C. Pantelides (1991). The Modeling and Simulation of Combined Discrete/Continuous Processes. Proc., PSE'91, Montebello, Canada.
5. Benayoune, M. and P. E. Preece (1987). Review of Information Management Systems in Computer-Aided Engineering. Computers. Chem. Engng., 11, pp. 1-6.
6. Birewar, D.B. and IE. Grossman (1990). Simultaneous Synthesis, Sizing, and Scheduling of Multiproduct Batch Plants. Ind. Eng. Chem. Res., 22, 2242-2251.
7. Bretelle, D., E. S. Chua and S. Macchietto (1994). Simulation and on-line operation of a PVC Process for Operator, Computers Chem Engng, 18S, pp. 547-551
8. Clark, S. M. and G. S. Joglekar (1992). General and Special Purpose Simulation Software for Batch Process Engineering. NATO Advanced Study Institute on Batch Processing Systems Engineering, Antalya, Turkey.
9. Cherry et al. (1985). Proc. PSE '85, (F. A. Perris, ed.), IChemE Symp. Series No. 92.
10. Cott, B. J. (1989). An Integrated Management System for the Operation of Multipurpose Batch Plants. PhD Thesis, Imperial College, University of London.
11. Cott, B.J. and S. Macchietto (1989a). Minimising the effects of batch process variability using on-line schedule modification Computers. Chem. Engng., 13, ½, pp. 105-113.
12. Cott, BJ. and S. Macchietto (1989b). An Integrated Approach to Computer-Aided Operation of Batch Chemical Plants. Computers. Chem. Engng., 13, 11/12, 1263-1271.
13. Cott, BJ. and S. Macchietto (1989c). A General Completion Time Determination Algorithm for Batch Processes. presented at AIChE Annual Meeting, San Francisco.
14. Craft, J. (1985). The impact of CAD and Database Techniques in Process Engineering. IChemE Symp. Series No. 92, Pergamon Press, Oxford.
15. Crooks, C. A. (1992). Synthesis of Operating Procedures for Chemical Plants. PhD Thesis, Imperial College, University of London.
16. Crooks, C.A., K. Kuriyan and S. Macchietto (1992). Integration of Batch Plant Design, Automation and Operation Software Tools, Computers Chem. Engng., 16S, pp 289-296.
17. Crooks, C.A. and S. Macchietto (1992). A Combined MILP and Logic-Based Approach to the Synthesis of Operating Procedures for Batch Plants. Chem. Eng. Comm.,11, pp. 117-144.
18. Crooks, C.A. and S. Macchietto (1993a). The Synthesis of Operating Procedures for Batch and Semi-continuous Chemical Plants - I. The Method. Submitted for Publication.
19. Crooks, C.A. and S. Macchietto (1993b). The Synthesis of Operating Procedures for Batch and Semi-continuous Chemical Plants - II. Applications. Submitted for Publication.
20. Crooks, C.A., N. Shah, C. C. Pantelides and S. Macchietto (1993). Detailed Scheduling of General Batch and Semi-continuous Operations. Submitted for Publication.
21. Czulek, A. J. (1988). An Experimental Simulator for Batch Chemical Processes. Computers. Chem. Engng., 12, 2/3, pp. 253-259.
22. Halasz, L., M. Hofmeister and D. W. T. Rippin (1992). Gantt-Kit - An Interactive Scheduling Tool. NATO Advanced Study Institute on Batch Processing Systems Engineering, This volume pp. 706.

23. Hanish, H. M. (1992). Coordination Control Modeling in Batch Production Systems by Means of Petri Nets. Computers Chem. Engng., 16, 1, pp. 1-10.
24. Hasebe, S. and I. Hashimoto (1992). Present Status of Batch Process Systems Engineering in Japan, NATO Advanced Study Institute on Batch Processing Systems Engineering, This volume pp. 49.
25. Huang, Y. W. and L. T. Fan (1988). Designing an Object-Relational Hybrid Database for Chemical Process Engineering. Computers. Chem. Engng., 12, 9/10, pp. 973-983.
26. ISA-dS88.01 (1992). Batch Control Systems: Models and Terminology, Draft 5, ISA, Triangle Park, NC, USA.
27. Jayakumar and G. V. Reklaitis (1991). Graph Partitioning with multiple Property Constraints in Multifloor Batch Chemical Plant Layout. Paper presented at AIChE National Meeting, Los Angeles.
28. Kassianides, S. (1991). An Integrated System for Computer-Based Training of Process Operators. PhD Thesis, Imperial College, University of London.
29. Kim, W. (1990). An Introduction to Object-Oriented Databases, MIT Press, Cambridge.
30. Kondili, E., C.C. Pantelides, R.W.H. Sargent (1988). A General Algorithm for Scheduling of Batch Operations. Proceedings 3rd Intl. Symp. on Process Systems Engineering, pp. 62-75. Sydney, Australia.
31. Macchietto, S. (1992a). Automation Research on a Food Processing Pilot Plant, IChemE Symp. Series No. 126, pp.179-189.
32. Macchietto, S. (1992b). Interactions between Design and Operation of Batch Plants. pp. 113-126 in Interactions Between Process Design and Process Control, IFAC Workshop, J.D. Perkins ed.,Pergamon Press (1992)
33. Macchietto, S., M. Matzopoulos and G. Stuart (1988). On-line simulation and optimization as an aid to plant operation-the easy way", pp. 669-677, in Computer Aided Process Operations, (G.V.Reklaitis and H.D. Spriggs eds.), Elsevier.
34. Marquardt, W. (1992). An Object-Oriented Representation of Structured Process Models. Computers. Chem. Engng., 16.
35. Moon, I. G. J. Powers, J. R. Burch and E. M. Clarke (1992). Automatic Verification of Sequential Control Systems using Temporal Logic. AIChE J. 38, 1, p. 67.
36. Motard, R. L. (1989). Integrated Computer-Aided Process Engineering. Computers. Chem. Engng., 13, 11/12, pp. 1199-1206.
37. Mujtaba, I. M. and S. Macchietto (1992). An Optimal Recycle Policy for Multicomponent Batch Distillation. Computers Chem. Engng., ICS, pp.273-280.
38. Papageorgiu, L. G. and C. C. Pantelides (1992). A Hierarchical Approach for Campaign Planning of Multipurpose Batch Plants, Computers Chem. Engng., 17S.
39. Pekny, J., V. Venkatasubramanian and G. V. Reklaitis (1991). Prospects for Computer-Aided Process Operations in the Process Industries. Proc. COPE '91, Elsevier, pp. 435-446.
40. Perkins, J. D. (1992). Computer-Integrated Process Design - Status and Trends. Computers Chem. Engng., 16S.
41. Piela, P. C., T. G. Epperly, K. M. Westerberg and A. W. Westerberg (1991). ASCEND: An Object Oriented Computer Environment for Modeling and Analysis. The Modeling Language. Computers. Chem. Engng., 15, 1, pp.53-72
42. Preston, M. L. (1990). Integrated Process Plant Design. Proc. FOCAPD III, CACHE-Elsevier.
43. Puigjaner, L., A. Espuna, I. Palou and J. Torres (1991). A prototype Computer Integrated Manufacturing Modular Unit for Education and Training in Batch Process Operations. Proc. COPE '91, Elsevier, pp. 427-432.
44. Raman, R. and I. E. Grossmann (1991). Relation between MILP Modeling and Logical Inferencing for Chemical Process Synthesis. Computers Chem. Engng., 15, 2, pp. 73-84.
45. Reklaitis, G.V. (1991). Perspectives on Scheduling and Planning of Process Operations. Proceedings 4th Intl. Symp. on Process Systems Engineering, Montebello, Quebec, Canada
46. Rijnsdorp, J. E. (1991). Integrated Process Control and Automation. Elsevier.
47. Rippin, D. W. T. (1991). Control of Batch Processes. Proc. IEAC Symp. Dycord+ '89, Maastricht, The Netherlands, pp. 115-125.
48. Rosenof, P. H. and A. Gosh (1987). Batch Process Automation - Theory and Practice. Van Nostrand Reinhold Co., New York.
49. Sawyer, P. Computer-Controlled Batch Processing. to be published by IChemE, Rugby, UK (Personal Communication) .
50. Shah, N. (1992). Efficient Scheduling, Planning and Design of Multipurpose Batch Plants. PhD Thesis, Imperial College, University of London.
51. Shah, N., C.C. Pantelides and R.W.H. Sargent (1992). A General Algorithm for Short-Term Scheduling of Batch Operations - II. Computational issues. Computers. Chem. Engng., 17, 2, pp 229-244.

52. Stephanopoulos, G., G. Henning and H. Leone (1990a). MODEL.LA. A Modeling Language for Process Engineering - I. The formal Framework. Computers Chem. Engng., 14, 8, pp. 813-846.
53. Stephanopoulos, G., G. Henning and H. Leone (1990b). MODEL.LA. A Modeling Language for Process Engineering - II. Multifaceted Modeling of Processing Systems. Computers Chem. Engng., 14, 8, pp.847-869.
54. Stephanopoulos, G. (1991). Towards the Intelligent Controller: Formal Integration of Pattern Recognition with Control Theory. Proc. CPC IV.
55. Venkatasubramanian, V. (1991). Recall and Generalization Performances of Neural Networks for Process Fault Diagnosis. Proc. CPC IV., pp. 647-664.
56. Williams, TJ., (1989) (editor). A Reference Model for Computer Integrated Manufacturing, A Description from the Viewpoint of Industrial Automation. Instrument Society of America, Research Triangle park, NC.
57. Yamalidou, E. C. and J. Kantor (1990). Modeling discrete-Event Dynamical Systems for Chemical Process Control - A survey of Several New Techniques. Computers Chem. Engng., 14, 3, pp.281-299.
58. Ziegler, B. P. (1984). Multifaceted Modeling and Discrete Event Simulation. Academic Press, London.

An Interval-Based Mathematical Model for the Scheduling of Resource-Constrained Batch Chemical Processes

M.G. Zentner and Gintaras V. Reklaitis

School of Chemical Engineering, Purdue University, West Lafayette, IN 47907, USA

Abstract: This paper is an extension of previously presented work concerned with the use of interval analysis and search techniques for the exact modelling of the solution space of resource-constrained batch chemical process scheduling problems. A mathematical programming formulation is presented which is based on the notion of interval analysis and which uses a non-uniform discretization of time derived only from real process events. No additional discretization variables are introduced. While conventional discretization models rely on the rounding of problem data in order to reduce the size of the formulation, the base formulation proposed here does not require such data modification. The importance of this difference is that, although a conventional discretization model can be solved exactly, an exact solution is found to an approximate problem; whereas the model proposed here is not sensitive to the timing of events and therefore always gives an exact representation of the problem data with minimal discretization. However, added expense is incurred by the introduction of sequencing variables and constraints. Preliminary comparison with existing mixed integer formulations indicates promise for drastic reduction in model size for certain types of problems.

Keywords: Mathematical Programming, Scheduling, Batch Process, Resource Constraints, Time Discretization, Time Interval, Interval Analysis, Chemical Engineering.

Introduction

Over the past decade, research in the scheduling of chemical processes has evolved from the treatment of simple, easily understandable problems which were closely related to many classical problems in operations research (OR) to the treatment of problems of industrial importance which are more realistic in terms of problem detail. Both the sophistication of solution techniques being applied to these problems and the level of problem detail treated have grown in parallel. However, the basic elements of problem formulation still draw largely upon the ideas set forth by earlier researchers in OR.

In this paper, the concepts behind a different exact modelling scheme for resource constrained processes which is based on the notions of time interval analysis are discussed. Not only can interval analysis be used as the underlying logic for a mathematical formulation of the problem, but also as a means of determining which parts of the problem need not be included in the formulation.

The next section of the paper will concentrate on previous research in the area. This will be followed by a discussion of the structure of the specific problem being treated here. Next, some of the ideas behind interval analysis will be briefly discussed. The details of the formulation will then be presented. Finally, the size of the formulation will be examined relative to the MILP size required by conventional formulations.

Background

In the area of OR, process scheduling has been studied extensively for several decades. Scheduling research in the chemical industry can no longer be considered a new field, although most research progress has occurred in the past ten to fifteen years. Only relatively recently have chemical engineers been studying problems with resource constraints. In this section of the paper, OR contributions to scheduling will be briefly discussed. Next, some research in chemical engineering toward problems without resource constraints will be summarized. The focus then will shift to a more detailed discussion of specific work in the area of resource-constrained batch chemical processes. Finally, some conclusions about the current state of research will be given.

A large portion of the scheduling research in OR has concentrated on special cases of various problems. The focus has been toward developing special purpose algorithms which result in optimal solutions. Problems for which such algorithms are not found have been formulated mathematically. There has also been a significant amount of effort directed toward the construction of heuristics which perform well on many classes of problems. Most heuristic work is based on job-prioritization and list scheduling, often coupled with neighborhood search techniques. A general overview of this area is provided by Baker in [1] and a further examination by French in [7]. Surveys of various approaches are given in [8,9,10,14]. In [2], Blazewicz, et al, provide an extensive summary of OR approaches to several special cases of resource-constrained problems. For problems of very simple structure, exact algorithms are presented. For problems of a more complex nature, mathematical formulations are provided. Finally, an important line of research in the area of resource-constrained problems was initiated by Fox [6], who presented one of the first approaches which attempted to reason about the problem in terms of its constraints.

Much of the early work in scheduling in the area of chemical engineering focussed on various special cases of batch processing problem structures in the absence of resource constraints. Overviews of the various structural features present in batch processes as well as of some of the approaches to problems of these various structures are given in [16,18,22]. Most recently, Reklaitis [17] revisits earlier reviews in order to provide not only more precise classifications of currently treated problems, but also to outline current needs for further research.

More recently, researchers have recognized the importance of resource constraints in chemical processes. Both heuristic and exact modelling approaches have been presented, however, the exact approaches have yielded the most productive results toward better understanding of the problem. Lazaro and Puigjaner [12] allowed for the incorporation of resource constraints in problems with unlimited intermediate storage. A more realistic case of a general batch process was treated by Egli and Rippin [5]. An enumeration scheme was employed in order to solve relatively small sized resource-constrained problems. More recently, a stream of research based on a formulation proposed by Kondili, et al [11], has resulted in solution approaches to larger, more realistic problems. The formulation is based on the uniform discretization of time [3,13,15], and thus is limited in its capability to exactly represent problems without the rounding of problem data. Sahinidis and Grossman [19] recognized sub-structures of the formulation in [11] which resembled lot-sizing problems. By disaggregating some of the continuous variables in the formulation, they were able to propose a larger formulation which had tighter LP relaxations than the original formulation, resulting in reductions in solution time ranging from 10% to 81%. However, the structure of the binary variable set was not changed. In addition to exact approaches, some research has been focussed toward heuristic algorithms. Cott and Machietto [4] and Tongo [20] propose schemes which resemble simulation type logic, using a "first available" time slot heuristic. The multiproduct batch plant with resource constraints was treated by Tsirukis and Reklaitis [21] using a decomposition approach. The planning problem MINLP was solved using a generalized Hopfield network. Subsequently, the resource allocation problem was solved by restricting the search space with a feature extraction algorithm prior to detailed optimization. Finally, work presented by the authors [23,24,25] studied the resource-constrained problem from a different perspective. A framework was constructed which was based on a non-uniform discretization of time defined by real process events. The decisions were handled by a search framework which produced a time interval for each task and a set of precedence constraints. The interval/constraint problem was then solved using an LP formulation.

The common theme throughout the modelling of resource-constrained scheduling problems is the uniform discretization of time. The difficulty with such a scheme is that exact modelling of processes is quite dependent on the timing of events. In order to accurately model a process, the greatest common divisor of all event times must be used as the width of the discretization intervals. Clearly, when problem data is expressed in terms of real numbers, rounding must be done in order to obtain a reasonable interval width. Furthermore, in practice the rounding is often more severe than simply converting real numbers into integers. For example, since an hourly discretization of a week would result in a large formulation, often the discretization must be more coarse. This effectively limits the scheduling of many tasks which would require an hour or less. As a result, for certain scheduling problems, a formulation which is not sensitive to the timing of events may prove advantageous.

Problem Definition

The formulation presented in this paper addresses the modelling of resource constrained batch process scheduling problems which are quite general in structure. However, since the level of the scheduling problem treated here is that of the detailed timing of batches, some restrictive assumptions are made. It is assumed that input information will be known about the structure of the processing equipment network, resource availabilities, product recipes, assignments of units to tasks, and production requirements. Other aspects of a more general problem will be treated in future work on this formulation. The discussion in this section will address plant structure, recipe structure, and production requirements.

Plant Structure

The proposed plant structure allows for the most general form of equipment connectivity. Equipment items may be connected to each other or to storage vessels in any fashion. In addition, the dynamic nature of connections can be modelled through the use of resources, that is, in order for a transfer to occur, the units must be connectable and a resource which represents a transfer line must be reserved. The connectivity of the various resources to the processing equipment can also be specified.

Recipe Structure

The recipe structure will be discussed in terms of three issues. First, the basics of product recipes will be defined. Then, the allowed storage modes will be covered. Finally, the interaction between resources and product recipes will be outlined.

The recipe structure considered encompasses all of the customary attributes of chemical product recipes. Specifically, the following attributes of each recipe step - processing unit pair are accommodated: sequence dependent setup times, transfer times both into and out of the unit from and to other specific units, processing times, and clean out times.

However, this treatment goes beyond the conventional level of detail in that each recipe task can be subdivided into multiple sub-tasks, called steps. For example, a recipe task may have two distinct phases, heating and stabilizing, which could have vastly different resource requirements and rate expressions. In general terms, the decomposition is first into recipe tasks, each of which is assigned to one processing unit and which are related by precedence constraints. Each recipe task can be further divided into operations. The operations may include set-up, transfer-in, processing, transfer-out, and clean-out. Each operation can more finely be composed of individual recipe steps, as in the case of the multiple processing steps for a recipe task which was mentioned above. Depending on the processing unit chosen for a task, the structure of the steps necessary for the task may be different.

Recipe steps can have several attributes. Each recipe step can have its own resource requirements, which may be dependent on which processing unit is chosen for the recipe task. Each recipe step can have a finite allowed wait time which may follow the step. Such wait times can be used to represent product instabilities. It may also be specified that during the processing of all steps of a given task, an aggregate amount of idle time may be incurred. This is an amount of idle time that may be distributed among the individual steps such that the total idle time incurred during the processing of the steps does not exceed the aggregate allowed idle time. The first step of the recipe of a product may specify several necessary precursor products, thus representing the combination of intermediate materials during processing.

The handling of storage in this approach is more comprehensive than most of the past treatments, although it still is not completely general. The aggregation of batches in storage, for example, where three batches of G enter the storage vessel and exit as two larger batches of G, is not determined, rather it must be specified by the user as problem input. However, several aspects of storage are modelled here that have not been extensively studied in past approaches.

The first issue deals with the fed batch mode. If material in an upstream unit is ready to transfer out of the unit and into the next stage, which is operated in fed batch mode, two scenarios are possible. The slow (fed batch) transfer can take place directly from the unit, occupying the unit until the fed batch phase of the downstream recipe step is complete. Another option is that the material can be quickly transferred to storage, clearing the upstream processing unit, and allowing the batch fed operation to proceed from the storage tank. Such options can be described in the framework presented here.

The second feature is the allowance for finite wait times for a recipe task. The current accepted description language for specifying storage is insufficient for representing batches that are only stable for a limited amount of time. A decomposition of the storage availability and the allowed storage time into separate vectors is proposed as an expansion of the present terminology. Specifically, these two vectors can be "multiplied" in order to define complete storage policies as shown below:

$$\begin{bmatrix} \text{ZW} \\ \text{FW} \\ \text{UW} \end{bmatrix} \times [\text{NIS} \quad \text{FIS} \quad \text{MIS} \quad \text{UIS}]$$

$$= \begin{bmatrix} \text{ZW} & \text{ZW} & \text{ZW} & \text{ZW} \\ \text{FW*NIS} & \text{FW*FIS} & \text{FW*MIS} & \text{FW*UIS} \\ \text{UW*NIS} & \text{UW*FIS} & \text{UW*MIS} & \text{UW*UIS} \end{bmatrix}$$

The letters ZW, FW, and UW represent the amount of time a product may stay in storage: Zero Wait, Finite Wait, or Unlimited Wait, respectively. Of the three, FW has not been extensively treated in previous approaches. The letters in the other vector represent storage conditions. NIS means that while no storage vessels may be used, the batch may remain in the processing unit for storage. FIS indicates that the batch must leave the processing unit but may be stored in one of a finite number of storage vessels. MIS indicates that both NIS and FIS apply. Finally, UIS implies that there is an unlimited number of storage vessels. The meaning of the product of a storage time and availability attribute should be self evident: FW*NIS means that the product may be stored in the processing unit for only a finite amount of time, and so forth. For obvious reasons, any availability multiplied by ZW reduces to ZW. The above types of storage policies are important when dealing with sensitive intermediate products with limited stability, for example, biochemical intermediates.

Also related to storage is the use of idle time, which may be incurred in a processing unit prior to the initiation of processing, between the steps associated with a processing operation, and following the completion of processing. This allows a process to be interrupted between phases when different resources are required but are not yet available as well as to be delayed prior to processing and/or held after processing.

The final aspect of the recipe description outlines how resources are to be treated. Resources may be of two types, renewable (RR) or non-renewable (NRR). The renewable resource is one which is replenished after use, for example, manpower, a transfer pipe, or electricity. Every level change of a renewable resource is treated as an absolute change and is not cumulative. Non-renewable resources are those that are consumed, for example, raw materials. Each specified level change of a non-renewable resource has a cumulative effect. That is, if there was 1 ton of raw material A and the level is changed by the arrival of 3 more tons, the level of raw material A would be 4 tons, whereas in the renewable case, the new level would have been 3 tons. The availability (or arrival) patterns of resources can be specified to be following a cyclic pattern or an acyclic pattern, as in weekly manpower levels or raw material levels, respectively.

Each recipe step may require several resources which serve different roles. As previously discussed, each task may be split into several steps, which allows for the specification of different resource demands for different phases of the task. For example, transfer into a unit may be divided into initial rapid charging of the unit followed by a slower fed batch transfer, each requiring different levels of electric power.

In the above discussion, it was shown that the recipe structure defined in this approach encompasses not only the consideration of the precedence of recipe steps, but also consideration of a variety of storage policies involved for each recipe task and treatment of detailed resource demands for each recipe step. The recipe model is sufficiently general to represent the vast majority of product recipes.

Production Requests

Each required product may have a due date and a release time. The release time of a product is the earliest point in time at which production of the product may be initiated. While both due dates and release times seem to impose the same type of constraint at opposite ends of a schedule, they are distinct. Specifically, a due date may not be met but a satisfactory schedule may still be obtained, whereas a release time is a hard constraint and in most cases can not be violated.

A restrictive assumption made in the current model is that the batch sizes will be known in advance, since this decision is somewhat higher in abstraction than the notion of detailed scheduling. This assumption implies, then, that batch aggregation will not be determined by the model presented here since the logic for determining when to aggregate and split batches would be nearly the same as that required in order to size batches at the onset. However, the assumption does not restrict the specification of aggregation and splitting by the user in the description of the product recipe through the renaming of products after aggregations and splits. This assumption also implies that the user must specify the number of batches to be produced in addition to their sizes in order to describe the overall production scheme. The amount of a product specified in a product order is assumed to be one batch.

A final ramification of assuming that batch sizes are fixed is that task durations and resource requirements are fixed as well. These stipulations are fairly restrictive, but one should note that using a constant level of resource requirement by a task is at best an approximation anyway. As a task is processed, the level of resource required can vary significantly over the duration of the task due to external influences and controller actions.

Summary of Problem Structure

The proposed scheduling problem includes a number of scheduling features that have not been treated before, many of which are specific to chemical processes and thus typically are not regarded in operations research studies. To recapitulate, the new ideas are the treatment of complex recipes including mixing of intermediate products, an allowance for splitting major tasks into many steps, the intelligent accommodation of storage with fed-batch processing mode, a new definition of storage modes allowing for finite stability times of intermediates, and a complete treatment of resources. The salient features of the problem can thus be outlined as follows.

- General equipment connectivity
- Division of recipes into tasks
- Division of tasks into operations
- Division of operations into steps
- Finite allowed idle times
- Aggregate idle times
- Merging of intermediates
- Complex storage policies
- Non-renewable and renewable resource constraints
- Fixed batch sizes
- Fixed unit and resource assignments
- Product release times
- Product due dates

Many of these individual features have seen little or no attention in the literature. Some of the assumptions with regard to fixed data in the problem description are restrictive, however, these aspects of the problem can be modelled by the addition of more variables and constraints to the formulation proposed here.

Definitions

Before developing the formulation, it is necessary to introduce some additional concepts. Specifically, the concepts of resource confluence intervals (RCI's), RCI elimination criteria, and event free intervals (EFI's) will be used in order to pre-analyze the problem so that only pertinent parts of the problem are formulated.

A resource confluence interval is an interval of time for a step such that all resources required for that step are present in sufficient amounts so that the step can be processed. In addition, an RCI spans as wide a time interval as possible. This means that, for a given step,

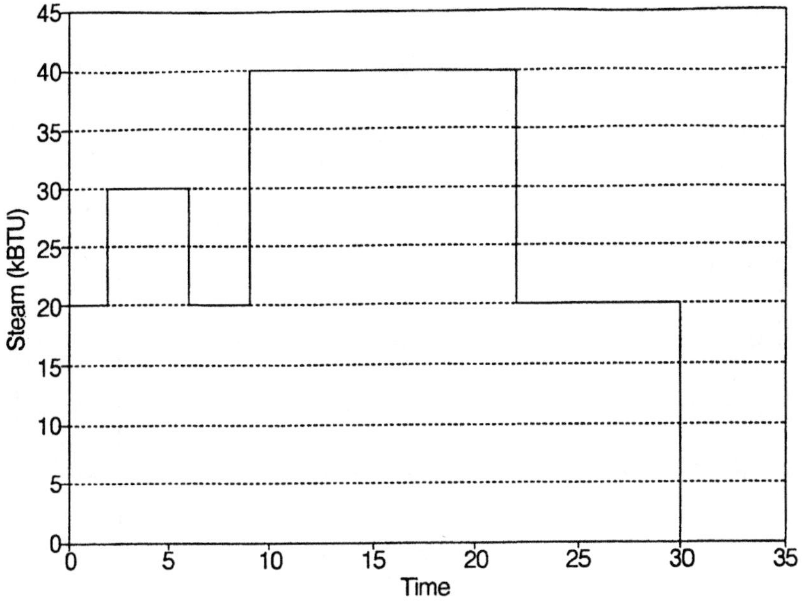

Figure 1 - Resource Profile for Steam

two RCI's never will exist end-to-end. For example, consider the resource profiles in Figures 1 and 2. If a step required 25 kW of electricity and 25 kBTU of steam, RCI's for that step would exist over the intervals [2,6], [10.5,17.5], and [20,22]. During all other times, at least one of the two resources is present in insufficient quantity. Also, note that none of the above intervals overlaps with any of the others, or touches end-to-end. This is because the intervals are created over as large a time interval as possible. Intervals which are shorter than the duration of the step are not considered valid RCI's and receive no further consideration. Further details of the construction of RCI's are developed in more detail in [23].

After preliminary analysis, there can be several RCI's for each step. Based on the set of RCI's and on pair-wise consideration of steps, some RCI's may be shortened or altogether eliminated. For a given pair of steps i and j, where step i precedes step j, the duration parameters $D_{ij,min}$ and $D_{ij,max}$ can be computed from recipe data. The durations of all steps which must occur between the start times of steps i and j indicate a minimum amount of time which must elapse between the start times of the two steps, $D_{ij,min}$. The durations of the steps between i and j coupled with their allowed idle times define the maximum time which may

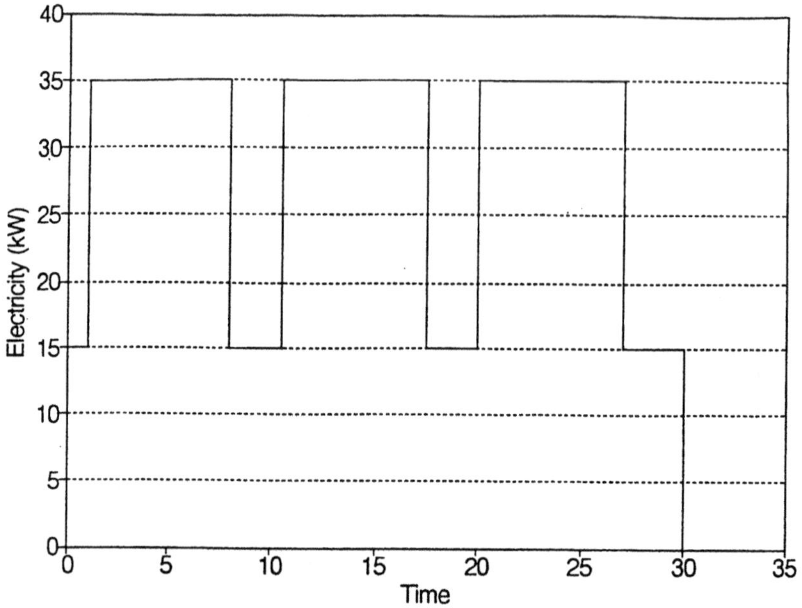

Figure 2 - Resource Profile for Electricity

elapse between the start times of the two steps, $D_{ij,max}$. These two parameters are used in Figure 3, which illustrates one criterion for eliminating an RCI. Consider the pair of steps i and j where i precedes j. Further, examine one fixed RCI for i, RCI_i, against several RCI's for j, as shown in Figure 3. The point labelled A on the axis is the earliest time step j could end if RCI_i was used for step i. Similarly, the point labelled B is the latest time step j could start if RCI_i was used for step i. If there is no RCI for j with end time $RCI_{j,ET} \geq A$ and start time $RCI_{j,ST} \leq B$, then no portion of RCI_i is useable, since its use will preclude a feasible RCI assignment for j. Several other such criteria for shortening and eliminating RCI's exist and are detailed in [23].

From the discussion of RCI's, it is apparent that several changes in resource level can take place during an RCI, as long as the resource level is not exceeded by the demands of its step at any point during the RCI. This implies that the discretization of time must be more detailed for the summation of resource consumption. For this purpose, the notion of an event free interval (EFI) is introduced. An EFI is an interval of time for a step (during one of its

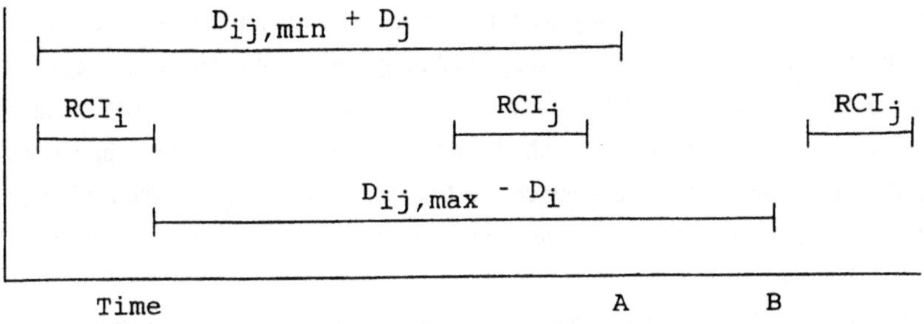

Figure 3 - An RCI Elimination Criterion

RCI's) during which no resource required by that step changes level. Depending on resource requirements, the set of EFI's may be different for the various steps in the problem.

The tools for resource interval creation and processing presented in this section serve as valuable pre-analysis tools prior to the generation of a detailed formulation. Through the generation of RCI's and subsequent elimination and shortening, the overall set of EFI's which must be considered in the final formulation can be considerably reduced.

Mathematical Formulation

The discussion will now be focussed on the set of mathematical formulae used for the representation of the non-uniformly discretized formulation. First, the representation of time and the constraints associated with it will be presented. Then, the concepts and constraints involved with the sequencing variables will be discussed. Next, the manner in which renewable resource constraints are handled will be given. These three aspects of the formulation are those which allow it to be different in structure than past approaches. Following the discussion of these three key aspects, the rest of the formulation, which largely consists of constraints for the incorporation of additional problem features such as linkage with a continuous formulation, optional use of storage vessels, non-renewable resource constraints, and sequence dependent set-up times, will be developed.

The Formulation of Time

The mathematical description of the time domain for resource-constrained scheduling problems, which is the simplest concept to represent using conventional approaches, typically is the root of the intractability in problem solution. A large number of decision variables is usually necessary in order to divide the scheduling horizon into small, uniform increments. Although time discretization will be used in this work, that discretization will be based upon a non-uniform set of time intervals, which allows a significant reduction in the number of variables. In the interest of following convention, the variable X_{ij} is introduced where $X_{ij} = 1$ if an interval is assigned to step j which encompasses the ith event free interval (EFI) for step j.

The number of variables introduced by such a definition is not as obvious as it might at first seem. Depending on recipe structure, the discretization for each step may not be the same as that for every other step. For example, if there is a transfer step which uses electricity and a pre-heating step which uses steam and if the sets of EFI's defined by steam and electricity are different, the discretization for the transfer and pre-heating steps would be different. If, however, the pre-heating step uses both steam and electricity, the discretization for the two steps must be identical (taken from the combination of the two resource profiles) in order to sum up resource usage. Even though the transfer step does not use steam, it inherits the discretization from all steps which share electricity since it must, at some point, be considered in conjunction with steam utilizing steps. In addition, application of interval narrowing and elimination criteria may make some X_{ij} variables unnecessary.

Example: Consider a case where two steps, A1 and B1, which use the resources steam and electricity from Figures 1 and 2, are to be processed. Step B1 uses steam, and step A1 uses both steam and electricity. Considered by itself, step B1 would result in discretization points at times 0, 2, 6, 9, 22, and 30. However, since both A1 and B1 use steam and since A1 also uses electricity, both A1 and B1 will inherit a discretization based on both resources. All points of discretization will then be 0, 1, 2, 6, 8, 9, 10.5, 17.5, 20, 22, 27, and 30.

The qualitative description above will now be cast into a more precise form. Let $\mathcal{S}$ be a set of steps which share at least one resource with step j and which are not otherwise constrained with respect to step j because of recipe considerations. Let $\mathcal{R}$ be the set of all resources required by every step $\in \mathcal{S}$. Every step $\notin \mathcal{S}$, which uses at least one resource in $\mathcal{R}$ and which is not constrained with respect to j by the recipe, is added to the set $\mathcal{S}$. The resources required by the new members of $\mathcal{S}$ are then used in updating $\mathcal{R}$. The definition of $\mathcal{S}$ continues in a recursive fashion until the set can no longer be expanded. Every step j $\in \mathcal{S}$ derives its discretization from the resources in $\mathcal{R}$.

In the worst case, all steps will consider every resource in the time discretization. Let s be the number of steps, r the number of resources, and e the average number of events per resource. Then, the number of X_{ij} variables will be $s(er - 1)$. Through pre-analysis, some of these variables can be eliminated. For example, if there is a release time specified for a product, all X_{ij} variables where the ith EFI occurs prior to the release time are not considered.

One can further envision several elimination criteria based on the values of the parameters $D_{i,j,min}$ and $D_{i,j,max}$.

In many cases, an interval assigned to a step may span several EFI's within a given resource confluence interval (RCI). Only certain pairs of EFI's can be used as the end points of an interval assigned to a step. As discussed earlier, each RCI for a step can be composed of several sequential EFI's. Each pair of EFI's allowed to serve as end points must satisfy two criteria. First, the time spanned between the start time of the first EFI and the end time of the last EFI must be greater than the duration of the step. It is also important that the time spanned by the EFI pair is not larger than necessary. The allowed pairs of EFI's for step j can be determined by sliding a time interval which represents the duration of step j along the time horizon. Initially, the start of the interval will coincide with the start of the first EFI for step j. One allowed pair is defined which starts with the first EFI and ends with the EFI in which the duration interval ends. Then, the interval is moved forward in time. Each time either the end or start of the interval crosses the boundary between two X_{ij}'s, a new pair of allowed EFI's is discovered. In order to estimate the worst case number of these pairs, recall that there are $(er - 1)$ X_{ij} variables (intervals) for each step. This means that the start and end of the duration interval can each cross at most $(er - 1)$ interval boundaries. There are also the initial position and final position of the duration interval, which results in a total possible of $s(2(er - 1) + 2) = 2ser$ pairs in the worst case. Many pairs will never be necessary because the step duration is wider than the duration of the single EFI's. Further, considerations of $D_{ij,min}$ and $D_{ij,max}$ can be used, as discussed earlier, to restrict the number of allowed EFI pairs by using interval elimination ideas.

It is important to be sure that all X_{ij}'s within the assigned interval have a value of 1 (an assigned interval for a step must be continuous). When the two EFI's, EFI_a and EFI_h, are chosen as the end points for the interval, for all i where $a \leq i \leq h$, then $X_{ij} = 1$. The constraint below enforces this condition.

$$h - a + 1 - (h - a + 1)[(1 - X_{hj}) + (1 - X_{aj})] \leq \sum_{i=a}^{h} X_{ij} \tag{1}$$

Note that this constraint need not be created when a = h or when a = h - 1. When a given pair of X variables is chosen for the end-points of the interval for a step, not only must all of the X's between the end points be set to 1, but the X's outside of the chosen interval must be set to 0. This includes all X's in the RCI in which the interval is chosen and all X's outside of the chosen RCI. The following constraint accomplishes this.

$$(a + z - 1 - h)[(1 - X_{hj}) + (1 - X_{aj}) + X_{a-1,j} + X_{h+1,j}] \geq \sum_{i=1}^{a-1} X_{ij} + \sum_{i=h+1}^{z} X_{ij} \tag{2}$$

Since one instance of constraints 1 and 2 must be created for every allowed EFI pair for a step, the worst case estimate for the number of each of these constraints is $2ser$.

It is also necessary to ensure that each step is assigned a time slot. For this purpose, the following constraint is created for each step.

$$1 \leq \sum_{i=1}^{z} X_{ij} \tag{3}$$

When choices for steps are present, as in the case of multiple options for the use of storage or sequence dependent set-up steps, constraint type 3 will be modified to include X's for all options. The size of constraint set 3 is s.

It is also necessary to ensure that enough consecutive EFI's are reserved in order to accommodate the duration of each step. Allow TB_{ij} and TE_{mj} to represent the start and end times of the ith and m^{th} EFI, respectively. For all X_{ij} and X_{mj} of every step j where $TE_{mj} - TB_{ij} < D_j$, $i \leq m$, the following constraint is created.

$$X_{i-1,j} + X_{m+1,j} \geq i - m + \sum_{a=i}^{m} X_{aj} \tag{4}$$

Essentially, when all EFI's between i and m are being used, the constraint forces an additional EFI to be used. Otherwise, if only part or none of the EFI's between i and m are used, the constraint has no effect. In the worst case, all EFI's will exist within the same RCI for each step. For a single step, then, the worst case number of constraints from type 4 is the summation from 1 to $(er - 1)$. For all steps, the number of constraints is $ser(er - 1)/2$. However, if this is the case, note that the number of constraints of types 1 and 2 will be much less than the worst case since there would have been only one allowed pair variable in the worst case for type 4.

Finally, since the notion of allowed pairs is being used for the definition of constraint types 1 and 2, it is necessary make sure that no additional start/end pairs of EFI's are chosen which span wider time intervals than the allowed pairs. In other words, it is either necessary to enforce constraint types 1 and 2 for all EFI pairs or create a constraint which ensures that only allowed pairs of EFI's will be chosen as interval end-points. For every allowed pair of EFI's h and m such that there is no allowed pair $\{i,l:(i \geq h \wedge l < m) \vee (i > h \wedge l \leq m)\}$, the following constraint states that if h and m are the chosen allowed pair, no EFI's outside of the time spanned by h and m will be chosen.

$$X_{h-1,j} + X_{m+1,j} \leq 2[2 - (X_{hj} + X_{mj})] \qquad (5)$$

The size of constraint set 5 is equal to the number of allowed EFI pairs in the worst case, which results in a possible $2ser$ constraints.

Sequencing Variables and Constraints

In order to ensure that resource levels will not be violated, it is necessary to know whether or not two steps are sequenced with respect to each other. Further, it is important to propagate the effects of sequencing two steps to the rest of the steps in the problem. For these purposes, several constraint types are needed.

First, the variables s_{ij} and s_{ji} are introduced for every pair of steps i and j which use at least one common resource and which are not already constrained because of recipe considerations. If step x follows step y, then $s_{yx} = 1$, $s_{yx} = 0$ otherwise. It follows that:

$$s_{ij} + s_{ji} \leq 1 \qquad (6)$$

which states that the two steps can only be constrained in at most one direction with respect to each other. More strongly, if steps i and j share a processing unit:

$$s_{ij} + s_{ji} = 1 \qquad (7)$$

In the worst case, every step could be constrained with respect to every other step except itself, resulting in $s(s-1)$ variables and $s(s-1)/2$ constraints.

Constraints are also necessary in order to propagate the effects of a sequence decision. For example, suppose that two steps, j and l, are constrained such that j precedes l ($s_{jl} = 1$). The implication is that all steps which precede j also precede l ($s_{xl} = 1$ if $s_{xj} = 1$). Similarly, but in the opposite direction, $s_{jx} = 1$ if $s_{lx} = 1$. In order to model this situation, for every unique triplet of steps x, j, and l (note that {1,2,3} and {1,3,2} are unique triplets) which share some common resource, the following non-linear constraint represents the propagation of constraints:

$$s_{xl} = s_{jl}s_{xj} \qquad (8)$$

Using a simple linearization construction, the above constraint translates into the following pair of linear constraints:

$$s_{xl} \leq 0.5(s_{jl} + s_{xj}) \qquad (9)$$

$$s_{xl} \geq s_{jl} + s_{xj} - 1 \qquad (10)$$

In the worst case, every step will have the potential to overlap every other step, though in practicality this would never happen. There are $s(s-1)(s-2)$ unique triplets, giving a total of $2s(s-1)(s-2)$ linear constraints in the worst scenario.

Finally, additional constraints arise from recipe concerns. Specifically, no task can be pre-empted on its processing unit in preference to the completion of another task on the same processing unit. Recalling that a task is constructed of individual steps, this implies additional constraints among the steps of tasks assigned to the same processing unit. Let T_A and T_B be tasks assigned to the same processing unit. Further, allow $a \in T_A$ and $b \in T_B$ to represent steps which belong to tasks A and B, respectively. The symbol $|T_X|$ is used to indicate the number of steps involved with task X. With these definitions, constraints can be formulated which state that if at least one step of task A is related to one step of task B, all other steps of both tasks must be similarly related. For every s_{ba} where $b \in T_B$ and $a \in T_A$:

$$s_{ba} \geq (1/|T_A||T_B|) \sum_{a \in T_A} \sum_{b \in T_B} s_{ba} \qquad (11)$$

In order to get an estimate on the number of constraints of type 11 necessary in the worst case, let t represent the number of tasks and assume all tasks are assigned to the same processing unit. In such a case, there are $t(t-1)$ possible unique task pairs. With each pair of tasks, there are potentially $(s/t)^2$ step sequencing variables, s_{ba}, resulting in a total of $s^2(t-1)/t$ possible constraints. Again, it should be noted that this would rarely be the case in a real process environment.

Of the three types of step sequencing constraints proposed, the first seems obvious and is conventional, while the second and third are more specialized to this approach and work in conjunction with more of the constraints presented in subsequent sections. It was also shown that if one is not careful in deciding which pairs of steps do not need sequencing variables during the formulation of a problem, much extra information can contaminate what might be an otherwise concise formulation.

Renewable Resource Constraints

A major cause of complexity in mathematical formulations of scheduling problems is the consideration of shared resources. Conventionally, with the fine, uniform discretization of time, the levels of resource usage during each interval are summed and compared to the available level of resource. A similar approach applies to the non-uniform grid presented here; however, it is important to remember that in this formulation, constraints apply only to the potential for various steps to share resources over an interval, not the absolute fact of resource sharing.

In order to represent the potential for steps to overlap, an expression can be written in terms of the variables presented so far. If there are n steps which share some resource and none of which have any recipe-derived constraints with respect to each other, they have the potential to overlap during a given EFI_e if:

$$\prod_{i=1}^{n} X_{ei} \prod_{j=i+1}^{n} (1 - (s_{ij} + s_{ji})) = 1 \tag{12}$$

Equation 12 states that steps i through n can overlap if they are assigned during the given EFI and if none of the steps have any constraints with respect to any of the other steps. In order to constrain the actual problem, it would be easy to enumerate all possible overlapping sets for each EFI and state which ones are allowed and which are not; however, this would produce an undesirable number of constraints.

Further examination of the problem suggests a logical point for constraint building based upon the concept of maximal overlapping sets. A maximal overlapping set for an EFI is a set of steps which all could overlap in time and not cause any resource level violation such that there exists a step that can be added to the set which would cause a resource level violation. Specifically, since a maximal overlapping set will cause no violation, no constraint is necessary for any maximal or sub-maximal overlapping set. An algorithm for the generation of maximal overlapping sets is given in [23]. A constraint is necessary, however, for each combination of a maximal overlapping set and a step which is not a member of the set and which would cause a resource level violation. These constraints will represent the boundaries of allowable resource allocation, making additional constraints unnecessary. The format of the constraints will be that of equation 12, except that, since it is desired that the set of steps not overlap, the right hand side of the constraint will be 0, not 1. This form can subsequently be linearized.

Example: Consider a case with 4 steps (1, 2, 3, and 4) which share some resource and can exist during EFI_e. If the level of resource during EFI_e is 2 and the demands of steps 1, 2, 3, and 4 are 0.5, 0.5, 0.75 and 1.0 respectively, then the maximal overlapping sets are {1,2,3}, {3,4}, and {1,2,4}. The set of constraints that arises is then:

$$X_{1e}(1-(s_{12}+s_{21}))(1-(s_{13}+s_{31}))(1-(s_{14}+s_{41}))X_{2e}$$
$$(1-(s_{23}+s_{32}))(1-(s_{24}+s_{42}))X_{3e}(1-(s_{34}+s_{43}))X_{4e} = 0 \tag{13}$$
$$X_{1e}(1-(s_{13}+s_{31}))(1-(s_{14}+s_{41}))X_{3e}(1-(s_{34}+s_{43}))X_{4e} = 0 \tag{14}$$
$$X_{2e}(1-(s_{23}+s_{32}))(1-(s_{24}+s_{42}))X_{3e}(1-(s_{34}+s_{43}))X_{4e} = 0 \tag{15}$$

Constraint 13 arises from the first maximal overlapping set combined with step 4. Constraints 14 and 15 arise from the second set combined with steps 1 and 2. The third maximal overlapping set combined with step 1 would yield a constraint identical to 13. The set of constraints is complete in that it disallows all resource violating combinations but does not eliminate any valid overlapping combinations. The set above can be transformed into the following linear set:

$$X_{1e}+(1-(s_{12}+s_{21}))+(1-(s_{13}+s_{31}))+(1-(s_{14}+s_{41}))+X_{2e}$$
$$+(1-(s_{23}+s_{32}))+(1-(s_{24}+s_{42}))+X_{3e}+(1-(s_{34}+s_{43}))+X_{4e} \leq 9 \quad (16)$$
$$X_{1e}+(1-(s_{13}+s_{31}))+(1-(s_{14}+s_{41}))+X_{3e}+(1-(s_{34}+s_{43}))+X_{4e} \leq 5 \quad (17)$$
$$X_{2e}+(1-(s_{23}+s_{32}))+(1-(s_{24}+s_{42}))+X_{3e}+(1-(s_{34}+s_{43}))+X_{4e} \leq 5 \quad (18)$$

The number on the right hand side of constraints 16-18 represents an integer 1 less than the number of 0-1 terms on the left hand side. Effectively, the constraints indicate that all of the events associated with each of the left hand side terms cannot simultaneously take place.

Constraints of the type 16-18 can be expressed in a general form. First, let an augmented maximal overlapping set (AMOS) be formed by the addition of one step, which is not a member of a maximal overlapping set (MOS) and which causes a resource level violation, to the MOS. Therefore, from each MOS, several AMOS's can arise. For each unique AMOS, the following constraint is written:

$$\sum_{i \in AMOS} (X_{ei} + \sum_{\substack{j \in AMOS \\ j > i}} (1 - (s_{ij} + s_{ji}))) \leq |AMOS| - 1 \quad (19)$$

The term |AMOS| indicates the number of 0-1 terms on the left hand side of the inequality. Within one EFI, the worst case would be when, if there are n total steps which could overlap, n/2 steps belong to each MOS. This would give rise to $n!/((n/2)!)^2$ MOS's. Let the number of MOS's then be K. To form an AMOS, there are n/2 steps possible to add to each MOS. However, since each AMOS would be produced twice, this number is divided by two, resulting in Kn/4 constraints (AMOS's).

The number of resource violation constraints introduced by constraint type 19 is seen to be quite problem dependent. Although the constraint was originally a highly non-linear decision constraint, exploiting the structure of the product of terms and the physical meaning of the constraint results in a rather simple linear form of inequality. No additional variables were necessary in order to account for resource usage. However, some pre-analysis of the type alluded to concerning the generation of MOS's is necessary for the construction of a minimal set of resource usage constraints.

Linkage with the Continuous Formulation

The discussion of the time horizon has been centered on a wide, non-uniform discretization of time. In contrast to the conventional form of discretization, the selection of an interval ($X_{ij} = 1$) in this formulation does not necessarily dictate a start time for a step. Instead, the selection of an interval indicates that the start time of a step may take place somewhere within the interval. The variable B_j and the parameter D_j respectively represent the start time and duration of step j. A continuous formulation which relates the various B_j's is discussed in [25]. The formulation includes straight-forward constraints on idle times and step durations which are based on recipe concerns. Also included were constraints which dealt with the use of storage vessels in order to ensure that a series of fill and empty steps did not cause the vessel to run dry. Additional constraints which can be used to link the continuous formulation in [25] to the discrete portion of the formulation presented here.

The relationship between the continuous variables and the binary interval variables is given by the following two constraints.

$$B_j \geq (X_{ij} - X_{i-1,j})TB_{ij} \tag{20}$$
$$B_j + D_j \leq (X_{ij} - X_{i+1,j})TE_{ij} + H(1 - (X_{ij} - X_{i+1,j})) \tag{21}$$

The first constraint forces the start time of a step to be greater than the start time of the first chosen interval, TB_{ij}. The second constraint forces the start time of a step to be less than the end time of the last chosen interval, TE_{ij}, minus the duration of the step. The symbol H represents the time horizon. The size of this constraint set will be the same as the number of X variables, $s(\epsilon r - 1)$.

More constraints are needed for every possible sequencing decision. Specifically, for every s_{ij} variable, the following constraint is important:

$$B_j - B_i + (1 - s_{ij})(H + D_i) \geq D_i \tag{22}$$

Constraint 22 indicates that if two steps are sequenced, a space of at least the step duration must be left between the start times. These sequence constraints are in addition to those implied by product recipes, which would already be included in the balance of the continuous formulation.

The constraints presented in this section are intended only to link the discrete formulation presented here to the continuous formulation for exact time assignment. All of the constraint types presented in [25] apply to this expanded version of the continuous formulation in order to form a complete mathematical formulation.

Optional Use of Storage Vessels

Several constraints are necessary in order to model the optional use of storage. The first constraint is a modified form of constraint 3, given below:

$$\sum_i X_{i,T1(j\to k)} + \sum_s \sum_i X_{i,T1(j\to s)} \geq 1 \qquad (23)$$

The arrow notation indicates that transfer is from a task to its successor task, k, or to a storage vessel, s. The T1 indicates that it is only the first transfer step of a transfer operation. The constraint indicates that at least one mode of transfer between tasks must be chosen, using storage or a direct transfer. The first summation covers the direct transfer while the second, double summation covers all possible storage units. The number of such constraints in the worst case is *t*, where *t* is the number of tasks in the problem.

Constraints of the type of constraint 1 remain applicable. However, constraint type 2 must be modified in order to exclude all other storage vessels once a mode of transfer has been chosen. For each allowed pair of start and end EFI's $X_{i,T1(j\to x)}$ and $X_{m,T1(j\to x)}$, where x represents k or s (regular transfer or storage), for every task which can use storage, the following constraint is necessary.

$$\sum_{a=1}^{i-1} X_{a,T1(j\to x)} + \sum_{a=m+1}^{z} X_{a,T1(j\to x)} + \sum_{y\neq x} \sum_{a=1}^{z} X_{a,T1(j\to y)} \leq$$

$$NT[(1-X_{i,T1(j\to x)}) + (1-X_{m,T1(j\to x)}) + X_{i-1,T1(j\to x)} + X_{m+1,T1(j\to x)}] \qquad (24)$$

The symbol NT represents the number of terms on the left hand side of the constraint. The size of this constraint set is dependent on the number of transfer options (allowed storage vessels) and on the number of resource events, giving $2tyer$ constraints, where y represents the average number of storage choices per task.

Once a transfer mode is selected by choosing an interval for the first transfer step of that transfer mode, the rest of the steps associated with that transfer operation (additional transfers and storage vessel clean-outs, indicated by (T,C)x) must be selected as well. If a transfer mode is not selected, then none of the subsequent steps of the operation should be selected. The following two constraints ensure these two properties.

$$\sum_i X_{i,T1} \leq NT \sum_i X_{i,(T,C)x} \qquad (25)$$

$$\sum_{(T,C)x} \sum_i X_{i,T1} \leq NT \sum_i X_{i,T1} \qquad (26)$$

An instance of constraint 25 will be created for every (T,C)x for every mode of transfer, giving a total of *tzy* constraints, where *z* represents the average number of transfer and clean-out steps associated with the use of storage. An instance of constraint 26 will be created for each mode of transfer between two tasks, giving a total worst case number of constraints of *ty*.

The combination of the above constraints effectively handles the use of optional storage vessels. Storage of material within a processing unit is handled by the idle time constraints of the continuous linear program presented in [25].

Non-Renewable Resource Constraints

It is necessary to introduce an extra binary variable for handling non-renewable resources. The variable Z_{ij} is used to represent the starting point of an interval for step j. It is only necessary to introduce the Z variables when a step uses a non-renewable resource. In the worst case, every step would use non-renewable resources, giving $s(er - 1)$ possible variables. In practice, a small percentage of the total number of steps will actually require non-renewable resources.

If $X_{ij} = 1$ and $X_{i-1,j} = 0$ then $Z_{ij} = 1$, otherwise $Z_{ij} = 0$. In order to enforce this rule, the following two constraints are necessary for each Z_{ij}.

$$Z_{ij} \geq X_{ij} - X_{i-1,j} \tag{27}$$

$$Z_{ij} \leq (1 - X_{i-1,j} + X_{ij})/2 \tag{28}$$

In the worst case, there would be $s(er - 1)$ instances of each constraint. In practice, this will rarely be the case.

Over the scheduling horizon, non-renewable resource arrivals define events in time. Each non-renewable resource may have its own event calendar, where an event in non-renewable resource N will be referred to by e_N. For each non-renewable resource N, a constraint is created for every event e_N of the form:

$$\sum_{s \text{ uses } N} \sum_{TB_a < e_N} Z_{as} RU_{sN} \leq RL_{e-\delta,N} \tag{29}$$

The term RU_{sN} refers to the required level of resource N by step s. The term $RL_{e-\delta,N}$ refers to the level of resource N just prior to the event e_N. The constraint accounts for a cumulative sum of all tentative reservations for resource N prior to the event e_N. The size of this constraint set in the worst case is *er*. As mentioned earlier, this will rarely be the case.

Sequence-Dependent Set-Up Times

In order to consider the effect of sequence dependent set-up steps on each processing unit, an extra set of constraints in place of constraint type 3 is necessary. This is due to the fact that s_{ij} is used not only to represent a direct precedence relationship here, but indirect relations as well. Also, since, in some cases, there may be no set-up required because of production sequence, it is not as simple as stating that at least one X = 1 for some set-up operation.

For each task, there are several potential set-up operations, the choice of which depends on the task that immediately preceded the task on the processing unit. Each of these potential operations for the set-up of the unit has an initial set-up step, SU1, and may have subsequent set-up steps, SUx. Let $X_{a,SU1(G \to H)}$ and $X_{b,SU1(G \to H)}$ represent an allowed pair of start and end EFI's for a string of EFI's from EFI_a to EFI_b for the first set-up step required if task G immediately precedes task H (where G and H are not necessarily of the same batch) on the processing unit. Further, allow the subscript X-P1 to represent the first processing step of task X. Then for every potential initial sequence dependent set-up step on a unit U, the following pair of constraints are written (where T represents the number of tasks assigned to U):

$$1 + \sum_{X \neq J, X \neq I} s_{J\text{-}P1,X\text{-}P1} - s_{I\text{-}P1,X\text{-}P1} \leq \sum_{a=1}^{n} X_{a,SU1(I \to J)} + (1 - s_{I\text{-}P1,J\text{-}P1})(T - 1) \qquad (30)$$

$$1 + (1/|X|) \sum_{X \neq J, X \neq I} s_{J\text{-}P1,X\text{-}P1} - s_{I\text{-}P1,X\text{-}P1} \geq X_{a,SU1(I \to J)} \qquad (31)$$

The first constraint ensures that the proper set-up operation will be initiated when task I immediately precedes task J on the same processing unit. The second constraint states that if task I precedes task J, but not immediately, the sequence dependent set-up operation between I and J is not chosen. When task I precedes task J, the constraints are switched on, otherwise, they contribute no effect on determining set-up decisions. The summation subscript X is used to indicate all tasks except I and J assigned to the same unit to which tasks I and J are assigned. If task I does precede task J, the term $\Sigma\, s_{J\text{-}P1,X\text{-}P1} - s_{I\text{-}P1,X\text{-}P1}$ will take on a value of 0 if no tasks are between I and J, otherwise, it will be negative. The term |X| indicates the number of tasks assigned to the processing unit, excluding I and J. If not every pair of steps requires a set-up, the value of |X| must be suitably altered in order to match the number of terms in the summation. Constraints 30 and 31 will then indicate that an interval for the appropriate set-up must be chosen when I immediately precedes J, otherwise no interval should be chosen for the set-up. In the worst case, every task is assigned to the same unit. The result would be $t(t-1)$ instances of constraint 30. Also in the worst case, there will be

$er-1$ X_{ij} variables for each step. This, multiplied by the possible number of sequence-dependent set-ups gives $t(t-1)(er-1)$ instances of constraint 31.

It is also necessary to restrict the binary variables when task J precedes task I, as in the following constraint which disallows sequence dependent set-ups when the precedence is in reverse order.

$$\sum_{a=1}^{n} X_{a,SU1(I\rightarrow J)} \leq ns_{I-P1,J-P1} \tag{32}$$

There are as many constraints of type 32 as there are possible worst case sequences, $t(t-1)$.

When the initial set-up step for a sequence dependent operation is selected, the rest of the steps in the operation must be carried through. Similarly, if the initial step of an operation is not selected, none of the subsequent steps should be selected. The following two constraints enforce these rules.

$$\sum_{a=1}^{n} X_{a,SU1(I\rightarrow J)} \leq n \sum_{a=1}^{m} X_{a,SUx(I\rightarrow J)} \tag{33}$$

$$\sum_{x=1}^{q} \sum_{i=1}^{n} X_{i,SUx(I\rightarrow J)} \leq qn \sum_{i=1}^{m} X_{i,SU1(I\rightarrow J)} \tag{34}$$

The first constraint is created for each set-up step, SUx, where $x > 1$. In the worst case, this will give $qt(t-1)$ constraints, where q is the average number of set-up steps in a set-up operation. The second constraint is created for each possible sequence, resulting in $t(t-1)$ constraints.

As with other formulations, the consideration of sequence dependent operations in this formulation requires additional constraints. However, each of the constraints is strictly an integer constraint, and the number of such constraints is polynomial in size.

Evaluation of the Formulation

Although the dimensions of the formulation presented here have been discussed in terms of various problem attributes, it is helpful to view the relative size of an instance of the formulation in contrast conventional formulations. Specifically, our formulation will be compared to that of Kondili, et al [11]. First, a small example problem will be discussed. Then there will be brief comparisons of the sizes of the formulation of [11], the formulation of [11] with intelligent interval analysis, and the formulation presented here.

Table 1 - Example Problem Data

Task	Step	Duration	Electricity required (kW)	Steam required (kBTU)
A1	Process	5.0	25.0	10.0
A2	Process	4.0	12.5	10.0
C1	Process	10.0	7.5	25.0
C2	Process	6.0	20.0	7.5

The example to be considered involves four processing steps for the production of two products. Each of the steps requires various amounts of the resources steam and electricity from Figures 1 and 2 in the amounts shown in Table 1 along with the step durations. No transfer steps or due-dates are considered.

The first element of problem size in the formulation of [11] is the discretization of time. In order to exactly model the solution space, it is necessary to use a discretization width equal to the greatest common factor of the step durations and resource event times. In this case, that interval is of width 0.5. Since there are 4 tasks with processing units specified, the scheduling horizon of 30 time units results in a total of 240 binary variables. Since the formulation of [11] also determines batch size, and since there are 6 states of material in the example, there are a total of 360 continuous state variables. An additional 120 continuous variables are necessary in order to model the storage of material in a processing unit during idle time. In order to ensure that no unit begins more than one task during any interval, 120 binary constraints are necessary. In order to ensure that no step begins processing on a unit while another step is still in process, 240 binary constraints are needed. In order to force the amount of material undergoing a task to 0 when the task is not in process, 240 mixed binary/continuous constraints are used. Material balances result in 60 continuous constraints. Since there are two resources and 60 time periods, 120 mixed binary/continuous constraints are necessary in order to assure that resource levels are not violated. The consideration of storage of material in an equipment item adds 240 mixed binary/continuous constraints which ensure that no material is stored in an equipment item while processing is taking place, and 120 more mixed constraints which ensure the forward flow of stored material.

Using intelligent interval analysis as presented in the earlier portion of this paper, the size of the formulation suggested in [11] can be drastically reduced. The ideas of the creation of RCI's and the RCI set elimination criteria can be used in order to compute intervals of the scheduling horizon during which some of the steps will never take place. The result is two-fold. First, the number of discretization variables is reduced for some of the steps. Second, the reduction in the number of variables implies that fewer constraints are necessary. The number of binary variables can be reduced from 240 to 98. In addition, the total number of continuous variables can be reduced from 720 to 496. Because of the reduction in the number of binary variables, the number of binary constraints is reduced from 360 to 148 and the

Table 2 - Formulation Size Comparisons

Item	Kondili, et al	Kondili, et al with Analysis	Reduction Factor	Proposed Formulation	Reduction Factor
Binary Variables	240	98	0.41	30	0.13
Continuous Variables	720	496	0.69	4	0.006
Binary Constraints	360	148	0.41	88	0.24
Continuous Constraints	60	58	0.97	2	0.03
Mixed Constraints	720	362	0.50	30	0.04

number of mixed constraints from 720 to 362. There is very little change in the number of continuous constraints.

The formulation presented in this paper coupled with intelligent interval analysis results in a significantly smaller formulation. Intelligent pre-processing of intervals results in 22 binary X_{ij} variables. Consideration of step sequencing results in 8 unspecified s_{ij} variables. A total of 20 binary constraints are introduced concerning step sequencing. Renewable resource level considerations give 6 binary constraints. Note, however, that as the number of steps increases, this number has the potential to rise quickly, depending on problem structure. There are 30 mixed constraints required for linking the discrete and continuous portions of the problem. In order to ensure correct use of the time horizon, 62 binary constraints are introduced. Only 4 continuous variables are introduced to model the start times of each of the steps. Finally, 2 continuous constraints are used in order to separate the start times of successor and predecessor steps as specified in the product recipes.

The resulting formulation sizes are summarized in Table 2. The column labelled, "Kondili, et al," contains estimates on the formulation size as presented in [11]. The column labelled, "Kondili, et al with Analysis," gives an estimate of the size of the formulation in [11] with the application of the interval elimination criteria alluded to in this paper and described in [23]. The first reduction factor column is a decimal which represents how large the Kondili, et al formulation is with interval analysis compared to the raw formulation. The next column presents the size of the formulation presented in this paper. The final reduction factor column gives the size of the proposed formulation as a fraction of the raw Kondili, et al formulation. When examining the reduction factors, it is important to remember that the reduction in solution complexity is exponentially, not linearly, related to the number of binary variables. Therefore, a reduction of 50% in the number of binary variables reduces the solution complexity by significantly more than 50%.

In fairness, it is important to mention that the Kondili, et al formulation includes considerations of batch sizing, which can not be eliminated from the formulation even when such decisions are not necessary. The result is an increased number of continuous variables and mixed constraints (in this example, 240 continuous variables and 240 constraints). Batch sizing decisions are an important consideration in practical problems, but are not included in the current form of the proposed formulation, which only involves the detailed scheduling of known tasks.

It is equally important to mention that no additional variables or constraints are added to the proposed formulation in order to model material storage in the processing unit, whereas the formulation of Kondili, et al required 120 extra continuous variables for this purpose. Further, the proposed formulation allows the storage of material to take place prior to a task, after a task, and between the various processing steps of a task. The formulation of Kondili, et al allows idle time only prior to the initiation of processing. Specifically, whatever material is being stored at some time in a unit must later be processed in that same unit. This implies that unless two successive recipe tasks are assigned to the same unit, material must vacate the unit as soon as processing is completed. The formulation proposed in this paper allows more flexibility in fitting the various resource demands under resource profiles through the use of idle time when external storage vessels are not available.

The question remains of which formulation is ultimately the best. The information in Table 2 seems to indicate that the formulation presented here is not just marginally better than previous work, but that it is a quantum leap. While this is clearly true for the example problem discussed here, it is not universally the case. When problems can be conveniently approximated with large, uniform time discretization intervals, and when there is a large number of steps which must be sequenced, the conventional formulation could be a better choice. However, the user of a conventional formulation must decide when the round-off of task durations and resource events (necessary in order to get a coarse time discretization) begins to represent a problem different than reality. The formulation presented here will exactly model the process regardless of the values of task durations and event times. The approximate conventional formulation may yield an exact solution, but the result may be an exact solution to an approximated problem and not an exact solution to the exact problem. This limitation becomes especially important when dealing with tight scheduling applications in which data rounding may exclude feasible solutions.

Conclusions

In this paper, several aspects of the resource-constrained batch chemical process scheduling problem were analyzed in an exact mathematical form. First, it was demonstrated that, by judicious choice of modelling variables through the use of RCI's and RCI elimination criteria, the size of a formulation can be drastically restricted. The formulation was then presented with respect to all aspects of the detailed scheduling problem discussed in the problem definition section. The flavour of the formulation was motivated by the scheduling method outlined in earlier work [23,24,25]. Finally, comparison with the currently accepted formulation on a small problem indicated promise for further study of enhanced formulations based on the notions of interval analysis.

All variables in this formulation can be mapped to some exact physical cause. No artificial constructs have been added to the problem in order to generate the model. In comparison with the traditional discretization model, the model presented here can be either better or worse in terms of solution complexity, depending on problem structure. A properly structured problem can always be used to demonstrate that one method is better than another, but such demonstrations seldom result in gaining any fundamental understanding about the problem. What is important here is that a new formulation is given which is, in essence, an exact formulation, whereas previous modelling approaches have effectively rounded the problem data and give an approximate model. While the solution of such approximate models produces an "exact" solution, it is important to realize that an "exact" solution to an inexact formulation does not necessarily give an optimal solution to the original problem. In terms of analogy, consider the area of process modelling. It is customary to generate an exact model of the process in terms of a set of differential equations and boundary conditions. Typically, this model cannot be analytically solved, so time and space are discretized and a numerical solution is generated. For practical use, an empirical model is as useful as the numerical solution; however, modelling is a valuable exercise in further understanding of the process. The same is true of building models for scheduling problems, except that here, sometimes even discretized models are not solvable. The model presented here can be useful both for further understanding of the process constraints as well as for obtaining practical solutions for small to medium sized problems. The reader should now be aware that there exists more than one means of mathematically modelling a problem, that the choice of the best paradigm for a given application requires some analysis, and that by investing time toward problem analysis, order of magnitude or greater savings may be realized.

Nomenclature

AMOS	augmented maximally overlapping set.
B_i	start time of step i.
$D_{ij,max}$	the maximum amount of time which can elapse between the start times of steps i and j.
$D_{ij,min}$	the minimum amount of time which must elapse between the start times of steps i and j.
D_K	Duration of step K, where K is one of the step notation subscripts.
D_x	duration of step x.
δ	a small decimal number.
EFI	event free interval.
e	average number of resource events per resource.
e_N	event time of non-renewable resource N.
H	length of the scheduling horizon.
K	the number of augmented maximally overlapping sets.
M	a large number.
MOS	maximally overlapping set.
NRR	non-renewable resource.
NT	number of terms on a given side of a constraint.
p	number of RCI's per step.
q	average number of set-up steps in a set-up operation.
$\mathcal{R}$	a set of resources used by potentially interacting steps.
RCI	resource confluence interval.
RL_{eR}	resource level of resource R available at time e.
RR	renewable resource.
RU_{sR}	resource level required of resource R by step s.
r	number of resources in a problem.
$\mathcal{S}$	a set of potentially interacting steps.
s	number of steps in a problem.
S_{ij}	sequence variable for step i preceding step j.
T	number of tasks assigned to a given unit.
TB_{xy}	the start time of an EFI.
TE_{xy}	the end time of an EFI.
t	number of tasks in a problem.
X_{ij}	a binary variable indicating the use of EFI_i by step j.
y	number of storage options following the completion of a task.
Z_{ij}	a binary variable indicating that X_{ij} represents the first EFI used by step j.
z	average number of transfer and clean-out steps associated with a use of storage.

References

1. Baker, K.R. Introduction to Sequencing and Scheduling. Wiley, New York, 1974.
2. Blazewicz, Jacek, Wojciech Cellary, Roman Slowinski, and Jan Weglarz. Scheduling Under Resource Constraints - Deterministic Models. Annals of Operations Research, Ed. Peter L. Hammer, J.C. Basltzer AG Scientific Publishing Company, Switzerland, Vol. 7, No. 1-4, December, 1986, pp. 1-226.
3. Bowman, E.H. "The Schedule Sequencing Problem." Operations Research, Vol. 7, 1959, pp. 621-624.
4. Cott, B.J. and S. Machietto. "A General Completion Time Determination Algorithm for Batch Processes." AIChE Annual Meeting, San Francisco, November, 1989.
5. Egli, U.M. and D.W.T. Rippin. "Short-Term Scheduling for Multiproduct Batch Chemical Plants." Computers and Chemical Engineering, Vol. 10, No. 4, 1986, pp. 303-325.
6. Fox, Mark S. and Stephen F. Smith. "ISIS - a Knowledge-Based System for Factory Scheduling." Expert Systems Journal, Vol. 1, No. 1, 1984, pp. 25-49.
7. French, S. Sequencing and Scheduling: An Introduction to the Mathematics of the Job-Shop. Horwood, Chichester, 1982.
8. Graves, S.C. "A Review of Production Scheduling." Operations Research, Vol. 29, No. 4, 1981, pp. 646-675.
9. Gupta, S.K. and J. Kyparsis. "Single Machine Scheduling Research." Omega International Jounal of Management Science, Vol. 15, No. 3, 1987, pp. 207-227.
10. King, J.R. and A.S. Spachis. "Heuristics for Flow-Shop Scheduling." International Journal of Production Research, Vol. 18, No. 3, 1980, pp. 345-357.
11. Kondili, E., C.C. Pantelides, and R.W.H. Sargent. "A General Algorithm for Scheduling Batch Operations." Third International Symposium on Process Systems Engineering, Sydney, August, 1988, pp. 62-75.
12. Lazaro, M. and L. Puijaner. "Simulation and Optimization of Multi-Product Plants for Batch and Semi-Batch Operation." I. Chem. Symp. Series, 92, 1985, pp. 209-222.
13. Manne, A.S. "On the Job-Shop Scheduling Problem." Operations Research, Vol. 8, 1960, pp. 219-223.
14. Park, Yang Byung, C. Dennis Pegden, and E. Emory Enscore. "A Survey and Evaluation of Static Flowshop Scheduling Heuristics." International Journal of Production Research, Vol. 22, No. 1, 1984, 127-141.
15. Pritsker, A.A.B., L.J. Watters, and P.M. Wolfe. "Multiproject Scheduling with Limited Resources: A Zero-One Programming Approach." Management Science, Vol. 16, 1969, pp. 93-108.
16. Reklaitis, G.V. "Review of Scheduling of Process Operations." AICHE Symposium Series, Vol. 78, No. 214, 1982, pp. 119-133.
17. Reklaitis, G.V. "Perspectives on Scheduling and Planning of Process Operations." Proceedings of the Fourth International Symposium of Process Systems Engineering, Montebello, Canada, August, 1991.
18. Rippin, D.W.T. "Design and Operation of Multiproduct and Multipurpose Batch Chemical Plants - An Analysis of Problem Structure." Computers and Chemical Engineering, Vol. 7, No. 4, 1983, pp. 463-481.
19. Sahinidis, N.V. and I. Grossman. "Reformulation of Multiperiod MILP Models for Planning and Scheduling of Chemical Processes." Computers and Chemical Engineering, Vol. 15, No. 4, 1991, pp. 255-272.
20. Tongo, G.O. Scheduling of Batch Chemical Processes with Resource Constraints. Diss. Purdue University, West Lafayette, IN, August, 1990.
21. Tsirukis, T. and Reklaitis, G.V. "A Comprehensive Framework for the Scheduling of Resource Constrained Multipurpose Batch Plants." Proceedings of the Fourth International Symposium of Process Systems Engineering, Montebello, Canada, August, 1991.
22. Wellons, M.C. and G.V. Reklaitis. "Problems in the Scheduling of Batch Chemical Processes." TIMS/ORSA Joint National Meeting, Paper no. MA35.3, April, 1988.
23. Zentner, Michael Gerard. "An Interval-Based Framework for the Scheduling of Resource-Constrained Batch Chemical Processes." Diss. Purdue University, May, 1992.
24. Zentner, M.G. and G.V. Reklaitis. An Interval-Based Approach for Resource Constrained Batch Process Scheduling Part I: Interval Processing Framework. Computer Oriented Process Engineering Proceedings of COPE-91, L. Puigjaner and A. Espuna Eds., Elsevier, Amsterdam, October, 1991, pp. 151-157.
25. Zentner, M.G. and G.V. Reklaitis. "An Interval-Based Approach for Resource Constrained Batch Process Scheduling Part II: Assignment and Adaptive Storage Retrofitting." AIChE Annual Meeting, Los Angeles, Paper No. 140d, November, 1991.

Batch Processing in Textile and Leather Industries

L. Puigjaner, A. Espuña, G. Santos, and M. Graells

Dept. Enginyeria Química, Universitat Politècnica de Catalunya, Avda. Diagonal, 647, E-08028 Spain

Abstract: Incorporating new technologies to the batch process industries constitutes a difficult task essentially dependent on the specific characteristics of the type of industry contemplated. Present trends in the methodology used to solve the efficient production scheduling problem can be grouped into two categories: the general purpose scheduler that employs the rigorous Mixed Integer Linear Programming (MILP) optimization strategies approach and the development of a heuristic-based customized approach which focusses on particular industrial sectors.

This paper describes how the combination of optimization and expert systems has been used to solve the production scheduling problem in the textile and leather industrial sectors. Detailed plant simulation is essential for production plant evaluation and eventual validation, which is facilitated by the data base management system. Mixed Integer Linear Programming is used to solve the optimization problem in the case of the leather industry, while the textile manufacturing scheduling problem is solved using ad-hoc hierarchical expert rules together with the plant manager supervision in a highly interactive mode. The reported industrial experience in both cases will be presented and results from specific case studies will be analyzed.

Keywords: Batch processing, multiproduct plant, production scheduling

Introduction

Batch processing is receiving increasing attention from a wide variety of industrial sectors due to the potential offered by this kind of production network: flexibility in the assignment of resources to multiple products, quick adaptation of capabilities to product market demand fluctuations, easy acceptance of alternate raw materials and/or manufacturing recipes. These potential attributes must be consolidated by the integration of adequate management tools and techniques that will enhance the productivity of batch operations, minimize idle times and ensure uniform product quality.

Computer powered batch processing optimization tools have for some time been developed at academic institutions with little or no impact on the industrial world. They attempt to solve partial aspects of the problem with varying degree of realism, thus generating an adverse gap between promise and the practice. On the other hand, software houses offer a variety of batch management commercial packages mainly devoted to material resource planning or production enhancement through simulation, without considering to actual process optimization.

This situation is not alleviated by the fact that batch processing cannot be easily generalized and that the level of detail required to describe industrial scale operations may become very complex and often problem-specific, which may involve significant amounts of speciality products [8].

To obviate the present circumstances two basic approaches are currently underway. On the one hand, the rigorous treatment of batch process operations optimization is contemplated, even for the most complex case of multipurpose network configuration [9,5]. On the other hand, problem specific solutions are attempted within a general framework [1,2]. The rigorous approach to the comprehensive treatment of the general resource constrained scheduling problem involves the solution to very large scale non-linear mixed integer programming models that can only be honestly contemplated in the perspective of developing parallel and distributed algorithms using highly parallel and distributed computers [7]. This assumes that present expectations of progress in computing technology will become a reality and a new generation of low cost supercomputers will be accessible to medium/small sized enterprises.

In the meantime, and even in the long range, the second approach may prove to be a better alternative in many cases, provided that heuristic formulations of specific problems are constructed from more informal problem descriptions and/or user intervention using AI methods. In this paper, this second approach has been used to solve the production scheduling problem in the textile industry. The strategy used to find good but suboptimal solutions is compared with another industrial case study leather manufacturing where strict branch and bound techniques are implemented to reach the optimum production plan.

Modelling Framework

Computer aided production scheduling of batch plants requires a comprehensive modeling framework representing the detailed batch/semicontinuous individual process operations, the connectivity aspects of the process units and the available resources. An open representation framework has been devised that facilitates the treatment of the general resource constrained scheduling problem and its application to specific environments [3].

Modeling of plant operations considers mixed product campaigns made-up of the tasks carried out on each of the m=1,2,3...,M available units. Such modeling includes the identification of task j=1,2,3,...,J_i belonging to batch n=1,2,...,N being processed in equipment m=1,2,..., M that is used for the lth time. The binary variable X_{jlmn} describes the assignment of tasks to individual equipment, and the binary variable Y_{in} is used to associate product i=1,2...,I and batch number. In turn, any task is made up of five elementary subtasks s=1,2,...,S (S=5), which may include (Fig. 1) waiting time (TW) only after the operation subtask.

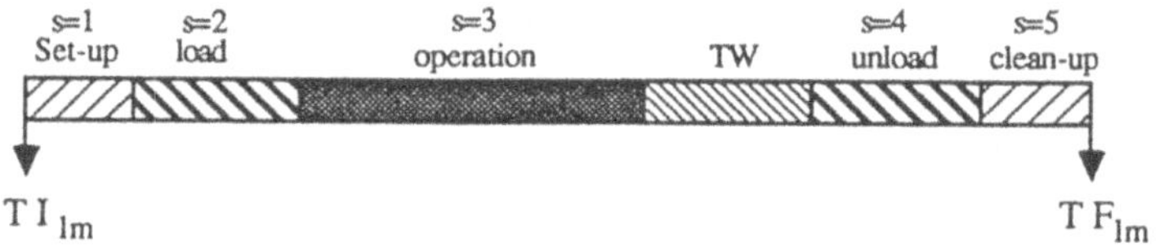

Fig. 1. Set of subtasks considered in every task starting at time TI_{lm} and finishing at TF_{lm}

Once calculated the t_{lms}, the initial and final times for the operation of equipment m for the lth time are related by:

$$TF_{lm} = TI_{lm} + \sum_{s=1}^{S=5} t_{lms} + TW_{lm} \tag{1}$$

The batch or semicontinuous mode of operation of the unit is described by (see Fig. 2):

$$z_m \in \{0,1\} \tag{2}$$

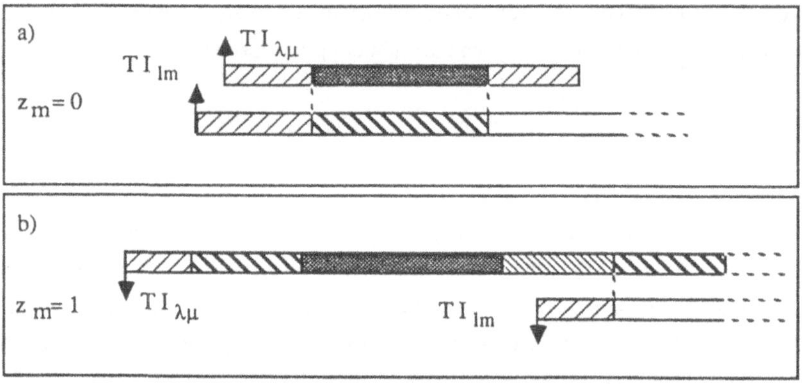

Fig. 2. Unit operation: a) semicontinuous operation; b) batch operation

and the time relation between consecutive stages of the same batch n is given by:

$$\sum_{m=1}^{M}\sum_{l=1}^{L_m} X_{j+1,lmn}\,(TI_{lm} - t_{lm1}) =$$

$$\sum_{\mu=1}^{M}\sum_{\lambda=1}^{L_m} X_{j\lambda\mu n}\left[TI_{\lambda\mu} + (1-z_\mu)\left\{\sum_{s=1}^{3} t_{\lambda\mu s} + TW_{\lambda\mu}\right\} + z_\mu\, t_{\lambda\mu 1}\right] \qquad (3)$$

The relationship between l and l+1 successive jobs is given by:

$$TI_{l+1,m} \geq TF_{lm} \qquad m = 1, ..., M; \;\; l=1, ..., L_m-1 \qquad (4)$$

The modeling framework just described considers the plant operation for a certain short-term period $q=1,2,...Q_p$ of a specific mid-term period $p=1,2,...P$. Total demand $\mathbf{D}_i$ to be covered during the available time horizon **H** is specified for each mid-term period p (D_{ip}).

The objective function and constraints

The objective function f considers the penalty costs per product i and production period (p) associated to production (PT_{ip}), inventory (S_{ip}), delays (R_{ip}) and unfilled demand (U_{ip}) [3].

$$f = \sum_{i=1}^{I}\sum_{p=1}^{P} PT_{ip} + S_{ip} + R_{ip} + U_{ip} \qquad (5)$$

to be minimized under the "Finite Wait" (FW) constraint, which expands the concept of mixed intermediate storage (MIS) to all possible policies, form Zero Wait (ZW) to No Intermediate Storage (NIS):

$$\sum_{n=1}^{N}\sum_{i=1}^{I}\sum_{j=1}^{J_i} Y_{in}\, X_{jlmn}\, TW_{ijm}^{max} \geq TW_{lm} \geq 0 \qquad \forall\, l,m \qquad (6)$$

and the time horizon constraint:

$$T_{pq} = \max_{l,m}\left\{TF_{lm}\right\} \quad ; \quad T_{pq} \leq H_{pq} \qquad (7)$$

Utilities constraints are given by:

$$\mathbf{W}_u(t) = \sum_{j=1}^{M}\sum_{n=1}^{N}\sum_{s=1}^{S=5}\sum_{k=1}^{K_{iju}} W_{njsku}(t) \leq \mathbf{W}_u^{max} \quad ; \; -\infty < t < \infty \;\; \forall u \qquad (8)$$

where the contribution of consumption njsk to general utility **u** pattern can be expressed by equations:

$$W_{klmsu}(t) = \sum_{n=1}^{N} \sum_{i=1}^{I} \sum_{i=1}^{J_i} X_{jlmn} \cdot Y_{in} \cdot \omega_{ijsku} \left\{ \theta(t - TI^{w}_{klmsu}) - \theta(t - TF^{w}_{klmsu}) \right\}$$

$$\forall\, k,l,m,s,u \quad (9)$$

$$TI^{w}_{klmsu} = TI_{lm} + \sum_{\sigma=1}^{s-1} t_{lm\sigma} + TW_{lm} \cdot \theta(s-4) + t_{lms} \sum_{n=1}^{N} \sum_{i=1}^{I} \sum_{j=1}^{J_i} X_{jlmn} Y_{in}\, \delta t^{o}_{ijkmsu} \quad (10)$$

$$TF^{w}_{klmsu} = TI_{lm} + \sum_{\sigma=1}^{s-1} t_{lm\sigma} + TW_{lm} \cdot \theta(s-4) + t_{lms} \sum_{n=1}^{N} \sum_{i=1}^{I} \sum_{j=1}^{J_i} X_{jlmn} Y_{in}\, \delta t_{ijkmsu} \quad (11)$$

being:

$$\theta(t) = \begin{cases} 0 & t \leq 0 \\ 1 & t > 0 \end{cases}$$

K_{ijsu} = times utility **u** is used during subtask ijs; $k = 1, \ldots, K_{ijsu}$
ω_{ijsku} = utility **u** power used the kth time during subtask ijs
δt^{o}_{ijkmsu} = starting time, related to subtask time, of consumption W_{klmsu}
δt_{ijkmsu} = finishing time, related to subtask time, of consumption W_{klmsu}

Additionally, intermediate storage considerations are also being incorporated in the actual formulation and will be presented elsewhere [10].

The preceding formulation of the problem of optimum production scheduling of a general processing network is a mixed integer non-linear programming problem (MINLP). Solutions to this problem can be obtained in reasonable computing time using rigorous approaches like reformulation techniques applied to linearize it in the space of the integer variables [12], and specific problem decomposition techniques [11,6] when the number of products is relatively low (i. e. less than 50).

Case study

The rigorous approach has been used to solve the optimization of process operations in the leather industry [4]. Detailed simulation has been carried out, covering the different manufacturing stages from pickled skins to "finflex" operation, and totalling 25 steps that are differently shared or bypassed according to the individual recipes of the final products (Fig. 3). The general modeling framework has been accommodated to accept the specificities of this type of industry, leading to the following working hypotheses taken from industrial practice:

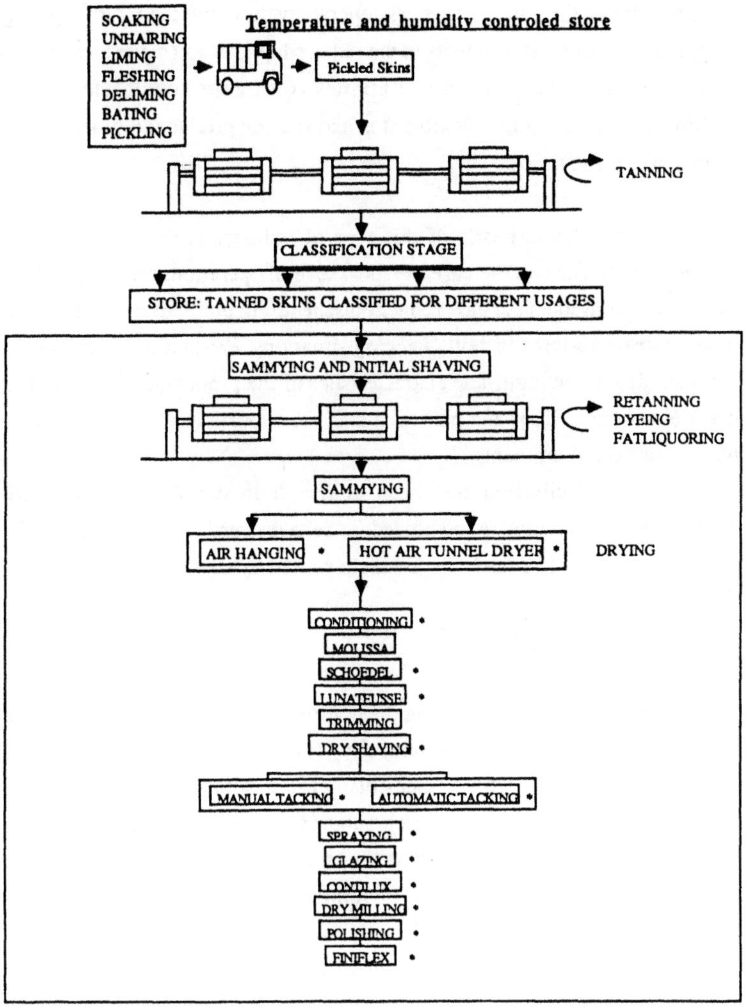

(*) Product dependent optional operation

Fig. 3. Flowsheet of process operations in the leather industry

- Every production step is carried out in batch/semicontinuous equipment.
- Set-up and clean-up times are considered for each unit.
- Task transfer mixed policy, is permitted (FW).
- Sharing manpower is the main constrained resource.
- The jobshop network configuration is considered.
- Batch processing advancement or postponement is allowed.

Then, the optimization of process operations is attempted by minimizing the total elapsed time required to produce all products in the required quantities (makespan) for a specified time horizon and under particular demand and utilities constraints. While the time horizon provides for a continuous period of time, flexible stop and starting production times can be introduced according to manufacturing needs.

A specific characteristic of this type of industry is the shared use of limited resources at discrete levels. Such is the case of labor: several operations may share different tasks at different times. As a consequence, the production scheduling problem becomes acrylic (aperiodic), leading to an increased source of bottlenecking situations. Production planning also becomes especially complex due to the multi-site characteristics of the production facilities that require the adequate management of utilities at a particular site, the allocation of production to tasks, and the inter-site transportation requirements.

Manpower limitation has been analyzed in detail in the overall process operation optimization. The optimization procedure behaves very well, finding satisfactory solutions to the aperiodic constrained scheduling problem caused by shared limiting resources [4].

The heuristic approach

An alternative approach to solving the constrained scheduling problem is implicit enumeration. Branch and Bound methods are generally very useful in discrete enumeration. But the bounding procedure depends on two key elements: a lower bound on the value of the performance index associated with the solution of each subproblem and a trial solution to the original problem. Heuristic rules are useful for providing the trial solution or upper bound in the procedure. General heuristic procedures are of limited use in industrial practice when the number of products and units increases significantly (over 50 jobs and 50 machines).

A user-guided heuristic methodology has been developed for specific scenarios. Basically, a branch and bound methodology is used to solve the optimization problem but a set of ad-hoc heuristic rules are provided for upper bounding and lower bounding and to conduct the search tree towards good solutions in less computing time. Experience-based heuristics have proven to be very helpful in finding realistic industrially valid solutions with reasonable computing effort even in complex production networks, as will be discussed later. A catalog of priorized rules can be automatically used to construct the tree, but a semi-manual mode is also possible, making use of the knowledge and experience of the expert engineer in two ways: by adding and/or modifying the existing catalog and by influencing the tree search using qualitative information to decide which variables should be fixed at any level during the tree branching (Fig. 4). The algorithm includes "manual optimization tools" of graphic interactive characteristics that can be actuated via mouse using the Gantt chart representation of the occupancy of processing and

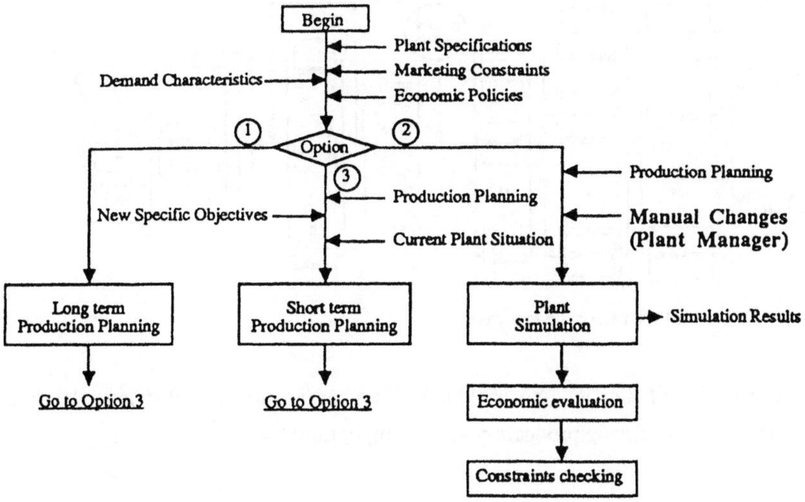

Fig. 4. Simplified flowchart of the heuristic approach

storage units. The level of branching and the number of alternative solutions to be kept or fathomed -if wished- are under user control, but if left under the automatic mode they will be handled in such a way as to minimize computer memory storage.

Case study

Heuristic procedures are especially attractive for finding problem specific solutions, mainly when the complexity involved in the manufacturing task representation and the very large variety of interrelated products and other organizational characteristics require tailor-made strategies that cannot be dealt with in general purpose algorithms, even with sophisticated knowledge representation and high performance optimization solvers.

Such is the textile industry. Producing textiles requires a complex manipulation of a large number of variables related to fabric properties such as twist, sizing, dying, printing and finishing characteristics. In order to study the complexity of the problem we will concentrate on the weaving stage, one of the most often difficult and economically penalizing steps in the textile manufacturing process (Fig. 5). Here, different fabrics and designs are produced using diverse types of machines which can be shared by some products.

This situation leads to a very complex problem of task assignment and optimization of production lines for the best use of existing equipment in order to meet some specified demand. Typical figures (Table 1) indicate that even for qualified and experienced personnel making the

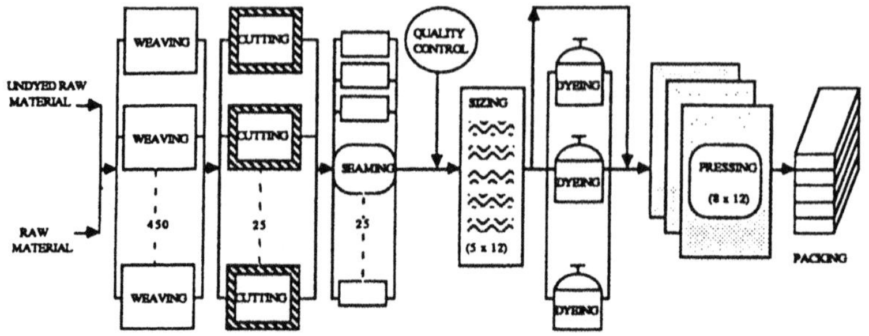

Fig. 5. Production recipe for the test case

production plan of such an industrial facility is a burdensome task if it has to be done manually, even without considering production scheduling optimization.

Table 1. Flowshop characteristics

Problem Dimensions	
Total Number of Processing Stages	8
Total Number of Processing Units in the Main Stage	450
Number of Types of Processing Units in the Main Stage	21
Total Number of Different Products	34000
Number of Product Families	800
Total Number of Compatibility Restrictions	12000
Number of Compatibility Aspects	30
Total Number of Production Priority Aspects	10

An ad-hoc solution strategy has been developed using two independent modules. The first module deals with:

- Simulation of the plant under diverse production conditions.
- Economic evaluation of such production plans according to an objective function that includes all elements allowing quantification, raw materials costs, operation time costs for each production line, utilities and manpower requirement costs, penalty costs for delivery time delays, etc. This evaluation implies the execution of alternate production plans that meet the desired demand.
- Checking for all production restrictions. The availability of the plant resources indicated above will determine the flexibility of each production plan.
- Verifying the discrepancies against the objectives described above.

The second module generates alternate long and short term production plans according to the general objectives described above and additional information coming from the plant status at the current time. The optimization procedure is indicated in Figure 6, the subplant calculation is shown in Figure 7 and the short term algorithm flowsheet is given in Figure 8.

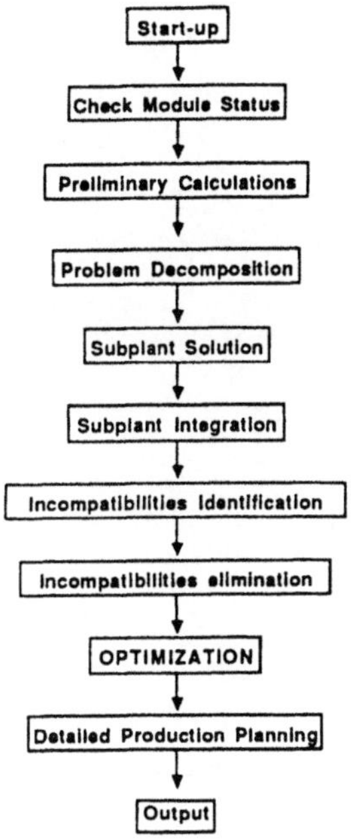

Fig. 6. The optimization procedure.

As it is essentially an order-driven system, an additional list of general priorities taken from practical experience, which can be conventionally weighed up by the plant manager, has been included in the objective function.

1. First serve children's size demand (as the market requires).
2. Some articles must be produced first in order to permit subsequent specific processes.
3. Assorted lots must be produced simultaneously.
4. There is a set of preferred customers that must be served first.
5. Due dates must be guaranteed for each product.
6. Individual product campaigns are preferred over alternating products in each unit. There are machine dependent penalty costs associated with product changeover.
7. Assign priorities to each unit, taking into account its production characteristics.
8. Establish the appropriate products to be produced for each unit set-up.

Some of these constraints may be found to be contradictory, and then appropriate penalty functions are evaluated to choose the prevalent one, and finally assign the production order to the specific unit. As the optimization procedure only considers single variable objective functions, all constraints and penalties have to be expressed in economic terms.

During day-to-day work, the short-term production planning module provides information necessary to modify current production plans, which are manually generated or suggested by the long-term optimization module, and conveniently adapts them to changing production patterns not considered in the long-term planning.

Finally, the simulation module allows the production manager to easily modify long and short term production plans, and make an objective evaluation of the consequences of such modifications (or of the ones suggested by the short-term production planning module). This is very useful when market pressures introduce unexpected changes in a long term production planning policy.

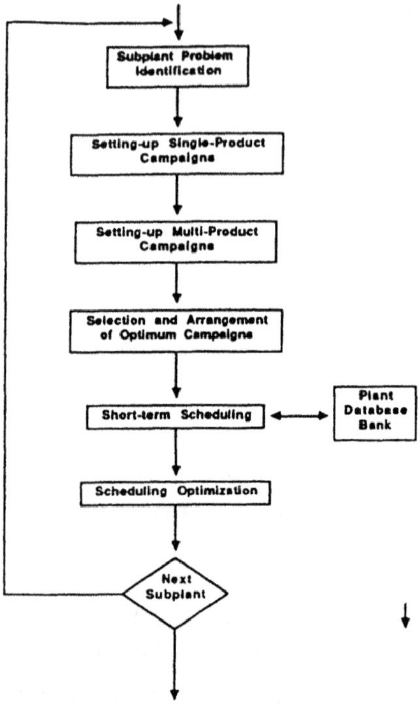

Fig. 7. The subplant calculation specified flowchart

The software package has been implemented in a workstation and provided with over 100 custom-mode colour windows to facilitate input and user interaction, thus obtaining high versatility and ease of training in the use of the program. Extensive use of graphics provides useful information to the user via Gantt charts, histograms and the like. It is worth note that although the input/output interfaces and the solution structure itself, are specific to the textile sector, this software framework can be easily extended to cover other industrial applications that may have common similarities.

Final considerations

For firstly thing, at the process topological level, it would be desirable to have an interactive scheduling system provided with an adequate interface to input the user's model that accurately represents the process plant. Detailed deterministic simulations can be carried out using existing simulators [13] that provide the expected performance of the plant for a given schedule and alternative configurations. Unfortunately, a general purpose simulator should be able to model a broad range of process industries, making it too costly in hardware and software for the small/medium enterprise due to the variety and specificity of this type of production facilities.

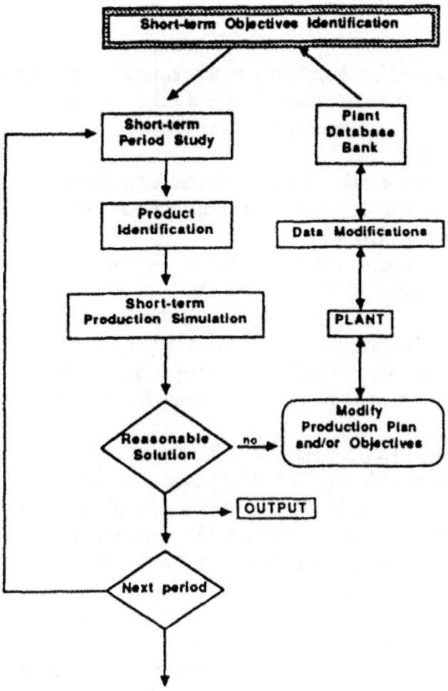

Fig . 8. The short term algorithm flowchart

However, it should be understood that one way or another plant modeling must be performed at the highest possible level of accuracy. This is a hard reality that requires further investigation, since it is not clear either on economical or on technical grounds that more powerful interfaces incorporating object-oriented structures capable of representing any plant configuration would be a better alternative to specialized process engineering houses making specific plant applications of standard low-cost scheduling optimizers ready for interfacing with the specific application.

Two industrial applications have been presented that show how both, exact solution methodologies and heuristic implementation can be valid for solving the constrained scheduling optimization problem, provided that detailed framework modeling is used. But at the same time they point out the need for custom-made solutions oriented towards a limited range of applications with common characteristics. Diversification of commercial schedule optimizers may be a good policy both in terms of economy and quality.

Acknowledgements

Part of this work was carried out under the auspices of the CIRIT (Comissió Interdepartamental de Recerca i Investigació Tecnològica, projecte QFN 89-4006). Support received from the CEC (JOULE 43) is also thankfully appreciated.

References

1. Abad, A., Espuña, A., Puiganer, L.: Computer Simulation and Optimization of Textile Fibers Manufacturing Process Operations. Computer Oriented Process Eng. (ed. by L. Puiganer and A. Espuña), pp. 177-184 (1991)
2. Espuña, A., Puigjaner, L.: Solving the Production Planning Problem for Parallel Multiproduct Plants. Chem. Eng. Res. Dev., 67, pp. 589-592, 1990
3. Graells, M.: Working Paper. Universitat Politecnica de Catalunya, Barcelona 1992
4. Graells, M., Espuña, A., Puiganer, L.: Optimization of Process Operations in the Leather Industry. In: Proc. ESCAPE-1 Symposium. Elsinore: Denmark 1992
5. Lazaro, M., Espuña, A., Puigjaner, L.,: A Comprehensive approach to Multipurpose Batch Plants Production Planning. Comput. & Chem. Engng. 13, 1031-1047 (1989)
6. Lazaro, M., Puigjaner, L.,: Solution of Integer Optimization Problems subjected to Non Linear Restrictions: An improved algorithm, Comput. & Chem. Engng. 12, 443-448 (1988)
7. Pekny, J., Venkatasubramanian, V., Reklaitis, G.V.: Prospects for Computer-Aided Process Operations in the Process Industries. Computer Oriented Process Engineering (L. Puigjaner and A. Espuna, eds.), pp. 435-447. Amsterdam: Elsevier 1991.
8. Puigjaner, L.: Improving the Design and Management of Production Chains in the Process Industry. Multi Suppliers Operation (W. van Puyembrocek, ed.), pp. 10.1- 10.20. Bruxelles: IOS Press 1992
9. Rapacoulias, C., Shah, N. and Pantelides, C.C.: Optimal Scheduling of Order-Driven Batch Chemical Plants. Computer Oriented Process Eng. (L. Puigjaner and A. Espuña, eds.), pp. 145-150. Amsterdam: Elsevier 1991
10. Santos, G., Espuña, A., Graells, M., Puigjaner, L.: Improving the Design Strategy of Multiproduct Batch Plants with Intermediate Storage. AIChE Annual Meeting, Florida, 1992.
11. Shah, N.,Pantelides, C.C., Sargent, R.W.H.: Efficient Solution Techniques for Optimal Scheduling of Batch Operators. Submitted to Comput. & Chem. Engng (1991)
12. Shainidis, N.V., Grossmann, I.E.: Reformulation of Multiperiod MILP Models for Planning and Scheduling of Chemical Processes. Comput. & Chem. Engng. 15, 255-272 (1991)

Baker's Yeast Plant Scheduling for Wastewater Equalization

Neyyire (Renda) Tümsen[1], S. Giray Velioğlu[2] and Öner Hortaçsu[1]

[1] Department of Chemical Engineering, Boğaziçi University, Istanbul, Turkey

[2] Department of Civil Engineering, Boğaziçi University, Istanbul, Turkey
(Present Address: Halk Sigorta A.S., Istanbul, Turkey)

Abstract: A Baker's yeast plant was modeled using data obtained from the normal operation of the plant via Monte Carlo analysis. The model which was first validated with respect to the normal plant operations was then used to predict wastewater discharges from two scheduling scenarios. The results showed that the plant wastewater discharge may be equalized if a 25 % reduction may be considered to be economically acceptable since wastewater equalization will not necessitate additional facilities for wastewater equalization external to the existing. However, if only one fermentation vat may be added to the process then wastewater equalization would be realized with 24 % increase in the capacity. The additional cost of the new vat is considered to be off-set by the increased plant capacity.

Keywords: Scheduling, wastewater equalization, baker's yeast

1. Introduction

Equalization is among one of the most widely used practices for volume and strength reduction of wastewater. Normally wastewater equalization is achieved by directing the wastewater into tanks where they are held for a specified period of time. However, it can also be achieved by good timing and scheduling of operations, especially when wastewater is discharged from a series of successive activities. In this paper, an approach aimed at the realization of the latter goal is presented, through the development of a Monte Carlo simulation model as applied to a baker's yeast production plant.

Monte Carlo methods are well established tools of numerical analysis that aid the decision making involving the design and operation of complex systems subject to uncertainty. The name Monte Carlo is derived from the fact that the method includes random sampling from statistical distributions. In the Monte Carlo approach random numbers are used to represent the status of each component in the system at an instant of time. As numbers are generated through random sampling, the entire set of numbers is modified, according to the logic of the model, in order to represent the new status of the system [3 and 7]. In carrying out a Monte Carlo simulation, the values of the variables are obtained by random sampling from specified distributions defined by the characteristics of these variables.

2. Simulation Model

As previously stated, this study aims to present an approach to achieve wastewater equalization via mathematical simulation. The wastewater discharged from a batch process show variations in concentrations of pollutants and quantities of wastewater discharged from batch to batch. Moreover, time and duration of discharges are generally not regular, therefore are not predictable except in a probabilistic sense. Hence, for industrial activities involving a series of batch processes Monte Carlo simulation is well-suited, for the analysis and evaluation of the effects of different scheduling schemes. Clearly, the simulation of wastewater discharges require the determination of the activities that contribute to the wastewater streams along with their precedence relationships, followed by data collection. The data should then be examined and evaluated to define the probabilistic nature of the important attributes of the activities (e.g. duration and idle time of the activities, quantities of wastewater, etc.)

As Monte Carlo simulation experiments require "data generation" mechanisms by producing a large number of random variates obeying the desired probability distributions, the use of Monte Carlo simulation has become more restricted to computers. Most computers, today, are equipped with a random number generating algorithm. The most commonly used algorithm, in fact, produces a nonrandom sequence of numbers (pseudo-random numbers) each being completely determined by its predecessors and consequently all numbers being determined by the initial (seed) number [4].

Upon completion of the evaluation of the random behavior of the components, the validity of the assessed probability distributions has to be checked. This can best be done by comparing the observed and simulated behavior of the system. Means of achieving wastewater equalization can be initiated only after it is shown that the probabilistic nature of the activities, of the system, are adequately described by the derived distributions and the system is well-represented by the simulation model. Different scenarios for timing and scheduling of activities can be experimented with and their "degrees of equalization" can be quantified. Degrees of equalization stand for the extent of peak load smoothing, which can be quantified by an "equalization metric", such as the maximum absolute deviation from the mean. It must be noted that the evaluation of the scenarios should not solely be based on the degree of equalization. Other important factors, such as the production level, should also be taken into account. Hence, the simulation experiment should be able to lead to a trade-off analysis between the degree of equalization and other important factors (i.e. the production level).

3. Application of the Model

Application of Monte Carlo simulation to chemical and environmental engineering problems is not too common. Tayyabkhan and Richardson [8] have analyzed a complex batch blending operation by the Monte Carlo technique. Berthouex and Brown [1] used Monte Carlo simulation to investigate the patterns and quantities of wastes discharged from a tanning industry. Peak load smoothing by altering the schedules of waste discharges from specific tanning process batches have been evaluated. However, in their study the Monte Carlo analysis is focused on evaluation of wastewater characteristic only. They considered the duration of operations to be deterministic which greatly simplifies the timing and scheduling of activities. In this study, the duration and the idle time between activities (operations) are viewed within the perspective of Monte Carlo analysis in addition to the evaluation of wastewater characteristics. In doing so, Berthouex and Brown's [1] approach is significantly extended. Although the resulting system is more difficult to deal with, its selection over that of the system considered by Berthouex and Brown [1] is justified, as the duration and the idle time of the activities significantly affect the discharge pattern and characteristics of the wastewater.

The simulation model outlined in the previous section, is applied to a baker's yeast production plant. It should be noted that most simulation models are system-specific, therefore, the emphasis should be on the concepts and the logic of the model. Hence, within this framework; the application below should be viewed as a specific illustration of the concepts and logic of the model rather than a unique solution to the problem of wastewater equalization.

3.1 Process Description

Baker's yeast production process consists of rapid multiplication of the yeast microorganisms, in a controlled medium of nutrients, followed by separation and washing and filtration of the fermented wort. The flow diagram of a typical baker's yeast production plants is shown in

Fig.1. Yeast plant wastewaters are highly contaminated with organic compounds and thus have a specific yeasty odor, high color and turbidity with a pH of 4.3 - 6.5 [6].

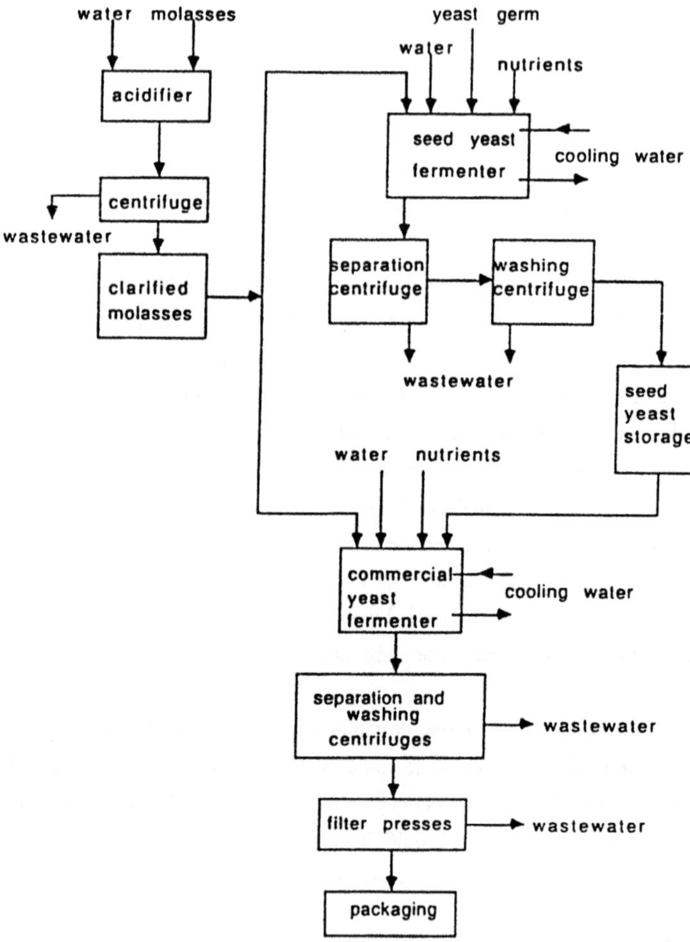

Figure 1. Flow Diagram of a Typical Baker's Yeast Manufacturing Process

3.2 Data Collection and Evaluation

Data collection was carried out in a yeast plant which had five fermentation vats. In the plant Vat No 1 and 2 operate alternately, following one another, to produce the seed yeast while the other three (Vats no. 3,4 and 5) are used in manufacturing the commercial product. (The first two vats will be referred to as "seed vats", and the last three vats will be referred to as "commercial vats" from hereon.). The seed vats were of the same size and specifications. A complete cycle of seed yeast fermentation (i.e. fermentation, separation and washing of the fermented wort, equipment washing and preparation for the next cycle) was observed to take about 25 to 26 hours. Fermentation in a seed yeast vat starts approximately when the cycle is completed in the other seed vat. The total amount of seed yeast required for one batch of the commercial vats could be produced in a single batch in either of Vats No: 1. and 2. The commercial yeast vats were not of the same size. Vat No: 5 was somewhat larger and its fermentation duration is about sixteen hours as opposed to 13-15 hours for Vats No: 3 and 4.

Once the fermentation is completed in one of the vats the fermented wort is separated and washed. Two centrifuges are used for this purpose, with Centrifuge No. 1 used for the seed yeast and Centrifuge No. 2 for the commercial yeast. Since the seed vats operate alternately (one at a time) their cycle completion times do not coincide, and thus the fermented wort can be directed to Centrifuge No.1 without delay. However, the commercial vats can operate concurrently, therefore, fermentation may be completed in any one of them while Centrifuge No.2 being used for separating fermented wort from another vat. In such a case, the separation of the fermented wort from the most recent cycle of the commercial vats would be delayed until Centrifuge No.2 is free.

As the separation and washing operation continues, the yeast milk, the concentrated liquor of the fermented wort is directed to cool in the storage tanks. It is gradually drawn from the storage tanks to the filter presses. If any yeast milk is already present from previous batches, then the most recently washed yeast milk is held in the vessel until the ongoing filtration operation ends. Otherwise, filtration starts within 10-20 minutes following completion of fermentation. Finally, the yeast is compressed and packed as the final product for sale.

In the application of the model only the discharges from separation and washing and filtration operations were considered. In the plant, the cooling waters from seed vats were discharged into the sewers, while the cooling waters of the commercial vats were reused. The cooling waters, together with the effluents from equipment washing, though large in quantities, are not highly contaminated. As suggested by Koziorowski and Kucharski [5], dilute effluents from fermentation plants are isolated and treated separately, thus concern was focused on the discharge behavior, patterns and equalization of concentrated effluents. We will note here that should the importance of wastewater streams not considered in this study become apparent, they can also be included in the analysis with no difficulty.

Throughout the study, COD (chemical oxygen demand) was used to characterize the organic strength of wastewater. BOD (biochemical oxygen demand) tests were also carried out to cross-check the consistency of the values with those reported in the literature. Grab samples which were collected from the effluents of the centrifuges approximately at the middle of the discharge periods, proved to be satisfactory as evidenced by running COD tests for two different samples collected from the same discharge at different time periods. The 15% difference in the results was considered insignificant in view of the possible experimental errors for the COD range in question. Composite samples were collected from filtration effluents. All samples were preserved according to the methods and recommendations given by the EPA [3] until the COD tests were carried out. Weekly averages of the COD and BOD_5 concentrations are summarized in Table 1. As can be observed from Table 1, the obtained COD and BOD_5 concentrations are in generally good agreement with those reported in the literature.

Flow measurements from the filter presses were made manually, by collecting a certain volume of liquid waste for a known period of time. On the other hand, separation and washing centrifuges discharged large quantities of wastewater, prohibiting manual measurement. For the latter case, the height of the fermented wort was measured at the end of the fermentation period to compute its volume. The volume of yeast milk was computed in similar manner. The volume difference between the two volumes was assumed to quantify the centrifuge effluent, and the average flow rate was estimated by dividing the volume difference by the discharge duration.

Data obtained from the survey, covering a period of two months are given in Renda [7]. Probability distributions for the relevant variables were determined empirically from the collected data.

Frequency diagrams were constructed from the data for each variable. The probabilistic behavior of the variables were then visually determined to be described by either normal or log-normal models by comparison of the frequency diagrams with the respective density functions. In addition, the data were plotted on normal and log-normal probability papers. When both distributions appeared to be plausible the preference was made on the basis of chi-square goodness of fit test. For the data sets that did not fit into either of the above distributions or into any other theoretical distribution, a "table look-up" method was used. In the "table look-up" method the cumulative probabilities are computed from the data and the value of the variate is determined in accordance with its relative frequency. A detailed discussion of data evaluation is given in Renda [7]. Table 2 summarizes the relevant process variables and their associated probability distributions. In Table 2 normal and log-normal distributions are designated by N (m,s) and LN(l,§) where m and s are the mean and standard deviation of the variate and l and § are the mean and the standard deviation of the natural logarithm of the variate, respectively.

Table 1: COD and BOD_5 Concentrations for Filtration and Separation and Washing Effluents.

Process	COD	BOD_5
Filtration	6880	5525
	7200	4100
	10560	7250
(Nemerov [6]; Çiler et al.[2])	7000	5000
Separation - Washing	20000	14050
	19200	13800
	21600	14250
(Nemerov [6]; Çiler et al.[2])	21500	16800

3.3 Assumptions

In delineating the time horizon for the analysis, the start and completion times of the activities are of interest. The 24-hour operating day is divided into 15 minute intervals to retain the detail and sensitivity of the overall process. Such a time interval was chosen because the shortest activity duration (i.e. separation and washing were assumed to be adequately covered by 15 minute increments.

To begin a simulation run, the start time of fermentation vats are drawn from a uniform distribution U (0,96). In other words, it was assumed that any vat could start operation, independent of others, at any 15 minute interval during the day, noting that there are ninety-six 15 minute intervals in a 24-hour day. The operational scheme of any activity was assumed to be independent of the duration of other activities. For example, the separation and washing begins right after the completion of the fermentation regardless of the duration of the fermentation. Each vat was assumed to operate once a day, therefore, every seven simulation runs describe a working week. That is, the number of weeks was increased by one after every seven cycles (i.e. group of batches) and simulation is stopped when the desired simulation time in terms of working weeks is reached.

Table 2. Relevant Process Variables and Their Associated Probability Distributions

COD(mg / lt)			
Filtration (SFI)			N (8290, 3670)
Commercial Yeast Separation and Washing		(SSE)	N (19930, 2990)
Seed Yeast Separation			N (23800, 6390)
Seed Yeast Washing			N (8400, 3640)
Flow, Q (m^3/15 min)			
Filtration (QFI)		(QSE)	LN (-1.5966, 0.2090)
Commercial Yeast Separation and Washing			LN (2.2819, 0.2055)
Seed Yeast Separation			LN (1.9659, 0.0773)
Seed Yeast Washing			LN (1.3942, 0.0902)
Duration(hrs)			
Separation-Commercial Yeast & Seed Yeast (TSE) - Table look-up method			
Filtration - Table look-up method			
Fermentation			
- Seed Yeast Vats No.1 and No.2 - Table look-up method			
- Commercial Yeast	Vat No.3		LN (2.7391, 0.0627)
	Vat No.4		LN (2.7272, 0.0555)
	Vat No.5		LN (2.8203, 0.0356)
Idle Time Till the Next Batch (hrs)			
Seed Yeast Vats No.1 and No.2		(TID)	N(0,1)
Commercial Yeast	Vat No.3		LN (1.9636, 0.4425)
	Vat No.4		LN(2.1088, 0.4515)
	Vat No.5		LN(2.4801, 0.1641)
Yeast Produced (kg/ batch)			
Vat No.3			N (4940, 313)
Vat No.4			N (6150, 572)
Vat No.5			N (11800,1170)

$TFE_{i,j} \sim LN(l_{TFE_j}, TFE_j)$

N - Normal Distribution

LN - Log Normal Distribution

As mentioned in Section 3.2, filtration starts 10-20 minutes after the completion of the fermentation operation. Since variations within 10-20 minutes can not be differentiated within the time scales used in this study, the start times of filter presses were assumed to be 15 minutes following fermentation (i.e. one time unit).

3.4 Mathematical Formulation

The first activity of the process is fermentation. A fermentation activity for the i th batch(cycle) in the i th vat is defined as $AFE_{i,j}$. The completion time of $AFE_{i,j}$, denoted by $FEC_{i,j}$ is simply obtained by adding the duration of $AFE_{i,j}$ (i.e. $TFE_{i,j}$) to the start time (i.e. $FES_{i,j}$) of $AFE_{i,j}$:

$$FEC_{i,j} = FES_{i,j} + TFE_{i,j} \qquad (\forall\, i, j) \tag{1}$$

The completion of the activity $AFE_{i,j}$ calls for the separation and washing operation (i.e., activity $ASE_{i,j}$). Thus, the start time of $ASE_{i,j}$ (i.e., $SES_{i,j}$) is set equal to the completion time of $AFE_{i,j}$:

$$SES_{i,j} = FEC_{i,j} \qquad (\forall\, i, j) \tag{2}$$

As mentioned in Section 3.2, it must be noted that for commercial vats, j = 3, 4 and 5, a check has to be made regarding Centrifuge No.2 . If Centrifuge No. 2 is busy, $SES_{i,j}$ has to be delayed accordingly. Therefore, the completion time of $ASE_{i,j}$, i.e., $TSE_{i,j}$, is set equal to the start time of the previous separation . Thus, in general

$$SEC_{i,j} = SES_{i,j} + TSE_{i,j} \qquad (\forall\, i, j) \tag{3}$$

where, the duration of separation $TSE_{i,j}$ is computed by the table look-up method.

As the separation process continues, the yeast milk is directed to the filter presses, the start time of which (i,e., $FIS_{i,j}$) is one time unit after the completion of $AFE_{i,j}$ (i.e. $FEC_{i,j}$):

$$FIS_{i,j} = FEC_{i,j} + 1 \qquad (\forall\, i, j = j^x) \tag{4}$$

It must be noted that Equation (4), as well as any other expression for the filtration operation, is only valid for commercial vats, since seed yeast is not filtered. This is reflected in Equation (4) by index j^x (j^x is a subset of j denoting commercial vats only). The completion time of the filtration operation (i.e. $FIC_{i,j}$) is obtained by adding the duration of filtration (i.e., $TFI_{i,j}$) to the start time of filtration (i.e., $FIS_{i,j}$) :

$$FIC_{i,j} = FIS_{i,j} + TFI_{i,j} \qquad (\forall\ i,j=j^x) \qquad (5)$$

The above series of activities define a complete process cycle (batch). The computations for the start of the next cycle is initiated by addition of the idle time (i.e., $TID_{i,j}$) to the completion time of fermentation (i.e., $FEC_{i,j}$):

$$FIS_{i,j} = FEC_{i,j} + TID_{i,j} \qquad (\forall\ i,j) \qquad (6)$$

and the computations are carried out until the stopping criteria is met.

Throughout the above computations, whenever separation and washing ($ASE_{i,j}$) and filtration ($AFI_{i,j}$) activities are ongoing, a wastewater stream is generated having a flow rate (i.e., $QSE_{i,j}$ and $QFI_{i,j}$) and a COD (i.e., $SES_{i,j}$ and $SFI_{i,j}$) concentration in accordance with their probability distributions.

3.5 Computational Scheme

The average period for the restart of a batch being over 24 hours suggested that a daily simulation run would not be meaningful. Therefore, a week was chosen as the minimum simulation period. Since the processes were not continuous waste discharges were tracked by means of clock index numbers. The flow rates and COD values, from all operations that fall into the same interval (clock index number) were superimposed. A computer code was developed following the formulae of Section 2.4. The probabilities and the associated values for the variables for which table look-up method is used, are included in the program statements. All other relevant information, such as the mean and standard deviation for a specific variable obeying normal or log-normal distributions including the total simulation time are specified within the READ command. For convenience in keeping track of the clock index numbers, the computations for the commercial vats and seed vats are carried out separately but the commercial vats (i.e. $j=j^x$) are handled first in the calculations. Details of the computational scheme and the computer code are given in Renda [7].

Computations are initiated with the generation of three uniformly distributed numbers between 0 and 96, designating the fermentation start times of commercial vats (i.e., $j= j^x$). As it is practical and sometimes necessary to delineate the activities in progressive order, computations are first carried out for the commercial vat having the minimum start time. For example, if the generated numbers turned out to be 57, 6 and 94, corresponding to the start times of Vats No 3, 4 and 5 in terms of 15-minute time units, respectively, then the computations proceed with Vat No.4 since its start time is the minimum. Negative values for the normal variates were not allowed throughout the computations.

3.6 Validation of the Model

The model is validated by testing the simulation outputs with the observed behavior of the system. This is often quite difficult since both the simulation results and observed behavior of the system are of probabilistic nature. Thus, a single simulation outcome, being a random sample, should not be expected to agree with survey data in every detail. However, it should agree with the important aspects if the system is properly simulated and if the input data are realistic.

To validate the model, simulated patterns of the fermentation durations for all the vats are compared with corresponding survey data in Figure 2. Since the fermentation operation is the keystone of the overall process an acceptable degree of agreement between the simulated and observed data is necessary. Figure 3 gives the volumetric flow variations with time as determined from the model and the flow behavior of the real system. The figure is plotted discontinuously because the observed data were available only for the day time hours.

Figures 2 and 3 show that the derived distributions adequately describe the probabilistic nature of the system on the whole. The good agreement between the weekly averages of the amount of yeast produced, flow rates and COD values obtained from a single simulation run for seven weeks as given in Table 3 indicate the consistency of the model on a weekly basis.

Actually, the outcome of a single simulation run should be viewed as a single observation of an experiment. Since the analysis in a simulation model is carried through using random numbers, different simulation outputs are expected to show variations. The sensitivity and the implicit variability of the model mainly depends upon:

i. the initial state (start-up) of the system

ii. the sequence of random numbers used in simulation, that is, the initializers (seed numbers) used in the random number generation process.

Figure 4 assesses the effect of the former issue. As can be observed from Figure 4, the simulated and the observed fermentation durations generally agree well after a day. Hence, the conclusions drawn from the very first day of the simulation run were disregarded. In other words, the start up period was considered to be one day and the system was considered to be in its normal state of operations after the first day of the simulation run.

To evaluate the effect of the initializers, 20 simulation runs were made, each with a drastically different initializer. The 95% confidence intervals for the important variables of the system, obtained from these runs, were narrow enough to conclude that the system behavior was virtually independent of the initializers used. Hence it was believed that the system is satisfactorily simulated and attention was focused on the appropriate timing and scheduling of activities to achieve wastewater equalization as discussed in the next section.

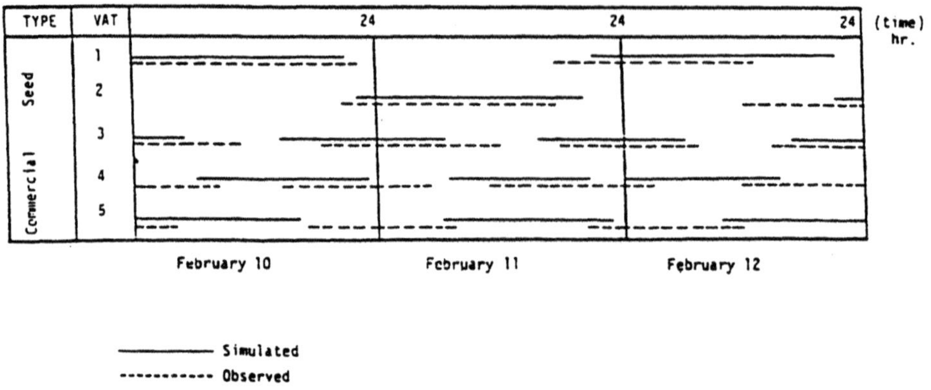

Figure 2. Fermentation Durations

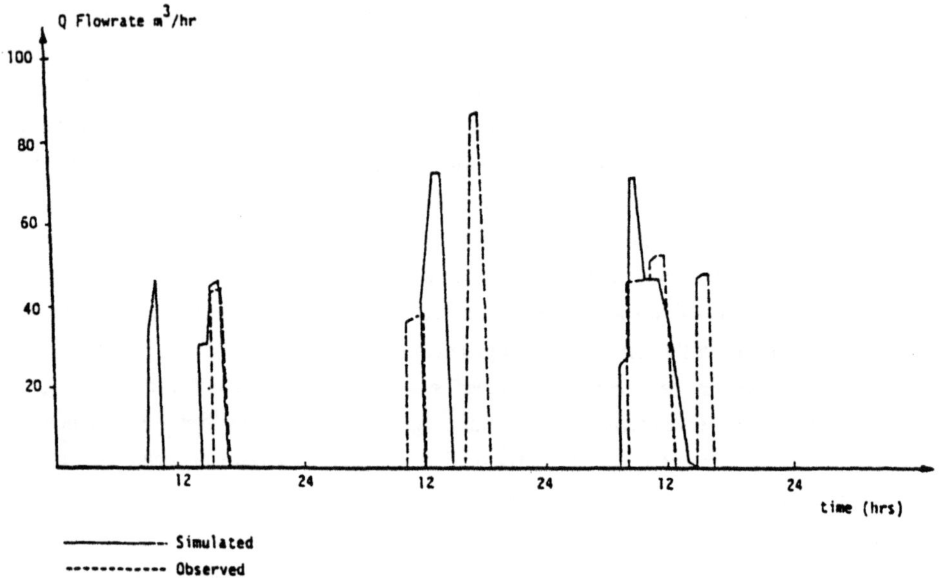

Figure 3. Variations in Wastewater Flow Rates with time

Table 3: Simulated Weekly Averages for the Amount of Yeast Produced, Flow Rate and COD

	Week	Yeast Production (kg/week)	Flow rate (m^3 /week)	COD * (kg/week)
	1	136425	2321	43658
	2	132628	2575	49587
	3	138554	2517	51310
	4	136261	2374	43596
	5	141886	2714	50287
	6	136334	2512	48336
	7	139960	2137	36717
Mean		137535	2450	46213

* COD kg/week - COD mg/lt x Flow m^3/hr x Conversion factor

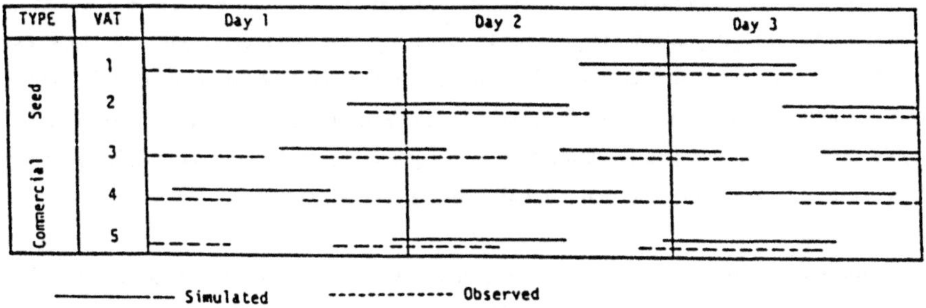

Figure 4. Duration of Fermentations Following Process Start-up

3.7 Timing and Scheduling of Successive Activities

Apparently the problem of timing and scheduling of activities to achieve wastewater equalization is very complex because of the large number of possible combinations. Thus, the solution should be sought through an optimization analysis which is difficult to implement. In this study since the probabilistic behavior of the operation start and end times do not yield to the deterministic scheduling algorithms available an intuitive approach was attempted. This mode of solution will more easily satisfy the system-specific observations.

It was observed that separation and washing operations were largely responsible for the large deviations in flow and consequently in the stream concentrations. Therefore, it was thought that distributing separation and washing periods throughout the week would somewhat result in equalization. The wastes discharged from separation and washing can be regarded as slug discharges due to their short discharge durations. Thus, if the discharge durations could be extended then the volumetric discharge rate would necessarily decrease. Starting from this point, the main aim should be to obtain sequential discharge patterns for the vats within a given period of time. In other words, the discharges should be arranged such that when discharging from one of the vats ends, the next vat should have completed fermentation and start discharging its own waste. Within this framework, the suggested operational scheme requires attention to the timing and scheduling of fermentation operations in all five vats in order to fix the start and completion times for separation and washing processes.

From the simulation runs the daily discharge of the wastewater was observed to be around 350 m^3. Therefore equalization schemes should aim to distribute the 350 m^3 of wastewater uniformly throughout the day (i.e. at a rate of 15 m^3/hr). In view of the wastewater volumes a discharge pattern is foreseen at a rate of 15 m^3/hr. However, the complete cycle of Vat No.5, namely fermentation, separation and washing and preparation for the next cycle, takes at least 26 hours. Thus it is physically impossible to discharge wastewater from Vat No.5 every 22-24 hours. Therefore, it was concluded that either the operational schemes of other vats have to be arranged so that Vat No.5 would be able to discharge wastewater after completing its cycle, i.e. 26-27 hours, (Equalization Scheme I) or another vat would have to be introduced into the system to absorb the burden (Equalization Scheme II).

Equalization Scheme I In the present operational schedule, when the seed yeast is being prepared in Vat No.1, the other seed vat, Vat No.2, is idle and vice versa. This equalization scheme is based on the idea of making beneficial use of the idle seed vat by allocating it for commercial yeast production when idle. However, if only one vat is used for seed yeast production, the daily seed yeast production would not be adequate for the remaining four commercial vats. Therefore, it is proposed that vats No. 1, 2 and 3 are to be scheduled in such a manner that they could be used interchangeably for both seed yeast and commercial yeast production. Although such on arrangement may seem to be operationally difficult, it is thought to be manageable provided that sufficient time is allowed for the interchange between batches.

Equalization Scheme II: In this scheme installation of a new vat (Vat No. 6), having the specifications of Vat No.5 is proposed. This leads to a discharge sequence of vats in order of

1, 3, 5, 2, 4 and 6. The cost of adding a new vat is expected to be offset by the increase in the overall yeast production.

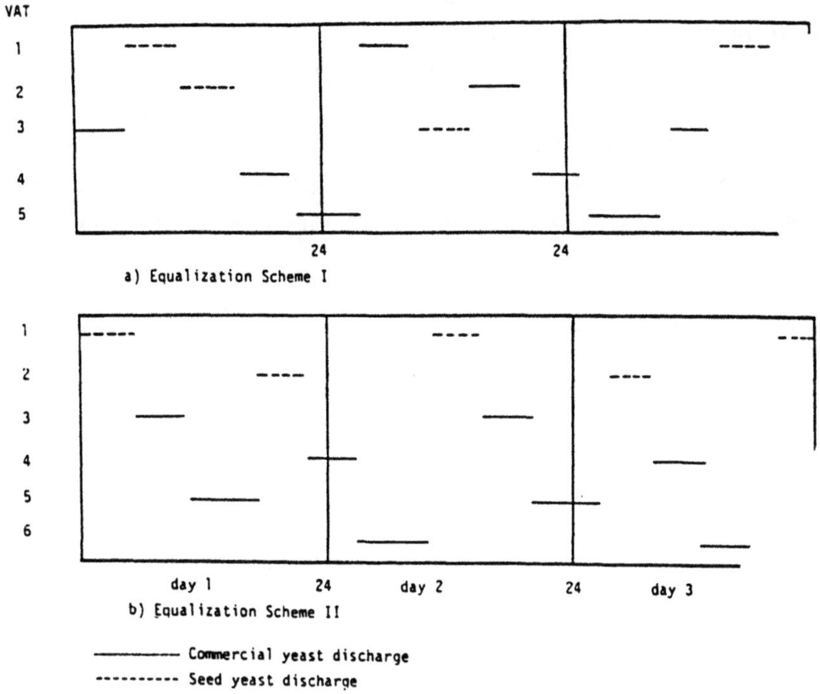

Figure 5. Fixed Discharge Periods to Provide a Relatively Continuc.

When the discharge periods, that is, the start and completion times, for the separation and washing operation are fixed as depicted in Figure 5 such that they follow one another the resulting flow and COD variations appear as given in Figure 6 and 7, respectively. As can be observed from these figures both equalization schemes definitely achieve a noticeable degree of equalization, both in flow and COD when compared with the present operational practice. It should be observed that the mean values of flow in Figure 6 and COD in Figure 7 for Equalization Scheme I, Equalization Scheme II, and the present operational practice, although close to each other, are different. This is due to the random behavior of the output variables as well as due to the fact that different operational schemes produce different quantities of yeast and consequently discharge wastewaters having varying characteristics.

To quantify the degree of equalization in a given operational scheme, three different expressions (i.e., equalization metrics) were used:

1) average of the absolute deviations from the mean
2) average of the squares of the deviations from the mean,
3) maximum absolute deviation from the mean

The three equalization metrics for the different schemes are summarized in Table 4, as obtained from a seven week simulation run. Apparently, Equalization Scheme I results in a satisfactory wastewater equalization when compared to the present operational practice at the expense of a reduction of about 35 tons/week, i.e. 25%, in yeast production. On the other hand, Equalization Scheme II achieves a better degree of equalization while increasing the yeast production by about 33 tons/week , i.e. 24%, at the expense of the installation of a new vat.

Table 4. Yeast Production and degree of equalization

Scheme	Yeast	Flow			COD		
	Produced (ton/week)	Avg. D* (m^3/hr)	Avg. D^2 (m^3/hr)^2	Max. D (m^3/hr)	Avg. D* (m^3/hr)	Avg. D^2 (m^3/hr)^2	Max. D (m^3/hr)
Present Scheme	137.5	72.9	41.3	77.5	1264	739	1487
Eq'tion Sch'm I	103.8	16.9	12.0	17.6	397	253	331
Eq'tion Sch'm II	170.1	5.0	3.1	4.3	210	127	163

* D = absolute deviation from the mean

4. Conclusions

A system consisting of a series of successive activities each possessing random behavior is shown to be satisfactorily simulated by a Monte Carlo model. Such a model was used with the aim of achieving flow and strength equalization of wastewater. The proposed approach involves appropriate timing and scheduling of the activities, and thus the discharges, based on system-specific observations. An application of the proposed methodology to a baker's yeast production plant revealed that the method is indeed feasible in achieving a satisfactory degree of equalization.

The smallest time increment considered in the analysis was 15 minutes. It is thought that shorter time increments would lengthen computations with unnecessary details. Apparently, the computations with 15 minute increments resulted in simulation clock index numbers in the orders of thousands. This would mean more computational load and would not improve the precision of the method appreciably.

It should be emphasized that the proposed approach can be adapted for other equalization purposes, such as idle machine hours, storage area for products, process stream, etc.. In all cases the simulation results can lead to a trade-off analysis between degree of equalization and product level and the decision maker can choose his/her operational policy in view of his/her personal preference.

References

1. Berthouex, P.M. and L. C. Brown, 'Monte Carlo Simulation of Industrial Waste Discharges', ASCE, Journal of Sanitary Engineering Division, V. 95, Oct. 1969.
2. Çiler, M., Çetinkaya, N., Dumlu, G., Ilhan, R., Kinayyiğit, G., Mutlu, H., Öktem, O. Timur, A., and Topal, H., "Pakmaya Fabrikasi Aritma Tesisleri Proje Çalismalari Ara Raporu ", TÜBITAK, Gebze (Turkey),1981 *(in Turkish)*
3. EPA-U.S. Environmental Protection Agency, Manual of Methods for Chemical Analysis of Water and Wastes, Technology Transfer Series, National Environmental Research Center, Cincinnati, Ohio, 1974.
4. Fishman, G.S., Concepts and Methods in Discrete Event Digital Simulation, John Wiley and Sons, Inc., New York, 1973.
5. Koziorowski, B. and J.Kuchorski, Industrial Waste Disposal, Wydawnictwa Naukowo-Techniczne, Warsaw, 1972.
6. Nemerow, N.L., "Liquid Waste of Industry: Theories, Practices and Treatment", Addison-Wesley Pub. Co., Reading, Mass., 1971.
7. Renda, A.N., "Wastewater Equalization by Monte Carlo Simulation", Masters Thesis, Dept. of Chemical Engineering, Boğaziçi University, Istanbul, Turkey, 1981.
8. Tayyabkhan, M.T. and T.C. Richardson, 'Monte Carlo Techniques', Chemical Engineering Progress, V.61-1, Jan. 1965.

Simple Model Predictive Control Studies on a Batch Polymerization Reactor

Ali Karaduman and Rıdvan Berber

Department of Chemical Engineering, Ankara University, Tandogan, Ankara , 06100 - Turkey

Abstract : A single step control algorithm has been developed and implemented for the temperature control of a batch styrene polymerization reactor. The algorithm is based on the energy balance of the system with assumed linear dependency of the overall heat transfer coefficient on feed flow rate through the cooling jacket. The heat generation term in the energy equation is treated as an inferentially measured process disturbance and predicted by on-line process measurements. The algorithm was tested in real-time runs and resulted in a good performance for maintaining the reactor temperature at its set point during the isothermal reaction stage.

Keywords : Process control, model predictive control, batch polymerization

Introduction

Batch chemical reactors are extensively used in polymer, chemical and pharmaceutical industries because of kinetic advantages in some systems. For radical chain polymerization, for example, they are often preferred for high conversion compared to moderate conversion that may be obtained with continuous systems with recycle of monomer and solvent. However, because of the fact that the batch system has no steady state, the dynamic nature is much more complicated and this makes temperature control difficult. There are still cases in which a batch reactor is simply controlled by human operator.

It has been well established that there exists close relationships between the operating conditions of reactors and the quality of the polymer produced. Therefore, the first problem of the batch polymerization control is to determine the set of time dependent controls, the most important being the optimum temperature profile. This has been subject for a great deal of research work [l, 5, 14, 18, 19].

In polymerization systems, the primary control objective is to control the polymer quality. This implies some structural properties of the polymer such as the average molecular weight, the molecular weight distribution and the chain length distribution which depend on the reaction conditions and affect the end-use of the polymer. The second objective of the control problem is to maintain the polymerization temperature as close as possible to its desired profile in spite of time variant and often nonlinear changes. A servo control strategy is, thus, required to move the system along the predetermined temperature vs. time trajectory like the one shown in Figure 1. In fact, these two control objectives (MWD and temperature) are correlated.

Temperature control in batch reactors, therefore, can be divided into three phases: a heating phase to raise the reactor temperature to its target as fast as possible without overshoot, and a

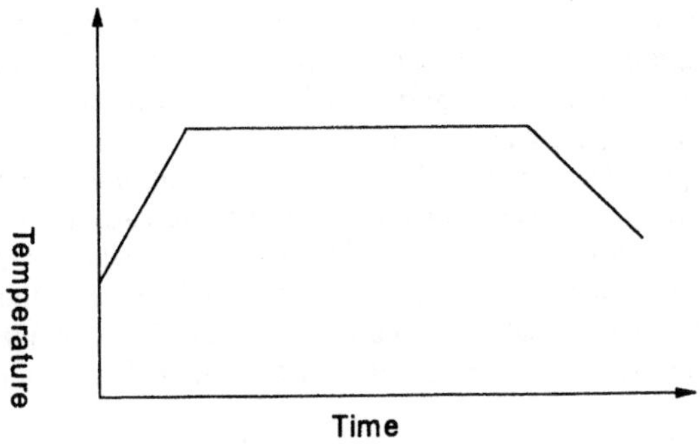

Figure 1. Desired Temperature Trajectory of a Batch Polymerization Reactor

stabilizing phase to maintain the reactor temperature at a set point or set point trajectory, and a cooling down period. Different control algorithms are normally required for the heating and cooling process. Implementing the change in control mode is critical.

For the heating phase, the control policies are bang-bang in nature. The problem during stabilizing control is the very uneven heat load on the cooling system because the polymerization reaction exhibits an auto-acceleration rate in the heat generation phase. The change in this phase occurs so quickly that maintaining a constant temperature in the reactor by an automatic controller becomes very difficult. As the reaction progresses, the heat generation rate decreases as a result of the monomer consumption.

The number of studies on control of batch polymerization reactors relative to that of continuous reactors is limited in the literature. Therefore, there are still unresolved problems and the area provides an interesting challenge.

MacGregor [11] gave a literature survey on temperature control of polymerization reactors. Despite of the fact that the most widely used controllers are PID type, the complicated control requirements for batch reactors may not be fulfilled by classical controllers, because of the time varying characteristics and other nonlinearities in the process. The standard PID algorithms may be sufficient only when the polymerization time is large and polymerization heat effects are small. Kiparissides and Shah [10] used, what they called, an extremely well tuned PID controller in their simulation run and observed more than two degrees overshoot during the transition from the heating to the stationary phase under PID control. PID controller tuned to operate in a stochastic environment gave an oscillatory performance when deterministic measurements were used. They concluded that a well-tuned PID controller would only perform satisfactorily if the signal to noise ratio did not change significantly. On the other hand, the evolution of digital control computers allowed one to produce much better designs without considerations for hardware realizability. What is required, therefore, for batch reactors is an adaptive control technique which can easily be implemented in digital computer environment. Kiparissides and Shah [10] evaluated two adaptive control techniques for self-tuning control of a batch PVC reactor and reported better performance compared to classical PID case. MacGregor [11] later reported that all of various adaptive controllers appeared to be robust and superior to fixed parameter PID controllers.

An alternative approach to the problem was to investigate the control strategies employed by human operator and design "fuzzy" logic controllers [17].

To overcome the difficulty associated with sharp increase in the rate of polymerization at reduced polymerization time, feedforward control was used by Amrehn [1] to take action in advance by predicting the amount of polymerization heat in a given reaction stage. Later Jutan and Uppal [9] employed a feedforward- feedback control algorithm.

"Parametric control" strategy was employed by Jutan and Rodrigez [8] for control of a batch reactor heating system which involved expressing the manipulated variables of the process in terms of a new set of variables, called parametric variables. They established the parametric relationships experimentally to satisfy the requirements. In Ponnuswamy et al. [14]'s work, different open loop policies for the batch reactor were derived and optimal temperature policies calculated from the optimal feedback control law were implemented on an experimental reactor system through a PI controller.

In recent years, "predictive control" techniques have been proposed based on a nonparametric model, namely an impulse response or discrete convolution model [12]. Predictive control techniques are advantageous because of usefulness for processes with unusual dynamics where specifying the model structure for a parametric model would be difficult as well as inherent time delay compensation and ability to accommodate process constraints. This approach is applicable theoretically to any system which can be described by a set of linear differential equations. Two predictive control techniques have evoked a great deal of interest recently : Model algorithmic control (MAC) and dynamic matrix control (DMC). Model predictive control approach and its comparison to IMC systems is well described by Cheng and Brosilow [4] and, Hidalgo and Brosilow [7]. As for the polymerization systems, MPC combined with coordinated control strategy, has been applied to a continuous styrene polymerization. They simulated the styrene reactor and control system using the complete fourth order model for the process and the reduced third-order system (ignoring the equation for initiator concentration) for the model and, have demonstrated that stable control and exact tracking of the set point can be accomplished.

Most recently, Westerholt et al. [20] described a time-optimal algorithm for second order plus dead time systems and tested it via simulation for the startup of an exothermic batch reactor as well as for some other systems. They reported that the algorithm had performed well on all of physical systems tested.

This paper aims to develop a simple, easy-to-implement control algorithm and establish its feasibility for temperature control of a batch solution polymerization reactor at the isothermal stage.

Process Description

The process model for the free radical solution polymerization of styrene in a jacketed continuous stirred tank reactor has been given in early investigations [3,5,7,19].

Reaction steps for free radical polymerization are represented by the following equations:

$$I \xrightarrow{k_d} 2R \qquad (\textit{initiator decomposition}) \tag{1}$$

$$M + R \xrightarrow{k_i} P_1 \qquad (initiation) \tag{2}$$

$$P_n + M \xrightarrow{k_p} P_{n+1} \qquad (propagation) \tag{3}$$

$$P_n + P_m \xrightarrow{k_{tc}} T_n + T_m \qquad (termination) \tag{4}$$

$$P_n + P_m \xrightarrow{k_{td}} T_n + T_m \qquad (termination) \tag{5}$$

Following these kinetics, the model for a batch process, is composed of material balances for initiator and monomer, and energy balances for the reactor and jacket as given by equations 6-9. Perfect mixing, constant holdup and constant physical properties are assumed in the reactor.

$$\frac{d[I]}{dt} = -k_d\,[I] \tag{6}$$

$$\frac{d[M]}{dt} = -k_p\,[M]\,[P] \tag{7}$$

$$[P] = [\frac{2fk_d[I]}{k_t}]^{\frac{1}{2}} \tag{7a}$$

$$\rho c_p v \frac{dT}{dt} = (-\Delta H) v k_p [M][P] - UA(T - T_c) \tag{8}$$

$$\rho c_{pc} v \frac{dT_c}{dt} = f_c c_{pc}(T_{cf} - T_c) + UA(T - T_c) \qquad (9)$$

It is known that, polymerization of certain monomers in concentrated solution is accompanied by a marked deviation from first order kinetics in the direction of an increase in reaction rate and molecular weight, termed "gel effect". However, benzoyl peroxide-initiated polymerization of styrene is reported to be accurately first order up to quite high conversions [2]. In virtue of this fact, the viscosity effects have not been included in the above model at the present stage of the work.

In the energy balances, the assumption that for a sufficiently rapid coolant flow rate the inlet and outlet temperatures would not vary by much holds and, therefore, an average temperature for the cooling jacket was used. Literature values were used for process parameters [3].

Control Algorithm

The approach used here for temperature control of the batch polymerization process in general is a simple model predictive control. MPC is explained in terms of a sequence of tasks [4]. The tasks define the MPC structure just as block diagrams define the structure of IMC [6].

Task 1. Use process model and current measurement to estimate the disturbances entering the process. If there is a process dead time, then disturbance estimates are extrapolated forward in time by the amount of the dead time. This task explicitly estimates the disturbance so that the model can be used to predict the future process outputs after the dead time.

Task 2. Predict the process state one dead time into the future based on current measurement, current and past controls and extrapolated disturbances.

Task 3. Compute the desired future values of the process state beyond the dead time starting with the process state predicted in Task 2. Desired future state should evolve as a dynamical system whose relative order is the same or greater than that of the process model. Modeling errors in MPC will often be compensated by assuming that the errors between the model and process outputs were caused by unmeasured disturbances.

Task 4. Compute the controls which force the model to track the desired future states computed in Task 3 as closely as possible over some time horizon into the future.

In some versions of MPC such as Model Predictive Heuristic Control [16] and Quadratic Dynamic Matrix Control [15], control is calculated by minimizing the integral of the square deviation between the desired trajectory and the model trajectory over a preset time horizon subject to control effort and process constraints.

An alternative approach, is to select the current control so that the output or the state of the model matches that of the reference dynamical system at some future time. This is computationally much less intensive than the previous method. We, therefore, want to use this approach here in order to develop an easy-to-implement and effective control algorithm for the isothermal stage of the batch polymerization reactor.

The overall objective of polymerization control can be accomplished by controlling the reactor temperature and monomer and initiator concentrations. However, because of its large effect on the structural characteristics of the polymers, the temperature control bears probably most importance. This was particularly true for solution polymerization where no gel effect was present because of relatively low viscosity polymerization. Therefore, we have chosen to control the reactor temperature by manipulating the flowrate of cooling water into the jacket. We assume that the only disturbance to the system is the heat of polymerization, represented by the first term on the right hand side of reactor energy balance, eqn. (8). We treat this nonlinear heat generation term as an inferential disturbance (to be predicted from process measurements) and, denote it by D. Thus, the energy equation becomes

$$\rho c_p v \frac{dT}{dt} = D - UA(T - T_c) \tag{10}$$

The exothermic reaction enthalpy, ΔH is a complicated function of the concentrations. If one wishes to obtain a theoretical expression for it, the necessary detailed information is often not available. As we do require information about ΔH in order to control the process, our suggestion that its value be predicted from on-line process measurements would be a practical and realistic approach. Such an approach was, in fact, previously used by Jutan and Uppal [9] in their feedforward control studies via simulation, but was not implemented in a real-time control algorithm.

However, what we end up with is still a nonlinear equation because of the overall heat transfer coefficient. U is the essential term in the above equation that relates the manipulated variable to the process variable. We calculated U by evaluating the film coefficients h_i and h_o for the reactor side and jacket side respectively. For h_i the correlation given by Perry [13] for heat transfer from agitated liquid contents of vessels to jacketed walls was used while h_o was calculated for laminar flow in annuli according to a correlation again provided by Perry [13]. The resultant equation was highly nonlinear as follows,

$$U = \frac{1}{336.44 + 1310.1\, f_c^{-1/3}} \tag{11}$$

However, if U is plotted against f_c for the range of operation in our cooling water feeder in the experimental system, one notices that after a flowrate of 20 cm^3 /s the graph exhibits a linear behavior as can be seen in Figure 2. As the experimental results later demonstrated, the linearized portion of this relationship mostly fell into the actual operating range of the controller. Therefore, we suggest that the following linear equation, thus obtained, may be used for simplicity

$$U = 10.968 * 10^{-4} + 7.083 * 10^{-6} f_c \tag{12}$$

If we substitute this equation into equation (10), then we get a linear first order system on which we can establish a model predictive control algorithm fairly easily. We, consequently, suggest the following control algorithm:

1) Using measured values of T and T_C, solve the suggested form of the energy equation (eqn. 10 coupled with 12) for D, the disturbance estimate. For a small enough sampling time, we treat this disturbance estimate constant.

2) Calculate the desired value of the temperature in the control horizon as follows.

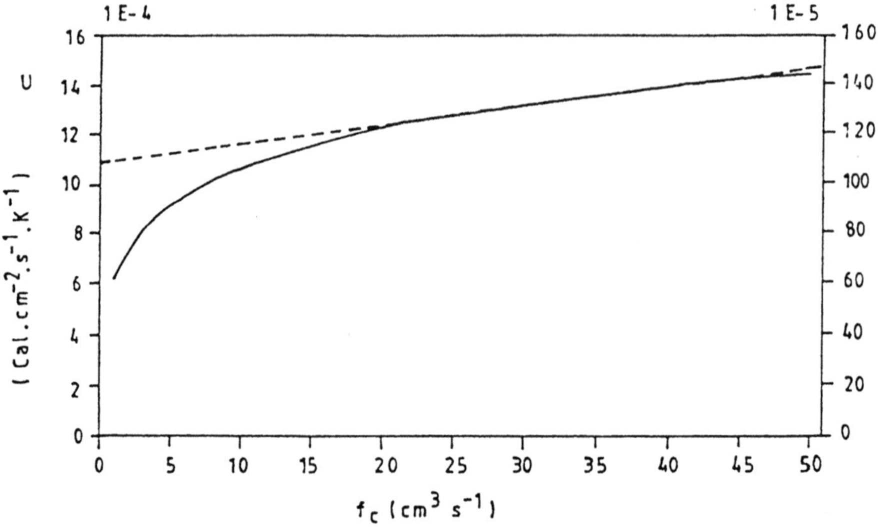

Figure 2. Dependency of Overall Heat Transfer Coefficient on Cooling Water Flow Rate

$$T_d\,(t_k+H) = e^{\frac{-H}{\epsilon}}\,T(t_k) + (1-e^{\frac{-H}{\epsilon}})T_s \tag{13}$$

This reference trajectory was obtained by applying a first order filter on the current reactor temperature to reach the desired value. The sampling time for the temperature measurement was 5 seconds. The control horizon was chosen two times the sampling time for stability reasons, as suggested by Hidalgo and Brosilow [7] as a consequence of their numerical studies.

3) Find the value, $T(t_k+H)$, that process variable would reach if there were no control starting from $T(t_k)$ again by making use of the linearized model above. Here, we assumed that there were no dead time.

4) Determine the step response of the system for a unit step change in control, S(H)

5) Calculate the control effort from the control law such that

$$u(t_k) = \frac{T_d\,(t_k+H) - T(t_k+H)}{S(H)} \tag{14}$$

As the first implementation of this control calculation resulted in some oscillations, we used a first order digital filter on this calculated control effort to stabilize the control action and smooth the operation. The final form of the actually applied control effort, therefore, was

$$c(t_k) = \alpha u(t_k) + (1-\alpha)C(t_k-\Delta t) \tag{15}$$

As for the filtering constant, first experiments indicated that $\alpha = 0.3$ would be a good choice in our case.

This algorithm was implemented in an interactive real-time monitoring and control software programmed in Turbo Pascal version 5.0.

Experimental

The bench scale experimental system under investigation in this work consists mainly of a 1-liter batch stirred tank reactor with cooling jacket and necessary auxiliary equipment for on-line microcomputer control. A diagram of the process is shown in Figure 3.

The reactor is made out of glass and surrounded by a metal jacket which is insulated by a thick layer of glasswool. To avoid any pressure build-up in the system, the reactor was operated under a condenser that was fitted at the top. The control strategy developed in this paper manipulates the flowrate of cooling water through the jacket to maintain the reactor temperature at desired level during the isothermal, heat-generating phase of the reaction.

Prior to the experiments, styrene monomer was distilled under vacuum and stored in a refrigerator until use. As for the initiator, pure benzoyl peroxide (Bz_2O_2) crystals freshly prepared by crystallizing from a solution in chloroform were used. Toluene was chosen as solvent for the reaction medium.

Experiments were started by loading the styrene monomer (6 mol/l) and solvent into the reactor. Initially, the reactor was purged with nitrogen gas and was kept under nitrogen atmosphere until the end of the experiment. The initial temperature to start the operation was first

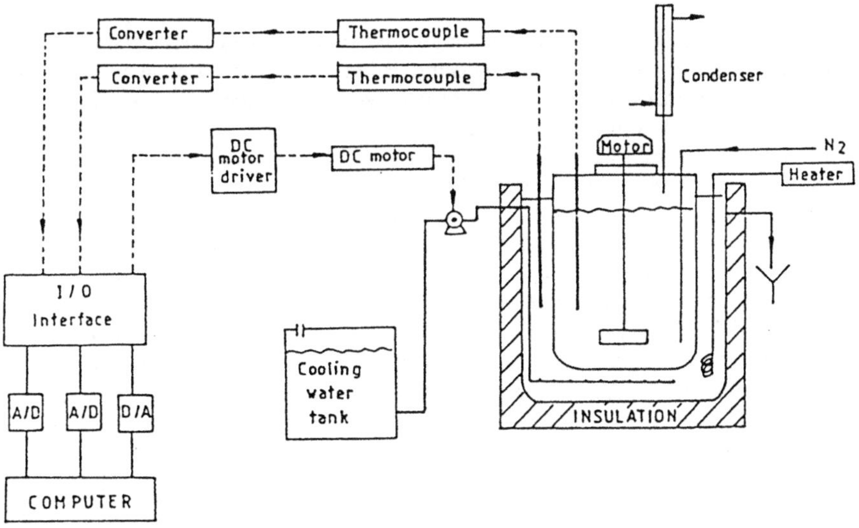

Figure 3. Experimental System

reached by heating the contents of the reactor with an external electric heater under manual control. This was a few degrees below the set point temperature for the reaction. The initiator could not be used for the initial warming up period because of the fact that the reaction rate of the initiator (Bz_2O_2) was very low at temperatures below 70 °C. This implies that an external source for heating was needed.

In industrial applications, the polymerization of styrene is usually conducted at 70 °C where the completion of reaction takes long and the rate of heat generation is relatively low. As was noted on the literature survey section, the classical control algorithms work well under these conditions. We have chosen a slightly higher set point temperature to have a higher rate of heat generation. This would be advantageous in two aspects; (i) to test the algorithm under severe conditions to see the performance so that it would be supposed to work under mild conditions, (ii) to be able to decrease the time required for completion of polymerization. Thus, the set point, in all experiments, was 90.2 °C. When the reactor reached that value, desired amount of initiator was dissolved in toluene and was charged to the reactor. Automatic measurement, control and data collection by microcomputer was started simultaneously. Control algorithm used the current values of the reactor and cooling jacket temperatures, and previous value of the control. The current control signal was then calculated and sent at each sampling interval to the DC motor driver through a D/A converter. This driver then regulated the 12V DC motor to manipulate the cooling water flowrate passing through. In all of the experiments, we used a cooling water supply at 70 °C.

After control experiments, dilute solution viscosity of polystyrene was measured in a capillary viscometer of Ubbelohde type to asses the molecular properties of the polymer. The average molecular weights calculated from the intrinsic viscosity measurements by Mark-Houwink-Sakurada equation were around 70,000 for three experiments reported here.

Results and Discussion

The experiments were exploratory in nature with the aim of determining if temperature of a batch polymerization reactor could be effectively controlled by the proposed simple model predictive control algorithm. At this stage of the work, only the isothermal stationary phase of the reaction was attempted to be controlled by microcomputer.

Our preliminary results had revealed that when a cooling water at room temperature were used, the continuing heat transfer-between the contents of the reactor and jacket even when the cooling water flow was zero, would bring the temperature further down below the set point, therefore limiting the time that the reactor would have been maintained at the set point. To avoid this, we used a coolant supply of 70 °C. Results from the experiments conducted thus far under specified conditions gave temperature tracking profiles as illustrated in Figures 4 to 6.

Figure 4 and 5 indicate two attempts to bring the reactor to the desired set point and hold it there. In both experiments, the temperature rises as the reaction starts at the time of addition of the initiator as described in the experimental part. During the early phase of this heat-up period, the algorithm applies no control. However, as the temperature approaches the set point, the control effort senses that the estimated disturbance may drive the process above the set point and, takes corrective action to compensate for its possible future effects. The closer the reactor temperature is to the set point, the more the control is applied. Then, consequently, with a slight overshoot, the effect of exothermic heat generation is compensated by the control action and the rate of increase in temperature slows down, finally settling the temperature at the desired value.

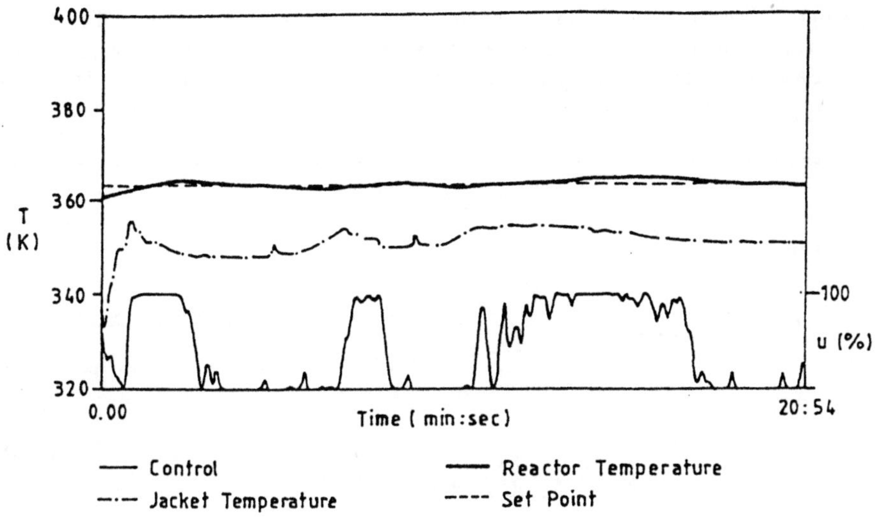

Figure 4. Control Experiment, Run 1, [I_o=0.084 mol/l, M_i=6 mol/l, T_{Co}=343 K]

Meanwhile, the control action also decreases and comes down to zero. Filter time constant (s) in these two experiments was 100 seconds.

Figure 6 shows the closed loop behavior of the control system when a filter time constant of 200 seconds was used. In that case, the control system takes a rather sluggish action to bring the

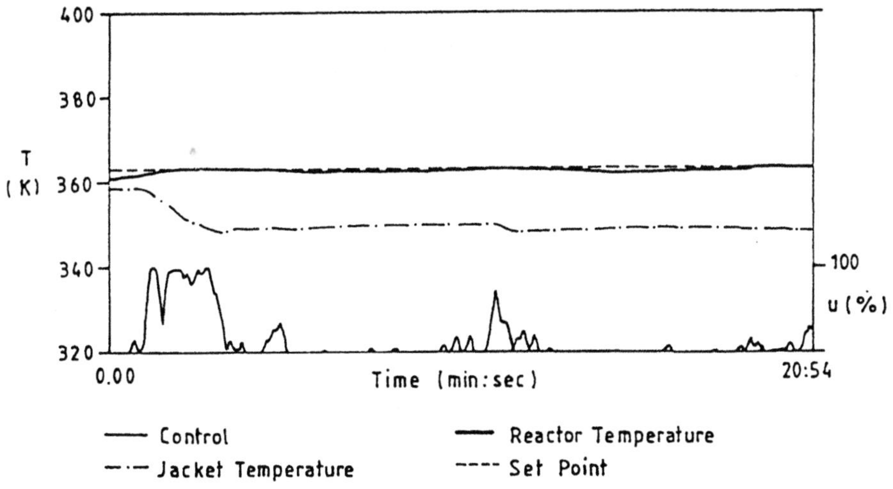

Figure 5. Control Experiment, Run 2, [I_o=0.084 mol/l, M_o= 6 mol/l, T_{Co}=343 K]

temperature to the set point. The applied control effort is remarkably less compared to the case observed in experiments shown in Figures 4 and 5. When filter time constant was small, immediate corrective action was taken to bring the process rapidly to the set point. Large s on the other hand, results in rather small changes in the control effort as reflected in Figure 6.

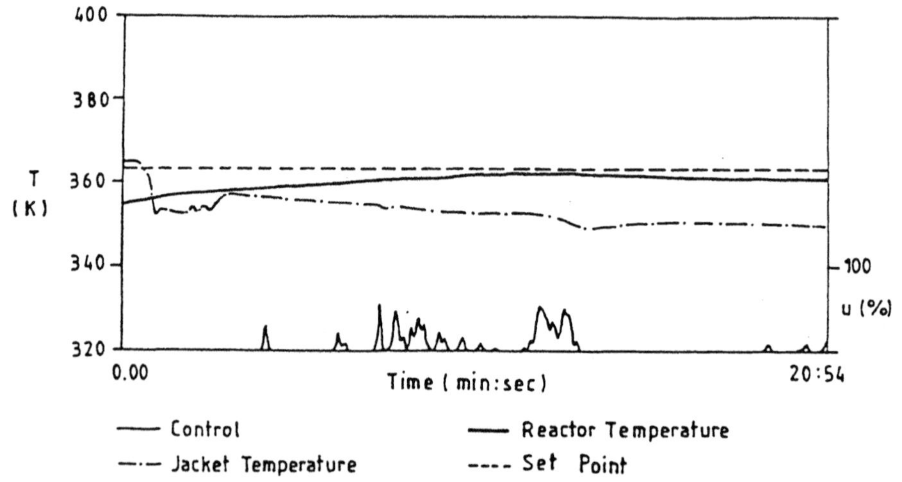

Figure 6. Control Experiment, Run 3, [I_o=0.084 mol/l, M_o= 6 mol/l, T_{Co}=343 K]

We suppose that the linearized relationship, between the overall heat transfer coefficient and the coolant flow rate, as illustrated in Figure 2, can be assumed to hold down to a flow rate of 15 ml/s. It is, therefore, to be noted that, the control effort, i.e. actual operating range of the coolant flowrate falls mostly within this linearized region (i.e. 30 to 100%) that defined by eqn. (12).

From the above discussions, we can conclude that the proposed algorithm holds promise for further investigation and for practical applications on a larger scale, because it provides the ability to minimize the effects of disturbances.

The suggested algorithm offers two advantages. First, it is simple to design so that it can be easily implemented in a microcomputer without resorting to excessive computation. Second, it has only one tuning parameter, namely ϵ, that can be either theoretically determined for simple linear first order systems or can be found experimentally. This is obviously an important superiority over the conventional, three parameter PID control. No need to mention that, in addition, the inherent advantages of MPC, like minimization of the effect of control effort saturation, also hold.

Besides, the simple methodology we suggest here can be applied to the temperature control of similar exothermic batch reaction systems, because we do not actually use kinetic model for control, we rather make an online inferential measurement to sense the instantaneous reaction conditions. Therefore, the algorithm can be expected to work even if changes in kinetics occur at high conversions. In different polymerization conditions, however, one should take the fact into account that controlling the reactor temperature may not be suitable to control the polymer properties which is the ultimate goal of the control problem in polymerization systems. In the next phase of this work, we would like to take the nonlinearities of the system into consideration and use the kinetic model to calculate the control effort.

The difficulties associated with the use of low temperature coolant that we had observed during our preliminary experiments indicated the need for a heating medium computationally coupled with this algorithm. In this case, if the temperature at the jacket side dropped sharply with applied control effort leading to a decrease in the reactor temperature below the set point, heat could be applied to bring the reactor back to set point. Such a medium could have also been used for the initial heating up of the reactor to the desired reaction temperature.

Conclusion

We can conclude, from the exploratory studies reported here, that the suggested simple MPC control algorithm can be useful in applications particularly where an easy-to-implement and practical control algorithm is needed.

However, the more usual solution to the problem would be to live with the process nonlinearities and use more complex algorithms such as computation of the control by a Newton algorithm while considering the full model without any simplification. We believe that this would lead to a more robust algorithm. In this case, one would be able to use both heating and cooling modes by probably manipulating the external heat input into the system together with the cooling flowrate. Therefore, a predetermined optimal time-temperature trajectory for the whole sequence of a batch operation would be controlled.

Acknowledgment: This work was supported by a research grant from Ankara University research fund under contract number 91-25-00-93 which is gratefully acknowledged. We also acknowledge fruitful discussions with Prof. C. B. Brosilow throughout this work.

Notation

A	: heat transfer area of reactor, cm^2
C_p	: mean heat capacity, cal g^{-1} K^{-1}
$C(t_k - \Delta t)$	: actual control applied at previous sampling period
f	: initiator efficiency
f_c	: flowrate of cooling jacket fluid, s^{-1}
H	: control horizon
I	: initiator
$k_d, k_i, k_p, k_{tc}, k_{td}$	: Arrhenius rate constants for dissociation, initiation, propagation, combination termination and disproportionation termination respectively
M	: monomer
P_k	: growing species of length k units
R	: initiated radical
T	: temperature of reactor, K
T_c	: mixing cup temperature of cooling jacket fluid, K
T_s	: temperature set point, K
t	: time, s
t_k	: current time, s
U	: overall heat transfer coefficient, cal cm^{-2} s^{-1} K^{-1}
u	: control effort
V	: reactor volume, l
[ ]	: represents concentration
o	: represents initial conditions

Greek Letters

ρ	: mean density of reactor fluid, g l^{-1}
ρ_c	: density of cooling jacket fluid, g l^{-1}
α	: a constant such that $\alpha = 1$ corresponds to no filtering while approaching zero means ignorance of measurement
$-\Delta H$	: heat of polymerization reaction, cal mol^{-1}
Δt	: sampling time, s
ϵ	: adjustable tuning parameter

References

1. Amrehn. H.: Computer Control in Polymerization Industry. Automatica, 13. 533?-545 (1977)
2. Billmeyer. F.W.: Textbook of Polymer Science. 2nd ed. Wiley- Interscience, New York (1971)
3. Brooks, B.W.: Dynamic Behavior of a Continuous Flow Polymerization Reactor . Chem. Eng. Sci. 36. 589-593 (1981)
4. Cheng. C.M. and Brosilow. C.B.: Model Predictive Control of Unstable Processes. AIChE 1987 Annual Meeting. New York (1987)
5. Chen, S.A. and Jeng. W.F.: Minimum End Time Policies for Batchwise Radical Chain Polymerization. Chem. Eng. Sci. 33. 735-743 (1978)
6. Garcia. C. and Morari. M.: Internal Model Control 1. A Unifying Review and Some New Results. Ind. Eng. Chem. Process Des. Dev. 21, 308-323 (1982)
7. Hidalgo. P.M and Brosilow. C.B.: Nonlinear Model Predictive Control of Styrene Polymerization at Unstable Operating Points. Computers chem. Engng. 14, 481-494 (1990)
8. Jutan. A. and Rodrigez. E.S.: Application of Parametric Control Concepts to Decoupler Design for a Batch Reactor. Can. J. Chem. Eng. 65, 858-866 (1987)
9. Jutan, A. and Uppal, A.: Combined Feedforward-Feedback Servo Control Scheme for an Exothermic Batch Reactor. Ind. Eng. Chem. Proc. Des. Dev. 23. 597-502 (1984)
10. Kiparissides. C. and Shah. S.L.: Self-tuning and Stable Adaptive Control of a Batch Polymerization Reactor, Automatica, 19,225-235 (1983)
11. MacGregor, J.F.. Penlidis, A.. Hamielec. A.E.: Control of Polymerization Reactors: A Review. Polym. Proc. Eng. 2. 179-206 (1984)
12. Marchelti. J.L.; Mellichamp, D.A.; Seborg, D.E.: Predictive Control Based on Discrete Convolution Models, Ind. Eng. Chem. Proc. Des. Dev. 22, 488-495 (1983)
13. Perry. R.H. and Chilton. C.H. (editors): Chemical Engineers' Handbook, 5th ed. pp. 10-13, 10-16, McGraw Hill Inc. N.Y. (1974)
14. Ponnuswamy, S.R.; Shah, S.L. and Kiparissides, C.A.: Computer Optimal Control of Batch Polymerization Reactors, Ind. Eng. Chem. Res. 26, 2229-2236 (1987)
15. Prett. D.M. and Garcia, C.E.: Fundamental Process Control. Butterworths. Boston (1988)
16. Richalet. J.. Rault, A., Testud, J.L. and Papon, J.: Model Predictive Heuristic Control: Applications to Industrial Processes. Automatica, 14, 413-428 (1978)
17. Rong, G.: Microcomputer Control of a Pilot Batch Polymerization Reactor and Investigation of Fuzzy Control Algorithms, Master Thesis, Khejiang University, China (1986)
18. Sacks. M.E.; Lee. S. and Biesenberger, J.A.: Effect of Temperature Variations on Molecular Weight Distributions: -Batch Chain Addition Polymerizations, Chem. Eng. Sci. 28. 241- 257 (1973)
19. Villermaux, J. and Blavier, L.: A New Method for Modeling Free Radical Homogenous Polymerization Reactions. Chem. Eng. Sci. 39. 87-99 (1984)
20. Westerholt, E. von; Beard, J.N.; Melsheimer. S.S.: Time- Optimal Startup Control Algorithm for Batch Processes, Ind. Eng. Chem. Res. 30. 1205-1212 (1991)

Retrofit Design and Energy Integration of Brewery Operations

Denis J. Mignon

Belgian National Fund for Scientific Research (F.N.R.S.), Rue d'Egmont 5, B-1050 Brussels and Process Engineering Department, Université Catholique de Louvain, B-1348 Louvain-la-Neuve BELGIUM

Abstract: A detailed simulation model of the four brewhouses of an industrial brewery has been developed with the help of the BATCHES simulator. The use of this model has allowed a thorough study of the utilization rates of the pieces of equipment. The study has shown that these utilization rates are generally quite low and that some pieces of equipment are largely oversized. Two new configurations have been proposed for the brewhouses, reducing by up to one third the number of pieces of equipment.

On the side of the energy integration, the use of the model has shown that the combination of the proposed modifications of the process, namely a better standardization of the brewing plant, an adequate production planning and a deliberate limitation of the steam availability level, permits a considerable reduction of the steam consumption peaks. From the most unfavourable situation to the most favourable one encountered during this simulation study, the global reduction amounts to 55%.

Keywords: BATCHES / batch processes / brewery / brewing process / energy integration / optimization / retrofitting / simulation.

1. Using BATCHES for the Simulation of the Brewing Process

While following the general objective of modelling and optimizing the brewing process, we have developed a detailed simulation model of the brewhouses of a test brewery having a capacity of two million hectolitres per year.

In its actual configuration, the "hot part" of this brewery consists of four brewhouses grouped two by two. The brewhouses of group A are equipped with classical filter-presses whereas the brewhouses of group B use the new mash filter 2001. Each group of two brewhouses comprises only one pair of mills and one chain of wort treatment down-stream of the boiling vessels. This configuration is represented in Fig. 1. As the four brewhouses have not been built at the same time and do not consist of the same pieces of equipment, they also do not work with the same load and with the same cycle time.

Realized on the basis of BATCHES [1-3], the simulation model now allows us to follow the consumption of the raw materials, the utilization of utilities, the status of all pieces of equipment, and so on. It can also be used to study the bottlenecks of the process or the influence of the production planning and of the synchronization of the four brewhouses on the parameters previously cited. Finally, it could be used to look at the way any disturbance, an equipment failure or a lack of operators for example, will spread into the whole process, in order do determine the time available to the operators to restore the system back to normal operating conditions before being obliged to stop the production completely.

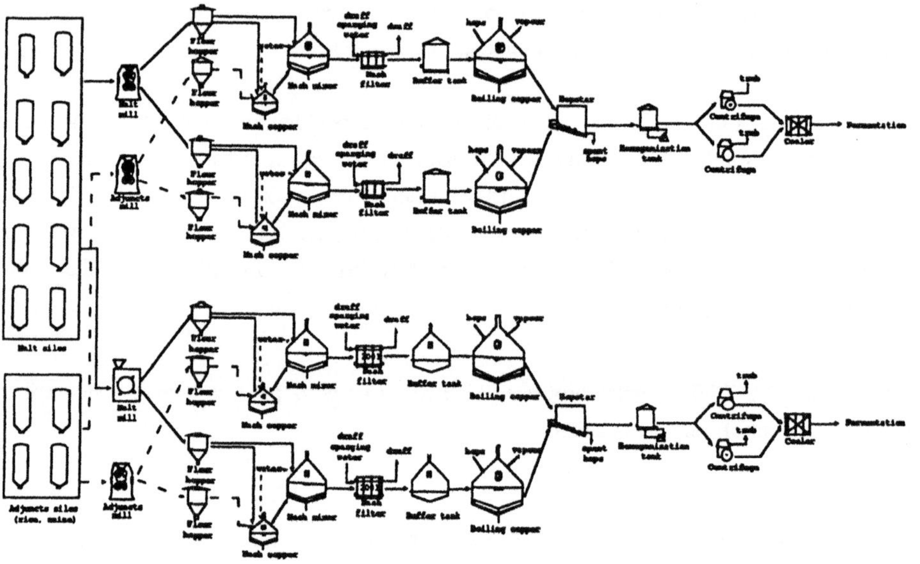

Fig. 1. Present configuration of the brewhouses

More specifically, the use of the model has allowed a thorough study of the utilization rates of the pieces of equipment of the brewhouses as well as an analysis of the influence of the production planning on the energy consumption profiles.

2. Study of the Utilization Rates of the Pieces of Equipment

The realization of a series of simulations of the functioning of the test brewery brewhouses has allowed us to study the utilization rate of their equipment. It clearly appears, as shown in the left part of Table 1, that these utilization rates are generally quite low. This is mainly the case for the mills, the buffer tanks, the mash coppers, the homogenization tanks upstream of the centrifuges and, to a lesser extent, for the mash filters.

	Utilization rate (%) of the equipment in the case of configuration			
Name of the piece of equipment	present group A	present group B	modified #1	modified #2
Malt mill	**39.3**	50.5	79.1	81.7
Adjuncts mill	**9.8**	**13.4**	20.8	21.5
Malt flour hopper #1	100.0	100.0	100.0	100.0
Malt flour hopper #2	100.0	100.0	100.0	100.0
Adjuncts flour hopper #1	100.0	100.0	100.0	100.0
Adjuncts flour hopper #2	100.0	100.0	---	---
Mash mixer #1	64.5	68.1	56.0	58.5
Mash mixer #2	64.5	69.0	56.2	58.6
Mash copper #1	**30.5**	**32.7**	55.0	56.8
Mash copper #2	**30.4**	**34.2**	---	---
Mash filter #1	**42.0**	**46.6**	81.3	85.2
Mash filter #2	**42.0**	**47.6**	---	---
Buffer tank #1	**18.2**	**4.7**	---	---
Buffer tank #2	**18.4**	**4.8**	---	---
Boiling copper #1	69.2	91.8	78.5	81.1
Boiling copper #2	69.2	91.9	79.7	81.5
Hopstar	**20.3**	**12.4**	19.4	10.0
Homogenization tank	47.8	39.3	67.1	31.4
Centrifuges	51.8	46.7	72.5	38.3
Cooler	51.8	46.7	72.5	38.3

Table 1. Equipment utilization statistics: comparison of the modified configurations to the present configuration

This study has also shown that some pieces of equipment are largely oversized. This is the case for the buffer tanks located between the mash filters and the boiling coppers, as well as for the homogenization tanks.

Taking into account all these observations, we have proposed two new configurations for the brewhouses, quite different from those in operation. In order to be as rational as possible, these configurations consist of four identical brewhouses and should be operated with a constant load value. Such a restriction could be considered as an important limitation of the operating flexibility of the industrial plant. However, it represents a totally acceptable constraint for large dedicated breweries like the one that has been dealt with in our study. As illustrated by Fig. 2 and summarized in Table 2, the suggested configurations represent a notable reduction of the number of pieces of equipment required. This reduction amounts to one third in the most favourable case. As can be easily noticed, the only difference between both modified configurations is the presence of one chain of wort treatment downstream of the boiling vessels in the case of the first one, whereas the second one comprises two such chains, which results in a larger operating flexibility.

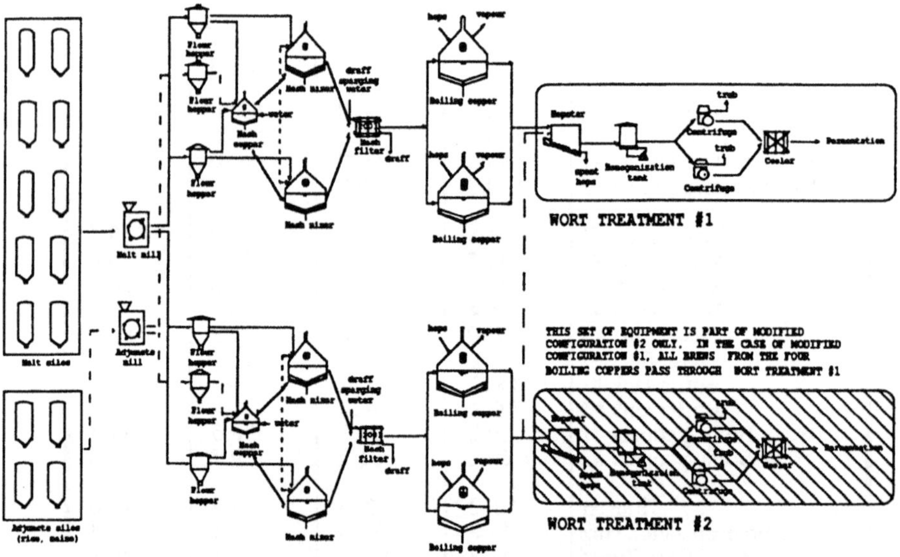

Fig. 2. Modified configurations of the brewhouses

A series of simulations have been performed on the basis of these two modified configurations. They have proven their feasibility, provided that a strict and quite rigid control of the production planning be enforced in the case of the first one, whereas the second one has the same operating flexibility as the present configuration. At the present time, two new breweries whose lay-outs are based on this second modified configuration are being put into operation in Belgium. The main difference between the configuration that we have proposed and the one adopted for these new plants is that they are composed of three brewing groups in parallel, instead of two. The other differences are just design details.

The right part of Table 1 shows the utilization rates of the equipment of both modified configurations. It illustrates very well the obtainable gain in comparison with the basic configuration (left part of the table). The absence of figure in the right part of the table means that these particular pieces of equipment are no longer included in the proposed new configurations.

Type of equipment	Present configuration	Modified configurations	
		#1	#2
malt & adjuncts mills	4	2	2
mash coppers	4	2	2
adjuncts flour hoppers	4	2	2
mash filters	4	2	2
buffer tanks (before boiling)	4	0	0
wort treatment chains (after boiling)	2	1	2
TOTAL NUMBER OF PIECES OF EQUIPMENT	**42**	**25**	**30**

Table 2. Number of pieces of equipment of the brewhouses

3. Energy Integration

The simulations that have been performed have allowed the determination of the time profiles of the process utilities consumption and, in particular, of the steam consumption. It has been possible to show that these profiles exhibit some peak demands of very short duration which greatly exceed the mean demand. Figure 3 shows an example of such a profile.

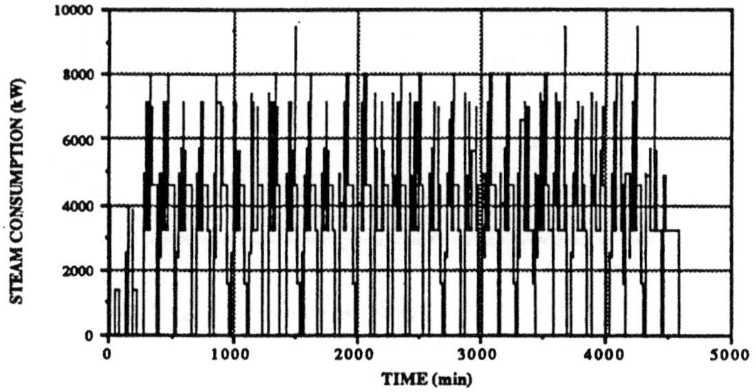

Fig. 3. Steam consumption profile for the realization of 15 brews in each of the four brewhouses. Cycle time: 145 min. Offset between brewing groups: 0 min.

3.1. Standardization of the brewhouses

The standardization of the brewhouses that can be reached by adopting one of the modified configurations gives rise to a reduction of some 20% of the maximum peak value of the steam demand. Indeed, by using the four brewhouses with constant values for the load and for the cycle time, one avoids having an offset between the brewhouses that varies with time, thus avoiding the steam consumption profile exhibiting a quasi unforeseeable behaviour.

3.2. Influence of the production planning

It also appears that the production planning exerts an important influence on the steam demand profile. Indeed, given the design of the brewhouses (one single wort treatment downstream of the boiling operation per group of two brewhouses), the two brewhouses of a same group are necessarily synchronized. On the other hand, in the case of the present configuration and of the second modified configuration, there is no synchronization constraint between the two brewing groups. In such a context, it was opportune to study the influence of the offset between these two groups on the various parameters of the steam consumption. Table 3 presents the results of the simulation runs for different values of the offset between the two brewing groups. It shows that an adequate offset of the two brewing groups starting times gives rise to a considerable reduction, once again about 20%, of the maximum level of the steam peak demands.

Cycle time [min]	Steam availability level [kW]	Offset between the brewing groups [min]	Highest steam consumption level [kW]
145	infinite	0	9500
145	infinite	10	8756
145	infinite	20	7583
145	infinite	30	8223
145	infinite	40	8273
145	infinite	50	**7547**
145	infinite	60	9500
145	infinite	70	8419

Table 3. Influence of the offset between brewing groups on the peak values of the steam consumption profile

However, the approach used so far is not satisfactory. Indeed, if we wanted to study the steam consumption profile for all the possible values of the offset, we would have to make as many simulation runs as there are such values. Even on a fast computing platform, such a study would be extremely time-consuming. On the other hand, if we do not consider all the possible offset values, we may miss the real optimal value which, it is quite obvious and was proved by the simulation study, can not be predicted by any means.

For this reason, we have developed a systematic approach of the problem which still rests on the use of the simulation model, but not so heavily. This approach can be split up into the following stages:

a) A simulation run is used to generate the steam consumption profile for 50 consecutive brews in the same brewhouse. These brews are processed with such a cycle time that their respective consumption profiles do not interfere with each other and do not overlap. Figure 4.a presents a part (10 brews) of the resulting profile. We then use these data to feed computer routines that are independent from the simulator. These routines, written in C programming language, execute the next four steps (b to e) of the method.

b) On the basis of the 50 individual profiles generated by the simulation run, a mean profile is computed. If we call *P(t)* the profile for *n* consecutive brews generated with a cycle time value *p*, the mean profile *M(t)* is given by the formula

$$M(t) = \frac{\sum_{i=1}^{n} P[t + (i-1).p]}{n} \qquad \forall\, t \in [0,p[\qquad (1)$$

whose application requires that the suitable values of *P(t)* be computed by interpolation since *P(t)* is defined in a discrete way. The mean profile for one individual brew that has been computed by applying formula 1 is shown at Fig. 4.b.

c) The mean profile is then used to compute the resulting profile *S(t)* of consecutive overlapping brews in one brewhouse, as defined by the formula

$$S(t) = M(t) + M(t-T) \qquad (2)$$

where T is the brewing cycle time. The periodic part of *S(t)* is represented at Fig. 4.c.

d) We can then generate the steam consumption profile *F(t)* of one brewing group consisting of two brewhouses with an offset *D*, following the formula

$$F(t) = S(t) + S(t-D) \qquad (3)$$

Fig. 4.a

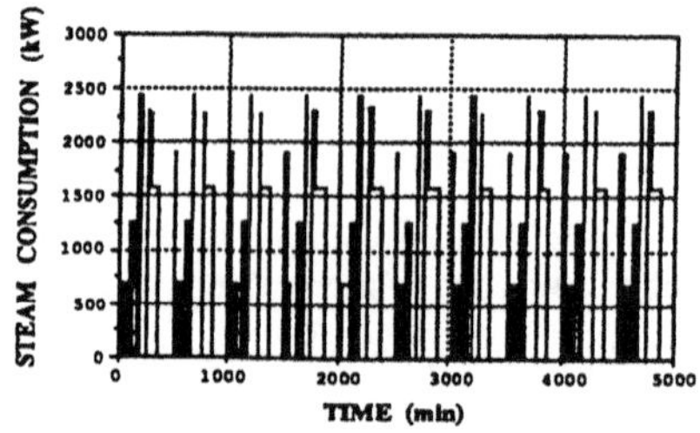

Fig. 4.b

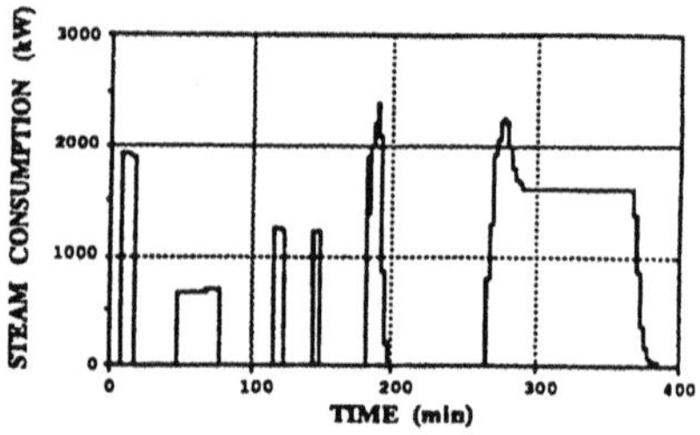

Fig. 4.c

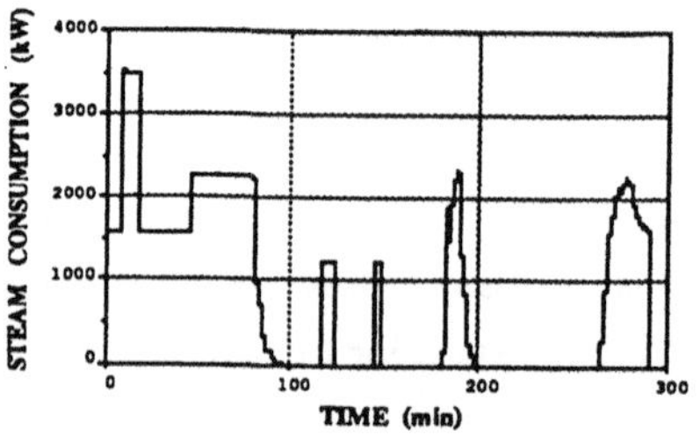

Fig. 4.d

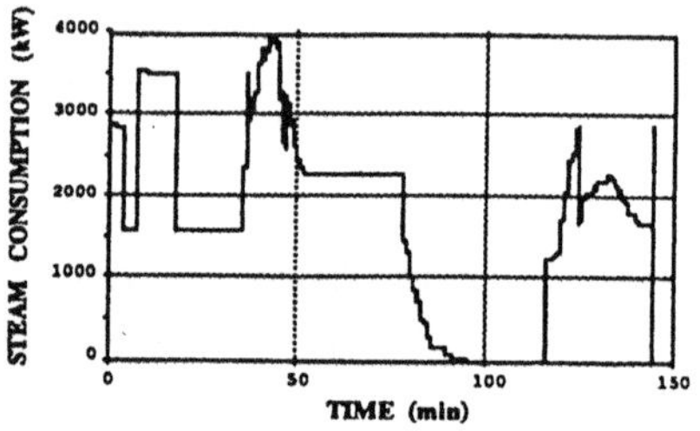

Fig. 4. a) Steam consumption profile of consecutive, non-overlapping brews in one brewhouse. Profile generated by a simulation run with BATCHES.
b) Computed mean steam consumption profile of one isolated brew.
c) Computed periodic part of the steam consumption profile of consecutive, overlapping brews in one brewhouse. Cycle time = 290 min.
d) Computed periodic part of the steam consumption profile of a brewing group consisting of two brewhouses operating in parallel. Offset = 145 min.

The periodic part of *F(t)* is represented at Fig. 4.d in the particular case where $D = T / 2$ which means that the two brewhouses are perfectly synchronized. Moreover, this would be the normal operating way for a configuration such as the one studied here.

e) The global steam consumption profiles of the brewing plant consisting of two (or three) groups operating in parallel with an offset X (or with offsets X and Y) are given by

$$R_{2,X}(t) = F(t) + F(t-X) \triangleq R_2(t,X) \quad (4)$$

$$R_{3,X,Y}(t) = F(t) + F(t-X) + F(t-Y) \triangleq R_3(t,X,Y) \quad (4')$$

and the maximum consumption peak, as a function of the offset(s) between the brewing groups, is defined by the formulae

$$\max_t \left[R_2(t,X)\right] \triangleq MAX(X) \quad (5)$$

$$\max_t \left[R_3(t,X,Y)\right] \triangleq MAX(X,Y) \quad (5')$$

Fig. 5.a

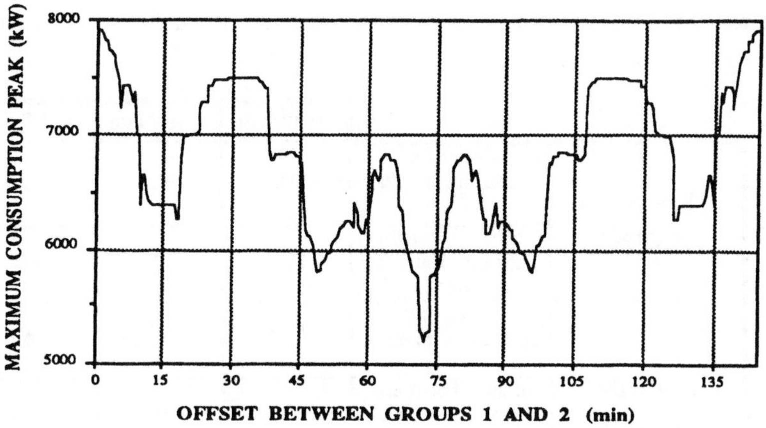

Fig. 5.b

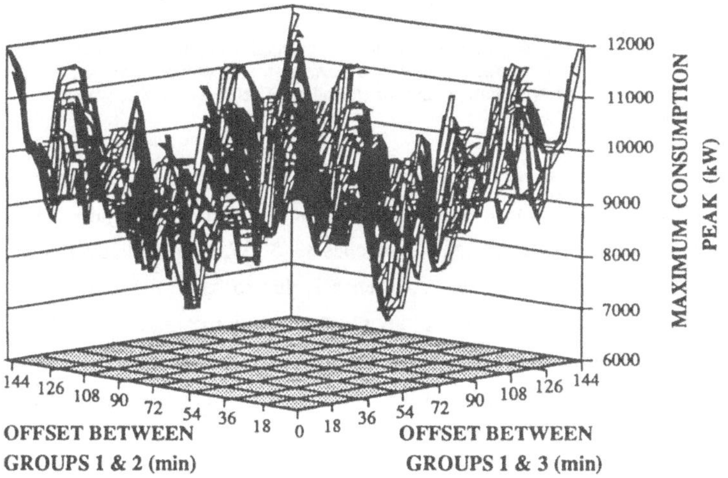

Fig. 5. Influence of the offset(s) between brewing groups on the maximum consumption peak. a) Two brewing groups in parallel. b) Three brewing groups in parallel

Figure 5 represents the function MAX (X) and the response surface MAX (X,Y) whereas Table 4 summarizes the most important information of these two diagrams, namely the highest and the lowest values of the maximum consumption peak, as well as the corresponding offsets between the brewing groups. As can be also seen, the difference between the situations with the most unfavourable offset and with the most favourable one amounts to 35% for two brewing groups, and reaches 41% in the case of three groups.

		minimum	maximum	% difference
2 BREWING GROUPS	maximum consumption peak [kW]	5182	7969	**35**
	offset between groups 1 and 2 [min]	72.5	0	--
3 BREWING GROUPS	maximum consumption peak [kW]	7007	11953	**41**
	offset between groups 1 and 2 [min]	48	0	--
	offset between groups 1 and 3 [min]	96	0	--

Table 4. Highest and lowest values of the maximum peak of the global steam consumption profile of a brewing plant composed of two or three brewing groups, and corresponding offsets between these groups.

f) In order to check the validity of the optimum offset computed above in the case of two brewing groups (72.5 min), we called upon the simulator once again. In this instance the simulated highest consumption peak reaches 8345 kW, a value that is much higher than the one predicted. Obviously, this difference is due to some parameters of the simulation model whose stochastic nature could not be taken into account in our method. However, as will be shown in the next paragraph, the information that we have obtained through its application will prove very useful since it will be possible to come very near to the computed optimum.

3.3. Deliberate limitation of the steam availability level

Actually, a statistical analysis of the steam consumption in the most favourable case computed by the method described above (second modified configuration; cycle time: 145 min; offset: 72.5 min; computed steam peak demand: 5182 kW; real steam peak demand: 8345 kW) reveals, as illustrated by Fig. 6.b, that only 1.7% of the total steam consumption consists of demand peaks higher than 6000 kW, and only 3.7% of peaks higher than 5500 kW. As a consequence, we have studied the behaviour of the process when the steam availability level is voluntarily limited to 6000 kW, or even to 5500 kW, all other parameters being the same as in the former simulation runs.

Figures 6.c and 6.d show the steam consumption profile and its statistical analysis in terms of power classes for simulation runs with the most severe of these two constraints. The analysis of the results of these runs shows that a more regular steam consumption profile has been obtained, as could be expected, but also that the additional waiting times generated by the

Fig. 6.a

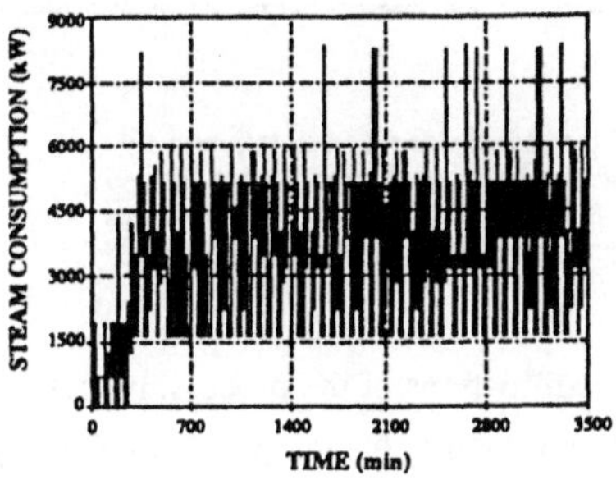

Fig. 6.b

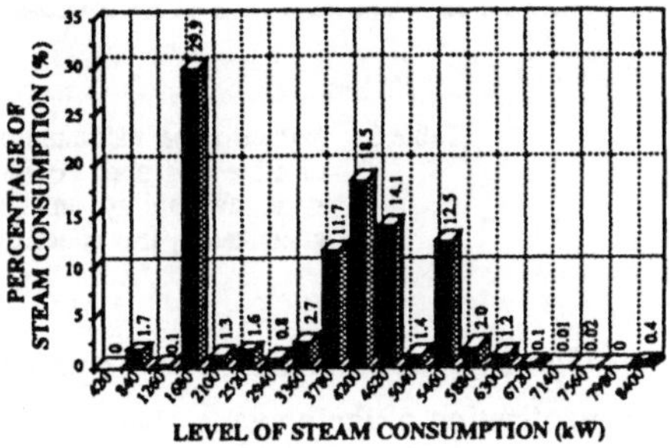

Fig. 6.c

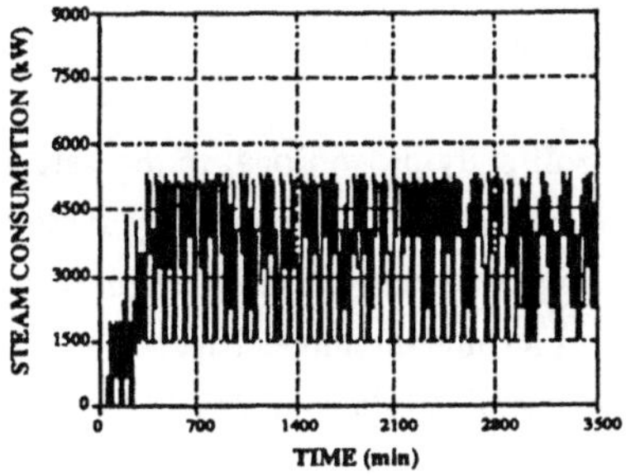

Fig. 6.d

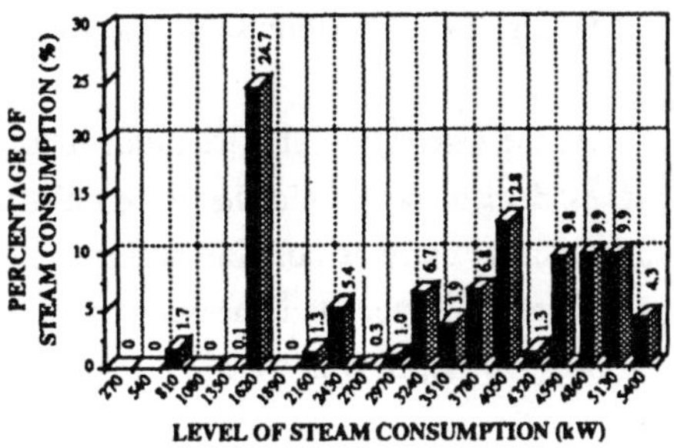

Fig. 6. Influence of the availability level of steam on its consumption profile. Cycle time: 145 min. Offset: 72.5 min. (6.a and 6.b: consumption profile and statistical analysis for infinite availability level; 6.c and 6.d: consumption profile and statistical analysis for availability level limited to 5500 kW)

new constraint are almost negligible. They can be completely neglected in the case of a steam availability limited to 6000 kW whereas, in the case of a limitation to 5500 kW, they are of the same order of magnitude as the stochastic variations of the operation times of the tasks they affect. Even in this last case, they are acceptable since they do not cumulate, thus lengthening the overall cycle time only in a small extent and leaving unaffected the quality of the end product. These supplementary waiting times are summarized in Table 5, together with the values of the maximum demand peaks obtained during these last simulation runs.

Steam availability level [kW]	Highest steam consumption level [kW]	MEAN ADDITIONAL WAITING TIME AT [min]			
		Mash mixing	Mash cooking	Filtration	Wort boiling
infinite	8345	--	--	--	--
6000	5986	0.1	0.0	0.8	1.2
5500	**5296**	6.6	4.8	3.3	3.7

Table 5. Values of the maximum peak of the global steam consumption profile of a brewing plant composed of two brewing groups, with a limited steam availability, and corresponding mean additional waiting times generated by this constraint. (Cycle time: 145 min; Offset: 72.5 min)

Globally, the combination of the proposed modifications of the process, namely a better standardization of the brewing plant, an adequate production planning and a voluntary limitation of the steam availability level, allows us to reach a considerable reduction of the steam consumption peaks. From the most unfavourable situation to the most favourable one encountered during our simulation study, the global reduction amounts to 55%.

As the size of the utilities production plants is directly proportional to the value of the highest demand peak which has to be satisfied by these plants, it is essential to try to minimize this value as early as possible during the design stage of a new industrial plant in order to reduce its investment and operating costs. The use of a simulation model such as the one that has been developed with BATCHES has proved very useful in following this objective.

References

1. Batch Process Technologies:BATCHES Reference Manual. Batch Process Technologies Inc., West-Lafayette, USA 1989-1990
2. Batch Process Technologies: BATCHES User's Manual. Batch Process Technologies Inc., West-Lafayette, USA 1989-1990
3. Clark, S.M. and Kuriyan, K.: BATCHES - Simulation software for managing semicontinuous and batch processes. Batch Process Technologies Inc., West-Lafayette, USA April 1989 (unpublished)

List of Participants

NATO Countries

Belgium

Christine Bernot, Chemical Eng. Dep., U. of Massachusetts, Amherst, MA USA, Presently at Solvay, Belgium.

Denis J. Mignon, Unite des Procedes - Universite Catholique de Louvain 1, Voie Minckelers B-1348 Louvain-la-neuve, Belgium

Canada

John F. MacGregor, McMaster University, Chemical Eng. Dept., Hamilton, Ontario L8S 4L7, Canada

Germany

Henner Schmitt, LSGC ENSIC BP 451 54001, Nancy Cedex, France

Greece

Christos Georgakis, Lehigh U., Chemical Engineering Department., Bethlehem, PA 18015, USA

Savoula Papageorgaki, Purdue University, 50 S. Meridith Av. Apt. 1, Pasadena, CA 91106, USA

Lazarus Papageorgio, University of London (Imperial College), Dept. of ChE & Chem. Tech., Prince Consort Road, London, SW7 2BY, England

Athanasios G. Tsirukis, California Inst. of Technology, Chemical Eng. 210-41, Pasadena, CA 91125, USA

Vasilius Voudouris, Dept. Chem. Eng., Carnegie Mellon Univ., Pittsburgh, PA 15213, USA

Italy

Sandro Macchietto, University of London (Imperial College), Dept. of ChE & Chem. Tech., Prince Consort Road, London, SW7 2BY, England

France

Michel Lucet, Associate Director, Rhone-Poulenc Industrialisation, Centre DE Decines 24, Avenue Jean Jaures, 69151 Decines Charpieu Cedex, B.p.16, France

Norway

Kristian M. Lien, Universitetet I Trondheim, Norges Tekniske Hogskole, Institutt for Kjemiteknikk, N-7034 Trondheim, Norway

Eva Sorensen, Dept. of Chemical Eng., University of Trondheim-NTH, Sem Salandsvei 4, N-7034 Trondheim, Norway

Portugal

Ana Cristina Santos Amaro, I.S.C.A.C. (Polytechnic of Coimbra), Rua Luis De Camoes, 3000 Coimbra, Portugal

Jose Almiro Castro, Universidade De Coimbra, Departamento de Engenharia Quimica, Largo Marques de Pombal, 3000 Coimbra, Portugal

M. Lucelinda Alcantara da Cunha, Departamento de Ciencia dos Materials, FCT/UNL Quinta Da Torre, 2825 Monte De Caparica, Portugal

Paulo Falcao Ferreira, Dept. Chem. Eng., Univ. of Coimbra, Rua Cidade de Salamanca 93, 3000 Coimbra, Portugal

Jorge M.S.S. Martins, Escola Superior De Tecnologia De Viseu, Estr. Circunvalao, 3500 Viseu, Portugal

Alirio Rodriguez, Universidad de Porto, Departamento de Engenharia Quimica, Faculdade de Engenharia, Rua dos Bragas, 4099 Porto Codex, Portugal

Spain

Estonia Espuna, Universitat Politecnica De Katalunya, Departament d'engineria Quimica, ETSEIB, Diagonal, 647, E-08028 Barcelona, Spain

Raimon Grav, Universitat Politecnica De Katalunya, Departament d'engineria Quimica ETSEIB, Diagonal, 647, E-08028 Barcelona, Spain

Luis Puigjaner, Universitat Politecnica De Katalunya, Departament d'engineria Quimica, ETSEIB, Diagonal, 647, E-08028 Barcelona, Spain

Turkey

Uğur Akman, Boğaziçi U., Chemical Engineering Dep., 80815 Bebek, Istanbul, Türkiye

Rıdvan Berber, U. of Ankara, Kimya Mühendisliği Bölümü, Tandoğan Ankara 06100,Türkiye

Esen Bolat, Yıldız Universitesi, Kimya Mühendisliği Bölümü, Şişli Istanbul, Türkiye

Vildan Dinçbaş, Boğaziçi U., Chemical Engineering Dep., 80815 Bebek, Istanbul, Türkiye

Türker Gürkan, METU Chem. Eng. Dept., 06531 Ankara, Türkiye

Öner Hortaçsu, Boğaziçi U., Chemical Engineering Dep., 80815 Bebek, Istanbul, Türkiye

Ersan Kalafatoğlu, Marmara Araştırma Merkezi, Kimya Mühendisliği Araştırma Bölümü, PK 21, Gebze Kocaeli, Türkiye

Ali Karaduman, U. of Ankara, Kimya Mühendisliği Bölümü, Tandoğan Ankara 06100,Türkiye

Derya Kibar, Petkim Araştırma Merkezi, Kocaeli, Türkiye

Bilgin Kısakürek, Chemical Eng. Dep., METU., 06531 Ankara, Türkiye

İlsen Önsan, Boğaziçi U., Chemical Engineering Dep., 80815 Bebek, Istanbul, Türkiye

Nuran Örs, Marmara Araştırma Merkezi, Kimya Mühendisliği Araştırma Bölümü, PK 21, Gebze Kocaeli, Türkiye

Canan Özgen, Chem. Eng. Dept., METU 06531 Ankara, Türkiye

Candan Tamerler, Boğaziçi U., Chemical Engineering Dep., 80815 Bebek, Istanbul, Türkiye

Ertan Taşkın, Chemical Eng. Dep., METU., 06531 Ankara, Türkiye

Metin Türkay, Chemical Eng. Dep., METU., 06531 Ankara, Türkiye

United Kingdom

Kamal Kuriyan, University of London (Imperial College), Dept. of ChE & Chem. Tech., Prince Consort Road, London, SW7 2BY, England

Phillip Law, ETH Technisch-Chemisches Lab., Systems group, CH-8092 Zurich, Switzerland

Jack W. Ponton, Chemical Engineering Department,University of Edinburgh, Edinburgh, Scotland

David W.T. Rippin, ETH, Technisch-Chemisches Lab., Systems group CH-8092, Zurich, Switzerland

Rodger.W.H. Sargent, University of London (Imperial College), Dept. of ChE & Chem. Tech., Prince Consort Road, London, SW7 2BY, England

Nilay Shah, University of London (Imperial College) Dept. of ChE & Chem. Tech., Prince Consort Road, London, SW7 2BY, England

Zhang Xueya, University of London (Imperial College), Dept. of ChE & Chem. Tech. Prince Consort Road, London, SW7 2BY, England

United States of America

Mukul Agarwal, ETH, Technisch-Chemisches Lab. Systems group, CH-8092 Zurich, Switzerland

Yaman Arkun, Georgia Institute of Tech., School of Chemical Eng., Atlanta, GA 30332-0100, USA

Ali Çınar, Illinois Institute of Technology, Chemical Engineering Department, 10 W. 33rd St., Perlstein Hall, Rm. 105, Chicago, IL 60616, USA

Luis Garcia-Rubio, Chemical Engineering Dep., U. of South Florida, Tampa Fl 33620, USA

Alicia Garcia, Civil Engineering Department, University of South Florida, Tampa Fl 33620, USA

Ignacio E. Grossmann, Department of Chemical Engineering, Carnegie Mellon University, Pittsburgh, Pennsylvania 15213-3890, USA

Muzaffer Kapanoğlu, Industrial Eng. Dep., University of South Florida, Tampa Fl 33620, USA

Girish Joglekar, Batch Process Technologies, 1291 E. Cumberland Avenue, West Lafayette, IN 47906, USA

Joseph F. Pekny, Purdue University, School of Chemical Eng., West Lafayette, IN 47907, USA

Gintaras V. Reklaitis, Purdue U., School of Chemical Eng., West Lafayette, IN 47907, USA

Aydın K. Sunol, Chemical Engineering Dep., University of South Florida, Tampa Fl 33620 USA

Venkat Venkatasubramanian, Purdue U., School of Chemical Eng. West Lafayette, IN 47907, USA

Arthur W. Westerberg, Carnegie-Mellon U., Chemical Eng. Dep. Pittsburgh, PA 15213, USA

Mike Zentner, Purdue University, School of Chemical Eng., West Lafayette, IN 47907, USA

Non NATO Countries

Czechoslovakia

Miroslav Hofmeister, Technisch-Chemisches Lab., Systems group, ETH, CH-8092 Zurich, Switzerland

Hungary

Bela Csukas, Dept. of Chem. Eng. Cybernetics, U. of Veszprem, H-8201 POB 158, Hungary

Gyula Kortvelyessy, Research Manager, SZEVIKI R&D Institute, Stahly Str. 13, Budapest, H-1085 Hungary

Laszlo Halasz, Technisch-Chemisches Lab., Systems group, ETH-Zentrum, CH-8092 Zurich, Switzerland

Japan

Shinji Hasebe, Kyoto University, Dep. of Chemical Engineering, Process Systems Engineering Group, Sakyo-ku, Kyoto, 606-01, Japan

Sweden

Dag E. Ravemark, Technisch-Chemisches Laboratorium, Systems group, ETH 8092 Zurich, Switzerland

Index

NATO ASI Series F

Including Special Programmes on Sensory Systems for Robotic Control (ROB) and on Advanced Educational Technology (AET)

Vol. 137: Technology-Based Learning Environments. Psychological and Educational Foundations. Edited by S. Vosniadou, E. De Corte and H. Mandl. X, 302 pages. 1994. *(AET)*

Vol. 138: Exploiting Mental Imagery with Computers in Mathematics Education. Edited by R. Sutherland and J. Mason. VIII, 326 pages. 1995. *(AET)*

Vol. 139: Proof and Computation. Edited by H. Schwichtenberg. VII, 470 pages. 1995.

Vol. 140: Automating Instructional Design: Computer-Based Development and Delivery Tools. Edited by R. D. Tennyson and A. E. Barron. IX, 618 pages. 1995. *(AET)*

Vol. 141: Organizational Learning and Technological Change. Edited by C. Zucchermaglio, S. Bagnara and S. U. Stucky. X, 368 pages. 1995. *(AET)*

Vol. 142: Dialogue and Instruction. Modeling Interaction in Intelligent Tutoring Systems. Edited by R.-J. Beun, M. Baker and M. Reiner. IX, 368 pages. 1995. *(AET)*

Vol. 143: Batch Processing Systems Engineering. Fundamentals of Chemical Engineering. Edited by G. V. Reklaitis, A. K. Sunol, D. W. T. Rippin, and Ö. Hortaçsu. XIV, 868 pages. 1996.

Vol. 144: The Biology and Technology of Intelligent Autonomous Agents. Edited by Luc Steels. VIII, 517 pages. 1995.

Vol. 145: Advanced Educational Technology: Research Issues and Future Potential. Edited by T. T. Liao. VIII, 219 pages. 1996. *(AET)*

Vol. 146: Computers and Exploratory Learning. Edited by A. A. diSessa, C. Hoyles and R. Noss. VIII, 482 pages. 1995. *(AET)*

Vol. 147: Speech Recognition and Coding. New Advances and Trends. Edited by A. J. Rubio Ayuso and J. M. López Soler. XI, 505 pages. 1995.

Vol. 148: Knowledge Acquisition, Organization, and Use in Biology. Edited by K. M. Fisher and M. R. Kibby. X, 246 pages. 1996. *(AET)*

Vol. 149: Emergent Computing Methods in Engineering Design. Applications of Genetic Algorithms and Neural Networks. Edited by D.E. Grierson and P. Hajela. VIII, 350 pages. 1996.

Vol. 150: Speechreading by Humans and Machines. Edited by D. G. Stork and M. E. Hennecke. XV, 686 pages. 1996.

Vol. 151: Computational and Conversational Discourse. Burning Issues – An Interdisciplinary Account. Edited by E. H. Hovy and D. R. Scott. XII, 202 pages. 1996.

Vol. 152: Deductive Program Design. Edited by M. Broy. IX, 467 pages. 1996.

Vol. 153: Identification, Adaptation, Learning. Edited by S. Bittanti and G. Picci. XIV, 553 pages. 1996.

Vol. 154: Reliability and Maintenance of Complex Systems. Edited by S. Özekici. XI, 589 pages. 1996.

Vol. 156: Microcomputer-Based Labs: Educational Research and Standards. Edited by R.F. Tinker. XIV, 398 pages. 1996.

NATO ASI Series F

Including Special Programmes on Sensory Systems for Robotic Control (ROB) and on Advanced Educational Technology (AET)

Vol. 114: Intelligent Systems: Safety, Reliability and Maintainability Issues. Edited by O. Kaynak, G. Honderd and E. Grant. XI, 340 pages. 1993.

Vol. 115: Learning Electricity and Electronics with Advanced Educational Technology. Edited by M. Caillot. VII, 329 pages. 1993. *(AET)*

Vol. 116: Control Technology in Elementary Education. Edited by B. Denis. IX, 311 pages. 1993. *(AET)*

Vol. 117: Intelligent Learning Environments: The Case of Geometry. Edited by J.-M. Laborde. VIII, 267 pages. 1996. *(AET)*

Vol. 118: Program Design Calculi. Edited by M. Broy. VIII, 409 pages. 1993.

Vol. 119: Automating Instructional Design, Development, and Delivery. Edited by. R. D. Tennyson. VIII, 266 pages. 1994. *(AET)*

Vol. 120: Reliability and Safety Assessment of Dynamic Process Systems. Edited by T. Aldemir, N. O. Siu, A. Mosleh, P. C. Cacciabue and B. G. Göktepe. X, 242 pages. 1994.

Vol. 121: Learning from Computers: Mathematics Education and Technology. Edited by C. Keitel and K. Ruthven. XIII, 332 pages. 1993. *(AET)*

Vol. 122: Simulation-Based Experiential Learning. Edited by D. M. Towne, T. de Jong and H. Spada. XIV, 274 pages. 1993. *(AET)*

Vol. 123: User-Centred Requirements for Software Engineering Environments. Edited by D. J. Gilmore, R. L. Winder and F. Détienne. VII, 377 pages. 1994.

Vol. 124: Fundamentals in Handwriting Recognition. Edited by S. Impedovo. IX, 496 pages. 1994.

Vol. 125: Student Modelling: The Key to Individualized Knowledge-Based Instruction. Edited by J. E. Greer and G. I. McCalla. X, 383 pages. 1994. *(AET)*

Vol. 126: Shape in Picture. Mathematical Description of Shape in Grey-level Images. Edited by Y.-L. O, A. Toet, D. Foster, H. J. A. M. Heijmans and P. Meer. XI, 676 pages. 1994.

Vol. 127: Real Time Computing. Edited by W. A. Halang and A. D. Stoyenko. XXII, 762 pages. 1994.

Vol. 128: Computer Supported Collaborative Learning. Edited by C. O'Malley. X, 303 pages. 1994. *(AET)*

Vol. 129: Human-Machine Communication for Educational Systems Design. Edited by M. D. Brouwer-Janse and T. L. Harrington. X, 342 pages. 1994. *(AET)*

Vol. 130: Advances in Object-Oriented Database Systems. Edited by A. Dogac, M. T. Özsu, A. Biliris and T. Sellis. XI, 515 pages. 1994.

Vol. 131: Constraint Programming. Edited by B. Mayoh, E. Tyugu and J. Penjam. VII, 452 pages. 1994.

Vol. 132: Mathematical Modelling Courses for Engineering Education. Edited by Y. Ersoy and A. O. Moscardini. X, 246 pages. 1994. *(AET)*

Vol. 133: Collaborative Dialogue Technologies in Distance Learning. Edited by M. F. Verdejo and S. A. Cerri. XIV, 296 pages. 1994. *(AET)*

Vol. 134: Computer Integrated Production Systems and Organizations. The Human-Centred Approach. Edited by F. Schmid, S. Evans, A. W. S. Ainger and R. J. Grieve. X, 347 pages. 1994.

Vol. 135: Technology Education in School and Industry. Emerging Didactics for Human Resource Development. Edited by D. Blandow and M. J. Dyrenfurth. XI, 367 pages. 1994. *(AET)*

Vol. 136: From Statistics to Neural Networks. Theory and Pattern Recognition Applications. Edited by V. Cherkassky, J. H. Friedman and H. Wechsler. XII, 394 pages. 1994.

NATO ASI Series F

Including Special Programmes on Sensory Systems for Robotic Control (ROB) and on Advanced Educational Technology (AET)

Vol. 92: Hypermedia Courseware: Structures of Communication and Intelligent Help. Edited by A. Oliveira. X, 241 pages. 1992. *(AET)*

Vol. 93: Interactive Multimedia Learning Environments. Human Factors and Technical Considerations on Design Issues. Edited by M. Giardina. VIII, 254 pages. 1992. *(AET)*

Vol. 94: Logic and Algebra of Specification. Edited by F. L. Bauer, W. Brauer, and H. Schwichtenberg. VII, 442 pages. 1993.

Vol. 95: Comprehensive Systems Design: A New Educational Technology. Edited by C. M. Reigeluth, B. H. Banathy, and J. R. Olson. IX, 437 pages. 1993. *(AET)*

Vol. 96: New Directions in Educational Technology. Edited by E. Scanlon and T. O'Shea. VIII, 251 pages. 1992. *(AET)*

Vol. 97: Advanced Models of Cognition for Medical Training and Practice. Edited by D. A. Evans and V. L. Patel. XI, 372 pages. 1992. *(AET)*

Vol. 98: Medical Images: Formation, Handling and Evaluation. Edited by A. E. Todd-Pokropek and M. A. Viergever. IX, 700 pages. 1992.

Vol. 99: Multisensor Fusion for Computer Vision. Edited by J. K. Aggarwal. XI, 456 pages. 1993. *(ROB)*

Vol. 100: Communication from an Artificial Intelligence Perspective. Theoretical and Applied Issues. Edited by A. Ortony, J. Slack and O. Stock. XII, 260 pages. 1992.

Vol. 101: Recent Developments in Decision Support Systems. Edited by C. W. Holsapple and A. B. Whinston. XI, 618 pages. 1993.

Vol. 102: Robots and Biological Systems: Towards a New Bionics? Edited by P. Dario, G. Sandini and P. Aebischer. XII, 786 pages. 1993.

Vol. 103: Parallel Computing on Distributed Memory Multiprocessors. Edited by F. Özgüner and F. Erçal. VIII, 332 pages. 1993.

Vol. 104: Instructional Models in Computer-Based Learning Environments. Edited by S. Dijkstra, H. P. M. Krammer and J. J. G. van Merriënboer. X, 510 pages. 1993. *(AET)*

Vol. 105: Designing Environments for Constructive Learning. Edited by T. M. Duffy, J. Lowyck and D. H. Jonassen. VIII, 374 pages. 1993. *(AET)*

Vol. 106: Software for Parallel Computation. Edited by J. S. Kowalik and L. Grandinetti. IX, 363 pages. 1993.

Vol. 107: Advanced Educational Technologies for Mathematics and Science. Edited by D. L. Ferguson. XII, 749 pages. 1993. *(AET)*

Vol. 108: Concurrent Engineering: Tools and Technologies for Mechanical System Design. Edited by E. J. Haug. XIII, 998 pages. 1993.

Vol. 109: Advanced Educational Technology in Technology Education. Edited by A. Gordon, M. Hacker and M. de Vries. VIII, 253 pages. 1993. *(AET)*

Vol. 110: Verification and Validation of Complex Systems: Human Factors Issues. Edited by J. A. Wise, V. D. Hopkin and P. Stager. XIII, 704 pages. 1993.

Vol. 111: Cognitive Models and Intelligent Environments for Learning Programming. Edited by E. Lemut, B. du Boulay and G. Dettori. VIII, 305 pages. 1993. *(AET)*

Vol. 112: Item Banking: Interactive Testing and Self-Assessment. Edited by D. A. Leclercq and J. E. Bruno. VIII, 261 pages. 1993. *(AET)*

Vol. 113: Interactive Learning Technology for the Deaf. Edited by B. A. G. Elsendoorn and F. Coninx. XIII, 285 pages. 1993. *(AET)*

NATO ASI Series F

Including Special Programmes on Sensory Systems for Robotic Control (ROB) and on Advanced Educational Technology (AET)

Vol. 70: Numerical Linear Algebra, Digital Signal Processing and Parallel Algorithms. Edited by G. H. Golub and P. Van Dooren. XIII, 729 pages. 1991.

Vol. 71: Expert Systems and Robotics. Edited by T. Jordanides and B.Torby. XII, 744 pages. 1991.

Vol. 72: High-Capacity Local and Metropolitan Area Networks. Architecture and Performance Issues. Edited by G. Pujolle. X, 536 pages. 1991.

Vol. 73: Automation and Systems Issues in Air Traffic Control. Edited by J. A. Wise, V. D. Hopkin and M. L. Smith. XIX, 594 pages. 1991.

Vol. 74: Picture Archiving and Communication Systems (PACS) in Medicine. Edited by H. K. Huang, O. Ratib, A. R. Bakker and G. Witte. XI, 438 pages. 1991.

Vol. 75: Speech Recognition and Understanding. Recent Advances, Trends and Applications. Edited by P. Laface and Renato De Mori. XI, 559 pages. 1991.

Vol. 76: Multimedia Interface Design in Education. Edited by A. D. N. Edwards and S. Holland. XIV, 216 pages. 1992. *(AET)*

Vol. 77: Computer Algorithms for Solving Linear Algebraic Equations. The State of the Art. Edited by E. Spedicato. VIII, 352 pages. 1991.

Vol. 78: Integrating Advanced Technology into Technology Education. Edited by M. Hacker, A. Gordon and M. de Vries. VIII, 185 pages. 1991. *(AET)*

Vol. 79: Logic, Algebra, and Computation. Edited by F. L. Bauer. VII, 485 pages. 1991.

Vol. 80: Intelligent Tutoring Systems for Foreign Language Learning. Edited by M. L. Swartz and M. Yazdani. IX, 347 pages. 1992. *(AET)*

Vol. 81: Cognitive Tools for Learning. Edited by P. A. M. Kommers, D. H. Jonassen, and J. T. Mayes. X, 278 pages. 1992. *(AET)*

Vol. 82: Combinatorial Optimization. New Frontiers in Theory and Practice. Edited by M. Akgül, H. W. Hamacher, and S. Tüfekçi. XI, 334 pages. 1992.

Vol. 83: Active Perception and Robot Vision. Edited by A. K. Sood and H. Wechsler. IX, 756 pages. 1992.

Vol. 84: Computer-Based Learning Environments and Problem Solving. Edited by E. De Corte, M. C. Linn, H. Mandl, and L. Verschaffel. XVI, 488 pages. 1992. *(AET)*

Vol. 85: Adaptive Learning Environments. Foundations and Frontiers. Edited by M. Jones and P. H. Winne. VIII, 408 pages. 1992. *(AET)*

Vol. 86: Intelligent Learning Environments and Knowledge Acquisition in Physics. Edited by A. Tiberghien and H. Mandl. VIII, 285 pages. 1992. *(AET)*

Vol. 87: Cognitive Modelling and Interactive Environments. With demo diskettes (Apple and IBM compatible). Edited by F. L. Engel, D. G. Bouwhuis, T. Bösser, and G. d'Ydewalle. IX, 311 pages. 1992. *(AET)*

Vol. 88: Programming and Mathematical Method. Edited by M. Broy. VIII, 428 pages. 1992.

Vol. 89: Mathematical Problem Solving and New Information Technologies. Edited by J. P. Ponte, J. F. Matos, J. M. Matos, and D. Fernandes. XV, 346 pages. 1992. *(AET)*

Vol. 90: Collaborative Learning Through Computer Conferencing. Edited by A. R. Kaye. X, 260 pages. 1992. *(AET)*

Vol. 91: New Directions for Intelligent Tutoring Systems. Edited by E. Costa. X, 296 pages. 1992. *(AET)*

GPSR Compliance

The European Union's (EU) General Product Safety Regulation (GPSR) is a set of rules that requires consumer products to be safe and our obligations to ensure this.

If you have any concerns about our products, you can contact us on ProductSafety@springernature.com

In case Publisher is established outside the EU, the EU authorized representative is:

Springer Nature Customer Service Center GmbH
Europaplatz 3
69115 Heidelberg, Germany

Batch number: 10682269

Printed by Printforce, the Netherlands